BRITISH STANDARD

BS 7671:2018

incorporating
Corrigendum 2018
Amendment 1:2020
Corrigendum 2020
Amendment 2:2022

AF711819

Requirements for Electrical Installations

IET Wiring Regulations
Eighteenth Edition

The Institution of Engineering and Technology

bsi.

Publication Information

Published by the Institution of Engineering and Technology, London, United Kingdom in agreement with BSI.

The Institution of Engineering and Technology is registered as a Charity in England & Wales (no. 211014) and Scotland (no. SC038698).

The Institution of Engineering and Technology is the institution formed in 2006 by the joining together of the IEE (the Institution of Electrical Engineers) and the IIE (the Institution of Incorporated Engineers).

British Standards Institution is the independent national body responsible for preparing British Standards. It presents the UK view on standards in Europe and at the international level. It is incorporated by Royal Charter.

This British Standard (BS 7671:2018+A2:2022) was published under the joint authority of the IET and of the Standards Policy and Strategy Committee on 28 March 2022.

The UK participation in its preparation was entrusted to Joint Technical Committee JPEL/64 Electrical Installations. A list of organizations represented on this committee can be obtained on request to its Committee Manager.

This publication does not purport to include all the necessary provisions of a contract. Users are responsible for its correct application. Compliance with a British Standard cannot confer immunity from legal obligations.

The IET is taking the issues found with dangerous counterfeits of its titles very seriously. The hologram on the inside cover of this book provides an additional measure to help our customers identify counterfeit versions. For more information on counterfeits, the hologram and how to identify genuine IET publications, visit https://electrical.theiet.org/bs-7671/fake-awareness/

It is the constant aim of the IET and BSI to improve the quality of our products and services. We should be grateful if anyone finding an inaccuracy or ambiguity while using this British Standard would inform the Committee Manager, G D Cronshaw {gcronshaw@theiet.org}, The IET, Six Hills Way, Stevenage, SG1 2AY, United Kingdom.

IET Technical Regulations staff

as at August 2021

JPEL/64	Joint IET/BSI Technical Committee Electrical Installations	G D Cronshaw CEng FIET
JPEL/A/64	Verification	M Peace CEng MIET MCIBSE
JPEL/B/64	Thermal Effects	Eur Ing L D Markwell MSc BSc(Hons) CEng MIET MCIBSE LCGI
JPEL/C/64	Shock Protection	S G Devine IEng MIET
JPEL/D/64	External Influences	M Coles BEng(Hons) MIET
Associate Committee Manager		R H Giddings IEng MIET ACIBSE

ISBN 978-1-83953-218-4 (paperback)

ISBN 978-1-83953-220-7 (vitalsource)

Proofread by M Doughton IEng MIET LCGI and G Gundry MIET

Typeset in the UK by the Institution of Engineering and Technology, Stevenage

Printed and bound in the UK by CPI Group (UK) Ltd, Croydon, CR40 4YY

CONTENTS

Foreword

This British Standard is published under the direction of the British Standards Institution and the Institution of Engineering and Technology (IET).

Following a full review, this Standard replaced the 17th Edition of the IEE Wiring Regulations BS 7671:2008 as amended. Copyright is held jointly by the IET and BSI.

Technical authority for this Standard is vested in the Joint IET/BSI Technical Committee JPEL/64. This Joint Technical Committee, which is responsible for the work previously undertaken by the IEE Wiring Regulations Committee and the BSI Technical Committee PEL/64, meets the constitutional and operational requirements of both parent bodies. JPEL/64 has the responsibility for the content of this British Standard under the joint authority of the IET and the BSI Standards Board.

All references in this text to the IET Wiring Regulations or the Regulation(s), where not otherwise specifically identified, shall be taken to refer to BS 7671:2018 Requirements for Electrical Installations, as amended by Amendment 1:2020 and Amendment 2:2022.

Additions or alterations to regulations owing to the issue of BS 7671:2018:A2:2022 are indicated by a side bar in the margin.

Introduction to BS 7671:2018

BS 7671:2018 *Requirements for Electrical Installations* was issued on 1st July 2018 and is intended to come into effect on 1st January 2019. Installations designed after 31st December 2018 are to comply with BS 7671:2018.

The Regulations apply to the design, erection and verification of electrical installations, also additions and alterations to existing installations. Existing installations that have been installed in accordance with earlier editions of the Regulations may not comply with this edition in every respect. This does not necessarily mean that they are unsafe for continued use or require upgrading.

A summary of the main changes is given below.

NOTE: This is not an exhaustive list.

Part 1 Scope, object and fundamental principles

Regulation 133.1.3 (Selection of equipment) has been modified and now requires certain usage of equipment to be recorded on the appropriate electrical certification specified in Part 6.

Part 2 Definitions

Definitions have been expanded and modified.

Chapter 41 Protection against electric shock

Section 411 contains a number of significant changes. Some of the main ones are mentioned below.

Metallic pipes entering the building having an insulating section at their point of entry need not be connected to the protective equipotential bonding (Regulation 411.3.1.2).

The maximum disconnection times stated in Table 41.1 now apply to final circuits rated up to 63 A with one or more socket-outlets and final circuits rated up to 32 A supplying only fixed connected current-using equipment (Regulation 411.3.2.2).

Regulation 411.3.3 has been revised and now applies to socket-outlets with a rated current not exceeding 32 A. There is an exception to omit RCD protection where, other than for a dwelling, a documented risk assessment determines that RCD protection is not necessary.

A new Regulation 411.3.4 requires that, within domestic (household) premises, additional protection by an RCD with a rated residual operating current not exceeding 30 mA shall be provided for AC final circuits supplying luminaires.

Regulation 411.4.3 has an addition that states that no switching or isolating device shall be inserted in a PEN conductor.

Regulations 411.4.4 and 411.4.5 have been redrafted.

The regulations concerning IT systems (411.6) have been reorganized. Regulations 411.6.3.1 and 411.6.3.2 have been deleted and 411.6.4 redrafted and a new Regulation 411.6.5 inserted.

A new Regulation group (419) has been inserted where automatic disconnection according to Regulation 411.3.2 is not feasible, such as electronic equipment with limited short-circuit current.

Chapter 42 Protection against thermal effects

A new Regulation 421.1.7 has been introduced recommending the installation of arc fault detection devices (AFDDs) to mitigate the risk of fire in AC final circuits of a fixed installation due to the effects of arc fault currents.

Regulation 422.2.1 has been redrafted. Reference to conditions BD2, BD3 and BD4 has been deleted. A note has been added stating that cables need to satisfy the requirements of the CPR in respect of their reaction to fire and making reference to Appendix 2, item 17. Requirements have also been included for cables that are supplying safety circuits.

Chapter 44 Protection against voltage disturbances and electromagnetic disturbances

Section 443, which deals with protection against overvoltages of atmospheric origin or due to switching, has been redrafted.

The AQ criteria (conditions of external influence for lightning) for determining if protection against transient overvoltages is needed are no longer included. Instead, protection against transient overvoltages has to be provided where the consequence caused by overvoltage (see Regulation 443.4):

(i) could result in serious injury to, or loss of, human life, or

(ii) could result in interruption of public services/or damage to and cultural heritage, or

(iii) could result in interruption of commercial or industrial activity, or

(iv) could affect a large number of co-located individuals.

For all other cases, a risk assessment has to be performed in order to determine if protection against transient overvoltage is required.

There is an exception not to provide protection for single dwelling units in certain situations.

Chapter 46 Isolation and switching

A new Chapter 46 has been introduced. This deals with non-automatic local and remote isolation and switching measures for the prevention or removal of dangers associated with electrical installations or electrically powered equipment. Also, switching for the control of circuits or equipment. Where electrically powered equipment is within the scope of BS EN 60204, only the requirements of that standard apply.

Chapter 52 Selection and erection of wiring systems

Regulation 521.10.202, which gives requirements for the methods of support of wiring systems, has replaced Regulation 521.11.201. This is a significant change. Regulation 521.10.202 requires cables to be adequately supported against their premature collapse in the event of a fire and applies throughout the installation, not just in escape routes as previously.

Regulation 527.1.3 has also been modified, and a note added stating that cables also need to satisfy the requirements of the CPR in respect of their reaction to fire.

Chapter 53 Protection, isolation, switching, control and monitoring

This chapter has been completely revised and deals with general requirements for protection, isolation, switching, control and monitoring and with the requirements for selection and erection of the devices provided to fulfil such functions.

Section 534 Devices for protection against overvoltage

This section focuses mainly on the requirements for the selection and erection of SPDs for protection against transient overvoltages where required by Section 443, the BS EN 62305 series, or as otherwise stated. Section 534 has been completely revised. The most significant technical change refers to the selection requirements for the voltage protection level.

Chapter 54 Earthing arrangements and protective conductors

Two new regulations (542.2.3 and 542.2.8) have been introduced concerning earth electrodes.

Chapter 55 Other equipment

Regulation 550.1 introduces a new scope.

New Regulation 559.10 refers to ground-recessed luminaires, the selection and erection of which shall take account of the guidance given in Table A.1 of BS EN 60598-2-13.

Part 6 Inspection and testing

Part 6 has been completely restructured, including the regulation numbering to align with the CENELEC standard. Chapters 61, 62 and 63 have been deleted and their content now forms two new Chapters 64 and 65.

Section 704 Construction and demolition site installations

This section contains a number of small changes, including requirements for external influences (Regulation 704.512.2), and a modification to Regulation 704.410.3.6 to include the non-use of the protective measure of electrical separation.

Section 708 Electrical installations in caravan/camping parks and similar locations

This section contains a number of changes including requirements for socket-outlets, RCD protection, and operational conditions and external influences.

Section 710 Medical locations

This section contains a number of small changes including the removal of Table 710, and changes to Regulations 710.415.2.1 to 710.415.2.3 concerning equipotential bonding. In addition, a new Regulation 710.421.1.201 states requirements regarding the installation of AFDDs.

Section 715 Extra-low voltage lighting installations

This section contains only minor changes including modifications to Regulation 715.524.201.

Section 721 Electrical installations in caravans and motor caravans

This section contains a number of changes to the requirements concerning electrical separation, RCDs, proximity to non-electrical services and protective bonding conductors.

Section 722 Electric vehicle charging installations

This section contains significant changes to Regulation 722.411.4.1 concerning the use of a PME supply. The exception concerning reasonably practicable has been deleted. Changes have also been made to requirements for external influences, RCDs, socket-outlets and connectors.

Section 730 Onshore units of electrical shore connections for inland navigation vessels

This is an entirely new section. Section 730 applies to onshore installations dedicated to the supply of inland navigation vessels for commercial and administrative purposes, berthed in ports and berths. Most, if not all, of the measures used to reduce the risks in marinas apply equally to electrical shore connections for inland navigation vessels. One of the major differences between supplies to vessels in a typical marina and electrical shore connections for inland navigation vessels is the size of the supply needed.

Section 753 Heating cables and embedded heating systems

This retitled section has been completely revised. The scope of Section 753 has been extended to apply to embedded electric heating systems for surface heating. The requirements also apply to electric heating systems for de-icing, frost prevention and similar applications, and cover both indoor and outdoor systems. Heating systems for industrial and commercial applications complying with BS EN 60519, BS EN 62395 and BS EN 60079 are not covered.

Appendices

The following main changes have been made within the appendices:

Appendix 1 British Standards to which reference is made in the Regulations has been updated, as necessary.

Appendix 3 Time/current characteristics of overcurrent protective devices and RCDs

The previous content of Appendix 14 concerning earth fault loop impedance has been moved into Appendix 3.

Appendix 6 Model forms for certification and reporting

This appendix includes minor changes to the certificates, changes to the inspections (for new installation work only) for domestic and similar premises with up to 100 A supply, and examples of items requiring inspection for an electrical installation condition report.

Appendix 8 Current-carrying capacity and voltage drop for busbar trunking and powertrack systems

This appendix includes changes regarding rating factors for current-carrying capacity.

Appendix 14 Determination of prospective fault current

As stated earlier, the content of Appendix 14 concerning earth fault loop impedance has been moved into Appendix 3. Appendix 14 now contains information on the determination of prospective fault current.

Appendix 17 Energy efficiency

This is a new appendix that provides recommendations for the design and erection of electrical installations, including installations having local production and storage of energy, for optimizing the overall efficient use of electricity.

The recommendations within the scope of this appendix apply for new electrical installations and modification of existing electrical installations. Much of this appendix will not apply to domestic and similar installations.

It is intended that this appendix will be developed into Part 8 of BS 7671 in a future amendment.

Introduction to Amendment 1:2020

Amendment 1:2020 to BS 7671:2018 *Requirements for Electrical Installations* was issued on 1st February 2020 and may be implemented immediately. Electrical installations falling within the scope of Section 722, the erection of which is commenced after 31st July 2020, are to comply with BS 7671:2018 incorporating Amendment 1:2020.

Amendment 1:2020 amends Section 722 Electric Vehicle Charging Installations.

Part 2

Definitions relating to Section 722 have been included.

Section 700 General

This overarching section explains the need for Part 7.

Section 722

Electric vehicle charging installations, Section 722 amended, based on HD 60364-7-722:2018.

Section 722 applies to electric vehicle charging installations. It modifies the general requirements for protection against electric shock, and includes specific requirements with regard to PME systems, socket-outlets and connectors, external influences, isolation and switching and RCD protection.

Regulation 722.411.4.1 adds indent (iv) which adds an alternative solution requirement for charging installations.

The requirements of this section do not apply to wireless charging, such as inductive charging.

Appendix 1

References to British Standards for Section 722 have been included.

Index

Entries for Section 722 have been included.

Introduction to Amendment 2:2022

BS 7671:2018+A2:2022 *Requirements for Electrical Installations* was issued on 28th March 2022 and may be implemented immediately. BS 7671:2018+A1:2020 remains current and will be withdrawn on 27th September 2022.

The Regulations apply to the design, erection and verification of electrical installations, also additions and alterations to existing installations. Existing installations that have been installed in accordance with earlier editions of the Regulations may not comply with this edition in every respect. This does not necessarily mean that they are unsafe for continued use or require upgrading.

Contractual and legal considerations

This publication does not purport to include all the necessary provisions of a contract. Users are responsible for its correct application. Compliance with a British Standard cannot confer immunity from legal obligations.

NOTE 1: Completion of an electrical installation designed to the withdrawn standard can be subject to the contractual agreement between all parties involved.

NOTE 2: BSI publishes information on the uses of withdrawn British Standards.

A summary of the main changes is given below.

NOTE 3: This is not an exhaustive list.

Part 1 Scope, Object and Fundamental principles

Minor changes have been made including reference to prosumers electrical installations and installing fibre optic cables.

Part 2 Definitions

Definitions have been expanded and modified.

Chapter 41 Protection against electric shock

Section 411 contains a number of significant changes. Some of the main changes are mentioned below.

Regulation 411.3.1.2 has been redrafted. The regulation now requires that in each consumer's installation within a building extraneous-conductive-parts that are liable to introduce a dangerous potential difference be connected to the main earthing terminal.

A note has been added clarifying that where non-metallic pipes, e.g. plastic, enter the building and are then connected to metallic pipes within the building, the metallic pipes within the building do not normally require protective bonding as they are unlikely to be extraneous-conductive-parts. Reference to insulating sections has been removed.

Regulation 411.3.3 has been redrafted and now has three indents. There is still an exception to omit RCD protection where, other than BA1, BA3 or children (BA2, BA3) and mobile equipment (not exceeding 32 A) for use outdoors, a suitably documented risk assessment determines that RCD protection is not necessary.

Regulation 411.4.2 now recommends an additional connection to Earth, by means of an earth electrode in accordance with Chapter 54, is made to the main earthing terminal.

Chapter 42 Protection against thermal effects

Regulation 421.1.7 has been redrafted. It is now a requirement to protect final circuits supplying socket-outlets with a rated current not exceeding 32 A using arc fault detection devices (AFDD) in Higher Risk Residential Buildings, Houses in Multiple Occupation, Purpose-built student accommodation and Care homes. For all other premises, the regulation recommends AFDDs for single-phase AC final circuits supplying socket-outlets not exceeding 32 A. The use of AFDDs does not obviate the need to apply one or more measures provided in other clauses in this standard.

Regulation 422.1 (precautions where particular risks of fire exist) now introduces requirements for the fire safety design of the building(s) to be documented. Also requirements have been added where cables should have an improved fire performance.

Regulation 422.2 now introduces separate requirements for escape routes and a new concept of 'protected escape routes'. Reference to conditions BD2, BD3 and BD4 have been deleted.

Chapter 44 Protection against voltage disturbances and electromagnetic disturbances

Regulation 443.4 for determining if protection against transient overvoltages is needed has been redrafted. Regulation 443.4.1 now requires protection against transient overvoltages to be provided where the consequence caused by the overvoltage could result in:

(i) serious injury to, or loss of, human life

(ii) failure of a safety service, as defined in Part 2

(iii) significant financial or data loss.

For all other cases, protection against transient overvoltages shall be provided unless the owner of the installation declares it is not required due to any loss or damage being tolerable and they accept the risk of damage to equipment and any consequential loss.

Regulation 443.4.2 requires protection against overvoltages to be considered in the case of equipment likely to produce switching overvoltages or disturbances and gives conditions.

Regulation 443.5 (Risk assessment method) has been deleted and Annex A443 (examples of calculated risk level CRL for use of SPDs) has also been deleted.

Section 514 Identification and notices

Section 514 contains a number of significant changes (including illustrations of notices which have been removed and examples are provided in Appendix 11). Some of the main changes are mentioned below.

Table 51 has been modified slightly to take into account changes of colour identification for DC and abbreviations have been added.

Regulation 514.9.1 now contains an exception for domestic (household) premises.

Regulation 514.9.2 has been introduced to advise that all diagrams, charts, and information or instruction notices comply with the applicable standards specified.

Regulation 514.12 (Notices: periodic inspection and testing) now contains an exception for domestic (household) premises in certain situations.

Regulation 514.12.2, concerning the RCD notice, has been modified and now contains an exception for domestic (household) premises in certain situations.

Regulation 514.14 concerning non-standard colours has been deleted.

A new Regulation 514.16.1 has been introduced requiring a label to indicate the presence of SPDs. However, there is an exception for domestic (household) premises or similar.

Chapter 53 Protection, isolation, switching, control and monitoring

Regulation 531.3.2 indent (ii) highlights the use of RCBOs for individual final circuits in residential premises to minimise unwanted tripping.

Regulation 531.3.3 now states that RCD Type AC shall only be used to serve fixed equipment, where it is known that the load current contains no DC components.

Chapter 54 Earthing arrangements and protective conductors

Modifications have been made to Regulation Group 543.7 concerning earthing requirements for the installation of equipment having high protective conductor currents.

Regulation 544.1.1 concerning main protective bonding conductors has also been revised.

Chapter 55 Other equipment

551.7 Additional requirements have been added for installations where the generating set or sets may operate in parallel with other sources including systems for distribution of electricity to the public.

551.7.2 includes additional conditions.

554.4 Heating conductors and cables Regulations 554.4.2, 554.4.3, and 554.4.4 have been deleted.

Chapter 64 Initial verification

Regulation Group 643.3 has been redrafted. The requirements for testing insulation resistance where equipment is likely to influence the verification test or be damaged has been clarified and reference is made to a 250 V DC test following the connection of equipment.

The requirements for RCD testing have been changed and Table 3A (Time/current performance criteria for RCDs) in Appendix 3 has been deleted. Regardless of RCD Type, an alternating current test at rated residual operating current ($I_{\Delta n}$) is used to verify the effectiveness.

Section 702 Swimming pools and other basins

Regulation, 702.414.3, concerning SELV and PELV, has been introduced.

Section 703 Rooms and cabins containing sauna heaters

Regulation 703.414.3, concerning SELV and PELV, has been introduced.

Section 704 Construction and demolition site installations

Regulation 704.511.1 includes changes to plugs and socket-outlet requirements. Changes have also been made to requirements for wiring systems (704.52) and the requirements for devices for isolation have been redrafted (704.537).

Section 706 Conducting locations with restricted movement

The requirements for protection against electric shock have been completely redrafted. A new Regulation 706.415, requirements for additional protection, has also been introduced.

Section 710 Medical locations

Changes include requirements concerning socket-outlet circuits in the medical IT systems and minor changes to requirements concerning luminaires and lighting installations. Also, requirements concerning diagrams and documentation have been changed. Requirements for IT systems have also been redrafted. Changes also include requirements for fault protection and requirements for protective equipotential bonding.

Section 711 Exhibitions, shows and stands

A new Regulation 711.410.3.101 has been introduced. It sets out the requirements for all final circuits, other than those of safety services, having a rated current not exceeding 32 A supplying socket-outlets or handheld electrical equipment, and all final circuits for lighting to use either automatic disconnection of supply (with additional protection by means of an ≤30 mA RCD), SELV or PELV, or electrical separation. Regulation 711.42 concerning protection against thermal effects has been deleted. Requirements for isolation and switching have been redrafted. In addition, the regulations covering luminaires and lighting installations have been redrafted.

Section 712 Solar photovoltaic (PV) power supply systems

The technical content of this section has been extensively revised and expanded. Requirements of this section now apply to:

(i) PV installations not connected to a system for distribution of electricity to the public,
(ii) PV installations in parallel with a system for distribution of electricity to the public,
(iii) PV installations as an alternative to a system for distribution of electricity to the public.

Section 714 Outdoor lighting installations

This section includes changes to automatic disconnection of supply and additional protection.

Section 717 Mobile or transportable units

Regulation 717.411.4 has been expanded to take account of PME.

Section 722 Electric vehicle charging installations

This section incorporates BS 7671:2018+A1:2020 together with other changes, such as, taking account of prosumer's electrical installations (722.826.3.201). Indent (i) of Regulation 722.411.4.1 has been deleted, also changes have been made to the notes in particular the one concerning where equipment to be used is not covered by a British or Harmonized Standard. In addition, the Annex to 722 has been redrafted and includes guidance on indent (iii) of Regulation 722.411.4.1.

Section 753 Heating cables and embedded heating systems

Changes include the addition of new regulations (relocated from Chapter 53) covering impact and the installation of heating cables.

Chapter 82 Prosumer's low-voltage electrical installations

This is an entirely new chapter. Historically, utility companies have managed the public transmission and distribution network from the point of view of having central production adapted to demand variation. The objective of this chapter is to provide requirements, such that, low-voltage electrical installations are compatible with the current and future ways to deliver safely the electrical energy to current-using equipment either from the public network or from other local sources.

Chapter 82 provides additional requirements, measures and recommendations for design, erection and verification of all types of low voltage electrical installations. This includes local production and/or storage of energy in order to achieve compatibility with existing and future ways to deliver electrical energy to current-using equipment or to the public network by means of local sources. Such electrical installations are designated as Prosumer's Electrical Installations (PEIs).

Appendices

The following main changes have been made within the Appendices:

Appendix 3 Table 3A, covering the tripping times of RCDs, has been deleted.

Appendix 4 includes minor changes concerning a cable surrounded by thermal insulation.

Appendix 6 Model forms for certification and reporting

Minor changes have been made to the model forms and guidance for recipients. These include the addition of fields for recording the details of SPDs and AFDDs.

The schedule of inspections has been simplified for initial verification. A new example checklist of items requiring inspection during initial verification has been added to Appendix 6 but is not required to be provided with the certificate.

The single page generic schedule of test results used for EIC and EICR has been redrafted. There is now a separate page for the schedule of circuit details and a separate page for the schedule of test results.

Appendix 7 Deleted by BS 7671:2018+A2:2022

Appendix 11 Warning and user instruction labels

This is a new appendix that provides guidance for the types of safety signs, warning signs and instructions required to be applied to electrical installations.

Appendix 13 Escape routes and fire protection

The previous content of Appendix 13 concerning methods for measuring the insulation resistance/impedance of floors and walls to Earth has been removed and replaced by guidance on escape routes and fire protection. The new content of Appendix 13 provides detailed guidance on protected escape routes in order to afford a safe passage through the building to an exit.

Appendix 17 Energy efficiency

This appendix includes changes to the scope to cover dwellings and reference to the Building Regulations, changes to user decisions, also new design recommendations, new information on determination of meshes, new information on user requirements, new guidance covering impacts on the design of an installation, and new guidance on periodic assessment.

Editions

The following editions have been published:

FIRST EDITION	Entitled 'Rules and Regulations for the Prevention of Fire Risks Arising from Electric Lighting'. Issued in 1882.
SECOND EDITION	Issued in 1888.
THIRD EDITION	Entitled 'General Rules recommended for Wiring for the Supply of Electrical Energy'. Issued in 1897.
FOURTH EDITION	Issued in 1903.
FIFTH EDITION	Entitled 'Wiring Rules'. Issued in 1907.
SIXTH EDITION	Issued in 1911.
SEVENTH EDITION	Issued in 1916.
EIGHTH EDITION	Entitled 'Regulations for the Electrical Equipment of Buildings'. Issued in 1924.
NINTH EDITION	Issued in 1927.
TENTH EDITION	Issued in 1934.
ELEVENTH EDITION	Issued in 1939. Revised, issued in 1943. Reprinted with minor Amendments, 1945. Supplement issued, 1946. Revised Section 8 issued, 1948.
TWELFTH EDITION	Issued in 1950. Supplement issued, 1954.
THIRTEENTH EDITION	Issued in 1955. Reprinted 1958, 1961, 1962 and 1964.
FOURTEENTH EDITION	Issued in 1966. Reprinted incorporating Amendments, 1968. Reprinted incorporating Amendments, 1969. Supplement on use in metric terms issued, 1969. Amendments issued, 1970. Reprinted in metric units incorporating Amendments, 1970. Reprinted 1972. Reprinted 1973. Amendments issued, 1974. Reprinted incorporating Amendments, 1974. Amendments issued, 1976. Reprinted incorporating Amendments, 1976.
FIFTEENTH EDITION	Entitled 'Regulations for Electrical Installations'. Issued in 1981. (Red Cover) Amendments issued, 1 January 1983. Reprinted incorporating Amendments, 1983. (Green Cover) Amendments issued, 1 May 1984. Reprinted incorporating Amendments, 1984. (Yellow Cover) Amendments issued, 1 January 1985. Amendments issued, 1 January 1986. Reprinted incorporating Amendments, 1986. (Blue Cover) Amendments issued, 12 June 1987. Reprinted incorporating Amendments, 1987. (Brown Cover) Reprinted with minor corrections, 1988. (Brown Cover)
SIXTEENTH EDITION	Issued in 1991. (Red Cover) Reprinted with minor corrections, 1992. (Red Cover) Entitled and reprinted as 'Requirements for Electrical Installations BS 7671:1992'. (Red Cover) Amendment No 1 issued, December 1994. Reprinted incorporating Amendment No 1, 1994. (Green Cover) Amendment No 2 issued, December 1997. Reprinted incorporating Amendment No 2, 1997. (Yellow Cover) Amendment No 3 issued, April 2000. BS 7671:2001 issued, June 2001. (Blue Cover) Amendment No 1 issued, February 2002. Amendment No 2 issued, March 2004. Reprinted incorporating Amendments 1 and 2, 2004. (Brown Cover)
SEVENTEENTH EDITION	BS 7671:2008 issued, January 2008. (Red Cover) Reprinted incorporating Amendment No 1, 2011. (Green Cover) Amendment 2 issued, Aug 2013. Reprinted incorporating Amendments 2 and 3, 2015. (Yellow Cover)
EIGHTEENTH EDITION	BS 7671:2018 issued, July 2018. (Blue Cover) Amendment 1 issued, February 2020 (electronic) Amendment 2 issued, March 2022. (Brown Cover)

Joint IET/BSI Technical Committee JPEL/64 CONSTITUTION

as at August 2021

Eur Ing J T Bradley BSc CEng FIET (Chairman)

Eur Ing P Sayer CEng FIET GCGI (Deputy Chairman)

for IET

Eur Ing G Kenyon BEng(Hons) CEng MIET TechIOSH (Chair – JPEL/D/64)

H R Lovegrove IEng FIET

G G Willard DipEE CEng FIET JP (Deputy Chair – JPEL/D/64)

G Digilio IEng FIEE ACIBSE MSLL

G Gundry MIET

J Peckham IEng FIET LCGI

S Robinson CEng FIET FIMechE FCIBSE FSLL

and

K Hoare MSc CEng MCIBSE MIET (ACE - Association for Consultancy and Engineering)

S A MacConnacher BSc CEng MIEE MInstR (Association of Manufacturers of Domestic Appliances)

A Greer (Association of Manufacturers of Power Generating Systems)

M Froggatt (British Approvals Service for Cables)

Dr J Hodge CChem CSci FRSC (British Cables Association)

Eur Ing J Wright BEng CEng MIET (BEAMA Installation Ltd)

P Norris MIET (BEAMA Ltd)

G Brown BEng (Hons) IEng MIET (BEAMA Ltd)

C Smith (BEIS – Department for Business, Energy and Industrial Strategy)

J Devaney (BSI)

D Moore (BSI)

A Samad Khan MEng (Hons) CEng MIET MIEEE (Individual Capacity; UK Expert - CLC/TC64/WG 4; liaison GEL 81, PEL 37/1)

T Benstead CEng MSc BEd MCIBSE MIET (Chair – JPEL/B/64)

J Elliott BSc(Hons) PGCert CEng MIET (Certsure LLP; Chair – JPEL/A/64)

N Bird IEng MIET (Chair – JPEL/C/64)

D Cowburn BSc(Hons) (Competent Persons Forum)

Dr A C M Sung BSc(Hons) MSc PhD CEng FIET FCIBSE SrMIEEE MEI MHKIE(Ir) (CIBSE and EI)

K Siriwardhana BScEng (Hons), MSc, MBA, CEng, FIET, SMIEEE, MCIGRE (CIGRE-UK)

S Ogborn FIET MSET LCGI (City & Guilds of London Institute)

D Irwin (DC Users Forum)

B Dunn CEng BEng(Hons) (MoD)

K Sparrow IEng MIET CertEd (EAL)

S Khan BSc (Hons) MIET (ECA)

G Parker BEng (Hons) IEng MIET MCIBSE MEI LCGI (ECA)

R Cairney IEng MIET (Electrical Contractors' Association of Scotland t/a SELECT)

D Forrester IEng MIET MInstLM (Electrical Contractors' Association of Scotland t/a SELECT)

M Allen BEng(Hons) CEng MIET (Electrical Safety First)

D Spillett MIET MIOSH MIIRSM (Energy Networks Association and Liaison - GEL/8)

R Wardak MIET (The GAMBICA Association Ltd)

M Hadley (The GAMBICA Association Ltd)

S Mason BSc CEng FIET (Health and Safety Executive)

I Andrews (Heating Ventilating and Air Conditioning Manufacturers Association)

Eur Ing P Harris BEng(Hons) FIHEEM MIEE MCIBSE (Institute of Healthcare Engineering and Estate Management)

P Davies (Intertek-Group Plc)

G C T Pritchard BTech(Hons) CEng FILP MIET TechIOSH (Institution of Lighting Professionals and HEA)

Eur Ing L D Markwell MSc BSc (Hons) CEng MIET MCIBSE LCGI (Liaison – FSH/12)

M Cotterell (Liaison – GEL/82)

T Oldershaw (Liaison – TCT/7)

Eur Ing M Mullins BA CEng FIET (Liaison – JPEL/18)

M Peace CEng MIET MCIBSE (Liaison – PEL/69)

M Coles BEng(Hons) MIET (Liaison – GEL/82)

L C Barling (The Lighting Industry Association)

F Bertie MIET (NAPIT; Deputy Chair – JPEL/A/64)

I Storrar BSc (Hons) CEng FCIBSE FIHEEM MIET (NHS Scotland)

J Eade BEng(Hons) CEng MIET AMIMechE (Professional Lighting and Sound Association)

G Brindle BSc(Hons) CEng MIET CMILT (Railway Industry Association)

M Kenyon MIET (Safety Assessment Federation – SAFed)

Dr T Nicklin (Society of Motor Manufacturers and Traders Ltd)

N Lowe (Society of Motor Manufacturers and Traders Ltd)

C J Tanswell CEng MIEE (Society for Public Architecture, Construction, Engineering and Surveying – SPACES)

P Grinnall (SPATA Swimming Pool and Allied Trades Association Ltd)

Committee Manager

G D Cronshaw CEng FIET

Preface

BS 7671:2018, Requirements for Electrical Installations, takes account of the technical substance of agreements reached at CENELEC level in Harmonized Documents (HD). Where no CENELEC HD exists, IEC level documents are taken into account. In particular, the technical intent of the following CENELEC HDs and IEC documents is included:

HD or IEC	Year	Document reference	BS 7671
HD 60364-1	2008 (2017)	Fundamental principles, assessment of general ..., definitions	Parts 1, 2, 3
HD 60364-4-41	2007 (2017)	Protection against electric shock	Chapter 41
HD 60364-4-42	2011 (2015)	Protection against thermal effects	Chapter 42
HD 60364-4-43	2010	Protection against overcurrent	Chapter 43
IEC 60364-4-44	2007 (2015)	Protection against voltage disturbances and electromagnetic disturbances	Section 440
HD 60364-4-442	2012	Protection of low-voltage installations against temporary overvoltages due to earth faults in the high-voltage system and due to faults in the low voltage system	Section 442
HD 60364-4-443	2016	Protection against transient overvoltages of atmospheric origin or due to switching	Section 443
HD 60364-4-444	2010 (2012)	Protection against voltage and electromagnetic disturbances	Section 444
HD 60364-4-46	2016 (2017)	Isolation and switching	Chapter 46
HD 60364-5-51	2009 (2017)	Selection and erection of electrical equipment - Common rules	Chapter 51
HD 60364-5-52	2011 (2017)	Selection and erection of electrical equipment - Wiring systems	Chapter 52
HD 60364-5-53	2015 (2017)	Selection and erection of electrical equipment - Switchgear and controlgear	Chapter 53
HD 60364-5-534	2016	Devices for protection against transient overvoltages	Section 534
HD 60364-5-54	2011 (2017)	Earthing arrangements and protective conductors	Chapter 54
HD 60364-5-551	2010 (2016)	Low-voltage generating sets	Section 551
HD 60364-5-557	2013 (2016)	Auxiliary circuits	Section 557
HD 60364-5-559	2012 (2017)	Luminaires and lighting installations	Section 559
HD 60364-6-56	2018	Safety services	Chapter 56
HD 60364-6	2016 (2017)	Verification	Part 6
HD 60364-7-701	2007 (2017)	Locations containing a bath or shower	Section 701
HD 60364-7-702	2010	Swimming pools and fountains	Section 702
HD 60364-7-703	2005	Rooms and cabins containing sauna heaters	Section 703
HD 60364-7-704	2018	Construction and demolition site installations	Section 704
HD 60364-7-705	2007 (2017)	Agricultural and horticultural premises	Section 705
HD 60364-7-706	2007 (2020)	Conducting locations with restricted movement	Section 706
HD 60364-7-708	2017	Caravan parks, camping parks and similar locations	Section 708
HD 60364-7-709	2009 (2019)	Marinas and similar locations	Section 709
HD 60364-7-710	2012 (2013)	Medical locations	Section 710
HD 60364-7-711	2019	Exhibitions, shows and stands	Section 711
HD 60364-7-712	2016	Photovoltaic (PV) systems	Section 712
HD 60364-7-714	2012	External lighting installations	Section 714
HD 60364-7-715	2012 (2017)	Extra-low-voltage lighting installations	Section 715
HD 60364-7-717	2010 (2014)	Mobile or transportable units	Section 717
IEC 60364-7-721	2019	Electrical installations in caravans and motor caravans	Section 721
HD 60364-7-722	2018	Supplies for electric vehicles	Section 722
HD 60364-7-729	2009 (2017)	Operating or maintenance gangways	Section 729
HD 60364-7-730	2015	Onshore units of electrical shore connections for inland navigation vessels	Section 730
HD 60364-7-740	2006 (2017)	Temporary electrical installations for structures, amusement devices and booths at fairgrounds, amusement parks and circuses	Section 740
HD 60364-7-753	2014 (2014)	Heating cables and embedded heating systems	Section 753
HD 60364-8-1	2019 (2019)	Functional aspects - energy efficiency	Appx 17
HD 60364-8-2	2018	Prosumer's low-voltage electrical installations	Chapter 82

The dates in brackets refer to the year of issue of amendments to the documents.

Where the above documents contain UK special national conditions, those conditions have been incorporated within BS 7671. If BS 7671 is applied in other countries the above documents should be consulted to confirm the status of a particular regulation.

BS 7671 will continue to be amended from time to time to take account of the publication of new or amended CENELEC and IEC standards. The opportunity has been taken to revise regulations that experience has shown require clarification or to allow for new technology and methods.

Reference is made throughout BS 7671 to publications of the British Standards Institution, both specifications and codes of practice. Appendix 1 lists these publications and gives their full titles whereas throughout BS 7671 they are referred to only by their numbers.

Where reference is made in BS 7671 to a British Standard which takes account of a CENELEC Harmonization Document, European Norm (EN) or IEC standard, it is understood that the reference also relates to any European national standard similarly derived from the CENELEC standard, although account needs to be taken of any national exemptions.

Note by the Health and Safety Executive

The Health and Safety Executive (HSE) welcomes the publication of BS 7671:2018, Requirements for Electrical Installations, IET Wiring Regulations 18th Edition and its updating with the second amendment, published in 2022. BS 7671 and the IET/IEE Wiring Regulations have been extensively referred to in HSE guidance over the years. Installations which conform to the standards laid down in BS 7671:2018+A2:2022 are regarded by HSE as likely to achieve conformity with the relevant parts of the Electricity at Work Regulations 1989. Installations to which BS 7671 is relevant may have been designed and installed in accordance with an earlier edition, now superseded but then current. That, in itself, would not mean that the installation would fail to comply with the Electricity at Work Regulations, 1989.

Notes on the plan of the 18th Edition

This Edition is based on the plan agreed internationally for the arrangement of safety rules for electrical installations.

The regulation numbering follows the pattern and corresponding references of IEC HD 60364 series. The numbering does not, therefore, necessarily follow sequentially. The numbering system used in Part 7 is explained in Section 700.

In the numbering system used, the first digit signifies a Part, the second digit a Chapter, the third digit a Section and the subsequent digits the Regulation number. For example, Section number 413 is made up as follows:

PART 4 - PROTECTION FOR SAFETY

Chapter 41 (first chapter of Part 4) - PROTECTION AGAINST ELECTRIC SHOCK

Section 413 (third section of Chapter 41) - PROTECTIVE MEASURE: ELECTRICAL SEPARATION

Part 1 sets out the scope, object and fundamental principles.

Part 2 defines the sense in which certain terms are used throughout the Regulations, and provides a list of symbols used and a list of abbreviations used in the Standard.

The subjects of the subsequent parts are as indicated below:

Part	Subject
Part 3	Identification of the characteristics of the installation that will need to be taken into account in choosing and applying the requirements of the subsequent Parts. These characteristics may vary from one part of an installation to another and should be assessed for each location to be served by the installation.
Part 4	Description of the measures that are available for the protection of persons, livestock and property, and against the hazards that may arise from the use of electricity.
Part 5	Precautions to be taken in the selection and erection of the equipment of the installation.
Part 6	Inspection and testing.
Part 7	Special installations or locations - particular requirements.
Part 8	Functional requirements

The sequence of the plan should be followed in considering the application of any particular requirement of the Regulations. The general index provides a ready reference to particular regulations by subject, but in applying any one regulation the requirements of related regulations should be borne in mind. Cross-references are provided, and the index is arranged to facilitate this.

In many cases, a group of associated regulations is covered by a side heading which is identified by a two-part number, e.g. 544.2. Throughout the Regulations where reference is made to such a two-part number, that reference is to be taken to include all the individual regulation numbers which are covered by that side heading and include that two-part number.

The Regulations use the IEC decimal point numbering system to make it easier to embody future changes and additions resulting from ongoing international standards work within IEC and CENELEC. In order to identify and accommodate future IEC changes some regulations carry either a 100 or 200 number where 100 numbers represent CENELEC Harmonization Document reference numbers and 200 numbers represent UK-only regulations. Some regulations have not been 'updated' to indicate 100 and 200 numbers; that will only be done when those regulations require a significant rewrite.

Guidance on the language used within BS 7671

To help all users of BS 7671, the following definitions apply specifically to the language used within this Standard:

Guidance. Element containing informative text.

Informative annex. Annex that contains additional recommendations, guidance or statements intended to assist the understanding or use of the document.

Informative element:

- **Preliminary informative element.** Element that identifies the document, introduces its content and explains its background, its development and its relationship with other documents.
- **Supplementary informative element.** Element that provides additional information intended to assist the understanding or use of the document.

Normative annex. Annex that contains requirements or procedural instructions additional to those in the body of the document.

Normative element. Element that describes the scope of the document or sets out requirements.

Note. Element containing informative text.

Requirement. Expression in the content of a document conveying criteria to be fulfilled if compliance with the document is to be claimed and from which no deviation is permitted.

Recommendation. Expression in the content of a document conveying that among several possibilities one is recommended as particularly suitable, without mentioning or excluding others.

The following table describes verbal forms which may be used to express normative and informative elements:

Verbal forms used in BS 7671

Implication	Verbal form	Typical context
Requirement	Shall	Normative element
Recommendation	Should	Informative element
Permission	May	Informative element
Possibility and capability	Can	Informative element
Possibility	Might	Informative element
Description	Is	Informative element

The content of this table is based on *Rules for the structure and drafting of UK standards,* second edition published April 2017, published by BSI Standards Ltd.

PART 1

SCOPE, OBJECT AND FUNDAMENTAL PRINCIPLES

CONTENTS

PART 1

SCOPE, OBJECT AND FUNDAMENTAL PRINCIPLES

CHAPTER 11

SCOPE

110.1 GENERAL

110.1.1 The Regulations apply to the design, erection and verification of electrical installations such as those of:

(i) residential premises
(ii) commercial premises
(iii) public premises
(iv) industrial premises
(v) prefabricated buildings
(vi) low voltage generating sets
(vii) highway equipment and street furniture
(viii) locations containing a bath or shower
(ix) swimming pools and other basins
(x) rooms and cabins containing sauna heaters
(xi) construction and demolition sites
(xii) agricultural and horticultural premises
(xiii) conducting locations with restricted movement
(xiv) caravan / camping parks and similar locations
(xv) marinas and similar locations
(xvi) medical locations
(xvii) exhibitions, shows and stands
(xviii) solar photovoltaic (PV) power supply systems
(xix) outdoor lighting
(xx) extra-low voltage lighting
(xxi) mobile or transportable units
(xxii) caravans and motor caravans
(xxiii) electric vehicle charging
(xxiv) operating and maintenance gangways
(xxv) temporary installations for structures, amusement devices and booths at fairgrounds, amusement parks and circuses including professional stage and broadcast applications
(xxvi) floor and ceiling heating systems
(xxvii) onshore units of electrical shore connections for inland navigation vessels.

NOTE: 'Premises' covers the land and all facilities including buildings belonging to it.

110.1.2 The Regulations include requirements for:

(i) circuits supplied at nominal voltages up to and including 1000 V AC or 1500 V DC. For AC, the preferred frequencies which are taken into account in BS 7671 are 50 Hz, 60 Hz and 400 Hz. The use of other frequencies for special purposes is not excluded
(ii) circuits, other than the internal wiring of equipment, operating at voltages exceeding 1000 V and derived from an installation having a voltage not exceeding 1000 V AC, e.g. discharge lighting, electrostatic precipitators
(iii) wiring systems and cables not specifically covered by the standards for appliances
(iv) prosumer's low voltage electrical installations (PEI) including those located external to buildings
(v) all consumer installations external to buildings
(vi) fixed wiring for information and communication technology, signalling, control and the like, including installation and support of fibre optic cables (excluding internal wiring of equipment)
(vii) additions and alterations to installations and also parts of the existing installation affected by an addition or alteration.

110.1.3 The Regulations are intended to be applied to electrical installations generally but, in certain cases, they may need to be supplemented by the requirements or recommendations of other British or Harmonized Standards or by the requirements of the person ordering the work.

Such cases include the following:

(i) Electric signs and high voltage luminous discharge tube installations - BS 559 and BS EN 50107
(ii) Emergency lighting - BS 5266 and BS EN 1838
(iii) Explosive atmospheres - BS EN 60079 series
(iv) Electrical apparatus for use in the presence of combustible dust - BS EN 50281 and BS EN 60079
(v) Fire detection and fire alarm systems for buildings - BS 5839
(vi) Telecommunications systems - BS 6701
(vii) Electric surface heating systems - BS EN 60335-2-96
(viii) Electrical installations for open-cast mines and quarries - BS 6907
(ix) Temporary electrical systems for events, entertainment and related purposes - BS 7909
(x) Life safety and firefighting applications - BS 8519 and BS 9999.

110.2 EXCLUSIONS FROM SCOPE

The Regulations do not apply to the following installations:

(i) Systems for the distribution of electricity to the public other than prosumer's installations covered by Chapter 82
(ii) Railway traction equipment, rolling stock and signalling equipment
(iii) Equipment of motor vehicles, except those to which the requirements of the Regulations concerning caravans or mobile units are applicable
(iv) Equipment on board ships covered by BS 8450, BS EN 60092-507, BS EN ISO 13297 or BS EN ISO 10133
(v) Equipment of mobile and fixed offshore installations
(vi) Equipment of aircraft
(vii) Those aspects of mines specifically covered by Statutory Regulations
(viii) Radio interference suppression equipment, except so far as it affects safety of the electrical installation
(ix) Lightning protection systems for buildings and structures covered by BS EN 62305
(x) Those aspects of lift installations covered by relevant parts of BS 5655 and BS EN 81 and those aspects of escalator or moving walk installations covered by relevant parts of BS 5656 and BS EN 115
(xi) Electrical equipment of machines covered by BS EN 60204
(xii) Electric fences covered by BS EN 60335-2-76
(xiii) The DC side of cathodic protection systems complying with the relevant part(s) of BS EN ISO 12696, BS EN 12954, BS EN ISO 13174, BS EN 13636 and BS EN 14505.

111 NOT USED

112 NOT USED

113 EQUIPMENT

113.1 The Regulations apply to items of electrical equipment only so far as selection and application of the equipment in the installation are concerned. The Regulations do not deal with requirements for the construction of assemblies of electrical equipment, which are required to comply with appropriate standards.

114 RELATIONSHIP WITH STATUTORY REGULATIONS

114.1 The Regulations are non-statutory. They may, however, be used in a court of law in evidence to claim compliance with a statutory requirement. The relevant statutory provisions are listed in Appendix 2 and include Acts of Parliament and Regulations made thereunder. In some cases statutory Regulations may be accompanied by Codes of Practice approved under Section 16 of the Health and Safety at Work etc. Act 1974. The legal status of these Codes is explained in Section 17 of the 1974 Act.

For a supply given in accordance with the Electricity Safety, Quality and Continuity Regulations, it shall be deemed that the connection with Earth of the neutral of the supply is permanent. Outside England, Scotland, Wales and Northern Ireland, confirmation shall be sought from the distributor that the supply conforms to requirements corresponding to those of the Electricity Safety, Quality and Continuity Regulations (ESQCR), in this respect. Where the ESQCR does not apply, equipment for isolation and switching shall be selected accordingly as specified in Chapters 46 and 53.

115 INSTALLATIONS IN PREMISES SUBJECT TO LICENSING

115.1 For installations in premises over which a licensing or other authority exercises a statutory control, the requirements of that authority shall be ascertained and complied with in the design and execution of the installation.

CHAPTER 12

OBJECT AND EFFECTS

120 GENERAL

120.1 BS 7671 contains requirements for the design, erection and verification of electrical installations so as to provide for safety and proper functioning for the intended use.

120.2 Chapter 13 states the fundamental principles. It does not include detailed technical requirements, which may be subject to modification because of technical developments.

120.3 BS 7671 sets out technical requirements to enable electrical installations to conform with the fundamental principles of Chapter 13, as follows:

Part 3	Assessment of general characteristics
Part 4	Protection for safety
Part 5	Selection and erection of equipment
Part 6	Inspection and testing
Part 7	Special installations or locations
Part 8	Functional requirements.

Any intended departure from these Parts requires special consideration by the designer of the installation and shall be recorded on the appropriate electrical certification specified in Part 6. The resulting degree of safety of the installation shall be not less than that obtained by compliance with the Regulations.

CHAPTER 13

FUNDAMENTAL PRINCIPLES

131 PROTECTION FOR SAFETY

131.1 General

The requirements of this chapter are intended to provide for the safety of persons, livestock and property against dangers and damage which may arise in the reasonable use of electrical installations. The requirements to provide for the safety of livestock are applicable in locations intended for them.

In electrical installations, risk of injury may result from:

(i) shock currents

(ii) excessive temperatures likely to cause burns, fires and other injurious effects

(iii) ignition of a potentially explosive atmosphere

(iv) undervoltages, overvoltages and electromagnetic disturbances likely to cause or result in injury or damage

(v) mechanical movement of electrically actuated equipment, in so far as such injury is intended to be prevented by electrical emergency switching or by electrical switching for mechanical maintenance of non-electrical parts of such equipment

(vi) power supply interruptions and/or interruption of safety services

(vii) arcing or burning, likely to cause blinding effects, excessive pressure and/or toxic gases.

131.2 Protection against electric shock

131.2.1 Basic protection

Persons and livestock shall be protected against dangers that may arise from contact with live parts of the installation.

This protection can be achieved by one of the following methods:

(i) Preventing a current from passing through the body of any person or any livestock

(ii) Limiting the current which can pass through a body to an on-hazardous value.

131.2.2 Fault protection

Persons and livestock shall be protected against dangers that may arise from contact with exposed-conductive-parts during a fault.

This protection can be achieved by one of the following methods:

(i) Preventing a current resulting from a fault from passing through the body of any person or any livestock

(ii) Limiting the magnitude of a current resulting from a fault, which can pass through a body, to a non-hazardous value

(iii) Limiting the duration of a current resulting from a fault,which can pass through a body, to a non-hazardous time period.

In connection with fault protection, the application of the method of protective equipotential bonding is one of the important principles for safety.

131.3 Protection against thermal effects

131.3.1 The electrical installation shall be so arranged that the risk of ignition of flammable materials due to high temperature or electric arc is minimized. In addition, during normal operation of the electrical equipment, there shall be minimal risk of burns to persons or livestock.

131.3.2 Persons, livestock, fixed equipment and fixed materials adjacent to electrical equipment shall be protected against harmful effects of heat or thermal radiation emitted by electrical equipment, and in particular the following:

(i) Combustion, ignition, or degradation of materials

(ii) Risk of burns

(iii) Impairment of the safe function of installed equipment.

Electrical equipment shall not present a fire hazard to adjacent materials.

131.4 Protection against overcurrent

Persons and livestock shall be protected against injury, and property shall be protected against damage, due to excessive temperatures or electromechanical stresses caused by any overcurrents likely to arise in live conductors.

NOTE: Protection can be achieved by limiting the overcurrent to a safe value and/or duration.

131.5 Protection against fault current

Conductors other than live conductors, and any other parts intended to carry a fault current, shall be capable of carrying that current without attaining an excessive temperature. Electrical equipment, including conductors, shall be provided with mechanical protection against electromechanical stresses of fault currents as necessary to prevent injury or damage to persons, livestock or property.

131.6 Protection against voltage disturbances and measures against electromagnetic disturbances

131.6.1 Persons and livestock shall be protected against injury, and property shall be protected against any harmful effects, as a consequence of a fault between live parts of circuits supplied at different voltages, in accordance with Section 442.

131.6.2 Persons and livestock shall be protected against injury, and property shall be protected against damage, as a consequence of overvoltages such as those originating from atmospheric events or from switching, in accordance with Section 443.

NOTE: For protection against direct lightning strikes on the structure or nearby lightning strokes, refer to the BS EN 62305 series.

131.6.3 Persons and livestock shall be protected against injury, and property shall be protected against damage, as a consequence of undervoltage and any subsequent voltage recovery, in accordance with Section 445.

131.6.4 The installation shall have an adequate level of immunity against electromagnetic disturbances so as to function correctly in the specified environment, in accordance with Section 444. The installation design shall take into consideration the anticipated electromagnetic emissions, generated by the installation or the installed equipment, which shall be suitable for the current-using equipment used with, or connected to, the installation.

131.7 Protection against power supply interruption

Where danger or damage is expected to arise due to an interruption of supply, suitable provisions shall be made in the installation or installed equipment.

132 DESIGN

132.1 General

The electrical installation shall be designed by one or more skilled persons to provide for:

(i) the protection of persons, livestock and property in accordance with Section 131

(ii) the proper functioning of the electrical installation for the intended use.

The information required as a basis for design is stated in Regulations 132.2 to 132.5. The requirements with which the design shall comply are stated in Regulations 132.6 to 132.16.

132.2 Characteristics of available supply or supplies

Information on the characteristics of the available supply or supplies shall be determined by calculation, measurement, enquiry or inspection.

The following characteristics shall be included in the documentation referred to in Regulation 132.13 to show conformity with the Regulations:

(i) Nature of current: AC and/or DC

(ii) Purpose and number of conductors:

- for AC

  line conductor(s)

  neutral conductor

  protective conductor

  PEN conductor

- for DC

  conductors equivalent to those listed above (outer/middle/earthed live conductors, protective conductor, PEN conductor)

(iii) Values and tolerances:

- nominal voltage and voltage tolerances
- nominal frequency and frequency tolerances
- maximum current allowable
- prospective fault current
- external earth fault loop impedance

(iv) Protective measures inherent in the supply, e.g. earthed neutral or mid-wire

(v) Particular requirements of the distributor.

NOTE: If the distributor changes the characteristics of the power supply this may affect the safety of the installation.

132.3 Nature of demand

The number and type of circuits required for lighting, heating, power, control, signalling, communication and information technology, etc. shall be determined from knowledge of:

(i) location of points of power demand

(ii) loads to be expected on the various circuits

(iii) daily and yearly variation of demand

(iv) any special conditions, such as harmonics

(v) requirements for control, signalling, information and communications technology

(vi) anticipated future demand, if specified.

132.4 Electrical supply systems for safety services or standby electrical supply systems

Where a supply for safety services or standby electrical supply systems is specified the following shall be determined:

(i) Characteristics of the supply
(ii) Circuits to be supplied by the safety source.

132.5 Environmental conditions

132.5.1 The design of the electrical installation shall take into account the environmental conditions to which it will be subjected.

132.5.2 Equipment in surroundings susceptible to risk of fire or explosion shall be so constructed or protected, and such other special precautions shall be taken, as to prevent danger.

132.6 Cross-sectional area of conductors

The cross-sectional area of conductors shall be determined for both normal operating conditions and, where appropriate, for fault conditions according to:

(i) the admissible maximum temperature
(ii) the admissible voltage drop
(iii) the electromechanical stresses likely to occur due to short-circuit and earth fault currents
(iv) other mechanical stresses to which the conductors are likely to be exposed
(v) the maximum impedance for correct operation of short-circuit and earth fault protection
(vi) the method of installation
(vii) harmonics
(viii) thermal insulation.

132.7 Type of wiring and method of installation

The choice of the type of wiring system and the method of installation shall include consideration of the following:

(i) The nature of the location
(ii) The nature of the structure supporting the wiring
(iii) Accessibility of wiring to persons and livestock
(iv) Voltage
(v) The electromechanical stresses likely to occur due to short-circuit and earth fault currents
(vi) Electromagnetic interference
(vii) Other external influences (e.g. mechanical, thermal and those associated with fire) to which the wiring is likely to be exposed during the erection of the electrical installation or in service.

132.8 Protective equipment

The characteristics of protective equipment shall be determined with respect to their function, including protection against the effects of:

(i) overload
(ii) fault current
(iii) overvoltage
(iv) undervoltage and/or loss of voltage. |

The protective devices shall operate at values of current, voltage and time which are suitably related to the characteristics of the circuits and to the possibilities of danger.

132.9 Emergency control

An interrupting device shall be installed in such a way that it can be easily recognized and effectively and rapidly operated where, in the case of danger, there is a necessity for immediate interruption of the supply.

132.10 Disconnecting devices

Disconnecting devices shall be provided so as to permit switching and/or isolation of the electrical installation, circuits or individual items of equipment as required for operation, inspection, testing, fault detection, maintenance and repair.

132.11 Prevention of mutual detrimental influence

The electrical installation shall be arranged in such a way that no mutual detrimental influence will occur between electrical installations and non-electrical installations.

Electromagnetic interference shall be taken into account.

132.12 Accessibility of electrical equipment

Electrical equipment shall be arranged to provide:

(i) sufficient space for the initial installation and later replacement of individual items of electrical equipment

(ii) accessibility for operation, inspection, testing, fault detection, maintenance and repair.

132.13 Documentation for the electrical installation

Every electrical installation shall be provided with appropriate documentation, including that required by Regulations 313.1, 422.1, 514.9, 536.5, Part 6 and, where applicable, Part 7 and Part 8.

132.14 Protective devices and switches

132.14.1 A single-pole fuse, switch or circuit-breaker shall be inserted in the line conductor only.

132.14.2 No switch or circuit-breaker, except where linked, or fuse, shall be inserted in an earthed neutral conductor. Any linked switch or linked circuit-breaker inserted in an earthed neutral conductor shall be arranged to break all the related line conductors.

132.15 Isolation and switching

132.15.201 Effective means, suitably placed for ready operation, shall be provided so that all voltage may be cut off from every installation, from every circuit thereof and from all equipment, as may be necessary to prevent or remove danger.

132.15.202 Every fixed electric motor shall be provided with an efficient means of switching off, readily accessible, easily operated and so placed as to prevent danger.

132.16 Additions and alterations to an installation

No addition or alteration, temporary or permanent, shall be made to an existing installation, unless it has been ascertained that the rating and the condition of any existing equipment, including that of the distributor, will be adequate for the altered circumstances. Furthermore, the earthing and bonding arrangements, if necessary for the protective measure applied for the safety of the addition or alteration, shall be adequate.

133 SELECTION OF ELECTRICAL EQUIPMENT

133.1 General

133.1.1 Every item of equipment shall comply with the appropriate British or Harmonized Standard. In the absence of such a standard, reference shall be made to the appropriate International (IEC) standard or the appropriate standard of another country.

133.1.2 Where there are no applicable standards, the item of equipment concerned shall be selected by special agreement between the person specifying the installation and the installer.

133.1.3 Where equipment to be used is not in accordance with Regulation 133.1.1 or is used outside the scope of its standard, the designer or other person responsible for specifying the installation shall confirm that the equipment provides at least the same degree of safety as that afforded by compliance with the Regulations. Such use shall be recorded on the appropriate electrical certification specified in Part 6.

133.2 Characteristics

Every item of electrical equipment selected shall have suitable characteristics appropriate to the values and conditions on which the design of the electrical installation (see Section 132) is based and shall, in particular, fulfil the requirements of Regulations 133.2.1 to 4.

133.2.1 Voltage

Electrical equipment shall be suitable with respect to the maximum steady-state voltage (rms value for AC) likely to be applied, as well as overvoltages likely to occur.

NOTE: For certain equipment, it may also be necessary to take account of the lowest voltage likely to occur.

133.2.2 Current

Electrical equipment shall be selected with respect to the maximum steady current (rms value for AC) which it has to carry in normal service and with respect to the current likely to be carried in abnormal conditions and the period (e.g. operating time of protective devices, if any) during which it may be expected to flow.

133.2.3 Frequency

Equipment shall be suitable for the frequencies likely to occur in the circuit.

133.2.4 Power

Electrical equipment which is selected on the basis of its power characteristics shall be suitable for the duty demanded of the equipment, taking into account the load factor and the normal service conditions.

133.3 Conditions of installation

Electrical equipment shall be selected so as to withstand safely the stresses, the environmental conditions (see Regulation 132.5) and the characteristics of its location. An item of equipment which does not by design have the properties corresponding to its location may be used where adequate further protection is provided as part of the completed electrical installation.

133.4 Prevention of harmful effects

All electrical equipment shall be selected so that it will not cause harmful effects on other equipment or impair the supply during normal service, including switching operations.

NOTE: Examples of characteristics which are likely to have harmful effects are given in Chapter 33.

133.5 New materials and inventions

Where the use of a new material or invention leads to departures from the Regulations, the resulting degree of safety of the installation shall be not less than that obtained by compliance with the Regulations. Such use shall be recorded on the appropriate electrical certification specified in Part 6.

134 ERECTION AND INITIAL VERIFICATION OF ELECTRICAL INSTALLATIONS

134.1 Erection

134.1.1 Good workmanship by one or more skilled or instructed persons and proper materials shall be used in the erection of the electrical installation. The installation of electrical equipment shall take account of manufacturers' instructions.

134.1.2 The characteristics of the electrical equipment, as determined in accordance with Section 133, shall not be impaired by the process of erection.

134.1.3 Conductors shall be identified in accordance with Section 514. Where identification of terminals is necessary, they shall be identified in accordance with Section 514.

134.1.4 Every electrical joint and connection shall be of proper construction as regards conductance, insulation, mechanical strength and protection.

134.1.5 Electrical equipment shall be installed in such a manner that the design temperatures are not exceeded.

134.1.6 Electrical equipment likely to cause high temperatures or electric arcs shall be placed or guarded so as to minimize the risk of ignition of flammable materials.

Where the temperature of an exposed part of electrical equipment is likely to cause injury to persons or livestock that part shall be so located or guarded as to prevent accidental contact therewith.

134.1.7 Where necessary for safety purposes, suitable warning signs and/or notices shall be provided.

134.2 Initial verification

134.2.1 During erection and on completion of an installation or an addition or alteration to an installation, and before it is put into service, appropriate inspection and testing shall be carried out by one or more skilled persons competent to verify that the requirements of BS 7671 have been met.

Appropriate certification shall be issued in accordance with Chapter 64.

134.2.2 The designer of the installation shall make a recommendation for the interval to the first periodic inspection and test as detailed in Chapter 64.

NOTE: The requirements of Chapter 34 (Maintainability) should be taken into consideration.

135 PERIODIC INSPECTION AND TESTING

135.1 It is recommended that every electrical installation is subjected to periodic inspection and testing, in accordance with Chapter 65.

PART 2

DEFINITIONS

For the purposes of BS 7671, the following definitions shall apply. As far as practicable the definitions align with IEC 60050 - International Electrotechnical Vocabulary.

NOTE: Where a section number is listed, e.g. {444}, the definition only applies within that section.

8/20 Current impulse, {534}. A current impulse with a virtual front time of 8 μs and a time to half-value of 20 μs where:

(i) the front time is defined as $1.25(t_{90} - t_{10})$, where t_{90} and t_{10} are the 90 % and 10 % points on the leading edge of the waveform

(ii) the time to half-value is defined as the time between the virtual origin and the 50 % point on the tail. The virtual origin is the point where a straight line drawn through the 90 % and 10 % points on the leading edge of the waveform intersects the I = 0 line.

Accessory. A device, other than current-using equipment, associated with such equipment or with the wiring of an installation.

Agricultural and horticultural premises. Rooms, locations or areas where:

- livestock are kept, or
- feed, fertilizers, vegetable and animal products are produced, stored, prepared or processed, or
- plants are grown, such as greenhouses.

Ambient temperature. The temperature of the air or other medium where the equipment is to be used.

Amusement device. Ride, stand, textile or membrane building, side stall, side show, tent, booth or grandstand intended for the entertainment of the public.

Appliance. An item of current-using equipment other than a luminaire or an independent motor.

Arc fault detection device (AFDD). Device intended to mitigate the effects of arcing faults by disconnecting the circuit when an arc fault is detected.

NOTE: Such devices include:

(i) a single device having opening means able to open the protected circuit in specified conditions, or

(ii) a single device with an integrated protective device, or

(iii) a separate unit assembled on site with a declared protective device.

The integrated protective device (ii) or the declared protective device (iii) for assembly on site is either a circuit-breaker in accordance with BS EN 60898-1 or an RCD in accordance with BS EN 61008-1, BS EN 61009-1 or BS EN 62423.

Arm's reach. A zone of accessibility to touch, extending from any point on a surface where persons usually stand or move about to the limits which a person can reach with a hand in any direction without assistance. (See Figure 417.)

Arrangements for livestock keeping. Buildings and rooms (housing for animals), cages, runs or other containers used for continuous accommodation of livestock.

Auxiliary circuit. Circuit for transmission of signals intended for control, detection, supervision or measurement of the functional status of a main circuit.

Back-up protection. Protection of a device by overcurrent coordination between that device and an overcurrent protective device (OCPD) in series with it, generally but not necessarily on the supply side. This prevents any excessive stress on the electrical device.

NOTE: In BS 7671, back-up protection is not the same as combined short-circuit protection.

Barrier. A part providing a defined degree of protection against contact with live parts from any usual direction of access.

Basic insulation. Insulation applied to live parts to provide basic protection and which does not necessarily include insulation used exclusively for functional purposes.

Basic protection. Protection against electric shock under fault-free conditions.

NOTE: For low voltage installations, systems and equipment, basic protection generally corresponds to protection against 'contact of persons or livestock with live parts'.

Basin of fountain. A basin not intended to be occupied by persons and which cannot be accessed (reached by persons) without the use of ladders or similar means. For basins of fountains which may be occupied by persons, the requirements for swimming pools apply.

Bonding conductor. A protective conductor providing equipotential bonding.

Bonding network (BN), {444}. A set of interconnected conductive parts that provide a path for currents at frequencies from direct current (DC) to radio frequency (RF) intended to divert, block or impede the passage of electromagnetic energy.

Bonding ring conductor (BRC), {444}. A bus earthing conductor in the form of a closed ring.

NOTE: Normally the bonding ring conductor, as part of the bonding network, has multiple connections to the common bonding network (CBN) that improves its performance.

Booth. Non-stationary unit, intended to accommodate equipment generally for pleasure or demonstration purposes.

Building void, accessible. A space within the structure or the components of a building accessible only at certain points. Such voids include the space within partitions, suspended floors, ceilings and certain types of window frame, door frame and architrave.

Building void, non-accessible. A space within the structure or the components of a building which has no ready means of access.

Bunched. Cables are said to be bunched when two or more are contained within a single conduit, duct, ducting, or trunking or, if not enclosed, are not separated from each other by a specified distance.

Busbar trunking system. A type-tested assembly, in the form of an enclosed conductor system comprising solid conductors separated by insulating material. The assembly may consist of units such as:

- busbar trunking units, with or without tap-off facilities
- tap-off units where applicable
- phase-transposition, expansion, building-movement, flexible, end-feeder and adaptor units.

NOTE: Other system components may include tap-off units.

Bypass bonding conductor, {444}. Bonding conductor connected in parallel with the screens of cables.

Cable channel. An enclosure situated above or in the ground, ventilated or closed, and having dimensions which do not permit the access of persons but allow access to the conduits and/or cables throughout their length during and after installation. A cable channel may or may not form part of the building construction.

Cable cleat. A component of a support system, which consists of elements spaced at intervals along the length of the cable or conduit and which mechanically retains the cable or conduit.

Cable coupler. A means of enabling the connection or disconnection, at will, of two flexible cables. It consists of a connector and a plug.

Cable ducting. An enclosure of metal or insulating material, other than conduit or cable trunking, intended for the protection of cables which are drawn in after erection of the ducting.

Cable ladder. A cable support consisting of a series of transverse supporting elements rigidly fixed to main longitudinal supporting members.

Cable tray. A cable support consisting of a continuous base with raised edges and no covering. A cable tray may or may not be perforated.

Cable trunking. A closed enclosure normally of rectangular cross-section, of which one side is removable or hinged, used for the protection of cables and for the accommodation of other electrical equipment.

Cable tunnel. A corridor containing supporting structures for cables and joints and/or other elements of wiring systems and whose dimensions allow persons to pass freely throughout the entire length.

Caravan. A trailer leisure accommodation vehicle, used for touring, designed to meet the requirements for the construction and use of road vehicles (see also definitions of Motor caravan and Leisure accommodation vehicle).

Caravan park / camping park. Area of land that contains two or more caravan pitches and/or tents.

Caravan pitch. Plot of ground intended to be occupied by a leisure accommodation vehicle.

Caravan pitch electrical supply equipment. Equipment that provides means of connecting and disconnecting supply cables from leisure accommodation vehicles or tents with a mains electrical supply.

Cartridge fuse link. A device comprising a fuse element or two or more fuse elements connected in parallel enclosed in a cartridge usually filled with arc-extinguishing medium and connected to terminations (see fuse link).

Central power supply system. A system supplying the required emergency power to essential safety equipment.

Central power supply system (low power output). Central power supply system with a limitation of the power output of the system at 500 W for 3 h or 1 500 W for 1 h.

NOTE: A low power supply system normally comprises a maintenance-free battery and a charging and testing unit.

Circuit. An assembly of electrical equipment supplied from the same origin and protected against overcurrent by the same protective device(s).

Circuit-breaker. A device capable of making, carrying and breaking normal load currents and also making and automatically breaking, under predetermined conditions, abnormal currents such as short-circuit currents. It is usually required to operate infrequently although some types are suitable for frequent operation.

Circuit-breaker, Instantaneous trip (ICB). Circuit-breaker which only fulfils the short-circuit portion of overcurrent protection.

Circuit-breaker, linked. A circuit-breaker the contacts of which are so arranged as to make or break all poles simultaneously or in a definite sequence.

Circuit protective conductor (cpc). A protective conductor connecting exposed-conductive-parts of equipment to the main earthing terminal.

Class I equipment. Equipment in which protection against electric shock does not rely on basic insulation only, but which includes means for the connection of exposed-conductive-parts to a protective conductor in the fixed wiring of the installation (see BS EN 61140).

Class II equipment. Equipment in which protection against electric shock does not rely on basic insulation only, but in which additional safety precautions such as supplementary insulation are provided, there being no provision for the connection of exposed metalwork of the equipment to a protective conductor, and no reliance upon precautions to be taken in the fixed wiring of the installation (see BS EN 61140).

Class III equipment. Equipment in which protection against electric shock relies on supply at SELV and in which voltages higher than those of SELV are not generated (see BS EN 61140).

Cold tail. The interface between the fixed installation and a heating unit.

Combined short-circuit current capability. Maximum short-circuit current which can be handled by two short-circuit protective devices in series.

Combined short-circuit protection. Overcurrent coordination, in short-circuit conditions, of two OCPDs in series, resulting in a combined short-circuit current capability higher than one OCPD alone.

Common equipotential bonding system, common bonding network (CBN), {444}. Equipotential bonding system providing both protective equipotential bonding and functional equipotential bonding.

Complementary floor heating. Direct heating system integrated into the floor construction, for example, in the border zones close to outer walls, which complements the heat dissipation of a thermal storage floor heating system.

Conditional short-circuit current. Prospective current that a circuit or a switching device, protected by a specified short-circuit protective device, can satisfactorily withstand for the total operating time of that device under specified conditions of use and behaviour.

Conducting location with restricted movement. A location comprised mainly of metallic or conductive surrounding parts, within which it is likely that a person will come into contact through a substantial portion of their body with the conductive surrounding parts and where the possibility of preventing this contact is limited.

Conduit. A part of a closed wiring system for cables in electrical installations, allowing them to be drawn in and/or replaced, but not inserted laterally.

Connected mode. Operating mode which needs connection to the public network (direct feeding mode and reverse feeding mode).

Connector. The part of a cable coupler or of an appliance coupler which is provided with female contacts and is intended to be attached to the end of the flexible cable remote from the supply.

Consumer (of electricity). Entity or party which uses electricity for its own needs.

Consumer unit (may also be known as a consumer control unit or electricity control unit). A particular type of distribution board intended for operation by ordinary persons, comprising a type-tested coordinated assembly for the control and distribution of electrical energy, incorporating manual means of double-pole isolation on the incoming circuit(s) and an assembly of one or more protective devices, signalling and other devices proven during the type-test of the assembly as suitable for such use.

Continuity of service. The extent to which the operation of an electrical system approaches the intended state of freedom from supply interruption.

Continuous operating voltage (U_c), {534}. Maximum rms voltage which may be continuously applied to an SPD's mode of protection. This is equal to the rated voltage.

Control and protective switching device (CPS device). Switching device (or equipment) capable of operation other than by hand, but with or without local manual operating means. A CPS device provides both functions of contactor and OCPD.

Controlgear (see Switchgear).

Conventional impulse withstand voltage. The peak value of an impulse test voltage at which insulation does not show any disruptive discharge when subjected to a specified number of applications of impulses of this value, under specified conditions.

Coordination of electrical equipment. Correct way of selecting electrical devices in series to provide safety and continuity of service of the installation, taking into account short-circuit protection and/or overload protection and/or selectivity.

Current-carrying capacity of a conductor. The maximum current which can be carried by a conductor under specified conditions without its steady-state temperature exceeding a specified value.

Current-using equipment. Equipment which converts electrical energy into another form of energy, such as light, heat or motive power.

Danger. Risk of injury to persons (and livestock where expected to be present) from:

(i) fire, electric shock, burns, arcing and explosion arising from the use of electrical energy, and

(ii) mechanical movement of electrically controlled equipment, in so far as such danger is intended to be prevented by electrical emergency switching or by electrical switching for mechanical maintenance of non-electrical parts of such equipment.

DC system - see Appendix 9.

Departure. Deliberate decision not to comply fully with the requirements of BS 7671, for which the designer must declare that the resultant degree of safety is not less than that achievable by full compliance.

Design current (of a circuit). The magnitude of the current (rms value for AC) to be carried by the circuit in normal service.

Device for connecting a luminaire (DCL). System comprising an outlet and a connector providing a fixed luminaire with electrical connection to and disconnection from a fixed installation but not providing mechanical support for a luminaire.

Direct feeding mode. Operating mode in which the public network supplies the PEI *(see Reverse feeding mode)*.

NOTE: Local storage units can supply current-using equipment or be charged by local power supplies and/or the public distribution network.

Direct heating system. Heating system which generates heat from electrical energy and dissipates it to the room to be heated with a response time being as low as possible.

Direct lightning strike. Lightning strike impacting a component of the network directly.

NOTE 1: Examples of components of the network are: conductor, tower, substation equipment, etc.

NOTE 2: Consequences of a direct lightning strike include the thermal, magnetic and mechanical effects of the discharge current as well as the surges resulting from its voltage drop along the discharge path.

Disconnector. A mechanical switching device which, in the open position, complies with the requirements specified for the isolating function.

NOTE 1: A disconnector is otherwise known as an isolator.

NOTE 2: A disconnector is capable of opening and closing a circuit when either a negligible current is broken or made, or when no significant change in the voltage across the terminals of each pole of the disconnector occurs. It is also capable of carrying currents under normal circuit conditions and carrying for a specified time current under abnormal conditions such as those of short-circuit.

Discrimination (see Selectivity).

Distribution board. An assembly containing switching or protective devices (e.g. fuses, circuit-breakers, residual current operated devices) associated with one or more outgoing circuits fed from one or more incoming circuits, together with terminals for the neutral and circuit protective conductors. It may also include signalling and other control devices. Means of isolation may be included in the board or may be provided separately.

Distribution circuit. A circuit supplying a distribution board or switchgear.

A distribution circuit may also connect the origin of an installation to an outlying building or separate installation, when it is sometimes called a sub-main.

Distribution System Operator (DSO). Party operating a distribution system.

Distributor. A party that distributes electricity to consumers using electrical lines and equipment that they own or operate.

Double insulation. Insulation comprising both basic insulation and supplementary insulation.

Duct, Ducting (see Cable ducting).

Earth. The conductive mass of the Earth, whose electric potential at any point is conventionally taken as zero.

Earth electrode. Conductive part, which may be embedded in the soil or in a specific conductive medium, e.g. concrete or coke, in electrical contact with the Earth.

Earth electrode network, {444}. Part of an earthing arrangement comprising only the earth electrodes and their interconnections.

Earth electrode resistance. The resistance of an earth electrode to Earth.

Earth fault current. A current resulting from a fault of negligible impedance between a line conductor and an exposed-conductive-part or a protective conductor.

Earth fault loop impedance. The impedance of the earth fault current loop starting and ending at the point of earth fault. This impedance is denoted by the symbol Z_s.

The earth fault loop comprises the following, starting at the point of fault:

- the circuit protective conductor, and
- the consumer's earthing terminal and earthing conductor, and
- for TN systems, the metallic return path, and
- for TT and IT systems, the Earth return path, and
- the path through the earthed neutral point of the transformer, and
- the transformer winding, and
- the line conductor from the transformer to the point of fault.

Earth leakage current *(see Protective conductor current).*

Earthed concentric wiring. A wiring system in which one or more insulated conductors are completely surrounded throughout their length by a conductor, for example a metallic sheath, which acts as a PEN conductor.

Earthing. Connection of the exposed-conductive-parts of an installation to the main earthing terminal of that installation.

Earthing conductor. A protective conductor connecting the main earthing terminal of an installation to an earth electrode or to other means of earthing.

Efficiency measures (EM). Level of implementation of measures to improve energy efficiency of an electrical installation.

Electric shock. A dangerous physiological effect resulting from the passing of an electric current through a human body or livestock.

Electric vehicle (EV). Any vehicle propelled by an electric motor drawing current from a rechargeable storage battery or from other portable energy storage devices (rechargeable, using energy from a source off the vehicle such as a residential or public electricity service), which is manufactured primarily for use on public streets, roads or highways.

- **Charging equipment.** An assembly including one or more charging points.
- **Electric vehicle charging point.** The point where the electric vehicle is connected to the fixed installation.

  NOTE: The charging point is a socket-outlet where the charging cable belongs to the vehicle, or a connector, where the charging cable is a fixed part of the electric vehicle supply equipment.

- **Mode 1 charging.** Connection of the EV to the AC supply network utilizing standardized socket-outlets not exceeding 16 A and not exceeding 250 V AC single-phase or 480 V AC three-phase, at the supply side, and utilizing the power and protective earth conductors (according to BS EN 61851-1).
- **Mode 2 charging.** Connection of the EV to the AC supply network utilizing standardized socket-outlets not exceeding 32 A and not exceeding 250 V AC single-phase or 480 V AC three-phase, at the supply side, and utilizing the power and protective earth conductors together with a control pilot function and system of personnel protection against electric shock (RCD) between the EV and the plug or as part of the in-cable control box.
- **Mode 3 charging.** Connection of the EV to the AC supply network utilizing dedicated electric vehicle supply equipment where the control pilot function extends to control equipment in the electric vehicle supply equipment, permanently connected to the AC supply network.
- **Mode 4 charging.** Connection of the EV to the AC supply network utilizing an off-board charger where the control pilot function extends to equipment permanently connected to the AC supply.
- **Vehicle connector.** Part of a vehicle coupler integral with, or intended to be attached to, the flexible cable connected to the AC supply network (mains).
- **Vehicle coupler.** Means of enabling the manual connection of a flexible cable to an EV for the purpose of charging.

  NOTE: A vehicle coupler consists of two parts: a vehicle connector and a vehicle inlet.

Electrical circuit for safety services. Electrical circuit intended to be used as part of an electrical supply system for safety services.

Electrical Energy Management System (EEMS). System comprising different equipment and devices in the installation for the purpose of energy management.

NOTE: This equipment can be stand-alone or integrated in other larger equipment such as a home and building electronic system.

Electrical equipment (abbr: ***Equipment*****).** Any item for such purposes as generation, conversion, transmission, distribution or utilization of electrical energy, such as machines, transformers, apparatus, measuring instruments, protective devices, wiring systems, accessories, appliances and luminaires.

Electrical installation (abbr: ***Installation*****).** An assembly of associated electrical equipment having coordinated characteristics to fulfil specific purposes.

Electrical source for safety services. Electrical source intended to be used as part of an electrical supply system for safety services.

Electrical supply system for safety services. A supply system intended to maintain the operation of essential parts of an electrical installation and equipment:

(i) for the health and safety of persons and livestock, and

(ii) to avoid damage to the environment and to other equipment.

NOTE: The supply system includes the source and the circuit(s) up to the terminals of the electrical equipment.

Electrically independent earth electrodes. Earth electrodes located at such a distance from one another that the maximum current likely to flow through one of them does not significantly affect the potential of the other(s).

Electrode boiler (or electrode water heater). Equipment for the electrical heating of water or electrolyte by the passage of an electric current between electrodes immersed in the water or electrolyte.

Electronic convertor (static convertor). A convertor having no moving parts and notably using semiconductor rectifiers.

Emergency stopping. Emergency switching intended to stop an operation.

Emergency switching. An operation intended to remove, as quickly as possible, danger, which may have occurred unexpectedly.

Emergency switching off. Opening operation of a switching device intended to remove electric power from an electrical installation to avert or alleviate a hazardous situation.

Enclosure. A part providing protection of equipment against certain external influences and in any direction providing basic protection.

Equipment (see Electrical equipment).

Equipotential bonding. Electrical connection maintaining various exposed-conductive-parts and extraneous-conductive-parts at substantially the same potential. (See also *Protective equipotential bonding.*)

Escape route. Path to follow for access to a safe area in the event of an emergency. *(See also Protected escape route)*

Exhibition. Event intended for the purpose of displaying and/or selling products etc., which can take place in any suitable location, either a room, building or temporary structure.

Exposed-conductive-part. Conductive part of equipment which can be touched and which is not normally live, but which can become live under fault conditions.

External influence. Any influence external to an electrical installation which affects the design and safe operation of that installation.

Extra-low voltage (see Voltage, nominal).

Extraneous-conductive-part. A conductive part liable to introduce a potential, generally Earth potential, and not forming part of the electrical installation.

Fairground. Area where one or more stands, amusement devices or booths are erected for leisure use.

Fault. A circuit condition in which current flows through an abnormal or unintended path. This may result from an insulation failure or a bridging of insulation.

NOTE: Conventionally, the impedance between live conductors or between live conductors and exposed- or extraneous-conductive-parts at the fault position is considered negligible.

Fault current. A current resulting from a fault.

Fault protection. Protection against electric shock under single fault conditions.

NOTE: For low voltage installations, systems and equipment, fault protection generally corresponds to protection against 'contact of persons or livestock with exposed-conductive-parts which have become live under fault conditions'.

Final circuit. A circuit connected directly to current-using equipment, or to a socket-outlet or socket-outlets or other outlet points for the connection of such equipment.

Fixed equipment. Equipment designed to be fastened to a support or otherwise secured in a specific location.

Flexible cable. A cable whose structure and materials make it suitable to be flexed while in service.

Flexible sheet heating element. Heating element consisting of sheets of electrical insulation laminated with electrical resistance material, or a base material on which electrically insulated heating wires are fixed.

Flexible wiring system. A wiring system designed to provide mechanical flexibility in use without degradation of the electrical components.

Follow current interrupt rating, I_{fi} {534}. Prospective short-circuit current that an SPD is able to interrupt without operation of the OCPD.

Functional bonding conductor, {444}. Conductor provided for functional equipotential bonding.

Functional earth. Earthing of a point or points in a system or in an installation or in equipment, for purposes other than electrical safety, such as for proper functioning of electrical equipment.

Functional extra-low voltage (FELV). An extra-low voltage system in which not all of the protective measures required for SELV or PELV have been applied.

Functional switching. An operation intended to switch 'on' or 'off' or vary the supply of electrical energy to all or part of an installation for normal operating purposes.

Fuse. A device which, by the melting of one or more of its specially designed and proportioned components, opens the circuit in which it is inserted by breaking the current when this exceeds a given value for a sufficient time. The fuse comprises all the parts that form the complete device.

Fuse carrier. The movable part of a fuse designed to carry a fuse link.

Fuse element. A part of a fuse designed to melt when the fuse operates.

Fuse link. A part of a fuse, including the fuse element(s), which requires replacement by a new or renewable fuse link after the fuse has operated and before the fuse is put back into service.

Fused connection unit. A device associated with the fixed wiring of an installation by which appliances may be connected, and having provision for a replaceable cartridge fuse link.

Gas installation pipe. Any pipe, not being a service pipe (other than any part of a service pipe comprised in a primary meter installation) or a pipe comprised in a gas appliance, for conveying gas for a particular consumer and including any associated valve or other gas fitting.

Generating set. Equipment converting non-electrical energy into electricity.

Harmonized Standard. A standard which has been drawn up by common agreement between national standards bodies notified to the European Commission by all member states and published under national procedures.

Hazardous-live-part. A live part which can give, under certain conditions of external influence, an electric shock.

Heating cable. Cable with or without a shield or a metallic sheath, intended to give off heat for heating purposes.

Heating-free area. Unheated floor or ceiling area which is completely covered when placing pieces of furniture or kept free for built-in furniture.

Heating unit. Heating cable or flexible sheet heating element with rigidly fixed cold tails or terminal fittings which are connected to the terminals of the electrical installation.

High-density livestock rearing. Breeding and rearing of livestock for which the use of automatic systems for life support is necessary.

NOTE: Examples of automatic life support systems are those for ventilation, feeding and air-conditioning.

High voltage (see Voltage, nominal).

Highway. A highway means any way (other than a waterway) over which there is public passage and includes the highway verge and any bridge over which, or tunnel through which, the highway passes.

Highway distribution board. A fixed structure or underground chamber, located on a highway, used as a distribution point, for connecting more than one highway distribution circuit to a common origin. Street furniture which supplies more than one circuit is defined as a highway distribution board. The connection of a single temporary load to an item of street furniture shall not in itself make that item of street furniture into a highway distribution board.

Highway distribution circuit. A Band II circuit connecting the origin of the installation to a remote highway distribution board or items of street furniture. It may also connect a highway distribution board to street furniture.

Highway power supply. An electrical installation comprising an assembly of associated highway distribution circuits, highway distribution boards and street furniture, supplied from a common origin.

Houseboat. Floating decked structure which is designed or adapted for use as a place of permanent residence often kept in one place on inland water.

Impulse current (I_{imp}), {534}. A parameter used for the classification test for SPDs; it is defined by three elements, a current peak value, a charge Q and a specific energy W/R.

Impulse withstand voltage (U_W), {534}. The highest peak value of impulse voltage of prescribed form and polarity which does not cause breakdown of insulation under specified conditions.

Indirect lightning stroke. Lightning strike that does not impact directly any part of the network but that induces an overvoltage in that network.

NOTE: Consequences of an indirect lightning stroke for electrical installations are the associated overvoltages and the energy content of the incoming surge wave.

Inspection. Examination of an electrical installation using all the senses as appropriate.

Installation (see Electrical installation).

Instructed person (electrically). Person adequately advised or supervised by a skilled person (as defined) to enable that person to perceive risks and to avoid hazards which electricity can create.

NOTE 1: The term ‘(electrically)’ is assumed to be present where the term 'instructed person' is used throughout BS 7671.

NOTE 2: Regulation 16 of the Electricity at Work Regulations 1989 requires persons to be competent to prevent danger and injury. The HSE publication HSR25 provides guidance on this.

Insulation. Suitable non-conductive material enclosing, surrounding or supporting a conductor.

Insulation coordination, {534}. The selection of the electric strength of equipment in relation to the voltages which can appear on the system for which the equipment is intended, taking into account the service environment and the characteristics of the available protective devices.

Island Mode. Operating mode in which the PEI is disconnected from the public distribution system, but remains energized.

NOTE: An island mode can be either the result of an automatic process or the result of a deliberate action.

Isolation. Function intended to make dead for reasons of safety all or a discrete section of the electrical installation by separating the electrical installation, or section thereof, from every source of electrical energy.

Isolator. A mechanical switching device which, in the open position, complies with the requirements specified for the isolating function. An isolator is otherwise known as a disconnector.

Ladder (see Cable ladder).

Leakage current. Electric current in an unwanted conductive path under normal operating conditions.

Leisure accommodation vehicle. Unit of living accommodation for temporary or seasonal occupation which may meet requirements for construction and use of road vehicles.

Lightning protection zone (LPZ), {534}. Zone where the lightning electromagnetic environment is defined.

Line conductor. A conductor of an AC system for the transmission of electrical energy other than a neutral conductor, a protective conductor or a PEN conductor. The term also means the equivalent conductor of a DC system unless otherwise specified in BS 7671.

Live conductor (see Live part).

Live part. A conductor or conductive part intended to be energized in normal use, including a neutral conductor but, by convention, not a PEN conductor.

Load shedding. Approach where the electrical loads are switched off for variable periods of time to optimize demand.

Low voltage (see Voltage, nominal).

Low voltage switchgear and controlgear assembly. Combination of one or more low voltage switching devices together with associated control, measuring, signalling, protective, regulating equipment, with all the internal electrical and mechanical interconnection and structural parts (see BS EN 61439-1).

Luminaire. Equipment which distributes, filters or transforms the light transmitted from one or more lamps and which includes all the parts necessary for supporting, fixing and protecting the lamps, but not the lamps themselves, and where necessary, circuit auxiliaries together with the means for connecting them to the supply.

NOTE: Lamps includes devices such as light emitting diodes.

Luminaire supporting coupler (LSC). A means, comprising an LSC outlet and an LSC connector, providing mechanical support for a luminaire and the electrical connection to and disconnection from a fixed wiring installation.

Main earthing terminal. The terminal or bar provided for the connection of protective conductors, including protective bonding conductors, and conductors for functional earthing, if any, to the means of earthing.

Maintenance. Combination of all technical and administrative actions, including supervision actions, intended to retain an item in, or restore it to, a state in which it can perform a required function.

Marina. Facility for mooring and servicing of pleasure craft with fixed wharves, jetties, piers or pontoon arrangements capable of berthing one or more pleasure craft.

Mechanical maintenance. The replacement, refurbishment or cleaning of lamps and non-electrical parts of equipment, plant and machinery.

Medical location, {710}. Location intended for purposes of diagnosis, treatment including cosmetic treatment, monitoring and care of patients.

- **Applied part.** Part of medical electrical equipment that in normal use necessarily comes into physical contact with the patient for ME equipment or an ME system to perform its function.

- **Group 0.** Medical location where no applied parts are intended to be used and where discontinuity (failure) of the supply cannot cause danger to life.
- **Group 1.** Medical location where discontinuity (failure) of the supply does not represent a threat to the safety of the patient and applied parts are intended to be used:
  - externally
  - invasively to any part of the body except where Group 2 applies.
- **Group 2.** Medical location where applied parts are intended to be used, and where discontinuity (failure) of the supply can cause danger to life, in applications such as:
  - intracardiac procedures
  - vital treatments and surgical operations.

  NOTE: An intracardiac procedure is a procedure whereby an electrical conductor is placed within the heart of a patient or is likely to come into contact with the heart, such conductor being accessible outside the patient's body. In this context, an electrical conductor includes insulated wires such as cardiac pacing electrodes or intracardiac ECG electrodes, or insulated tubes filled with conducting fluid.
- **Medical electrical equipment (ME equipment).** Electrical equipment having an applied part or transferring energy to or from the patient or detecting such energy transfer to or from the patient and which is
  (a) provided with not more than one connection to a particular supply mains, and
  (b) intended by the manufacturer to be used
  - in the diagnosis, treatment or monitoring of a patient, or
  - for compensation or alleviation of disease, injury or disability.

  NOTE: ME equipment includes those accessories as defined by the manufacturer that are necessary to enable the normal use of the ME equipment.
- **Medical electrical system (ME system).** Combination, as specified by the manufacturer, of items of equipment, at least one of which is medical electrical equipment to be interconnected by functional connection or by use of a multiple socket-outlet.

  NOTE: The system includes those accessories which are needed for operating the system and are specified by the manufacturer.
- **Medical IT system.** IT electrical system fulfilling specific requirements for medical applications.

  NOTE: These supplies are also known as isolated power supply systems.
- **Patient.** Living being (person or animal) undergoing a medical, surgical or dental procedure.

  NOTE: A person under treatment for cosmetic purposes may be considered a patient.
- **Patient environment.** Any volume in which intentional or unintentional contact can occur between a patient and parts of the medical electrical equipment or medical electrical system or between a patient and other persons touching parts of the medical electrical equipment or medical electrical system.

  NOTE 1: For illustration see Figure 710.1.

  NOTE 2: This applies when the patient's position is predetermined; if not, all possible patient positions should be considered.

Meshed bonding network (MESH-BN), {444}. Bonding network in which all associated equipment frames, racks and cabinets and usually the DC power return conductor are bonded together as well as at multiple points to the CBN and may have the form of a mesh.

NOTE: A MESH-BN improves the performance of a common bonding network.

Metering. Application of a device measuring energy or other consumption.

Minimum illuminance. Illuminance for emergency lighting at the end of the rated operating time.

Minor works. Additions and alterations to an installation that do not extend to the provision of a new circuit.

NOTE: Examples include the addition of socket-outlets or lighting points to an existing circuit, the relocation of a light switch etc.

Mobile and offshore installations. Installations used for the exploration or development of liquid or gaseous hydrocarbon resources.

Mobile equipment (portable equipment (deprecated)). Electrical equipment which is moved while in operation or which can easily be moved from one place to another while connected to the supply.

Mobile home. A transportable leisure accommodation vehicle which includes means for mobility but does not meet the requirements for construction and use of road vehicles.

Motor caravan. Self-propelled leisure accommodation vehicle, used for touring, that meets the requirements for the construction and use of road vehicles.

NOTE: It is either adapted from a series production vehicle, or designed and built on an existing chassis, with or without the driving cab, the accommodation being either fixed or dismountable.

Neutral conductor. A conductor connected to the neutral point of a system and contributing to the transmission of electrical energy. The term also means the equivalent conductor of an IT or DC system unless otherwise specified in BS 7671 and also identifies either the mid-wire of a three-wire DC circuit or the earthed conductor of a two-wire earthed DC circuit.

Nominal discharge current (I_{nspd}), {534}. A parameter used for the classification test for Class I SPDs and for preconditioning of an SPD for Class I and Class II tests; it is defined by the crest value of current through an SPD, having a current waveform of 8/20.

Nominal voltage (see Voltage, nominal).

Non-compliance. A non-conformity that may give rise to danger.

Non-flame propagating. Liable to ignite as a result of an applied flame but, after the flame is removed, does not propagate further and extinguishes itself within a limited time.

Obstacle. A part preventing unintentional contact with live parts but not preventing deliberate contact.

Open-circuit voltage under standard test conditions $U_{oc\ STC}$. Voltage under standard test conditions across an unloaded (open) generator or on the DC side of the convertor.

Open-circuit maximum voltage. Maximum voltage across an unloaded (open) PV module, PV string, or PV array, or on the DC side of the PV PCE.

Operating and maintenance gangway, {729}. Gangway providing access to facilitate operations such as switching, controlling, setting, observation and maintenance of electrical equipment.

Operating mode. Operation of an installation with respect to the different sources of electrical energy and to energy flow.

Ordinary person. Person who is neither a skilled person nor an instructed person.

Origin of an installation. The position at which electrical energy is delivered to an electrical installation.

Origin of a temporary electrical installation. Point on the permanent installation or other source of supply from which electrical energy is delivered to the temporary electrical installation.

Overcurrent. A current exceeding the rated value. For conductors the rated value is the current-carrying capacity.

Overcurrent detection. A method of establishing that the value of current in a circuit exceeds a predetermined value for a specified length of time.

Overcurrent protective device (OCPD). Device provided to interrupt an electric circuit in case the conductor current in the electric circuit exceeds a predetermined value for a specified duration.

NOTE: Table A53.1 provides information regarding the different devices corresponding to the main generic function.

Overload current. An overcurrent occurring in a circuit which is electrically sound.

Partial selectivity. Selectivity up to a given overcurrent lower than the breaking capacity of the downstream device.

PEL. A conductor combining the functions of both a protective earthing conductor and a line conductor.

PELV (protective extra-low voltage). An extra-low voltage system which is not electrically separated from Earth, but which otherwise satisfies all the requirements for SELV.

PEM. A conductor combining the functions of both a protective earthing conductor and a midpoint conductor.

PEN conductor. A conductor combining the functions of both protective conductor and neutral conductor.

Person (see Skilled, Instructed, Ordinary).

Phase conductor (see Line conductor).

Pleasure craft. Any boat, vessel, yacht, motor launch, houseboat or other floating craft used exclusively for sport or leisure.

Plug. Accessory having pins designed to engage with the contacts of a socket-outlet, and incorporating means for the electrical connection and mechanical retention of a flexible cable.

Point (in wiring). A termination of the fixed wiring intended for the connection of current-using equipment.

Portable equipment (see Mobile equipment).

Powertrack. A system component, which is generally a linear assembly of spaced and supported busbars, providing electrical connection of accessories.

Powertrack system. An assembly of system components including a powertrack by which accessories may be connected to an electrical supply at one or more points (predetermined or otherwise) along the powertrack.

NOTE: The maximum current rating of a powertrack system is 63A.

Prefabricated wiring system. Wiring system consisting of wiring sections incorporating the means of interconnection designed to allow sections to be connected together to form a given system, and incorporating installation couplers conforming to BS EN 61535.

Producer (of electricity). Party generating electrical energy.

Prospective fault current (I_{pf}). The value of overcurrent at a given point in a circuit resulting from a fault of negligible impedance between live conductors having a difference of potential under normal operating conditions, or between a live conductor and an exposed-conductive-part.

Prosumer. Entity or party which can be both a producer and a consumer of electrical energy.

Prosumer's Electrical Installations (PEIs). Low voltage electrical installations connected or not to a public distribution network able to operate:

- with local power supplies, and/or
- with local storage units,

and that monitor and control the energy from the connected sources delivering it to:

- current-using equipment, and/or
- local storage units, and/or
- public distribution network.

PEI collective. Several consuming electrical installations connected to the same public distribution network and sharing one common set of local power supplies and energy storage equipment.

PEI individual. Single consuming and/or producing electrical installation.

PEI shared. Several consuming and/or producing electrical installations similar to an individual PEI connected to the same low-voltage public distribution network and sharing their individual power supplies and energy storage equipment between themselves.

Protected escape route. A route enclosed with specified fire-resisting construction designated for escape to a place of safety in the event of an emergency *(see also Escape route).*

Protective bonding conductor. Protective conductor provided for protective equipotential bonding.

Protective conductor (PE). A conductor used for some measures of protection against electric shock and intended for connecting together any of the following parts:

(i) Exposed-conductive-parts
(ii) Extraneous-conductive-parts
(iii) The main earthing terminal
(iv) Earth electrode(s)
(v) The earthed point of the source, or an artificial neutral.

Protective conductor current. Electric current appearing in a protective conductor, such as leakage current or electric current resulting from an insulation fault.

Protective earthing. Earthing of a point or points in a system or in an installation or in equipment for the purposes of safety.

Protective equipotential bonding. Equipotential bonding for the purposes of safety.

Protective multiple earthing (PME). An earthing arrangement, found in TN-C-S systems, in which the supply neutral conductor is used to connect the earthing conductor of an installation with Earth, in accordance with the Electricity Safety, Quality and Continuity Regulations (ESQCR) (see also Figure 3.9).

Protective separation. Separation of one electric circuit from another by means of:

(i) double insulation, or
(ii) basic insulation and electrically protective screening (shielding), or
(iii) reinforced insulation.

PV (Solar photovoltaic), {712}.

- **PV AC module.** Integrated module/convertor assembly where the electrical interface terminals are AC only. No access is provided to the DC side.
- **PV array.** Mechanically and electrically integrated assembly of PV modules, and other necessary components, to form a DC power supply unit.
- **PV array cable.** Output cable of a PV array.
- **PV array junction box.** Enclosure where PV strings of any PV array are electrically connected and where devices can be located.
- **PV cell.** Basic PV device which can generate electricity when exposed to light such as solar radiation.
- **PV convertor.** Device which converts DC voltage and DC current into AC voltage and AC current.
- **PV DC main cable.** Cable connecting the PV generator junction box to the DC terminals of the PV convertor.
- **PV generator.** Assembly of PV arrays.
- **PV generator junction box.** Enclosure where PV arrays are electrically connected and where devices can be located.
- **PV installation.** Erected equipment of a PV power supply system.
- **PV module.** Smallest completely environmentally protected assembly of interconnected PV cells.
- **PV string.** Circuit in which PV modules are connected in series, in order for a PV array to generate the required output voltage.
- **PV string cable.** Cable connecting PV modules to form a PV string.
- **PV supply cable.** Cable connecting the AC terminals of the PV convertor to a distribution circuit of the electrical installation.

Rated current. Value of current used for specification purposes, established for a specified set of operating conditions of a component, device, equipment or system.

Rated impulse withstand voltage level (U_w), {534}. The level of impulse withstand voltage assigned by the manufacturer to the equipment, or to part of it, characterizing the specified withstand capability of its insulation against overvoltages.

NOTE: For the purposes of BS 7671, only withstand voltage between live conductors and Earth is considered.

Reduced low voltage system. A system in which the nominal line-to-line voltage does not exceed 110 volts and the nominal line to Earth voltage does not exceed 63.5 volts.

Reinforced insulation. Single insulation applied to live parts, which provides a degree of protection against electric shock equivalent to double insulation under the conditions specified in the relevant standard. The term 'single insulation' does not imply that the insulation must be one homogeneous piece. It may comprise two or more layers which cannot be tested singly as supplementary or basic insulation.

Reporting. Communication of the results of periodic inspection and testing of an electrical installation to the person ordering the work.

Residences and other locations belonging to agricultural and horticultural premises. Residences and other locations which have a conductive connection to the agricultural and horticultural premises by either protective conductors of the same installation or by extraneous-conductive-parts.

NOTE: Examples of other locations include offices, social rooms, machine-halls, workrooms, garages and shops.

Residential park home. A factory produced relocatable dwelling designed for permanent residence which may be used for leisure purposes.

Residual current. Algebraic sum of the currents in the live conductors of a circuit at a point in the electrical installation.

Residual current device (RCD). Mechanical switching device designed to make, carry and break currents under normal service conditions and to cause the opening of the contacts when the residual current attains a given value under specified conditions.

NOTE 1: A residual current device can be a combination of various separate elements designed to detect and evaluate the residual current and to make and break current.

NOTE 2: RCD includes devices such as RCCB, RCBO, CBR and MRCD.

NOTE 3: MRCD is a modular residual current device.

Residual current operated circuit-breaker with integral overcurrent protection (RCBO). A residual current operated switching device designed to perform the functions of protection against overload and/or short-circuit.

Residual current operated circuit-breaker without integral overcurrent protection (RCCB). A residual current operated switching device not designed to perform the functions of protection against overload and/or short-circuit.

Residual operating current. Residual current which causes the RCD to operate under specified conditions.

Resistance area (for an earth electrode only). The surface area of ground (around an earth electrode) on which a significant voltage gradient may exist.

Response time. The time that elapses between the failure of the normal power supply and the ability of the auxiliary power supply to energize the equipment.

Restrictive conductive location (see Conducting location with restricted movement).

Reverse feeding mode. Operating mode in which the PEI supplies the public network. *(see Direct feeding mode).*

NOTE: Local storage units can supply current-using equipment and/or the public distribution network or be charged by local power supplies.

Ring final circuit. A final circuit arranged in the form of a ring and connected to a single point of supply.

Safety service. An electrical system for electrical equipment provided to protect or warn persons in the event of a hazard, or essential to their evacuation from a location.

Sauna. A room or location in which air is heated, in service, to high temperatures where the relative humidity is normally low, rising only for a short period of time when water is poured over the heater.

SELV (separated extra-low voltage). An extra-low voltage system which is electrically separated from Earth and from other systems in such a way that a single fault cannot give rise to the risk of electric shock.

Selectivity. Coordination of the operating characteristics of two or more protective devices such that, on the incidence of an overcurrent or residual current within stated limits, the device intended to operate within these limits does so, while the other(s) does (do) not.

NOTE: Fundamentally, selectivity is the ability of a protective device to operate in preference to another protective device in series.

- Partial: Selectivity up to a given overcurrent lower than the breaking capacity of the downstream device.
- Total: Selectivity for all overcurrents up to the value of the breaking capacity of the downstream device.

Shock (see Electric shock).

Shock current. A current passing through the body of a person or livestock such as to cause electric shock and having characteristics likely to cause dangerous effects.

Short-circuit current. An overcurrent resulting from a fault of negligible impedance between live conductors having a difference in potential under normal operating conditions.

Short-circuit current rating (I_{SCCR}), {534}. Maximum prospective short-circuit current from the power system for which the SPD, in conjunction with the OCPD specified, is rated.

Short-circuit current under standard test conditions ($I_{sc\ STC}$), {712}. Short-circuit current of a PV module, PV string, PV array or PV generator under standard test conditions.

Short-circuit maximum current. Maximum short-circuit current of a PV module, PV string or PV array.

Short-circuit protective device (SCPD). Device intended to protect a circuit or part of a circuit against short-circuit currents by interrupting them.

NOTE: Table A53.1 provides information regarding the different devices corresponding to the main generic function.

Show. Display or presentation in any suitable location, either a room, building or temporary structure.

Simple separation. Separation between circuits or between a circuit and Earth by means of basic insulation.

Simultaneously accessible parts. Conductors or conductive parts which can be touched simultaneously by a person or, in locations specifically intended for them, by livestock.

NOTE: Simultaneously accessible parts may be: live parts, exposed-conductive-parts, extraneous-conductive-parts, protective conductors or earth electrodes.

Skilled person (electrically). Person who possesses, as appropriate to the nature of the electrical work to be undertaken, adequate education, training and practical skills, and who is able to perceive risks and avoid hazards which electricity can create.

NOTE 1: The term '(electrically)' is assumed to be present where the term 'skilled person' is used throughout BS 7671.

NOTE 2: Regulation 16 of the Electricity at Work Regulations 1989 requires persons to be competent to prevent danger and injury. The HSE publication HSR25 provides guidance on this.

Socket-outlet. A device, provided with female contacts, which is intended to be installed with the fixed wiring, and intended to receive a plug. A luminaire track system is not regarded as a socket-outlet system.

SPD disconnector. Device for disconnecting an SPD, or part of an SPD, from the power system.

NOTE: This disconnecting device is not required to have isolating capability for safety purposes. It is to prevent a persistent fault on the system and is used to give an indication of an SPD's failure. Disconnectors can be internal (built in) or external (required by the manufacturer). There may be more than one disconnector function, for example an overcurrent protection function and a thermal protection function. These functions may be in separate units.

Spur. A branch from a ring or radial final circuit.

Stand. Area or temporary structure used for display, marketing or sales.

Standard test conditions (STC). Test conditions specified in BS EN 60904-3 for PV cells and PV modules.

Standby electrical source. Electrical source intended to maintain, for reasons other than safety, the supply to an electrical installation or a part or parts thereof, in case of interruption of the normal supply.

Standby electrical supply system. Supply system intended to maintain, for reasons other than safety, the functioning of an electrical installation or a part or parts thereof, in case of interruption of the normal supply.

Static convertor. A convertor having no moving parts and notably using semiconductor rectifiers.

Stationary equipment. Electrical equipment which is either fixed or which has a mass exceeding 18 kg and is not provided with a carrying handle.

Street furniture. Fixed equipment located on a highway.

Supplementary insulation. Independent insulation applied in addition to basic insulation for fault protection.

Supplier (see Distributor).

Surge current, {534}. A transient wave appearing as an overcurrent caused by a lightning electromagnetic impulse.

Surge protective device (SPD), {534}. A device that is intended to limit transient overvoltages and divert surge currents. It contains at least one non-linear component.

Switch, linked. A switch the contacts of which are so arranged as to make or break all poles simultaneously or in a definite sequence.

Switch. A device designed to make or break the current in one or more electric circuits.

Switch-disconnector. A switch which, in the open position, satisfies the isolating requirements specified for a disconnector.

NOTE: A switch-disconnector is otherwise known as an isolating switch.

Switchboard. An assembly of switchgear with or without instruments, but the term does not apply to groups of local switches in final circuits.

Switchgear. An assembly of main and auxiliary switching equipment for operation, regulation, protection or other control of an electrical installation.

System. An electrical system consisting of a single source or multiple sources running in parallel of electrical energy and an installation. See Part 3. For certain purposes of BS 7671, types of system are identified as follows, depending upon the relationship of the source, and of exposed-conductive-parts of the installation, to Earth:

- **TN system.** A system having one or more points of the source of energy directly earthed, the exposed-conductive-parts of the installation being connected to that point by protective conductors.
- **TN-C system.** A system in which neutral and protective functions are combined in a single conductor throughout the system.
- **TN-S system.** A system having separate neutral and protective conductors throughout the system (see Figure 3.8).
- **TN-C-S system.** A system in which neutral and protective functions are combined in a single conductor in part of the system (see Figure 3.9).
- **TT system.** A system having one point of the source of energy directly earthed, the exposed-conductive-parts of the installation being connected to earth electrodes electrically independent of the earth electrodes of the source (see Figure 3.10).
- **IT system.** A system having no direct connection between live parts and Earth, the exposed-conductive-parts of the electrical installation being earthed (see Appendix 9 Figure 9C).
- **Multiple source and DC systems** - see Appendix 9.

Temporary electrical installation. Electrical installation erected for a particular purpose and dismantled when no longer required for that purpose.

Temporary overvoltage (U_{TOV}), {534}. A fundamental frequency overvoltage occurring on the network at a given location, of relatively long duration.

NOTE 1: TOVs may be caused by faults inside the LV system ($U_{TOV,LV}$) or inside the HV system ($U_{TOV,HV}$)

NOTE 2: Temporary overvoltages, typically lasting up to several seconds, usually originate from switching operations or faults (for example, sudden load rejection, single-phase faults, etc.) and/or from non-linearity (ferroresonance effects, harmonics, etc.)

Temporary structure. A unit or part of a unit, including mobile portable units, situated indoors or outdoors, designed and intended to be assembled and dismantled.

Temporary supply unit. An enclosure containing equipment for the purpose of taking a temporary electrical supply safely from an item of street furniture.

Testing. Implementation of measures to assess an electrical installation by means of which its effectiveness is proved. This includes ascertaining values by means of appropriate measuring instruments, where measured values are not detectable by inspection.

Thermal storage floor heating system. Heating system in which, due to a limited charging period, a restricted availability of electrical energy is converted into heat and dissipated mainly through the surface of the floor to the room to be heated with an intended time delay.

Total selectivity. Selectivity for all overcurrents up to the value of the breaking capacity of the downstream device.

Triplen harmonics. The odd multiples of the 3rd harmonic of the fundamental frequency (e.g. 3rd, 9th, 15th, 21st)

Trunking (see Cable trunking).

Verification. All measures by means of which compliance of the electrical installation with the relevant requirements of BS 7671 are checked, comprising inspection, testing and certification.

Voltage, nominal. Voltage by which an installation (or part of an installation) is designated. The following ranges of nominal voltage (rms values for AC) are defined:

- **Extra-low.** Not exceeding 50 V AC or 120 V ripple-free DC, whether between conductors or to Earth.
- **Low.** Exceeding extra-low voltage but not exceeding 1000 V AC or 1500 V DC between conductors, or 600 V AC or 900 V DC between conductors and Earth.
- **High.** Normally exceeding low voltage.

NOTE: The actual voltage of the installation may differ from the nominal value by a quantity within normal tolerances, see Appendix 2.

Voltage, reduced (see Reduced low voltage system).

Voltage band

Band I

Band I covers:

- installations where protection against electric shock is provided under certain conditions by the value of voltage;
- installations where the voltage is limited for operational reasons (e.g. telecommunications, signalling, bell, control and alarm installations).

Extra-low voltage (ELV) will normally fall within voltage Band I.

Band II

Band II contains the voltages for supplies to household and most commercial and industrial installations. Low voltage (LV) will normally fall within voltage Band II.

NOTE: Band II voltages do not exceed 1000 V AC rms or 1500 V DC.

Voltage protection level (U_p), {534}. A parameter that characterizes the performance of an SPD in limiting the voltage across its terminals, which is selected from a list of preferred values; this value is greater than the highest value of the measured limiting voltages.

Wiring system. An assembly made up of cable or busbars and parts which secure and, if necessary, enclose the cable or busbars.

Fig 2.1 – Illustration of earthing and protective conductor terms (see Chapter 54)

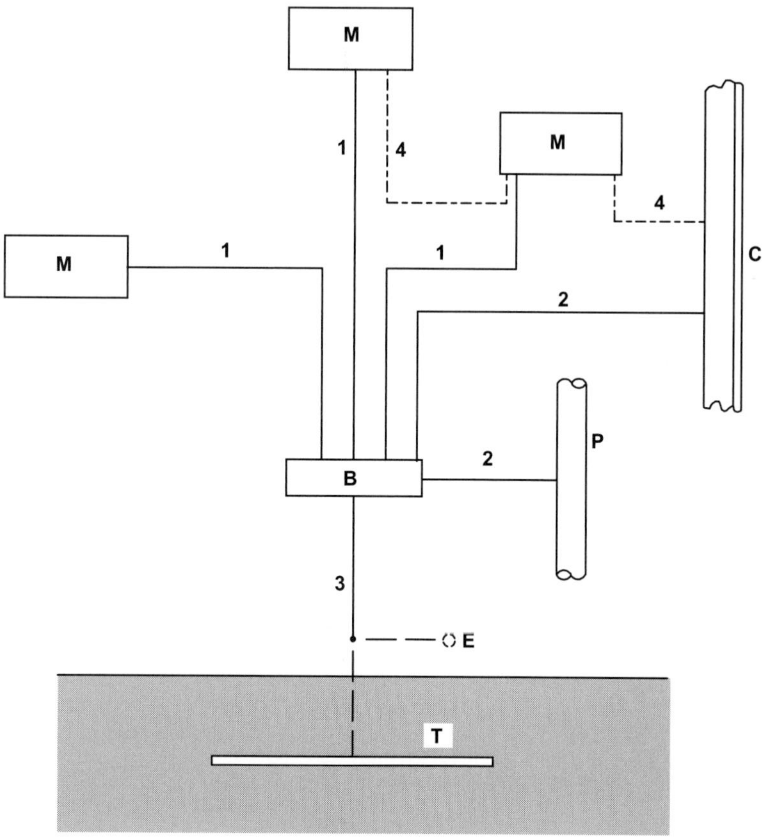

1,2,3,4	=	**protective conductors**
1	=	**circuit protective conductor**
2	=	**main protective bonding conductor**
3	=	**earthing conductor**
4	=	**supplementary protective bonding conductors (where required)**
B	=	**main earthing terminal (MET)**
M	=	**exposed-conductive-part**
C	=	**extraneous-conductive-part**
P	=	**main metallic water pipe (extraneous-conductive-part)**
T	=	**earth electrode (TT and IT systems)**
E	=	**other means of earthing (TN systems)**

SYMBOLS USED IN BS 7671

Symbol	Meaning		Example
C	rating factor - general		Appx 4 sec 3
c	battery capacity	Ah	A721.525
C_a	rating factor for ambient temperature		Appx 4 sec 3
C_c	rating factor for circuits buried in the ground		Appx 4 sec 3
C_d	rating factor for depth of burial		Appx 4 sec 3
C_f	rating factor for semi-enclosed fuse to *BS 3036*		Appx 4 sec 3
C_g	rating factor for grouping		Appx 4 sec 3
C_h	rating factor for higher harmonic currents in line conductors		Appx 4 sec 5.6
C_i	rating factor for conductors embedded in thermal insulation		Appx 4 sec 3
C_{max}	is the maximum voltage factor to take account of voltage variations depending on time and place, changing of transformer taps and other considerations. For a low voltage supply given in accordance with the ESQCR, C_{max} is given the value 1.1		A722.2
C_{min}	minimum voltage factor		411.4.4
C_s	rating factor for thermal resistivity of soil		Appx 4 sec 3
C_t	rating factor for operating temperature of conductor		Appx 4 sec 6.1
D_e	external cable diameter	mm	Appx 4, Table 4A2
ΣI_{zk}	the sum of the continuous current-carrying capacities of m conductors in parallel	A	Appx 10 sec 2
f	frequency in cycles per second	Hz	Appx 6
gG	class 'gG' utilization category of fuses to *BS 88-2* - general use		411.4.201
gM	class 'gM' utilization category of fuses to *BS 88-2* motor circuit application		411.4.201
I	current (general term)	A	
I_a	current causing operation of the protective device within the specified time.	A	411.4.4
I_b	design current of the circuit	A	433.1.1
I_{bh}	design current including the effect of third harmonic currents	A	Appx 4 sec 5.5.2
I_{bk}	design current for conductor k	A	Appx 10 sec 2
I_{bn}	neutral current due to third harmonic currents	A	App 4 sec 5.5.2
I_c	charging current	A	A721.525
I_{cc}	rated conditional short- circuit current	A	536.4.201
I_{cw}	rated short-time withstand current	A	434.5.3
I_d	fault current of first fault (IT system)	A	411.6.2
I_E	part of the earth fault current in the high voltage system that flows through the earthing arrangement of the transformer substation	A	442.1.2

Symbol	Meaning		Example
I_g	is the rated output current of the generating set		551.7.2
I_h	the fault current that flows through the earthing arrangement of the exposed-conductive-parts of the equipment of the low voltage installation during a period when there is a high voltage fault and a first fault in the low voltage installation	A	442.1.2
I_{hn}	nth harmonic current	A	Appx 4 sec 5.6
I_{imp}	selection of SPDs with regard to impulse discharge current		534.4.4.4
I_{inst}	is the rms maximum demand current of a single-phase installation (in amperes), including that of the electric vehicle charging load and any other loads, determined in accordance with Regulation 311.1		A722.3
I_{L1}, I_{L2} and I_{L3}	are the rms values of current (in amperes) in lines 1, 2 and 3, respectively, that were used when determining the value of I_m		A722.1
I_m	is the rms maximum neutral current of a three phase installation (in amperes), including that of the electric vehicle charging load and any other loads, determined in accordance with Regulation 311.1		A722.1
I_n	rated current or current setting of protective device	A	Table 41.3
$I_{\Delta n}$	rated residual operating current of RCD, in amperes	A	411.5.3
I_{nA}	the rated current (A) of an assembly which is the maximum load current that it is designed to manage and distribute		536.4.202
I_{nc}	the rated current (A) of a circuit as stated by the assembly manufacturer, taking into consideration the ratings of the devices within the circuits, their disposition and application		536.4.202
I_{nk}	rated current of the protective device for conductor k	A	Appx 10 sec 2
I_{nspd}	selection of SPDs with regard to discharge current		534.4.4.4
I_{pf}	prospective fault current	A	Appx 6
I_{pk}	rated peak withstand current		536.4.201
I_{SCCR}	selection of SPDs with regard to the short-circuit current rating		534.4.4.6
$I_{SC\ MAX}$	short-circuit maximum current		712.431.101
$I_{sc\ STC}$	short-circuit current under standard test conditions	A	712.433.1
I_t	tabulated current-carrying capacity of a cable	A	Appx 4 sec 3
I_z	current-carrying capacity of a cable for continuous service under the particular installation conditions concerned	A	433.1.1
I_{zk}	the continuous current-carrying capacity of conductor k	A	Appx 10 sec 2
I^2t	energy let-through rating of device	A^2s	434.5.2

Symbol	Meaning		Example
I_2	current causing effective operation of the overload protective device	A	433.1.1
λ	thermal conductivity	$W^{m-1}K^{-1}$	523.9
k	material factor taken from *Tables 43.1, 54.2* to *54.6*	$As^{½}mm^{-2}$	434.5.2
$k^2 S^2$	energy withstand of cable	A^2s	434.5.2
mV/A/m	voltage drop per ampere per metre	$mVA^{-1}m^{-1}$	Appx 4 sec 6
$(mV/A/m)_r$	resistive voltage drop per ampere per metre	$mVA^{-1}m^{-1}$	Appx 4 sec 6
$(mV/A/m)_x$	reactive voltage drop per ampere per metre	$mVA^{-1}m^{-1}$	Appx 4 sec 6
$(mV/A/m)_z$	impedance voltage drop per ampere per metre	$mVA^{-1}m^{-1}$	Appx 4 sec 6
n	number of circuits in a group		Appx 4 sec 2.3.3.1
R	resistance of supplementary bonding conductor	Ω	415.2.2
R_a	the response value of the IMD as described in BS EN 61557-8		538.1.1
R_A	the sum of the resistances of the earth electrode and the protective conductor connecting it to the exposed-conductive-parts	Ω	411.5.3
	also defined as the 'resistance of the earthing arrangement of the exposed-conductive-parts of the equipment of the low voltage installation'		442.1.2
$R_{A\ ev}$	is the sum of resistances of the earth electrode and the protective conductor connecting it to the main earthing terminal of the installation (in ohms)		A722.3
R_B	resistance of the earthing arrangement of the low voltage system neutral, for low voltage systems in which the earthing arrangements of the transformer substation and of the low voltage system neutral are electrically independent	Ω	442.1.2
R_E	resistance of the earthing arrangement of the transformer substation	Ω	442.1.2
R_f	the insulation resistance between the system to which it is connected and either the earth, the PE connection or another reference point for protective equipotential bonding		538.1.1
R_1	resistance of line conductor of a distribution or final circuit	Ω	Appx 6 Generic schedule of test results
R_2	resistance of circuit protective conductor (cpc) of a distribution or final circuit.	Ω	Appx 6 Generic schedule of test results
S	size (nominal cross-sectional area of conductor)	mm^2	543.1.3
$S_1,..S_m$	cross-sectional area of parallel conductors	mm^2	Appx 10 sec 2
S_k	cross-sectional area of conductor k	mm^2	Appx 10 sec 2
t	time (seconds)	s	434.5.2
t_p	maximum permitted normal operating conductor temperature	°C	Appx 4 sec 6.1
U	voltage between lines	V	411.6.4

Symbol	Meaning		Example
U_c	continuous operating voltage, *{534}*	V	534.4.4.3
U_f	power frequency fault voltage that appears in the low voltage system between exposed-conductive-parts and earth for the duration of the fault	V	442.1.2
U_{oc}	open-circuit voltage	V	Table 16A note 3
$U_{oc\ MAX}$	maximum voltage across an unloaded (open) PV module, PV string, or PV array, or on the DC side of the PV PCE		712.414.1.1
$U_{oc\ STC}$	open-circuit voltage under standard test conditions	V	712.433.101.1
U_p	voltage protection level of SPD, *{534}*	V	534.4.4.1
U_{TOV}	temporary overvoltage, *{534}*	V	Part 2 TOV
U_w	rated impulse withstand voltage level, *{534}*	V	Table 443.2
U_0	nominal AC rms or *ripple-free* DC line voltage to Earth	V	Table 41.1
U_1	power frequency stress voltage between the line conductor and the exposed-conductive-parts of the low voltage equipment of the transformer substation during the fault	V	442.1.2
U_2	power frequency stress voltage between the line conductor and the exposed-conductive-parts of the equipment of the low voltage installation during the fault	V	442.1.2
Z	the impedance between the low voltage system and an earthing arrangement	Ω	442.1.2
Z_l	the impedance of parallel conductor l	Ω	Appx 10 sec 2
Z_{db}	the earth fault loop impedance value measured at a distribution board	Ω	Appx 6 Model forms for certification and reporting
Z_e	that part of the earth fault loop impedance which is external to the installation	Ω	313.1
Z_k	the impedance of conductor k	Ω	Appx 10 sec 2
Z_m	the impedance of parallel conductor m	Ω	Appx 10 sec 2
Z_s	earth fault loop impedance	Ω	411.4.4
Z^1_s	neutral-earth loop impedance (IT systems with distributed neutral only)	Ω	411.6.5
cos Ø	power factor (sinusoidal)		Appx 4 sec 6.2

ABBREVIATIONS USED IN BS 7671

Abbreviation	Meaning	Example
ACS	Assembly for Construction Sites	704.537.2
AFDD	Arc fault detection devices	421.1.7
BN	Bonding network	Part 2
BRC	Bonding ring conductor	A444.1.4
BS	British Standard	110.1.3
BS EN	British Standard Euro Norm (BSI published version of European harmonized standard)	110.1.3
BS EN ISO	A BS EN which has the core text of an ISO standard	A721.55.2.6
BSI	British Standards Institution	
BSI IEC	British Standards Institution International Electrotechnical Commission. A BSI IEC means the UK has adopted an IEC standard that has not been put through from adoption in Europe	523.3
CBN	Common bonding network	444.5.2
CBR	Circuit-breaker incorporating residual current protection not suitable for use by ordinary persons	411.4.4 note 2
CENELEC	European Committee for Electrotechnical Standardization	Preface
cfl	compact fluorescent lamp	444.6.2
CLC/TS	European Committee for Electrotechnical Standardization, Technical Specification	534.1
cpc	circuit protective conductor	Part 2
DCL	Device for connecting a luminaire	411.7.5
DNO	Distribution network operator	560.6.9 note 1
DSO	Distribution System Operator	822.1
EMC	Electromagnetic compatibility	332.1

Abbreviation	Meaning	Example
EMI	Electromagnetic interference	332.2
EN	European Norm	Appx 1
ENA	Energy Networks Association	560.6.9 note 1
EPR	Earth potential rise – stress voltage	442.2
ESQCR	Electricity Safety, Quality and Continuity Regulations	114.1
EV	Electric vehicle	722.511.1
ELV	Extra-low voltage	Part 2
FELV	Functional extra-low voltage	411.7
HD	Harmonization Document	Preface
HF	High frequency	Appx 5 concise list
HSE	Health and Safety Executive	
HTM	Health Technical Memorandum	710.1 note 5
HV	High voltage	442.2
IEC	International Electrotechnical Commission	133.1.1
IEC/TS	International Electrotechnical Commission Technical Specification	706.1
IET	Institution of Engineering and Technology	
IMD	Insulation monitoring device	411.6.3
IP	International Protection Code	412.2.2.3
ISO	International Standards Organisation	A721.533.1.6
LPG	Liquefied petroleum gas	717.528.3.4
LPZ	Lightning protection zone	534.1
LSC	Luminaire supporting coupler	411.7.5
LV	Low voltage	
OCPD	Overcurrent protective device	411.6.3
PD	Published Document (IEC)	414.2

ABBREVIATIONS USED IN BS 7671

Abbreviation	Meaning	Example
PE	Protective conductor	Fig 3.8
PEI	Prosumer's Electrical Installation	Ch. 82
PELV	Protective extra-low voltage	410.3.3
PEN	Protective and neutral conductor (combined)	Fig 3.9
PME	Protective multiple earthing	Fig 3.9
PV	Photovoltaic	712.1
PVC	Polyvinyl chloride	709.521.1.4
RCBO	Residual current circuit-breaker with integral overcurrent protection	Table 41.3
RCCB	Residual current circuit-breaker without integral overcurrent protection	Appx 1
RCD	Residual current device (RCCB or RCBO)	
RCM	Residual current monitor	411.6.3
rms	root mean square	133.2.1
SELV	Separated extra-low voltage	410.3.3
SPD	Surge protective device	443.1.1
TSE	Transfer Switching Equipment	536.4.2.3

[END]

PART 3

ASSESSMENT OF GENERAL CHARACTERISTICS

CONTENTS

PART 3

ASSESSMENT OF GENERAL CHARACTERISTICS

CHAPTER 30

301 ASSESSMENT OF GENERAL CHARACTERISTICS

301.1 An assessment shall be made of the following characteristics of the installation in accordance with the chapters indicated:

(i) The purpose(s) for which the installation is intended to be used, its general structure and its supplies (Chapter 31)

(ii) The external influences to which it is to be exposed (Chapter 32)

(iii) The compatibility of its equipment (Chapter 33)

(iv) Its maintainability (Chapter 34)

(v) Recognized safety services (Chapter 35)

(vi) Assessment for continuity of service (Chapter 36).

These characteristics shall be taken into account in the choice of methods of protection for safety (see Chapters 41 to 44), isolation and switching (Chapter 46 and Section 537) and the selection and erection of equipment (Chapters 51 to 56).

CHAPTER 31

PURPOSES, SUPPLIES AND STRUCTURE

311 MAXIMUM DEMAND AND DIVERSITY

311.1 For economic and reliable design of an installation within thermal limits and admissible voltage drop, the maximum demand shall be determined. In determining the maximum demand of an installation or part thereof, diversity may be taken into account.

312 CONDUCTOR ARRANGEMENT AND SYSTEM EARTHING

The following characteristics shall be assessed:

(i) Arrangement of current-carrying conductors under normal operating conditions

(ii) Type of system earthing.

312.1 General

The following arrangements of current-carrying conductors under normal operating conditions are taken into account in BS 7671.

312.1.1 Current-carrying conductors in AC circuits

Fig 3.1 – Single-phase 2-wire

Fig 3.2 – Single-phase 3-wire

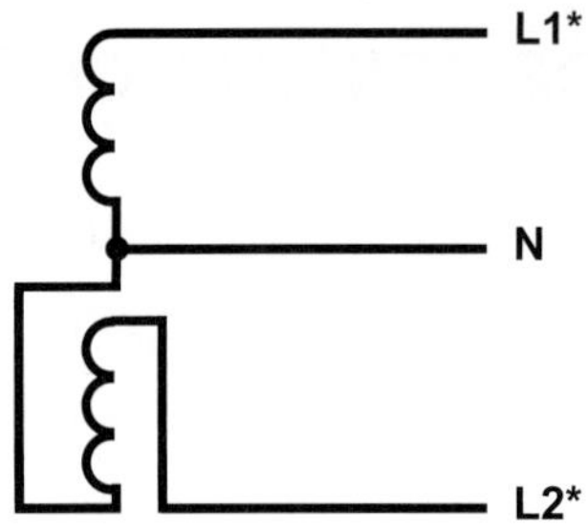

Phase angle 0°
* Numbering of conductors optional

Fig 3.3 – Two-phase 3-wire

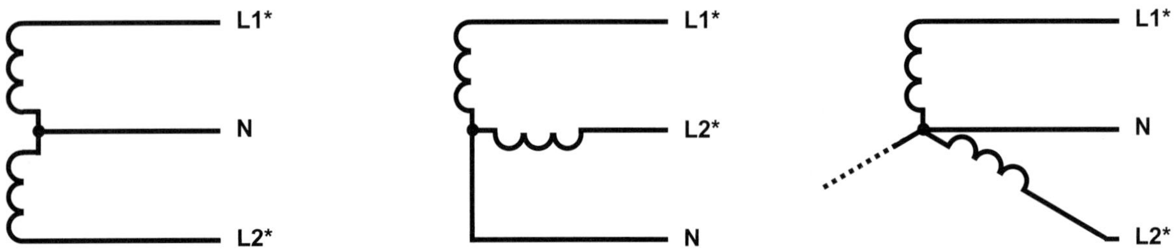

Phase angle 180° | Phase angle 90° | Phase angle 120°

* Numbering of conductors optional

Fig 3.4 – Three-phase 3-wire

Star connection | Delta connection

Fig 3.5 – Three-phase 4-wire

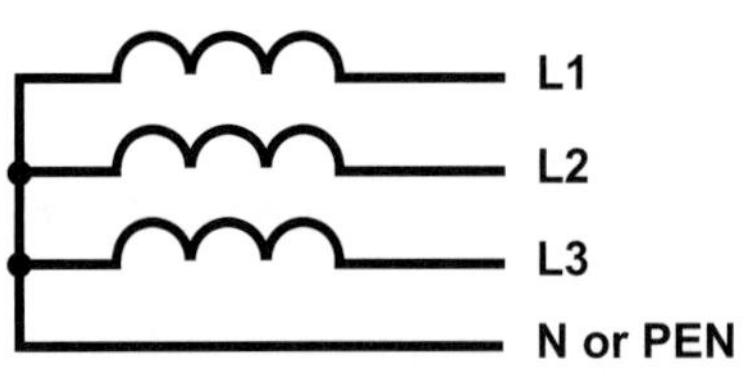

Three-phase, 4-wire with neutral conductor or PEN conductor. By definition, the PEN conductor is not a live conductor but a conductor carrying an operating current.

NOTE 1: In the case of a single-phase 2-wire arrangement which is derived from a three-phase 4-wire arrangement, the two conductors are either two line conductors or a line conductor and a neutral conductor or a line conductor and a PEN conductor.

NOTE 2: In installations with all loads connected between lines, the installation of the neutral conductor may not be necessary.

312.1.2 Current-carrying conductors in DC circuits

Fig 3.6 – 2-wire

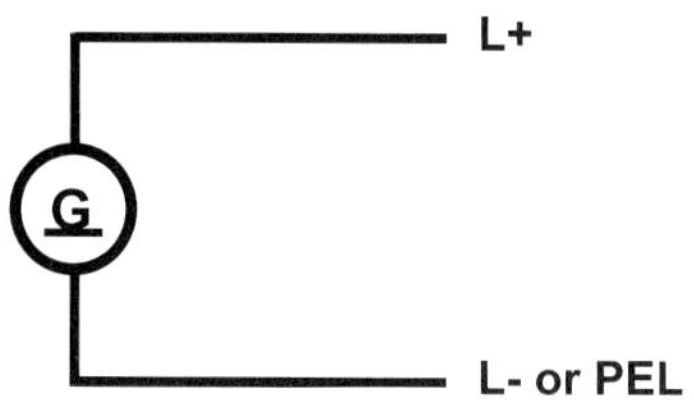

Fig 3.7 – 3-wire

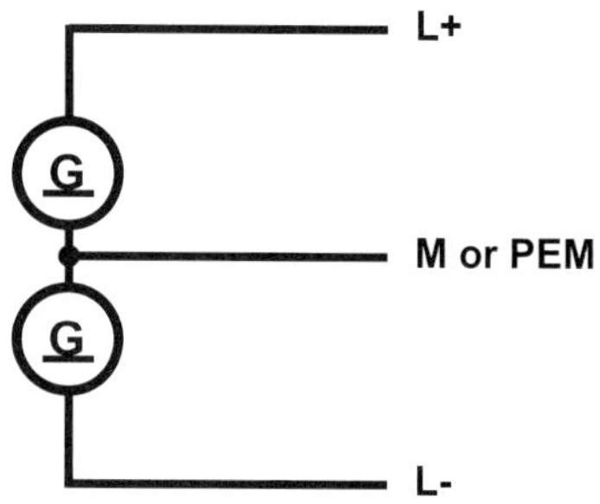

NOTE: PEL and PEM conductors are not live conductors although they carry operating current. Therefore, the designation 2-wire arrangement or 3-wire arrangement applies.

312.2 Types of system earthing

The following types of system earthing are taken into account in BS 7671.

NOTE 1: Figures 3.8 to 3.10 show examples of commonly used three-phase systems. For IT, multiple source, DC and other systems see Appendix 9.

NOTE 2: For private systems, the source and/or the distribution system may be considered as part of the installation within the meaning of BS 7671.

NOTE 3: The codes used have the following meanings:

First letter – Relationship of the power system to Earth:

T = direct connection of one point to Earth

I = all live parts isolated from Earth, or one point connected to Earth through a high impedance.

Second letter – Relationship of the exposed-conductive-parts of the installation to Earth:

T = direct electrical connection of exposed-conductive-parts to Earth, independently of the earthing of any point of the power system

N = direct electrical connection of the exposed-conductive-parts to the earthed point of the power system (in AC systems, the earthed point of the power system is normally the neutral point or, if a neutral point is not available, a line conductor).

Subsequent letter(s) (if any) – Arrangement of neutral and protective conductors:

S = protective function provided by a conductor separate from the neutral conductor or from the earthed line (or, in AC systems, earthed phase) conductor

C = neutral and protective functions combined in a single conductor (PEN conductor).

312.2.1 TN systems

312.2.1.1 Single-source systems

TN systems have one point directly earthed at the source, the exposed-conductive parts of the installation(s) being connected to that point by protective conductors. Two types of TN system are considered according to the arrangement of neutral and protective conductors, as follows:

Fig 3.8 – TN-S system

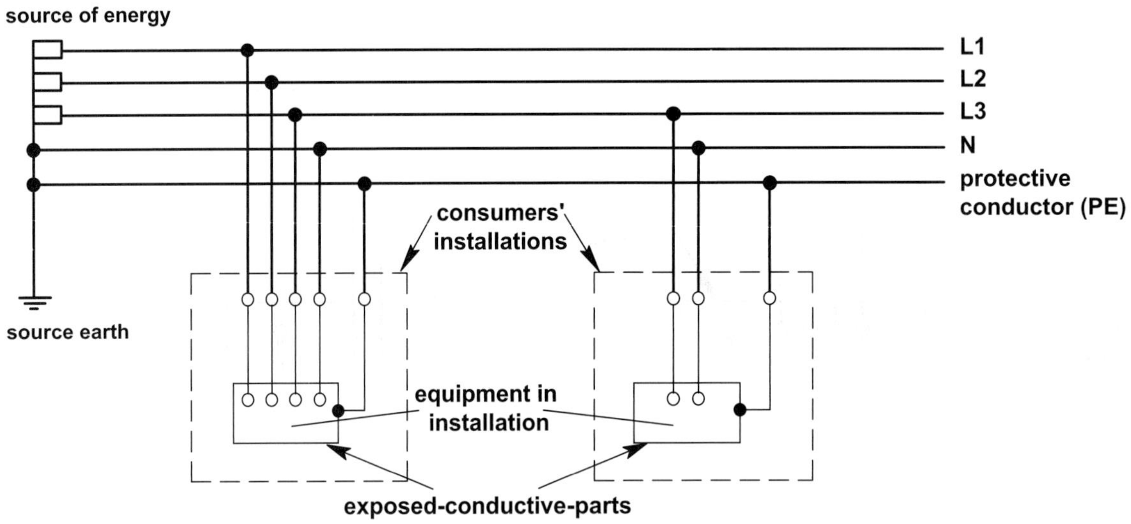

Separate neutral and protective conductors throughout the system.

The protective conductor (PE) is the metallic covering of the cable supplying the installations or a separate conductor.

All exposed-conductive-parts of an installation are connected to this protective conductor via the main earthing terminal of the installation.

Fig 3.9 – TN-C-S (PME) system

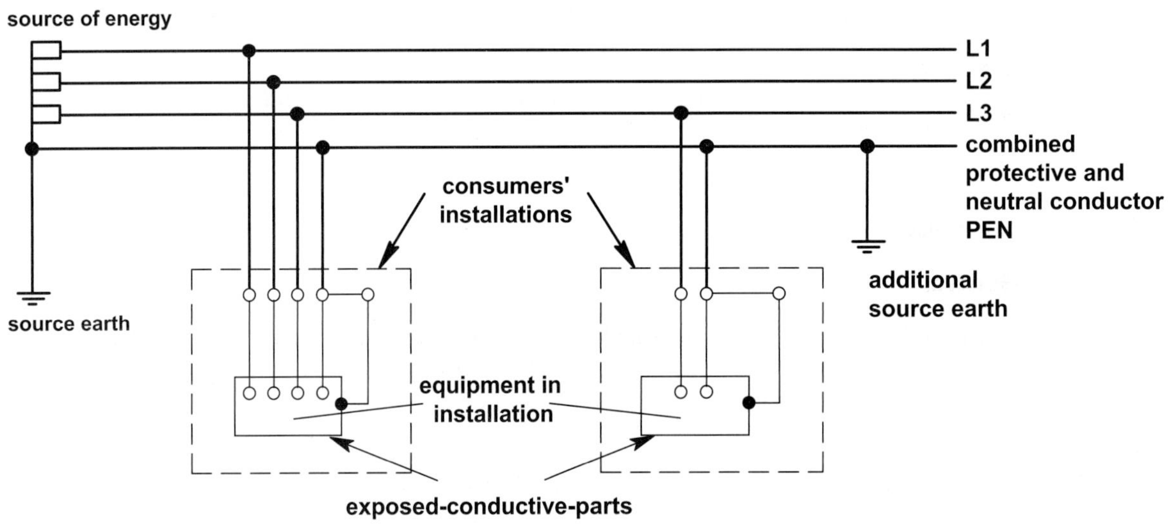

Neutral and protective functions combined in a single conductor (PEN) in a part of the system.

This type of distribution is known also as protective multiple earthing (PME).

The supply system PEN conductor is earthed at two or more points and an earth electrode may be necessary at or near a consumer's installation.

All exposed-conductive-parts of an installation are connected to the PEN conductor via the main earthing terminal and the neutral terminal, these terminals being linked together.

312.2.2 TT system

312.2.2.1 Single-source system

A TT system has only one point directly earthed at the source, the exposed-conductive-parts of the installation(s) being connected to earth electrodes electrically independent of the earth electrode of the supply system (the source earth).

Fig 3.10 – TT system

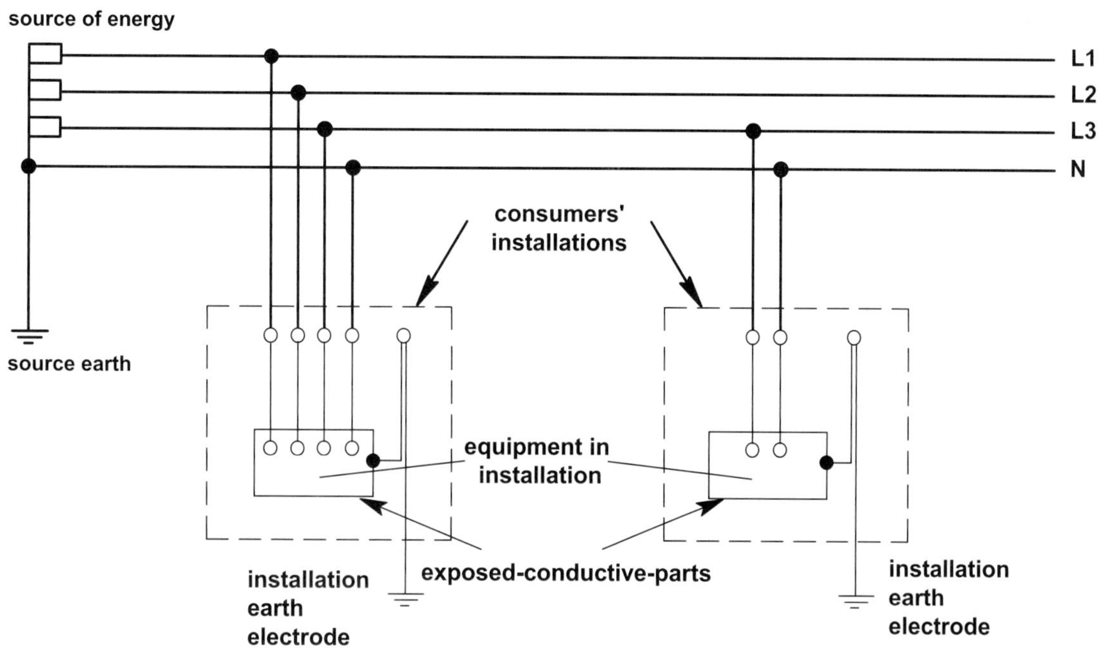

All exposed-conductive-parts of an installation are connected to an earth electrode which is electrically independent of the source earth.

Separate neutral and protective conductors throughout the system.

NOTE: Additional earthing of the PE in the installation may be provided.

312.3 Not used

312.4 IT, multiple source, DC and other systems

See Appendix 9.

313 SUPPLIES

313.1 General

The following characteristics of the supply or supplies, from whatever source, and the normal range of those characteristics where appropriate, shall be determined by calculation, measurement, enquiry or inspection:

(i) The nominal voltage(s) and its characteristics including harmonic distortion
(ii) The nature of the current and frequency
(iii) The prospective short-circuit current at the origin of the installation
(iv) The earth fault loop impedance of that part of the system external to the installation, Z_e
(v) The suitability for the requirements of the installation, including the maximum demand
(vi) The type and rating of the overcurrent protective device(s) acting at the origin of the installation.

These characteristics shall be ascertained for an external supply and shall be determined for a private source. These requirements are equally applicable to main supplies and to safety services and standby supplies.

NOTE: The above information should be provided by distributors on request (see Appendix 2 sec 2).

313.2 Supplies for safety services and standby systems

Where the provision of safety services is required, for example, by the authorities concerned with fire precautions and other conditions for emergency evacuation of the premises, and/or where the provision of standby supplies is required by the person specifying the installation, the characteristics of the source or sources of supply for safety services and/or standby systems shall be separately assessed. Such supplies shall have adequate capacity, reliability and rating and appropriate changeover time for the operation specified.

NOTE 1: For further requirements for supplies for safety services, see Chapter 35 hereafter and Chapter 56.

NOTE 2: For standby systems, there are no particular requirements in these Regulations.

314 DIVISION OF INSTALLATION

314.1 Every installation shall be divided into circuits, as necessary, to:

(i) avoid danger and minimize inconvenience in the event of a fault

(ii) facilitate safe inspection, testing and maintenance (see also Chapter 46 and Section 537)

(iii) take account of hazards that may arise from the failure of a single circuit such as a lighting circuit

(iv) reduce the possibility of unwanted tripping of RCDs due to excessive protective conductor (PE) currents not due to a fault

(v) mitigate the effects of electromagnetic disturbances (see also Chapter 44)

(vi) prevent the indirect energizing of a circuit intended to be isolated.

314.2 Separate circuits shall be provided for parts of the installation which need to be separately controlled, in such a way that those circuits are not affected by the failure of other circuits, and due account shall be taken of the consequences of the operation of any single protective device.

314.3 The number of final circuits required, and the number of points supplied by any final circuit, shall be such as to facilitate compliance with the requirements of Chapter 43 for overcurrent protection, Chapter 46 and Section 537 for isolation and switching and Chapter 52 as regards current-carrying capacities of conductors.

314.4 Where an installation comprises more than one final circuit, each final circuit shall be connected to a separate way in a distribution board. The wiring of each final circuit shall be electrically separate from that of every other final circuit, so as to prevent the indirect energizing of a final circuit intended to be isolated.

CHAPTER 32

CLASSIFICATION OF EXTERNAL INFLUENCES

Refer to Chapter 51 and Appendix 5.

CHAPTER 33

COMPATIBILITY

331 COMPATIBILITY OF CHARACTERISTICS

331.1 An assessment shall be made of any characteristics of equipment likely to have harmful effects upon other electrical equipment or other services or likely to impair the supply, for example, for coordination with concerned parties e.g. petrol stations, kiosks and shops within shops. Those characteristics include, for example:

(i) transient overvoltages
(ii) undervoltage
(iii) unbalanced loads
(iv) rapidly fluctuating loads
(v) starting currents
(vi) harmonic currents
(vii) earth leakage current
(viii) excessive PE conductor current not due to a fault
(ix) DC feedback
(x) high-frequency oscillations
(xi) necessity for additional connections to Earth
(xii) power factor.

For an external source of energy the distributor shall be consulted regarding any equipment of the installation having a characteristic likely to have significant influence on the supply.

332 ELECTROMAGNETIC COMPATIBILITY

332.1 All electrical equipment forming part of an electrical installation shall meet the appropriate electromagnetic compatibility (EMC) requirements and shall be in accordance with the relevant EMC standard.

332.2 Consideration shall be given by the designer of the electrical installation to measures reducing the effect of induced voltage disturbances and electromagnetic interferences (EMI). Measures are given in Chapter 44.

CHAPTER 34

MAINTAINABILITY

341 GENERAL

341.1 An assessment shall be made of the frequency and quality of maintenance the installation can reasonably be expected to receive during its intended life. The person or body responsible for the operation and/or maintenance of the installation shall be consulted. Those characteristics are to be taken into account in applying the requirements of Parts 4 to 8 so that, having regard to the frequency and quality of maintenance expected:

(i) any periodic inspection and testing, maintenance and repairs likely to be necessary during the intended life can be readily and safely carried out, and
(ii) the effectiveness of the protective measures for safety during the intended life shall not diminish, and
(iii) the reliability of equipment for proper functioning of the installation is appropriate to the intended life.

NOTE: There may be particular statutory requirements relating to maintenance.

CHAPTER 35

SAFETY SERVICES

351 GENERAL

NOTE 1: The need for safety services and their nature are frequently regulated by statutory authorities whose requirements have to be observed.

NOTE 2: Examples of safety services are: emergency escape lighting, fire detection and fire alarm systems, installations for fire pumps, firefighters lifts, smoke and heat extraction equipment.

351.1 The following electrical sources for safety services are recognized:

(i) Storage batteries
(ii) Primary cells
(iii) Generator sets independent of the normal supply
(iv) A separate feeder of the supply network that is effectively independent of the normal feeder (see Regulation 560.6.5).

352 CLASSIFICATION

Refer to Regulation 560.4.

CHAPTER 36

CONTINUITY OF SERVICE

361 GENERAL

361.1 An assessment shall be made for each circuit of any need for continuity of service considered necessary during the intended life of the installation, e.g. life-support systems. The following characteristics shall be considered:

(i) Selection of the system earthing
(ii) Selection of the protective device in order to achieves electivity
(iii) Number of circuits
(iv) Multiple power supplies
(v) Use of monitoring devices.

PART 4

PROTECTION FOR SAFETY

CONTENTS

CHAPTER 45 *NOT USED*

CHAPTER 46 ISOLATION AND SWITCHING

460	**SCOPE**
461	**GENERAL**
462	**ISOLATION**
463	**FUNCTIONAL SWITCHING (CONTROL)**
464	**SWITCHING OFF FOR MECHANICAL MAINTENANCE**
465	**EMERGENCY SWITCHING OFF**

CHAPTER 41

PROTECTION AGAINST ELECTRIC SHOCK

CONTENTS

CHAPTER 41

PROTECTION AGAINST ELECTRIC SHOCK

410 INTRODUCTION

This chapter deals with protection against electric shock as applied to electrical installations. It is based on BS EN 61140, which is a basic safety standard that applies to the protection of persons and livestock. BS EN 61140 is intended to give fundamental principles and requirements that are common to electrical installations and equipment or are necessary for their coordination.

The fundamental rule of protection against electric shock, according to BS EN 61140, is that hazardous-live-parts shall not be accessible and accessible conductive parts shall not be hazardous-live, both under normal conditions and under single fault conditions.

According to 4.2 of BS EN 61140, protection under normal conditions is provided by basic protective provisions and protection under single fault conditions is provided by fault protective provisions.

Alternatively, protection against electric shock is provided by an enhanced protective provision, which provides protection under normal conditions and under single fault conditions.

410.1 Scope

Chapter 41 specifies essential requirements regarding protection against electric shock, including basic protection and fault protection of persons and livestock. It deals also with the application and coordination of these requirements in relation to external influences.

Requirements are given for the application of additional protection in certain cases.

410.2 Not used

410.3 General requirements

410.3.1 In BS 7671, the following specification of voltages is intended unless stated otherwise:

- AC voltages are rms
- DC voltages are ripple-free.

Ripple-free is conventionally defined as an rms ripple voltage of not more than 10 % of the DC component.

410.3.2 A protective measure shall consist of:

(i) an appropriate combination of a provision for basic protection and an independent provision for fault protection, or

(ii) an enhanced protective provision which provides both basic protection and fault protection.

Additional protection is specified as part of a protective measure under certain conditions of external influence and in certain special locations (see the corresponding sections of Part 7).

NOTE 1: For special applications, protective measures which do not follow this concept are permitted (see Regulations 410.3.5 and 410.3.6).

NOTE 2: An example of an enhanced protective measure is reinforced insulation.

410.3.3 In each part of an installation one or more protective measures shall be applied, taking account of the conditions of external influence.

The following protective measures generally are permitted:

(i) automatic disconnection of supply (Section 411)

(ii) double or reinforced insulation (Section 412)

(iii) electrical separation for the supply to one item of current-using equipment (Section 413)

(iv) extra-low voltage (SELV and PELV) (Section 414).

The protective measures applied in the installation shall be considered in the selection and erection of equipment.

For particular installations see Regulations 410.3.4 to 410.3.9.

NOTE: In electrical installations the most commonly used protective measure is automatic disconnection of supply.

410.3.4 For special installations or locations, the particular protective measures specified in the corresponding section of Part 7 shall be applied.

410.3.5 The protective measures specified in Section 417, i.e. the use of obstacles and placing out of reach, shall only be used in installations restricted to:

(i) skilled persons, or

(ii) instructed persons under the supervision of skilled persons.

410.3.6 The protective measures specified in Section 418, i.e.

(i) non-conducting location

(ii) earth-free local equipotential bonding

(iii) electrical separation for the supply of more than one item of current-using equipment

shall be applied only where the installation is under the supervision of skilled or instructed persons so that unauthorized changes cannot be made.

410.3.7 If certain conditions of a protective measure cannot be met, supplementary provisions shall be applied so that the protective provisions together achieve the same degree of safety.

NOTE: An example of the application of this regulation is given in Regulation 411.7 (FELV).

410.3.8 Different protective measures applied to the same installation or part of an installation or within equipment shall have no influence on each other such that failure of one protective measure could impair the other protective measure or measures.

410.3.9 The provision for fault protection may be omitted for the following equipment:

(i) metal supports of overhead line insulators which are attached to the building and are placed out of arm's reach

(ii) steel reinforced concrete poles of overhead lines in which the steel reinforcement is not accessible

(iii) exposed-conductive-parts which, owing to their reduced dimensions (approximate maximum of 50 mm x 50 mm) or their disposition cannot be gripped or come into significant contact with a part of the human body and provided that connection with a protective conductor could only be made with difficulty or would be unreliable

NOTE: This exemption applies, for example, to bolts, rivets, nameplates, cable clips, screws and other fixings.

(iv) metal enclosures protecting equipment in accordance with Section 412

(v) unearthed street furniture supplied from an overhead line and inaccessible in normal use.

411 PROTECTIVE MEASURE: AUTOMATIC DISCONNECTION OF SUPPLY

411.1 General

Automatic disconnection of supply is a protective measure in which:

(i) basic protection is provided by basic insulation of live parts or by barriers or enclosures, in accordance with Section 416, and

(ii) fault protection is provided by protective earthing, protective equipotential bonding and automatic disconnection in case of a fault, in accordance with Regulations 411.3 to 411.6.

Where this protective measure is applied, Class II equipment may also be used.

Where specified, additional protection is provided by an RCD with rated residual operating current not exceeding 30 mA, in accordance with Regulation 415.1.

NOTE: Residual Current Monitors (RCMs) are not protective devices but they may be used to monitor residual currents in an electrical installation. RCMs produce an audible or audible and visual signal when a preselected value of residual current is reached.

411.2 Requirements for basic protection

All electrical equipment shall comply with one of the provisions for basic protection described in Section 416 or, where appropriate, Section 417.

411.3 Requirements for fault protection

411.3.1 Protective earthing and protective equipotential bonding

411.3.1.1 Protective earthing

Exposed-conductive-parts shall be connected to a protective conductor under the specific conditions for each type of system earthing as specified in Regulations 411.4 to 411.6.

Simultaneously accessible exposed-conductive-parts shall be connected to the same earthing system individually, in groups or collectively.

Conductors for protective earthing shall comply with Chapter 54.

A circuit protective conductor shall be run to and terminated at each point in wiring and at each accessory except a lampholder having no exposed-conductive-parts and suspended from such a point.

411.3.1.2 Protective equipotential bonding

In each consumer's installation within a building, extraneous-conductive-parts liable to introduce a dangerous potential difference shall be connected to the main earthing terminal by protective bonding conductors complying with Chapter 54. Examples of extraneous-conductive-parts may include:

(i) metallic water installation pipes

(ii) metallic gas installation pipes

(iii) other metallic installation pipework and ducting

(iv) central heating and air conditioning systems

(v) exposed metallic structural parts of the building.

NOTE: Where non-metallic pipes (e.g. plastic) enter a building and are then connected to metallic pipes within the building, the metallic pipes within the building do not normally require protective bonding as they are unlikely to be extraneous-conductive-parts.

Connection of a lightning protection system to the protective equipotential bonding shall be made in accordance with BS EN 62305.

Where an installation serves more than one building the above requirement shall be applied to each building.

Where an incoming telecommunication cable has a metallic sheath, equipotential bonding shall be applied. However, the consent of the owner or operator of the cable shall be obtained. Where consent is not granted, and as a consequence this requirement is not met, details shall be recorded in the description section of the appropriate electrical certification specified in Part 6.

411.3.2 Automatic disconnection in case of a fault

411.3.2.1 Except as provided by Regulation 411.3.2.5, a protective device shall automatically interrupt the supply to the line conductor of a circuit or equipment in the event of a fault of negligible impedance between the line conductor and an exposed-conductive-part or a protective conductor in the circuit or equipment within the disconnection time required by Regulation 411.3.2.2, 411.3.2.3 or 411.3.2.4.

The protective device shall be suitable for isolation of at least the line conductor.

NOTE: For IT systems, automatic disconnection is not necessarily required on the occurrence of a first fault (see Regulation 411.6.1). For the requirements for disconnection in the event of a second fault, occurring on a different live conductor, see Regulation 411.6.5.

411.3.2.2 Maximum disconnection times stated in Table 41.1 shall be applied to final circuits with a rated current not exceeding:

(i) 63 A with one or more socket-outlets, and

(ii) 32 A supplying only fixed connected current-using equipment.

TABLE 41.1 –
Maximum disconnection times

System	$50\ V < U_0 \leq 120\ V$ (s)		$120\ V < U_0 \leq 230\ V$ (s)		$230\ V < U_0 \leq 400\ V$ (s)		$U_0 > 400\ V$ (s)	
	AC	DC	AC	DC	AC	DC	AC	DC
TN	0.8	NOTE 1	0.4	1	0.2	0.4	0.1	0.1
TT	0.3	NOTE 1	0.2	0.4	0.07	0.2	0.04	0.1

Where in TT systems the disconnection is achieved by an overcurrent protective device and the protective equipotential bonding is connected with all extraneous-conductive-parts within the installation in accordance with Regulation 411.3.1.2, the maximum disconnection times applicable to TN systems may be used.

U_0 nominal AC rms or ripple-free DC line voltage to Earth.

Where compliance with this regulation is provided by an RCD, the disconnection times in accordance with Table 41.1 relate to prospective residual fault currents significantly higher than the rated residual operating current of the RCD.

NOTE 1: Disconnection is not required for protection against electric shock but may be required for other reasons, such as protection against thermal effects.

NOTE 2: Where disconnection is provided by an RCD, see Note 2 to Regulation 411.4.4 and Note 4 to Regulation 411.6.5.

411.3.2.3 In a TN system, a disconnection time not exceeding 5 s is permitted for a distribution circuit and for a circuit not covered by Regulation 411.3.2.2.

411.3.2.4 In a TT system, a disconnection time not exceeding 1 s is permitted for a distribution circuit and for a circuit not covered by Regulation 411.3.2.2.

411.3.2.5 Where it is not feasible for an overcurrent protective device to interrupt the supply in accordance with Regulation 411.3.2 or the use of an RCD for this purpose is not appropriate, see Section 419. However, disconnection may be required for reasons other than protection against electric shock.

411.3.3 Additional requirements for socket-outlets and for the supply of mobile equipment for use outdoors

In AC systems, additional protection by means of an RCD with a rated residual operating current not exceeding 30 mA shall be provided for:

(i) socket-outlets with a rated current not exceeding 32 A in locations where they are liable to be used by persons of capability BA1, BA3 or children (BA2, BA3),

(ii) socket-outlets with a rated current not exceeding 32 A in other locations, and

(iii) mobile equipment with a rated current not exceeding 32 A for use outdoors.

An exception to (ii) but not (i) or (iii) is permitted where a suitably documented risk assessment undertaken with the involvement of a skilled person (electrically) determines that RCD protection is not necessary.

NOTE 1: For the purpose of this exception, an ordinary person (BA1) instructed in the use of the installation does not become an instructed person (electrically) or cease to be an ordinary person.

The documented risk assessment shall be provided with the appropriate electrical installation certificate.

The requirements of Regulation 411.3.3 do not apply to FELV systems according to Regulation 411.7 or reduced low voltage systems according to Regulation 411.8.

NOTE 2: See also Regulations 314.1(iv) and 531.3.2 concerning the avoidance of unwanted tripping.

NOTE 3: RCD protection of all socket-outlets is recommended.

NOTE 4: See Appendix 2, item 11 in respect of risk assessment.

NOTE 5: A lighting distribution unit complying with BS 5733, shaver supply unit complying with BS EN 61558-2-5, luminaire track system, installation coupler, LSC or DCL is not regarded as a socket-outlet for the purposes of this regulation.

411.3.4 Additional requirements for circuits with luminaires

Within domestic (household) premises, additional protection by an RCD with a rated residual operating current not exceeding 30 mA shall be provided for AC final circuits supplying luminaires.

411.4 TN system

411.4.1 In a TN system, the integrity of the earthing of the installation depends on the reliable and effective connection of the PEN or PE conductors to Earth. Where the earthing is provided from a public or other supply system, compliance with the necessary conditions external to the installation is the responsibility of the distributor.

411.4.2 The neutral point or the midpoint of the power supply system shall be earthed. If a neutral point or midpoint is not available or not accessible, a line conductor shall be earthed.

Exposed-conductive-parts of the installation shall be connected by a protective conductor to the main earthing terminal of the installation, which shall be connected to the earthed point of the power supply system.

It is recommended that an additional connection to Earth, by means of an earth electrode in accordance with Chapter 54, is made to the main earthing terminal. This recommendation does not apply to outbuildings of dwellings served by the installation.

411.4.3 In a fixed installation, a single conductor may serve both as a protective conductor and as a neutral conductor (PEN conductor) provided that the requirements of Regulation 543.4 are satisfied. No switching or isolating device shall be inserted in the PEN conductor.

NOTE: Regulation 8(4) of the Electricity Safety, Quality and Continuity Regulations prohibits the use of PEN conductors in consumers' installations.

411.4.4 The characteristics of the protective devices (see Regulation 411.4.5) and the circuit impedances shall fulfil the following requirement:

$$Z_s \times I_a \leq U_0 \times C_{min}$$

where:

Z_s is the impedance in ohms (Ω) of the fault loop comprising:

- the source
- the line conductor up to the point of the fault, and
- the protective conductor between the point of the fault and the source

I_a is the current in amperes (A) causing the automatic operation of the protective device within the time specified in Regulation 411.3.2.2, or Regulation 411.3.2.3. When an RCD is used this current is the residual operating current providing disconnection in the time specified in Regulation 411.3.2.2, or Regulation 411.3.2.3

U_0 nominal AC rms or ripple-free DC line voltage to Earth

C_{min} is the minimum voltage factor to take account of voltage variations depending on time and place, changing of transformer taps and other considerations.

NOTE: For a low voltage supply given in accordance with the Electricity Safety, Quality and Continuity Regulations, C_{min} is given the value 0.95.

NOTE 1: Where compliance with this regulation is provided by an RCD, the disconnection times in accordance with Table 41.1 relate to prospective residual fault currents significantly higher than the rated residual operating current of the RCD.

NOTE 2: In TN systems the residual fault currents are significantly higher than $I_{\Delta n}$. Therefore, the disconnecting times in accordance with Table 41.1 are fulfilled where an RCD according to BS EN 61008, BS EN 61009 or BS EN 62423 is used. A circuit-breaker incorporating residual current protection not suitable for use by ordinary persons (CBR) according to BS EN 60947-2 can be used, provided the time delay is adjusted to afford compliance with Table 41.1.

411.4.5 The following types of protective device may be used for fault protection:

(i) An overcurrent protective device

(ii) An RCD.

Where an RCD is used for fault protection the circuit shall also incorporate an overcurrent protective device in accordance with Chapter 43.

An RCD shall not be used in a TN-C system.

Where an RCD is used in a TN-C-S system, a PEN conductor shall not be used on the load side. The connection of the protective conductor to the PEN conductor shall be made on the source side of the RCD.

NOTE: Where selectivity between RCDs is necessary, see Regulation 536.4.1.4.

411.4.201 Where a fuse is used to satisfy the requirements of Regulation 411.3.2.2, maximum values of earth fault loop impedance (Z_s) corresponding to a disconnection time of 0.4 s are stated in Table 41.2 for a nominal voltage (U_0) of 230 V. For types and rated currents of general purpose (gG) and motor circuit application (gM) fuses other than those mentioned in Table 41.2, reference should be made to the appropriate British or Harmonized Standard to determine the value of I_a for compliance with Regulation 411.4.4.

TABLE 41.2 –
Maximum earth fault loop impedance (Z_S) for fuses, for 0.4 s disconnection time with U_0 of 230 V (see Regulation 411.4.201)

(a) General purpose (gG) and motor circuit application (gM) fuses to BS 88-2 – fuse systems E (bolted) and G (clip-in)

Rating (amperes)	2	4	6	10	16	20	25	32	40	50	63
Z_s (ohms)	33.1	15.6	7.80	4.65	2.43	1.68	1.29	0.99	0.75	0.57	0.44

(b) Fuses to BS 88-3 fuse system C

Rating (amperes)	5	16	20	32	45	63
Z_s (ohms)	9.93	2.30	1.93	0.91	0.57	0.36

(c) Fuses to BS 3036

Rating (amperes)	5	15	20	30	45	60
Z_s (ohms)	9.10	2.43	1.68	1.04	0.56	0.40

(d) Fuses to BS 1362

Rating (amperes)	3	13
Z_s (ohms)	15.6	2.30

NOTE 1: The circuit loop impedances have been determined using a value for factor C_{min} of 0.95.

NOTE 2: The circuit loop impedances given in the table should not be exceeded when:

(i) the line conductors are at the appropriate maximum permitted operating temperature, as given in Table 52.1, and

(ii) the circuit protective conductors are at the appropriate assumed initial temperature, as given in Tables 54.2 to 54.5.

If the conductors are at a different temperature when tested, the reading should be adjusted accordingly. See Appendix 3.

NOTE 3: Where the line conductor insulation is of a type for which Table 52.1 gives a maximum permitted operating temperature exceeding 70 °C, such as thermosetting, but the conductor has been sized in accordance with Regulation 512.1.5:

(i) the maximum permitted operating temperature for the purpose of Note 2(i) is 70 °C, and

(ii) the assumed initial temperature for the purpose of Note 2(i) is that given in Tables 54.2 to 54.4 corresponding to an insulation material of 70 °C thermoplastic.

411.4.202 Where a circuit-breaker is used to satisfy the requirements of Regulation 411.3.2.2 or Regulation 411.3.2.3, the maximum value of earth fault loop impedance (Z_s) shall be determined by the formula in Regulation 411.4.4. Alternatively, for a nominal voltage (U_0) of 230 V and a disconnection time of 0.4 s in accordance with Regulation 411.3.2.2 or 5 s in accordance with Regulation 411.3.2.3, the values specified in Table 41.3 for the types and ratings of overcurrent devices listed may be used instead of calculation.

TABLE 41.3 –

Maximum earth fault loop impedance (Z_s) for circuit-breakers with U_0 of 230 V, for operation giving compliance with the 0.4 s disconnection time of Regulation 411.3.2.2 and 5 s disconnection time of Regulation 411.3.2.3 (for RCBOs see also Regulation 411.4.204)

(a) Type B circuit-breakers to BS EN 60898 and the overcurrent characteristics of RCBOs to BS EN 61009-1

Rating (amperes)	3	6	10	16	20	25	32	40	50	63	80	100	125	I_n
Z_s (ohms)	14.57	7.28	4.37	2.73	2.19	1.75	1.37	1.09	0.87	0.69	0.55	0.44	0.35	230 x 0.95/($5I_n$)

(b) Type C circuit-breakers to BS EN 60898 and the overcurrent characteristics of RCBOs to BS EN 61009-1

Rating (amperes)	6	10	16	20	25	32	40	50	63	80	100	125	I_n
Z_s (ohms)	3.64	2.19	1.37	1.09	0.87	0.68	0.55	0.44	0.35	0.27	0.22	0.17	230 x 0.95/($10I_n$)

(c) Type D circuit-breakers to BS EN 60898 and the overcurrent characteristics of RCBOs to BS EN 61009-1

Rating (amperes)	6	10	16	20	25	32	40	50	63	80	100	125	I_n
Z_s (ohms) 0.4 sec	1.82	1.09	0.68	0.55	0.44	0.34	0.27	0.22	0.17	0.14	0.11	0.09	230 x 0.95/($20I_n$)
Z_s (ohms) 5 secs	3.64	2.19	1.37	1.09	0.87	0.68	0.55	0.44	0.35	0.27	0.22	0.17	230 x 0.95/($10I_n$)

NOTE 1: The circuit loop impedances have been determined using a value for factor C_{min} of 0.95.

NOTE 2: The circuit loop impedances given in the table should not be exceeded when:

(i) the line conductors are at the appropriate maximum permitted operating temperature, as given in Table 52.1, and

(ii) the circuit protective conductors are at the appropriate assumed initial temperature, as given in Tables 54.2 to 54.5.

If the conductors are at a different temperature when tested, the reading should be adjusted accordingly. See Appendix 3.

NOTE 3: Where the line conductor insulation is of a type for which Table 52.1 gives a maximum permitted operating temperature exceeding 70 °C, such as thermosetting, but the conductor has been sized in accordance with Regulation 512.1.5:

(i) the maximum permitted operating temperature for the purpose of Note 2(i) is 70 °C, and

(ii) the assumed initial temperature for the purpose of Note 2(i) is that given in Tables 54.2 to 54.4 corresponding to an insulation material of 70 °C thermoplastic.

411.4.203 Where a fuse is used for a distribution circuit or a final circuit in accordance with Regulation 411.3.2.3, maximum values of earth fault loop impedance (Z_s) corresponding to a disconnection time of 5 s are stated in Table 41.4 for a nominal voltage (U_0) of 230 V. For types and rated currents of general purpose (gG) and motor circuit application (gM) fuses other than those mentioned in Table 41.4, reference should be made to the appropriate British or Harmonized Standard to determine the value of I_a for compliance with Regulation 411.4.4.

TABLE 41.4 –
Maximum earth fault loop impedance (Z_s) for fuses, for 5 s disconnection time with U_0 of 230 V (see Regulation 411.4.203)

(a) General purpose (gG) and motor circuit application (gM) fuses to BS 88-2 – fuse systems E (bolted) and G (clip in)								
Rating (amperes)	2	4	6	10	16	20	25	32
Z_s (ohms)	44	21	12	6.8	4.0	2.8	2.2	1.7
continued								
Rating (amperes)	40	50	63	80	100	125	160	200
Z_s (ohms)	1.3	0.99	0.78	0.55	0.42	0.32	0.27	0.18
(b) Fuses to BS 88-3 fuse system C								
Rating (amperes)	5	16	20	32	45	63	80	100
Z_s (ohms)	14.6	3.9	3.2	1.6	1.0	0.68	0.51	0.38
(c) Fuses to BS 3036								
Rating (amperes)	5	15	20	30	45	60	100	
Z_s (ohms)	16.8	5.08	3.64	2.51	1.51	1.07	0.51	
(d) Fuses to BS 1362								
Rating (amperes)	3	13						
Z_s (ohms)	22.0	3.64						

NOTE 1: The circuit loop impedances have been determined using a value for factor C_{min} of 0.95.

NOTE 2: The circuit loop impedances given in the table should not be exceeded when:

(i) the line conductors are at the appropriate maximum permitted operating temperature, as given in Table 52.1, and

(ii) the circuit protective conductors are at the appropriate assumed initial temperature, as given in Tables 54.2 to 54.4.

If the conductors are at a different temperature when tested, the reading should be adjusted accordingly. See Appendix 3.

NOTE 3: Where the line conductor insulation is of a type for which Table 52.1 gives a maximum permitted operating temperature exceeding 70 °C, such as thermosetting, but the conductor has been sized in accordance with Regulation 512.1.5:

(i) the maximum permitted operating temperature for the purpose of Note 2(i) is 70 °C, and

(ii) the assumed initial temperature for the purpose of Note 2(i) is that given in Tables 54.2 to 54.5 corresponding to an insulation material of 70 °C thermoplastic.

NOTE 4: Data for fuses of rating exceeding 200 A should be obtained from the manufacturer.

411.4.204 Where an RCD is used to satisfy the requirements of Regulation 411.3.2.2, 411.3.2.3 or 411.3.2.4 the maximum values of earth fault loop impedance in Table 41.5 may be applied for non-delayed RCDs to BS EN 61008-1 and BS EN 61009-1 for a nominal voltage U_0 of 230 V. In such cases, an overcurrent protective device shall provide protection against overload current and fault current in accordance with Chapter 43.

411.5 TT system

411.5.1 Every exposed-conductive-part which is to be protected by a single protective device shall be connected, via the main earthing terminal, to a common earth electrode. However, if two or more protective devices are in series, the exposed-conductive-parts may be connected to separate earth electrodes corresponding to each protective device.

The neutral point or the midpoint of the power supply system shall be earthed. If a neutral point or midpoint is not available or not accessible, a line conductor shall be earthed.

411.5.2 One or more of the following types of protective device shall be used, the former being preferred:

(i) An RCD

(ii) An overcurrent protective device.

NOTE 1: An appropriate overcurrent protective device may be used for fault protection provided a suitably low value of Z_S is permanently and reliably assured.

NOTE 2: Where an RCD is used for fault protection the circuit should also incorporate an overcurrent protective device in accordance with Chapter 43.

411.5.3 Where an RCD is used for fault protection, the following conditions shall be fulfilled:

(i) The disconnection time shall be that required by Regulation 411.3.2.2 or 411.3.2.4, and

(ii) $R_A \times I_{\Delta n} \leq 50$ V

where:

R_A is the sum of the resistances of the earth electrode and the protective conductor connecting it to the exposed-conductive-parts (in ohms)

$I_{\Delta n}$ is the rated residual operating current of the RCD.

The requirements of this regulation are met if the earth fault loop impedance of the circuit protected by the RCD meets the requirements of Table 41.5.

NOTE 1: Where selectivity between RCDs is necessary refer also to Regulation 536.4.1.4.

NOTE 2: Where R_A is not known, it may be replaced by Z_S.

TABLE 41.5 –
Maximum earth fault loop impedance (Z_S) for non-delayed and time delayed 'S' Type RCDs to BS EN 61008-1 and BS EN 61009-1 for U_0 of 230 V (see Regulation 411.5.3)

Rated residual operating current (mA)	Maximum earth fault loop impedance Z_s (ohms)
30	1667*
100	500*
300	167
500	100

Disconnection shall be within the times stated in Table 41.1.

NOTE 1: Figures for Z_s result from the application of Regulation 411.5.3(i) and (ii).

NOTE 2: * The resistance of the installation earth electrode should be as low as practicable. A value exceeding 200 ohms may not be stable. Refer to Regulation 542.2.4.

411.5.4 Where an overcurrent protective device is used the following condition shall be fulfilled:

$Z_s \times I_a \leq U_0 \times C_{min}$

where:

Z_s is the impedance in ohms (Ω) of the earth fault loop comprising:

- the source
- the line conductor up to the point of the fault
- the protective conductor from the exposed-conductive-parts
- the earthing conductor
- the earth electrode of the installation and
- the earth electrode of the source

I_a is the current in amperes (A) causing the automatic operation of the protective device within the time specified in Regulation 411.3.2.2 or 411.3.2.4

U_0 nominal AC rms or ripple-free DC line voltage to Earth

C_{min} is the minimum voltage factor to take account of voltage variations depending on time and place, changing of transformer taps and other considerations.

NOTE: For a low voltage supply given in accordance with the Electricity Safety, Quality and Continuity Regulations as amended, C_{min} is given the value 0.95.

411.6 IT system

411.6.1 In an IT system, live parts shall be insulated from Earth or connected to Earth through a sufficiently high impedance. This connection may be made either at the neutral point or midpoint of the system or at an artificial neutral point. The latter may be connected directly to Earth if the resulting impedance to Earth is sufficiently high at the system frequency. Where no neutral point or midpoint exists, a line conductor may be connected to Earth through a high impedance.

Where the above conditions apply, the fault current is then low in the event of a single fault to an exposed-conductive-part or to Earth, so that automatic disconnection in accordance with Regulation 411.3.2 is not imperative provided that the condition in Regulation 411.6.2 is fulfilled. Precautions shall be taken, however, to avoid the risk of harmful effects on a person in contact with simultaneously accessible exposed-conductive-parts in the event of two faults existing simultaneously.

NOTE 1: It is strongly recommended that IT systems with distributed neutrals should not be employed.

NOTE 2: To reduce overvoltage or to damp voltage oscillation, it may be necessary to provide earthing through impedances or artificial neutral points, and the characteristics of these should be appropriate to the requirements of the installation.

411.6.2 Exposed-conductive-parts shall be earthed individually, in groups, or collectively.

In AC systems the following condition shall be fulfilled to limit the touch voltage to:

$R_A \times I_d \leq 50\ \text{V}$

where:

R_A is the sum of the resistances in ohms of the earth electrode and protective conductor for the exposed-conductive-parts.

I_d is the fault current in amperes (A) of the first fault of negligible impedance between a line conductor and an exposed-conductive-part. The value of I_d takes account of leakage currents and the total earthing impedance of the electrical installation.

NOTE: No touch voltage limitation is considered in DC systems as the value of I_d can be considered to be negligibly low.

411.6.3 The following monitoring devices and protective devices may be used:

(i) insulation monitoring devices (IMDs)
(ii) residual current monitoring devices (RCMs)
(iii) insulation fault location systems (IFLS)
(iv) overcurrent protective devices (OCPD)
(v) residual current devices (RCDs).

NOTE 1: Where an RCD is used, tripping of the RCD in the event of a first fault cannot be excluded, due to capacitive leakage currents.

NOTE 2: In case of faults in two different Class I current-using devices supplied by different line conductors, the operation of an RCD is only likely to be achieved if every single item of current-using equipment is protected by an individual RCD. In such a case, the use of an overcurrent protective device is more suitable.

411.6.4 Where an IT system is designed not to disconnect in the event of a first fault, the occurrence of the first fault shall be indicated by either:

(i) an insulation monitoring device (IMD), which may be combined with an insulation fault location system (IFLS), or
(ii) a residual current monitor (RCM), provided the residual current is sufficiently high to be detected.

NOTE: RCMs are not able to detect symmetrical insulation faults.

The device shall initiate an audible and/or visual signal which shall continue as long as the fault persists. The signal can be initiated via a relay contact output, an electronic switching output or a communication protocol.

A visual and/or an audible alarm system shall be arranged at a suitable place, so that it is perceived by responsible persons.

If there are both audible and visible signals, it is permissible for the audible signal to be cancelled.

In addition, an insulation fault location system according to BS EN 61557-9 may be provided to indicate the location of a first fault from a live part to exposed-conductive-parts or Earth or another reference point.

411.6.5 After the occurrence of a first fault, conditions for automatic disconnection of supply in the event of a second fault occurring on a different live conductor shall be as follows:

(i) Where exposed-conductive-parts are interconnected by a protective conductor collectively earthed to the same earthing system, the conditions similar to a TN system apply and the following conditions shall be fulfilled where the neutral conductor is not distributed in AC systems and in DC systems where the midpoint conductor is not distributed:

$$Z_s \leq \frac{U \times C_{min}}{2I_a}$$

or where the neutral conductor or midpoint conductor respectively is distributed:

$$Z_s^1 \leq \frac{U_0 \times C_{min}}{2I_a}$$

where:

U	is the nominal AC or DC voltage, in volts, between line conductors
U_0	is the nominal AC or DC voltage, in volts, between line conductor and neutral conductor or midpoint conductor, as appropriate
Z_s	is the impedance in ohms of the fault loop comprising the line conductor and the protective conductor of the circuit
$Z^1{}_s$	is the impedance in ohms of the fault loop comprising the neutral conductor and the protective conductor of the circuit
I_a	is the current in amperes (A) causing operation of the protective device within the time required in Regulation 411.3.2.2 for TN systems or Regulation 411.3.2.3
C_{min}	is the minimum voltage factor to take account of voltage variations depending on time and place, changing of transformer taps and other considerations.

NOTE 1: The time stated in Table 41.1 of Regulation 411.3.2.2 for the TN system is applicable to IT systems with a distributed or non-distributed neutral conductor or midpoint conductor.

NOTE 2: The factor 2 in both formulae takes into account that in the event of the simultaneous occurrence of two faults, the faults may exist in different circuits.

NOTE 3: For fault loop impedance, the most severe case should be taken into account, e.g. a fault on the line conductor at the source and simultaneously another fault on the neutral conductor of current-using equipment of the circuit considered.

(ii) Where the exposed-conductive-parts are earthed in groups or individually, the following condition applies:

$$R_A \times I_a \leq 50 \text{ V}$$

where:

R_A	is the sum of the resistances, in ohms, of the earth electrode and the protective conductor to the exposed-conductive-parts
I_a	is the current in amperes causing automatic disconnection of the protective device in a time complying with that for TT systems in Table 41.1 of Regulation 411.3.2.2 or in a time complying with Regulation 411.3.2.4.

NOTE 4: If compliance to the requirements of (ii) is provided by an RCD, compliance with the disconnection times required for TT systems in Table 41.1 may require residual currents significantly higher than the rated residual operating current $I_{\Delta n}$ of the RCD applied.

411.7 Functional extra-low voltage (FELV)

411.7.1 General

Where, for functional reasons, a nominal voltage not exceeding 50 V AC or 120 V DC is used but not all the requirements of Section 414 relating to SELV or to PELV are fulfilled, and where SELV or PELV is not necessary, the supplementary provisions described in Regulation 411.7.2 and 411.7.3 shall be applied to provide basic protection and fault protection.

This combination of provisions is known as FELV.

NOTE: Such conditions may, for example, be encountered where the circuit contains equipment (such as transformers, relays, remote-control switches, contactors) insufficiently insulated with respect to circuits at higher voltage.

411.7.2 Requirements for basic protection

Basic protection shall be provided by either:

(i) basic insulation according to Regulation 416.1 corresponding to the nominal voltage of the primary circuit of the source, or

(ii) barriers or enclosures in accordance with Regulation 416.2.

411.7.3 Requirements for fault protection

The exposed-conductive-parts of the equipment of the FELV circuit shall be connected to the protective conductor of the primary circuit of the source, provided that the primary circuit is subject to protection by automatic disconnection of supply as described in Regulations 411.3 to 411.6.

411.7.4 Sources

The source of the FELV system shall either be a transformer with at least simple separation between windings or shall comply with Regulation 414.3.

If an extra-low voltage system is supplied from a higher voltage system by equipment which does not provide at least simple separation between that system and the extra-low voltage system, such as an autotransformer, a potentiometer or a semiconductor device, the output circuit is not part of a FELV system and is deemed to be an extension of the input circuit and shall be protected by the protective measure applied to the input circuit.

NOTE: This does not preclude connecting a conductor of the FELV circuit to the protective conductor of the primary circuit.

411.7.5 Plugs, socket-outlets, LSCs, DCLs and cable couplers

Every plug, socket-outlet, luminaire supporting coupler (LSC), device for connecting a luminaire (DCL) and cable coupler in a FELV system shall have a protective conductor contact and shall not be dimensionally compatible with those used for any other system in use in the same premises.

411.8 Reduced low voltage systems

411.8.1 General

411.8.1.1 Where, for functional reasons, the use of extra-low voltage is impracticable and there is no requirement for the use of SELV or PELV, a reduced low voltage system may be used, for which the provisions described in Regulations 411.8.2 to 411.8.5 shall be made to provide basic protection and fault protection. |

411.8.1.2 The nominal voltage of the reduced low voltage circuits shall not exceed 110 V AC rms between lines (three-phase 63.5 V to earthed neutral, single-phase 55 V to earthed midpoint).

411.8.2 Requirements for basic protection

Basic protection shall be provided by either:

(i) basic insulation according to Regulation 416.1 corresponding to the maximum nominal voltage of the reduced low voltage system given in Regulation 411.8.1.2, or

(ii) barriers or enclosures in accordance with Regulation 416.2.

411.8.3 Requirements for fault protection

Fault protection by automatic disconnection of supply shall be provided by means of an overcurrent protective device in each line conductor or by an RCD, and all exposed-conductive-parts of the reduced low voltage system shall be connected to Earth. The earth fault loop impedance at every point of utilization, including socket-outlets, shall be such that the disconnection time does not exceed 5 s.

Where a circuit-breaker is used, the maximum value of earth fault loop impedance (Z_s) shall be determined by the formula in Regulation 411.4.4. Alternatively, the values specified in Table 41.6 may be used instead of calculation for the nominal voltages (U_0) and the types and ratings of overcurrent device listed therein.

Where a fuse is used, the maximum values of earth fault loop impedance (Z_s) corresponding to a disconnection time of 5 s are stated in Table 41.6 for nominal voltages (U_0) of 55 V and 63.5 V.

For types and rated currents of fuses other than those mentioned in Table 41.6, reference should be made to the appropriate British or Harmonized Standard to determine the value of I_a for compliance with Regulation 411.4.4, according to the appropriate value of the nominal voltage (U_0).

Where fault protection is provided by an RCD, the product of the rated residual operating current ($I_{\Delta n}$) in amperes and the earth fault loop impedance in ohms shall not exceed 50 V.

TABLE 41.6 –
Maximum earth fault loop impedance (Z_s) for 5 s disconnection time and U_0 of 55 V (single-phase) and 63.5 V (three-phase)
(see Regulations 411.8.1.2 and 411.8.3)

	Circuit-breakers to BS EN 60898 and the overcurrent characteristics of RCBOs to BS EN 61009-1 Type				General purpose (gG) fuses to BS 88-2 – fuse systems E and G	
	B		C and D			
U_0 (Volts)	55	63.5	55	63.5	55	63.5
Rating amperes	Z_s ohms					
3	3.48	4.02	1.74	2.01		
6	1.74	2.01	0.87	1.01	2.90	3.35
10	1.05	1.21	0.52	0.60	1.63	1.89
16	0.65	0.75	0.33	0.38	0.95	1.10
20	0.52	0.60	0.26	0.30	0.67	0.77
25	0.42	0.48	0.21	0.24	0.52	0.60
					-	-
32	0.33	0.38	0.16	0.19	0.42	0.48
40	0.26	0.30	0.13	0.15	0.31	0.35
50	0.21	0.24	0.10	0.12	0.24	0.27
63	0.17	0.19	0.08	0.10	0.19	0.22
80	0.13	0.15	0.07	0.08	0.13	0.15
100	0.10	0.12	0.05	0.06	0.12	0.14
125	0.08	0.10	0.04	0.05	0.08	0.09
I_n	$10.4/I_n$	$12.1/I_n$	$5.2/I_n$	$6.1/I_n$		

NOTE 1: The circuit loop impedances have been determined using a value for factor C_{min} of 0.95.

NOTE 2: The circuit loop impedances given in the table should not be exceeded when:

(i) the line conductors are at the appropriate maximum permitted operating temperature, as given in Table 52.1, and

(ii) the circuit protective conductors are at the appropriate assumed initial temperature, as given in Tables 54.2 to 54.6.

If the conductors are at a different temperature when tested, the reading should be adjusted accordingly. See Appendix 3.

NOTE 3: Where the line conductor insulation is of a type for which Table 52.1 gives a maximum permitted operating temperature exceeding 70 °C, such as thermosetting, but the conductor has been sized in accordance with Regulation 512.1.5:

(i) the maximum permitted operating temperature for the purpose of Note 2(i) is 70 °C, and

(ii) the assumed initial temperature for the purpose of Note 2(ii) is that given in Tables 54.2 to 54.4 corresponding to an insulation material of 70 °C thermoplastic.

NOTE 4: Data for fuses of rating exceeding 200 A should be obtained from the manufacturer.

411.8.4 Sources

411.8.4.1 The source of supply to a reduced low voltage circuit shall be one of the following:

(i) A double-wound isolating transformer complying with BS EN 61558-1 and BS EN 61558-2-23

(ii) A motor-generator set having windings providing isolation equivalent to that provided by the windings of an isolating transformer

(iii) A source independent of other supplies, e.g. an engine-driven generator.

411.8.4.2 The neutral (star) point of the secondary windings of three-phase transformers and generators, or the midpoint of the secondary windings of single-phase transformers and generators, shall be connected to Earth.

411.8.5 Requirements for circuits

Every plug, socket-outlet, luminaire supporting coupler (LSC), device for connecting a luminaire (DCL) and cable coupler of a reduced low voltage system shall have a protective conductor contact and shall not be dimensionally compatible with those used for any other system in use in the same premises.

412 PROTECTIVE MEASURE: DOUBLE OR REINFORCED INSULATION

412.1 General

412.1.1 Double or reinforced insulation is a protective measure in which:

(i) basic protection is provided by basic insulation and fault protection is provided by supplementary insulation, or

(ii) basic and fault protection is provided by reinforced insulation between live parts and accessible parts.

NOTE: This protective measure is intended to prevent the appearance of a dangerous voltage on the accessible parts of electrical equipment through a fault in the basic insulation.

The protective measure of double or reinforced insulation is applicable in all situations, unless some limitations are given in the corresponding section of Part 7.

412.1.2 Where this protective measure is to be used as the sole protective measure (i.e. where a whole installation or circuit is intended to consist entirely of equipment with double insulation or reinforced insulation), it shall be verified that effective measures, for example by adequate supervision, are in place so that no change can be made that would impair the effectiveness of the protective measure. Therefore this protective measure shall not be applied to any circuit that includes, for example, a socket-outlet with an earthing contact, luminaire supporting coupler (LSC), device for connecting a luminaire (DCL) or cable coupler, or where a user may change items of equipment without authorization.

412.2 Requirements for basic protection and fault protection

412.2.1 Electrical equipment

Where the protective measure double or reinforced insulation is used for the complete installation or part thereof, electrical equipment shall comply with one of the following:

(i) Regulation 412.2.1.1, or

(ii) Regulations 412.2.1.2 and 412.2.2, or

(iii) Regulations 412.2.1.3 and 412.2.2.

412.2.1.1 Electrical equipment shall be of the following types, type-tested and marked to the relevant standards:

(i) electrical equipment having double or reinforced insulation (Class II equipment)

(ii) electrical equipment declared in the relevant product standard as equivalent to Class II, such as assemblies of electrical equipment having total insulation (see BS EN 61439 series).

NOTE: This equipment is identified by the symbol  refer to BS EN 60417: Class II equipment.

412.2.1.2 Electrical equipment having basic insulation only shall have supplementary insulation applied in the process of erecting the electrical installation, providing a degree of safety equivalent to electrical equipment according to Regulation 412.2.1.1 and complying with Regulations 412.2.2.1 to 412.2.2.3.

NOTE: The symbol  should be fixed in a visible position both on the exterior and interior of the enclosure.

412.2.1.3 Electrical equipment having uninsulated live parts shall have reinforced insulation applied in the process of erecting the electrical installation, providing a degree of safety equivalent to electrical equipment according to Regulation 412.2.1.1 and complying with Regulations 412.2.2.2 and 412.2.2.3. Such insulation is recognized only where constructional features prevent the application of double insulation.

NOTE: The symbol  should be fixed in a visible position both on the exterior and interior of the enclosure.

412.2.2 Enclosures

412.2.2.1 The electrical equipment being ready for operation, all conductive parts separated from live parts by basic insulation only, shall be contained in an insulating enclosure affording at least the degree of protection IPXXB or IP2X.

412.2.2.2 The following requirements apply as specified:

(i) the insulating enclosure shall not be traversed by conductive parts likely to transmit a potential; and

(ii) the insulating enclosure shall not contain any screws or other fixing means of insulating material which might need to be removed, or are likely to be removed, during installation and maintenance and whose replacement by metallic screws or other fixing means could impair the enclosure's insulation.

Where the insulating enclosure must be traversed by mechanical joints or connections (e.g. for operating handles of built-in equipment), these should be arranged in such a way that protection against shock in case of a fault is not impaired.

412.2.2.3 Where a lid or door in the insulating enclosure can be opened without the use of a tool or key, all conductive parts which are accessible if the lid or door is open shall be behind an insulating barrier (providing a degree of protection not less than IPXXB or IP2X) preventing persons from coming unintentionally into contact with those conductive parts. This insulating barrier shall be removable only by the use of a tool or key.

412.2.2.4 Conductive parts enclosed in the insulating enclosure shall not be connected to a protective conductor. However, provision may be made for connecting protective conductors which necessarily run through the enclosure in order to serve other items of electrical equipment whose supply circuit also runs through the enclosure. Inside the enclosure, any such conductors and their terminals shall be insulated as though they were live parts, and their terminals shall be marked as protective conductor (PE) terminals.

No exposed-conductive-parts or intermediate part shall be connected to a protective conductor unless specific provision for this is made in the specification for the equipment concerned.

412.2.2.5 The enclosure shall not adversely affect the operation of the equipment protected in this way.

412.2.3 Installation

412.2.3.1 The installation of equipment mentioned in Regulation 412.2.1 (fixing, connection of conductors, etc.) shall be effected in such a way as not to impair the protection afforded in compliance with the equipment specification.

412.2.3.2 Except where Regulation 412.1.2 applies, a circuit supplying one or more items of Class II equipment shall have a circuit protective conductor run to and terminated at each point in wiring and at each accessory.

NOTE: This requirement is intended to take account of the replacement by the user of Class II equipment by Class I equipment.

412.2.4 Wiring systems

412.2.4.1 Wiring systems installed in accordance with Chapter 52 are considered to meet the requirements of Regulation 412.2 if:

(i) the rated voltage of the cable(s) is not less than the nominal voltage of the system and at least 300/500 V, and

(ii) adequate mechanical protection of the basic insulation is provided by one or more of the following:

(a) The non-metallic sheath of the cable

(b) Non-metallic trunking or ducting complying with the BS EN 50085 series of standards, or non-metallic conduit complying with the BS EN 61386 series of standards.

NOTE 1: Cable product standards do not specify impulse withstand capability. However, it is considered that the insulation of the cabling system is at least equivalent to the requirement in BS EN 61140 for reinforced insulation.

NOTE 2: A wiring system should not be identified by the symbol  or by the symbol .

413 PROTECTIVE MEASURE: ELECTRICAL SEPARATION

413.1 General

413.1.1 Electrical separation is a protective measure in which:

(i) basic protection is provided by basic insulation of live parts or by barriers or enclosures in accordance with Section 416, and

(ii) fault protection is provided by simple separation of the separated circuit from other circuits and from Earth.

413.1.2 Except as permitted by Regulation 413.1.3, this protective measure shall be limited to the supply of one item of current-using equipment supplied from one unearthed source with simple separation.

NOTE: When this protective measure is used, it is particularly important that the basic insulation complies with the relevant product standard.

413.1.3 Where more than one item of current-using equipment is supplied from an unearthed source with simple separation, the requirements of Regulation 418.3 shall be met.

413.2 Requirements for basic protection

All electrical equipment shall be subject to one of the basic protective provisions in Section 416 or to the protective measures in Section 412.

413.3 Requirements for fault protection

413.3.1 Protection by electrical separation shall comply with Regulations 413.3.2 to 413.3.6.

413.3.2 The separated circuit shall be supplied through a source with at least simple separation, and the voltage of the separated circuit shall not exceed 500 V.

413.3.3 Live parts of the separated circuit shall not be connected at any point to another circuit or to Earth or to a protective conductor.

For electrical separation, arrangements shall be such that basic insulation is achieved between circuits in compliance with Regulation 416.1.

413.3.4 Flexible cables shall be visible throughout any part of their length liable to mechanical damage.

413.3.5 For separated circuits the use of separate wiring systems is recommended. If separated circuits and other circuits are in the same wiring system, multi-conductor cables without metallic covering, insulated conductors in insulating conduit, insulating ducting or insulating trunking shall be used, provided that:

(i) the rated voltage is not less than the highest nominal voltage, and

(ii) each circuit is protected against overcurrent.

413.3.6 The exposed-conductive-parts of the separated circuit shall not be connected either to the protective conductor or exposed-conductive-parts of other circuits, or to Earth.

NOTE: If the exposed-conductive-parts of the separated circuit are liable to come into contact, either intentionally or fortuitously, with the exposed-conductive-parts of other circuits, protection against electric shock no longer depends solely on protection by electrical separation but also on the protective provisions to which the latter exposed-conductive-parts are subject.

414 PROTECTIVE MEASURE: EXTRA-LOW VOLTAGE PROVIDED BY SELV OR PELV

414.1 General

414.1.1 Protection by extra-low voltage is a protective measure which consists of either of two different extra-low voltage systems:

(i) SELV, or

(ii) PELV.

Protection by extra-low voltage provided by SELV or PELV requires:

(iii) limitation of voltage in the SELV or PELV system to the upper limit of voltage Band I, 50 V AC or 120 V DC (see BS EN 61140), and

(iv) protective separation of the SELV or PELV system from all circuits other than SELV and PELV circuits, and basic insulation between the SELV or PELV system and other SELV or PELV systems, and

(v) for SELV systems only, basic insulation between the SELV system and Earth.

414.1.2 The use of SELV or PELV according to Section 414 is considered as a protective measure in all situations.

NOTE: In certain cases Part 7 limits the value of the extra-low voltage to a value lower than 50 V AC or 120 V DC.

414.2 Requirements for basic protection and fault protection

Both basic protection and fault protection are deemed to be provided where:

(i) the nominal voltage cannot exceed the upper limit of voltage Band I, and

(ii) the supply is from one of the sources listed in Regulation 414.3, and

(iii) the conditions of Regulation 414.4 are fulfilled.

NOTE 1: If the system is supplied from a higher voltage system by equipment which provides at least simple separation between that system and the extra-low voltage system but which does not meet the requirements for SELV and PELV sources in Regulation 414.3, the requirements for FELV may be applicable, see Regulation 411.7.

NOTE 2: DC voltages for ELV circuits generated by a semiconductor convertor (see BS EN 60146-2) require an internal AC voltage circuit to supply the rectifier stack. This internal AC voltage exceeds the DC voltage. The internal AC circuit is not to be considered as a higher voltage circuit within the meaning of this regulation. Between internal circuits and external higher voltage circuits, protective separation is required.

NOTE 3: In DC systems with batteries, the battery charging and floating voltages exceed the battery nominal voltage, depending on the type of battery. This does not require any protective provisions in addition to those specified in this regulation. The charging voltage should not exceed a maximum value of 75 V AC or 150 V DC as appropriate according to the environmental situation as given in Table 1 of PD 6536 (IEC 61201).

414.3 Sources for SELV and PELV

The following sources may be used for SELV and PELV systems:

(i) A safety isolating transformer in accordance with BS EN 61558-2-6 or BS EN 61558-2-8

(ii) A source of current providing a degree of safety equivalent to that of the safety isolating transformer specified in (i) (e.g. motor-generator with windings providing equivalent isolation)

(iii) An electrochemical source (e.g. a battery) or another source independent of a higher voltage circuit (e.g. a diesel-driven generator)

(iv) Certain electronic devices complying with appropriate standards, where provisions have been taken such that, even in the case of an internal fault, the voltage at the outgoing terminals cannot exceed the values specified in Regulation 414.1.1. Higher voltages at the outgoing terminals are, however, permitted where, in case of contact with a live part or in the event of a fault between a live part and an exposed-conductive-part, the voltage at the output terminals is immediately reduced to the value specified in Regulation 414.1.1 or less.

NOTE 1: Examples of such devices include insulation testing equipment and monitoring devices.

NOTE 2: Where higher voltages exist at the outgoing terminals, compliance with this regulation may be assumed if the voltage at the outgoing terminals is within the limits specified in Regulation 414.1.1 when measured with a voltmeter having an internal resistance of at least 3 000 ohms.

A mobile source supplied at low voltage, e.g. a safety isolating transformer or a motor-generator, shall be selected and erected in accordance with the requirements for protection by the use of double or reinforced insulation (see Section 412).

414.4 Requirements for SELV and PELV circuits

414.4.1 SELV and PELV circuits shall have:

(i) basic insulation between live parts and other SELV or PELV circuits, and

(ii) protective separation from live parts of circuits not being SELV or PELV, provided by double or reinforced insulation or by basic insulation and protective screening for the highest voltage present.

SELV circuits shall have basic insulation between live parts and Earth.

The PELV circuits and/or exposed-conductive-parts of equipment supplied by the PELV circuits may be earthed.

NOTE 1: In particular, protective separation is necessary between the live parts of electrical equipment such as relays, contactors and auxiliary switches, and any part of a higher voltage circuit or a FELV circuit.

NOTE 2: The earthing of PELV circuits may be achieved by a connection to Earth or to an earthed protective conductor within the source itself.

414.4.2 Protective separation of wiring systems of SELV or PELV circuits from the live parts of other circuits, which have at least basic insulation, shall be achieved by one of the following arrangements:

(i) SELV and PELV circuit conductors enclosed in a non-metallic sheath or insulating enclosure in addition to basic insulation

(ii) SELV and PELV circuit conductors separated from conductors of circuits at voltages higher than Band I by an earthed metallic sheath or earthed metallic screen

(iii) Circuit conductors at voltages higher than Band I may be contained in a multi-conductor cable or other grouping of conductors if the SELV and PELV conductors are insulated for the highest voltage present

(iv) The wiring systems of other circuits are in compliance with Regulation 412.2.4.1

(v) Physical separation.

414.4.3 Every socket-outlet and luminaire supporting coupler in a SELV or PELV system shall require the use of a plug which is not dimensionally compatible with those used for any other system in use in the same premises.

Plugs and socket-outlets in a SELV system shall not have a protective conductor contact.

414.4.4 Exposed-conductive-parts of SELV circuits shall not be connected to Earth, or to protective conductors or exposed-conductive-parts of another circuit.

NOTE: If the exposed-conductive-parts of SELV circuits are liable to come into contact, either fortuitously or intentionally, with the exposed-conductive-parts of other circuits, protection against electric shock no longer depends solely on protection by SELV, but also on the protective provisions to which the latter exposed-conductive-parts are subject.

414.4.5 If the nominal voltage exceeds 25 V AC or 60 V DC or if the equipment is immersed, basic protection shall be provided for SELV and PELV circuits by:

(i) insulation in accordance with Regulation 416.1

(ii) barriers or enclosures in accordance with Regulation 416.2.

Basic protection is generally unnecessary in normal dry conditions for:

(i) SELV circuits where the nominal voltage does not exceed 25 V AC or 60 V DC

(ii) PELV circuits where the nominal voltage does not exceed 25 V AC or 60 V DC and exposed-conductive-parts and/or the live parts are connected by a protective conductor to the main earthing terminal.

In all other cases, basic protection is not required if the nominal voltage of the SELV or PELV system does not exceed 12 V AC or 30 V DC.

415 ADDITIONAL PROTECTION

NOTE: Additional protection in accordance with Section 415 may be specified with the protective measure. In particular, additional protection may be required with the protective measure under certain conditions of external influence and in certain special locations (see the corresponding section of Part 7).

415.1 Additional protection: RCDs

415.1.1 The use of RCDs with a rated residual operating current not exceeding 30 mA is recognized in AC systems as additional protection in the event of failure of the provision for basic protection and/or the provision for fault protection or carelessness by users.

415.1.2 The use of RCDs is not recognized as a sole means of protection and does not obviate the need to apply one of the protective measures specified in Sections 411 to 414.

415.2 Additional protection: supplementary protective equipotential bonding

NOTE 1: Supplementary protective equipotential bonding is considered as an addition to fault protection.

NOTE 2: The use of supplementary protective equipotential bonding does not exclude the need to disconnect the supply for other reasons, for example protection against fire, thermal stresses in equipment, etc.

NOTE 3: Supplementary protective equipotential bonding may involve the entire installation, a part of the installation, an item of equipment, or a location.

NOTE 4: Additional requirements may be necessary for special locations (see the corresponding section of Part 7), or for other reasons.

415.2.1 Supplementary protective equipotential bonding shall include all simultaneously accessible exposed-conductive-parts of fixed equipment and extraneous-conductive-parts including where practicable the main metallic reinforcement of constructional reinforced concrete. The supplementary protective equipotential bonding system shall be connected to the protective conductors of all equipment including those of socket-outlets.

415.2.2 The resistance R between simultaneously accessible exposed-conductive-parts and extraneous-conductive-parts shall fulfil the following condition:

$R \leq 50 \text{ V}/I_a$ in AC systems

$R \leq 120 \text{ V}/I_a$ in DC systems

where I_a is the operating current in amperes (A) of the protective device or:

(i) for RCDs, $I_{\Delta n}$

(ii) for overcurrent devices, the 5 s operating current.

416 PROVISIONS FOR BASIC PROTECTION

NOTE: Provisions for basic protection provide protection under normal conditions and are applied where specified as a part of the chosen protective measure.

416.1 Basic insulation of live parts

Live parts shall be completely covered with insulation which can only be removed by destruction. For equipment, the insulation shall comply with the relevant standard for such electrical equipment.

NOTE: The insulation is intended to prevent contact with live parts. Paint, varnish, lacquer or similar products are generally not considered to provide adequate insulation for basic protection in normal service.

416.2 Barriers or enclosures

NOTE: Barriers or enclosures are intended to prevent contact with live parts.

416.2.1 Live parts shall be inside enclosures or behind barriers providing at least the degree of protection IPXXB or IP2X except that, where larger openings occur during the replacement of parts, such as certain lampholders or fuses, or where larger openings are necessary to allow the proper functioning of equipment according to the relevant requirements for the equipment:

(i) suitable precautions shall be taken to prevent persons or livestock from unintentionally touching live parts, and

(ii) as far as is reasonably practicable, persons will be aware that live parts can be touched through the opening and should not be touched intentionally, and

(iii) the opening shall be as small as is consistent with the requirement for proper functioning and for replacement of a part.

416.2.2 A horizontal top surface of a barrier or enclosure which is readily accessible shall provide a degree of protection of at least IPXXD or IP4X.

416.2.3 A barrier or enclosure shall be firmly secured in place and have sufficient stability and durability to maintain the required degree of protection and appropriate separation from live parts in the known conditions of normal service, taking account of relevant external influences.

416.2.4 Where it is necessary to remove a barrier or open an enclosure or remove parts of enclosures, this shall be possible only:

(i) by the use of a key or tool, or

(ii) after disconnection of the supply to live parts against which the barriers or enclosures afford protection, restoration of the supply being possible only after replacement or reclosure of the barrier or enclosure, or

(iii) where an intermediate barrier providing a degree of protection of at least IPXXB or IP2X prevents contact with live parts, by the use of a key or tool to remove the intermediate barrier.

NOTE: This regulation does not apply to:

- a ceiling rose complying with BS 67
- a cord operated switch complying with BS EN 60669-1
- a bayonet lampholder complying with BS EN 61184
- an Edison screw lampholder complying with BS EN 60238.

416.2.5 If, behind a barrier or in an enclosure, an item of equipment such as a capacitor is installed which may retain a dangerous electrical charge after it has been switched off, a warning label shall be provided. Small capacitors such as those used for arc extinction and for delaying the response of relays, etc shall not be considered dangerous.

NOTE: Unintentional contact is not considered dangerous if the voltage resulting from static charge falls below 120 V DC in less than 5 s after disconnection from the power supply.

417 OBSTACLES AND PLACING OUT OF REACH

417.1 Application

The protective measures of obstacles and placing out of reach provide basic protection only. They are for application in installations, with or without fault protection, that are controlled or supervised by skilled persons.

The conditions of supervision under which the basic protective provisions of Section 417 may be applied as part of the protective measure are given in Regulation 410.3.5.

417.2 Obstacles

NOTE: Obstacles are intended to prevent unintentional contact with live parts but not intentional contact by deliberate circumvention of the obstacle.

417.2.1 Obstacles shall prevent:

(i) unintentional bodily approach to live parts, and

(ii) unintentional contact with live parts during the operation of live equipment in normal service.

417.2.2 An obstacle may be removed without the use of a key or tool but shall be secured so as to prevent unintentional removal.

417.3 Placing out of reach

NOTE: Protection by placing out of reach is intended only to prevent unintentional contact with live parts.

A bare or insulated overhead line for distribution between buildings and structures shall be installed to the standard required by the Electricity Safety, Quality and Continuity Regulations.

417.3.1 Simultaneously accessible parts at different potentials shall not be within arm's reach.

A bare live part other than an overhead line shall not be within arm's reach or within 2.50 m of the following:

(i) An exposed-conductive-part

(ii) An extraneous-conductive-part

(iii) A bare live part of any other circuit.

NOTE: Two parts are deemed to be simultaneously accessible if they are not more than 2.50 m apart (see Figure 417).

417.3.2 If a normally occupied position is restricted in the horizontal direction by an obstacle (e.g. handrail, mesh screen) affording a degree of protection less than IPXXB or IP2X, arm's reach shall extend from that obstacle. In the overhead direction, arm's reach is 2.50 m from the surface, S, not taking into account any intermediate obstacle providing a degree of protection less than IPXXB.

NOTE: The values of arm's reach apply to contact directly with bare hands without assistance (e.g. tools or ladder).

Fig 417 – Arm's reach

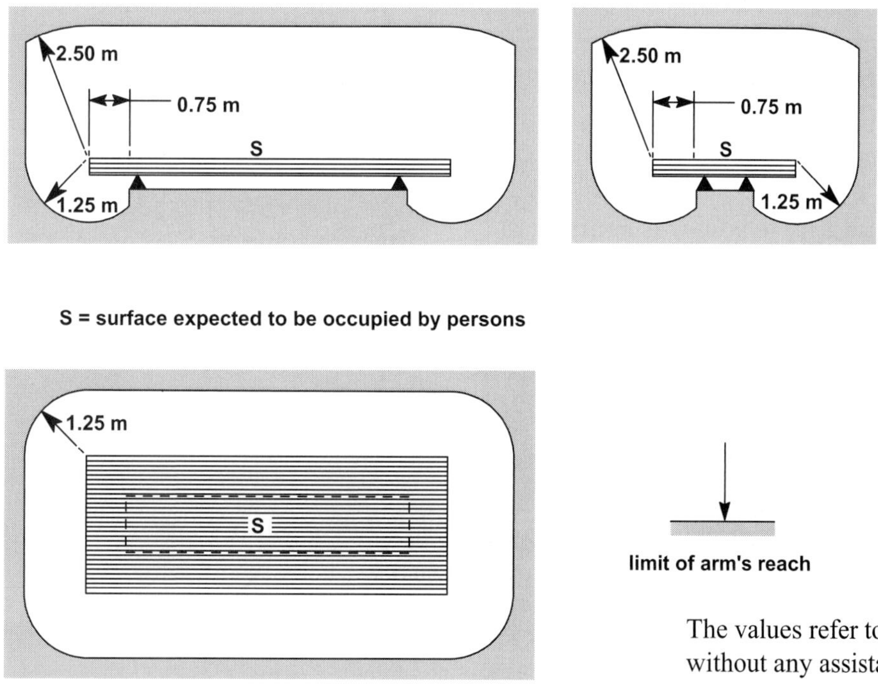

The values refer to bare hands without any assistance, e.g. from tools or a ladder.

417.3.3 In places where bulky or long conductive objects are normally handled, the distances required by Regulations 417.3.1 and 417.3.2 shall be increased, taking account of the relevant dimensions of those objects.

418 PROTECTIVE MEASURES FOR APPLICATION ONLY WHERE THE INSTALLATION IS CONTROLLED OR UNDER THE SUPERVISION OF SKILLED OR INSTRUCTED PERSONS

NOTE: The conditions of supervision under which the fault protective provisions of Section 418 may be applied as part of the protective measure are given in Regulation 410.3.6.

418.1 Non-conducting location

This method of protection is not recognized for general application.

NOTE: This protective measure is intended to prevent simultaneous contact with parts which may be at different potentials through failure of the basic insulation of live parts.

418.1.1 All electrical equipment shall comply with one of the provisions for basic protection described in Section 416.

418.1.2 Exposed-conductive-parts shall be arranged so that under ordinary circumstances persons will not come into simultaneous contact with:

(i) two exposed-conductive-parts, or

(ii) an exposed-conductive-part and any extraneous-conductive-part

if these parts are liable to be at different potentials through failure of the basic insulation of a live part.

418.1.3 In a non-conducting location there shall be no protective conductor.

418.1.4 Regulation 418.1.2 is fulfilled if the location has an insulating floor and walls and one or more of the following arrangements applies:

(i) Relative spacing of exposed-conductive-parts and of extraneous-conductive-parts as well as spacing of exposed-conductive-parts

This spacing is sufficient if the distance between two parts is not less than 2.5 m; this distance may be reduced to 1.25 m outside the zone of arm's reach.

(ii) Interposition of effective obstacles between exposed-conductive-parts and extraneous-conductive-parts

Such obstacles are sufficiently effective if they extend the distances to be surmounted to the values stated in point (i) above. They shall not be connected to Earth or to exposed-conductive-parts; as far as possible they shall be of insulating material.

(iii) Insulation or insulating arrangements of extraneous-conductive-parts.

The insulation shall be of sufficient mechanical strength and be able to withstand a test voltage of at least 2 000 V. Leakage current shall not exceed 1 mA in normal conditions of use.

418.1.5 The resistance of insulating floors and walls at every point of measurement under the conditions specified in Part 6 shall be not less than:

(i) 50 kΩ, where the nominal voltage of the installation does not exceed 500 V, or

(ii) 100 kΩ, where the nominal voltage of the installation exceeds 500 V.

NOTE: If at any point the resistance is less than the specified value, the floors and walls are deemed to be extraneous-conductive-parts for the purposes of protection against electric shock.

418.1.6 The arrangements made shall be permanent and it shall not be possible to make them ineffective. The arrangements shall also provide protection where the use of mobile equipment is envisaged.

NOTE 1: Attention is drawn to the risk that, where electrical installations are not under effective supervision, further conductive parts may be introduced at a later date (e.g. mobile Class I equipment, or extraneous-conductive-parts such as metallic water pipes), which may invalidate compliance with Regulation 418.1.6.

NOTE 2: It is essential that the insulation of floors and walls cannot be affected by humidity.

418.1.7 Precautions shall be taken so that extraneous-conductive-parts cannot cause a potential to appear within the vicinity of the location concerned.

418.2 Protection by earth-free local equipotential bonding

This method of protection shall be used only in special circumstances.

NOTE: Earth-free local equipotential bonding is intended to prevent the appearance of a dangerous touch voltage.

418.2.1 All electrical equipment shall comply with one of the provisions for basic protection described in Section 416.

418.2.2 Protective bonding conductors shall interconnect every simultaneously accessible exposed-conductive-part and extraneous-conductive-part.

418.2.3 The local protective bonding conductors shall neither be in electrical contact with Earth directly, nor through exposed-conductive-parts, nor through extraneous-conductive-parts.

NOTE: Where this requirement cannot be fulfilled, protection by automatic disconnection of supply is applicable (see Section 411).

418.2.4 Precautions shall be taken so that persons entering the equipotential location cannot be exposed to a dangerous potential difference, in particular, where a conductive floor insulated from Earth is connected to the earth-free protective bonding conductors.

418.2.5 Where this measure is applied, a warning notice complying with Regulation 514.13.2 shall be fixed in a prominent position adjacent to every point of access to the location concerned.

418.3 Electrical separation for the supply to more than one item of current-using equipment

Where the measure is used to supply two or more items of equipment from a single source, a warning notice complying with Regulation 514.13.2 shall be fixed in a prominent position adjacent to every point of access to the location concerned.

NOTE: Electrical separation of an individual circuit is intended to prevent shock currents through contact with exposed-conductive-parts that may be energized by a fault in the basic insulation of the circuit.

418.3.1 All electrical equipment shall comply with one of the provisions for basic protection described in Section 416.

418.3.2 Protection by electrical separation for the supply to more than one item of equipment shall be achieved by compliance with all the requirements of Section 413 except Regulation 413.1.2, and with the requirements in Regulations 418.3.3 to 418.3.8. |

418.3.3 Precautions shall be taken to protect the separated circuit from damage and insulation failure.

418.3.4 The exposed-conductive-parts of the separated circuit shall be connected together by insulated, non-earthed protective bonding conductors. Such conductors shall not be connected to the protective conductor or exposed-conductive-parts of any other circuit or to any extraneous-conductive-parts.

NOTE: See Note to Regulation 413.3.6.

418.3.5 Every socket-outlet shall be provided with a protective conductor contact which shall be connected to the equipotential bonding system provided in accordance with Regulation 418.3.4.

418.3.6 Except where supplying equipment with double or reinforced insulation, all flexible cables shall embody a protective conductor for use as a protective bonding conductor in accordance with Regulation 418.3.4.

418.3.7 If two faults affecting two exposed-conductive-parts occur and these are fed by conductors of different polarity, a protective device shall disconnect the supply in a disconnection time conforming with Table 41.1.

418.3.8 The product of the nominal voltage of the circuit in volts and length of the wiring system in metres shall not exceed 100 000 Vm, and the length of the wiring system shall not exceed 500 m.

419 PROVISIONS WHERE AUTOMATIC DISCONNECTION ACCORDING TO REGULATION 411.3.2 IS NOT FEASIBLE

419.1 Where automatic disconnection is not feasible in circumstances where:

(i) electronic equipment with limited short-circuit current is installed, or

(ii) the required disconnection times cannot be achieved by a protective device

the provisions of Regulations 419.2 and 419.3 are applicable.

419.2 For installations with power electronic convertors with nominal voltage U_0 greater than 50 V AC or 120 V DC and where automatic disconnection is not feasible, the output voltage of the source shall be reduced to 50 V AC or 120 V DC or less in the event of a fault between a live conductor and the protective conductor or Earth in a time as given in Regulation 411.3.2.2, 411.3.2.3 or 411.3.2.4, as appropriate (see BS EN 62477-1).

The power electronic convertor shall be one for which the manufacturer gives adequate methods for the initial verification and periodic inspection and testing of the installation.

419.3 Except where Regulation 419.2 applies, if automatic disconnection cannot be achieved in the time required by Regulation 411.3.2.2, 411.3.2.3 or 411.3.2.4 as appropriate, supplementary protective equipotential bonding shall be provided in accordance with Regulation 415.2 and the voltage between simultaneously accessible exposed-conductive-parts and/or extraneous-conductive-parts shall not exceed 50 V AC or 120 V DC.

CHAPTER 42

PROTECTION AGAINST THERMAL EFFECTS

CONTENTS

CHAPTER 42

420.1 Scope

This chapter applies to electrical installations and equipment with regard to measures for the protection of persons, livestock and property against:

(i) the harmful effects of heat or thermal radiation developed by electrical equipment

(ii) the ignition, combustion or degradation of materials

(iii) flames and smoke where a fire hazard could be propagated from an electrical installation to other nearby fire compartments, and

(iv) safety services being cut off by the failure of electrical equipment.

NOTE 1: For protection against thermal effects and fire, statutory requirements may be applicable. Refer to Appendix 2.

NOTE 2: Protection against overcurrent is dealt with in Chapter 43 of these Regulations.

421 PROTECTION AGAINST FIRE CAUSED BY ELECTRICAL EQUIPMENT

421.1 General requirements

421.1.1 Persons, livestock and property shall be protected against harmful effects of heat or fire which may be generated or propagated in electrical installations.

Manufacturers' instructions shall be taken into account in addition to the requirements of BS 7671.

NOTE 1: Harmful effects of heat or fire may be caused by:

- heat accumulation, heat radiation, hot components or equipment
- failure of electrical equipment such as protective devices, switchgear, thermostats, temperature limiters, seals of cable penetrations and wiring systems
- overcurrent
- insulation faults or arcs, sparks and high temperature particles
- harmonic currents
- external influences such as lightning surge
- inappropriate selection or erection of equipment.

NOTE 2: Lightning strikes and overvoltages are covered in BS EN 62305 and Section 443 of these Regulations.

421.1.2 Fixed electrical equipment shall be selected and erected such that its temperature in normal operation will not cause a fire. This shall be achieved by the construction of the equipment or by additional protective measures taken during erection.

The heat generated by electrical equipment shall not cause danger or harmful effects to adjacent fixed material or to material which may foreseeably be in proximity to such equipment.

Where fixed equipment may attain surface temperatures which could cause a fire hazard to adjacent materials, one or more of the following installation methods shall be adopted. The equipment shall:

(i) be mounted on a support which has low thermal conductance or within an enclosure which will withstand, with minimal risk of fire or harmful thermal effect, such temperatures as may be generated, or

(ii) be screened by materials of low thermal conductance which can withstand, with minimal risk of fire or harmful thermal effect, the heat emitted by the electrical equipment, or

(iii) be mounted so as to allow safe dissipation of heat and at a sufficient distance from adjacent material on which such temperatures could have deleterious effects. Any means of support shall be of low thermal conductance.

421.1.3 Where arcs, sparks or particles at high temperature may be emitted by fixed equipment in normal service, the equipment shall meet one or more of the following requirements. It shall be:

(i) totally enclosed in arc-resistant material

(ii) screened by arc-resistant material from materials upon which the emissions could have harmful effects

(iii) mounted so as to allow safe extinction of the emissions at a sufficient distance from materials upon which the emissions could have harmful effects

(iv) in compliance with its standard.

Arc-resistant material used for this protective measure shall be of low ignitability in accordance with BS EN ISO 11925-2, of low thermal conductivity and of adequate thickness to provide mechanical stability.

421.1.4 Fixed equipment causing a concentration and focusing of heat shall be at a sufficient distance from any fixed object or building element so that the object or element is not subjected to a dangerous temperature in normal conditions.

421.1.5 Where electrical equipment in a single location contains flammable liquid in significant quantity, adequate precautions shall be taken to prevent the spread of liquid, flame and the products of combustion.

NOTE 1: Examples of such precautions are:

(a) a retention pit of sufficient size to collect any leakage of liquid and provide extinction in the event of fire

(b) installation of the equipment in a chamber of adequate fire-resistance and the provision of sills or other means of preventing burning liquid spreading to other parts of the building, such a chamber being ventilated solely to the external atmosphere.

NOTE 2: The generally accepted lower limit for a significant quantity is 25 litres.

NOTE 3: For quantities less than 25 litres, it is sufficient to take precautions to prevent the escape of liquid.

NOTE 4: Products of combustion of liquid are considered to be smoke and gas.

421.1.201 Within domestic (household) premises, consumer units and similar switchgear assemblies shall comply with BS EN 61439-3 and shall:

(i) have their enclosure manufactured from non-combustible material, or

(ii) be enclosed in a cabinet or enclosure constructed of non-combustible material and complying with Regulation 132.12.

NOTE: Ferrous metal, e.g. steel, is deemed to be an example of a non-combustible material.

421.1.6 Materials used for the construction of enclosures of electrical equipment shall comply with the resistance to heat and fire requirements in an appropriate product standard.

Where no product standard exists, the materials of an enclosure shall withstand the highest temperature likely to be produced by the electrical equipment in normal use.

421.1.7 Arc fault detection devices (AFDD) conforming to BS EN 62606 shall be provided for single-phase AC final circuits supplying socket-outlets with a rated current not exceeding 32 A in:

- Higher Risk Residential Buildings (HRRB)
- Houses in Multiple Occupation (HMO)
- Purpose-built student accommodation
- Care homes.

NOTE 1: Higher Risk Residential Buildings are assumed to be residential buildings over 18 m in height or in excess of six storeys, whichever is met first. It is anticipated that in many areas higher risk residential buildings will be defined in legislation which can be subject to change over time, as well as in risk management procedures adopted by fire and rescue services. Current legislation should be applied.

For all other premises, the use of AFDDs conforming to BS EN 62606 is recommended for single-phase AC final circuits supplying socket-outlets not exceeding 32 A.

Where used, AFDDs shall be placed at the origin of the circuit to be protected.

The use of AFDDs does not obviate the need to apply one or more measures provided in other clauses in BS 7671.

NOTE 2: For busbar systems conforming to BS EN 61439-6 and Powertrack systems to BS EN 61534, the AFDD may be placed at a location other than the origin of the circuit.

422 PRECAUTIONS WHERE PARTICULAR RISKS OF FIRE EXIST

422.1 General

The requirements of this regulation shall be applied in addition to those of Section 421 for installations in locations where any of the conditions of external influence described in Regulations 422.2 to 422.6 exist.

For such locations the fire safety design of the building(s) should be documented. This information should be included in a fire safety manual produced by or for the person responsible for the building (the 'responsible person') as detailed in local and/or national fire safety legislation or guidance, including, for example BS 9999.

It is recommended that the electrical system designer/installer provides the person responsible for the building with details of the electrical system, setting out the basis of the design with respect to fire safety including routine maintenance requirements.

Where it is determined that cables should have an improved fire performance but are not covered by Regulations 422.2 to 422.6, this may be achieved by using cables with a minimum light transmittance of 60 % when tested in accordance with BS EN 61034-2 and,

(i) limited flame propagation according to the minimum requirements of the relevant part of BS EN 60332-3 series or

(ii) where cable management systems according to 422.2.1 (iii) or (iv) are used, cables shall conform to the minimum requirements of BS EN 60332-1-2.

422.1.1 Except for wiring systems meeting the requirements of Regulation 422.3.5, electrical equipment shall be restricted to that necessary to the use of the locations given in Regulation 422.1.

422.1.2 Electrical equipment shall be so selected and erected that its normal temperature rise and foreseeable temperature rise during a fault cannot cause a fire. This shall be achieved by the construction of the equipment or by additional protective measures taken during erection.

Special measures are not necessary where the temperature of surfaces is unlikely to cause combustion of nearby substances.

422.1.3 A temperature cut-out device shall have manual reset only.

422.2 Protected escape routes

Cables or other electrical equipment shall not be installed in a protected escape route unless part of:

(i) an essential fire safety or related safety system

(ii) general needs lighting

(iii) socket-outlets provided for cleaning or maintenance.

NOTE 1: Guidance is provided in Appendix 13.

NOTE 2: Generally, this means cables in a protected escape route should be limited to lighting and associated accessories, emergency lighting and fire detection and alarm systems, although cables for other safety systems may be necessary. Hospitals may have special requirements as detailed in Section 710.

422.2.1 Only cables that meet the following requirements shall be installed in protected escape routes:

(i) cables meeting the requirements of 560.8.1 or cables with resistance to flame propagation according to the recommended requirements of the relevant part of BS EN 60332-3 series or, where cable management systems according to (iii) or (iv) below are used, to the recommended requirements of BS EN 60332-1-2 and

(ii) a minimum of 60 % light transmittance when tested in accordance with BS EN 61034-2.

Cables in protected escape routes shall be as short as practicable. Cables encroaching on escape routes shall not be installed within arm's reach unless they are provided with protection against mechanical damage likely to occur during an evacuation.

Where used in protected escape routes, cable management systems shall be one or more of the following types and shall be of limited smoke production so as not to inhibit escape:

(iii) conduit systems classified as non-flame propagating according to BS EN 61386

(iv) cable trunking systems and cable ducting systems classified as non-flame propagating according to BS EN 50085

(v) cable tray and cable ladder systems classified as non-flame propagating according to BS EN 61537

(vi) powertrack systems meeting the requirements of BS EN 61534.

NOTE 1: Non-fire resisting cables need to satisfy the requirements of the CPR in respect of their reaction to fire. See Appendix 2, item 17. There are no requirements in the CPR for fire resisting cables described in Regulation 560.8.

NOTE 2: Ferrous metal, e.g. steel, is deemed to be an example of a material having limited smoke production.

Cables that are supplying safety circuits shall have a resistance to fire rating of either the time authorized by regulations for building elements or British Standards for the circuits or one hour in the absence of such a regulation or standard. Refer to Regulation 560.8 for such cables.

422.2.2 Switchgear or controlgear shall be accessible only to authorized persons. If switchgear or controlgear is placed in an escape route, it shall be enclosed in a cabinet or an enclosure constructed of non-combustible or not readily combustible material.

These requirements do not apply to items of switchgear or controlgear installed to facilitate evacuation, such as fire alarm call points.

422.2.3 In protected escape routes the use of electrical equipment containing flammable liquids is not permitted.

This requirement does not apply to individual capacitors incorporated in equipment, such as a capacitor installed in a discharge luminaire or a motor starter.

422.3 Locations with risks of fire due to the nature of processed or stored materials

The requirements of this regulation shall be applied in addition to those of Section 421 in locations where BE2 conditions exist.

This regulation does not apply to selection and erection of installations in locations with explosion risks, see BS EN 60079-14 and BS EN 61241-14.

NOTE 1: BE2 conditions exist where there is a risk of fire due to the manufacture, processing or storage of flammable materials including the presence of dust (see Appendix 5).

NOTE 2: Examples of locations presenting BE2 conditions include barns (due to the accumulation of dust and fibres), woodworking facilities, paper mills and textile factories (due to the storage and processing of combustible materials).

NOTE 3: Quantities of flammable materials or the surface or volume of the location may be regulated by national authorities.

422.3.1 Except for equipment for which an appropriate product standard specifies requirements, a luminaire shall be kept at an adequate distance from combustible materials. Unless otherwise recommended by the manufacturer, a small spotlight or projector shall be installed at the following minimum distance from combustible materials:

(i)	Rating up to 100 W	0.5 m
(ii)	Over 100 and up to 300 W	0.8 m
(iii)	Over 300 and up to 500 W	1.0 m.

Lamps and other components of luminaires shall be protected against foreseeable mechanical stresses. Such protective means shall not be fixed to lampholders unless they form an integral part of the luminaire or are fitted in accordance with the manufacturer's instructions.

A luminaire with a lamp that could eject flammable materials in case of failure shall be constructed with a safety protective shield for the lamp in accordance with the manufacturer's instructions.

422.3.2 Measures shall be taken to prevent an enclosure of electrical equipment such as a heater or resistor from exceeding the following temperatures:

(i) 90 °C under normal conditions, and

(ii) 115 °C under fault conditions.

Where materials such as dust or fibres sufficient to cause a fire hazard could accumulate on an enclosure of electrical equipment, adequate measures shall be taken to prevent the enclosure from exceeding the temperatures stated above.

NOTE: Luminaires marked ▽D in compliance with BS EN 60598-2-24 have limited surface temperature.

422.3.3 Switchgear or controlgear shall be installed outside the location unless:

(i) it is suitable for the location, or

(ii) it is installed in an enclosure providing a degree of protection of at least IP4X or, in the presence of dust, IP5X or, in the presence of electrically conductive dust, IP6X, except where Regulation 422.3.11 applies.

422.3.4 A cable shall as a minimum meet the requirements of BS EN 60332-1-2 unless it is fully embedded in non-combustible material such as plaster or concrete, or be installed in one or more of the following:

(i) A conduit system satisfying the test under fire conditions specified in BS EN 61386-1.

(ii) A cable trunking system or cable ducting system satisfying the test under fire conditions specified in the appropriate part of the BS EN 50085 series.

A cable tray system or cable ladder shall satisfy the test under fire conditions specified in BS EN 61537.

A powertrack system shall satisfy the test for resistance to flame propagation specified in the appropriate part of the BS EN 61534 series.

Wiring systems shall be selected and installed to minimize the propagation of flame.

Where the risk of flame propagation is high the cable shall meet the requirements of the appropriate part of BS EN 60332-3 series.

NOTE 1: The risk of flame propagation can be high where cables are bunched or installed in long vertical runs.

NOTE 2: Cables manufactured for the above application also need to satisfy the requirements of the CPR in respect of their reaction to fire. See Appendix 2, item 17.

NOTE 3: Where non-metallic cable management systems are proposed to be used in locations where a particular risk of fire exists, the manufacturer should be consulted regarding the suitability for use with cables not meeting BS EN 60332-3 series.

422.3.5 A wiring system which passes through the location but is not intended to supply electrical equipment in the location shall:

(i) meet the requirements of Regulation 422.3.4, and

(ii) have no connection or joint within the location, unless the connection or joint is installed within an enclosure that does not adversely affect the flame propagation characteristics of the wiring system, and

(iii) be protected against overcurrent in accordance with the requirements of Regulation 422.3.10, and

(iv) not employ bare live conductors.

422.3.6 *Not used*

422.3.7 A motor which is automatically or remotely controlled or which is not continuously supervised shall be protected against excessive temperature by a protective device with manual reset. A motor shall be protected against overtemperature in all operational modes.

422.3.8 Every luminaire shall:

(i) be appropriate for the location, and

(ii) be provided with an enclosure providing a degree of protection of at least IP4X or, in the presence of dust, IP5X or, in the presence of electrically conductive dust, IP6X, and

(iii) have a limited surface temperature in accordance with BS EN 60598-2-24, and

(iv) be of a type that prevents lamp components from falling from the luminaire.

In locations where there may be fire hazards due to dust or fibres, luminaires shall be installed so that dust or fibres cannot accumulate in dangerous amounts.

422.3.9 Wiring systems, other than mineral insulated cables, busbar trunking systems or powertrack systems, shall be protected against insulation faults:

(i) in a TN or TT system, by an RCD having a rated residual operating current ($I_{\Delta n}$) not exceeding 300 mA according with Regulation 531.3.2 and to relevant product standards.

Where a resistive fault may cause a fire, e.g. for overhead heating with heating film elements, the rated residual operating current shall not exceed 30 mA.

(ii) in an IT system, by an insulation monitoring device with audible and visual signals provided in accordance with Regulation 538.1. Disconnection times in the event of a second fault are given in Chapter 41. Alternatively, RCDs with a rated residual operating current as specified in (i) may be used. In the event of a second fault, see Chapter 41 for disconnection times.

422.3.10 Circuits supplying or traversing locations where BE2 conditions exist shall be protected against overload and against fault current by protective devices located outside and on the supply side of these locations. Circuits originating inside these locations shall be protected against overcurrent by protective devices located at their origin.

422.3.11 Regardless of the nominal voltage of a circuit supplied at extra-low voltage, live parts shall be either:

(i) contained in enclosures affording a degree of protection of at least IPXXB or IP2X, or

(ii) provided with insulation capable of withstanding a test voltage of 500 V DC for 1 minute.

These requirements are in addition to those of Section 414.

422.3.12 A PEN conductor shall not be used. This requirement does not apply to a circuit traversing the location.

422.3.13 Except as permitted by Regulation 461.2, every circuit shall be provided with a means of isolation from all live supply conductors by a linked switch or a linked circuit-breaker.

NOTE: Provision may be made for isolation of a group of circuits by a common means, if the service conditions allow this.

422.3.201 Flexible cables shall be of the following construction:

(i) Heavy duty type having a voltage rating of not less than 450/750 V, or

(ii) suitably protected against mechanical damage.

NOTE 1: Descriptions of light, ordinary and heavy duty types are given in BS EN 50565-1.

NOTE 2: Suitable flexible heavy duty cables can be found in BS EN 50525-2-21 (conventional elastomeric types) and BS EN 50525-3-21 (low smoke, halogen-free types).

422.3.202 A heating appliance shall be fixed.

422.3.203 A heat storage appliance shall be of a type which prevents the ignition of combustible dusts or fibres by the heat storing core.

422.4 Combustible constructional materials

The requirements of this regulation shall be applied in addition to those of Section 421 in locations where CA2 conditions exist.

NOTE: CA2 conditions exist where a building is mainly constructed of combustible materials, such as wood (see Appendix 5).

422.4.1 Precautions shall be taken so that electrical equipment does not ignite walls, floors or ceilings. In prefabricated hollow walls containing a pre-installed wiring system including accessories, all boxes and enclosures shall have a degree of protection of at least IP3X where the wall is liable to be drilled during erection.

422.4.2 Except for equipment for which an appropriate product standard specifies requirements, a luminaire shall be kept at an adequate distance from combustible materials. Unless otherwise recommended by the manufacturer, a small spotlight or projector shall be installed at the following minimum distance from combustible materials:

(i) Rating up to 100 W 0.5 m
(ii) Over 100 and up to 300 W 0.8 m
(iii) Over 300 and up to 500 W 1.0 m.

Lamps and other components of luminaires shall be protected against foreseeable mechanical stresses. Such protective means shall not be fixed to lampholders unless they form an integral part of the luminaire or are fitted in accordance with the manufacturer's instructions.

A luminaire with a lamp that could eject flammable materials in case of failure shall be constructed with a safety protective shield for the lamp in accordance with the manufacturer's instructions.

NOTE: Refer to Table 55.3 regarding the marking of luminaires and their installation or mounting on normally flammable surfaces.

422.4.201 Electrical equipment, e.g. installation boxes and distribution boards, installed on or in a combustible wall shall comply with the relevant standard for enclosure temperature rise.

422.4.202 Electrical equipment that does not comply with Regulation 422.4.201 shall be enclosed with a suitable thickness of non-flammable material. The effect of the material on the heat dissipation from electrical equipment shall be taken into account.

422.4.203 Cables shall comply with the requirements of BS EN 60332-1-2.

NOTE: Cables also need to satisfy the requirements of the CPR in respect of their reaction to fire. See Appendix 2, item 17.

422.4.204 Conduit and trunking systems shall be in accordance with BS EN 61386-1 and BS EN 50085-1 respectively and shall meet the fire-resistance tests within these standards.

422.5 Fire propagating structures

The requirements of this regulation shall be applied in addition to those of Section 421 in locations where CB2 conditions exist. Where the risk of flame propagation is high the cable shall meet the requirements of the appropriate part of BS EN 60332-3 series.

NOTE 1: CB2 conditions relate to the propagation of fire and exist where a building has a shape and dimensions which facilitate the spread of fire (e.g. chimney effect), such as high-rise buildings or where a building has a forced ventilation system (see Appendix 5).

NOTE 2: Fire detectors may be provided to activate measures for preventing propagation of fire, for example, the closing of fireproof shutters in ducts, troughs or trunking.

NOTE 3: Boxes and enclosures according to BS EN 60670-1 and BS EN 61439 series for use in hollow walls can be used.

NOTE 4: Cables also need to satisfy the requirements of the CPR in respect of their reaction to fire. See Appendix 2, item 17.

422.5.1 In structures where the shape and dimensions are such as will facilitate the spread of fire, precautions shall be taken so that the electrical installation does not propagate a fire (e.g. chimney effect).

422.6 Selection and erection of installations in locations of national, commercial, industrial or public significance

The requirements of Regulation 422.1 shall apply to locations that include buildings or rooms with assets of significant value. Examples include national monuments, museums and other public buildings. Buildings such as railway stations and airports are generally considered to be of public significance. Buildings or facilities such as laboratories, computer centres and certain industrial and storage facilities can be of commercial or industrial significance.

The following measures may be considered:

(i) Installation of mineral insulated cables according to BS EN 60702
(ii) Installation of cables with improved fire-resisting characteristics in case of a fire hazard, such as those complying with BS 7629-1, BS 7846 or BS 8592
(iii) Installation of cables in non-combustible solid walls, ceilings and floors
(iv) Installation of cables in areas with constructional partitions having a fire-resisting capability for a time of 30 minutes or 90 minutes, the latter in locations housing staircases and needed for an emergency escape.

Where these measures are not practicable improved fire protection may be possible by the use of reactive fire protection systems.

423 PROTECTION AGAINST BURNS

423.1 Excepting equipment for which a Harmonized Standard specifies a limiting temperature, an accessible part of fixed electrical equipment within arm's reach shall not attain a temperature in excess of the appropriate limit stated in Table 42.1. Each such part of the fixed installation likely to attain under normal load conditions, even for a short period, a temperature exceeding the appropriate limit in Table 42.1 shall be guarded so as to prevent accidental contact.

**TABLE 42.1 –
Temperature limit under normal load conditions for an accessible part of equipment within arm's reach**

Accessible part	Material of accessible surfaces	Maximum temperature (°C)
A hand-held part	Metallic	55
	Non-metallic	65
A part intended to be touched but not hand-held	Metallic	70
	Non-metallic	80
A part which need not be touched for normal operation	Metallic	80
	Non-metallic	90

424 PROTECTION AGAINST OVERHEATING

424.1 Forced air heating systems, appliances producing hot water or steam, and space heating appliances

Locations containing forced air heating systems, appliances producing hot water or steam, and space heating appliances, must comply with the appropriate parts of the Building Regulations.

CHAPTER 43

PROTECTION AGAINST OVERCURRENT

CONTENTS

CHAPTER 43

PROTECTION AGAINST OVERCURRENT

430 INTRODUCTION

430.1 Scope

This chapter provides requirements for the protection of live conductors from the effects of overcurrent.

This chapter describes how live conductors are protected by one or more devices for the automatic disconnection of the supply in the event of overload current (Section 433) and fault current (Section 434), except in cases where the overcurrent is limited in accordance with Section 436 or where the conditions described in Regulation 433.3 (omission of devices for protection against overload) or Regulation 434.3 (omission of devices for protection against fault current) are met. Coordination of overload current protection and fault current protection is also covered (Section 435).

NOTE 1: Live conductors protected against overload current in accordance with Section 433 are also considered to be protected against faults likely to cause overcurrents of a magnitude similar to overload currents.

NOTE 2: The requirements of this chapter do not take account of external influences.

NOTE 3: Protection of conductors according to these regulations does not necessarily protect the equipment connected to the conductors.

NOTE 4: Disconnection does not necessarily mean isolation in this chapter.

430.2 Not used

430.3 General requirement

A protective device shall be provided to break any overcurrent in the circuit conductors before such a current could cause a danger due to thermal or mechanical effects detrimental to insulation, connections, joints, terminations or the surroundings of the conductors.

The protection against overload current and the protection against fault current shall be coordinated in accordance with Section 435.

NOTE: An overcurrent may be an overload current or a fault current.

431 PROTECTION ACCORDING TO THE NATURE OF THE CIRCUITS AND THE DISTRIBUTION SYSTEM

431.1 Protection of line conductors

431.1.1 Except where Regulation 431.1.2 applies, detection of overcurrent shall be provided for all line conductors and shall cause the disconnection of the conductor in which the overcurrent is detected, but not necessarily the disconnection of the other line conductors except where the disconnection of one line conductor could cause damage or danger.

If disconnection of a single line conductor may cause danger, for example in the case of a three-phase motor, appropriate precautions shall be taken.

431.1.2 In a TN or TT system, for a circuit supplied between line conductors and in which the neutral conductor is not distributed, overcurrent detection need not be provided for one of the line conductors, provided that both the following conditions are simultaneously fulfilled:

(i) There exists, in the same circuit or on the supply side, differential protection intended to detect unbalanced loads and cause disconnection of all the line conductors, and

(ii) the neutral conductor is not distributed from an artificial neutral point of the circuits situated on the load side of the differential protective device mentioned in (i).

431.2 Protection of the neutral conductor

431.2.1 TN or TT system

The neutral conductor shall be protected against short-circuit current.

Where the cross-sectional area of the neutral conductor is at least equivalent to that of the line conductors, and the current in the neutral is not expected to exceed the value in the line conductors, it is not necessary to provide overcurrent detection for the neutral conductor or a disconnecting device for that conductor.

Where the cross-sectional area of the neutral conductor is less than that of the line conductors, it is necessary to provide overcurrent detection for the neutral conductor, appropriate to the cross-sectional area of the conductor. The overcurrent detection shall cause the disconnection of the line conductors, but not necessarily of the neutral conductor.

Except for disconnection complying with Regulation 461.2 the requirements for a neutral conductor apply to a PEN conductor.

Where the current in the neutral conductor is expected to exceed that in the line conductors refer to Regulation 431.2.3.

431.2.2 IT system

The neutral conductor shall not be distributed unless one of the following is met:

(i) Overcurrent detection is provided for the neutral conductor of every circuit. The overcurrent detection shall cause the disconnection of all the live conductors of the corresponding circuit, including the neutral conductor

(ii) The particular neutral conductor is effectively protected against short-circuit by a protective device installed on the supply side, for example at the origin of the installation, in accordance with Regulation 434.5

(iii) The particular circuit is protected by an RCD with a rated residual operating current ($I_{\Delta n}$) not exceeding 0.2 times the current-carrying capacity of the corresponding neutral conductor. The RCD shall disconnect all the live conductors of the corresponding circuit, including the neutral conductor. The device shall have sufficient breaking capacity for all poles.

431.2.3 Harmonic currents

Overcurrent detection shall be provided for the neutral conductor in a polyphase circuit where the harmonic content of the line currents is such that the current in the neutral conductor may exceed the current-carrying capacity of that conductor. The overcurrent detection shall cause disconnection of the line conductors but not necessarily the neutral conductor. Where the neutral is disconnected the requirements of Regulation 431.3 are applicable.

431.3 Disconnection and reconnection of the neutral conductor

Where a switch is placed in a neutral conductor, disconnection and reconnection shall be such that the neutral conductor shall not be disconnected before the line conductors and shall be reconnected at the same time as or before the line conductors.

432 NATURE OF PROTECTIVE DEVICES

A protective device shall be of the appropriate type indicated in Regulations 432.1 to 432.3.

432.1 Protection against both overload current and fault current

Except as permitted by Regulation 434.5.1, a device providing protection against both overload and fault current shall be capable of breaking, and for a circuit-breaker making, any overcurrent up to and including the maximum prospective fault current at the point where the device is installed.

432.2 Protection against overload current only

A device providing protection against overload current is generally an inverse-time-lag protective device whose rated short-circuit breaking capacity may be below the value of the maximum prospective fault current at the point where the device is installed. Such a device shall satisfy the relevant requirements of Section 433.

432.3 Protection against fault current only

A device providing protection against fault current only shall be installed where overload protection is achieved by other means or where Section 433 permits overload protection to be dispensed with. Except as permitted by Regulation 434.5.1, a device shall be capable of breaking, and for a circuit-breaker making, the fault current up to and including the prospective fault current. Such a device shall satisfy the relevant requirements of Section 434.

NOTE: Such a device may be:

(i) a circuit-breaker with a short-circuit release, or

(ii) a fuse.

432.4 Characteristics of protective devices

The time/current characteristics of an overcurrent protective device shall comply with those specified in BS 88 series, BS 3036, BS EN 60898, BS EN 60947-2 or BS EN 61009-1.

The use of another device is not precluded provided that its time/current characteristics provide a level of protection not less than that given by the devices listed above.

433 PROTECTION AGAINST OVERLOAD CURRENT

433.1 Coordination between conductor and overload protective device

Every circuit shall be designed so that a small overload of long duration is unlikely to occur.

433.1.1 The operating characteristics of a device protecting a conductor against overload shall satisfy the following conditions:

(i) The rated current or current setting of the protective device (I_n) is not less than the design current (I_b) of the circuit, and

(ii) the rated current or current setting of the protective device (I_n) does not exceed the lowest of the current-carrying capacities (I_z) of any of the conductors of the circuit, and

(iii) the current (I_2) causing effective operation of the protective device does not exceed 1.45 times the lowest of the current-carrying capacities (I_z) of any of the conductors of the circuit.

For adjustable protective devices, the rated current (I_n) is the current setting selected.

The current (I_2) causing effective operation of the protective device is given in the product standard or may be provided by the manufacturer.

NOTE 1: Where overload protection is provided by BS 3036 fuses, refer to Regulation 433.1.202.

NOTE 2: Protection in accordance with this regulation may not provide protection in all cases, for example, where sustained overcurrents less than I_2 occur.

433.1.201 Where the protective device is a general-purpose type (gG) fuse toBS 88-2, a fuse to BS 88-3, a circuit-breaker to BS EN 60898, a circuit-breaker to BS EN 60947-2 or a residual current circuit-breaker with integral overcurrent protection (RCBO) to BS EN 61009-1, compliance with conditions (i) and (ii) also results in compliance with condition (iii) of Regulation 433.1.1.

433.1.202 Where the protective device is a semi-enclosed fuse to BS 3036 compliance with condition (iii) of Regulation 433.1.1 is afforded if its rated current (I_n) does not exceed 0.725 times the current-carrying capacity (I_z) of the lowest rated conductor in the circuit protected.

433.1.203 For direct buried cables or cables in buried ducts where the tabulated current-carrying capacity is based on an ambient temperature of 20 °C compliance with condition (iii) of Regulation 433.1.1 is afforded if the rated current or current setting of the protective device (I_n) does not exceed 0.9 times the current-carrying capacity (I_z) of the lowest rated conductor in the circuit protected.

433.1.204 Accessories to BS 1363 may be supplied through a ring final circuit, with or without unfused spurs, protected by a 30 A or 32 A protective device complying with BS 88 series, BS 3036, BS EN 60898, BS EN 60947-2 or BS EN 61009-1 (RCBO). The circuit shall be wired with copper conductors having line and neutral conductors with a minimum cross-sectional area of 2.5 mm^2 except for two-core mineral insulated cables complying with BS EN 60702-1, for which the minimum cross-sectional area is 1.5 mm^2. Such circuits are deemed to meet the requirements of Regulation 433.1.1 if the current-carrying capacity (I_z) of the cable is not less than 20 A and if, under the intended conditions of use, the load current in any part of the circuit is unlikely to exceed for long periods the current-carrying capacity (I_z) of the cable.

433.2 Position of devices for protection against overload

433.2.1 Except where Regulation 433.2.2 or 433.3 applies, a device for protection against overload shall be installed at the point where a reduction occurs in the value of the current-carrying capacity of the conductors of the installation.

NOTE: A reduction in current-carrying capacity may be due to a change in cross-sectional area, method of installation, type of cable or conductor, or in environmental conditions.

433.2.2 The device protecting a conductor against overload may be installed along the run of that conductor if the part of the run between the point where a change occurs (in cross-sectional area, method of installation, type of cable or conductor, or in environmental conditions) and the position of the protective device has neither branch circuits nor outlets for connection of current-using equipment and fulfils at least one of the following conditions:

(i) It is protected against fault current in accordance with the requirements stated in Section 434

(ii) Its length does not exceed 3 m, it is installed in such a manner as to reduce the risk of fault to a minimum, and it is installed in such a manner as to reduce to a minimum the risk of fire or danger to persons (see also Regulation 434.2.1).

433.3 Omission of devices for protection against overload

This regulation shall not be applied to installations situated in locations presenting a fire risk or risk of explosion or where the requirements for special installations and locations specify different conditions.

433.3.1 General

A device for protection against overload need not be provided:

(i) for a conductor situated on the load side of the point where a reduction occurs in the value of current-carrying capacity, where the conductor is effectively protected against overload by a protective device installed on the supply side of that point

(ii) for a conductor which, because of the characteristics of the load or the supply, is not likely to carry overload current, provided that the conductor is protected against fault current in accordance with the requirements of Section 434

(iii) at the origin of an installation where the distributor provides an overload device and agrees that it affords protection to the part of the installation between the origin and the main distribution point of the installation where further overload protection is provided.

433.3.2 Position or omission of devices for protection against overload in IT systems

433.3.2.1 The provisions in Regulations 433.2.2 and 433.3 for an alternative position or omission of devices for protection against overload are not applicable to IT systems unless each circuit not protected against overload is protected by one of the following means:

(i) Use of the protective measures described in Regulation 413.2

(ii) An RCD that will operate immediately on the second fault

(iii) For permanently supervised systems only, the use of an insulation monitoring device which either:

(a) causes the disconnection of the circuit when the first fault occurs, or

(b) gives a signal indicating the presence of a fault. The fault shall be corrected in accordance with operational requirements and recognition of the consequences of a second fault.

433.3.2.2 In an IT system without a neutral conductor it is permitted to omit the overload protective device in one of the line conductors if an RCD is installed in each circuit.

433.3.3 Omission of devices for protection against overload for safety reasons

The omission of devices for protection against overload is permitted for circuits supplying current-using equipment where unexpected disconnection of the circuit could cause danger or damage.

Examples of such circuits are:

(i) the exciter circuit of a rotating machine

(ii) the supply circuit of a lifting magnet

(iii) the secondary circuit of a current transformer

(iv) a circuit supplying a fire extinguishing device

(v) a circuit supplying a safety service, such as a fire alarm or a gas alarm

(vi) a circuit supplying medical equipment used for life support in specific medical locations where an IT system is incorporated.

NOTE: In such situations consideration should be given to the provision of an overload alarm.

433.4 Overload protection of conductors in parallel

Where a single protective device protects two or more conductors in parallel there shall be no branch circuits or devices for isolation or switching in the parallel conductors.

This regulation does not preclude the use of ring final circuits with or without spur connections.

433.4.1 Equal current sharing between parallel conductors

Except for a ring final circuit, where spurs are permitted, where a single device protects conductors in parallel and the conductors are sharing currents equally, the value of I_z to be used in Regulation 433.1.1 is the sum of the current-carrying capacities of the parallel conductors.

The measures to be taken to achieve equal current sharing are given in Regulation 523.7 (i).

433.4.2 Unequal current sharing between parallel conductors

Where the use of a single conductor is impractical and the currents in the parallel conductors are unequal, the design current and requirements for overload protection for each conductor shall be considered individually.

NOTE: Currents in parallel conductors are considered to be unequal if the difference between the currents is more than 10 % of the design current for each conductor. Refer to paragraph 2 of Appendix 10.

434 PROTECTION AGAINST FAULT CURRENT

This section only considers the case of a fault between conductors belonging to the same circuit.

434.1 Determination of prospective fault current

The prospective fault current shall be determined at every relevant point of the installation. This shall be done by calculation, measurement or enquiry.

434.2 Position of devices for protection against fault current

A device providing protection against fault current shall be installed at the point where a reduction in the cross-sectional area or other change causes a reduction in the current-carrying capacity of the conductors, except where Regulation 434.2.1, 434.2.2 or 434.3 applies.

The requirements in Regulations 434.2.1 and 434.2.2 shall not be applied to installations situated in locations presenting a fire risk or risk of explosion or where special requirements for certain locations specify different conditions.

434.2.1 Except where Regulation 434.2.2 or 434.3 applies, a device for protection against fault current may be installed other than as specified in Regulation 434.2, under the following conditions:

In the part of the conductor between the point of reduction of cross-sectional area or other change and the position of the protective device there shall be no branch circuits or socket-outlets and that part of the conductor shall:

(i) not exceed 3 m in length, and

(ii) be installed in such a manner as to reduce the risk of fault to a minimum, and

NOTE: This condition may be obtained, for example, by reinforcing the protection of the wiring against external influences.

(iii) be installed in such a manner as to reduce to a minimum the risk of fire or danger to persons.

434.2.2 The device protecting a conductor may be installed on the supply side of the point where a change occurs (in cross-sectional area, method of installation, type of cable or conductor, or in environmental conditions) provided that it possesses an operating characteristic such that it protects the wiring situated on the load side against fault current, in accordance with Regulation 434.5.2.

434.3 Omission of devices for protection against fault current

A device for protection against fault current need not be provided for:

(i) a conductor connecting a generator, transformer, rectifier or an accumulator battery to the associated control panel where the protective device is placed in the panel

(ii) a circuit where disconnection could cause danger for the operation of the installation concerned, such as those quoted in Regulation 433.3.3

(iii) certain measuring circuits

(iv) the origin of an installation where the distributor installs one or more devices providing protection against fault current and agrees that such a device affords protection to the part of the installation between the origin and the main distribution point of the installation where further protection against fault current is provided,

provided that both of the following conditions are simultaneously fulfilled:

(a) The wiring is carried out in such a way as to reduce the risk of fault to a minimum (see item (ii) of Regulation 434.2.1), and

(b) the wiring is installed in such a manner as to reduce to a minimum the risk of fire or danger to persons.

434.4 Fault current protection of conductors in parallel

A single protective device may protect conductors in parallel against the effects of fault currents provided that the operating characteristic of the device results in its effective operation should a fault occur at the most onerous position in one of the parallel conductors. Account shall be taken of the sharing of the fault currents between the parallel conductors. A fault can be fed from both ends of a parallel conductor.

If operation of a single protective device may not be effective then one or more of the following measures shall be taken:

(i) The wiring shall be installed in such a manner as to reduce to a minimum the risk of a fault in any parallel conductor, for example, by the provision of protection against mechanical damage. In addition, conductors shall be installed in such a manner as to reduce to a minimum the risk of fire or danger to persons

(ii) For two conductors in parallel, a fault current protective device shall be provided at the supply end of each parallel conductor

(iii) For more than two conductors in parallel, a fault current protective device shall be provided at the supply and load ends of each parallel conductor.

NOTE: Further information is given in paragraph 3 of Appendix 10.

434.5 Characteristics of a fault current protective device

Every fault current protective device shall meet the requirements of this regulation.

434.5.1 Except where the following paragraph applies, the rated short-circuit breaking capacity of each device shall be not less than the maximum prospective fault current at the point at which the device is installed.

A lower breaking capacity is permitted if another protective device or devices having the necessary rated short-circuit breaking capacity is installed on the supply side. In this situation, the characteristics of the devices shall be co-ordinated so that the energy let-through of these devices does not exceed that which can be withstood, without damage, by the device(s) on the load side.

NOTE: Technical data for the selection of protective devices can be requested from the manufacturer.

434.5.2 A fault occurring at any point in a circuit shall be interrupted within a time such that the fault current does not cause the permitted limiting temperature of any conductor or cable to be exceeded.

For a fault of very short duration (less than 0.1 sec), for current limiting devices k^2S^2 shall be greater than the value of let-through energy (I^2t) quoted for the Class of protective device to BS EN 60898-1, BS EN 60898-2 or BS EN 61009-1, or as quoted by the manufacturer.

The time, t, in which a given fault current will raise the live conductors from the highest permissible temperature in normal duty to the limiting temperature, can, as an approximation, be calculated from the formula:

$$t = \frac{k^2\, S^2}{I^2}$$

where:

t is the duration in seconds

S is the cross-sectional area of conductor in mm^2

I is the effective fault current, in amperes, expressed for AC as the rms value, due account being taken of the current limiting effect of the circuit impedances

k is a factor taking account of the resistivity, temperature coefficient and heat capacity of the conductor material, and the appropriate initial and final temperatures. For common materials, the values of k are shown in Table 43.1.

TABLE 43.1 –
Values of k for common materials, for calculation of the effects of fault current for disconnection times up to 5 seconds

	Conductor insulation							
	Thermoplastic				**Thermosetting**		**Mineral insulated**	
	90 °C		70 °C		90 °C	60 °C	Thermoplastic sheath	Bare (unsheathed)
Conductor cross-sectional area	$\leq 300\ mm^2$	$> 300\ mm^2$	$\leq 300\ mm^2$	$> 300\ mm^2$				
Initial temperature	90 °C		70 °C		90 °C	60 °C	70 °C	105 °C
Final temperature	160 °C	140 °C	160 °C	140 °C	250 °C	200 °C	160 °C	250 °C
Copper conductor	k = 100	k = 86	k = 115	k = 103	k = 143	k = 141	k = 115	k = 135/115[a]
Aluminium conductor	k = 66	k = 57	k = 76	k = 68	k = 94	k = 93		
Tin soldered joints in copper conductors	k = 100	k = 86	k = 115	k = 103	k = 100	k = 122		

[a] This value shall be used for bare cables exposed to touch.

NOTE 1: The rated current or current setting of the fault current protective device may be greater than the current-carrying capacity of the cable.

NOTE 2: Other values of k can be determined by reference to BS 7454.

434.5.3 For a busbar trunking system complying with BS EN 61439-6 or a powertrack system complying with BS EN 61534, one of the following requirements shall apply:

(i) The rated short-time withstand current (I_{cw}) and the rated peak withstand current of a busbar trunking system or powertrack system shall be not lower than the rms value of the prospective fault current and the prospective fault peak current value, respectively. The maximum time for which the I_{cw} is defined for the busbar trunking system shall be greater than the maximum operating time of the protective device

(ii) The rated conditional short-circuit current of the busbar trunking system or powertrack system associated with a specific protective device shall be not lower than the prospective fault current.

435 COORDINATION OF OVERLOAD CURRENT AND FAULT CURRENT PROTECTION

435.1 Protection afforded by one device

A protective device providing protection against both overload current and fault current shall fulfil the requirements of the relevant regulations in Sections 433 and 434.

Except as required by Regulation 434.4 or 434.5.2, where an overload protective device complying with Regulation 433.1 is to provide fault current protection and has a rated short-circuit breaking capacity not less than the value of the maximum prospective fault current at its point of installation, it may be assumed that the requirements of this section are satisfied as regards fault current protection of the conductors on the load side of that point.

The validity of the assumption shall be checked, where there is doubt, for conductors in parallel and for certain types of circuit-breaker e.g. non-current-limiting types.

435.2 Protection afforded by separate devices

The requirements of Sections 433 and 434 apply, respectively, to the overload current protective device and the fault current protective device.

The characteristics of the devices shall be co-ordinated so that the energy let-through by the fault current protective device does not exceed that which can be withstood without damage by the overload protective device (see Section 536). This requirement does not exclude the type of coordination specified in BS EN 60947-4-1. For a circuit incorporating a motor starter, this requirement does not preclude the type of coordination described in BS EN 60947-4-1, in respect of which the advice of the manufacturer of the starter shall be sought.

436 LIMITATION OF OVERCURRENT BY THE CHARACTERISTICS OF THE SUPPLY

Conductors are considered to be protected against overload current and fault current where they are supplied from a source incapable of supplying a current exceeding the current-carrying capacity of the conductors (e.g. certain bell transformers, certain welding transformers and certain types of thermoelectric generating set).

CHAPTER 44

PROTECTION AGAINST VOLTAGE DISTURBANCES AND ELECTROMAGNETIC DISTURBANCES

CONTENTS

CHAPTER 44

PROTECTION AGAINST VOLTAGE DISTURBANCES

440 INTRODUCTION

440.1 Scope

These requirements are intended to provide for the safety of electrical installations in the event of voltage disturbances and electromagnetic disturbances generated due to different specified causes.

The requirements are not intended to apply to systems for distribution of energy to the public, or to power generation and transmission for such systems, although such disturbances may be conducted into or between electrical installations via these supply systems. The requirements of this chapter are in addition to those of Chapter 43.

440.2 General

This chapter covers the protection of electrical installations and measures against voltage disturbances and electromagnetic disturbances. The requirements are arranged into four sections as follows:

(i)	Section 442	Protection of low voltage installations against temporary overvoltages due to earth faults in the high voltage system and due to faults in the low voltage system
(ii)	Section 443	Protection against overvoltages of atmospheric origin or due to switching
(iii)	Section 444	Measures against electromagnetic disturbances
(iv)	Section 445	Protection against undervoltage.

441 NOT USED

442 PROTECTION OF LOW VOLTAGE INSTALLATIONS AGAINST TEMPORARY OVERVOLTAGES DUE TO EARTH FAULTS IN THE HIGH VOLTAGE SYSTEM AND DUE TO FAULTS IN THE LOW VOLTAGE SYSTEM

442.1 Scope and object

This regulation provides requirements for the safety of the low voltage installation in the event of:

(i) a fault between the high voltage system and Earth in the transformer substation that supplies the low voltage installation

(ii) loss of the supply neutral in the low voltage system

(iii) short-circuit between a line conductor and neutral in the low voltage installation

(iv) accidental earthing of a line conductor of a low voltage IT system.

NOTE: The requirements for the earthing of transformers that provide a supply from a system for distribution of electricity in accordance with the Electricity Safety, Quality and Continuity Regulations (ESQCR) are addressed in the Distribution Code.

442.1.1 General

Section 442 gives rules for the designer and installer of the substation. It is necessary to have the following information on the high voltage system:

(i) Quality of the system earthing

(ii) Maximum level of earth fault current

(iii) Resistance of the earthing arrangement.

The following regulations consider four situations which generally cause the most severe temporary overvoltages:

(iv) Fault between the high voltage system(s) and Earth (see Regulation 442.2)

(v) Loss of the neutral in a low voltage system (see Regulation 442.3)

(vi) Accidental earthing of a low voltage IT system (see Regulation 442.4)

(vii) Short-circuit in the low voltage installation (see Regulation 442.5).

442.1.2 Symbols

In Section 442 the following symbols are used (see Figure 44.1):

I_E part of the earth fault current in the high voltage system that flows through the earthing arrangement of the transformer substation

R_A resistance of the earthing arrangement of the exposed-conductive-parts of the equipment of the low voltage installation

R_B resistance of the earthing arrangement of the low voltage system neutral, for low voltage systems in which the earthing arrangements of the transformer substation and of the low voltage system neutral are electrically independent

R_E resistance of the earthing arrangement of the transformer substation

U_0 in TN and TT systems: nominal AC rms line voltage to Earth

in IT systems: nominal AC rms voltage between line conductor and neutral conductor or midpoint conductor, as appropriate

U_1 power frequency stress voltage between the line conductor and the exposed-conductive-parts of the low voltage equipment of the transformer substation during the fault

U_2 power frequency stress voltage between the line conductor and the exposed-conductive-parts of the equipment of the low voltage installation during the fault

U_f power frequency fault voltage that appears in the low voltage system between exposed-conductive-parts and Earth for the duration of the fault.

NOTE 1: The power frequency stress voltages (U_1 and U_2) are the voltages that appear across the insulation of low voltage equipment and across surge protective devices connected to the low voltage system.

The following additional symbols are used in respect of IT systems in which the exposed-conductive-parts of the equipment of the low voltage installation are connected to an earthing arrangement that is electrically independent of the earthing arrangement of the transformer substation.

I_h the fault current that flows through the earthing arrangement of the exposed-conductive-parts of the equipment of the low voltage installation during a period when there is a high voltage fault and a first fault in the low voltage installation (see Table 44.1)

I_d the fault current, in accordance with Regulation 411.6.2, that flows through the earthing arrangement of the exposed-conductive-parts of the low voltage installation during the first fault in a low voltage system (see Table 44.1)

Z is the impedance (for example, the IMD internal impedance or the artificial neutral impedance) between the low voltage system and an earthing arrangement.

NOTE 2: An earthing arrangement may be considered electrically independent of another earthing arrangement if a rise of potential with respect to Earth in one earthing arrangement does not cause an unacceptable rise of potential with respect to Earth in the other earthing arrangement.

442.2 Overvoltages in low voltage (LV) systems during a high voltage (HV) earth fault

In case of a fault to Earth in the HV side of the substation the following types of overvoltage may affect the LV installation:

(i) Power frequency fault voltage (U_f)

(ii) Power frequency stress voltages (U_1 and U_2).

Table 44.1 provides the relevant methods of calculation for the different types of overvoltage.

Fig 44.1 – Representative diagram for connections to Earth in the substation and the LV installation and the overvoltages occurring in case of faults

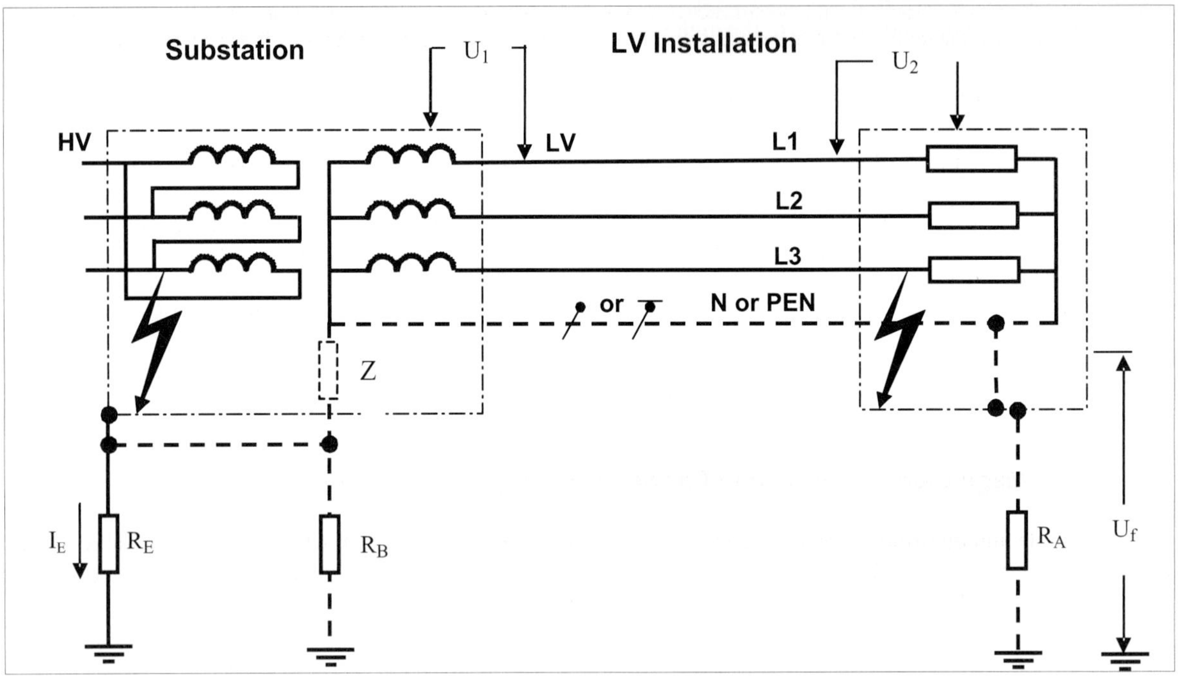

Where high and low voltage earthing systems exist in proximity to each other, two practices are presently used:

- interconnection of all high voltage (R_E) and low voltage (R_B) earthing systems
- separation of high voltage (R_E) from low voltage (R_B) earthing systems.

The general method used is interconnection. The high and low voltage earthing systems shall be interconnected if the low voltage system is totally confined within the area covered by the high voltage earthing system (see BS EN 50522).

NOTE 1: Details of the different types of system earthing are shown in Part 3 (TN, TT) and Appendix 9 (IT).

NOTE 2: The requirements for the earthing of transformers that provide a supply from a system for distribution of electricity in accordance with the Electricity Safety, Quality and Continuity Regulations (ESQCR) are addressed in the Distribution Code.

NOTE 3: BS EN 50522, Table 2, gives minimum requirements for determining whether it is feasible to interconnect high voltage and low voltage earthing arrangements. Feasibility is dependent on the earthing design for the substation achieving minimum safe touch and step voltages and a tolerable 'earth potential rise' (EPR - stress voltage).

TABLE 44.1 – Power frequency stress voltages and power frequency fault voltage in the low voltage system

Type of system earthing	Type of earth connections	U_1	U_2	U_f
TT	R_E and R_B connected	U_0	$R_E \times I_E + U_0$	0 *
	R_E and R_B separated	$R_E \times I_E + U_0$	U_0 *	0 *
TN	R_E and R_B connected	U_0	U_0 *	$R_E \times I_E$ **
	R_E and R_B separated	$R_E \cdot I_E + U_0$	U_0 *	0 *
IT	R_E and Z connected	U_0	$R_E \times I_E + U_0$	0 *
	R_E and R_A separated †	$U_0 \times \sqrt{3}$	$R_E \times I_E + U_0 \times \sqrt{3}$	$R_A \times I_h$
	R_E and Z connected	U_0	U_0	$R_E \times I_E$
	R_E and R_A interconnected †	$U_0 \times \sqrt{3}$	$U_0 \times \sqrt{3}$	$R_E \times I_E$
	R_E and Z separated	$R_E \times I_E + U_0$	U_0 *	0 *
	R_E and R_A separated †	$R_E \times I_E + U_0 \times \sqrt{3}$	$U_0 \times \sqrt{3}$	$R_A \times I_d$

† With existing earth fault in the installation.

* No consideration need be given

** See Regulation 442.2.1

NOTE 1: The requirements for U_1 and U_2 are derived from design criteria for insulation of low voltage equipment with regard to temporary power frequency overvoltage (see also Table 44.2).

NOTE 2: In a system whose neutral is connected to the earthing arrangement of the transformer substation, such temporary power frequency overvoltage is also to be expected across insulation which is not in an earthed enclosure where the equipment is outside a building.

NOTE 3: In TT and TN systems the terms 'connected' and 'separated' refer to the electrical connection between R_E and R_B. For IT systems the terms refer to the electrical connection between R_E and Z and the connection between R_E and R_A.

442.2.1 Power frequency fault voltage

The fault voltage U_f as calculated in Table 44.1, which appears in the low voltage installation between exposed-conductive-parts and Earth, shall not exceed a dangerous level.

NOTE 1: In a TN system where R_E and R_B are connected together (see Table 44.1), their connection to a low voltage global earthing system as described in BS EN 50522:2010, clause 3.4.19, can be considered to be a safety measure against dangerous fault voltages.

NOTE 2: A global earthing system is an earthing system created by the interconnection of local earthing systems that provides, by the proximity of these earthing systems, that there are no dangerous touch voltages.

In installations outside a global earthing system, additional connection shall be made between the PEN conductor and Earth.

442.2.2 Magnitude and duration of power frequency stress voltages

The magnitude and duration of the power frequency stress voltages (U_1 and U_2), where specified in Table 44.1, on the equipment in the low voltage installation due to an earth fault in the high voltage system, shall not exceed the requirements given in Table 44.2.

TABLE 44.2 – Permissible power frequency stress voltage

Duration of the earth fault in the high voltage system t	Permissible power frequency stress voltage on equipment in low voltage installations U
> 5 s	U_0 + 250 V
≤ 5 s	U_0 + 1200 V

In systems without a neutral conductor, U_0 shall be the line-to-line voltage.

NOTE 1: The first line of the table relates to high voltage systems having long disconnection times, for example, isolated neutral and resonant earthed high voltage systems. The second line relates to high voltage systems having short disconnection times, for example, low-impedance earthed high voltage systems. Both lines together are relevant design criteria for insulation of low voltage equipment with regard to temporary power frequency overvoltage, BS EN IEC 60664-1.

NOTE 2: In a system whose neutral is connected to the earthing arrangement of the transformer substation, such temporary power frequency overvoltage is also to be expected across insulation which is not in an earthed enclosure where the equipment is outside a building.

442.2.3 Requirements for calculation of limits

The requirements of Regulations 442.2.1 and 442.2.2 are deemed to be fulfilled for installations receiving a supply at low voltage from a system for distribution of electricity to the public.

To fulfil the above requirements coordination between the HV system operator and the LV system installer is necessary. Compliance with the above requirements mainly falls into the responsibility of the substation installer/owner/operator who needs also to fulfil requirements provided by BS EN 61936-1. Therefore, the calculation for U_1, U_2 and U_f is normally not necessary for the LV system installer.

Possible measures to fulfil the above requirements are, for example:

(i) separation of HV and LV earthing arrangements

(ii) change of LV system earthing

(iii) reduction of earth resistance, R_E.

442.3 Power frequency stress voltage in the event of loss of the neutral conductor in TN or TT system

Consideration shall be given to the fact that, if the neutral conductor in a three-phase TN or TT system is interrupted, basic, double and reinforced insulation as well as components rated for the voltage between line and neutral conductors can be temporarily stressed with the line-to-line voltage. The stress voltage can reach up to $U = \sqrt{3}\,U_0$.

442.4 Power frequency stress voltage in the event of an earth fault in a IT system with distributed neutral

Consideration shall be given to the fact that, if a line conductor of an IT system is earthed accidentally, insulation or components rated for the voltage between line and neutral conductors can be temporarily stressed with the line-to-line voltage. The stress voltage can reach up to $U = \sqrt{3}\ U_0$.

442.5 Power frequency stress voltage in the event of short-circuit between a line conductor and the neutral conductor

Consideration shall be given to the fact that, if a short-circuit occurs in the low voltage installation between a line conductor and the neutral conductor, the voltage between the other line conductors and the neutral conductor can reach the value of $1.45 \times U_0$ for a time up to 5s.

443 PROTECTION AGAINST TRANSIENT OVERVOLTAGES OF ATMOSPHERIC ORIGIN OR DUE TO SWITCHING

443.1 Scope and object

443.1.1 This section deals with protection of electrical installations against transient overvoltages of atmospheric origin transmitted by the supply distribution system, including direct strikes to the supply system, and against switching overvoltages generated by the equipment within the installation.

This section does not specify requirements for protection against transient overvoltages due to direct or nearby lightning strokes on the structure.

NOTE 1: For risk management for protection against transient overvoltages due to direct or nearby lightning strokes on the structure, see BS EN 62305-2.

In general, switching overvoltages have lower amplitude than transient overvoltages of atmospheric origin and therefore the requirements regarding protection against transient overvoltages of atmospheric origin normally cover protection against switching overvoltages.

If no transient overvoltage protection against disturbances of atmospheric origin is installed, protection against switching overvoltages may need to be provided.

NOTE 2: Overvoltages due to switching can be longer in duration and can contain more energy than transient overvoltages of atmospheric origin (see Regulation 443.4.2).

NOTE 3: The characteristics of transient overvoltages of atmospheric origin depend on factors such as:

- the nature of the supply distribution system (underground or overhead)
- the possible existence of at least one surge protective device (SPD) upstream of the installation
- the voltage level of the supply system.

NOTE 4: Transient overvoltages transmitted by the supply distribution system are not significantly attenuated downstream in most installations.

This section does not cover overvoltages transmitted by information and communications technology systems. See BS EN 50174, BS EN 62305-4 and PD CLC/TS 61643-22.

NOTE 5: As regards transient overvoltages of atmospheric origin, no distinction is made between earthed and unearthed systems.

Where protection against overvoltages is by the use of surge protective devices (SPDs) they shall be selected and erected in accordance with Section 534.

NOTE 6: Examples of equipment with various rated impulse voltages are given in Table 443.2.

NOTE 7: Some electronic equipment may have protection levels lower than Category I of Table 443.2.

This section does not apply to installations where the consequences of overvoltage are:

(a) explosion

(b) chemical or radioactive emissions.

NOTE 8: BS EN 62305-2 applies for such high risk installations.

443.2 Not used

443.3 Not used

443.4 Overvoltage control

443.4.1 Transient overvoltages due to the effects of indirect lightning strokes

Protection against transient overvoltages shall be provided where the consequence caused by the overvoltage could result in:

(i) serious injury to, or loss of, human life
(ii) failure of a safety service, as defined in Part 2
(iii) significant financial or data loss.

For all other cases, protection against transient overvoltages shall be provided unless the owner of the installation declares it is not required due to any loss or damage being tolerable and they accept the risk of damage to equipment and any consequential loss.

443.4.2 Transient overvoltages caused by equipment

Protection against overvoltages shall be considered in the case of equipment likely to produce switching overvoltages or disturbances exceeding the applicable rated impulse voltage of equipment according to Table 443.2, e.g. where inductive or capacitive equipment, such as motors, transformers, capacitor banks, storage units or high current loads are installed.

443.5 *Deleted by BS 7671:2018+A2:2022*

Fig 44.2 – *Deleted by BS 7671:2018+A2:2022*

TABLE 443.1 – *Deleted by BS 7671:2018+A2:2022*

Fig 44.3 – *Deleted by BS 7671:2018+A2:2022*

443.6 Classification of rated impulse voltages (overvoltage categories)

443.6.1 Purpose of classification of rated impulse voltages (overvoltage categories)

Overvoltage categories are defined within electrical installations for the purpose of insulation coordination and a related classification of equipment with rated impulse voltages is provided (see Table 443.2).

The rated impulse voltage is used to classify equipment energized directly from the low voltage electrical installation into overvoltage categories.

Rated impulse voltages for equipment selected according to the nominal voltage are provided to distinguish different levels of availability of equipment with regard to continuity of service and an acceptable risk of failure.

Inherent overvoltage control for insulation coordination, based only on the rated impulse voltage of the equipment in accordance with BS EN IEC 60664-1, is not sufficient where:

- transient overvoltages transmitted via the supply distribution system are not significantly attenuated downstream

  NOTE: Insulation coordination can be achieved in the whole installation by transient overvoltage protection of the equipment corresponding to the classified rated impulse voltage, thereby reducing the risk of failure to an acceptable level.

- surge currents and partial lightning currents are distributed via underground cables
- equipment is connected to multiple services, e.g. power, telecommunications and data lines.

It is necessary to consider the rated impulse voltage U_W (see BS EN IEC 60664-1) of the most sensitive equipment to be protected in the system or, in cases where a temporary loss of function of equipment is unacceptable, the impulse immunity voltage (see BS EN 61000-4-5).

443.6.2 Rated impulse voltages of equipment and overvoltage categories

Category IV equipment is suitable for use at, or in the proximity of, the origin of the electrical installation, for example, upstream of the main distribution board. Equipment of category IV has a very high impulse withstand capability providing the required high degree of reliability, and shall have a rated impulse voltage not less than the value specified in Table 443.2.

Category III equipment is suitable for use in the fixed installation downstream of and including the main distribution board, providing a high degree of availability, and shall have a rated impulse voltage not less than the value specified in Table 443.2.

Category II equipment is suitable for connection to the fixed installation, providing a degree of availability normally required for current-using equipment, and shall have a rated impulse voltage not less than the value specified in Table 443.2.

Category I equipment is only suitable for use in the fixed installation where SPDs are installed outside the equipment to limit transient overvoltages to the specified level, and shall have a rated impulse voltage not less than the value specified in Table 443.2. Therefore, equipment with a rated impulse voltage corresponding to overvoltage category I should, preferably, not be installed at or near the origin of the installation.

TABLE 443.2 – Required rated impulse voltage of equipment (U_W)

Nominal voltage of the installation V[a]	Voltage line to neutral derived from nominal voltages AC or DC up to and including V	Required rated impulse voltage of equipment[b] kV			
		Overvoltage category IV (equipment with very high rated impulse voltage)	Overvoltage category III (equipment with high rated impulse voltage)	Overvoltage category II (equipment with normal rated impulse voltage)	Overvoltage category I (equipment with reduced rated impulse voltage)
		For example, energy meter, telecontrol systems	For example, distribution boards, switches socket-outlets	For example, domestic appliances, tools	For example, sensitive electronic equipment such as alarm panels, computers and home electronics
120/208	150	4	2.5	1.5	0.8
230/400[c] 277/480	300	6	4	2.5	1.5
400/690	600	8	6	4	2.5
1000	1000	12	8	6	4
1500 DC	1500 DC	-	-	-	-

a According to BS EN 60038.

b This rated impulse voltage is applied between live conductors and PE.

c For IT systems operating at 220-240 V, the 230/400 row should be used, due to the voltage to earth at the earth fault on one line.

ANNEX A443 Deleted by BS 7671:2018+A2:2022

ANNEX B443 (Informative)
GUIDANCE ON OVERVOLTAGE CONTROL BY SPDs APPLIED TO OVERHEAD LINES

Where an installation is supplied by, or includes, an overhead line and SPDs are required according to Regulation Group 443.4, the protective control of the overvoltage level may be obtained either by installing surge protective devices directly in the installation close to the origin of the installation or, with the consent of the network operator, in the overhead lines of the supply distribution network.

As an example, the following measures may be applied:

(a) in the case of overhead supply distribution networks, overvoltage protection is erected at network junction points and especially at the end of each feeder longer than 0.5 km. Surge protective devices should be erected at every 0.5 km distance along the supply distribution lines. Nevertheless, the distance between surge protective devices should in no case exceed 1 km

(b) if a supply distribution network is erected partly as an overhead network and partly as an underground network, overvoltage protection in the overhead lines should be applied in accordance with (a) at each transition point from an overhead line to an underground cable

(c) in a TN distribution network supplying electrical installations, where the protective measure automatic disconnection of supply is applied, the earthing conductors of the surge protective devices connected to the line conductors are connected to the PEN conductor or to the PE conductor

(d) in a TT distribution network supplying electrical installations, where the protective measure automatic disconnection of supply is applied, surge protective devices are provided for the line conductors and for the neutral conductor. At the place where the neutral conductor of the supply network is effectively earthed, a surge protective device for the neutral conductor is not necessary.

444 MEASURES AGAINST ELECTROMAGNETIC DISTURBANCES

444.1 Scope

This section provides basic requirements and recommendations to enable the avoidance and reduction of electromagnetic disturbances.

Those involved in the design, installation and maintenance of, and alterations to, electrical installations shall give due consideration to the measures described in this section.

Electromagnetic disturbances can disturb or damage information and communications technology systems or information and communications technology equipment as well as equipment with electronic components or circuits. Currents due to lightning, switching operations, short-circuits and other electromagnetic phenomena might cause overvoltages and electromagnetic interference.

These effects are potentially more severe:

(i) where large metal loops exist

(ii) where different electrical wiring systems are installed in common routes, e.g. for power supply and for signalling and/or data communication cables connecting information and communications technology equipment within a building.

The value of the induced voltage depends on the rate of change (dI/dt) of the interference current and on the size of the loop.

Power cables carrying large currents with a high rate of change of current (dI/dt) (e.g. the starting current of lifts or currents controlled by rectifiers) can induce overvoltages in cables of information and communications technology systems, which can influence or damage information and communications technology equipment or similar electrical equipment.

In or near rooms for medical use, electromagnetic disturbances associated with electrical installations can interfere with medical electrical equipment.

The requirements and recommendations given in this section can have an influence on the overall design of the building including its structural aspects.

The requirements of the following standards shall be applied where appropriate:

(iii) BS 6701: Telecommunications equipment and telecommunications cabling. Specification for installation, operation and maintenance

(iv) BS EN 50310: Telecommunications bonding networks for buildings and other structures

(v) BS EN 50174 series: Information technology. Cabling installation

(vi) BS IEC 61000-5-2: Electromagnetic compatibility (EMC). Installation and mitigation guidelines. Earthing and cabling.

444.2 Not used

444.3 Not used

444.4 Electromagnetic disturbances

444.4.1 Sources of electromagnetic disturbances

Consideration shall be given to the location of the sources of electromagnetic disturbances relative to the positioning of other equipment. Potential sources of electromagnetic disturbances within an installation typically include:

(i) switching devices for inductive loads

(ii) electric motors

(iii) fluorescent lighting

(iv) welding machines

(v) rectifiers

(vi) choppers

(vii) frequency convertors/regulators including Variable Speed Drives (VSDs)

(viii) lifts

(ix) transformers

(x) switchgear

(xi) power distribution busbars.

NOTE: For further information refer to the BS EN 50174 series of standards.

444.4.2 Measures to reduce electromagnetic interference (EMI)

444.4.2.1 The following measures shall be considered, where appropriate, in order to reduce the effects of electromagnetic interference:

(i) Where screened signal or data cables are used, care should be taken to limit the fault current from power systems flowing through the screens and cores of signal cables, or data cables, which are earthed. Additional conductors may be necessary, e.g. a bypass conductor for screen reinforcement, see Figure 44.4

Fig 44.4 – Bypass conductor for screen reinforcement to provide a common equipotential bonding system

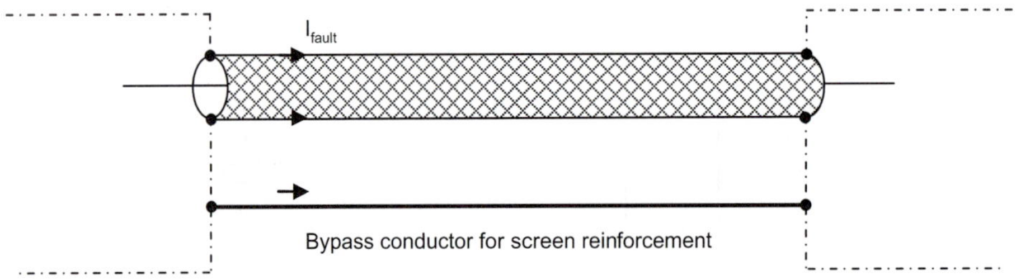

(ii) The use of surge protective devices and/or filters to improve electromagnetic compatibility with regard to conducted electromagnetic phenomena for electrical equipment sensitive to electromagnetic disturbances

(iii) The installation of power cables (i.e. line, neutral and any protective earth conductors) close together in order to minimize cable loop areas

(iv) The separation of power and signal cables

(v) The installation of an equipotential bonding network, see Regulation 444.5.2.

444.4.3 TN system

To minimize electromagnetic disturbances, the following requirements shall be met.

444.4.3.1 A PEN conductor shall not be used downstream of the origin of the installation.

NOTE: Regulation 8(4) of the Electricity Safety, Quality and Continuity Regulations (ESQCR) prohibits the use of PEN conductors in consumers' installations.

444.4.3.2 The installation shall have separate neutral and protective conductors downstream of the origin of the installation; see Figure 44.5.

Fig 44.5 – Avoidance of neutral conductor currents in a bonded structure by using an installation forming part of a TN-C-S system from the origin of the public supply up to and including the final circuit within a building

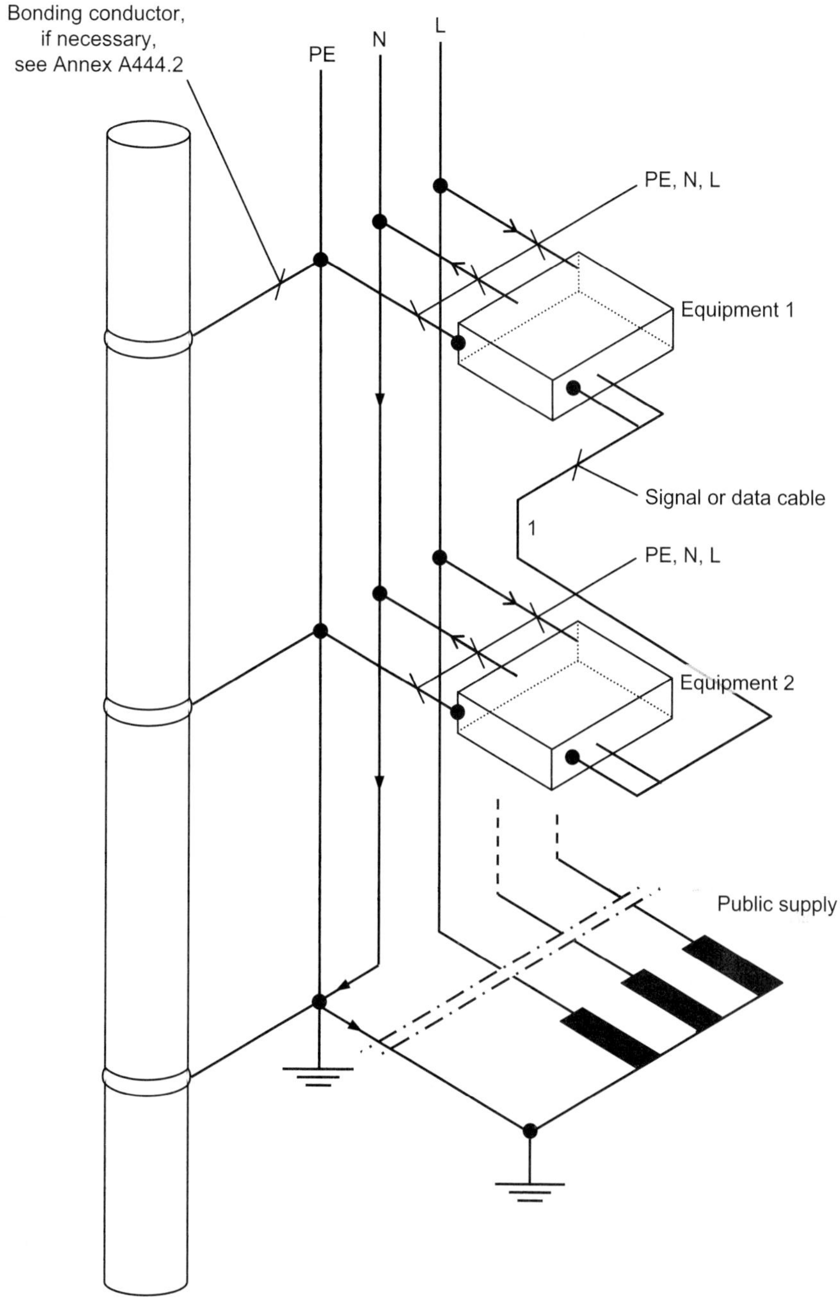

Key :

1 Loops of limited area formed by signal or data cables.

444.4.3.3 Where the complete low voltage installation including the transformer is operated only by the user, an installation forming part of a TN-S system shall be installed; see Figure 44.6.

Fig 44.6 – Avoidance of neutral conductor currents in a bonded structure by using an installation forming part of a TN-S system downstream of a consumer's private supply transformer

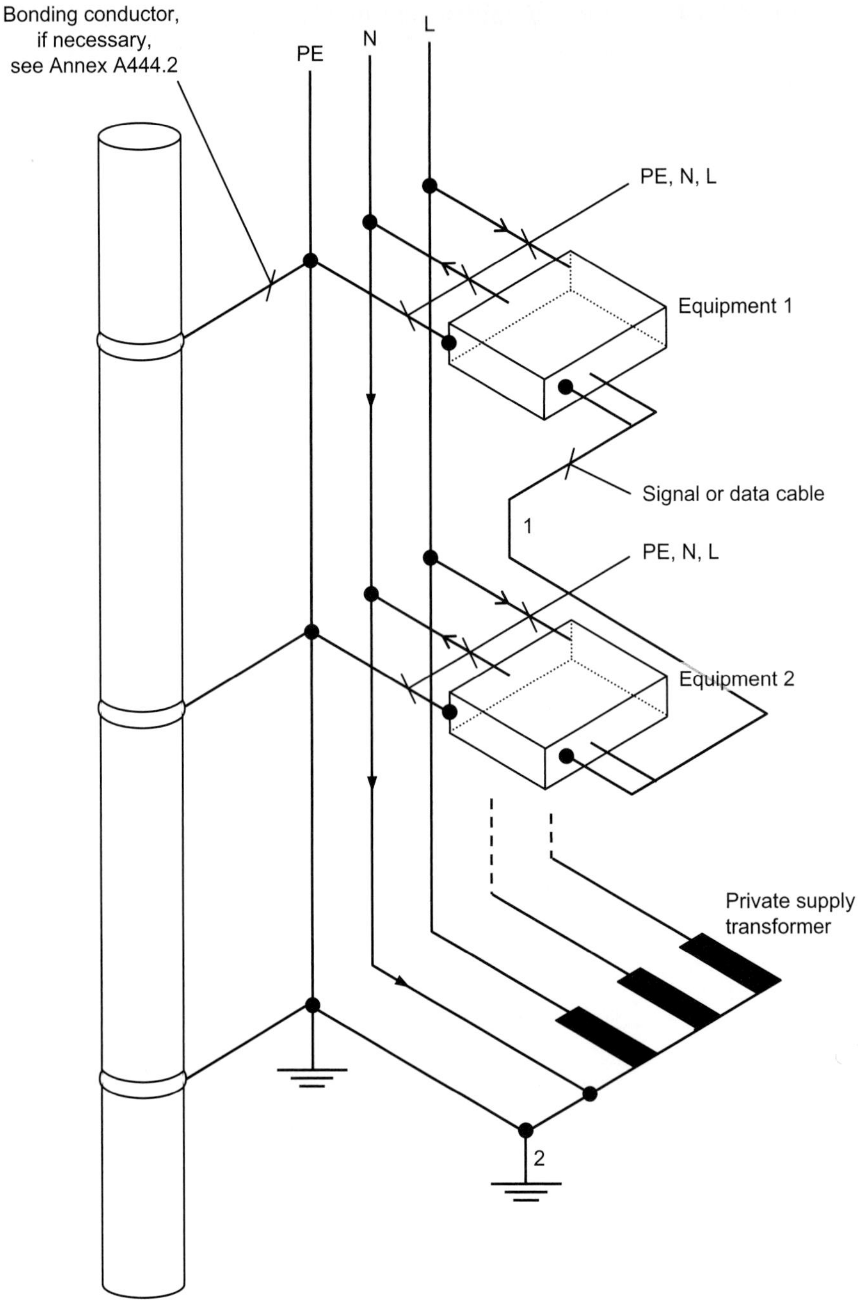

Key :

1 Loops of limited area formed by signal or data cables.

2 The point of neutral earthing may be made at the transformer or the main LV switchgear.

444.4.4 TT system

In an installation forming part of a TT system, such as that shown in Figure 44.7, consideration shall be given to overvoltages which might exist between live parts and extraneous-conductive-parts where the extraneous-conductive-parts of different buildings are connected to different earth electrodes.

The use of an isolating transformer to provide a TN-S system shall be considered.

Fig 44.7 – Installation forming part of a TT system within a building installation

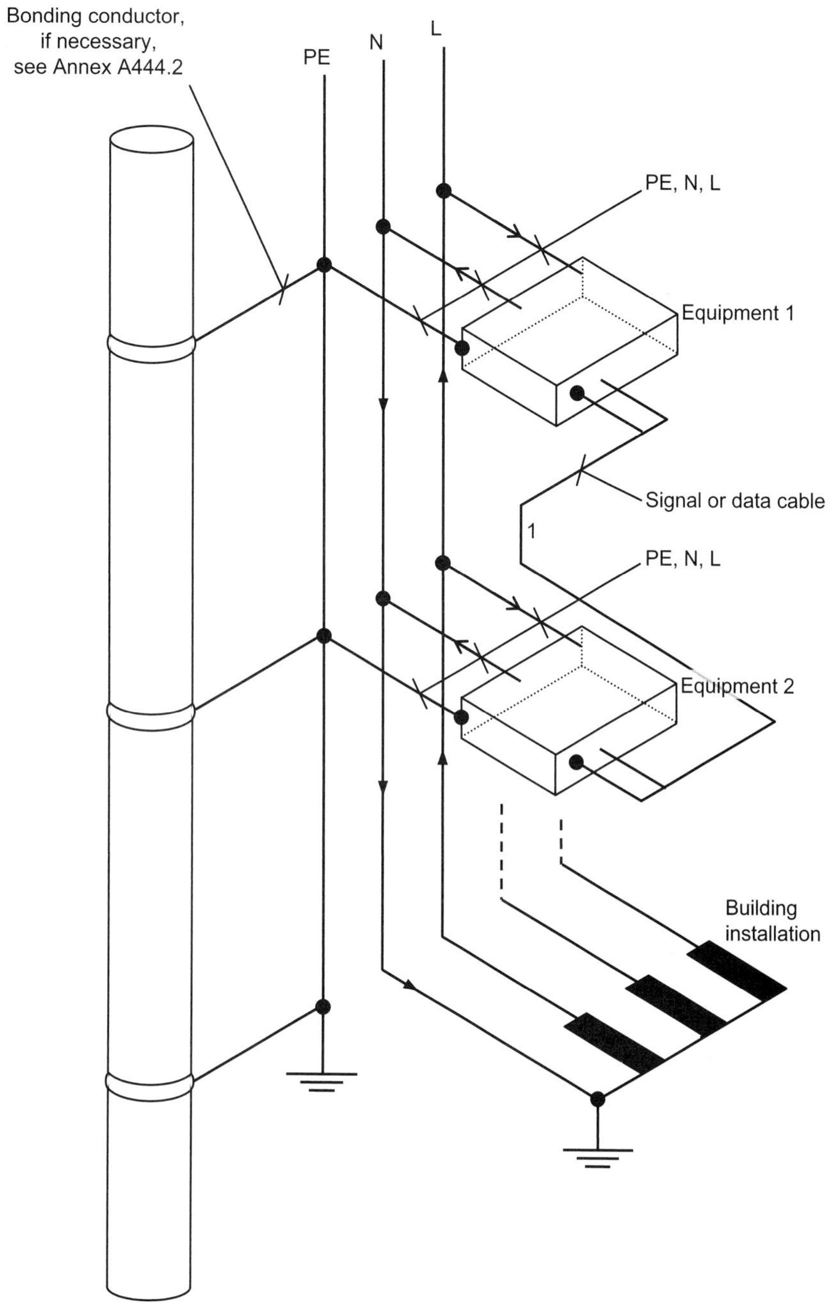

Key :

1 Loops of limited area formed by signal or data cables.

Where screened signal cables or data cables are common to several buildings supplied from an installation forming part of a TT system, the use of a bypass bonding conductor (see Figure 44.8) or single-point bonding shall be considered. The bypass conductor shall have a minimum cross-sectional area of 16 mm^2 copper or equivalent, the equivalent cross-sectional area being selected in accordance with Regulation 544.1.

Where the live conductors of the supply into any of the buildings exceed 35 mm^2 in cross-sectional area the bypass conductor shall have a minimum cross-sectional area in accordance with Table 54.8.

Fig 44.8 – Example of a substitute or bypass bonding conductor in an installation forming part of a TT system

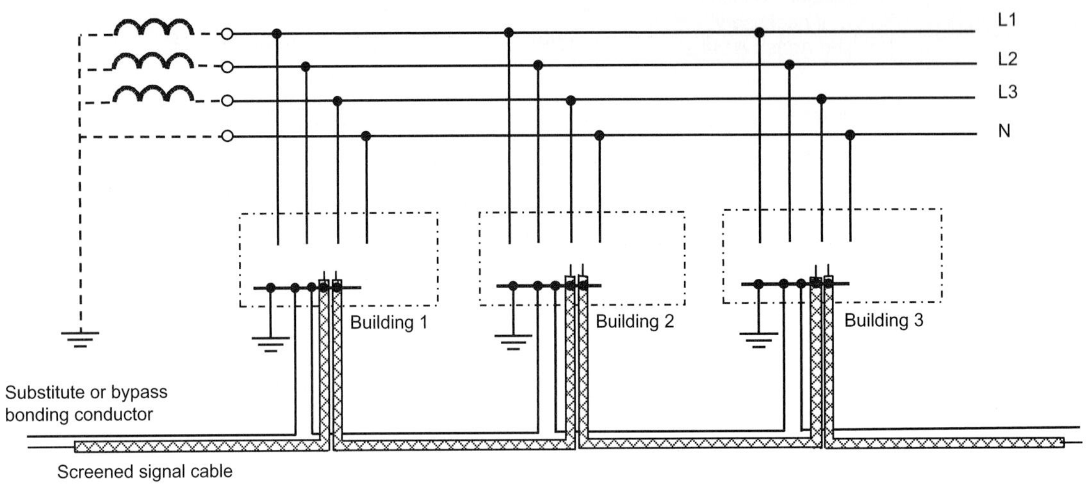

If consent according to the last paragraph of Regulation 411.3.1.2 (telecommunication cables) cannot be obtained, it is the responsibility of the owner or operator of the cable to avoid any danger due to the exclusion of those cables from the connection to the main equipotential bonding.

444.4.5 *Not used*

444.4.6 Multiple-source TN or TT power supplies

For TN or TT multiple-source power supplies to an installation, the system shall be earthed at one point only.

For a TN system, to avoid having the neutral current flowing through the protective conductor, a single point of connection only shall be made as illustrated in Figure 44.9.

Fig 44.9 – TN multiple-source power supply with single connection between PEN and earth

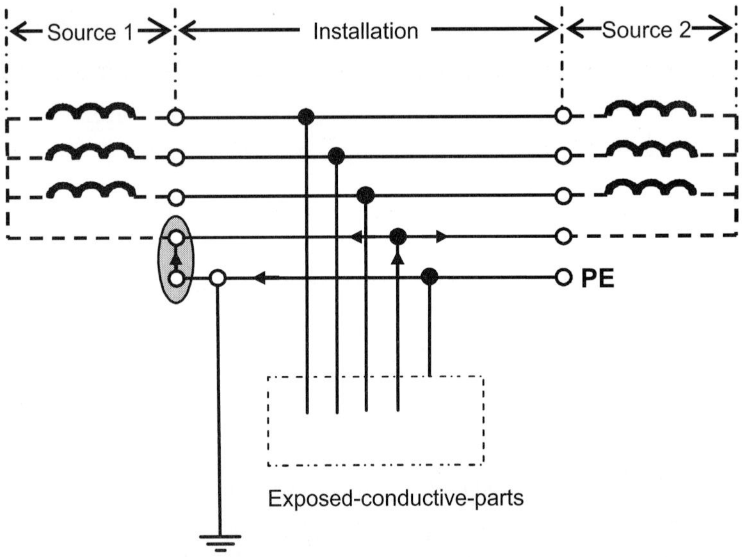

NOTE 1: Where multiple earthing of the star points of the sources of supply is applied, neutral conductor currents might flow back to the relevant star point, not only via the neutral conductor, but also via the protective conductor as shown in Figure 44.10. For this reason the sum of the partial currents flowing in the installation is no longer zero and a stray magnetic field is created, similar to that of a single conductor cable.

NOTE 2: In the case of a single conductor cable carrying AC current, a circular electromagnetic field is generated around the conductor that might interfere with electronic equipment. Harmonic currents produce similar electromagnetic fields but they attenuate more rapidly than those produced by the fundamental current.

Fig 44.10 – TN multiple-source power supply with unsuitable multiple connection between PEN and earth

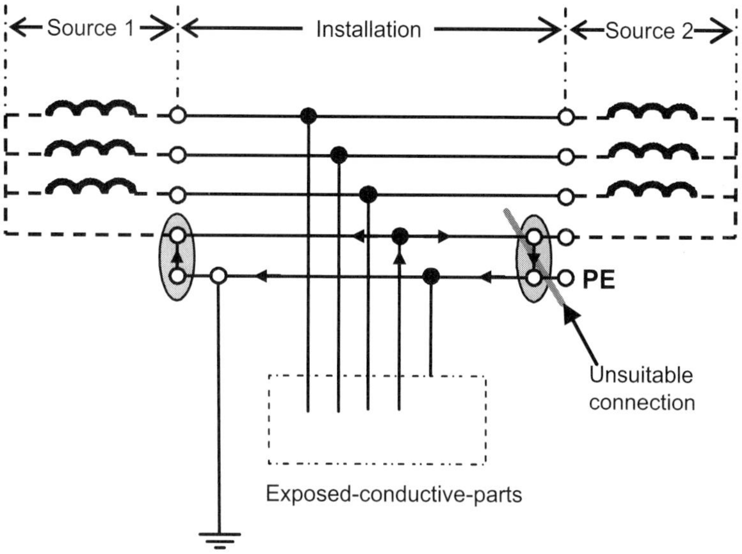

444.4.7 Transfer of supply

In an installation forming part of a TN system the transfer from one supply to an alternative supply shall be by means of a multipole switching device which switches the line conductors and the neutral conductor, if any.

NOTE: This method prevents electromagnetic fields due to stray currents in the main supply system of an installation.

444.4.8 Not used

444.4.9 Separate buildings

Where different buildings have separate equipotential bonding systems, metal-free optical fibre cables or other non-conducting systems are preferred for signal and data transmission, e.g. microwave signal transformer for isolation in accordance with BS EN 61558-2-1, 2-4, 2-6, 2-15 and either BS EN 60950-1 or BS EN 62368-1.

444.4.10 Inside buildings

Within a building, the requirements and recommendations of the following standards shall be applied for control, signalling and communication circuits:

(i) BS EN 50174-1: Information technology – Cabling installation: Installation specification and quality assurance

(ii) BS EN 50174-2: Information technology – Cabling installation: Installation planning and practices inside buildings

(iii) BS EN 50310: Telecommunications bonding networks for buildings and other structures.

444.5 Earthing and equipotential bonding

444.5.1 Interconnection of earth electrodes

444.5.1.1 Within a single building

All protective and functional earthing conductors of an installation within a building shall be connected to the main earthing terminal, as required by Regulation 542.4.1, except where this is precluded by the requirements of legislation or Part 7.

444.5.1.2 Between buildings

For communication and data exchange between several buildings, the requirements of Regulation 542.1.3.3 apply to both the protective and functional earthing requirements.

NOTE: Where interconnection of the earth electrodes is not possible or practicable, it is recommended that separation of communications networks is applied, for example, by using optical or radio links.

444.5.2 Equipotential bonding networks

The structure selected for these conductors shall be appropriate for the installation:

(i) Metal sheaths, screens or armouring of cables shall be bonded to the common bonding network (CBN) unless such bonding is required to be omitted for safety reasons

(ii) Where screened signal or data cables are earthed, care shall be taken to limit the fault current from power systems flowing through the screens and cores of signal cables or data cables

(iii) The impedance of equipotential bonding connections intended to carry functional earth currents having high frequency components shall be as low as practicable and this should be achieved by the use of multiple, separated bonds that are as short as possible

NOTE: Where bonds of up to 1 metre long are used, their inductive reactance and impedance of route can be reduced by choosing a conductive braid or a bonding strap/strip (with a width to thickness ratio of at least 5:1 and a length to width ratio no greater than 5:1).

(iv) Where a lightning protection system is installed, reference shall be made to BS EN 62305.

444.5.3 Sizing and installation of copper bonding ring network conductors

Equipotential bonding designed as a bonding ring network shall have the following minimum nominal dimensions:

(i) Flat cross-section: 25 mm × 3 mm

(ii) Round diameter: 8 mm.

Bare conductors shall be protected against corrosion at their supports and on their passage through walls.

444.5.3.1 Parts to be connected to the equipotential bonding network

The following parts shall be connected to the equipotential bonding network:

(i) Metallic containment, conductive screens, conductive sheaths or armouring of data transmission cables or of information and communications technology equipment

(ii) Functional earthing conductors of antenna systems

(iii) Conductors of the earthed pole of a DC supply for information and communications technology equipment

(iv) Functional earthing conductors

(v) Protective conductors.

444.5.4 *Not used*

444.5.5 *Not used*

444.5.6 *Not used*

444.5.7 Earthing arrangements and equipotential bonding of information and communications technology installations for functional purposes

444.5.7.1 Earthing busbar

Where an earthing busbar is required for functional purposes, consideration shall be given to extending the main earthing terminal of the building by using one or more earthing busbars. This enables information and communications technology installations to be connected to the main earthing terminal by the shortest practicable route from any point in the building. Where the earthing busbar is erected to support the equipotential bonding network of a significant amount of information and communications technology equipment in a building, consideration shall be given to the installation of a bonding ring conductor or common mesh bonding network; see Annex A444 Figure A444.2.

Consideration shall be given to the need for accessibility of the earthing busbar throughout its length and to the protection of bare conductors to prevent corrosion.

444.5.7.2 Cross-sectional area of the earthing busbar

For installations connected to a supply having a capacity of 200 A per phase or more, the cross-sectional area of the earthing busbar shall be not less than 50 mm^2 copper and shall be selected in accordance with Regulation 444.5.2(iii).

For supplies having a capacity of less than 200 A per phase the earthing busbar shall be selected in accordance with Table 54.8.

Where the earthing busbar is used as part of a DC return current path, its cross-sectional area shall be selected according to the expected DC return currents.

444.6 Segregation of circuits

444.6.1 General

Cables that are used at voltage Band II (low voltage) and cables that are used at voltage Band I (extra-low voltage) which share the same cable management system or the same route, shall be installed according to the requirements of Regulations 528.1 and 528.2. Circuits of the same voltage band might also require segregation or separation.

Electrical safety and electromagnetic compatibility might produce different segregation or separation requirements. The design shall meet both requirements.

444.6.2 Equipment

The minimum distance between information and communications technology cables and discharge, neon and mercury vapour (or other high-intensity discharge) lamps shall be 130 mm. In this regard, low energy lamps (cfl) are to be considered as gas discharge sources. Data wiring racks and electrical equipment shall always be separated.

ANNEX A444 (Informative)
MEASURES AGAINST ELECTROMAGNETIC DISTURBANCES

A444.1 Structures for the network of bonding conductors and earthing conductors

For dwellings, where normally a limited amount of electronic equipment is in use, a protective conductor network in the form of a star network might be acceptable.

For commercial and industrial buildings and similar buildings containing multiple electronic applications, a common equipotential bonding system is useful in order to comply with the EMC requirements of different types of equipment.

The four basic structures described as follows might be used, depending on the importance and vulnerability of equipment.

NOTE: For further information, the methodology referred to in BS EN 50310 (Telecommunications bonding networks for buildings and other structures) is generally applicable.

A444.1.1 Protective conductors in a star network

This type of network is applicable to small installations associated with dwellings, small commercial buildings, etc., and from a general point of view to equipment that is not interconnected by signal cables; see Figure A444.1.

Fig A444.1 – Example of protective conductors in star network

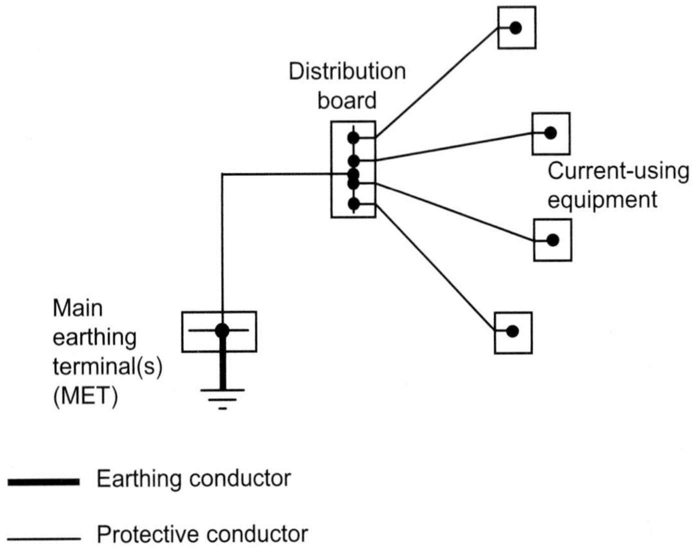

A444.1.2 Multiple meshed bonding star network

This type of network is applicable to small installations with different small groups of interconnected communicating equipment. It enables the local dispersion of currents caused by electromagnetic interference; see Figure A444.2.

Fig A444.2 – Example of multiple meshed bonding star network

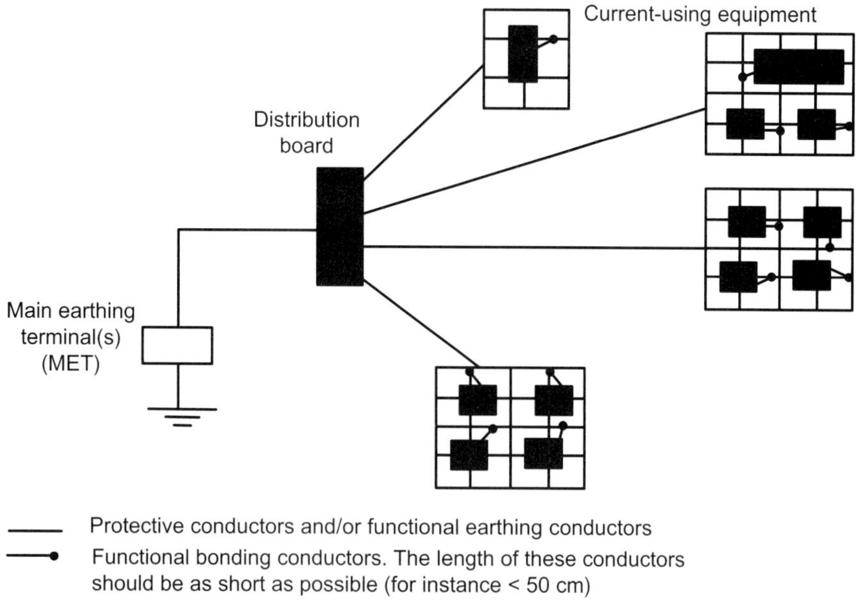

A444.1.3 Common meshed bonding star network

This type of network is applicable to installations with a high density of communicating equipment corresponding to critical applications; see Figure A444.3. It is suitable for protection of private automatic branch exchange equipment (PABX) and centralized data processing systems.

A meshed equipotential bonding network is enhanced by the existing metallic structure of the building. It is supplemented by conductors forming the square mesh.

The mesh size depends on the selected level of protection against lightning, on the immunity level of the equipment and on the frequencies used for data transmission.

Mesh size should be adapted to the dimensions of the installation to be protected and should be in accordance with the recommendations of BS EN 50310. Where concerns exist, the mesh size should be adapted to the dimensions of the installation to be protected, but should not exceed 2 m × 2 m in areas where equipment susceptible to electromagnetic environmental interferences is installed.

NOTE: The mesh size refers to the dimensions of square spaces enclosed by the conductors forming the mesh.

In some cases, parts of this network may be meshed more closely in order to meet specific requirements.

Fig A444.3 – Example of a common meshed bonding star network

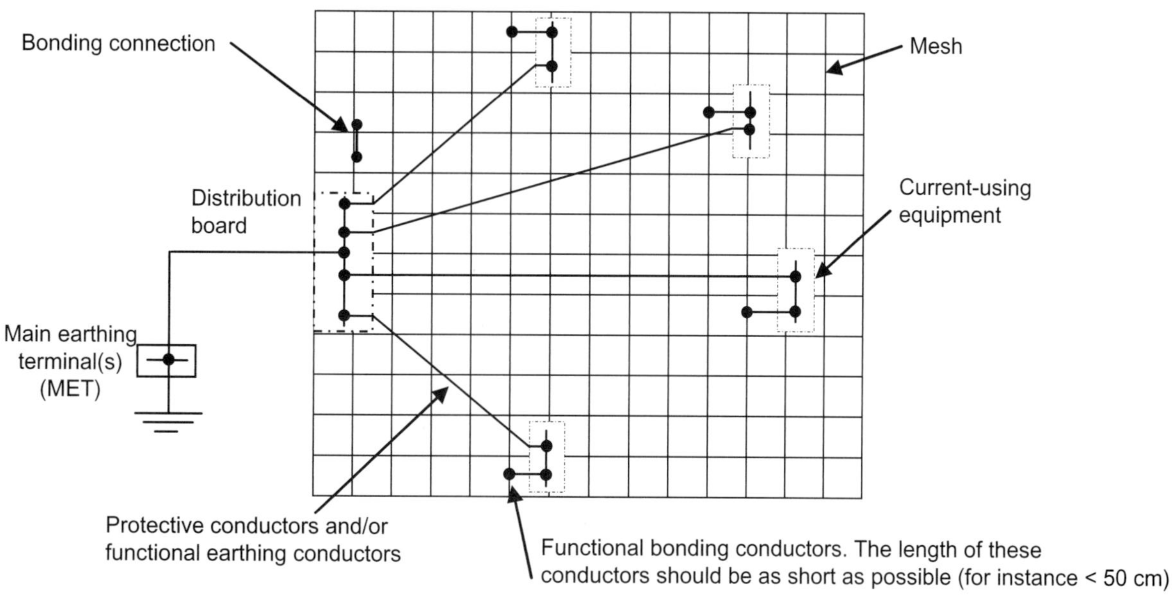

A444.1.4 Protective conductors connected to a bonding ring conductor

An equipotential bonding network in the form of a bonding ring conductor (BRC) is shown in Figure A444.4 on the top floor of the structure. The BRC should preferably be made of copper, bare or insulated, and installed in such a manner that it remains accessible everywhere, e.g. by mounting on a cable tray, in a metallic conduit (see BS EN 61386 series), employing a surface mounted method of installation or cable trunking. All protective and functional earthing conductors may be connected to the BRC.

A444.2 Equipotential bonding networks in buildings with several floors

For buildings with several floors, it is recommended that, on each floor, an equipotential bonding system be installed; see Figure A444.4 for examples of bonding networks in common use; each floor is a type of network. The bonding systems of the different floors should be interconnected, at least twice, by conductors selected in accordance with the requirements of Chapter 54.

Fig A444.4 – Example of equipotential bonding networks in a structure without a lightning protection system

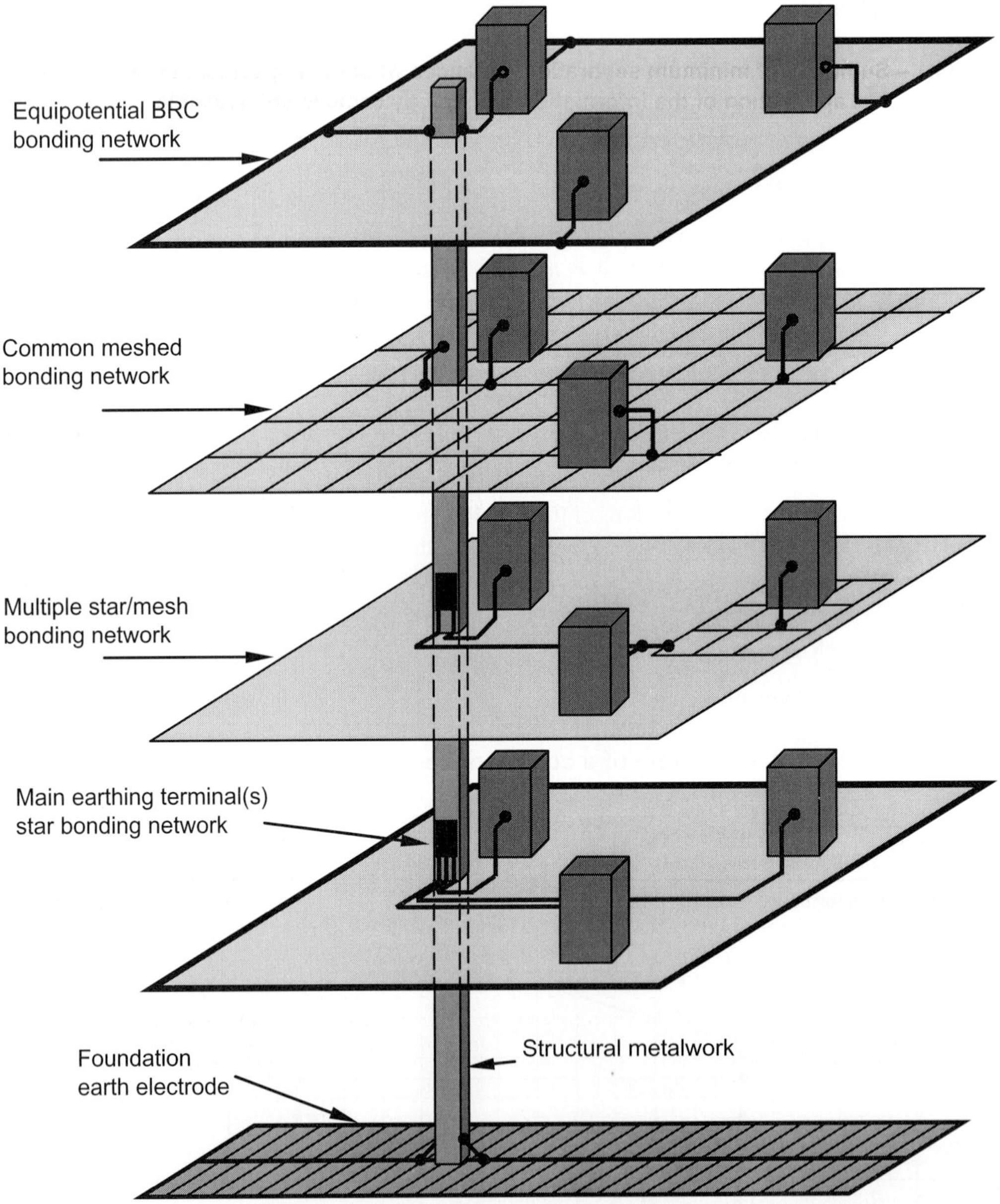

A444.3 Installations containing a high density of interconnected equipment

In severe electromagnetic environments, it is recommended that the common meshed bonding star network described in A444.1.3 be adopted.

A444.4 Design guidelines for segregation of circuits

Where both the specification of the information technology cable and its intended application is known, the requirements of BS EN 50174-2 and BS EN 50174-3 are appropriate.

BS EN 50174 series standards contain requirements and recommendations for the installation of information technology cabling which support a range of applications delivering the following services:

(i) ICT (information and communications technology) e.g. local area networks

(ii) BCT (broadcast communication technologies) e.g. audio-visual, television

(iii) CCCB (command control and communications in buildings) e.g. building automation

(iv) PMCA (process monitoring, control and automation) e.g. industrial networks (Fieldbus).

Where the specification and/or intended application of the information technology cable is not available, then the cable separation distance between the power and information technology cables should be a minimum of 200 mm in free air.

This distance can be reduced if a screened power cable, a metallic barrier, or containment system is used as described in Table A444.1.

TABLE A444.1 – Summary of minimum separation distances where the specification and/or the intended application of the information technology cable is not available

These recommendations of segregation are based upon the following assumptions:

(i) The electromagnetic environment complies with the levels defined in the BS EN 61000-6 series of standards for conducted and radiated disturbances (e.g. mains power cabling)

(ii) The LV supply is non-deformed but has high-frequency content consistent with the switching and operation of connected equipment in accordance with the BS EN 61000-6 series of standards

NOTE: "Deformed" LV power supplies and the use of other equipment lie outside the scope of BS 7671 and might require additional engineering practices.

(iii) The total design current in the LV circuits does not exceed 600 A

(iv) Balanced information technology/telecommunications cables have electromagnetic immunity performance in accordance with BS EN 50288 series standards for Category 5 and above

(v) Coaxial information technology/telecommunications cables have electromagnetic immunity performance in accordance with BS EN 50117-4-1 standard for Category BCT-C

(vi) The applications supported by the cabling are designed to operate using the information technology cabling installed or to be installed.

Containment applied to the mains power cabling		
No containment or open metallic containment A[1]	Perforated open metallic containment B[2]	Solid metallic containment C[3]
200 mm	150 mm	Note 4

NOTE 1: Screening performance (DC-100MHz) equivalent to welded mesh steel basket of mesh size 50 mm × 100 mm (excluding ladders). This screening performance is also achieved with steel tray (duct without cover) of less than 1.0 mm wall thickness and more than 20 % equally distributed perforated area. No part of the cable within the containment should be less than 10 mm below the top of the barrier.

NOTE 2: Screening performance (DC-100 MHz) equivalent to steel tray (duct without cover) of 1.0 mm wall thickness and no more than 20 % equally distributed perforated area. This screening performance is also achieved with screened power cables that do not meet the performance defined in Note 1. No part of the cable within the containment should be less than 10 mm below the top of the barrier.

NOTE 3: Screening performance (DC-100 MHz) equivalent to a fully enclosed steel containment system having a minimum wall thickness of 1.5 mm. Separation specified is in addition to that provided by any divider/barrier.

NOTE 4: No physical separation other than that provided by the containment.

NOTE: Zero segregation in the Table references additional segregation/separation for EMC over and above the requirements for safety. Safety considerations must always take precedence over EMC requirements.

Where the above conditions do not apply, see Table A444.2.

TABLE A444.2 – Minimum separation between power and signal cables (m)

Power Cable Voltage (V)	Minimum Separation between Power and Signal Cables (m)	Power Cable Current (A)	Minimum Separation between Power and Signal Cables (m)
115	0.25	5	0.24
240	0.45	15	0.35
415	0.58	50	0.5
3300	1.1	100	0.6
6600	1.25	300	0.85
11000	1.4	600	1.05

NOTE 1: The values in Table A444.2 can be used specifically for long parallel runs of cables.

NOTE 2: The worst-case separation based on voltage or current should be used.

Additional areas of concern are expressed in Regulation 444.4.1.

The minimum separation between the information technology cables and mains power cables includes all allowances for cable movement between their fixing points or other restraints (see example in Figure A444.5).

Fig A444.5 – Example of cable separation distance

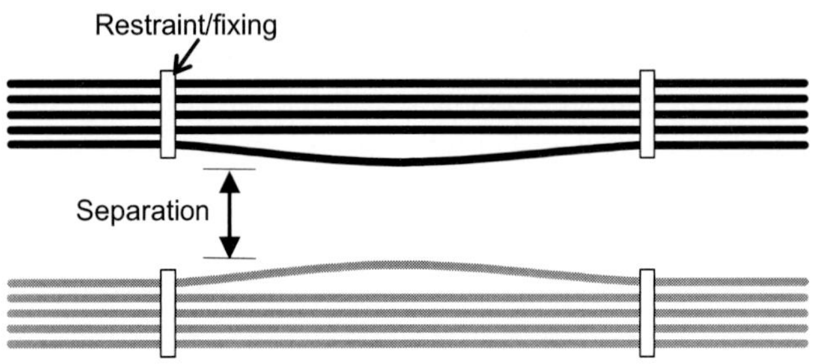

The minimum separation requirement applies in three dimensions. However, where information technology cables and mains power cables are required to cross and required minimum separation cannot be maintained then maintaining the angle of their crossing at 90 degrees on either side of the crossing for a distance no less than the applicable minimum separation requirement will minimize any electromagnetic disturbances.

A444.5 Conditions for zero segregation

See BS EN 50174 series.

445 PROTECTION AGAINST UNDERVOLTAGE

445.1 General requirements

445.1.1 Suitable precautions shall be taken where a reduction in voltage, or loss and subsequent restoration of voltage, could cause danger. Provisions for a circuit supplying a motor shall comply with Regulation 552.1.3.

Where current-using equipment or any other part of the installation may be damaged by a drop in voltage and it is verified that such damage is unlikely to cause danger, one of the following arrangements shall be adopted:

(i) Suitable precautions against the damage foreseen shall be provided

(ii) It shall be verified, in consultation with the person or body responsible for the operation and maintenance of the installation, that the damage foreseen is an acceptable risk.

445.1.2 A suitable time delay may be incorporated in the operation of an undervoltage protective device if the operation of the equipment to which the protection relates allows without danger a brief reduction or loss of voltage.

445.1.3 Any delay in the opening or reclosing of a contactor shall not impede instantaneous disconnection by a control device or a protective device.

445.1.4 The characteristics of an undervoltage protective device shall be compatible with the requirements for starting and use of the equipment to which the protection relates, as stated in the appropriate British or Harmonized Standard.

445.1.5 Where the reclosure of a protective device is likely to cause danger, the reclosure shall not be automatic.

CHAPTER 45 *NOT USED*

CHAPTER 46

ISOLATION AND SWITCHING

CONTENTS

CHAPTER 46

ISOLATION AND SWITCHING

460 SCOPE

Chapter 46 deals with:

(i) non-automatic local and remote isolation and switching measures for the prevention or removal of dangers associated with electrical installations or electrically powered equipment, and

(ii) switching for the control of circuits or equipment.

Where electrically powered equipment is within the scope of BS EN 60204, only the requirements of that standard apply.

461 GENERAL

461.1 According to the intended function(s), every device provided for isolation or switching shall comply with the relevant requirements of Chapter 53.

461.2 In TN-C systems and TN-C-S systems, the PEN conductor shall not be isolated or switched.

In TN-C-S and TN-S systems, isolation or switching of the neutral conductor is not required if protective equipotential bonding is installed and either:

(i) the neutral conductor is reliably connected to Earth by a low resistance to meet the disconnection times of the protective devices according to the requirements of Chapter 41, or

(ii) the distributor declares that either the PEN or the neutral conductor of the supply is reliably connected to Earth by a low resistance to meet the disconnection times of the protective devices according to the requirements of Chapter 41.

462 ISOLATION

462.1 Each electrical installation shall have provisions for isolation from each supply.

462.1.201 A main linked switch or linked circuit-breaker shall be provided as near as practicable to the origin of every installation as a means of switching the supply on load and as a means of isolation.

A main switch intended for operation by ordinary persons, e.g. of a household or similar installation, shall interrupt both live conductors of a single-phase supply.

462.2 Every circuit shall be provided with isolation means for all live conductors, except as detailed in Regulation 461.2.

Provision may be made for isolating a group of circuits by a common means, if the service conditions allow this.

462.3 Devices for isolation shall be designed and/or installed so as to prevent unintentional or inadvertent closure.

Examples of precautions are as follows:

- Located within a lockable space or lockable enclosure
- Padlocking
- Located adjacent to the associated equipment.

462.4 Where residual electrical energy is potentially present, suitable means shall be provided for its discharge.

Where relevant, a warning label indicating the discharge time required before the enclosure can be safely opened shall be provided.

NOTE: Storage units do not need to be discharged since they are considered to be supply sources.

463 FUNCTIONAL SWITCHING (CONTROL)

463.1 General

463.1.1 Functional switching shall be provided for each part of a circuit which may require to be controlled independently of other parts of the installation.

463.1.2 Functional switching devices need not necessarily switch off all live conductors of a circuit. A single-pole switching device shall not be placed in the neutral conductor except for the connection of the control device for lighting circuits as shown in Figure 46.1.

Fig 46.1 – Lamp control circuit with switching in the neutral conductor

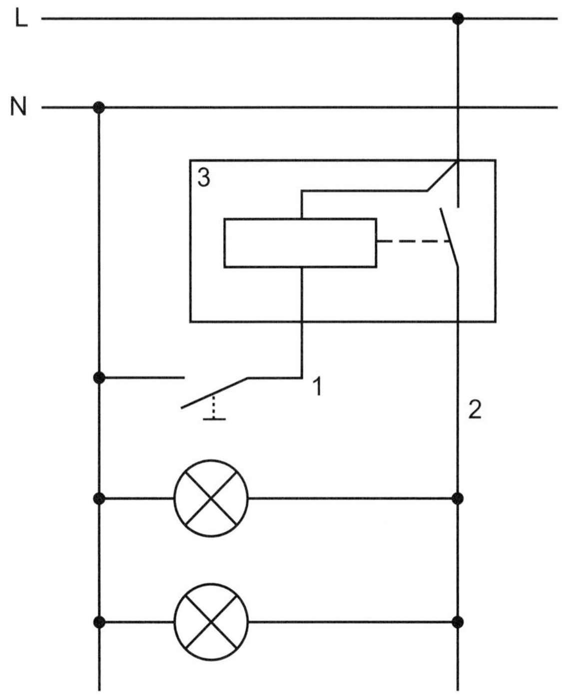

Key

1 connection of the control device for lighting circuits
2 circuit supplying the lamps
3 control device.

Switching of the neutral shall be in compliance with Regulation 530.3.3.

463.1.3 In general, all current-using equipment requiring control shall be controlled by an appropriate functional switching device.

A single functional switching device may control several items of current-using equipment intended to operate simultaneously.

463.1.4 Functional switching devices ensuring the changeover of supply from alternative sources shall switch off all live conductors and shall not be capable of putting the sources in parallel, unless the installation is specifically designed for this condition.

463.2 Auxiliary circuits

Auxiliary circuits shall be designed, arranged and protected to limit dangers resulting from a fault in the auxiliary circuit or an insulation fault between the auxiliary circuit and other conductive parts liable to cause malfunction (e.g. inadvertent operation) of the controlled apparatus; see Section 557.

463.3 Motor control

463.3.1 Motor control circuits shall be designed so as to prevent any motor from restarting automatically after a stoppage due to a fall in or loss of voltage, if such starting is liable to cause danger.

Earth faults in control circuits should not cause unintentional starting, potentially hazardous motion, or prevent stopping of the motor.

463.3.2 Where reverse-current braking of a motor is provided, provision shall be made for the avoidance of reversal of the direction of rotation at the end of braking if such reversal may cause danger.

463.3.3 Where safety depends on the direction of rotation of a motor, provision shall be made for the prevention of reverse operation due to a reversal of phases.

NOTE: Attention is drawn to danger which may arise from the loss of one phase.

464 SWITCHING OFF FOR MECHANICAL MAINTENANCE

464.1 Means for switching off shall be provided where mechanical maintenance may involve a risk of physical injury.

The switching off shall cause the disconnection of all live conductors, except as provided in Regulation 461.2, by a device suitable for isolation.

NOTE 1: Electrically powered mechanical equipment may include rotating machines as well as heating elements and electromagnetic equipment.

NOTE 2: Systems powered by other means, e.g. pneumatic, hydraulic or steam, are not covered by these regulations. In such cases, switching off any associated supply of electricity may not be a sufficient measure.

NOTE 3: Where electrically powered equipment is within the scope of BS EN 60204, the requirements for switching off for mechanical maintenance of that standard apply.

464.2 Suitable means shall be provided to prevent electrically powered equipment from inadvertently or unintentionally reactivating during mechanical maintenance, unless the means of switching off is continuously under the control of any person performing such maintenance.

465 EMERGENCY SWITCHING OFF

465.1 Means shall be provided for emergency switching off of any part of an installation where it may be necessary to control the supply to remove an unexpected danger.

465.2 Where a risk of electric shock or another risk of electrical origin is involved, the emergency switching off shall cause the disconnection of all live conductors, except as provided in Regulation 461.2, by a device suitable for isolation.

465.3 Means for emergency switching off shall act as directly as possible on the appropriate supply conductors.

The arrangement for emergency switching off shall be such that one single action only will interrupt the appropriate supply.

465.4 The arrangement of the emergency switching shall be such that its operation does not introduce a further danger or interfere with the complete operation necessary to remove the danger.

NOTE: The operation of the switching device is to be understood as switching off in case of emergency and switching on to re-energize the relevant circuit.

The emergency operation function shall not impair the effectiveness of protective devices or of devices with other safety functions.

PART 5
SELECTION AND ERECTION OF EQUIPMENT
CONTENTS

CHAPTER 55 OTHER EQUIPMENT

CHAPTER 56 SAFTEY SERVICES

CHAPTER 51

COMMON RULES

CONTENTS

CHAPTER 51

COMMON RULES

510 INTRODUCTION

510.1 General

This chapter deals with the selection of equipment and its erection. It provides common rules for compliance with measures of protection for safety, requirements for proper functioning for intended use of the installation, and requirements appropriate to the external influences.

510.2 Not used

510.3 Every item of equipment shall be selected and erected so as to allow compliance with the regulations stated in this chapter and the relevant regulations in other parts of BS 7671 and shall take account of manufacturers' instructions.

511 COMPLIANCE WITH STANDARDS

511.1 Every item of equipment shall comply with the relevant requirements of the applicable British or Harmonized Standard, appropriate to the intended use of the equipment. The edition of the standard shall be the current edition, with those amendments pertaining at a date to be agreed by the parties to the contract concerned (see Appendix 1).

Alternatively, if equipment complying with a foreign national standard based on an IEC Standard is to be used, the designer or other person responsible for specifying the installation shall verify that any differences between that standard and the corresponding British or Harmonized Standard will not result in a lesser degree of safety than that afforded by compliance with the British or Harmonized Standard. Such use shall be recorded on the appropriate electrical certification specified in Part 6.

NOTE 1: As a minimum, an appropriate foreign national standard should define relevant safety parameters, together with test criteria and requirements. For some products, for instance cables, these include the constructional details.

NOTE 2: The use of foreign national standards not based on an IEC Standard, and/or not compatible with the safety provisions of BS 7671, is not recommended.

511.2 Where equipment to be used is not covered by a British or Harmonized Standard or is to be used outside the scope of its standard, the designer or other person responsible for specifying the installation shall confirm that the equipment provides at least the same degree of safety as that afforded by compliance with BS 7671. Such use shall be recorded on the appropriate electrical certification specified in Part 6.

NOTE: An assessment of compliance with BS 7671 in respect of the degree of safety afforded would be expected to consider all the same criteria as in Note 1 to Regulation 511.1. This also applies to private specifications not containing the same criteria as in Note 1 to Regulation 511.1.

512 OPERATIONAL CONDITIONS AND EXTERNAL INFLUENCES

512.1 Operational conditions

512.1.1 Voltage

Every item of equipment shall be suitable for the nominal voltage (U_0) of the installation or the part of the installation concerned, where necessary taking account of the highest and/or lowest voltage likely to occur in normal service. In an IT system, equipment shall be insulated for the nominal voltage between lines.

512.1.2 Current

Every item of equipment shall be suitable for:

(i) the design current, taking into account any capacitive and inductive effects, and

(ii) the current likely to flow in abnormal conditions for such periods of time as are determined by the characteristics of the protective devices concerned.

512.1.3 Frequency

If frequency has an influence on the characteristics of the equipment, the rated frequency of the equipment shall correspond to the nominal frequency of the supply to the circuit concerned.

512.1.4 Power

Every item of equipment selected on the basis of its power characteristics shall be suitable for the duty demanded of the equipment.

512.1.5 Compatibility

Every item of equipment shall be selected and erected so that it will neither cause harmful effects to other equipment nor impair the supply during normal service including switching operations.

Switchgear, protective devices, accessories and other types of equipment shall not be connected to conductors intended to operate at a temperature exceeding 70 °C at the equipment in normal service unless the equipment manufacturer has confirmed that the equipment is suitable for such conditions. The 70 °C limit is met if the live conductor size is chosen based on the current rating for 70 °C cables of similar construction.

See also Regulation 526.2 concerning electrical connections, Regulations 523.1 and 536.4.202 and Table 4A3.

The designer of the fixed installation shall verify that the installed fixed equipment, where relevant, is designed and manufactured in accordance with the Electromagnetic Compatibility Regulations 2016 and, upon request, the responsible person for the fixed installation shall provide the required documentation as specified by the Electromagnetic Compatibility Regulations.

NOTE 1: Information on the parameters to be considered is given in Section 444. The level of detail of the documentation may vary from very simple information to much more detailed documentation for complex installations involving important potential EMC aspects.

NOTE 2: The responsible person referred to in this regulation is as defined in the Electromagnetic Compatibility Regulations 2016 as the installer.

NOTE 3: Where the current rating is to be based on 70 °C, current-carrying capacities given in Tables 4D1 to 4D5 or 4H1 to 4H4 of Appendix 4 may be used for 90 °C thermosetting insulated cables.

512.1.6 Impulse withstand voltage

Equipment shall be selected so that its impulse withstand voltage is at least equal to the required minimum impulse withstand voltage according to the overvoltage category at the point of installation as defined in Section 443.

512.2 External influences

512.2.1 Equipment shall be of a design appropriate to the situation in which it is to be used or its mode of installation shall take account of the conditions likely to be encountered.

512.2.2 If the equipment does not, by its construction, have the characteristics relevant to the external influences of its location, it may nevertheless be used on condition that it is provided with appropriate additional protection in the erection of the installation. Such protection shall not adversely affect the operation of the equipment thus protected.

512.2.3 Where different external influences occur simultaneously, they may have independent or mutual effects and the degree of protection shall be provided accordingly.

512.2.4 The selection of equipment according to external influences is necessary not only for proper functioning, but also for the reliability of the measures of protection for safety complying with BS 7671 generally. Measures of protection afforded by the construction of equipment are valid only for the given conditions of external influence if the corresponding equipment specification tests are made in these conditions of external influence.

NOTE: For the purpose of BS 7671, the following classes of external influence are conventionally regarded as normal:

AA Ambient temperature	AA4
AB Atmospheric humidity	AB4
Other environmental conditions (AC to AS)	XX1 of each parameter
Utilization and construction of buildings (B and C)	XX1 of each parameter, except XX2 for the parameter BC

513 ACCESSIBILITY

513.1 Except for a joint in cables where Section 526 allows such a joint to be inaccessible, every item of equipment shall be arranged so as to facilitate its operation, inspection and maintenance and access to each connection. Such facility shall not be significantly impaired by mounting equipment in an enclosure or a compartment.

514 IDENTIFICATION AND NOTICES

514.1 General

514.1.1 Except where there is no possibility of confusion, a label or other suitable means of identification shall be provided to indicate the purpose of each item of switchgear and controlgear. Where the operator cannot observe the operation of switchgear and controlgear and where this might lead to danger, a suitable indicator complying, where applicable, with BS EN 60073 and BS EN 60447, shall be fixed in a position visible to the operator.

514.1.2 So far as is reasonably practicable, wiring shall be so arranged or marked that it can be identified for inspection, testing, repair or alteration of the installation.

514.1.3 Except where there is no possibility of confusion, unambiguous marking shall be provided at the interface between conductors identified in accordance with BS 7671 and conductors identified to previous versions of BS 7671.

514.2 Not used

514.3 Identification of conductors

514.3.1 Except where identification is not required by Regulation 514.6, cores of cables shall be identified by:

(i) colour as required by Regulation 514.4 and/or
(ii) letters and/or numbers as required by Regulation 514.5.

514.3.2 Every core of a cable shall be identifiable at its terminations and preferably throughout its length. Binding and sleeves for identification purposes shall comply with BS 3858 where appropriate.

514.4 Identification of conductors by colour

514.4.1 Neutral or midpoint conductor

Where a circuit includes a neutral or midpoint conductor identified by colour, the colour used shall be blue.

514.4.2 Protective conductor

The bi-colour combination green-and-yellow shall be used exclusively for identification of a protective conductor and this combination shall not be used for any other purpose. In this combination one of the colours shall cover at least 30 % and at most 70 % of the surface being coloured, while the other colour shall cover the remainder of the surface.

Single-core cables and conductors in multicore cables identified by green-and-yellow throughout their length shall only be used as a protective conductor and shall not be overmarked at their terminations, except as permitted by Regulation 514.4.3.

A bare conductor or busbar used as a protective conductor shall be identified, where necessary, by equal green-and-yellow stripes, each not less than 15 mm and not more than 100 mm wide, close together, either throughout the length of the conductor or in each compartment and unit and at each accessible position.

514.4.3 PEN conductor

A PEN conductor shall, when insulated, be marked by one of the following methods:

(i) Green-and-yellow throughout its length with, in addition, blue markings at the terminations
(ii) Blue throughout its length, with green-and-yellow markings at the terminations.

514.4.4 Other conductors

Other conductors shall be identified by colour in accordance with Table 51.

514.4.5 The single colour green shall not be used for the identification of:

(i) live conductors in power circuits
(ii) protective conductors
(iii) functional earthing and bonding conductors.

514.4.6 Bare conductors

A bare conductor shall be identified, where necessary, by the application of tape, sleeve or disc of the appropriate colour prescribed in Table 51 or by painting with such a colour.

514.5 Identification of conductors by letters and/or numbers

514.5.1 The lettering or numbering system applies to identification of individual conductors and of conductors in a group. The identification shall be clearly legible and durable. All characters shall be in strong contrast to the colour of the insulation. The identification shall be given in letters and/or Arabic numerals. In order to avoid confusion, unattached numerals 6 and 9 shall be underlined.

514.5.2 Protective conductor

Conductors with green-and-yellow colour identification shall not be numbered other than for the purpose of circuit identification.

514.5.3 Alphanumeric

The preferred alphanumeric system is described in Table 51.

514.5.4 Numeric

Conductors may be identified by numbers, the number 0 being reserved for the neutral or midpoint conductor.

514.6 Omission of identification by colour or marking

514.6.1 Identification by colour or marking is not required for:

(i) concentric conductors of cables
(ii) metal sheath or armour of cables when used as a protective conductor
(iii) bare conductors where permanent identification is not practicable
(iv) extraneous-conductive-parts used as a protective conductor
(v) exposed-conductive-parts used as a protective conductor.

TABLE 51 – Identification of conductors

Function	Alphanumeric	Colour
Protective conductor	PE	Green-and-yellow (GNYE)
Protective bonding conductors	PB	Green-and-yellow (GNYE)
Functional earthing conductor	FE	Pink (PK)
AC power circuit[1]		
Line of single-phase circuit	L	Brown (BN)
Neutral of single- or three-phase circuit	N	Blue (BU)
Line 1 of three-phase AC circuit	L1	Brown (BN)
Line 2 of three-phase AC circuit	L2	Black (BK)
Line 3 of three-phase AC circuit	L3	Grey (GY)
Two-wire unearthed DC power circuit		
Positive of two-wire circuit	L+	Red (RD)
Negative of two-wire circuit	L-	White (WH)
Two-wire earthed DC power circuit		
Positive (of negative earthed) circuit	L+	Red (RD)
Negative (of negative earthed) circuit[2]	M	Blue (BU)
Positive (of positive earthed) circuit[2]	M	Blue (BU)
Negative (of positive earthed) circuit	L-	White (WH)
Three-wire DC power circuit		
Outer positive of two-wire circuit derived from three-wire system	L+	Red (RD)
Outer negative of two-wire circuit derived from three-wire system	L-	White (WH)
Positive of three-wire circuit	L+	Red (RD)
Mid-wire of three-wire circuit[2][3]	M	Blue (BU)
Negative of three-wire circuit	L-	White (WH)
Control circuits, ELV and other applications		
Line conductor	L	Black (BK), Brown (BN), Red (RD), Orange (OG), Yellow (YE), Green (GN), Violet (VT), Grey (GY), White (WH), Pink (PK) or Turquoise (TQ)
Neutral or mid-wire[4]	N or M	Blue (BU)

NOTE: [1] Power circuits include lighting circuits.

[2] M identifies either the mid-wire of a three-wire DC circuit, or the earthed conductor of a two-wire earthed DC circuit.

[3] Only the middle wire of three-wire circuits may be earthed.

[4] An earthed PELV conductor is blue.

514.7 Not used

514.8 Identification of a protective device

514.8.1 A protective device shall be arranged and identified so that the circuit protected may be easily recognized.

514.9 Diagrams and documentation

514.9.1 A diagram, chart or table or equivalent form of information shall be provided indicating in particular:

(i) the type and composition of each circuit (points of utilization served, number and size of conductors, type of wiring), and

(ii) the method used for compliance with Regulation 410.3.2, and

(iii) the information necessary for the identification of each device performing the functions of protection, isolation and switching, and its location, and

(iv) any circuit or equipment vulnerable to the electrical tests as required by Part 6.

For simple installations the foregoing information may be given in a schedule. A durable copy of the schedule relating to a distribution board shall be provided within or adjacent to each distribution board.

Any symbol used shall comply with IEC 60617.

The requirements of this regulation need not be applied for domestic (household) premises or similar installations where certification for initial verification, complete with Guidance for Recipients as detailed in Appendix 6, has been issued to the person ordering the work.

514.9.2 All diagrams, charts, and information or instruction notices shall comply with BS EN 61082-1, BS EN IEC/IEEE 82079-1, and, where appropriate, BS EN 81346-1.

All warning notices and other relevant safety signs shall comply with BS ISO 3864, BS EN ISO 7010, and BS EN IEC/IEEE 82079-1.

514.10 Warning notice: voltage

514.10.1 Every item of equipment or enclosure within which a nominal voltage exceeding 230 volts to earth exists and where the presence of such a voltage would not normally be expected, shall be so arranged that before access is gained to a live part, a notice warning of the maximum voltage to earth present is clearly visible.

514.11 Warning notice: isolation

514.11.1 A warning notice of such durable material as to be likely to remain easily legible throughout the life of the installation shall be fixed in each position where there are live parts which are not capable of being isolated by a single device. The location of each disconnector (isolator) shall be indicated unless there is no possibility of confusion.

514.12 Notices: periodic inspection and testing

514.12.1 An instruction notice of such durable material as to be likely to remain easily legible throughout the life of the installation shall be fixed in a prominent position at or near the relevant distribution board or boards upon completion of the work carried out in accordance with Chapter 64 or 65. The notice shall be inscribed in indelible characters and shall read as follows:

Important
This installation should be periodically inspected and tested and a report on its condition obtained, as prescribed in BS 7671 Requirements for Electrical Installations. Date of last inspection .... Recommended date of next inspection ....

The requirements of this regulation need not be applied for domestic (household) premises or similar installations where certification for initial verification, complete with Guidance for Recipients as detailed in Appendix 6, has been issued to the person ordering the work.

NOTE: Examples are provided in Figure 11A of Appendix 11.

514.12.2 Where an installation incorporates an RCD an instruction notice shall be fixed in a prominent position at or near each RCD in the installation. The notice shall be inscribed in indelible characters and shall read as follows:

> This installation, or part of it, is protected by a device which automatically switches off the supply if a fault develops. Test six-monthly by pressing the relevant test button(s) which should operate the device. Afterwards, manually switch on the device. If the device does not operate, or indicates a fault, seek expert advice.

NOTE 1: Testing frequencies of RCDs in temporary installations may need increasing.

The requirements of this regulation need not be applied for domestic (household) premises or similar installations where certification for initial verification, complete with Guidance for Recipients as detailed in Appendix 6, has been issued to the person ordering the work.

NOTE 2: Examples are provided in Figure 11B of Appendix 11.

NOTE 3: The instruction notice can be applied equally for other devices incorporating a user-test facility for operation by ordinary persons.

514.13 Warning notices: earthing and bonding connections

514.13.1 A durable warning notice with the words 'Safety Electrical Connection – Do Not Remove' shall be securely fixed in a visible position at or near:

(i) the point of connection of every earthing conductor to an earth electrode, and
(ii) the point of connection of every bonding conductor to an extraneous-conductive-part, and
(iii) the main earthing terminal, where separate from main switchgear.

The warning notice may be provided by a clamp according to BS 951.

NOTE: An example of a suitable warning notice is provided in Figure 11C of Appendix 11.

514.13.2 Where Regulation 418.2.5 or 418.3 applies, the warning notice specified shall be durably marked and shall incorporate the following wording:

> 'The protective bonding conductors associated with the electrical installation in this location MUST NOT BE CONNECTED TO EARTH.
>
> Equipment having exposed-conductive-parts connected to earth must not be brought into this location.'

NOTE: An example is provided in Figure 11D of Appendix 11.

514.14 Deleted by BS 7671:2018+A2:2022

514.15 Warning notice: alternative or additional supplies

514.15.1 Where an installation includes alternative or additional sources of supply, warning notices shall be affixed at the following locations in the installation:

(i) At the origin of the installation
(ii) At the meter position, if remote from the origin
(iii) At the consumer unit or distribution board to which the alternative or additional sources are connected
(iv) At all points of isolation of all sources of supply.

The warning notice shall be durably marked and identify the relevant point(s) of isolation.

NOTE: An example is provided in Figure 11E of Appendix 11.

514.16 Notice: presence of SPDs in an installation

514.16.1 The presence of SPDs in an installation shall be indicated by an information notice at or near the relevant distribution board(s).

The requirements of this regulation need not be applied for domestic (household) premises or similar installations where the information is recorded on the appropriate certification for initial verification, complete with Guidance for Recipients as detailed in Appendix 6, and issued to the person ordering the work.

514.17 Notice: high protective conductor current

A warning notice shall be provided where required by Regulation 543.7.1.205.

515 PREVENTION OF MUTUAL DETRIMENTAL INFLUENCE

515.1 Prevention of mutual detrimental influence

Electrical equipment shall be selected and erected so as to avoid any harmful influence between the electrical installation and any non-electrical installations envisaged.

NOTE: For EMC see Sections 332 and 444.

515.2 Where equipment carrying current of different types or at different voltages is grouped in a common assembly (such as a switchboard, a cubicle or a control desk or box), all the equipment belonging to any one type of current or any one voltage shall be effectively segregated wherever necessary to avoid mutual detrimental influence.

The immunity levels of equipment shall be chosen taking into account the electromagnetic disturbances that can occur when connected and erected as for normal use, and taking into account the intended level of continuity of service necessary for the application. See the specific equipment standard or the relevant part of BS EN 61000 series.

CHAPTER 52

SELECTION AND ERECTION OF WIRING SYSTEMS

CONTENTS

CHAPTER 52

SELECTION AND ERECTION OF WIRING SYSTEMS

520 INTRODUCTION

520.1 Scope

This chapter deals with the selection and erection of wiring systems.

NOTE: These regulations also apply in general to protective conductors. Chapter 54 contains further requirements for those conductors.

520.2 Not used

520.3 Not used

520.4 General

Consideration shall be given to the application of the fundamental principles of Chapter 13 as it applies to:

(i) cables and conductors
(ii) their connections, terminations and/or jointing
(iii) their associated supports or suspensions, and
(iv) their enclosure or methods of protection against external influences.

521 TYPES OF WIRING SYSTEM

The requirements of Regulations 521.1 to 521.201 do not apply to busbar and powertrack systems covered by Regulation 521.4.

521.1 The installation method of a wiring system in relation to the type of conductor or cable used shall be in accordance with Table 4A1 of Appendix 4, provided the external influences are taken into account according to Section 522.

521.2 The installation method of a wiring system in relation to the situation concerned shall be in accordance with Table 4A2 of Appendix 4. Other methods of installation of cables and conductors not included in Table 4A2 are permitted provided that they fulfil the requirements of this chapter.

521.3 Examples of wiring systems, excluding systems covered by Regulation 521.4, are shown in Table 4A2. Table 4A2 gives examples of installation methods of cables including reference method for obtaining current-carrying capacity where it is considered that the same current-carrying capacities can safely be used. It is not implied that such methods must be employed or that other methods are prohibited.

521.201 Prefabricated wiring systems intended for permanent connection in fixed installations incorporating installation couplers conforming to BS EN 61535, shall comply with BS 8488.

521.4 Busbar trunking systems and powertrack systems

A busbar trunking system shall comply with BS EN 61439-6 and a powertrack system shall comply with the appropriate part of the BS EN 61534 series. A busbar trunking system or a powertrack system shall be installed in accordance with the manufacturer's instructions, taking account of external influences. See also Appendix 8.

521.5 AC circuits: electromagnetic effects

521.5.1 Ferromagnetic enclosures: electromagnetic effects

The conductors of an AC circuit installed in a ferromagnetic enclosure shall be arranged so that all line conductors and the neutral conductor, if any, and the appropriate protective conductor are contained within the same enclosure.

Where such conductors enter a ferrous enclosure, they shall be arranged such that the conductors are only collectively surrounded by ferromagnetic material.

These requirements do not preclude the use of an additional protective conductor in parallel with the steel wire armouring of a cable where such is required to comply with the requirements of the appropriate regulations in Chapters 41 and 54. It is permitted for such an additional protective conductor to enter the ferrous enclosure individually.

521.5.2 Single-core cables armoured with steel wire or steel tape shall not be used for an AC circuit.

NOTE: The steel wire or steel tape armour of a single-core cable is regarded as a ferromagnetic enclosure. For single-core armoured cables, the use of aluminium armour may be considered.

521.5.201 Electromechanical stresses

Every conductor or cable shall have adequate strength and be so installed as to withstand the electromechanical forces that may be caused by any current, including fault current, it may have to carry in service.

521.6 Conduit, ducting, trunking, tray and ladder systems

Two or more circuits are allowed in the same conduit, ducting or trunking system provided the requirements of Section 528 are met.

Cable conduits shall comply with the appropriate part of the BS EN 61386 series, cable trunking or ducting shall comply with the appropriate part of the BS EN 50085 series and cable tray and ladder systems shall comply with BS EN 61537.

521.7 Multicore cables: two or more circuits

Two or more circuits are allowed in the same cable provided the requirements of Section 528 are met.

521.8 Circuit arrangements

521.8.1 Each part of a circuit shall be arranged such that the conductors are not distributed over different multicore cables, conduits, ducting systems, trunking systems or tray or ladder systems.

This requirement need not be met where a number of multicore cables, forming one circuit, are installed in parallel. Where multicore cables are installed in parallel each cable shall contain one conductor of each line.

521.8.2 The line and neutral conductors of each final circuit shall be electrically separate from those of every other final circuit, so as to prevent the indirect energizing of a final circuit intended to be isolated.

521.8.3 Where two or more circuits are terminated in a single junction box this shall comply with BS EN 60670-22.

521.9 Use of flexible cables

521.9.1 A flexible cable shall be used for fixed wiring only where the relevant provisions of BS 7671 are met. |
Flexible cables used for fixed wiring shall be of the heavy duty type unless the risk of damage during installation and service, due to impact or other mechanical stresses, is low or has been minimized or protection against mechanical damage is provided.

NOTE: Descriptions of light, ordinary and heavy duty types are given in BS EN 50565-1.

521.9.2 Equipment that is intended to be moved in use shall be connected by flexible cables, except equipment supplied by contact rails.

521.9.3 Stationary equipment which is moved temporarily for the purposes of connecting, cleaning etc., e.g. cookers or flush-mounting units for installations in false floors, shall be connected with flexible cable. If the equipment is not subject to vibration then non-flexible cables may be used.

521.10 Installation of cables

521.10.1 Non-sheathed cables for fixed wiring shall be enclosed in conduit, ducting or trunking. This requirement does not apply to a protective conductor complying with Section 543.

Non-sheathed cables are permitted if the cable trunking system provides at least the degree of protection IPXXD or IP4X, and if the cover can only be removed by means of a tool or a deliberate action.

NOTE: For a cable trunking system to meet IP4X requirements, IP4X trunking and related system components would need to be installed. If a system includes site-fabricated joints the installer must confirm the completed item meets at least the degree of protection IPXXD.

521.10.201 A bare live conductor shall be installed on insulators.

521.10.202 Wiring systems shall be supported such that they will not be liable to premature collapse in the event of a fire.

NOTE 1: Wiring systems hanging across access or egress routes may hinder evacuation and firefighting activities.

NOTE 2: Cables installed in or on steel cable containment systems are deemed to meet the requirements of this regulation.

NOTE 3: This regulation precludes, for example, the use of non-metallic cable clips or cable ties as the sole means of support where cables are clipped direct to exposed surfaces or suspended under cable tray, and the use of non-metallic cable trunking as the sole means of support of the cables therein.

NOTE 4: Suitably spaced steel or copper clips, saddles or ties are examples that will meet the requirements of this regulation.

522 SELECTION AND ERECTION OF WIRING SYSTEMS IN RELATION TO EXTERNAL INFLUENCES

The installation method selected shall be such that protection against the expected external influences is provided in all appropriate parts of the wiring system. Particular care shall be taken at changes in direction and where wiring enters into equipment.

NOTE: The external influences categorized in Appendix 5 which are of significance to wiring systems are included in this section.

522.1 Ambient temperature (AA)

522.1.1 A wiring system shall be selected and erected so as to be suitable for the highest and lowest local ambient temperatures and so that the limiting temperature in normal operation (see Table 52.1) and the limiting temperature in case of a fault (see Table 43.1) will not be exceeded.

522.1.2 Wiring system components, including cables and wiring accessories, shall only be installed or handled at temperatures within the limits stated in the relevant product specification or as given by the manufacturer.

522.2 External heat sources

522.2.1 In order to avoid the effects of heat from external sources, one or more of the following methods or an equally effective method shall be used to protect a wiring system:

(i) Shielding
(ii) Placing sufficiently far from the source of heat
(iii) Selecting a system with due regard for the additional temperature rise which may occur
(iv) Local reinforcement or substitution of insulating material.

NOTE: Heat from external sources may be radiated, conducted or convected, e.g.:

- from hot water systems
- from plant, appliances and luminaires
- from a manufacturing process
- through heat conducting materials
- from solar gain of the wiring system or its surrounding medium.

522.2.201 Parts of a cable within an accessory, appliance or luminaire shall be suitable for the temperatures likely to be encountered, as determined in accordance with Regulation 522.1.1, or shall be provided with additional insulation suitable for those temperatures.

522.3 Presence of water (AD) or high humidity (AB)

522.3.1 A wiring system shall be selected and erected so that no damage is caused by condensation or ingress of water during installation, use and maintenance. The completed wiring system shall comply with the IP degree of protection (see BS EN 60529) relevant to the particular location.

NOTE: Special considerations apply to wiring systems liable to frequent splashing, immersion or submersion.

522.3.2 Where water may collect or condensation may form in a wiring system, provision shall be made for its escape.

522.3.3 Where a wiring system may be subjected to waves (AD6), protection against mechanical damage shall be afforded by one or more of the methods of Regulations 522.6 to 522.8.

522.4 Presence of solid foreign bodies (AE)

522.4.1 A wiring system shall be selected and erected so as to minimize the danger arising from the ingress of solid foreign bodies. The completed wiring system shall comply with the IP degree of protection (see BS EN 60529) relevant to the particular location.

522.4.2 In a location where dust in significant quantity is present (AE4), additional precautions shall be taken to prevent the accumulation of dust or other substances in quantities which could adversely affect heat dissipation from the wiring system.

NOTE: A wiring system which facilitates the removal of dust may be necessary (see Section 529).

522.5 Presence of corrosive or polluting substances (AF)

522.5.1 Where the presence of corrosive or polluting substances, including water, is likely to give rise to corrosion or deterioration, parts of the wiring system likely to be affected shall be suitably protected or manufactured from a material resistant to such substances.

NOTE: Suitable protection for application during erection may include protective tapes, paints or grease.

522.5.2 Dissimilar metals liable to initiate electrolytic action shall not be placed in contact with each other, unless special arrangements are made to avoid the consequences of such contact.

522.5.3 Materials liable to cause mutual or individual deterioration or hazardous degradation shall not be placed in contact with each other.

522.6 Impact (AG)

522.6.1 Wiring systems shall be selected and erected so as to minimize the damage arising from mechanical stress, e.g. by impact, abrasion, penetration, tension or compression during installation, use or maintenance.

522.6.2 In a fixed installation where impacts of medium severity (AG2) or high severity (AG3) can occur protection shall be afforded by:

(i) the mechanical characteristics of the wiring system, or
(ii) the location selected, or
(iii) the provision of additional local or general protection against mechanical damage, or
(iv) any combination of the above.

NOTE: Examples are areas where the floor is likely to be penetrated and areas used by forklift trucks.

522.6.3 *Not used*

522.6.4 The degree of protection of electrical equipment shall be maintained after installation of the cables and conductors.

522.6.201 A cable installed under a floor or above a ceiling shall be run in such a position that it is not liable to be damaged by contact with the floor or ceiling or their fixings.

A cable passing through a joist within a floor or ceiling construction or through a ceiling support (e.g. under floorboards), shall:

(i) be installed at least 50 mm measured vertically from the top, or bottom as appropriate, of the joist or batten, or
(ii) comply with Regulation 522.6.204.

522.6.202 A cable installed in a wall or partition at a depth of less than 50 mm from a surface of the wall or partition shall:

(i) be installed in a zone within 150 mm from the top of the wall or partition or within 150 mm of an angle formed by two adjoining walls or partitions. Where the cable is connected to a point, accessory or switchgear on any surface of the wall or partition, the cable may be installed in a zone either horizontally or vertically, to the point, accessory or switchgear. Where the location of the accessory, point or switchgear can be determined from the reverse side, a zone formed on one side of a wall of 100 mm thickness or less or partition of 100 mm thickness or less extends to the reverse side, or
(ii) comply with Regulation 522.6.204.

Where indent (i) but not indent (ii) applies, the cable shall be provided with additional protection by means of an RCD having the characteristics specified in Regulation 415.1.1.

522.6.203 Irrespective of its buried depth, a cable concealed in a wall or partition, the internal construction of which includes metallic parts, other than metallic fixings such as nails, screws and the like, shall:

(i) be provided with additional protection by means of an RCD having the characteristics specified in Regulation 415.1.1, or
(ii) comply with Regulation 522.6.204.

For a cable installed at a depth of less than 50 mm from the surface of a wall or partition the requirements of Regulation 522.6.202(i) shall also apply.

522.6.204 For the purposes of Regulation 522.6.201(ii), Regulation 522.6.202(ii) and Regulation 522.6.203(ii), a cable shall:

(i) incorporate an earthed metallic covering which complies with the requirements of BS 7671 for a protective conductor of the circuit concerned, the cable complying with BS 5467, BS 6724, BS 7846, BS 8436 or BS EN 60702-1, or

(ii) be installed in earthed conduit complying with BS EN 61386-21 and satisfying the requirements of BS 7671 for a protective conductor, or

(iii) be enclosed in earthed trunking or ducting complying with BS EN 50085-2-1 and satisfying the requirements of BS 7671 for a protective conductor, or

(iv) be provided with mechanical protection against damage sufficient to prevent penetration of the cable by nails, screws and the like, or

(v) form part of a SELV or PELV circuit meeting the requirements of Regulation 414.4.

522.7 Vibration (AH)

522.7.1 A wiring system supported by or fixed to a structure or equipment subject to vibration of medium severity (AH2) or high severity (AH3) shall be suitable for such conditions, particularly where cables and cable connections are concerned.

522.7.2 For the fixed installation of suspended current-using equipment, e.g. luminaires, connection shall be made by cable with flexible cores. Where no vibration or movement can be expected, cable with non-flexible cores may be used.

522.8 Other mechanical stresses (AJ)

522.8.1 A wiring system shall be selected and erected to avoid during installation, use or maintenance, damage to the sheath or insulation of cables and their terminations. The use of any lubricants that can have a detrimental effect on the cable or wiring system are not permitted.

522.8.2 Where buried in the structure, a conduit system or cable ducting system, other than a pre-wired conduit assembly specifically designed for the installation, shall be completely erected between access points before any cable is drawn in.

522.8.3 The radius of every bend in a wiring system shall be such that conductors or cables do not suffer damage and terminations are not stressed.

522.8.4 Where conductors or cables are not supported continuously due to the method of installation, they shall be supported by suitable means at appropriate intervals in such a manner that the conductors or cables do not suffer damage by their own weight.

522.8.5 Every cable or conductor shall be supported in such a way that it is not exposed to undue mechanical strain and so that there is no appreciable mechanical strain on the terminations of the conductors, account being taken of mechanical strain imposed by the supported weight of the cable or conductor itself.

NOTE: Consumer unit meter tails are included in the requirements of this regulation.

522.8.6 A wiring system intended for the drawing in or out of conductors or cables shall have adequate means of access to allow this operation.

522.8.7 A wiring system buried in a floor shall be sufficiently protected to prevent damage caused by the intended use of the floor.

522.8.8 *Not used*

522.8.9 *Not used*

522.8.10 Except where installed in a conduit or duct which provides equivalent protection against mechanical damage, a cable buried in the ground shall incorporate an earthed armour or metal sheath or both, suitable for use as a protective conductor. The location of buried cables shall be marked by cable covers or a suitable marker tape. Buried conduits and ducts shall be suitably identified. Buried cables, conduits and ducts shall be at a sufficient depth to avoid being damaged by any reasonably foreseeable disturbance of the ground.

NOTE: BS EN 61386-24 is the standard for underground conduits.

522.8.11 Cable supports and enclosures shall not have sharp edges liable to damage the wiring system.

522.8.12 A cable or conductors shall not be damaged by the means of fixing.

522.8.13 Cables, busbars and other electrical conductors which pass across expansion joints shall be so selected or erected that anticipated movement does not cause damage to the electrical equipment.

522.8.14 No wiring system shall penetrate an element of building construction which is intended to be load bearing unless the integrity of the load-bearing element can be assured after such penetration.

522.9 Presence of flora and/or mould growth (AK)

522.9.1 Where the conditions experienced or expected constitute a hazard (AK2), the wiring system shall be selected accordingly or special protective measures shall be adopted.

NOTE 1: An installation method which facilitates the removal of such growths may be necessary (see Section 529).

NOTE 2: Possible preventive measures are closed types of installation (conduit or channel), maintaining distances to plants and regular cleaning of the relevant wiring system.

522.10 Presence of fauna (AL)

522.10.1 Where conditions experienced or expected constitute a hazard (AL2), the wiring system shall be selected accordingly or special protective measures shall be adopted, for example, by:

(i) the mechanical characteristics of the wiring system, or

(ii) the location selected, or

(iii) the provision of additional local or general protection against mechanical damage, or

(iv) any combination of the above.

522.11 Solar radiation (AN) and ultraviolet radiation

522.11.1 Where significant solar radiation (AN2) or ultraviolet radiation is experienced or expected, a wiring system suitable for the conditions shall be selected and erected or adequate shielding shall be provided. Special precautions may need to be taken for equipment subject to ionising radiation.

NOTE: See also Regulation 522.2.1 dealing with temperature rise.

522.12 Seismic effects (AP)

522.12.1 The wiring system shall be selected and erected with due regard to the seismic hazards of the location of the installation.

522.12.2 Where the seismic hazards experienced are low severity (AP2) or higher, particular attention shall be paid to the following:

(i) The fixing of wiring systems to the building structure

(ii) The connections between the fixed wiring and all items of essential equipment, e.g. safety services, shall be selected for their flexible quality.

522.13 Movement of air (AR)

522.13.1 See Regulation 522.7, Vibration (AH), and Regulation 522.8, Other mechanical stresses (AJ).

522.14 Nature of processed or stored materials (BE)

522.14.1 See Section 527, Selection and erection of wiring systems to minimize the spread of fire and Section 422, Precautions where particular risks of fire exist.

522.15 Building design (CB)

522.15.1 Where risks due to structural movement exist (CB3), the cable support and protection system employed shall be capable of permitting relative movement so that conductors and cables are not subjected to excessive mechanical stress.

522.15.2 For a flexible structure or a structure intended to move (CB4), a flexible wiring system shall be used.

523 CURRENT-CARRYING CAPACITIES OF CABLES

523.1 The current, including any harmonic current, to be carried by any conductor for sustained periods during normal operation shall be such that the appropriate temperature limit specified in Table 52.1 is not exceeded. The value of current shall be selected in accordance with Regulation 523.2, or determined in accordance with Regulation 523.3.

TABLE 52.1 – Maximum operating temperatures for types of cable insulation

Type of insulation	Temperature limit[a]
Thermoplastic	70 °C at the conductor
Thermosetting	90 °C at the conductor[b]
Mineral (Thermoplastic covered or bare exposed to touch)	70 °C at the sheath
Mineral (bare not exposed to touch and not in contact with combustible material)	105 °C at the sheath[b, c]

[a] The maximum permissible conductor temperatures given in Table 52.1 on which the tabulated current-carrying capacities given in Appendix 4 are based, have been taken from IEC 60502-1 and BS EN 60702-1 and are shown on these tables in Appendix 4.

[b] Where a conductor operates at a temperature exceeding 70 °C it shall be ascertained that the equipment connected to the conductor is suitable for the resulting temperature at the connection.

[c] For mineral insulated cables, higher operating temperatures may be permissible dependent upon the temperature rating of the cable, its terminations, the environmental conditions and other external influences.

NOTE: For temperature limits applicable to specific types of cable or insulation, refer to the cable specification or manufacturer.

523.2 The requirement of Regulation 523.1 is considered to be satisfied if the current for non-sheathed and sheathed cables does not exceed the appropriate values selected from the tables of current-carrying capacity given in Appendix 4 with reference to Table 4A2, subject to any necessary rating factors.

NOTE: The current-carrying capacities given in the tables are provided for guidance. It is recognized that there will be some tolerance in the current-carrying capacities depending on the environmental conditions and the precise construction of the cables.

523.3 The appropriate value of current-carrying capacity may also be determined as described in BS 7769 series (some parts of the BS 7769 series are now numbered BS IEC 60287 series, eventually all parts will be renumbered), or by test, or by calculation using a recognized method, provided that the method is stated. Where appropriate, account shall be taken of the characteristics of the load and, for buried cables, the effective thermal resistance of the soil.

523.4 The ambient temperature shall be considered to be the temperature of the surrounding medium when the non-sheathed or sheathed cable(s) under consideration are not loaded.

523.5 Groups containing more than one circuit

The group rating factors, see Tables 4C1 to 4C6 of Appendix 4, are applicable to groups of non-sheathed or sheathed cables having the same maximum operating temperature.

For groups containing non-sheathed or sheathed cables having different maximum operating temperatures, the current-carrying capacity of all the non-sheathed or sheathed cables in the group shall be based on the lowest maximum operating temperature of any cable in the group together with the appropriate group rating factor.

If, due to known operating conditions, a non-sheathed or sheathed cable is expected to carry a current not greater than 30 % of its grouped current-carrying capacity, it may be ignored for the purpose of obtaining the rating factor for the rest of the group.

523.6 Number of loaded conductors

523.6.1 The number of conductors to be considered in a circuit are those carrying load current. Where conductors in polyphase circuits carry balanced currents, the associated neutral conductor need not be taken into consideration. Under these conditions a four-core cable is given the same current-carrying capacity as a three-core cable having the same conductor cross-sectional area for each line conductor. The neutral conductor shall be considered as a loaded conductor in the case of the presence of third harmonic current or multiples of the third harmonic presenting a total harmonic distortion greater than 15 % of the fundamental line current.

523.6.2 Where the neutral conductor in a multicore cable carries current as a result of an imbalance in the line currents, the temperature rise due to the neutral current is offset by the reduction in the heat generated by one or more of the line conductors. In this case the conductor size shall be chosen on the basis of the highest line current.

In all cases the neutral conductor shall have a cross-sectional area adequate to afford compliance with Regulation 523.1.

523.6.3 Where the neutral conductor carries current without a corresponding reduction in load of the line conductors, the neutral conductor shall be taken into account in ascertaining the current-carrying capacity of the circuit. Such currents may be caused by a significant harmonic current in three-phase circuits. If the total harmonic distortion due to third harmonic current or multiples of the third harmonic is greater than 15 % of the fundamental line current the neutral conductor shall not be smaller than the line conductors. Thermal effects due to the presence of third harmonic or multiples of third harmonic currents and the corresponding rating factors for higher harmonic currents are given in Appendix 4, section 5.5.

523.6.4 Conductors which serve the purpose of protective conductors only are not to be taken into consideration. PEN conductors shall be taken into consideration in the same way as neutral conductors.

523.6.201 The tabulated current-carrying capacities in Appendix 4 are based on the fundamental frequency only and do not take account of the effect of harmonics.

523.7 Conductors in parallel

Where two or more live conductors or PEN conductors are connected in parallel in a system, either:

(i) measures shall be taken to achieve equal load current sharing between them

This requirement is considered to be fulfilled if the conductors are of the same material, have the same cross-sectional area, are approximately the same length and have no branch circuits along their length, and either:

(a) the conductors in parallel are multicore cables or twisted single-core cables or non-sheathed cables, or

(b) the conductors in parallel are non-twisted single-core cables or non-sheathed cables in trefoil or flat formation and where the cross-sectional area is greater than 50 mm^2 in copper or 70 mm^2 in aluminium, the special configuration necessary for such formations is adopted. These configurations consist of suitable groupings and spacings of the different lines or poles

or

(ii) special consideration shall be given to the load current sharing to meet the requirements of Regulation 523.1.

This regulation does not preclude the use of ring final circuits with or without spur connections.

Where adequate current sharing is not possible or where four or more conductors have to be connected in parallel consideration shall be given to the use of busbar trunking.

523.8 Variation of installation conditions along a route

Where the heat dissipation differs from one part of a route to another, the current-carrying capacity of cables at each part of the route shall be appropriate for that part of the route.

523.9 Cables in thermal insulation

A cable should preferably not be installed in a location where it is liable to be covered by thermal insulation. Where a cable is to be run in a space to which thermal insulation is likely to be applied it shall, wherever practicable, be fixed in a position such that it will not be covered by the thermal insulation. Where fixing in such a position is impracticable the cross-sectional area of the cable shall be selected to meet the requirements of Chapter 43. Where necessary, the nature of the load (e.g. cyclic) and diversity may be taken into account.

For a cable installed in a thermally insulated wall or above a thermally insulated ceiling, the cable being in contact with a thermally conductive surface on one side, current-carrying capacities are tabulated in Appendix 4.

For a single cable likely to be totally surrounded by thermally insulating material over a length of 0.5 m or more, the current-carrying capacity shall be taken, in the absence of more precise information, as 0.5 times the current-carrying capacity for that cable clipped direct to a surface and open (Reference Method C).

Where a cable is to be totally surrounded by thermal insulation for less than 0.5 m the current-carrying capacity of the cable shall be reduced appropriately depending on the size of cable, length in insulation and thermal properties of the insulation. The derating factors in item 2.6 of Appendix 4 are appropriate to conductor sizes up to 10 mm^2 in thermal insulation having a thermal conductivity (λ) greater than 0.04 Wm^{-1}K^{-1}.

TABLE 52.2 – *Moved to Appendix 4, item 2.6*

523.201 Armoured single-core cables

The metallic sheaths and/or non-magnetic armour of single-core cables in the same circuit shall normally be bonded together at both ends of their run (solid bonding). Alternatively, the sheaths or armour of such cables having conductors of cross-sectional area exceeding 50 mm^2 and a non-conducting outer sheath may be bonded together at one point in their run (single point bonding) with suitable insulation at the unbonded ends, in which case the length of the cables from the bonding point shall be limited so that voltages from sheaths and/or armour to Earth:

(i) do not cause corrosion when the cables are carrying their full load current, for example by limiting the voltage to 25 V, and

(ii) do not cause danger or damage to property when the cables are carrying short-circuit current.

524 CROSS-SECTIONAL AREAS OF CONDUCTORS

524.1 The cross-sectional area of each conductor in a circuit shall be not less than the values given in Table 52.3, except as provided for extra-low voltage lighting installations according to Regulation 715.524.201.

524.2 Neutral conductors

524.2.1 The neutral conductor, if any, shall have a cross-sectional area not less than that of the line conductor:

(i) in single-phase, two-wire circuits, whatever the cross-sectional area

(ii) in polyphase and single-phase three-wire circuits, where the size of the line conductors is less than or equal to 16 mm^2 for copper, or 25 mm^2 for aluminium

(iii) in circuits where it is required according to Regulation 523.6.3.

524.2.2 If the total harmonic content due to triplen harmonics is greater than 33 % of the fundamental line current, an increase in the cross-sectional area of the neutral conductor may be required (see Regulation 523.6.3 and Appendix 4, section 5.5).

524.2.3 For a polyphase circuit where each line conductor has a cross-sectional area greater than 16 mm^2 for copper or 25 mm^2 for aluminium, the neutral conductor is permitted to have a smaller cross-sectional area than that of the line conductors provided that the following conditions are simultaneously fulfilled:

(i) The expected maximum current including harmonics, if any, in the neutral conductor during normal service is not greater than the current-carrying capacity of the reduced cross-sectional area of the neutral conductor, and

NOTE: The load carried by the circuit under normal service conditions should be practically equally distributed between the lines.

(ii) the neutral conductor is protected against overcurrents according to Regulation 431.2, and

(iii) the size of the neutral conductor is at least equal to 16 mm^2 for copper or 25 mm^2 for aluminium, account being taken of Regulation 523.6.3.

TABLE 52.3 – Minimum cross-sectional area of conductors

Type of wiring system	Use of the circuit	Conductor	
		Material	Cross-sectional area mm²
Non-sheathed and sheathed cables	Lighting and power circuits	Copper Aluminium	1.0 16 (see Note 3)
	Signalling and control circuits	Copper	0.5 (see Note 1)
Bare conductors	Power circuits	Copper Aluminium	10 16 (see Note 3)
	Signalling and control circuits	Copper	4
Non-sheathed and sheathed flexible cables	For a specific appliance	Copper	As specified in the product standard
	For any other application		0.75[a]
	Extra-low voltage circuits for special applications (see Note 2)		0.75

NOTE 1: In information technology, signalling and control circuits intended for electronic equipment a minimum cross-sectional area of 0.1 mm² is permitted.

NOTE 2: For special requirements for ELV lighting see Section 715.

NOTE 3: Connectors used to terminate aluminium conductors shall be tested and approved for this specific use.

[a] In multicore flexible cables containing seven or more cores, Note 1 applies.

525 VOLTAGE DROP IN CONSUMERS' INSTALLATIONS

525.1 In the absence of any other consideration, under normal service conditions the voltage at the terminals of any fixed current-using equipment shall be greater than the lower limit corresponding to the product standard relevant to the equipment.

525.201 Where fixed current-using equipment is not the subject of a product standard the voltage at the terminals shall be such as not to impair the safe functioning of that equipment.

525.202 The above requirements are deemed to be satisfied if the voltage drop between the origin of the installation (usually the supply terminals) and a socket-outlet or the terminals of fixed current-using equipment does not exceed that stated in Appendix 4, section 6.4.

525.203 A greater voltage drop than stated in Appendix 4, section 6.4 may be accepted for a motor during starting periods and for other equipment with high inrush currents, provided that it is verified that the voltage variations are within the limits specified in the relevant product standard for the equipment or, in the absence of a product standard, in accordance with the manufacturer's recommendations.

526 ELECTRICAL CONNECTIONS

526.1 Every connection between conductors or between a conductor and other equipment shall provide durable electrical continuity and adequate mechanical strength and protection.

NOTE: See Regulation 522.8 – Other mechanical stresses.

526.2 The selection of the means of connection shall take account of, as appropriate:

(i) the material of the conductor and its insulation
(ii) the number and shape of the wires forming the conductor
(iii) the cross-sectional area of the conductor
(iv) the number of conductors to be connected together
(v) the temperature attained at the terminals in normal service such that the effectiveness of the insulation of the conductors connected to them is not impaired
(vi) the provision of adequate locking arrangements in situations subject to vibration or thermal cycling.

Where a soldered connection is used the design shall take account of creep, mechanical stress and temperature rise under fault conditions.

NOTE 1: Applicable standards include BS EN 60947-7, the BS EN 60998 series and BS EN 61535.

NOTE 2: Terminals without the marking 'r' (only rigid conductor), 'f' (only flexible conductor), 's' or 'sol' (only solid conductor) are suitable for the connection of all types of conductors.

526.3 Every connection shall be accessible for inspection, testing and maintenance, except for the following:

(i) A joint designed to be buried in the ground
(ii) A compound-filled or encapsulated joint
(iii) A connection between a cold tail and the heating element as in ceiling heating, floor heating or a trace heating system
(iv) A joint made by welding, soldering, brazing or appropriate compression tool
(v) Joints or connections made in equipment by the manufacturer of the product and not intended to be inspected or maintained
(vi) Equipment complying with BS 5733 for a maintenance-free accessory and marked with the symbol (MF) and installed in accordance with the manufacturer's instructions.

526.4 Where necessary, precautions shall be taken so that the temperature attained by a connection in normal service shall not impair the effectiveness of the insulation of the conductors connected to it or any insulating material used to support the connection. Where a cable is to be connected to a bare conductor or busbar its type of insulation and/or sheath shall be suitable for the maximum operating temperature of the bare conductor or busbar.

526.5 Every termination and joint in a live conductor or a PEN conductor shall be made within one of the following or a combination thereof:

(i) A suitable accessory complying with the appropriate product standard
(ii) An equipment enclosure complying with the appropriate product standard
(iii) An enclosure partially formed or completed with building material which is non-combustible when tested to BS 476-4.

526.6 There shall be no appreciable mechanical strain on the connections of conductors.

526.7 Where a connection is made in an enclosure the enclosure shall provide adequate mechanical protection and protection against relevant external influences.

526.8 Cores of sheathed cables from which the sheath has been removed and non-sheathed cables at the termination of conduit, ducting or trunking shall be enclosed as required by Regulation 526.5.

526.9 Connection of multiwire, fine wire and very fine wire conductors

526.9.1 In order to avoid inappropriate separation or spreading of individual wires of multiwire, fine wire or very fine wire conductors, suitable terminals shall be used or the conductor ends shall be suitably treated.

526.9.2 Soldering (tinning) of the whole conductor end of multiwire, fine wire and very fine wire conductors is not permitted if screw terminals are used.

526.9.3 Soldered (tinned) conductor ends on fine wire and very fine wire conductors are not permissible at connection and junction points which are subject in service to a relative movement between the soldered and the non-soldered part of the conductor.

527 SELECTION AND ERECTION OF WIRING SYSTEMS TO MINIMIZE THE SPREAD OF FIRE

527.1 Precautions within a fire-segregated compartment

527.1.1 The risk of spread of fire shall be minimized by the selection of appropriate materials and erection in accordance with Section 527.

NOTE: A fire-segregated compartment (fire compartment) is considered to be an enclosed space, which may be subdivided, separated from adjoining spaces within a building by elements of construction having a specified fire resistance.

527.1.2 A wiring system shall be installed so that the general building structural performance and fire safety are not reduced.

527.1.3 Cables complying with, at least, the requirements BS EN 60332-1-2 may be installed without special precautions.

Where the fire-segregated compartment provides a means of evacuation in an emergency then the cable shall meet the requirements of the appropriate part of BS EN 60332-3 series.

NOTE: Cables manufactured for the above application also need to satisfy the requirements of the CPR in respect of their reaction to fire. See Appendix 2, item 17.

527.1.4 Cables not complying with the cable requirements of Regulation 527.1.3 shall be limited to short lengths for connection of appliances to the permanent wiring system and shall not pass from one fire-segregated compartment to another.

527.1.5 Products having the necessary resistance to flame propagation as specified in the BS EN 61386 series, the appropriate part of BS EN 50085 series, BS EN 61439-6, BS EN 61534 series, BS EN 61537 or BS EN 60570 may be installed without special precautions. Other products complying with standards having similar requirements for resistance to flame propagation may be installed without special precautions.

527.1.6 Parts of wiring systems other than cables which do not comply, as a minimum, with the flame propagation requirements as specified in the BS EN 61386 series, the appropriate part of BS EN 50085 series, BS EN 61439-6, BS EN 61534 series or BS EN 61537 but which comply in all other respects with the requirements of their respective product standard shall, if used, be completely enclosed in suitable non-combustible building materials.

527.2 Sealing of wiring system penetrations

527.2.1 Where a wiring system passes through elements of building construction such as floors, walls, roofs, ceilings, partitions or cavity barriers, the openings remaining after passage of the wiring system shall be sealed according to the degree of fire-resistance (if any) prescribed for the respective element of building construction before penetration.

This requirement is satisfied if the sealing of the wiring system concerned has passed a relevant type test meeting the requirements of Regulation 527.2.3.

527.2.1.1 During the erection of a wiring system temporary sealing arrangements shall be provided as appropriate.

527.2.1.2 During alteration work, sealing which has been disturbed shall be reinstated as soon as practicable.

527.2.2 A wiring system such as a conduit system, cable ducting system, cable trunking system, busbar or busbar trunking system which penetrates elements of building construction having specified fire-resistance shall be internally sealed to the degree of fire-resistance of the respective element before penetration as well as being externally sealed as required by Regulation 527.2.1.

This requirement is satisfied if the sealing of the wiring system concerned has passed a relevant type test meeting the requirements of Regulation 527.2.3.

NOTE: Reference should be made to the intended manufacturers' guidance on the specific fire sealing product to be used with regard to the size and arrangement of trunking or containment.

527.2.3 A conduit system, cable trunking system or cable ducting system classified as non-flame propagating according to the relevant product standard and having a maximum internal cross-sectional area of 710 mm^2 need not be internally sealed provided that:

(i) the system satisfies the test of BS EN 60529 for IP33, and

(ii) any termination of the system in one of the compartments, separated by the building construction being penetrated, satisfies the test of BS EN 60529 for IP33.

527.2.4 Any sealing arrangement intended to satisfy Regulation 527.2.1 or 527.2.1.1 shall resist external influences to the same degree as the wiring system with which it is used and, in addition, it shall meet all of the following requirements:

(i) It shall be resistant to the products of combustion to the same extent as the elements of building construction which have been penetrated

(ii) It shall provide the same degree of protection from water penetration as that required for the building construction element in which it has been installed

(iii) It shall be compatible with the material of the wiring system with which it is in contact

(iv) It shall permit thermal movement of the wiring system without reduction of the sealing quality

(v) It shall be of adequate mechanical stability to withstand the stresses which may arise through damage to the support of the wiring system due to fire.

The seal and the wiring system shall be protected from dripping water which may travel along the wiring system or which may otherwise collect around the seal unless the materials used in the seal are all resistant to moisture when finally assembled for use.

NOTE: This regulation may be satisfied if:

- either cable cleats, cable ties or cable supports are installed within 750 mm of the seal, and are able to withstand the mechanical loads expected following the collapse of the supports on the fire side of the seal to the extent that no strain is transferred to the seal, or
- the design of the sealing system itself provides adequate support.

528 PROXIMITY OF WIRING SYSTEMS TO OTHER SERVICES

528.1 Proximity to electrical services

Except where one of the following methods is adopted, neither a voltage Band I nor a voltage Band II circuit shall be contained in the same wiring system as a circuit of nominal voltage exceeding that of low voltage, and a Band I circuit shall not be contained in the same wiring system as a Band II circuit:

(i) Every cable or conductor is insulated for the highest voltage present

(ii) Each conductor of a multicore cable is insulated for the highest voltage present in the cable

(iii) The cables are insulated for their system voltage and installed in a separate compartment of a cable ducting or cable trunking system

(iv) The cables are installed on a cable tray system where physical separation is provided by a partition

(v) A separate conduit, trunking or ducting system is employed

(vi) For a multicore cable, the cores of the Band I circuit are separated from the cores of the Band II circuit by an earthed metal screen of equivalent current-carrying capacity to that of the largest core of a Band II circuit.

For SELV and PELV systems the requirements of Regulation 414.4 shall apply.

NOTE 1: In the case of proximity of wiring systems and lightning protection systems, BS EN 62305 should be considered.

NOTE 2: Recommendations for separation and segregation in relation to safety services are given in BS 5266, BS 5839 and BS 8519.

528.2 Proximity of communications cables

In the event of crossing or proximity of underground telecommunication cables and underground power cables, a minimum clearance of 100 mm shall be maintained, or the requirements according to (i) or (ii) shall be fulfilled:

(i) A fire-retardant partition shall be provided between the cables, e.g. bricks, cable protecting caps (clay, concrete), shaped blocks (concrete), protective cable conduit or troughs made of fire-retardant materials

(ii) For crossings, mechanical protection between the cables shall be provided, e.g. cable conduit, concrete cable protecting caps or shaped blocks.

NOTE 1: Special considerations of electrical interference, both electromagnetic and electrostatic, may apply to telecommunication circuits, data transfer circuits and the like.

NOTE 2: Segregation requirements for communications services are given in BS 6701 and BS EN 50174 series.

528.3 Proximity to non-electrical services

528.3.1 A wiring system shall not be installed in the vicinity of services which produce heat, smoke or fumes likely to be detrimental to the wiring, unless it is protected from harmful effects by shielding arranged so as not to affect the dissipation of heat from the wiring.

In areas not specifically designed for the installation of cables, e.g. service shafts and cavities, the cables shall be laid so that they are not exposed to any harmful influence by the normal operation of adjacent installations (e.g. gas, water or steam lines).

528.3.2 Where a wiring system is routed below services liable to cause condensation (such as water, steam or gas services), precautions shall be taken to protect the wiring system from deleterious effects.

528.3.3 Where an electrical service is to be installed in proximity to one or more non-electrical services it shall be so arranged that any foreseeable operation carried out on the other services will not cause damage to the electrical service or the converse.

NOTE: This may be achieved by:

(i) suitable spacing between the services, or

(ii) the use of mechanical or thermal shielding.

528.3.4 Where an electrical service is located in close proximity to one or more non-electrical services, both the following conditions shall be met:

(i) The wiring system shall be suitably protected against the hazards likely to arise from the presence of the other services in normal use

(ii) Fault protection shall be afforded in accordance with the requirements of Section 411.

NOTE: The requirements for segregation between low pressure gas systems and electrical equipment are given in BS 6891.

528.3.5 No cable shall be run in a lift or hoist well unless it forms part of the lift installation as defined in BS EN 81 series.

529 SELECTION AND ERECTION OF WIRING SYSTEMS IN RELATION TO MAINTAINABILITY, INCLUDING CLEANING

529.1 With regard to maintainability, reference shall be made to Regulation 132.12.

529.2 Where it is necessary to remove any protective measure in order to carry out maintenance, provision shall be made so that the protective measure can be reinstated without reduction of the degree of protection originally intended.

529.3 Provision shall be made for safe and adequate access to all parts of a wiring system which may require maintenance.

NOTE: In some situations, it may be necessary to provide permanent means of access by ladders, walkways, etc.

CHAPTER 53

PROTECTION, ISOLATION, SWITCHING, CONTROL AND MONITORING

CONTENTS

CHAPTER 53

PROTECTION, ISOLATION, SWITCHING, CONTROL AND MONITORING

530 INTRODUCTION

530.1 Scope

This chapter deals with general requirements for protection, isolation, switching, control and monitoring and with the requirements for selection and erection of the devices provided to fulfil such functions.

NOTE: In Chapter 53, reference to short-circuit, where relevant, includes earth fault. Short-circuit Protective Device (SCPD), where relevant, also means earth fault current protective device.

530.2 Not used

530.3 General and common requirements

530.3.1 Every item of equipment shall be selected and erected so as to allow compliance with the requirements stated in the following regulations.

530.3.2 The moving contacts of multipole switching devices marked for the connection of the neutral or midpoint may close before and open after the other contacts.

530.3.3 A switching device shall not be inserted in the neutral conductor alone.

530.3.4 Devices embodying more than one function, as defined in the following regulations, shall comply with the relevant requirements of this chapter appropriate to each separate function.

530.3.5 Equipment intended for protection only shall not be provided for functional switching of circuits.

NOTE: See Table 537.4 – Guidance on the selection of protective, isolation and switching devices.

530.4 Fixing of equipment

530.4.1 Taking into account the manufacturer's instructions, if any, equipment shall be erected in such a way that connections between wiring and equipment shall not be subject to undue stress or strain resulting from the normal use of the equipment.

530.4.2 Unenclosed equipment shall be mounted in a suitable mounting box or enclosure in compliance with the relevant part of BS EN 60670, BS EN 62208 or other relevant standards such as BS EN 61439 series. Socket-outlets, connection units, plate switches and similar accessories shall be fitted to a mounting box complying with BS 4662 or BS 5733 and with the relevant part of BS EN 60670.

530.4.3 Equipment such as circuit-breakers, switches, socket-outlets, control equipment, etc. may be installed on or in a cable trunking system complying with BS EN 50085 series. Wherever equipment is fixed on or in cable trunking, skirting trunking or in mouldings it shall not be fixed on covers which can be removed inadvertently.

531 DEVICES FOR PROTECTION AGAINST ELECTRIC SHOCK BY AUTOMATIC DISCONNECTION OF SUPPLY

531.1 General

531.1.1 Devices for protection against electric shock by automatic disconnection of supply shall be suitable for isolation in accordance with Chapter 46 and Section 537.

Automatic reclosing of devices for protection against electric shock by automatic disconnection of supply shall only be installed in the associated part of the installation where access is restricted to instructed persons or skilled persons only.

A warning notice shall be clearly displayed near the point of access to the associated part of the installation controlled by the auto-reclosing device, indicating its automatic reclose function must be disengaged prior to entry.

NOTE: Automatic Reclosing Devices are intended to reclose circuit-breakers, RCBOs and RCCBs after tripping, in order to re-establish continuity of service.

Requirements for the selection of devices for protection against electric shock by automatic disconnection of supply are given in the regulations listed below.

In TN, TT and IT systems the following protective devices may be used:

(i) overcurrent protective devices, in accordance with Regulation 531.2

(ii) residual current devices (RCDs), in accordance with Regulation 531.3.

Devices according to BS EN 60947-2 marked with the voltage value followed by the symbol (IT) shall not be used in IT systems for such voltage.

In addition, in IT systems the following monitoring devices may be used to detect insulation fault conditions:

(iii) insulation monitoring devices (IMDs), in accordance with Regulation 538.1

(iv) equipment for insulation fault location, in accordance with Regulation 538.2

(v) residual current monitors (RCMs), in accordance with Regulation 538.4.

531.2 Overcurrent protective devices

531.2.1 General

Where overcurrent protective devices are used for protection against electric shock by automatic disconnection of supply they shall be selected in accordance with Section 533.

531.2.2 TN systems

In TN systems, overcurrent protective devices when used as devices for fault protection shall be selected and erected in order to comply with the requirements specified in Chapter 41 (see in particular Regulation 411.4.4).

If, for certain equipment or for certain parts of the installation, the maximum disconnection times in Table 41.1 cannot be fulfilled by the overcurrent protective devices, those parts shall be protected by an RCD in compliance with Regulation 531.3.5.2.

In TN-S systems, the neutral need not be disconnected if the supply conditions are such that the neutral conductor can be considered to be reliably at earth potential.

In TN-C systems, the PEN conductor shall not be disconnected.

531.2.2.201 Except in certain special installations or locations (Part 7), there is no requirement under overcurrent conditions to disconnect/switch the neutral in TT or TN systems.

531.2.3 TT systems

In TT systems, overcurrent protective devices may be used for fault protection provided that a suitably low value of Z_s is permanently and reliably assured (see Regulation 411.5.4) so that in case of a fault, tripping of the overcurrent protective device in compliance with the required disconnection times is achieved.

531.2.4 IT systems

Overcurrent protective devices, when used as devices for fault protection in the event of a second fault, shall comply with:

(i) Regulation 531.2.2, taking into account the requirements of Regulation 411.6.5(i), where exposed-conductive-parts are interconnected, or

(ii) Regulation 531.2.3, taking into account the requirements of Regulation 411.6.5(ii), where exposed-conductive-parts are earthed in groups or individually.

In IT systems, if disconnection required by Chapter 41 in the event of the second fault to earth cannot be achieved by an overcurrent protective device, one or more RCDs shall be used to provide the required fault protection within the installation.

NOTE: Reference is also made to Regulation 419.3 where supplementary protective equipotential bonding is required in those cases where automatic disconnection according to Regulation 411.3.2.1 cannot be achieved.

Overcurrent protective devices used in IT systems shall have line poles suitable for line-to-line voltage applications and a neutral pole, if applicable, suitable for the line-to-neutral voltage for operation in case of a second insulation fault.

In IT systems, in the event of a second fault, the operation of the overcurrent protective device shall result in the disconnection of all corresponding live conductors, including the neutral conductor, if any (see also Regulation 431.2.2).

531.3 Residual current devices (RCDs)

531.3.1 General

Except where Regulation 531.3.1.201 applies, an RCD shall disconnect all live conductors of the circuit protected.

The protective conductor shall not pass through the sensor of the RCD except where this is unavoidable, e.g. in the case of armoured cables. In such exceptional cases, the protective conductor alone has to be passed again through the sensor but in the reverse direction. The protective conductor shall be insulated and shall not be earthed either at the first or at the second passing through the sensor.

A protective conductor current shall not contribute to the measurement of the residual current.

531.3.1.201 For protection against electric shock, there is no requirement to disconnect/switch the neutral in TT or TN systems.

531.3.1.202 It is not permissible to introduce an external connection for the purpose of intentionally creating a residual current to trip an RCD.

NOTE 1: This does not preclude the use of an RCD with a functional earth connection.

NOTE 2: This does not preclude the use of a test instrument in accordance with Part 6.

531.3.2 Unwanted tripping

Residual current protective devices shall be selected and erected such as to limit the risk of unwanted tripping. The following shall be considered:

(i) subdivision of circuits with individual associated RCDs. RCDs shall be selected and the circuits subdivided in such a way that any earth leakage current likely to occur during normal operation of the connected load will not cause unwanted tripping of the device. See also Section 314

(ii) the use of RCBOs for individual final circuits in residential premises. See also Section 314

(iii) in order to avoid unwanted tripping by protective conductor currents and/or earth leakage currents, the accumulation of such currents downstream of the RCD shall be not more than 30 % of the rated residual operating current

NOTE 1: This will also allow a better selection of the type of RCDs according to the nature of the circuit or the load.

NOTE 2: RCDs may operate at any value of residual current in excess of 50 % of the rated residual current.

(iv) use of short time-delayed RCDs, provided the applicable requirements of Chapter 41 are met

NOTE 3: In the case of transient effects, tripping of the RCD may occur by charging of bypass capacitors or by other electromagnetic disturbances.

(v) coordination of general type RCDs, selective type RCDs and time-delayed RCDs (CBRs according to BS EN 60947-2) as covered in Section 536

NOTE 4: CBR is a circuit-breaker incorporating residual current protection.

(vi) coordination of RCDs with surge protective devices (SPD) according to Regulation 534.4.7.

531.3.3 Types of RCD

Different types of RCD exist, depending on their behaviour in the presence of DC components and frequencies. The appropriate RCD shall be selected from the following:

(i) RCD Type AC: RCD tripping on alternating sinusoidal residual current, suddenly applied or smoothly increasing

(ii) RCD Type A: RCD tripping on alternating sinusoidal residual current and on residual pulsating direct current, suddenly applied or smoothly increasing

NOTE 1: For RCD Type A, tripping is achieved for residual pulsating direct currents superimposed on a smooth direct current up to 6 mA.

(iii) RCD Type F: RCD for which tripping is achieved as for Type A and in addition:

(a) for composite residual currents, whether suddenly applied or slowly rising, intended for circuit supplied between line and neutral or line and earthed middle conductor

(b) for residual pulsating direct currents superimposed on smooth direct current.

NOTE 2: For RCD Type F, tripping is achieved for residual pulsating direct currents superimposed on a smooth direct current up to 10 mA.

(iv) RCD Type B: RCD for which tripping is achieved as for Type F and in addition:

(a) for residual sinusoidal alternating currents up to 1 kHz

(b) for residual alternating currents superimposed on a smooth direct current

(c) for residual pulsating direct currents superimposed on a smooth direct current

(d) for residual pulsating rectified direct current which results from two or more phases

(e) for residual smooth direct currents, whether suddenly applied or slowly increased, independent of polarity.

NOTE 3: For RCD Type B, tripping is achieved for residual pulsating direct currents superimposed on a smooth direct current up to 0.4 times the rated residual current ($I_{\Delta n}$) or 10 mA, whichever is the highest value.

RCD Type AC shall only be used to serve fixed equipment, where it is known that the load current contains no DC components.

NOTE 1: Examples of fixed equipment with a load current containing no DC components can include but not be limited to electric heating appliances and/or simple filament lighting, neither containing electronic components.

NOTE 2: For guidance on the correct use of RCDs for household and similar use, see PD IEC/TR 62350.

NOTE 3: Some typical fault currents in circuits comprising semiconductors are given in Annex A53, Figure A53.1.

531.3.4 Selection according to the accessibility to the installation

531.3.4.1 In AC installations having RCDs that are intended to be operated by ordinary persons, the RCDs shall comply with:

- BS EN 61008 series for RCCBs, or
- BS EN 61009 series for RCBOs, or
- BS EN 62423 for Type F and Type B RCCBs and RCBOs, or
- BS 7288 for SRCDs and FCURCDs.

NOTE 1: RCCB is a Residual Current operated Circuit-Breaker without integral overcurrent protection. RCBO is a Residual Current operated Circuit-Breaker with integral overcurrent protection.

NOTE 2: SRCD is a socket-outlet incorporating a Residual Current Device and FCURCD is a Fused Connection Unit incorporating a Residual Current Device. Devices conforming to BS 7288 only provide 'additional protection'.

531.3.4.201 Where an RCD may be operated by an ordinary person, it shall be designed or installed so that it is not possible to modify or adjust the setting or the calibration of its rated residual operating current ($I_{\Delta n}$) or time delay mechanism without a deliberate act involving the use of either a key or a tool and resulting in a visible indication of its setting or calibration.

531.3.4.2 In AC installations having RCDs that are intended to be operated by instructed persons or skilled persons, the RCDs shall comply with:

- BS EN 61008 series for RCCBs, or
- BS EN 61009 series for RCBOs, or
- BS EN 62423 for Type F and Type B RCCBs and RCBOs, or
- BS EN 60947-2 for CBRs and MRCDs, or
- BS 7288 for SRCDs and FCURCDs.

NOTE: Devices conforming to BS 7288 only provide 'additional protection'.

531.3.5 RCDs for fault protection

531.3.5.1 General

The use of RCDs shall provide protection against faults in compliance with Regulation 411.3.

The selection of RCDs depends on the type of earthing system (see Regulations 531.3.5.2, 531.3.5.3 and 531.3.5.4, the presence of DC components and frequencies (Regulation 531.3.3).

531.3.5.2 TN system

RCDs shall be installed at the origin of that part of the installation to be protected. The requirements for unwanted tripping in accordance with Regulation 531.3.2 shall also be taken into account.

NOTE: Except where particular restriction for selectivity applies, several circuits may be protected by the same device.

The division of the PEN conductor into neutral conductor and protective conductor shall take place at the supply side of the RCDs.

On the load side of the RCD, connection between the protective and neutral conductors is not permitted.

An RCD shall not be used in a TN-C system.

531.3.5.3 TT system

531.3.5.3.1 Location of RCDs

RCDs shall be erected at the origin of that part of the installation to be protected. The requirements for unwanted tripping, in accordance with Regulation 531.3.2, shall also be taken into account.

NOTE: Where there is more than one origin, this requirement applies to each origin.

531.3.5.3.2 Selection of the rated residual operating current of the RCD

The rated residual operating current value ($I_{\Delta n}$) of an RCD shall not exceed the $I_{\Delta n}$ corresponding to the maximum value of the earth resistance R_A to the exposed-conductive-parts, taking into account the possible seasonal variations, including soil freezing and drying, of the part of the installation protected by this device, as shown in Table 53.1.

R_A is the sum of the resistances in ohms of the earth electrode and the protective conductor connecting it to the exposed-conductive-parts.

TABLE 53.1 – Correlation between the maximum value of earth resistance R_A and the maximum rated residual operating current $I_{\Delta n}$ of the RCD

Maximum value of R_A (Ω)	Maximum $I_{\Delta n}$ of the RCD
2.5	20 A
5	10 A
10	5 A
17	3 A
50	1 A
100	500 mA
167	300 mA
500	100 mA
1667	30 mA

531.3.5.3.2.201 For Class I enclosures in TT systems where RCD protection is used on outgoing circuits, double or reinforced insulation of all live conductors (incoming cables, extension terminals, etc.) on the supply side of the incoming device, e.g. main switch, shall be used. Insulated and non-metallic sheathed cables are deemed to meet the requirements of double or reinforced insulation.

NOTE 1: When selecting equipment, consideration should be given to the assembly manufacturer's internal line interconnecting cable links on the supply side of an RCD being insulated and non-metallic sheathed, or having reinforced insulation or equivalent mechanical protection.

NOTE 2: Only the assembly manufacturer's approved internal interconnecting cable links should be used.

531.3.5.4 IT system

531.3.5.4.1 General

In IT systems, protection of the neutral conductor by RCD is permitted provided that the requirements of Regulation 431.2.2 are fulfilled.

531.3.5.4.2 Case of second fault on another live conductor when exposed-conductive-parts are interconnected

Where RCDs are used according to Regulation 411.6.3(v), one RCD per circuit shall be used.

The operating characteristics of this RCD shall be selected according to Table 41.1 of Chapter 41.

531.3.5.4.3 Case of second fault on another live conductor when exposed-conductive-parts are not interconnected

Where, in an installation, all exposed-conductive-parts are not interconnected, one RCD shall protect each group of interconnected exposed-conductive-parts.

The conditions for determining the characteristics of the RCD shall be those for TT systems defined in Regulation 411.5:

- The rated residual operating current $I_{\Delta n}$ has to be selected according to Table 53.1.
- The disconnection time has to comply with the values given in Regulation 411.3.

In addition, fault protection for every circuit located downstream of this RCD shall be provided in accordance with the requirements of Regulation 411.6.3(v). In this case every final circuit shall be individually protected by an RCD.

531.3.6 RCDs for additional protection

The use of RCDs with a rated residual operating current not exceeding 30 mA is recognized as additional protection in compliance with Regulation 415.1. These RCDs shall be provided to comply with the requirements of Regulation 411.3.3.

RCDs for additional protection in AC installations shall comply with:

- BS EN 61008 series, or
- BS EN 61009 series, or
- BS EN 62423, or
- BS 7288.

Where installed at the origin of a final circuit or a group of final circuits, an RCD with a rated residual operating current not exceeding 30 mA may provide fault protection and additional protection simultaneously.

NOTE: Consideration shall be given to the division of the installation (see Regulations 531.3.2 and 314.2).

532 DEVICES FOR PROTECTION AGAINST THE RISK OF FIRE

532.1 General

In locations where, in accordance with Chapter 42, a particular risk of fire exists, preventive protection measures against the risk of fire are required. This may also apply to other locations of the electrical installation, depending on a risk analysis.

NOTE: A suitable evaluation of the risk should be carried out by one or more persons competent in fire risk assessments.

The selection of protective and monitoring devices shall take into account the nature of the load and the likelihood of the device to operate, e.g. fault currents of higher frequencies, DC fault currents or increased leakage currents. (See also Regulation 331.1.)

532.2 Residual current devices (RCDs) for protection against the risk of fire

RCDs shall comply with Regulations 531.3.1 to 531.3.4 and with the applicable requirements of Regulation 531.3.5.

RCDs with a rated residual operating current not exceeding 300 mA shall be used. RCDs shall be installed at the origin of the circuit to be protected.

532.3 Residual current monitoring devices (RCMs) for protection against the risk of fire in IT systems

In IT systems, RCMs may be used as an alternative to RCDs in accordance with Regulation 532.2, provided that the location is supervised by one or more skilled or instructed person(s).

RCMs shall be in accordance with BS EN 62020 and operate in conjunction with switchgear suitable for isolation.

RCMs shall be installed at the origin of final circuits. The rated residual operating current shall not exceed 300 mA. Audible and visual signals shall be provided by the RCMs.

532.4 Insulation monitoring devices (IMDs) for protection against the risk of fire in IT systems

Insulation monitoring devices applied in IT systems for protection against the risk of fire shall comply with the requirements of Section 538.

A fault location system in accordance with the requirement of Regulation 538.2 able to locate the faulty circuit, may also be used.

532.5 Internal arc fault protection in a switchgear and controlgear assembly

Where required for special applications, internal arc fault protection (e.g. optical detection system) detecting a fault arc together with a protection system can be selected in order to extinguish the arc.

Alternatively, an assembly with arc ignition protected zone(s) can be selected to minimise the risk of an arcing fault. (Arcing class I in accordance with PD IEC/TR 61641.)

NOTE: Internal arc fault protection is typically associated with a switchboard used in special applications. An assembly that has been subject to an internal arcing fault can require maintenance or replacement.

532.6 Arc fault detection devices (AFDDs)

Where specified, arc fault detection devices shall be installed:

(i) at the origin of the final circuits to be protected, and

(ii) in AC single-phase circuits not exceeding 230 V.

AFDDs shall comply with BS EN 62606. Coordination of AFDDs with overcurrent protective devices, if necessary, shall take account of the manufacturer's instructions.

533 DEVICES FOR PROTECTION AGAINST OVERCURRENT

533.1 General requirements

533.1.1 Compliance with standards

A device for protection against overcurrent shall comply with one or more of the following:

- BS 88 series
- BS 646
- BS 1362
- BS 3036
- BS EN 60898 series
- BS EN 60947 series
- BS EN 61009 series
- BS EN 62423.

The use of another device is not precluded provided that its time/current characteristics provide a level of protection not less than that given by the devices listed above.

The following protective devices shall be used only for protection against short-circuit current:

- instantaneous trip circuit-breakers (ICB) in accordance with Annex O of BS EN 60947-2
- aM and aR type fuses in accordance with BS HD 60269-2 or BS HD 60269-3.

For every fuse and circuit-breaker there shall be provided on or adjacent to it an indication of its intended rated current as appropriate to the circuit it protects. For a semi-enclosed fuse, the intended rated current to be indicated is the value to be selected in accordance with Regulation 533.1.2.3.

533.1.2 Fuses

533.1.2.1 A fuse base shall be arranged so as to exclude the possibility of the fuse carrier making contact between conductive parts belonging to two adjacent fuse bases.

A fuse base using screw-in fuses shall be connected so that the centre contact is connected to the conductor from the supply and the shell contact is connected to the conductor to the load.

533.1.2.2 Fuses having fuse links likely to be removed or replaced by an ordinary person shall be of a type which complies with BS 88-3, BS 3036 or BS 1362. Such a fuse link shall either:

(i) have marked on or adjacent to it an indication of the type of fuse link intended to be used, or

(ii) be of a type such that there is no possibility of inadvertent replacement by a fuse link having the intended rated current but a higher fusing factor than that intended.

NOTE: In polyphase systems additional measures may be needed, e.g. an all-pole switch on the supply side, in order to prevent the risk of unintentional contact with live parts on the load side.

Fuses or combination units having fuse links likely to be removed and replaced only by skilled or instructed person(s) shall be installed in such a manner that the fuse links can be removed or replaced without unintentional contact with live parts.

533.1.2.3 A fuse shall preferably be of the cartridge type. Where a semi-enclosed fuse is selected, it shall be fitted with an element in accordance with the manufacturer's instructions, if any. In the absence of such instructions, it shall be fitted with a single element of tinned copper wire of the appropriate diameter specified in Table 533.1.

TABLE 533.1 – Sizes of tinned copper wire for use in semi-enclosed fuses

Rated current of fuse element (A)	Nominal diameter of wire (mm)
3	0.15
5	0.2
10	0.35
15	0.5
20	0.6
25	0.75
30	0.85
45	1.25
60	1.53
80	1.8
100	2.0

533.1.3 Circuit-breakers

Where a circuit-breaker may be operated by an ordinary person, it shall be designed or installed so that it is not possible to modify the setting or the calibration of its overcurrent release without a deliberate act involving the use of either a key or a tool and resulting in a visible indication of its setting or calibration.

Where a screw-in type circuit-breaker is used in a fuse base, the requirements of Regulation 533.1.2.1 also apply.

According to the requirements of Annex H of BS EN 60947-2, devices marked with the voltage value followed by the symbol  shall not be used in IT systems for such voltage.

533.2 Selection of devices for overload protection of wiring systems

533.2.1 The rated current (or current setting) of the protective device shall be chosen in accordance with Regulation 433.1.

In certain cases, to avoid unintentional operation, the peak current values of the loads may have to be taken into consideration.

In the case of a cyclic load, the values of I_n and I_2 shall be chosen on the basis of values of I_b and I_z for the thermally equivalent constant load

where:

I_b is the current for which the circuit is designed

I_z is the current-carrying capacity of a cable for continuous service under the particular installation conditions concerned

I_n is the rated current or current setting of the protective device

I_2 is the current causing effective operation of the overload protective device within the conventional time as stated in the product standard.

NOTE: The current causing effective operation in the conventional time of protective devices may also be named I_t (for circuit-breakers) or I_f (for fuses) according to the product standards. Both I_t and I_f are multiples of I_n and attention should be given to the correct representation of values and indexes.

533.2.2 Additional requirements for protection against overload when harmonic currents are present

When selecting an overload protective device to comply with Regulation 433.1, account shall be taken of harmonic currents in accordance with Regulation 431.2.3.

NOTE: See also Appendix 4 for further details on harmonics.

533.3 Selection of devices for protection of wiring systems against fault current

The application of the regulations of Chapter 43 shall take into account both the minimum and maximum fault current conditions, so that the highest energy let-through is taken into account.

Where the standard covering a protective device specifies both a rated service short-circuit breaking capacity and a rated ultimate short-circuit breaking capacity, it is acceptable to select the protective device on the basis of the ultimate short-circuit breaking capacity for the maximum fault current conditions. Operational circumstances may, however, make it desirable to select the protective device on the service short-circuit breaking capacity, e.g. where a protective device is placed at the origin of the installation.

Where the short-circuit breaking capacity of the protective device is lower than the maximum prospective short-circuit or earth fault current that is expected at its point of installation, it is necessary to comply with the requirements of the last paragraph of Regulation 536.1 and Regulation 536.5.

NOTE: To calculate the maximum and minimum fault currents, see Technical Report PD CLC/TR 50480 'Determination of cross-sectional area of conductors and selection of protective devices'. See also Appendix 4, paragraph 2.5.

534 DEVICES FOR PROTECTION AGAINST OVERVOLTAGE

For further information see Appendix 16.

534.1 General

This section contains provisions for the application of voltage limitation in order to obtain insulation coordination in the cases described in Section 443, BS EN IEC 60664-1, BS EN 62305 series and BS EN 61643 series. This section focuses mainly on the requirements for the selection and erection of SPDs for protection against transient overvoltages where required by Section 443, the BS EN 62305 series, or as otherwise stated.

BS EN 62305-4 and DD/CLC/TS 61643-12 series deal with the protection against the effects of direct lightning strokes or strokes near to the supply system. Both documents describe the selection and the application of surge protective devices (SPDs) according to the Lightning Protection Zones (LPZ) concept. The LPZ concept describes the installation of Type 1, Type 2 and Type 3 SPDs. See Figure 534.1.

Fig 534.1 – Lightning Protection Zones (LPZ) concept

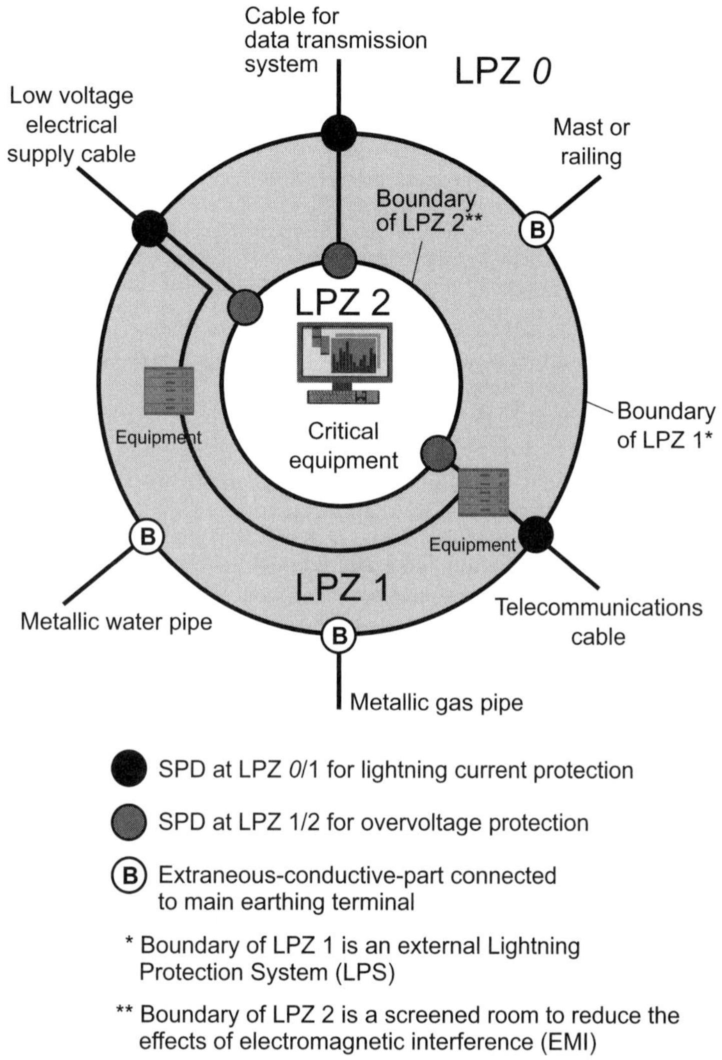

NOTE 1: SPDs with more than one Type of classification are available, e.g. combined Type 1+2 or combined Type 2+3.

Section 534 does not take into account:

- surge protective components incorporated in appliances connected to the installation
- portable surge protective devices (SPD).

NOTE 2: Further information can be found in DD CLC/TS 61643-12.

Section 534 applies to AC power circuits. As far as it is applicable, the requirements of Section 534 may be followed for DC power circuits.

NOTE 3: Overvoltages of atmospheric origin and electrical switching events can affect metallic data, signal and telecommunication lines. Protection measures for these systems are detailed within PD CLC/TS 61643-22.

534.2 Not used

534.3 Not used

534.4 Selection and erection of SPDs

534.4.1 SPD types and location

534.4.1.1 Where SPDs are required:

(i) SPDs installed at the origin of the electrical installation shall be Type 1 or Type 2

(ii) SPDs installed close to sensitive equipment to further protect against switching transients originating within the building shall be Type 2 or Type 3.

NOTE: Type 1 SPDs are often referred to as equipotential bonding SPDs and are fitted at the origin of the electrical installation to specifically prevent dangerous sparking which could lead to fire or electric shock hazards. In accordance with BS EN 62305-4, a lightning protection system which only employs equipotential bonding SPDs provides no effective protection against failure of sensitive electrical and electronic systems. Further SPDs (Type 2 and Type 3) are required to protect sensitive and critical equipment (for example, hospital equipment and fire/security alarm systems) downstream of the origin of the electrical installation.

534.4.1.2 In accordance with the LPZ concept, where a cable crosses the zonal interface, further SPDs shall be installed to preserve the zone integrity.

534.4.1.3 Where the installation of SPDs is required by Section 443 and the structure is equipped with an external lightning protection system or protection against the effects of direct lightning, Type 1 SPDs shall be installed as close as possible to the origin of the electrical installation.

534.4.1.4 Where the installation of SPDs is required by Section 443 and the structure is not equipped with an external lightning protection system or does not require protection against the effects of direct lightning, Type 2 SPDs shall be installed as close as possible to the origin of the electrical installation.

534.4.1.5 Type 2 or Type 3 SPDs shall be located in the fixed electrical installation, for example, in sub-distribution boards or close to the equipment to be protected, to achieve the required voltage protection level; see Figure 534.2. These SPDs shall be coordinated with Type 1 and/or Type 2 SPDs being installed at the origin of the electrical installation (see Regulation 534.4.4.5).

Fig 534.2 – Example of installation of Type 1, Type 2 and Type 3 SPDs

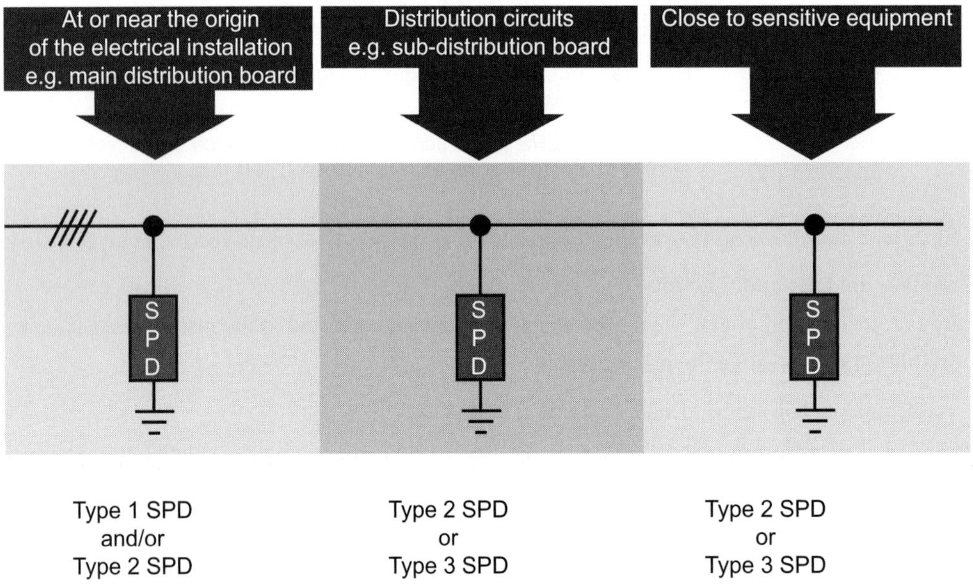

534.4.1.6 Consideration shall be given to the provision of SPDs to protect from other sources, such as:

(i) switching overvoltages produced by current-using equipment located within the installation

(ii) overvoltages on other incoming services such as metallic telecommunication and signalling services

(iii) overvoltages on other services feeding other structures such as secondary buildings, external installations/lighting, power lines feeding external sensors.

These SPDs shall be installed and located as close as possible to the origin of such events.

NOTE: For further information see DD CLC/TS 61643-12 series and BS EN 62305-4.

534.4.1.7 For surge protection labelling requirements, see Regulation 514.16.1.

534.4.2 Connection modes of SPDs

Protection against transient overvoltages shall be provided:

- between live conductors and PE (common mode protection), and/or
- between live conductors (differential mode protection).

NOTE: For further information see DD CLC/TS 61643-12.

534.4.3 Connection types

534.4.3.1 Connection Type CT1

Connection Type CT1 (e.g. 4+0 configuration):

- SPD assembly providing a mode of protection between each live conductor (line and neutral conductors, if available) and PE.

An example of connection Type CT1 for application in a three-phase system is given in Figure 534.3.

Fig 534.3 – Connection Type CT1 (4+0-configuration) for a three-phase system with neutral

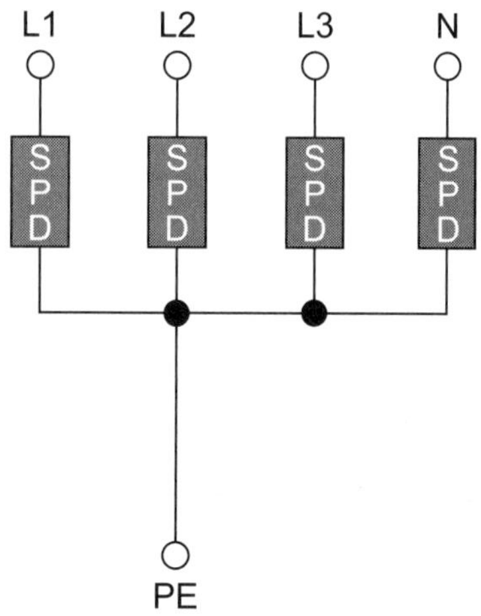

534.4.3.2 Connection Type CT2

Connection Type CT2 (e.g. 3+1-configuration):

- SPD assembly providing a mode of protection between each line conductor and the neutral conductor, and between the neutral conductor and PE.

An example of connection Type CT2 for application in a three-phase system is given in Figure 534.4.

Fig 534.4 – Connection Type CT2 (e.g. 3+1-configuration) for a three-phase system with neutral

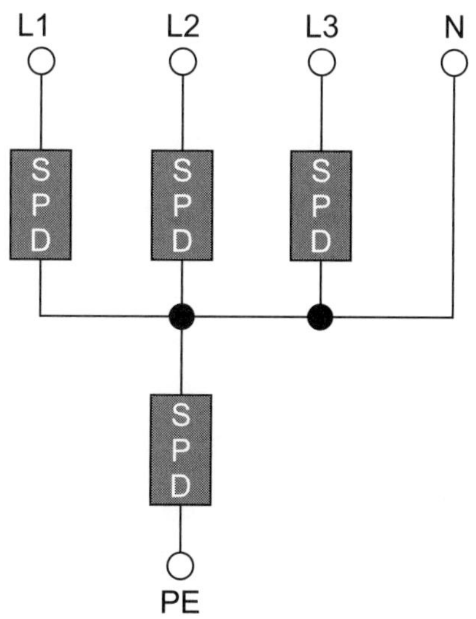

534.4.4 Selection of SPDs

534.4.4.1 General

The following parameters shall be considered for SPD protection:

- voltage protection level (U_p) of the SPD and rated impulse withstand voltage (U_W) of the equipment to be protected (see Regulation 534.4.4.2)
- continuous operating voltage (U_c) of the SPD, i.e. supply system (TT, TN, IT) (see Regulation 534.4.4.3)
- nominal discharge current (I_{nspd}) and/or impulse discharge current (I_{imp}) of the SPD (see Regulation 534.4.4.4)
- SPD coordination (see Regulation 534.4.4.5)
- expected short-circuit current (I_{SCCR}) (see Regulation 534.4.4.6)
- follow current interrupt rating (I_{fi}) (SPDs shall comply with the requirements of BS EN 61643-11).

NOTE: Additional information regarding selection and application is given in DD CLC/TS 61643-12. Annex D of DD CLC/TS 61643-12 provides application examples of selecting SPDs.

534.4.4.2 Selection with regard to voltage protection level (U_p)

The voltage protection level (U_p) of SPDs shall be selected in accordance with impulse withstand voltage Category II of Table 443.2 and in no case exceed the required rated impulse voltage of the equipment.

NOTE 1: In some cases, for example where the continuous operation of the equipment is critical, the voltage protection level (U_p) of SPDs can be selected to be lower than the impulse immunity of equipment. The impulse immunity voltage of equipment is lower than the impulse withstand of equipment and requires SPDs with lower voltage protection level U_p between live conductors (e.g. line to neutral) to avoid equipment malfunction, particularly against switching transients. For further information see DD CLC/TS 61643-12.

In installations operating at 230/400 V, the voltage protection level of the installed SPD assembly shall not exceed 2.5 kV, as the SPD's connecting leads have additional inductive voltage drop across them (see Regulation 534.4.8). It may, therefore, be necessary to select an SPD with a lower voltage protection level.

If the distance between the SPD and equipment to be protected (protective distance) is greater than 10 m, oscillations could lead to a voltage at the equipment terminals of up to twice the SPD's voltage protection level. Consideration shall be given to the provision of additional coordinated SPDs, closer to the equipment, or the selection of SPDs with a lower voltage protection level.

NOTE 2: It is recommended that the voltage protection level provided by SPDs does not exceed a safety margin of 80 % of the required rated impulse voltage for equipment according to Table 443.2 and corresponding to overvoltage category II.

This safety margin is not necessary where one of the following cases applies:

- where the equipment is connected directly to the SPD terminals
- where a protection scheme according to Figure 534.9 is already applied
- where the voltage drop across the overcurrent protection in the SPD branch circuit is already taken into account for the voltage protection level U_p
- where protection according to overvoltage category II is provided but only overvoltage category III or IV equipment is installed at this location.

NOTE 3: DD CLC/TS 61643-12 gives additional information about the rated impulse voltage of equipment and the given U_p for the SPD.

Table 534.1 Not used

534.4.4.3 Selection of SPDs with regard to continuous operating voltage (U_c)

In AC installations, the maximum continuous operating voltage U_c of SPDs shall be equal to or higher than required by Table 534.2.

TABLE 534.2 – Minimum required U_c of the SPD dependent on the supply system configuration

SPD connected between (as applicable)	System configuration of distribution network		
	TN system	**TT system**	**IT system**
Line conductor and neutral conductor	$\frac{1.1U}{\sqrt{3}}$ or (0.64 × U)	$\frac{1.1U}{\sqrt{3}}$ or (0.64 × U)	$\frac{1.1U}{\sqrt{3}}$ or (0.64 × U)
Line conductor and PE conductor	$\frac{1.1U}{\sqrt{3}}$ or (0.64 × U)	$\frac{1.1U}{\sqrt{3}}$ or (0.64 × U)	1.1 U
Neutral conductor and PE conductor	$\frac{U}{\sqrt{3}}$ (see NOTE 2)	$\frac{U}{\sqrt{3}}$ (see NOTE 2)	$\frac{1.1U}{\sqrt{3}}$ or (0.64 × U)
Line conductors	1.1 U	1.1 U	1.1 U
NOTE 1: U is the line-to-line voltage of the low voltage system.			
NOTE 2: These values are related to worst-case fault conditions, therefore, the tolerance of 10 % is not taken into account.			

534.4.4.4 Selection of SPDs with regard to discharge current (I_{nspd}) and impulse discharge current (I_{imp})

At or near the origin of the electrical installation, SPDs shall comply with one of the following cases, as applicable:

- where the building is protected against direct lightning strike, SPDs at the origin of the electrical installation shall be selected according to Regulation 534.4.4.4.2 and Table 534.4
- in other cases, SPDs shall be selected according to Regulation 534.4.4.4.1.

SPDs installed downstream of the SPDs at or near the origin of the electrical installation shall also comply with the coordination requirements in Regulation 534.4.4.5.

NOTE: Overvoltages due to switching can be longer in duration and can contain more energy than the transient overvoltages of atmospheric origin. This has to be considered for the selection of SPDs with regard to nominal discharge current and impulse discharge current. For further information see DD CLC/TS 61643-12.

534.4.4.4.1 Type 2 SPDs

Where Type 2 SPDs are required at or near the origin of the electrical installation, their nominal discharge current (I_{nspd}) shall be not less than that given in Table 534.3.

TABLE 534.3 – Nominal discharge current (I_{nspd}) depending on supply system and connection Type

Connection	I_{nspd} in kA			
	Supply system			
	Single-phase		Three-phase	
	CT1	CT2	CT1	CT2
L - N		5		5
L - PE	5		5	
N - PE	5	10	5	20

534.4.4.4.2 Type 1 SPDs

Where Type 1 SPDs are required at or near the origin of the electrical installation, one of the following cases applies:

(i) Where no risk analysis according to BS EN 62305-2 has been carried out, the impulse discharge current (I_{imp}) shall be not less than as given in Table 534.4.

TABLE 534.4 – Selection of impulse discharge current (I_{imp}) where the building is protected against direct lightning strike

Connection	I_{imp} in kA			
	Supply system			
	Single-phase		Three-phase	
	CT1	CT2	CT1	CT2
L - N		12.5		12.5
L - PE	12.5		12.5	
N - PE	12.5	25	12.5	50
NOTE: This table refers to lightning protection levels (LPL) III and IV.				

(ii) Where the risk analysis according to BS EN 62305-2 has been carried out, the impulse discharge current (I_{imp}) shall be determined according to the BS EN 62305 series.

534.4.4.5 Coordination of two or more SPDs

SPDs shall be selected and erected such as to provide coordination in operation by reference to the manufacturer's data.

534.4.4.6 Selection of SPDs with regard to the short-circuit current rating (I_{SCCR})

In general, the short-circuit current rating I_{SCCR} of an SPD, as stated by the manufacturer, shall not be lower than the maximum prospective short-circuit current at the connection points of the SPD assembly; see Figure 534.5.

NOTE 1: The SPD alone, or as an assembly in conjunction with its disconnector and/or overcurrent protective device (OCPD), is required to withstand the short-circuit current rating I_{SCCR} as stated by the manufacturer. This value is defined and tested according to BS EN 61643-11, in conjunction with the disconnector and/or OCPD as stated by the manufacturer.

NOTE 2: This requirement does not apply to SPDs connected between the neutral conductor and PE in TN or TT systems, for which this is already covered by the product standard BS EN 61643-11.

For SPDs connected between the neutral conductor and PE in IT systems, the short-circuit current rating I_{SCCR} of the SPD shall not be lower than the maximum prospective short-circuit current at the connection points of this SPD in case of a second fault.

534.4.4.7 Selection of SPDs with regard to the follow current interrupt rating (I_{fi})

In general, the follow current interrupt rating I_{fi} of the SPD, if declared by the manufacturer, shall not be lower than the maximum prospective short-circuit current at the connection points of the SPD assembly. See Figure 534.5.

For SPDs connected between the neutral conductor and PE in IT systems, the follow current interrupt rating I_{fi} of the SPD, if declared by the manufacturer, shall not be lower than the maximum prospective short-circuit current at the connection points of this SPD in case of a second fault.

Fig 534.5 – Connection points of an SPD assembly

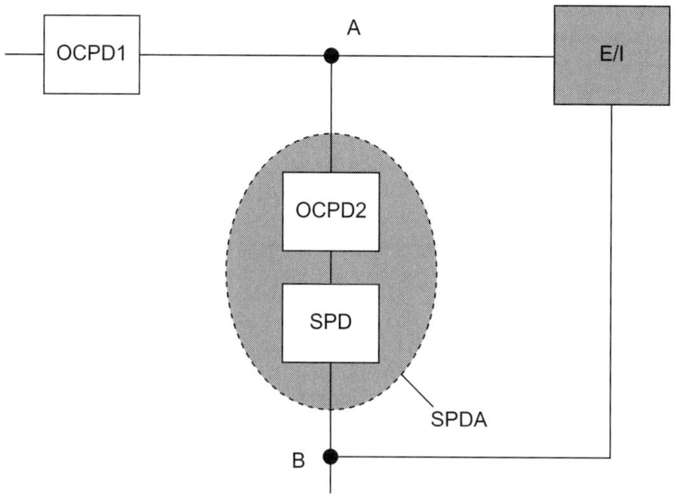

Key

OCPD 1	overcurrent protective device in the installation
OCPD 2	overcurrent protective device specified by the SPD manufacturer
SPD	surge protective device
SPDA	SPD assembly
A & B	connection points of SPD assembly
E/I	equipment or installation to be protected.

534.4.5 Protection of the SPD against overcurrent

534.4.5.1 General

SPD installations shall be protected against overcurrent with respect to short-circuit currents.

This protection may be internal and/or external to the SPD according to the manufacturer's instructions.

The ratings and characteristics of external OCPDs for protecting the SPD assembly shall be selected in accordance with Section 434 and be the highest permissible rating to provide a high surge current capability for the complete assembly whilst not exceeding the ratings and characteristics as required in the SPD manufacturer's installation instructions for the maximum overcurrent protection.

534.4.5.2 Arrangement of SPDs with relation to overcurrent protection

In case of OCPD operation arising from SPD failure, the continuity of the supply to the equipment is unaffected. However, neither the installation nor the equipment is protected against possible further overvoltages (see Figure 534.6). In such an arrangement, the effective voltage protection level within the installation is increased due to the additional voltage drop at the external OCPD connected in series with the SPD.

Fig 534.6 – Example of overcurrent protection in the SPD branch by using a dedicated external overcurrent protective device

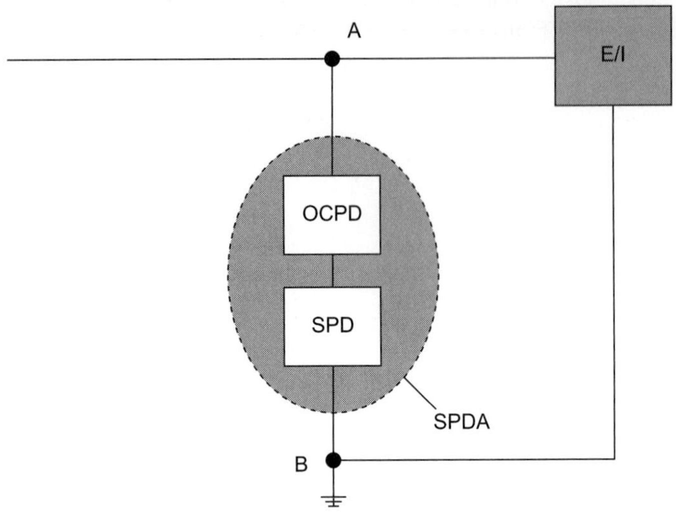

Key

OCPD	overcurrent protective device specified by the SPD manufacturer
SPD	surge protective device
SPDA	SPD assembly
A & B	connection points of SPD assembly
E/I	equipment or installation to be protected.

534.4.5.3 Selectivity between OCPDs

Where selectivity between OCPDs is necessary to prevent danger and where required for proper functioning of the installation, the manufacturer's instructions shall be taken into account; see Regulation 536.4.

534.4.6 Fault protection

Fault protection, as defined in Chapter 41, shall remain effective in the protected installation even in the event of SPD failure.

In the case of automatic disconnection of supply:

- in TN systems, this may generally be fulfilled by the OCPD on the supply side of the SPD (See Figure 16A1 of Appendix 16);
- in TT systems, this may be fulfilled by:
  (a) the installation of SPDs downstream (load side) of an RCD (See Figure 16A2 of Appendix 16), or
  (b) the installation of SPDs upstream (supply side) of the main RCD. Owing to the possibility of a failure of an SPD connected between neutral conductor and PE, the conditions of Regulation 411.4.1 shall be met and the SPDs shall be installed in accordance with connection Type CT2 (See Figure 16A3 of Appendix 16).
- in IT systems, no additional measure is needed (See Figure 16A4 of Appendix 16).

TABLE 534.5 – Connection of the SPD dependent on supply system

Supply system at the connection point of the SPD assembly	Connection Type	
	CT1	CT2
TN system	Y	Y
TT system	SPD only downstream of RCD	Y
IT system with neutral	Y	Y
IT system without neutral	Y	N/A
NOTE 1: Y = applicable		
NOTE 2: N/A = not applicable		

NOTE: Additional requirements might apply for SPDs installed in the area of influence of applications such as railway systems, HV power systems, mobile units, etc.

534.4.7 SPD installation in conjunction with RCDs

Where SPDs are installed in accordance with Regulation 534.4.1 and are on the load side of an RCD, an RCD having an immunity to surge currents of at least 3 kA 8/20 shall be used.

NOTE 1: Type S RCDs in accordance with BS EN 61008-1 and BS EN 61009-1 satisfy this requirement.

Installation of Type 1 SPDs downstream of an RCD is not recommended.

NOTE 2: In the case of a surge current higher than 3 kA 8/20, the RCD may trip causing interruption of the supply.

534.4.8 Connection of SPDs

All conductors and interconnections to the relevant line to be protected, together with the connections between the SPD and any external OCPD, shall be kept as short and as straight as possible and any unnecessary cable loops shall be avoided.

The length of the connecting conductors is defined as the sum of the path length of conductors used from the line conductor to the PE in between connection points A and B as shown in Figure 534.8 (for example, a + b + c). |

Fig 534.7 ***Not used***

Fig 534.8 – Connection of the SPD

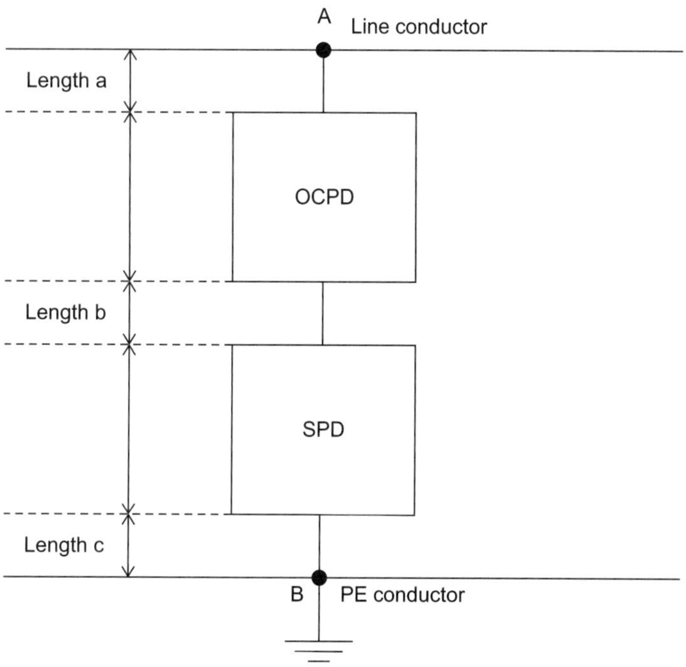

Key

OCPD	overcurrent protective device
SPD	surge protective device
PE conductor	protective earthing conductor
A and B	connection points of SPD assembly.

NOTE: When the OCPD is not present, length b is equal to 0.

Consideration shall be given to limit the total wiring length of conductors between connection points of the SPD assembly, which should preferably not exceed 0.5 m and in no case exceed 1.0 m.

To meet these requirements, the main protective conductor shall be connected to the earthing terminal located as near as possible to the SPD by adding, if necessary, an intermediate earthing terminal (see Figure 534.9).

To determine the total length of the connecting conductors according to Figure 534.9, the following conductor lengths shall not be taken into account:

- between the main earthing terminal and the intermediate earthing terminal
- between the intermediate earthing terminal and the PE conductor.

The length (and therefore inductance) of the conductors between the SPDs and the main earthing terminal shall be kept to a minimum. SPDs may be connected to the main earthing terminal or to the protective conductor via metallic parts, e.g. the metallic enclosures of the assembly (see Regulation 534.4.2), provided it is connected to PE and meets the requirements for a protective conductor in accordance with Chapter 54. Connection of the relevant SPD(s) to the main earthing terminal, and in addition to the main protective conductor, may improve the voltage protection level.

If the total wiring length (a+b+c) as defined in Figure 534.8 exceeds 0.5 m, at least one of the following options shall be chosen:

- select an SPD with a lower voltage protection level U_p (a 1 m length of rectilinear conductor carrying a discharge current of 10 kA (8/20) adds a voltage drop of about 1 000 V)
- install a second coordinated SPD close to the equipment to be protected so as to adapt the voltage protection level U_p to the rated impulse voltage of the equipment to be protected
- use the installation method as shown in Figure 534.9.

Fig 534.9 – Example of installation of an SPD in order to decrease lead length of SPD supply conductors

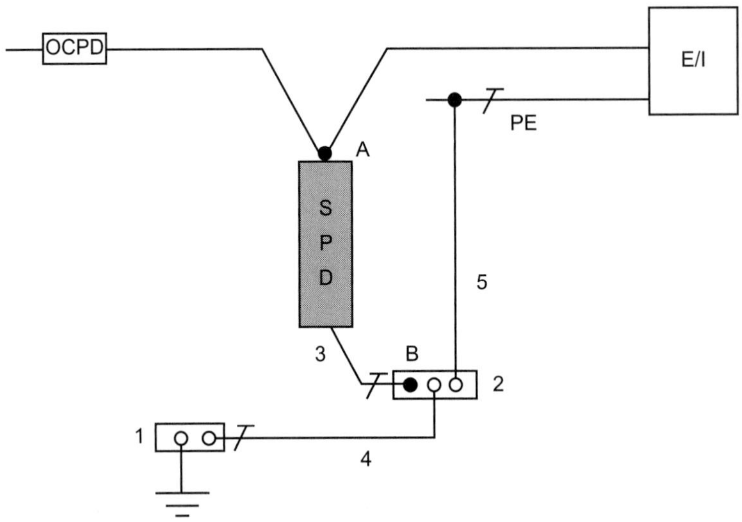

Key

OCPD	overcurrent protective device
SPD	surge protective device
PE conductor	protective earthing
E/I	equipment/installation to be protected
1	main earthing terminal
2	intermediate earthing terminal
3	length c (to be considered)
4	cable length need not be considered
5	cable length need not be considered
A and B	connection points of SPD assembly.

NOTE: For further information see DD CLC/TS 61643-12.

534.4.9 Not used

534.4.10 Connecting conductors of SPDs

Conductors between SPDs and the main earthing terminal or the protective conductor shall have a cross-sectional area not less than:

- 6 mm^2 copper or equivalent for Type 2 SPDs installed at or near the origin of the installation
- 16 mm^2 copper or equivalent for Type 1 SPDs installed at or near the origin of the installation.

Referring to Regulation 433.3.1(ii), conductors connecting SPDs and the OCPDs to live conductors shall be rated to withstand the prospective short-circuit current to be expected and shall have a cross-sectional area not less than:

- 2.5 mm^2 copper or equivalent for Type 2 SPDs installed at or near the origin of the installation
- 6 mm^2 copper or equivalent for Type 1 SPDs installed at or near the origin of the installation.

535 DEVICES FOR PROTECTION AGAINST UNDERVOLTAGE

Devices for protection against undervoltage shall comply with the relevant requirements of Section 445.

For protection of persons and property, devices for protection against undervoltage may be required.

Protective devices against undervoltage are selected as follows:

(i) direct operating undervoltage release:
  - lower value of the relay operating voltage
  - higher value of the relay operating voltage
  - time delay (if required)

(ii) indirect operating undervoltage release:
  - lower value of the operating voltage
  - higher value of the operating voltage
  - time delay (if required)

(iii) automatic reclosing when the voltage is restored:
  - with reclosing prevention
  - without reclosing prevention.

The characteristics of the protective devices against undervoltage shall be coordinated with the requirements in the relevant standards for switching on (inrush current) operation and switching off of the electrical equipment.

536 COORDINATION OF ELECTRICAL EQUIPMENT FOR PROTECTION, ISOLATION, SWITCHING AND CONTROL

536.1 General

This section covers coordination in the case of a fault and overload conditions and also takes into consideration aspects in Section 133 relevant to the coordination of electrical devices as follows:

- overcurrent protective device (OCPD)
- control and protective switching device (CPS)
- residual current device (RCD)
- contactor and starter
- switch and disconnector.

NOTE 1: Coordination of monitoring devices is under consideration.

NOTE 2: Reference to the meaning of some of the abbreviations used in this section may be found in Table A53.1 in Annex A53 located at the end of this chapter.

Section 536 does not provide requirements for the selection of an electrical device alone, but provides requirements for the selection of electrical devices to achieve electrical coordination between them.

The requirements also cover aspects of continuity of supply of the installation.

536.2 Electrical devices considered and function provided

NOTE: Table A53.1 in Annex A53 shows the functions provided by different electrical devices.

536.3 Coordination requirements

For selecting electrical devices as covered by the following regulations, the mutual interaction between those devices shall be considered so that they do not adversely affect the safety of the installation.

The coordination of electrical devices considers requirements in case of:

- short-circuit
- overload
- residual currents.

Aspects for coordination of electrical devices are:

- selectivity
- short-circuit protection
- overload protection.

Electrical devices shall be selected taking into account the coordination characteristics as given by the manufacturers.

536.4 Requirements for selectivity

536.4.1 General

Selectivity between OCPDs is defined in Regulation 536.4.1.2 for overload conditions and in Regulation 536.4.1.3 for short-circuit conditions. Selectivity between RCDs is defined in Regulation 536.4.1.4 and selectivity between OCPD and RCD is defined in Regulation 536.4.1.5.

In this regulation, the OCPD could be replaced by an SCPD.

536.4.1.2 Selectivity under overload conditions between OCPDs

536.4.1.2.1 General requirements

Where selectivity is required, as shown in Figure 536.1, the design shall be verified either by:

- desk study, taking into account the relevant product standards and the manufacturer's literature, or
- appropriate software tools where information is provided by the manufacturer for this specific use, or
- tests in accordance with the applicable product standard (in order to achieve the correct test performances and reproducibility), or
- manufacturer's declaration.

In the case of a desk study, when time/current characteristics are used to verify selectivity, account shall be taken of the reference ambient temperature for which the tripping curves are given and to load conditions before the overcurrent.

Fig 536.1 – Selectivity between OCPDs

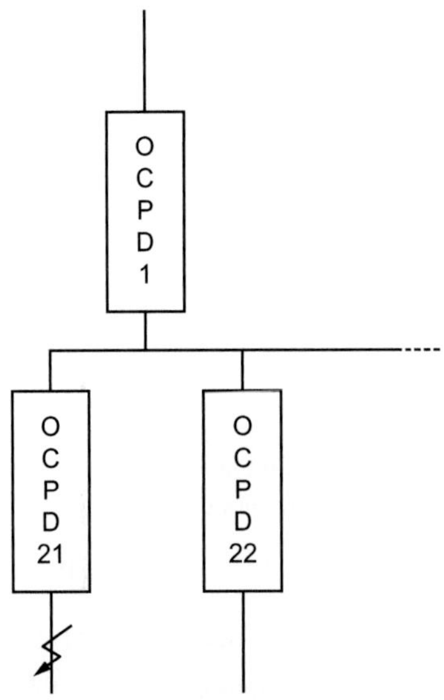

536.4.1.3 Selectivity under short-circuit conditions between OCPDs

536.4.1.3.1 General requirements

Where selectivity is required (see Figure 536.1), verification shall be made either by:

- desk study, taking into account the relevant product standard and the manufacturer's literature, or
- appropriate software tools where information is provided by the manufacturer for this specific use, or
- tests in accordance with the applicable product standard (in order to achieve the correct test performances and reproducibility), or
- manufacturer's declaration.

In the case of a desk study, when energy limitation curves are used to verify selectivity, account shall be taken of the voltage for which the curves are given.

In cases given in A53.7 and A53.10 in Annex A53, selectivity will be obtained for short-circuit currents up to a specific value, the selectivity limit current. The value of this limit will be given by the manufacturer. In a particular installation selectivity may be total or partial.

NOTE: Generally, manufacturers provide tables giving information on selectivity in case of short-circuit.

536.4.1.4 Selectivity between RCDs

(i) General requirements

Where selectivity is required (see Figure 536.2), verification shall be made either by:

- desk study, taking into account the relevant product standard and the manufacturer's literature, or
- appropriate software tools where information is provided by the manufacturer for this specific use, or
- tests in accordance with the applicable product standard (in order to achieve the correct test performances and reproducibility), or
- manufacturer's declaration.

NOTE 1: Generally, manufacturers provide information specifying selectivity between RCDs.

(ii) Selectivity in case of residual currents

Selectivity in case of residual currents, as shown in Figure 536.2, is given under the following conditions:

- the upstream RCD is of selective type (type S or time-delayed type with appropriate time delay setting), and
- the ratio of the rated residual operating current of the upstream RCD to that of the downstream RCD is at least 3:1.

In the case of RCDs with adjustable rated residual operating current and time delay, reference shall be made to manufacturer's instructions for selectivity.

NOTE 2: RCD type S is in accordance with BS EN 61008 series or BS EN 61009 series.

NOTE 3: A time-delay type RCD in accordance with BS EN 60947-2, Annex B or Annex M will be marked with the symbol Δt followed by the limiting non-actuating time in ms or marked with an [S].

Fig 536.2 – Selectivity between RCDs in case of residual current

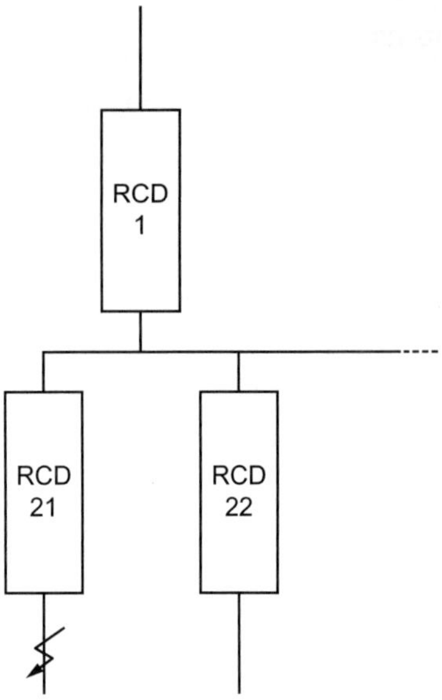

NOTE 4: RCD1 is type S or time-delayed type.

536.4.1.5 Selectivity between OCPDs and RCDs

536.4.1.5.1 Selectivity between RCD(s) and upstream OCPD

In the event of an earth fault, current may reach a high value that could exceed the instantaneous tripping current of the upstream OCPD. Therefore, when selectivity between RCD(s) and upstream OCPD is required, an RCBO or CBR shall be used and the requirements of selectivity according to Regulation 536.4.1.2 shall be applied.

Fig 536.3 – Selectivity between OCPD and RCD using RCBOs

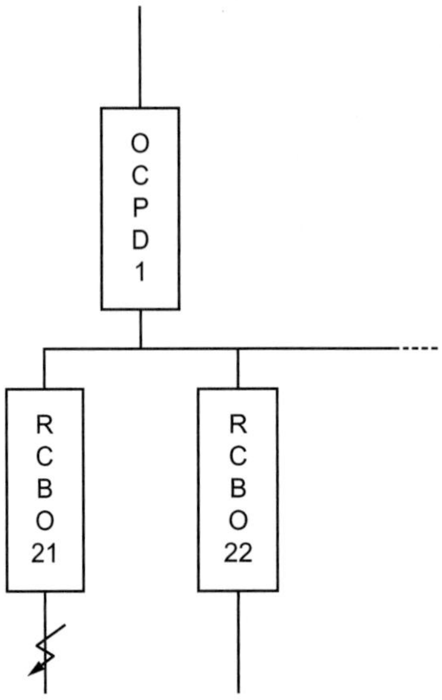

In Figure 536.3, if RCCBs are used instead of RCBOs, selectivity cannot be guaranteed. In this case, if there are OCPDs downstream of the RCCB, as shown in Figure 536.4, selectivity may be achieved provided that the requirements of Regulation 536.4.1.2 and Regulation 536.4.1.3 are fulfilled. In addition, the connection between the RCCB and the downstream OCPD shall be selected and erected so as to minimize the risk of earth faults.

NOTE: In order to minimize the risk of faults between RCCB and downstream OCPD, specific wiring accessories may be used (e.g. specific busbars).

Fig 536.4 – Selectivity between OCPD and RCD using RCCBs

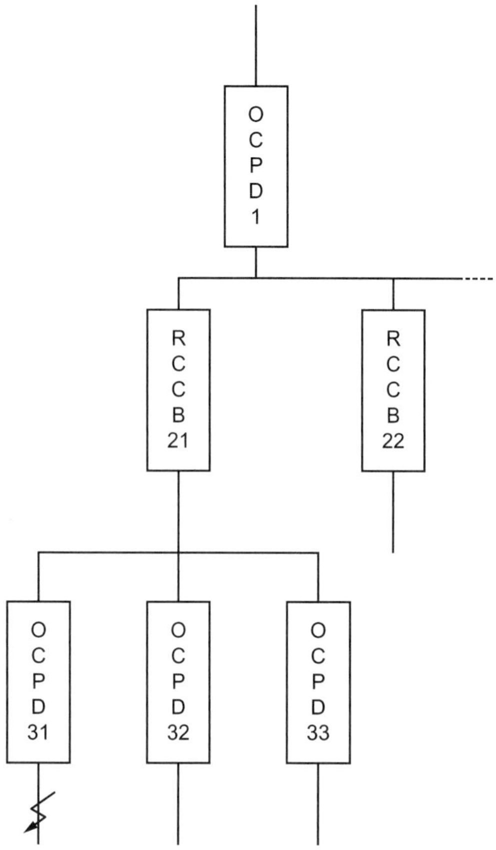

536.4.1.6 Selectivity between RCDs and downstream OCPDs

In the event of an earth fault, the earth fault current may be lower than the instantaneous tripping current of the downstream OCPD. In this case, the upstream RCD will trip and selectivity may not be achieved. Therefore, when selectivity between RCDs and downstream OCPD is required, RCBOs shall be used and selectivity requirements according to Regulation 536.4.1.4 shall be applied.

Fig 536.5 – Selectivity between upstream RCCB and RCBOs

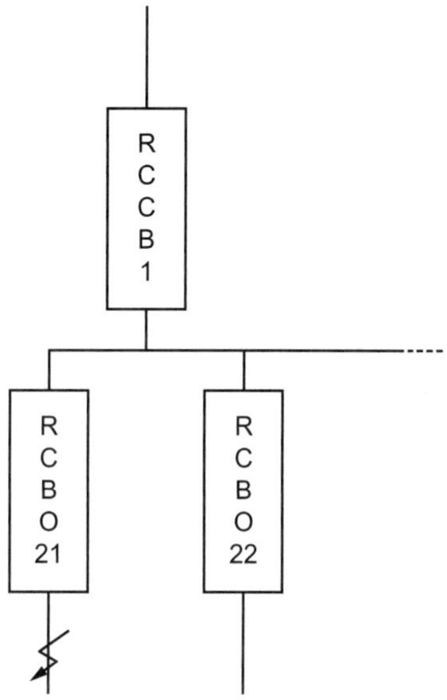

NOTE: RCCB1 is type S or time-delay type.

536.4.2 Requirements for protection in case of short-circuit

536.4.2.1 Combined short-circuit protection of OCPDs

In this regulation, the OCPD could be replaced by an SCPD.

This breaking technique allows the use of short-circuit protective devices with a lower breaking capacity than required in Chapter 43. It is only applicable to short-circuit protective devices.

Fig 536.6 – Typical configuration for combined short-circuit protection of OCPDs

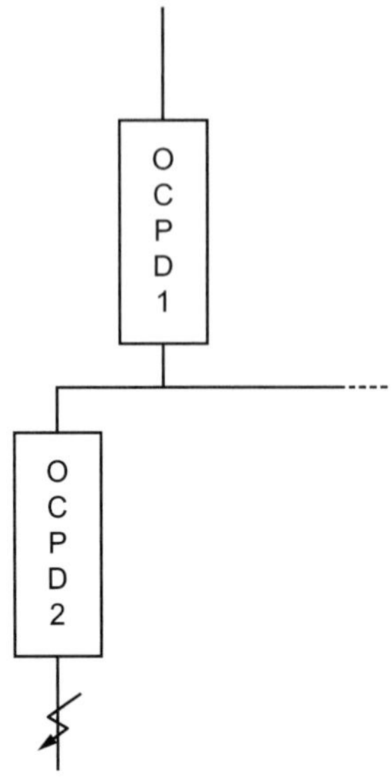

Regulation 434.5.1 permits a device with a lower rated breaking capacity than the prospective short-circuit current at its point of installation to be used in specific conditions.

When selecting two OCPDs for combined short-circuit protection, reference shall be made to the instructions of the manufacturer of the downstream OCPD. These instructions are derived from tests performed according to relevant product standards, as applicable (e.g. BS EN 60947-2 and BS EN 60898-1). Where no information is available from the manufacturer, combined short-circuit protection of OCPDs shall not be used, and each OCPD shall have the required short-circuit capability at the point of installation.

If declared by the manufacturer of both devices, the combined short-circuit capability of OCPD1 and OCPD2 may be higher than the breaking capacity of either OCPD. In this case, the connection between the two devices has to minimize the risk of short-circuits and there shall not be any short-circuit contributions by other active equipment in parallel to OCPD1.

NOTE 1: Coordination of an OCPD with a separate current limiter to increase the short-circuit breaking capacity of an OCPD may be used according to the manufacturer's instructions.

NOTE 2: Combined short-circuit protection may be used for manual motor starter having a short-circuit capability in combination with OCPD, when declared in the manufacturer documentation.

536.4.2.2 Back-up protection of contactors or overload relays

In this regulation, the OCPD could be replaced by an SCPD.

Contactors complying with BS EN 60947-4-1 or BS EN 61095 do not provide protection against short-circuit, thus they shall be protected by an upstream overcurrent protective device (OCPD).

NOTE 1: Control and Protective Switching Devices (CPS) provide protection against short-circuit, and therefore provide intrinsic coordination.

Figure 536.7 gives a typical schematic for coordination of a contactor with a short-circuit protective device (OCPD).

Fig 536.7 – Coordination between OCPD and contactor in case of short-circuit

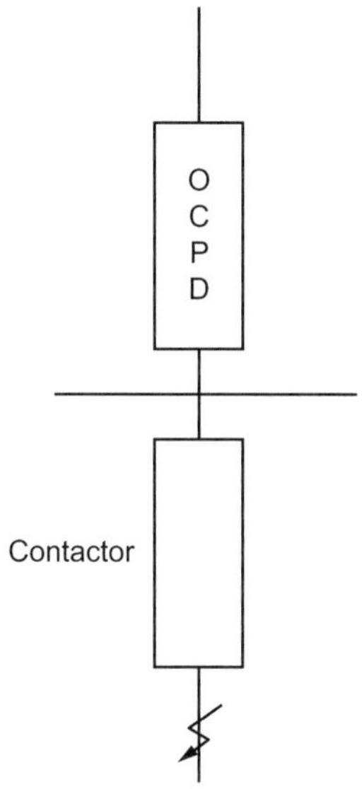

Regulation 512.1.2 requires equipment to be selected for the design current which it has to carry in normal service and the current likely to flow in abnormal conditions. In the event of short-circuit, the let-through energy and the peak current may cause the contactor to open its contacts at a level of current beyond its making and breaking capacity. Coordination of the contactor and the OCPD is needed to achieve safe operation in case of short-circuit.

Coordination between a contactor and an OCPD is verified by mandatory short-circuit tests according to BS EN 60947-4-1 or to BS EN 61095, as applicable.

Contactors shall be selected and erected in conjunction with the short-circuit protective device declared by the manufacturer, such that the rated conditional short-circuit current is higher than the prospective short-circuit current at the point of installation.

The rated conditional short-circuit current can only be obtained by type-testing and thus the data for the selection of the OCPD shall be obtained from the manufacturer of the contactor, taking into account the rated operational current, rated operational voltage and the corresponding utilization category.

NOTE 2: This information is generally provided in coordination tables which summarize the combination of devices (OCPD, contactor or motor starter) able to maintain a safe behaviour for a given rated conditional short-circuit.

In a motor circuit the overload current protection may be provided by separate overload relay electrically connected with the contactor. In such a case protection of the overload relay shall also be achieved, since the overload relay may be damaged by the let-through energy (I^2t) of the OCPD. This information is generally provided in the manufacturer's coordination tables.

536.4.2.3 Back-up protection of switches, Transfer Switching Equipment (TSE) or impulse relays

Switches complying with BS EN 60947-3 or BS EN 60669-2-4, Transfer Switching Equipment (TSE) complying with BS EN 60947-6-1 and impulse relays complying with BS EN 60669-2-2 are used to switch loads or circuits (e.g. distribution switchboard, lighting circuit, specific load). These devices do not provide protection against short-circuit, thus they shall be protected by an overcurrent protective device (OCPD).

NOTE: Fuse combination units to BS EN 60947-3 consist of a switch and integral fuse(s) and have a short-circuit capability declared by the manufacturer which does not require the use an upstream OCPD.

Should a short-circuit occur on the load side of a switch, the current will flow through both devices (OCPD and switch); therefore the let-through energy and the peak current limited by the OCPD need to be compatible with the short-circuit withstand of the switch, the TSE or the impulse relay.

Figure 536.9 gives a typical schematic for coordination of a switch with a short-circuit protective device (OCPD).

Fig 536.8 Not used

Fig 536.9 – Coordination between OCPD and switch

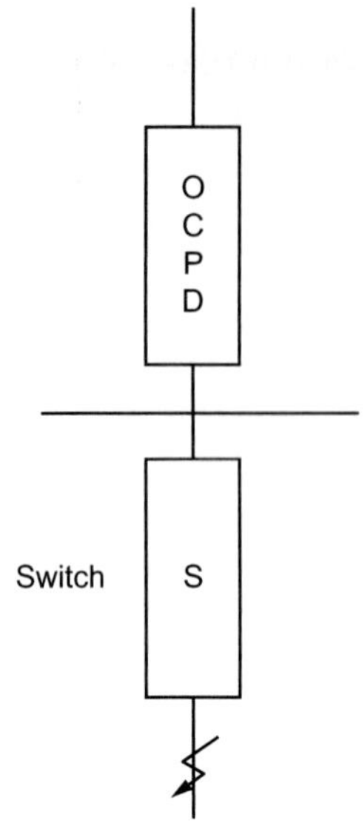

The OCPD may also be situated downstream of the switch, TSE or impulse relay provided that the connection between the switch and the downstream OCPD is selected and erected so as to minimize the risk of earth faults and short-circuits.

Switches, Transfer Switching Equipment and impulse relays shall be selected and erected in conjunction with the short-circuit protective device declared by the manufacturer in order that their rated conditional short-circuit current is equal to or higher than the prospective short-circuit current at the point of installation.

The rated conditional short-circuit current of the switch can only be obtained by type-testing and thus the data for the selection of the OCPD shall be obtained from the manufacturer of the switch, taking into account the rated operational current and the rated operational voltage.

For switches according to BS EN 60947-3, where the OCPD is not specified by the switch manufacturer, an alternative method for coordination between the OCPD and the switch is as follows:

- the rated short-circuit making capacity of the switch is higher than the peak value of the prospective short-circuit current at the point of installation, and
- the OCPD time-current characteristic is within the limits of I_{cw} of the switch, as stated by the manufacturer, or
- the rated short-circuit making capacity is higher than the peak value of the prospective short-circuit current at the point of installation, and
- the rated short time withstand current is higher than the prospective short-circuit current at the point of installation and the corresponding withstand time is longer than the operating time of the SCPD, if applicable, or
- if at the prospective short-circuit current, the protective device has an energy let-through (I^2t) and cut-off current not exceeding the values for the switch provided by the manufacturer.

536.4.2.4 Back-up protection of RCCBs

In this regulation, the OCPD could be replaced by an SCPD.

RCCBs complying with BS EN 61008 series are intended to protect persons against electric shocks. They may also be used to provide protection against fire hazards due to a persistent earth fault. These devices have a limited short-circuit current withstand capability, thus they shall be protected by an upstream overcurrent protective device.

Should a short-circuit occur on the load side of a RCCB, the current will flow through both devices (OCPD and RCCB); therefore the let-through energy and the peak current limited by the OCPD need to be compatible with the short-circuit withstand capability of the RCCB.

Figure 536.10 gives a typical schematic for coordination of a RCCB with a short-circuit protective device.

Fig 536.10 – Coordination between OCPD and RCCB

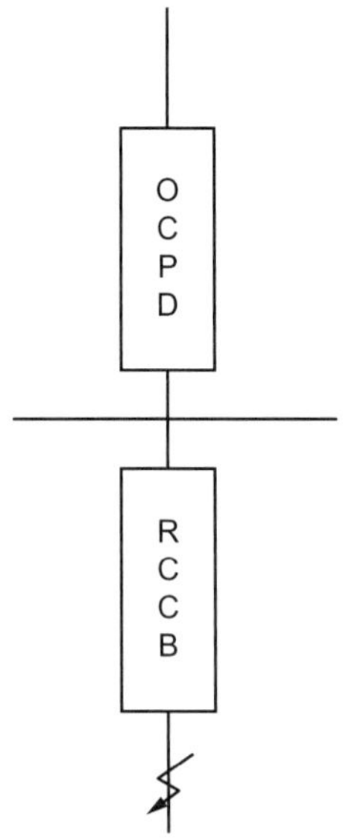

RCCBs shall be selected and erected in conjunction with the OCPD declared by the manufacturer in order that their rated conditional short-circuit current is higher than the prospective short-circuit current at the point of installation.

The rated conditional short-circuit current of the RCCB with the OCPD, related to the relevant rated operational current and rated operational voltage, is given by the manufacturer based on tests results according to BS EN 61008 series.

The OCPD may also be situated downstream of the RCCB provided that the connection between RCCB and downstream OCPD is selected and erected so as to minimize the risk of earth faults and short-circuits.

536.4.3 Requirements for protection in case of overload

536.4.3.1 Overload protection of contactor or SCPDs

Contactors complying with BS EN 60947-4-1 or BS EN 61095 and SCPDs without integral overload protection, such as ICBs complying with BS EN 60947-2, shall be protected by an overload protective device.

Devices for overload protection are selected for the protection of cables. For overload protection of contactors or SCPDs, the rated current of the OCPD shall be selected according to manufacturers' information.

NOTE: In cases where Regulation 433.3 permits omission of overload protection, coordination in case of overload does not apply.

536.4.3.2 Overload protection of RCCB, switch, Transfer Switching Equipment (TSE) or impulse relay

Residual current circuit-breakers (RCCB) complying with BS EN 61008 series, switches complying with BS EN 60947-3 or BS EN 60669-2-4, impulse relays complying with BS EN 60669-2-2 and TSEs complying with BS EN 60947-6-1 are used to switch loads or circuits. None of these devices provide protection against overload, therefore they shall be protected by an overcurrent protective device (OCPD).

Regulation 433.1 requires that devices for overload protection be selected for the protection of cables. For overload protection of RCCBs and switches, the rated current of the OCPD shall take account of the manufacturer's information; in general, the OCPD is installed upstream of the RCCB or the switch.

The rated current of a switch or RCCB may also be based on the application of diversity factors to the downstream circuits according to Regulation 311.1, and the rated current of the OCPD shall be selected according to the manufacturer's instructions. See also Regulation 536.4.202.

NOTE: When using an RCBO instead of an RCCB, no further consideration regarding overload protection of the RCBO is necessary.

Regulation 433.3 specifies conditions for omission of overload protection. In such cases overload coordination does not require verification.

536.4.4 Requirements for selectivity between OCPDs equipped with undervoltage relay

In case of fault (short-circuit or earth fault) a high fault current may generate a voltage drop through the installation. Voltage drop may also occur in the installation for other reasons (for example, switching and reclosing of HV switching device).

Where one undervoltage relay is installed in the OCPD on the supply side or elsewhere in the installation, possible undervoltage relay operation shall not impair selectivity achieved between OCPDs and/or RCDs in series.

Fig 536.11 – Selectivity with OCPD and undervoltage relays

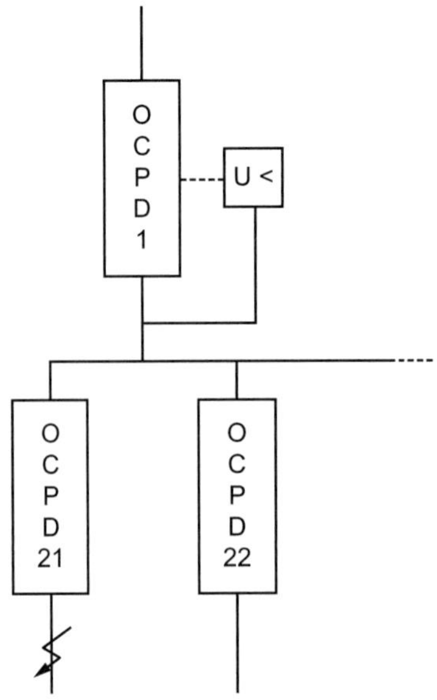

For selectivity, undervoltage relay operation shall be time-delayed according to the maximum short-circuit or earth fault clearance time. In all cases, instructions provided by the manufacturer shall be fulfilled in order not to impair the safety of the electrical installation.

536.4.5 Low voltage assemblies according to BS EN 61439 series

The interface characteristics of an assembly conforming to BS EN 61439 series shall be compatible with the ratings of the circuits to which it is connected and with the installation conditions. The characteristics of the assembly shall be declared by the manufacturer, taking into account the interface characteristics of the relevant BS EN 61439 product standard.

536.4.201 Fault current (short-circuit) ratings

The relevant fault current (short-circuit) rating of the assembly should be equal to or exceed the maximum prospective fault current at the point of connection to the system.

The terminology used to define the short-circuit rating of an assembly is given in the BS EN 61439 series as follows:

- rated short-time withstand current, I_{cw}
- rated peak withstand current, I_{pk}
- rated conditional short-circuit current, I_{cc}.

The assembly manufacturer's ratings and instructions shall be taken into account.

For an installation with a 230 V single-phase supply rated up to 100 A that is under the control of ordinary persons, switchgear and controlgear assemblies shall either comply with BS EN 61439-3 having a suitable fault current (short-circuit) rating for the maximum prospective fault current at the point of connection to the system or be a consumer unit incorporating components and protective devices specified by the manufacturer complying with BS EN 61439-3, including the 16 kA conditional short-circuit test described in Annex ZB of the standard.

NOTE: When the single-phase PSCC value of 19.6 kA is cited from Engineering Recommendation P25, the 16 kA conditional rating described in Annex ZB of BS EN 61439-3, for incoming service equipment, will satisfy design requirements where the service cable is at least 2 metres in length.

536.4.202 Current ratings

The relevant design current shall not exceed the rated current of an assembly (I_{nA}) or rated current of a circuit (I_{nc}) of the associated assembly, having taken any applicable diversity/loading factors into account.

The terminology used to define the rating of an assembly in relation to load/design current used in BS EN 61439 series can be summarized as follows:

- The rated current of an assembly (I_{nA}) (A) is the maximum load current that it is designed to manage and distribute.
- The rated current of a circuit (I_{nc}) (A) is stated by the assembly manufacturer, taking into consideration the ratings of the devices within the circuits, their disposition and application.

The current rating(s) of an assembly circuit may be lower than the rated current(s) of the device(s) according to their respective device standard, when installed in the assembly; therefore, the assembly manufacturer's ratings and instructions shall be taken into account.

Rated diversity (loading factor) can be stated by the assembly manufacturer, e.g. for groups of circuits.

The rated current of a switch or RCCB (I_{nA} and I_{nc}) shall be based upon:

- the sum of final circuit current demand after any applicable load diversity factors, or
- the sum of final circuit current demand after any applicable load diversity factors together with allowances for diversity between final circuits, or
- the sum of the downstream OCPDs/circuit rated current multiplied by a diversity factor.

However, overload protection shall not solely be based on the use of diversity factors of the downstream circuits. To achieve overload protection of RCCBs or switches, the rated current of the OCPD shall be selected according to the manufacturer's instructions.

536.4.203 Integration of devices and components

The relevant part of the BS EN 61439 series shall be applied to the integration of mechanical and electrical devices and components, e.g. circuit-breakers, control devices, busbars into an empty enclosure or existing low voltage assembly.

In low voltage assemblies to the BS EN 61439 series, e.g. consumer units, distribution boards, incorporated devices and components shall only be those declared suitable according to the assembly manufacturer's instructions or literature.

NOTE 1: The use of individual components complying with their respective product standard(s) does not indicate their compatibility when installed with other components in a low voltage switchgear and controlgear assembly.

NOTE 2: Incorporated components inside the assembly can be from different manufacturers. It is essential that all incorporated components should have had their compatibility for the final enclosed arrangements verified by the original manufacturer of the assembly and be assembled in accordance with their instructions e.g. the consumer unit, distribution board manufacturer. The original manufacturer is the organization that carried out the original design and the associated verification of the low voltage switchgear and controlgear assembly to the relevant part of the BS EN 61439 series. If an assembly deviates from its original manufacturer's instructions, or includes components not included in the original verification, the person introducing the deviation becomes the original manufacturer with the corresponding obligations.

536.5 Documentation

The information mentioned in the previous regulations for coordination of electrical devices may be found in manufacturers' documentation (e.g. catalogue, instruction sheets, calculation software).

When the initial verification is made, the documentation concerning the selection of devices for coordination shall be added to the design documentation in accordance with the requirements of Regulation 132.13.

537 ISOLATION AND SWITCHING

537.1 General

This section provides requirements for:

(i) non-automatic local and remote isolation and switching measures for the prevention or removal of dangers associated with electrical installations or electrically-powered equipment and machines, and

(ii) functional switching and control.

537.1.1 Any device for isolation and switching according to Sections 462 to 465 shall comply with the relevant requirements included in this section.

In certain instances, additional requirements may be necessary for combined functions.

NOTE 1: Table 537.4 summarizes the functions provided by the devices for isolation and switching, together with indication of the relevant product standards.

NOTE 2: For some applications such as motor control, the switching device needs to withstand the inrush current.

TABLE 537.4 – Guidance on the selection of protective, isolation and switching devices

Device	Standard	Isolation(4)	Emergency switching off(2)	Functional switching(5)
Switching device	BS EN 50428	No	No	Yes
	BS EN 60669-1	No	No	Yes
	BS EN 60669-2-1	No	No	Yes
	BS EN 60669-2-2	No	No	Yes
	BS EN 60669-2-3	No	No	Yes
	BS EN 60669-2-4	Yes(3)	Yes	Yes
	BS EN 60947-3	Yes(1,3)	Yes	Yes
	BS EN 60947-5-1	No	No	Yes
Contactor	BS EN 60947-4-1	Yes(1)	Yes	Yes
	BS EN 61095	No	No	Yes
Starters	BS EN 60947-4-1	Yes(1)	Yes	Yes
	BS EN 60947-4-2	No	No	Yes
	BS EN 60947-4-3	No	No	Yes
Circuit-breaker	BS EN 60898	Yes(3)	Yes	Yes
	BS EN 60947-2	Yes(1)	Yes	Yes
	BS EN 61009-1	Yes(3)	Yes	Yes
RCD	BS EN 60947-2	Yes(1)	Yes	Yes
	BS EN 61008 series	Yes(3)	Yes	Yes
	BS EN 61009 series	Yes(3)	Yes	Yes
	BS 7288	Yes(3)	No	Yes
Arc fault detection devices	BS EN 62606	Yes(3)	Yes	Yes
Isolating switch	BS EN 60669-2-4	Yes(3)	Yes	Yes
	BS EN 60947-3	Yes(1,3)	Yes	Yes
Plug and socket-outlet (≤ 20 A)	BS EN 60309	Yes(3)	No	Yes
Plug and socket-outlet (≥ 32 A)	BS EN 60309	Yes(3)	No	No
Device for the connection of luminaire	BS EN 61995-1	Yes(3)	No	No
Control and protective switching device for equipment (CPS)	BS EN 60947-6-1	Yes(1,3)	Yes	Yes
	BS EN 60947-6-2	Yes(1,3)	Yes	Yes
Fuse (removal of fuse link)	BS 88 series	Yes	No	No
	BS 3036	Yes	No	No
Device with semiconductors	BS EN 50428	No	No	Yes
	BS EN 60669-2-1	No	No	Yes
Luminaire Supporting Coupler	BS 6972	Yes(3)	No	No
Plug and unswitched socket-outlet	BS 1363-1	Yes(3)	No	Yes
	BS 1363-2	Yes(3)	No	Yes
Plug and switched socket-outlet(6)	BS 1363-1	Yes(3)	No	Yes
	BS 1363-2	Yes(3)	No	Yes
Plug and unswitched socket-outlet	BS 546	Yes(3)	No	Yes

Device	Standard	Isolation[4]	Emergency switching off[2]	Functional switching[5]
Plug and switched socket-outlet[6]	BS 546	Yes[3]	No	Yes
Plug and socket-outlet	BS 5733	Yes[3]	No	Yes
Switched fused connection unit	BS 1363-4	Yes[3]	Yes	Yes
Unswitched fused connection unit	BS 1363-4	Yes[3] (Removal of fuse link)	No	No
Cooker Control Unit switch	BS 4177	Yes[3]	Yes	Yes

Yes = Function provided, No = Function not provided

[1] Function provided if the device is suitable and marked with the symbol for isolation (see IEC 60617 identity number S00288)

[2] See Regulation 537.3.3.

[3] Device is suitable for on-load isolation, i.e. disconnection whilst carrying load current.

[4] In an installation forming part of a TT or IT system, isolation requires disconnection of all the live conductors. See Regulation 462.2.

[5] Circuit-breakers, AFDDs and RCDs are primarily circuit protective devices and, as such, they are not intended for frequent load switching. Infrequent switching of circuit-breakers on-load is admissible for the purposes of isolation or emergency switching. For a more frequent duty, the number of operations and load characteristics according to the manufacturer's instructions should be taken into account or an alternative device from those listed as suitable for functional switching in Table 537.4 should be employed.

[6] The switch of a socket-outlet is not required to be suitable for isolation, as isolation is achieved by withdrawal of the plug. See Regulation 537.1.3.

NOTE 1: An entry of (1,3) means that the device is suitable for on-load isolation only if it is marked with the symbol for **on-load** isolation

NOTE 2: In the above table, the functions provided by the devices for isolation and switching are summarized, together with an indication of the relevant product standards.

537.1.2 Where an installation or an item of equipment or enclosure contains live parts connected to more than one supply, a durable warning notice shall be placed in such a position that any person, before gaining access to live parts, will be warned of the need to isolate those parts from the various supplies unless an interlocking arrangement is provided to isolate all the circuits concerned.

537.1.3 Plugs and socket-outlets, connectors and devices for connection of luminaires may be used for providing the isolation and switching functions in accordance with Table 537.4.

The isolation and switching functions are provided by the disconnection of the plug from the outlet or connector from the inlet, as applicable.

537.1.4 *Not used*

537.1.5 Where an installation is supplied from more than one source of energy, one of which requires a means of earthing independent of the means of earthing of other sources and it is necessary to provide that not more than one means of earthing is applied at any time, a switching device may be inserted in the connection between the neutral point and the means of earthing, provided that the device is:

(i) a multipole, linked switching device arranged to disconnect and connect the earthing conductor for the appropriate source at substantially the same time as the related live conductors, or

(ii) a switching device interlocked with a multipole, linked switching device inserted in the related live conductors such that the earthing conductor for the appropriate source shall not be interrupted before the related live conductors and shall be re-established not later than when the live conductors are reconnected.

Switching devices provided in accordance with (i) and (ii) shall meet the requirements of Chapter 46 for a device for isolation.

537.2 Devices for isolation

537.2.1 Devices for isolation shall be of a type for which the isolation function is explicitly recognized by the relevant product standard or as identified in Table 537.4.

537.2.2 Semiconductor devices shall not be used as isolating devices.

537.2.3 Devices suitable for isolation shall be selected according to the requirements which are based on the overvoltage categories applicable at their point of installation.

Devices for isolation shall be designed for overvoltage category III or IV except the plug of a plug and socket-outlet combination identified in Table 537.4 as suitable for isolation.

NOTE: Where electrically powered equipment is within the scope of BS EN 60204, the requirements for isolation of that standard apply.

537.2.4 Devices for isolation shall be selected and/or installed so as to prevent unwanted or unintentional closure (see Regulation 462.3).

This may be achieved by locating the device in a lockable space or lockable enclosure or by padlocking or by other suitable means.

537.2.5 Provision shall be made for securing off-load isolating devices against unwanted or unintentional opening.

This may be achieved, for example, by locating the device in a lockable space or lockable enclosure or by padlocking. Alternatively, the off-load device may be interlocked with a load-breaking one.

537.2.6 Means of isolation shall preferably be provided by a multipole switching device which disconnects all applicable poles of the relevant supply but single-pole devices situated adjacent to each other are not excluded, subject to the provisions of Regulation 461.2.

537.2.7 Each device used for isolation shall be clearly identified by position or durable marking to indicate the installation or circuit it isolates.

537.2.8 Where a link is inserted in the neutral conductor for isolating purposes, the link shall comply with the following requirements:

- it cannot be removed without the use of a tool, and
- it is accessible to one or more skilled persons only.

537.3 Devices for switching

537.3.1 Functional switching and control devices

537.3.1.1 The devices for functional switching and control shall be selected in accordance with Table 537.4.

537.3.1.2 Functional switching devices shall be suitable for the most onerous duty they are intended to perform. The characteristic of the load to be switched shall be considered (e.g. utilization category).

537.3.1.3 Functional switching devices may control the current without necessarily opening the corresponding poles.

NOTE: Semiconductor switching devices and some control auxiliaries are examples of devices capable of interrupting the current in the circuit but not opening the corresponding poles.

537.3.2 Devices for switching off for mechanical maintenance

537.3.2.1 Selection and erection of devices for switching off for mechanical maintenance shall be in accordance with the following regulations and shall comply with Regulation group 537.2.

NOTE: Where electrically powered equipment is within the scope of BS EN 60204, the requirements for switching off for mechanical maintenance of that standard apply.

537.3.2.2 Devices for switching off for mechanical maintenance shall be inserted in the main supply circuit.

Where a switch is provided for this purpose, it shall be capable of cutting off the full load current of the relevant part of the installation.

Interruption of a circuit for the control of mechanical movement is permitted only where a condition equivalent to the direct interruption of the main supply is provided by one of the following:

(i) Supplementary safeguards, such as mechanical retainers

(ii) Compliance with the requirements of a British or Harmonized Standard specification for the control devices used.

537.3.2.3 Devices for switching off for mechanical maintenance or control switches for such devices shall require manual operation.

The open position of the contacts of the device shall be visible or be clearly and reliably indicated.

The indication required by this regulation may be achieved by the use of the symbols 'O' and 'I' to indicate the open and closed positions respectively.

537.3.2.4 Devices for switching off for mechanical maintenance shall be clearly identified by position or durable marking so as to be identifiable for their intended use.

537.3.3 Devices for emergency switching off

NOTE: Emergency switching off is an emergency operation intended to switch off the supply of electrical energy to all or part of an installation where a risk of electric shock or another risk of electrical origin is involved.

537.3.3.1 Selection and erection of devices for emergency switching off shall be in accordance with the following regulations and shall comply with Regulation group 537.2.

NOTE: Where electrically powered equipment is within the scope of BS EN 60204, the requirements for emergency switching off of that standard apply.

537.3.3.2 The devices for emergency switching off shall be capable of breaking the full load current of the relevant parts of the installation taking into account stalled motor currents where appropriate.

537.3.3.3 Means for emergency switching off may consist of:

- one switching device capable of directly cutting off the appropriate supply, or
- a combination of devices activated by a single action for the purpose of cutting off the appropriate supply.

Plugs and socket-outlets shall not be provided for use as means for emergency switching off.

537.3.3.4 Devices for emergency switching off shall provide the switching of the main circuit.

Hand-operated switching devices for direct interruption of the main circuit shall be selected where practicable.

Remote control switching of circuit-breakers, control and protective switching devices or residual current devices (RCD) shall be opened by de-energization of coils, or other equivalent failure-to-safety techniques/actuators.

537.3.3.5 The means of operating (handles, push-buttons, etc.) devices for emergency switching off shall be clearly identified, preferably by colour. If a colour is used for identification, this shall be RED with a contrasting background (e.g. yellow).

NOTE: The contrasting background may or may not include text.

537.3.3.6 The means of operating shall be readily accessible at places where a danger might occur and, where appropriate, at any additional remote position from which that danger can be removed.

Devices for emergency switching off shall be so placed as to be readily identifiable and convenient for their intended use.

537.3.3.7 The means of operation of a device for emergency switching off shall be capable of latching in the 'OFF' position, unless both the means of operation for emergency switching off and for re-energizing are under the control of the same person.

The release of an emergency switching device operated remotely shall not re-energize the relevant part of the installation.

The operation of the emergency switching device shall have priority over any other function relative to safety and shall not be inhibited by any other operation of the installation.

537.4 Firefighter's switches

537.4.1 Firefighter's switches shall comply with BS EN 60669-2-6 or BS EN 60947-3.

537.4.2 A firefighter's switch shall be provided in the low voltage circuit supplying:

(i) outdoor lighting installations operating at a voltage exceeding low voltage, and

(ii) indoor discharge lighting installations operating at a voltage exceeding low voltage.

This requirement does not apply to a portable discharge lighting luminaire or to a sign of rating not exceeding 100 W and fed from an accessible socket-outlet.

NOTE: Installations in certain premises subject to licensing conditions, such as petrol station forecourts, may require the installation of a firefighter's switch.

537.4.2.1 Every outdoor installation on each single premises should wherever practicable be controlled by a single firefighter's switch. Similarly, every internal installation in each single premises should be controlled by a single firefighter's switch independent of the switch for any outdoor installation.

537.4.2.2 Every firefighter's switch should comply with the following requirements, where applicable:

(i) for an outdoor installation, the switch should be outside the building and adjacent to the equipment, or alternatively a notice indicating the position of the switch should be placed adjacent to the equipment and a notice should be fixed near the switch so as to render it clearly distinguishable

(ii) for an indoor installation, the switch should be in the main entrance to the building or in another position to be agreed with the local fire authority

(iii) the switch should be placed in a conspicuous position, reasonably accessible to firefighters, at not more than 2.75 m from the ground or the standing beneath the switch.

537.4.3 A firefighter's switch shall be easily visible, accessible and marked to indicate the installation or part of the installation which it controls.

537.4.4 The following information shall be distinctly and durably marked on the firefighter's switch in a position where it can be seen clearly by a person standing on the ground at the intended site, without opening the enclosure and when the switch is installed:

- 'ON' and 'OFF' positions, in letters not less than 10 mm high;
- letters reading 'FIREFIGHTER'S SWITCH' or 'FIRE SWITCH' in letters not less than 10 mm high.

Once installed, the handle off position shall be up.

NOTE: The 'ON' position means powered and the 'OFF' position means unpowered.

538 MONITORING DEVICES

538.1 Insulation monitoring devices (IMDs) for IT systems

538.1.1 General

IMDs shall be in accordance with BS EN 61557-8.

An IMD is intended to permanently monitor the insulation resistance of an IT system and provides an alarm where the insulation resistance R_f is below the response value R_a.

R_a is the response value of the IMD as described in BS EN 61557-8.

R_f is the insulation resistance between the system to which it is connected and either the earth, the PE connection or another reference point for protective equipotential bonding.

Examples of these systems would be an electrical installation, a mobile generator or a safety service.

An IMD shall be installed in an IT system where selected to meet with the requirement of Regulation 411.6.4.

IMDs shall be installed as close as practicable to the origin of the part of the installation to be monitored.

Instructions shall be provided indicating that when the IMD detects an insulation fault to earth, the insulation fault shall be located and eliminated in order to restore normal operating conditions with the shortest practicable delay.

Where the IT system is used for continuity of supply, the occurrence of a first insulation fault shall be indicated at a suitable location so it is audible and/or visible by instructed or skilled person(s).

NOTE: This alarm may be relayed through a building management system (BMS).

It is recommended to use an IMD that signals an interruption of the measurement connections to the system conductors and earth.

538.1.2 Installation of insulation monitoring devices

Where a neutral conductor is distributed, an insulation monitoring device (IMD) may be connected to the neutral conductor. In this case, no overcurrent protective device shall be inserted in the conductor connecting the IMD to the neutral.

An IMD shall be connected symmetrically or unipolarly between the live conductors and earth or the PE connection or another reference point for protective equipotential bonding.

Where the neutral conductor is not distributed, the 'line' terminal of the IMD may be connected:

(i) either to an artificial neutral point with the three impedances connected to line conductors, or

(ii) to a line conductor.

Where, in a polyphase system, the IMD is connected between one line conductor and earth, it shall be suitable to withstand at least the line-to-line voltage between its 'line' terminal and its 'earth' terminal.

NOTE: This voltage appears across these two terminals in the case of a single insulation fault on another line conductor.

For DC installations, the 'line' terminal(s) of the IMD shall be connected either directly to the midpoint, if any, or to one or all of the supply conductors.

The supply circuit of IMD shall be connected either to the installation on the same circuit of the connecting point of the 'line' terminal and as close as possible to the origin of the system, or to an auxiliary source.

The connecting point to the installation shall be selected in such a way that the IMD is able to monitor the insulation of the installation in all operating conditions.

Where the installation is supplied from more than one power supply, connected in parallel, one IMD per supply shall be used, provided they are interlocked in such a way that only one IMD remains connected to the system. All other IMDs monitor the disconnected power supply, enabling the reconnection of this supply without any pre-existing insulation fault.

IMDs shall be able to measure the insulation resistance of the system if DC components caused by electronic equipment, e.g. rectifiers or convertors, are contained in the fault current.

538.1.3 Setting of the insulation monitoring device

The IMD shall have a selection of setting values and be adjusted to suit the respective installation.

When operating normally with the maximum of loads connected, the IMD shall be set to a lower value corresponding to the normal insulation of the system.

NOTE: A value of 100 Ω/V (300 Ω/V for pre-warning) of the rated system voltage is an example of typical setting values.

Where IMDs are installed in locations where ordinary persons have access to their use they shall be selected and/or installed in such a way that it shall be impossible to modify the settings except by the use of a key, a tool or a password.

538.2 Equipment for insulation fault location in IT systems

Equipment for insulation fault location shall be in accordance with BS EN 61557-9. Where an IT system has been selected for continuity of service, it is recommended to combine the IMD with devices enabling the fault location on load. Their function is to indicate the faulty circuit when the IMD has detected an insulation fault.

538.3 Monitoring of off-line circuits

The insulation monitoring of circuits switched off may be carried out in TN, TT and IT systems with insulation monitoring devices (IMDs) provided that the IMD is automatically deactivated whenever the safety equipment is activated. A prerequisite for it is that the monitored electrical circuits are isolated from all poles of the system.

NOTE: As an example, this can be applicable to a circuit comprising safety equipment which is normally de-energized such that the safety equipment is allowed to work without intervention of supply during the emergency.

The reduction of the insulation level shall be indicated locally by either a visual or an audible signal with the option of remote indication.

If the equipment is disconnected from the installation during the off-load insulation measuring process, the insulation levels to be measured are generally very high. The alarm threshold should be above 300 kΩ.

538.4 Residual current monitors (RCMs)

538.4.1 General

RCMs shall comply with BS EN 62020.

An RCM permanently monitors leakage and fault currents to earth of the downstream installation or part thereof and is intended to inform the user about the level of these currents of that part of the installation being monitored.

RCMs are not intended to provide protection against electric shock.

Where an RCD is installed upstream of the RCM, it is recommended that the RCM be set to a residual actuating current no higher than half of the rated residual operating current ($I_{\Delta n}$) of the RCD.

It is recommended that RCMs are installed at the origin of the outgoing circuits.

RCMs shall initiate an audible and/or visual signal, which shall continue as long as the fault persists.

538.4.2 RCMs installed in IT systems

In IT systems where interruption of the supply in case of a first insulation fault to earth is not required or not permitted, RCMs may be installed to indicate the occurrence of a first insulation fault from a live part to exposed-conductive-parts or to earth in accordance with Regulation 411.6.4.

Where used in IT systems, it is recommended to use directionally discriminating RCMs, in order to avoid unwanted signalling of leakage current when high leakage capacitances are liable to exist downstream from the point of installation of the RCM.

ANNEX A53
(Informative)
Device functions and coordination

TABLE A53.1 – Devices and associated functions

Devices					Functions			
Product	**OCPD[2]**	**SCPD[2]**	**RCD[2]**	**Standard**	**Overload protection**	**Short-circuit protection**	**Residual current protection**	**Switching only**
Circuit-breaker	X			BS EN 60947-2 BS EN 60898-1 BS EN 60898-2	Yes	Yes	No	No
RCCB			X	BS EN 61008 series BS EN 62423	No	No	Yes	No
RCBO	X		X	BS EN 61009 series BS EN 62423	Yes	Yes	Yes	No
CBR	X		X	BS EN 60947-2, Annex B	Yes	Yes	Yes	No
MRCD[3]	X		X	BS EN 60947-2, Annex M	Yes	Yes	Yes	No
ICB		X		BS EN 60947-2, Annex O	No	Yes	No	No
Fuse with full range breaking capacity (e.g. gG, gM)[1]	X			BS EN 60269 series	Yes	Yes	No	No
Fuse with partial range breaking capacity (e.g. aM)[1]		X		BS EN 60269 series	No	Yes	No	No
CPS	X			BS EN 60947-6-2	Yes	Yes	No	No
Contactor				BS EN 60947-4-1 BS EN 61095	No	No	No	Yes
Overload relay				BS EN 60947-4-1	Yes	No	No	No
Switch or switch -disconnector				BS EN 60947-3 BS EN 60669-2-2 BS EN 60669-2-4	No	No	No	Yes
TSE				BS EN 60947-6-1	No	No	No	Yes

[1] fuse combination units according to BS EN 60947-3 are considered in this row

[2] generic abbreviations used in this document for devices

[3] when associated with a circuit-breaker.

A53.1 Basis of correct coordination

The basis of the coordination between electrical equipment is to take advantage of the correct combination of their electrical characteristics in order not to impair:

- the safety of an installation (i.e. to avoid equipment combustion due to an electrical fault). Combined short-circuit protection of OCPDs and back-up protection by an OCPD relate to the safety of an installation
- the safety due to continuity of service, if needed (i.e. to restrict disconnection to the faulty circuit in case of overcurrent or fault to earth).

Selectivity between electrical devices provides continuity of service and thus avoids dangers linked to the loss of supply of specific circuits.

A53.2 Parameters

The parameters for the correct coordination between electrical devices may be:

- design current
- prospective short-circuit or fault current
- operating time of devices
- system voltage
- energy (let-through I^2t values)
- peak let-through current.

In addition, for fuses, the following parameters should be considered:

- pre-arcing I^2t and pre-arcing time
- operating I^2t and operating time.

Device coordination table

Table A53.2 shows the types of combination of devices and cells to show how various modes of coordination can affect safety.

TABLE A53.2 – Device coordination in a LV electrical installation

	Regulation	Modes of coordination	Devices involved	Impact on safety of installation	Impact on safety due to continuity of service
Current or time based conditions	536.4.1.2 536.4.1.3	Selectivity	Between OCPDs		Yes
	536.4.1.4		Between RCDs		Yes
	536.4.1.5		Between OCPD and RCD		Yes
	536.4.2.1	Combined short-circuit protection	OCPDs	Yes	
	536.4.2.2	Back-up protection in case of short-circuit	Between OCPD and contactors, overload relays	Yes	
	536.4.2.3		Between OCPD and switched TSE or impulse relays	Yes	
	536.4.2.4		Between OCPD and RCCB	Yes	
	536.4.3.1	Protection in case of overload	Between OCPD and contactor or SCPD	Yes	
	536.4.3.2		Between OCPD and switched TSE or impulse relays	Yes	
	536.4.3.2		Between OCPD and RCCB	Yes	
Voltage based conditions	536.4.4	Selectivity	Between OCPDs equipped with under-voltage relay		Yes

A53.3 Between circuit-breakers or circuit-breaker and CPS or circuit-breaker and overload relay or circuit-breaker and motor starter

Selectivity in case of overload is verified by the comparison of time/current characteristics of the devices involved. The maximum operating time of the device on the load side must be lower than the non-tripping time of the circuit-breaker for any overload current. Separation of the characteristics in both the time and current axes provides selective operation in this zone.

A53.4 Between fuses

Selectivity in case of overload is verified by the comparison of time/current characteristics of the fuses involved. The total operating time of the fuse on the load side must be lower than the pre-arcing time of the fuse on the supply side. Fuses to BS EN 60269-2 of the same utilization category (e.g. type gG) with rated current of 16 A and above will provide total selectivity if the ratio of the rated currents is 1.6:1 or greater. Separation of the characteristics in both the time and current axes provides selective operation in this zone.

A53.5 Between circuit-breaker (upstream) and fuse (downstream)

Selectivity in case of overload is verified by comparison of the time/current characteristics, taking into account the trip setting of the circuit-breaker where applicable. When using published time-current characteristics, the maximum operating time curve must be taken for the downstream device and the minimum operating time curve must be taken for the upstream device. Separation of the characteristics in both the time and current axes provides selective operation in this zone.

A53.6 Between fuse (upstream) and circuit-breaker (downstream)

Selectivity in case of overload is verified by comparison of the time/current characteristics, taking into account the trip setting of the circuit-breaker where applicable. The maximum operating time of the circuit-breaker as given by the manufacturer must be lower than the minimum pre-arcing time of the fuse as given by the product standard. Separation of the characteristics in both the time and current axes provides selective operation in this zone.

A53.7 Between circuit-breakers

Generally, reference should be made to the manufacturers' documentation.

In principle, selectivity is assured up to the fault current level at which the peak current let-through of the downstream circuit-breaker is less than the peak value corresponding to the instantaneous tripping level of the upstream circuit-breaker. Where the upstream circuit-breaker has dedicated selective behaviour, the selectivity limit might be higher.

A53.8 Between fuses

Selectivity in case of short-circuit is verified by comparison of the I^2t values. The maximum operating I^2t value of the fuse on the load side should be lower than the minimum pre-arcing I^2t value of the fuse on the supply side. Fuses to BS EN 60269-1 of the same utilization category (e.g. type gG) with rated current of 16 A and above will provide total selectivity if the ratio of the rated currents is 1.6:1 or greater.

A53.9 Between circuit-breaker (upstream) and fuse (downstream)

The peak let-through current of the fuse should be lower than the minimum instantaneous tripping current of the circuit-breaker.

Data for peak values of fuses should be taken from the relevant standard or the manufacturers' documentation. If data is taken from the manufacturer, this should be stated in the documentation of the installation.

A53.10 Between fuse (upstream) and circuit-breaker (downstream)

The minimum pre-arcing I^2t value of the fuse should be higher than the maximum let-through I^2t value of the circuit-breaker for any short-circuit current up to the maximum prospective short-circuit current to be considered. Data for I^2t values of fuses should be taken from the relevant standard/or the manufacturers' documentation. If data is taken from the manufacturer, this should be stated in the documentation of the installation. The maximum let-through I^2t value of the circuit-breaker should be obtained from the manufacturers' data.

Possible fault currents in systems with semiconductors

In the diagrams of the following Figure A53.1, circuits with most likely fault currents in connection with semiconductor devices are shown.

Fig A53.1 – Possible fault currents in systems with semiconductors

	Circuit diagram with fault location	Shape of load current I_L	Shape of earth fault current I_F	Protection provided by RCD tripping characteristic
1	Phase control 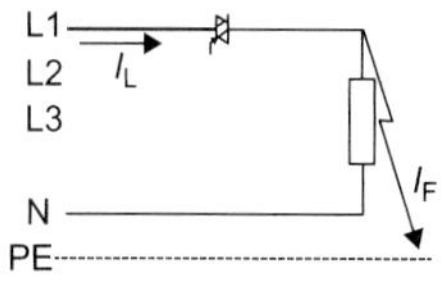	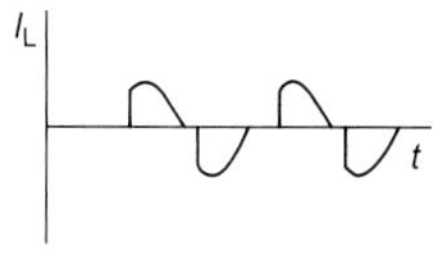	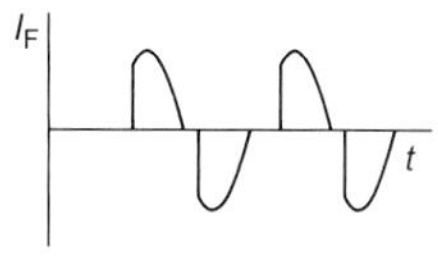	AC, A, F, B
2	Burst control 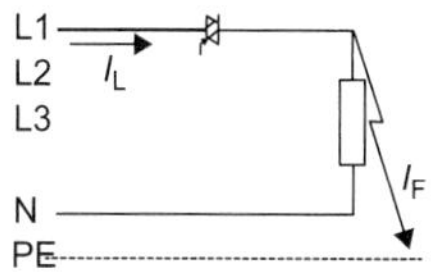	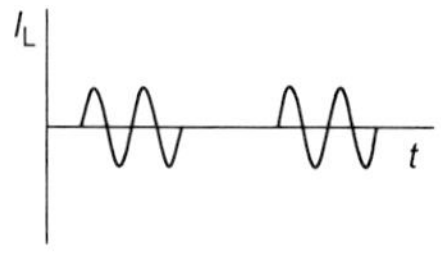	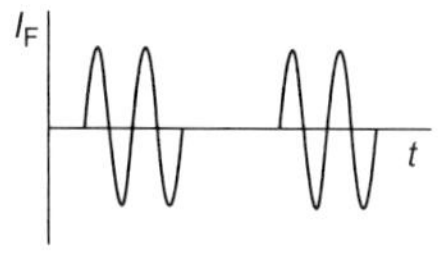	AC, A, F, B
3	Single phase 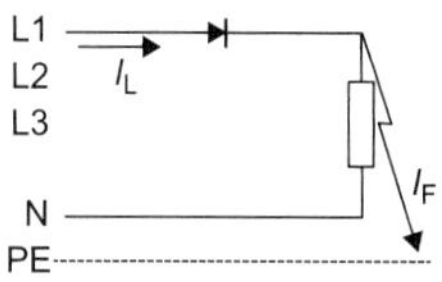	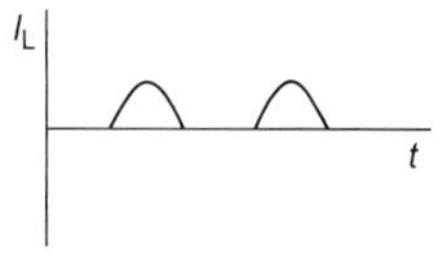	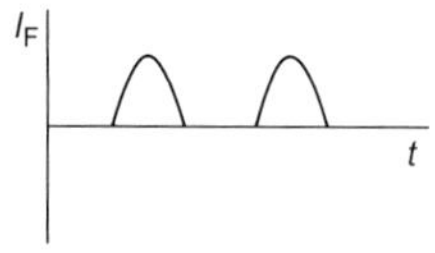	A, F, B
4	Two-pulse bridge 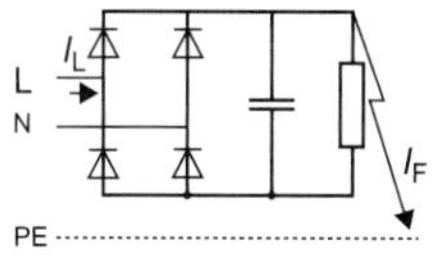	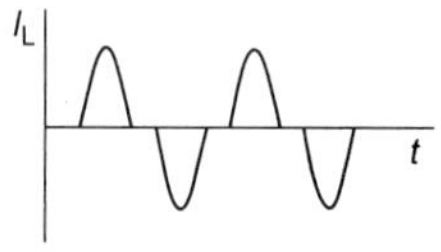	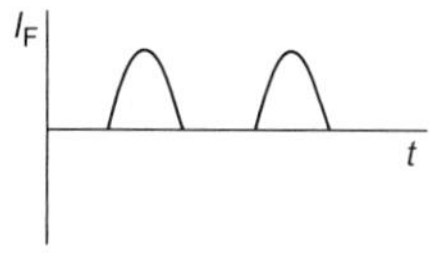	A, F, B
5	Two-pulse bridge, half controlled 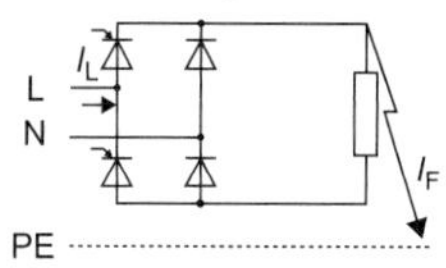	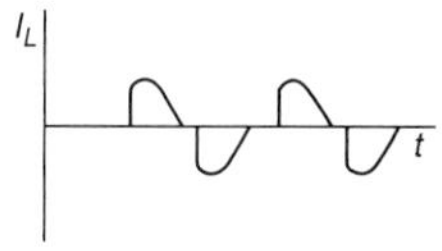	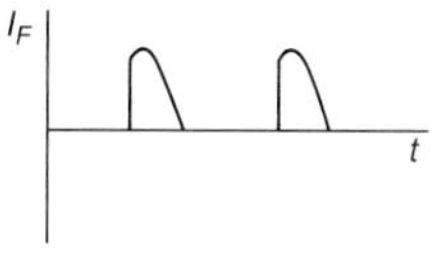	A, F, B
6	Single-phase with smoothing 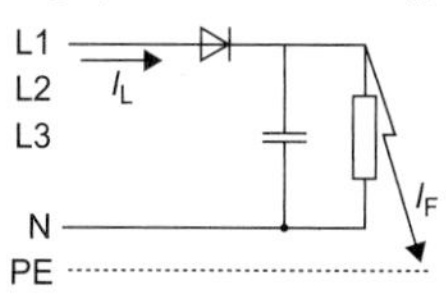	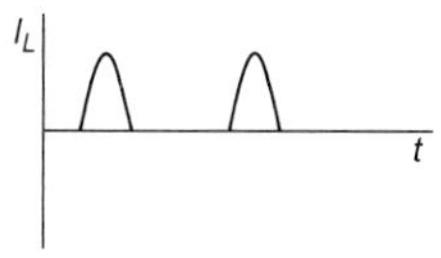	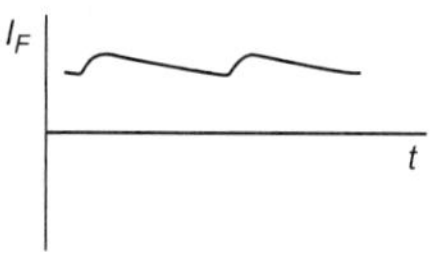	B
7	Two-Pulse bridge between phases 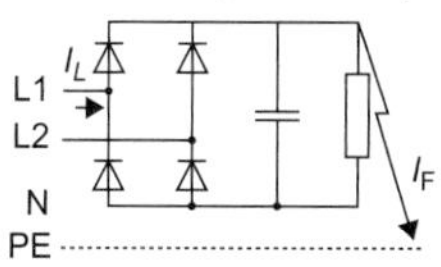	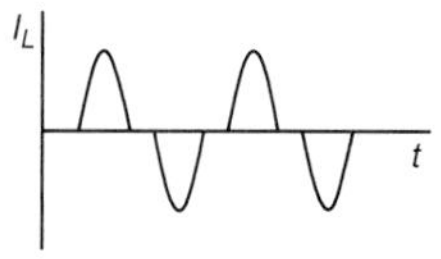	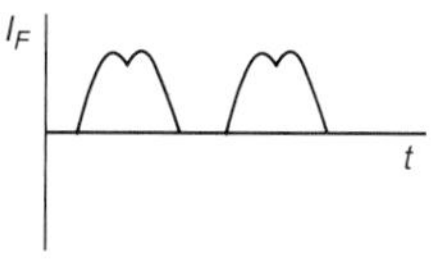	B
8	Three-phase star 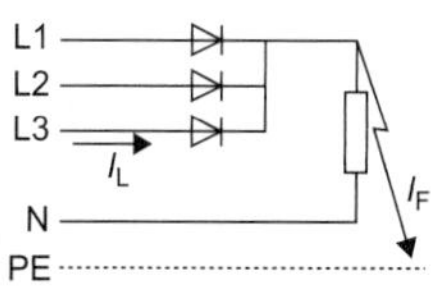	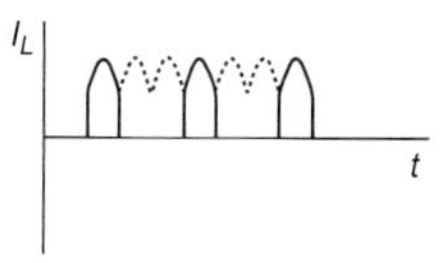	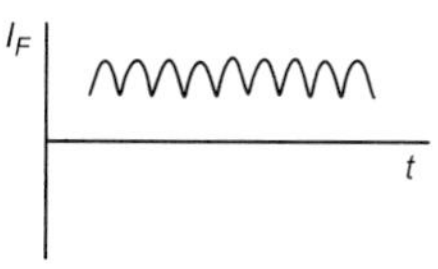	B
9	Six-pulse bridge 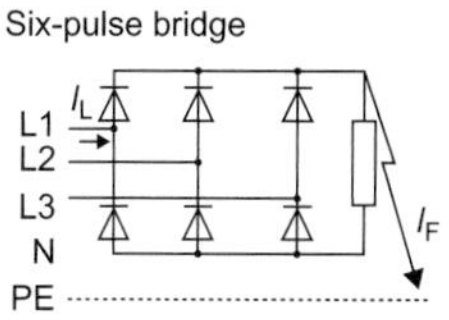	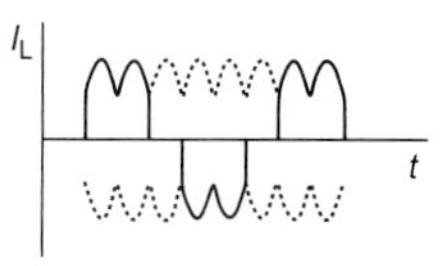	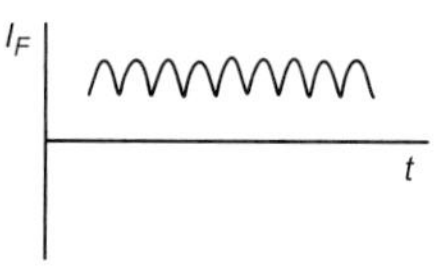	B

CHAPTER 54

EARTHING ARRANGEMENTS AND PROTECTIVE CONDUCTORS

CONTENTS

CHAPTER 54

EARTHING ARRANGEMENTS AND PROTECTIVE CONDUCTORS

541 GENERAL

541.1 Every means of earthing and every protective conductor shall be selected and erected so as to satisfy the requirements of the Regulations.

541.2 The earthing system of an installation may be subdivided, in which case each part thus divided shall comply with the requirements of this chapter.

541.3 Where there is also a lightning protection system, reference shall be made to BS EN 62305.

542 EARTHING ARRANGEMENTS

542.1 General requirements

542.1.1 The earthing arrangements may be used jointly or separately for protective and functional purposes, according to the requirements of the installation.

542.1.201 The main earthing terminal shall be connected with Earth by one of the methods described in Regulations 542.1.2.1 to 542.1.2.3, as appropriate to the type of system of which the installation is to form a part and in compliance with Regulations 542.1.3.1 and 542.1.3.2.

NOTE: Refer to Part 2 and Appendix 9 for definitions of systems.

542.1.2 Supply arrangements

542.1.2.1 For a TN-S system, means shall be provided for the main earthing terminal of the installation to be connected to the earthed point of the source of energy. Part of the connection may be formed by the distributor's lines and equipment.

542.1.2.2 For a TN-C-S system, where protective multiple earthing is provided, means shall be provided for the main earthing terminal of the installation to be connected by the distributor to the neutral of the source of energy.

542.1.2.3 For a TT or IT system, the main earthing terminal shall be connected via an earthing conductor to an earth electrode complying with Regulation 542.2.

542.1.2.4 Where the supply to an installation is at high voltage, protection against faults between the high voltage supply and earth shall be provided in accordance with Section 442.

542.1.3 Installation earthing arrangements

542.1.3.1 The earthing arrangements shall be such that:

(i) the value of impedance from the consumer's main earthing terminal to the earthed point of the supply for TN systems, or to Earth for TT and IT systems, is in accordance with the protective and functional requirements of the installation, and considered to be continuously effective, and

(ii) earth fault currents and protective conductor currents which may occur are carried without danger, particularly from thermal, thermomechanical and electromechanical stresses, and

(iii) they are adequately robust or have additional mechanical protection appropriate to the assessed conditions of external influence.

542.1.3.2 Precautions shall be taken against the risk of damage to other metallic parts through electrolysis.

542.1.3.3 Where a number of installations have separate earthing arrangements, any protective conductors common to any of these installations shall either be capable of carrying the maximum fault current likely to flow through them or be earthed within one installation only and insulated from the earthing arrangements of any other installation. In the latter circumstances, if the protective conductor forms part of a cable, the protective conductor shall be earthed only in the installation containing the associated protective device.

542.2 Earth electrodes

542.2.1 The design used for, and the construction of, an earth electrode shall be such as to withstand damage and to take account of possible increase in resistance due to corrosion.

542.2.2 Suitable earth electrodes shall be used. The following types of earth electrode are recognized for the purposes of the Regulations:

(i) Earth rods or pipes
(ii) Earth tapes or wires
(iii) Earth plates
(iv) Underground structural metalwork embedded in foundations or other metalwork installed in the foundations
(v) Welded metal reinforcement of concrete (except pre-stressed concrete) embedded in the ground
(vi) Lead sheaths and other metal coverings of cables, where not precluded by Regulation 542.2.5
(vii) other suitable underground metalwork.

NOTE: Further information on earth electrodes can be found in BS 7430.

542.2.3 Where foundation earth electrodes are installed, the materials and dimensions of the earth electrodes shall be selected to withstand corrosion and to have adequate mechanical strength.

NOTE: If a lightning protection system (LPS) is present, BS EN 62305-1 applies.

542.2.4 The type and embedded depth of an earth electrode shall be such that soil drying and freezing will not increase its resistance above the required value.

542.2.5 The use, as an earth electrode, of the lead sheath or other metal covering of a cable shall be subject to all of the following conditions:

(i) Adequate precautions shall be taken to prevent excessive deterioration by corrosion
(ii) The sheath or covering shall be in effective contact with Earth
(iii) The consent of the owner of the cable shall be obtained
(iv) Arrangements shall exist for the owner of the electrical installation to be warned of any proposed change to the cable which might affect its suitability as an earth electrode.

542.2.6 A metallic pipe for gases or flammable liquids shall not be used as an earth electrode. The metallic pipe of a water utility supply shall not be used as an earth electrode. Other metallic water supply pipework shall not be used as an earth electrode unless precautions are taken against its removal and it has been considered for such a use.

542.2.7 An earth electrode shall not consist of a metal object immersed in water.

542.2.8 Where an earth electrode consists of parts that must be connected together, connections shall be made by welding, pressure connectors, clamps or other suitable mechanical connectors.

542.3 Earthing conductors

542.3.1 Every earthing conductor shall comply with Section 543 and, where PME conditions apply, shall meet the requirements of Regulation 544.1.1 for the cross-sectional area of a main protective bonding conductor. In addition, where buried in the ground, the earthing conductor shall have a cross-sectional area not less than that stated in Table 54.1. For a tape or strip conductor, the thickness shall be such as to withstand mechanical damage and corrosion.

NOTE: For further information see BS 7430.

TABLE 54.1 –
Minimum cross-sectional area of a buried earthing conductor

	Protected against mechanical damage	Not protected against mechanical damage
Protected against corrosion by a sheath	$2.5\ mm^2$ copper $10\ mm^2$ steel	$16\ mm^2$ copper $16\ mm^2$ coated steel
Not protected against corrosion	$25\ mm^2$ copper $50\ mm^2$ steel	

542.3.2 The connection of an earthing conductor to an earth electrode or other means of earthing shall be soundly made and be electrically and mechanically satisfactory, and labelled in accordance with Regulation 514.13.1(i). It shall be suitably protected against corrosion.

542.4 Main earthing terminals or bars

542.4.1 In every installation a main earthing terminal shall be provided to connect the following to the earthing conductor:

(i) The circuit protective conductors

(ii) The protective bonding conductors

(iii) Functional earthing conductors (if required)

(iv) Lightning protection system bonding conductor, if any (see Regulation 411.3.1.2).

542.4.2 To facilitate measurement of the resistance of the earthing arrangements, means shall be provided in an accessible position for disconnecting the earthing conductor. Such means may conveniently be combined with the main earthing terminal or bar. Any joint shall be capable of disconnection only by means of a tool.

543 PROTECTIVE CONDUCTORS

543.1 Cross-sectional areas

543.1.1 The cross-sectional area of every protective conductor, other than a protective bonding conductor, shall be:

(i) calculated in accordance with Regulation 543.1.3, or

(ii) selected in accordance with Regulation 543.1.4.

Calculation in accordance with Regulation 543.1.3 is necessary if the choice of cross-sectional area of line conductors has been determined by considerations of short-circuit current and if the earth fault current is expected to be less than the short-circuit current.

If the protective conductor:

(iii) is not an integral part of a cable, or

(iv) is not formed by conduit, ducting or trunking, or

(v) is not contained in an enclosure formed by a wiring system,

the cross-sectional area shall be not less than 2.5 mm^2 copper equivalent if protection against mechanical damage is provided, and 4 mm^2 copper equivalent if mechanical protection is not provided (see also Regulation 543.3.1).

For a protective conductor buried in the ground Regulation 542.3.1 for earthing conductors also applies. The cross-sectional area of a protective bonding conductor shall comply with Section 544.

543.1.2 Where a protective conductor is common to two or more circuits, its cross-sectional area shall be:

(i) calculated in accordance with Regulation 543.1.3 for the most onerous of the values of fault current and operating time encountered in each of the various circuits, or

(ii) selected in accordance with Regulation 543.1.4 so as to correspond to the cross-sectional area of the largest line conductor of the circuits.

543.1.3 The cross-sectional area, where calculated, shall be not less than the value determined by the following formula or shall be obtained by reference to BS 7454:

$$S = \frac{\sqrt{I^2 t}}{k}$$

NOTE: This equation is an adiabatic equation and is applicable for disconnection times not exceeding 5s.

where:

S is the nominal cross-sectional area of the conductor in mm^2

I is the value in amperes (rms for AC) of fault current for a fault of negligible impedance, which can flow through the associated protective device, due account being taken of the current limiting effect of the circuit impedances and the limiting capability (I^2t) of that protective device

t is the operating time of the protective device in seconds corresponding to the fault current I amperes

k is a factor taking account of the resistivity, temperature coefficient and heat capacity of the conductor material, and the appropriate initial and final temperatures.

Values of k for protective conductors in various use or service are as given in Tables 54.2 to 54.6. The values are based on the initial and final temperatures indicated in each table.

Where the application of the formula produces a non-standard size, a conductor having a larger standard cross-sectional area shall be used.

TABLE 54.2 –
Values of k for insulated protective conductor not incorporated in a cable and not bunched with cables, or for seperate bare protective conductor in contact with cable covering but not bunched with cables, where the assumed initial temperature is 30 °C

Material of conductor	Insulation of protective conductor or cable covering		
	70 °C thermoplastic	90 °C thermoplastic	90 °C thermosetting
Copper	143/133*	143/133*	176
Aluminium	95/88*	95/88*	116
Steel	52	52	64
Assumed initial temperature	30 °C	30 °C	30 °C
Final temperature	160 °C/140 °C*	160 °C/140 °C*	250 °C

* Above 300 mm^2

TABLE 54.3 –
Values of k for protective conductor incorporated in a cable or bunched with cables, where the assumed initial temperature is 70 °C or greater

Material of conductor	Insulation material		
	70 °C thermoplastic	90 °C thermoplastic	90 °C thermosetting
Copper	115/103*	100/86*	143
Aluminium	76/68*	66/57*	94
Assumed initial temperature	70 °C	90 °C	90 °C
Final temperature	160 °C/140 °C*	160 °C/140 °C*	250 °C

* Above 300 mm^2

TABLE 54.4 –
Values of k for the sheath or armour of a cable as the protective conductor

Material of conductor	Insulation material		
	70 °C thermoplastic	90 °C thermoplastic	90 °C thermosetting
Aluminium	93	85	85
Steel	51	46	46
Lead	26	23	23
Assumed initial temperature	60 °C	80 °C	80 °C
Final temperature	200 °C	200 °C	200 °C

TABLE 54.5 –
Values of k for steel conduit, ducting and trunking as the protective conductor

Material of protective conductor	Insulation material		
	70 °C thermoplastic	90 °C thermoplastic	90 °C thermosetting
Steel conduit, ducting and trunking	47	44	58
Assumed initial temperature	50 °C	60 °C	60 °C
Final temperature	160 °C	160 °C	250 °C

TABLE 54.6 –
Values of k for bare conductor where there is no risk of damage to any neighbouring material by the temperatures indicated

NOTE: The temperatures indicated are valid only where they do not impair the quality of the connections

Material of conductor	Conditions		
	Visible and in restricted areas	Normal conditions	Fire risk
Copper	228	159	138
Aluminium	125	105	91
Steel	82	58	50
Assumed initial temperature	30 °C	30 °C	30 °C
Final temperature			
Copper conductor	500 °C	200 °C	150 °C
Aluminium conductor	300 °C	200 °C	150 °C
Steel conductor	500 °C	200 °C	150 °C

543.1.4 Where it is desired not to calculate the minimum cross-sectional area of a protective conductor in accordance with Regulation 543.1.3, the cross-sectional area may be determined in accordance with Table 54.7. Where the application of Table 54.7 produces a non-standard size, a conductor having a larger standard cross-sectional area shall be used.

TABLE 54.7 –
Minimum cross-sectional area of protective conductor in relation to the cross-sectional area of associated line conductor

Cross-sectional area of line conductor S	Minimum cross-sectional area of the corresponding protective conductor	
	If the protective conductor is of the same material as the line conductor	If the protective conductor is not of the same material as the line conductor
(mm^2)	(mm^2)	(mm^2)
$S \leq 16$	S	$\frac{k_1}{k_2} \times S$
$16 < S \leq 35$	16	$\frac{k_1}{k_2} \times 16$
$S > 35$	$\frac{S}{2}$	$\frac{k_1}{k_2} \times \frac{S}{2}$

where:

k_1 is the value of k for the line conductor, selected from Table 43.1 in Chapter 43 according to the materials of both conductor and insulation.

k_2 is the value of k for the protective conductor, selected from Tables 54.2 to 54.6, as applicable.

543.2 Types of protective conductor

543.2.1 A protective conductor may consist of one or more of the following:

(i) A single-core cable
(ii) A conductor in a cable
(iii) An insulated or bare conductor in a common enclosure with insulated live conductors
(iv) A fixed bare or insulated conductor
(v) A metal covering, for example, the sheath, screen or armouring of a cable
(vi) A metal conduit, metallic cable management system or other enclosure or electrically continuous support system for conductors
(vii) an extraneous-conductive-part complying with Regulation 543.2.6.

543.2.2 Where a metal enclosure or frame of a low voltage switchgear or controlgear assembly or busbar trunking system is used as a protective conductor, it shall satisfy the following three requirements:

(i) Its electrical continuity shall be assured, either by construction or by suitable connection, in such a way as to be protected against mechanical, chemical or electrochemical deterioration
(ii) Its cross-sectional area shall be at least equal to that resulting from the application of Regulation 543.1, or verified by test in accordance with the appropriate part of BS EN 61439 series
(iii) It shall permit the connection of other protective conductors at every predetermined tap-off point.

543.2.3 A gas pipe, an oil pipe, flexible or pliable conduit, support wires or other flexible metallic parts, or constructional parts subject to mechanical stress in normal service, shall not be selected as a protective conductor.

543.2.4 A protective conductor of the types described in items (i) to (iv) of Regulation 543.2.1 and of cross-sectional area 10 mm^2 or less, shall be of copper.

543.2.5 The metal covering including the sheath (bare or insulated) of a cable, in particular the sheath of a mineral insulated cable, trunking and ducting for electrical purposes and metal conduit, may be used as a protective conductor for the associated circuit, if it satisfies both requirements of items (i) and (ii) of Regulation 543.2.2.

543.2.6 Except as prohibited by Regulation 543.2.3, an extraneous-conductive-part may be used as a protective conductor if it satisfies all the following requirements:

(i) Electrical continuity shall be assured, either by construction or by suitable connection, in such a way as to be protected against mechanical, chemical or electrochemical deterioration
(ii) The cross-sectional area shall be at least equal to that resulting from the application of Regulation 543.1.1
(iii) Unless compensatory measures are provided, precautions shall be taken against its removal
(iv) It has been considered for such a use and, if necessary, suitably adapted.

543.2.7 Where the protective conductor is formed by metal conduit, trunking or ducting or the metal sheath and/or armour of a cable, the earthing terminal of each accessory shall be connected by a separate protective conductor to an earthing terminal incorporated in the associated box or other enclosure.

543.2.8 *Not used*

543.2.9 Except where the circuit protective conductor is formed by a metal covering or enclosure containing all of the conductors of the ring, the circuit protective conductor of every ring final circuit shall also be run in the form of a ring having both ends connected to the earthing terminal at the origin of the circuit.

543.2.10 A separate metal enclosure for cable shall not be used as a PEN conductor.

543.3 Preservation of electrical continuity of protective conductors

543.3.1 A protective conductor shall be suitably protected against mechanical and chemical deterioration and electrodynamic effects.

543.3.2 Every connection and joint shall be accessible for inspection, testing and maintenance except as provided by Regulation 526.3.

543.3.201 A protective conductor having a cross-sectional area up to and including 6 mm^2 shall be protected throughout by a covering at least equivalent to that provided by the insulation of a single-core non-sheathed cable of appropriate size having a voltage rating of at least 450/750 V, except for the following:

(i) A protective conductor forming part of a multicore cable

(ii) A metal conduit, metallic cable management system or other enclosure or electrically continuous support system for conductors, where used as a protective conductor.

Where the sheath of a cable incorporating an uninsulated protective conductor of cross-sectional area up to and including 6 mm^2 is removed adjacent to joints and terminations, the protective conductor shall be protected by insulating sleeving complying with BS EN 60684 series.

543.3.3.101 No switching device shall be inserted in a protective conductor, except:

(i) as permitted by Regulation 537.1.5

(ii) a multipole, linked switching device in which the protective conductor circuit is not interrupted before the live conductors and is re-established not later than when the live conductors are reconnected

(iii) a switching device interlocked with a multipole, linked switching device inserted in the live conductors such that the protective conductor circuit shall not be interrupted before the live conductors and shall be re-established not later than when the live conductors are reconnected, or

(iv) a multipole plug-in device in which the protective conductor circuit shall not be interrupted before the live conductors and shall be re-established not later than when the live conductors are reconnected.

Switching devices provided in accordance with (i), (ii), (iii) and (iv) shall meet the requirements of Chapter 46 and Section 537 for a device for isolation.

Joints for test purposes that can be disconnected only by the use of a tool may be inserted in a protective conductor.

543.3.4 Where electrical earth monitoring is used, no dedicated devices (e.g. operating sensors, coils) shall be connected in series with the protective conductor (see BS 4444).

543.3.5 An exposed-conductive-part of equipment shall not be used to form a protective conductor for other equipment except as provided by Regulations 543.2.1, 543.2.2 and 543.2.5.

543.3.6 Every joint in metallic conduit shall be mechanically and electrically continuous.

543.4 Combined protective and neutral (PEN) conductors

543.4.1 PEN conductors shall not be used within an installation except as permitted by Regulation 543.4.2.

NOTE: Regulation 8(4) of the Electricity Safety, Quality and Continuity Regulations (ESQCR) prohibits the use of PEN conductors in consumers' installations.

543.4.2 The provisions of Regulations 543.4.3 to 543.4.8 may be applied only:

(i) where any necessary authorization for use of a PEN conductor has been obtained and where the installation complies with the conditions for that authorization, or

(ii) where the installation is supplied by a privately owned transformer or convertor in such a way that there is no metallic connection (except for the earthing connection) with the distributor's network, or

(iii) where the supply is obtained from a private generating plant.

543.4.3 If, from any point of the installation, the neutral and protective functions are provided by separate conductors, those conductors shall not then be reconnected together beyond that point. At the point of separation, separate terminals or bars shall be provided for the protective and neutral conductors. The PEN conductor shall be connected to the terminals or bar intended for the protective earthing conductor and the neutral conductor. The conductance of the terminal link or bar shall be not less than that specified in Regulation 543.4.5.

543.4.4 The outer conductor of a concentric cable shall not be common to more than one circuit. This requirement does not preclude the use of a twin or multicore cable to serve a number of points contained within one final circuit.

543.4.5 The conductance of the outer conductor of a concentric cable (measured at a temperature of 20 °C) shall:

(i) for a single-core cable, be not less than that of the internal conductor

(ii) for a multicore cable serving a number of points contained within one final circuit or having the internal conductors connected in parallel, be not less than that of the internal conductors connected in parallel.

543.4.6 At every joint in the outer conductor of a concentric cable and at a termination, the continuity of that joint shall be supplemented by a conductor additional to any means used for sealing and clamping the outer conductor. The conductance of the additional conductor shall be not less than that specified in Regulation 543.4.5 for the outer conductor.

543.4.7 No means of isolation or switching shall be inserted in the outer conductor of a concentric cable.

543.4.8 Excepting a cable to BS EN 60702-1 installed in accordance with the manufacturer's instructions, the PEN conductor of every cable shall be insulated or have an insulating covering suitable for the highest voltage to which it may be subjected.

543.4.201 For a fixed installation, a conductor of a cable not subject to flexing and having a cross-sectional area not less than 10 mm^2 for copper or 16 mm^2 for aluminium may serve as a PEN conductor provided that the part of the installation concerned is not supplied through an RCD.

543.5 Earthing arrangements for combined protective and functional purposes

543.5.1 Where earthing for combined protective and functional purposes is required, the requirements for protective measures shall take precedence.

543.6 Earthing arrangements for protective purposes

543.6.1 Where overcurrent protective devices are used for fault protection, the protective conductor shall be incorporated in the same wiring system as the live conductors or in their immediate proximity.

543.7 Earthing requirements for the installation of equipment having high protective conductor currents

543.7.1 General

543.7.1.201 *Deleted by BS 7671:2018+A2:2022*

543.7.1.202 Equipment having a protective conductor current exceeding 10 mA shall be connected to the supply by one of the following methods:

(i) Permanently connected to the wiring of the installation, with the protective conductor selected in accordance with Regulation 543.7.1.203.

(ii) A flexible cable with a plug and socket-outlet complying with BS EN 60309-2, provided that the protective conductor of the associated flexible cable is of a cross-sectional area not less than 2.5 mm^2 for plugs rated at 16 A and not less than 4 mm^2 for plugs rated above 16 A.

(iii) A protective conductor complying with Section 543 with an earth monitoring system to BS 4444 installed which, in the event of a continuity fault occurring in the protective conductor, automatically disconnects the supply to the equipment.

543.7.1.203 The wiring of every final circuit and distribution circuit intended to supply one or more items of equipment, such that the total protective conductor current is likely to exceed 10 mA, shall have a high integrity protective connection complying with one or more of the following:

(i) A single protective conductor having a cross-sectional area of not less than 10 mm^2, complying with the requirements of Regulations 543.2 and 543.3

(ii) A single copper protective conductor having a cross-sectional area of not less than 4 mm^2, complying with the requirements of Regulations 543.2 and 543.3, the protective conductor being enclosed to provide additional protection against mechanical damage, for example, within a flexible conduit

(iii) Two individual protective conductors, each complying with the requirements of Section 543. The two protective conductors may be of different types, e.g. a metal conduit together with an additional conductor of a cable enclosed in the same conduit.

Where the two individual protective conductors are both incorporated in a multicore cable, the total cross-sectional area of all the conductors including the live conductors shall be not less than 10 mm^2. One of the protective conductors may be formed by the metallic sheath, armour or wire braid screen incorporated in the construction of the cable and complying with Regulation 543.2.5

(iv) An earth monitoring system to BS 4444 may be installed which, in the event of a continuity fault occurring in the protective conductor, automatically disconnects the supply to the equipment

(v) Connection of the equipment to the supply by means of a double-wound transformer or equivalent unit, such as a motor-alternator set, the protective conductor of the incoming supply being connected to the exposed-conductive-parts of the equipment and to a point of the secondary winding of the transformer or equivalent device. The protective conductor(s) between the equipment and the transformer or equivalent device shall comply with one of the arrangements described in (i) to (iv) above.

543.7.1.204 *Deleted by BS 7671:2018+A2:2022*

543.7.1.205 At the distribution board information shall be provided indicating those circuits having a high protective conductor current. This information shall be positioned so as to be visible to a person who is modifying or extending the circuit.

543.7.2 Socket-outlet final circuits

543.7.2.201 For a final circuit with a number of socket-outlets or connection units intended to supply two or more items of equipment, where it is known or reasonably to be expected that the total protective conductor current in normal service will exceed 10 mA, the circuit shall be provided with a high integrity protective conductor connection complying with the requirements of Regulation 543.7.1. The following arrangements of the final circuit are acceptable:

(i) A ring final circuit with a ring protective conductor. Spurs, if provided, require high integrity protective conductor connections complying with the requirements of Regulation 543.7.1

(ii) A radial final circuit with a single protective conductor:

(a) the protective conductor being connected as a ring, or

(b) a separate protective conductor being provided at the final socket-outlet by connection to the metal conduit or ducting.

(iii) Other circuits complying with the requirements of Regulation 543.7.1.

544 PROTECTIVE BONDING CONDUCTORS

544.1 Main protective bonding conductors

544.1.1 Except where PME conditions apply, a main protective bonding conductor shall have a cross-sectional area not less than half the cross-sectional area required for the earthing conductor of the installation. Where an installation serves more than one building, a main protective bonding conductor shall be selected in accordance with the characteristics of the distribution circuit protective conductor for that particular building. The cross-sectional area shall be not less than 6 mm^2, and need not exceed 25 mm^2 if the bonding conductor is of copper or a cross-sectional area affording equivalent conductance in other metals.

Except for highway power supplies and street furniture, where PME conditions apply the main protective bonding conductor shall be selected in accordance with the PEN conductor of the supply and Table 54.8.

Where an installation has more than one source of supply to which PME conditions apply, a main protective bonding conductor shall be selected according to the largest PEN conductor of the supply.

TABLE 54.8 –
Minimum cross-sectional area of the main protective bonding conductor in relation to the PEN conductor of the supply

NOTE: Local distributor's network conditions may require a larger conductor.

Copper equivalent cross-sectional area of the PEN conductor	Minimum copper equivalent* cross-sectional area of the main protective bonding conductor
35 mm^2 or less	10 mm^2
over 35 mm^2 up to 50 mm^2	16 mm^2
over 50 mm^2 up to 95 mm^2	25 mm^2
over 95 mm^2 up to 150 mm^2	35 mm^2
over 150 mm^2	50 mm^2

* The minimum copper equivalent cross-sectional area is given by a copper bonding conductor of the tabulated cross-sectional area or a bonding conductor of another metal affording equivalent conductance.

544.1.2 The main protective bonding connection to any extraneous-conductive-part such as gas, water or other metallic pipework or service shall be made as near as practicable to the point of entry of that part into the premises. Where there is a meter, isolation point or union, the connection shall be made to the consumer's hard metal pipework and before any branch pipework. Where practicable the connection shall be made within 600 mm of the meter outlet union or at the point of entry to the building if the meter is external.

544.2 Supplementary bonding conductors

544.2.1 A supplementary bonding conductor connecting two exposed-conductive-parts shall have a conductance, if sheathed or otherwise provided with mechanical protection, not less than that of the smaller protective conductor connected to the exposed-conductive-parts. If mechanical protection is not provided, its cross-sectional area shall be not less than 4 mm^2.

544.2.2 A supplementary bonding conductor connecting an exposed-conductive-part to an extraneous-conductive-part shall have a conductance, if sheathed or otherwise provided with mechanical protection, not less than half that of the protective conductor connected to the exposed-conductive-part. If mechanical protection is not provided, its cross-sectional area shall be not less than 4 mm^2.

544.2.3 A supplementary bonding conductor connecting two extraneous-conductive-parts shall have a cross-sectional area not less than 2.5 mm^2 if sheathed or otherwise provided with mechanical protection or 4 mm^2 if mechanical protection is not provided, except that where one of the extraneous-conductive-parts is connected to an exposed-conductive-part in compliance with Regulation 544.2.2, that regulation shall apply also to the conductor connecting the two extraneous-conductive-parts.

544.2.4 Except where Regulation 544.2.5 applies, supplementary bonding shall be provided by a supplementary conductor, a conductive part of a permanent and reliable nature, or by a combination of these.

544.2.5 Where supplementary bonding is to be applied to a fixed appliance which is supplied via a short length of flexible cable from an adjacent connection unit or other accessory, incorporating a flex outlet, the circuit protective conductor within the flexible cable shall be deemed to provide the supplementary bonding connection to the exposed-conductive-parts of the appliance, from the earthing terminal in the connection unit or other accessory.

CHAPTER 55

OTHER EQUIPMENT

CONTENTS

CHAPTER 55

OTHER EQUIPMENT

550.1 Scope

Chapter 55 specifies requirements and recommendations for the selection and erection of low voltage electrical equipment not covered by other parts of BS 7671 except Part 7 and Chapter 82, intended to be part of the fixed installation.

551 LOW VOLTAGE GENERATING SETS

551.1 Scope

This section applies to low voltage and extra-low voltage installations which incorporate generating sets intended to supply, either continuously or occasionally, all or part of the installation. Requirements are included for installations with the following supply arrangements:

(i) Supply to an installation which is not connected to a system for distribution of electricity to the public
(ii) Supply to an installation as an alternative to a system for distribution of electricity to the public
(iii) Supply to an installation in parallel with a system for distribution of electricity to the public
(iv) Appropriate combinations of the above.

This section does not apply to self-contained items of extra-low voltage electrical equipment which incorporate both the source of energy and the energy-using load and for which a specific product standard exists that includes the requirements for electrical safety.

NOTE: Where a generating set with an output not exceeding 16 A is to be connected in parallel with a system for distribution of electricity to the public, procedures for informing the electricity distributor are given in the Electricity Safety, Quality and Continuity Regulations (ESQCR). In addition to the ESQCR requirements, where a generating set with an output exceeding 16 A is to be connected in parallel with a system for distribution of electricity to the public, requirements of the electricity distributor should be ascertained before the generating set is connected. Requirements of the distributor for the connection of units rated up to 16 A are given in BS EN 50549-1.

551.1.1 Generating sets with the following power sources are considered:

(i) Combustion engines
(ii) Turbines
(iii) Electric motors
(iv) Photovoltaic cells
(v) Batteries
(vi) Other suitable sources.

551.1.2 Generating sets with the following electrical characteristics are considered:

(i) Mains-excited and separately excited synchronous generators
(ii) Mains-excited and self-excited asynchronous generators
(iii) Mains-commutated and self-commutated static convertors with or without bypass facilities.

551.1.3 The use of generating sets for the following purposes is considered:

(i) Supply to permanent installations
(ii) Supply to temporary installations
(iii) Supply to mobile equipment which is not connected to a permanent fixed installation
(iv) Supply to mobile units (Section 717 also applies).

551.2 General requirements

551.2.1 The means of excitation and commutation shall be appropriate for the intended use of the generating set and the safety and proper functioning of other sources of supply shall not be impaired by the generating set.

551.2.2 The prospective short-circuit current and prospective earth fault current shall be assessed for each source of supply or combination of sources which can operate independently of other sources or combinations. The short-circuit rating of protective devices within the installation and, where appropriate, connected to a system for distribution of electricity to the public, shall not be exceeded for any of the intended methods of operation of the sources.

551.2.3 Where the generating set is intended to provide a supply to an installation which is not connected to a system for distribution of electricity to the public or to provide a supply as a switched alternative to such a system, the capacity and operating characteristics of the generating set shall be such that danger or damage to equipment does not arise after the connection or disconnection of any intended load as a result of the deviation of the voltage or frequency from the intended operating range. Means shall be provided to automatically disconnect such parts of the installation as may be necessary if the capacity of the generating set is exceeded.

NOTE 1: Consideration should be given to the intended duty cycle and size of individual connected loads as a proportion of the capacity of the generating set and to the starting characteristics of any connected electric motors.

NOTE 2: Consideration should be given to the power factor specified for protective devices in the installation.

NOTE 3: The installation of a generating set within an existing building or installation may change the conditions of external influence for the installation (see Part 3), for example by the introduction of moving parts, parts at high temperature or by the presence of flammable fluids and noxious gases, etc.

551.2.4 Provision for isolation shall meet the requirements of Chapter 46 and Section 537 for each source or combination of sources of supply.

551.3 Protective measure: Extra-low voltage provided by SELV or PELV

551.3.1 Additional requirements for SELV and PELV where the installation is supplied from more than one source

Where a SELV or PELV system may be supplied by more than one source, the requirements of Regulation 414.3 shall apply to each source. Where one or more of the sources is earthed, the requirements for PELV systems in Regulation 414.4 shall apply.

If one or more of the sources does not meet the requirements of Regulation 414.3, the system shall be treated as a FELV system and the requirements of Regulation 411.7 shall apply.

551.3.2 Where it is necessary to maintain the supply to an extra-low voltage system following the loss of one or more sources of supply, each source of supply or combination of sources of supply which can operate independently of other sources or combinations shall be capable of supplying the intended load of the extra-low voltage system. Provisions shall be made so that loss of the low voltage supply to an extra-low voltage source does not lead to danger or damage to other extra-low voltage equipment.

NOTE: Such provisions may be necessary in supplies for safety services (see Chapter 56).

551.4 Fault protection

551.4.1 Fault protection shall be provided for the installation in respect of each source of supply or combination of sources of supply that can operate independently of other sources or combinations of sources.

The fault protective provisions shall be selected or precautions shall be taken so that where fault protective provisions are achieved in different ways within the same installation or part of an installation according to the active sources of supply, no influence shall occur or conditions arise that could impair the effectiveness of the fault protective provisions.

NOTE: This might, for example, require the use of a transformer providing electrical separation between parts of the installation using different earthing systems.

551.4.2 The generating set shall be connected so that any provision within the installation for protection by RCDs in accordance with Chapter 41 remains effective for every intended combination of sources of supply.

NOTE: Connection of live parts of the generator with Earth may affect the protective measure.

551.4.3 Protection by automatic disconnection of supply

551.4.3.1 Protection by automatic disconnection of supply shall be provided in accordance with Section 411, except as modified for particular cases by Regulation 551.4.3.2, 551.4.3.3 or 551.4.4.

551.4.3.2 Additional requirements for installations where the generating set provides a switched alternative to the system for distribution of electricity to the public (standby systems)

551.4.3.2.1 Protection by automatic disconnection of supply shall not rely upon the connection to the earthed point of the system for distribution of electricity to the public when the generator is operating as a switched alternative to a TN system. A suitable means of earthing shall be provided.

551.4.3.3 Additional requirements for installations incorporating static convertors

551.4.3.3.1 Where fault protection for parts of the installation supplied by the static convertor relies upon the automatic closure of the bypass switch and the operation of protective devices on the supply side of the bypass switch is not within the time required by Section 411, supplementary equipotential bonding shall be provided between simultaneously accessible exposed-conductive-parts and extraneous-conductive-parts on the load side of the static convertor in accordance with Regulation 415.2.

The resistance (R) of the supplementary protective bonding conductor required between simultaneously accessible exposed-conductive-parts and extraneous-conductive-parts shall fulfil the following condition:

$$R \leq 50/I_a$$

where:

I_a is the maximum earth fault current which can be supplied by the static convertor when the bypass switch is closed.

NOTE: Where such equipment is intended to operate in parallel with a system for distribution of electricity to the public, the requirements of Regulation 551.7 also apply.

551.4.3.3.2 Precautions shall be taken or equipment shall be selected so that the correct operation of protective devices is not impaired by DC currents generated by a static convertor or by the presence of filters.

551.4.3.3.3 A means of isolation shall be installed on both sides of a static convertor. This requirement does not apply on the power source side of a static convertor which is integrated in the same enclosure as the power source.

551.4.4 Additional requirements for protection by automatic disconnection where the installation and generating set are not permanently fixed

This regulation applies to portable generating sets and to generating sets which are intended to be moved to unspecified locations for temporary or short-term use. Such generating sets may be part of an installation which is subject to similar use. This regulation does not apply to permanent fixed installations.

NOTE: For suitable connection arrangements see BS EN 60309 series.

551.4.4.1 Between separate items of equipment, protective conductors shall be provided which are part of a suitable cable and which comply with Table 54.7.

All protective conductors shall comply with the relevant requirements of Chapter 54. |

551.4.4.2 In a TN, TT or IT system, every final circuit shall be provided with additional protection by means of an RCD having the characteristics specified in Regulation 415.1.

NOTE: In an IT system, an RCD may not operate unless one of the earth faults is on a part of the system on the supply side of the device.

551.5 Protection against overcurrent

551.5.1 Where overcurrent protection of the generating set is required, it shall be located as near as practicable to the generator terminals.

NOTE: The contribution to the prospective short-circuit current by a generating set may be time-dependent and may be much less than the contribution made by a system for distribution of electricity to the public.

551.5.2 Where a generating set is intended to operate in parallel with a system for distribution of electricity to the public, or where two or more generating sets may operate in parallel, circulating harmonic currents shall be limited so that the thermal rating of conductors is not exceeded.

The effects of circulating harmonic currents may be limited by one or more of the following:

(i) The selection of generating sets with compensated windings

(ii) The provision of a suitable impedance in the connection to the generator star points

(iii) The provision of switches which interrupt the circulatory circuit but which are interlocked so that at all times fault protection is not impaired

(iv) The provision of filtering equipment

(v) Other suitable means.

NOTE: Consideration should be given to the maximum voltage which may be produced across an impedance connected to limit circulating harmonic currents.

551.6 Additional requirements for installations where the generating set provides a supply as a switched alternative to the system for distribution of electricity to the public (standby systems)

551.6.1 Precautions complying with the relevant requirements of Chapter 46 and Section 537 for isolation shall be taken so that the generator cannot operate in parallel with the system for distribution of electricity to the public. Suitable precautions may include one or more of the following:

(i) An electrical, mechanical or electromechanical interlock between the operating mechanisms or control circuits of the changeover switching devices

(ii) A system of locks with a single transferable key

(iii) A three-position break-before-make changeover switch

(iv) An automatic changeover switching device with a suitable interlock

(v) Other means providing equivalent security of operation.

551.6.2 For a TN-S system where the neutral is not isolated, any RCD shall be positioned to avoid incorrect operation due to the existence of any parallel neutral-earth path.

NOTE: It may be desirable in a TN system to disconnect the neutral of the installation from the neutral or PEN of the system for distribution of electricity to the public to avoid disturbances such as induced voltage surges caused by lightning.

551.7 Additional requirements for installations where the generating set may operate in parallel with other sources including systems for distribution of electricity to the public

551.7.1 When a generating set is used as an additional source of supply in parallel with another source, both of the following conditions shall be fulfilled:

(i) Protection against thermal effects in accordance with Chapter 42 and protection against overcurrent in accordance with Chapter 43 shall remain effective in all situations

(ii) Where an RCD is providing additional protection in accordance with Regulation 415.1 for a circuit connecting the generator set to the installation, the RCD shall disconnect all live conductors, including the neutral conductor.

551.7.2 A generating set used as an additional source of supply in parallel with another source shall be installed:

- on the supply side of all the overcurrent protective devices for the final circuits of the installation, or
- on the load side of all the overcurrent protective devices for a final circuit of the installation, but in this case all the following additional requirements shall be fulfilled:

  (i) The conductors of the final circuit shall meet the following requirement:

  $I_z \geq I_n + I_g$

  where:

  I_z is the current-carrying capacity of the final circuit conductors

  I_n is the rated current or current setting of the protective device of the final circuit

  I_g is the rated output current of the generating set

  (ii) A generating set shall not be connected to a final circuit by means of a plug and socket-outlet

  (iii) The line and neutral conductors of the final circuit and of the generating set shall not be connected to Earth

  (iv) Unless the device providing automatic disconnection of the final circuit in accordance with Regulation 411.3.2 disconnects the line and neutral conductors, it shall be verified that the combination of the disconnection time of the protective device for the final circuit and the time taken for the output voltage of the generating set to reduce to 50 V or less is not greater than the disconnection time required by Regulation 411.3.2 for the final circuit.

Where the generating set is connected via a low voltage switchgear and controlgear assembly then:

(v) The assembly shall be selected such that:

$I_{nA} \geq I_n + I_{g(s)}$

where:

I_{nA} is the rated current of the assembly

I_n is the rated current or current setting of the incoming circuit overcurrent protective device either incorporated within the low voltage switchgear and controlgear assembly or upstream of it

$I_{g(s)}$ is the rated output current of the generating set or sets.

This regulation does not apply to a static convertor provided to supply specific items of current-using equipment within the final circuit to which it is connected.

551.7.3 In selecting and using a generating set to run in parallel with the system for distribution of electricity to the public, care shall be taken to avoid adverse effects to that system and to other installations in respect of power factor, voltage changes, harmonic distortion, unbalance, starting, synchronizing or voltage fluctuation effects. Where synchronization is necessary, the use of an automatic synchronizing system which considers frequency, phase sequence and voltage is to be preferred.

551.7.4 Means of automatic switching shall be provided to disconnect the generating set from the system for distribution of electricity to the public in the event of loss of that supply or deviation of the voltage or frequency at the supply terminals from declared values.

For a generating set with an output exceeding 16 A, the type of protection and the sensitivity and operating times depend upon the protection of the system for distribution of electricity to the public and the number of generating sets connected and shall be agreed by the distributor. For a generating set with an output not exceeding 16 A, the settings shall comply with BS EN 50549-1.

In the case of the presence of a static convertor, the means of switching shall be provided on the load side of the static convertor.

551.7.5 Means shall be provided to prevent the connection of a generating set to the system for distribution of electricity to the public in the event of loss of that supply or deviation of the voltage or frequency at the supply terminals from values required by Regulation 551.7.4.

NOTE: For a generating set with an output not exceeding 16 A intended to operate in parallel with a system for distribution of electricity to the public the requirements are given in BS EN 50549-1.

551.7.6 Means shall be provided to enable the generating set to be isolated from the system for distribution of electricity to the public. For a generating set with an output exceeding 16 A, the accessibility of this means of isolation shall comply with national rules and distribution system operator requirements. For a generating set with an output not exceeding 16 A, the accessibility of this means of isolation shall comply with BS EN 50549-1.

551.7.7 Where a generating set may operate as a switched alternative to the system for distribution of electricity to the public, the installation shall also comply with Regulation 551.6.

551.8 Requirements for installations incorporating stationary batteries

551.8.1 Stationary batteries shall be installed so that they are accessible only to skilled or instructed persons.

NOTE: This generally requires the battery to be installed in a secure location or, for smaller batteries, a secure enclosure.

The location or enclosure shall be adequately ventilated.

551.8.2 Battery connections shall have basic protection by insulation or enclosures or shall be arranged so that two bare conductive parts having between them a potential difference exceeding 120 volts cannot be inadvertently touched simultaneously.

552 ROTATING MACHINES

NOTE: See also Regulation 463.3 Motor control.

552.1 Rotating machines

552.1.1 All equipment, including cable, of every circuit carrying the starting, accelerating and load currents of a motor shall be suitable for a current at least equal to the full-load current rating of the motor when rated in accordance with the appropriate British or Harmonized Standard. Where the motor is intended for intermittent duty and for frequent starting and stopping, account shall be taken of any cumulative effects of the starting or braking currents upon the temperature rise of the equipment of the circuit.

552.1.2 Every electric motor having a rating exceeding 0.37 kW shall be provided with control equipment incorporating means of protection against overload of the motor. This requirement does not apply to a motor incorporated in an item of current-using equipment complying as a whole with an appropriate British or Harmonized Standard.

552.1.3 Except where failure to start after a brief interruption would be likely to cause greater danger, every motor shall be provided with means to prevent automatic restarting after a stoppage due to a drop in voltage or failure of supply, where unexpected restarting of the motor might cause danger. This requirement does not preclude arrangements for starting a motor at intervals by an automatic control device, where other adequate precautions are taken against danger from unexpected restarting.

553 ACCESSORIES

553.1 Plugs and socket-outlets

553.1.1 Every plug and socket-outlet shall comply with all the requirements of items (i) and (ii) below and, in addition, with the appropriate requirements of Regulations 553.1.2 to 553.2.2:

(i) Except for SELV circuits, it shall not be possible for any pin of a plug to make contact with any live contact of its associated socket-outlet while any other pin of the plug is completely exposed

(ii) It shall not be possible for any pin of a plug to make contact with any live contact of any socket-outlet within the same installation other than the type of socket-outlet for which the plug is designed.

553.1.2 Except for SELV or a special circuit from Regulation 553.1.5, every plug and socket-outlet shall be of the non-reversible type, with provision for the connection of a protective conductor.

553.1.3 Except where Regulation 553.1.5 applies, in a low voltage circuit every plug and socket-outlet shall conform with the applicable British Standard listed in Table 55.1.

TABLE 55.1 –
Plugs and socket-outlets for low voltage circuits

Type of plug and socket-outlet	Rating (amperes)	Applicable British Standard
Fused plugs and shuttered socket-outlets, 2-pole and earth, for AC	13	BS 1363 (fuses to BS 1362)
Plugs, fused or non-fused, and socket-outlets, 2-pole and earth	2, 5, 15, 30	BS 546 (fuses, if any, to BS 646)
Plugs and socket-outlets (industrial type)	16, 32, 63, 125	BS EN 60309-2

553.1.201 Every socket-outlet for household and similar use shall be of the shuttered type and, for an AC installation, shall preferably be of a type complying with BS 1363.

NOTE: USB charging ports, coaxial outlets, data outlets, HDMI and other AV outlets are not regarded as socket-outlets for the purposes of this regulation.

553.1.5 A plug and socket-outlet not complying with BS 1363, BS 546 or BS EN 60309-2, may be used in single-phase AC or two-wire DC circuits operating at a nominal voltage not exceeding 250 volts for:

(i) the connection of an electric clock, provided that the plug and socket-outlet are designed specifically for that purpose, and that each plug incorporates a fuse of rating not exceeding 3 amperes complying with BS 646 or BS 1362 as appropriate

(ii) the connection of an electric shaver, provided that the socket-outlet is either incorporated in a shaver supply unit complying with BS EN 61558-2-5 or is a type complying with BS 4573

(iii) a circuit having special characteristics such that danger would otherwise arise or it is necessary to distinguish the function of the circuit.

553.1.6 A socket-outlet on a wall or similar structure shall be mounted at a height above the floor or any working surface to minimize the risk of mechanical damage to the socket-outlet or to an associated plug and its flexible cable which might be caused during insertion, use or withdrawal of the plug.

553.1.7 Where mobile equipment is likely to be used, provision shall be made so that the equipment can be fed from an adjacent and conveniently accessible socket-outlet, taking account of the length of flexible cable normally fitted to portable appliances and luminaires.

553.2 Cable couplers

553.2.1 Except for a SELV or a Class II circuit, a cable coupler shall comply where appropriate with BS 6991, BS EN 61535, BS EN 60309-2 or BS EN 60320-1, shall be non-reversible and shall have provision for the connection of a protective conductor.

553.2.2 A cable coupler shall be arranged so that the connector of the coupler is fitted at the end of the cable remote from the supply.

554 CURRENT-USING EQUIPMENT

554.1 Electrode water heaters and boilers

554.1.1 Every electrode water heater and electrode boiler shall be connected to an AC system only, and shall be selected and erected in accordance with the appropriate requirements of this section.

554.1.2 The supply to the electrode water heater or electrode boiler shall be controlled by a linked circuit-breaker arranged to disconnect the supply from all electrodes simultaneously and provided with an overcurrent protective device in each conductor feeding an electrode.

554.1.3 The earthing of the electrode water heater or electrode boiler shall comply with the requirements of Chapter 54 and, in addition, the shell of the electrode water heater or electrode boiler shall be bonded to the metallic sheath and armour, if any, of the incoming supply cable. The protective conductor shall be connected to the shell of the electrode water heater or electrode boiler and shall comply with Regulation 543.1.1.

554.1.4 Where an electrode water heater or electrode boiler is directly connected to a supply at a voltage exceeding low voltage, the installation shall include an RCD arranged to disconnect the supply from the electrodes on the occurrence of a sustained earth leakage current in excess of 10 % of the rated current of the electrode water heater or electrode boiler under normal conditions of operation, except that if in any instance a higher value is essential for stability of operation of the electrode water heater or electrode boiler, the value may be increased to a maximum of 15 %. A time delay may be incorporated in the device to prevent unnecessary operation in the event of imbalance of short duration.

554.1.5 Where an electrode water heater or electrode boiler is connected to a three-phase low voltage supply, the shell of the electrode water heater or electrode boiler shall be connected to the neutral of the supply as well as to the earthing conductor. The current-carrying capacity of the neutral conductor shall be not less than that of the largest line conductor connected to the equipment.

554.1.6 Except as provided by Regulation 554.1.7, where the supply to an electrode water heater or electrode boiler is single-phase and one electrode is connected to a neutral conductor earthed by the distributor, the shell of the electrode water heater or electrode boiler shall be connected to the neutral of the supply as well as to the earthing conductor.

554.1.7 Where the electrode water heater or electrode boiler is not piped to a water supply or in physical contact with any earthed metal, and where the electrodes and the water in contact with the electrodes are so shielded in insulating material that they cannot be touched while the electrodes are live, a fuse in the line conductor may be substituted for the circuit-breaker required under Regulation 554.1.2 and the shell of the electrode water heater or electrode boiler need not be connected to the neutral of the supply.

554.2 Heaters for liquids or other substances having immersed heating elements

554.2.1 Every heater for liquid or other substance shall incorporate or be provided with an automatic device to prevent a dangerous rise in temperature.

554.3 Water heaters having immersed and uninsulated heating elements

554.3.1 Every single-phase water heater or boiler having an uninsulated heating element immersed in the water shall comply with the requirements of Regulations 554.3.2 and 554.3.3. This type of water heater or boiler is deemed not to be an electrode water heater or electrode boiler.

554.3.2 All metal parts of the heater or boiler which are in contact with the water (other than current-carrying parts) shall be solidly and metallically connected to a metal water pipe through which the water supply to the heater or boiler is provided, and that water pipe shall be connected to the main earthing terminal by means independent of the circuit protective conductor.

554.3.3 The heater or boiler shall be permanently connected to the electricity supply through a double-pole linked switch which is either separate from and within easy reach of the heater or boiler or is incorporated therein, and the wiring from the heater or boiler shall be connected directly to that switch without the use of a plug and socket-outlet.

554.3.4 Before a heater or boiler of the type referred to in Regulation 554.3.1 is connected, the installer shall confirm that no single-pole switch, non-linked circuit-breaker or fuse is fitted in the neutral conductor in any part of the circuit between the heater or boiler and the origin of the installation.

554.4 Heating conductors and cables

NOTE: For electric floor and ceiling heating systems in buildings the requirements of Section 753 must also be met.

554.4.1 Where a heating cable is required to pass through, or be in close proximity to, material which presents a fire hazard, the cable shall be enclosed in material having the ignitability characteristic 'P' as specified in BS 476-12 and shall be adequately protected from any mechanical damage reasonably foreseeable during installation and use.

555 TRANSFORMERS

555.1 Autotransformers and step-up transformers

555.1.1 Where an autotransformer is connected to a circuit having a neutral conductor, the common terminal of the winding shall be connected to the neutral conductor.

555.1.2 A step-up autotransformer shall not be connected to an IT system.

555.1.3 Where a step-up transformer is used, a linked switch shall be provided for disconnecting the transformer from all live conductors of the supply.

556 NOT USED

557 AUXILIARY CIRCUITS

557.1 Scope

This section applies to auxiliary circuits, except those covered by specific product or system standards, e.g. the construction of assemblies of electrical equipment to the appropriate part of the BS EN 61439 series.

557.2 Not used

557.3 Requirements for auxiliary circuits

557.3.1 General

The power supply, AC or DC, for an auxiliary circuit may be either dependent or independent of the main circuit according to its required function. If the status of the main circuit has to be signalled, then the signalling circuit shall be able to operate independently of that main circuit.

557.3.201 Control circuits

A control circuit shall be designed, arranged and protected to limit dangers resulting from a fault between the control circuit and other conductive parts liable to cause malfunction (e.g. inadvertent operation) of the controlled equipment.

557.3.2 Power supply for auxiliary circuits dependent on the main circuit

557.3.2.1 General

Auxiliary circuits with a power supply dependent on the main AC circuit shall be connected to the main circuit:

(i) directly (see Figure 55.1), or

(ii) via a rectifier (see Figure 55.2), or

(iii) via a transformer (see Figure 55.3).

It is recommended that auxiliary circuits supplying primarily electronic equipment or systems should not be supplied directly but at least via simple separation from the main circuit.

Fig 55.1 – Auxiliary circuit supplied directly from the main circuit

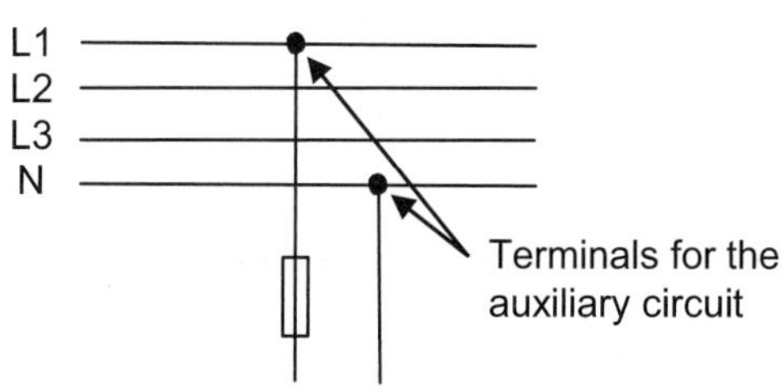

Fig 55.2 – Auxiliary circuit supplied from the main circuit via a rectifier

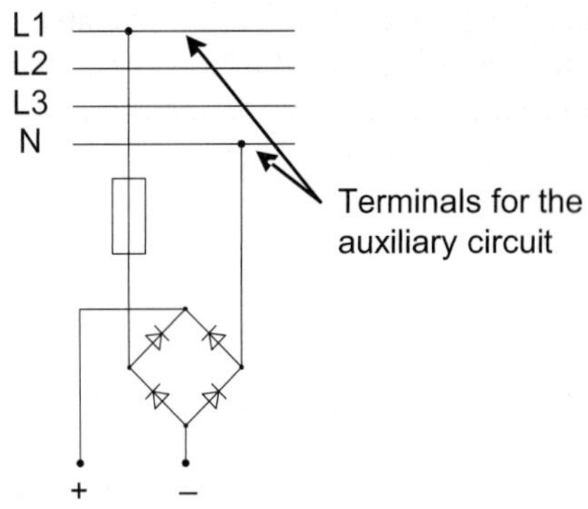

Fig 55.3 – Auxiliary circuit supplied from the main circuit via a transformer

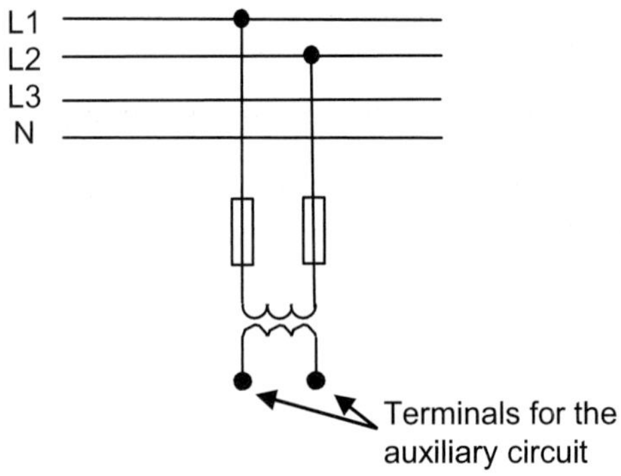

NOTE: For an auxiliary circuit supplied from the main circuit, either directly or via a rectifier, the auxiliary circuit begins at the connection point to the main circuit, see Figures 55.1 and 55.2. In the case of a transformer supply, the auxiliary circuit begins on the secondary side of the transformer, see Figure 55.3.

557.3.2.2 Auxiliary circuit supplied from the main circuit via transformer

Where an auxiliary circuit is supplied by more than one transformer, they shall be connected in parallel both on the primary and secondary sides.

557.3.3 Auxiliary circuit supplied by an independent source

Where an independent source is used a loss of supply or undervoltage of the main circuit source shall be detected. An independent auxiliary circuit shall not create a hazardous situation.

NOTE: Batteries and a power supply system independent of the mains are examples of independent sources.

557.3.4 Auxiliary circuits with or without connection to earth

557.3.4.1 General

An auxiliary circuit shall comply with the earthing requirements of BS 7671 except as modified by Regulation 557.3.4.2 or 557.3.4.3.

NOTE 1: It depends on the requirements for an auxiliary circuit as to whether it is operated earthed or unearthed. For example, in an earthed auxiliary circuit an earth fault in a non-earthed conductor leads to disconnection of the power supply of the auxiliary circuit; in an unearthed auxiliary circuit, an earth fault in a conductor leads only to a signal from the IMD (see Regulation 557.3.4.3).

NOTE 2: The use of unearthed auxiliary circuits should be considered where high reliability is required.

557.3.4.2 Earthed auxiliary circuit

An earthed auxiliary circuit supplied via a transformer shall be connected to earth only at one point on the secondary side of the transformer. The connection to earth shall be situated close to the transformer. The connection shall be easily accessible and capable of being isolated for insulation measurement.

557.3.4.3 Unearthed auxiliary circuit

Except in the case of a SELV or PELV circuit, if an auxiliary circuit is operated unearthed via a transformer, an insulation monitoring device (IMD) according to BS EN 61557-8 shall be installed on the secondary side. Consideration shall be given to the use of risk assessment to determine whether the signal from the IMD is to initiate an acoustic and/or a flash alarm, or be transmitted to a monitoring system.

557.3.5 Power supplies for auxiliary circuits

557.3.5.1 General

The rated voltage of an auxiliary circuit, and the components used in the circuit, shall be compatible with the supply to that circuit.

NOTE 1: If the supply voltage is too low for the design of the circuit then the operation will not be reliable for the proper function of relays.

Account shall be taken of the effects of voltage drop on the electrical equipment of the auxiliary circuit, including inrush and starting currents.

NOTE 2: Motors starting direct-on-line can draw a starting current from the main circuit that will reduce the supply voltage to an auxiliary circuit dependent on the main circuit, below the minimum operating voltage of the associated switchgear.

557.3.5.2 Standby power supply or power supply for safety services

Where a standby power supply or a power supply from a generating set is used to supply auxiliary circuits, the frequency variation shall be taken into account.

557.3.5.3 AC supply

The nominal voltage of control circuits shall preferably not exceed:

(i) 230 V for circuits with 50 Hz nominal frequency

(ii) 277 V for circuits with 60 Hz nominal frequency

respectively, taking into account voltage tolerances according to BS EN 60038.

NOTE: The dimensioning of cable length with respect to the conductor capacitances, e.g. connection to a limit switch, needs to be coordinated with the selected relays or solenoid valves. The standing voltage caused by high conductor capacitances may impair the switching off of relays or solenoid valves.

557.3.5.4 DC supply

557.3.5.4.1 Supply by a power system

The nominal voltage of control circuits shall preferably not exceed 220 V.

557.3.5.4.2 Supply by batteries

Where batteries are used as a power supply for auxiliary circuits, the voltage fluctuation due to charging or discharging shall not exceed voltage tolerances specified in BS EN 60038, unless the auxiliary circuit is specifically designed to compensate for such voltage fluctuation.

557.3.6 Protective measures

557.3.6.1 Protection of wiring systems

NOTE 1: In the case of extended auxiliary circuits it is necessary to check that the required operating current of the protective device will be achieved also at the far end of the respective cables or conductors, Regulation 433.1.

Single-phase earthed AC auxiliary circuits supplied on the secondary side of the transformer for an auxiliary supply and earthed DC auxiliary circuits are permitted to be protected by single-pole devices. The protective devices shall only be inserted in conductors which are not connected directly to earth.

In unearthed AC or DC auxiliary circuits, short-circuit protection shall be provided for all line conductors.

NOTE 2: The use of protective devices which disconnect all lines of an unearthed auxiliary circuit will aid fault diagnosis and maintenance activities.

If the short-circuit protective device on the primary side of the transformer for an auxiliary circuit is selected so that it also protects against short-circuit current on the secondary side, a protective device on the secondary side of the transformer may be omitted.

NOTE 3: For a fault on the transformer secondary side, the magnitude of the short-circuit current on the primary side depends also on the impedance of the transformer.

557.3.6.2 Protection against short-circuit

Switching contacts of electrical switching devices of the auxiliary circuit shall be protected against damage caused by short-circuit currents, according to the manufacturer's instructions.

557.4 Characteristics of cables and conductors

557.4.1 Minimum cross-sectional areas

In order to provide adequate mechanical strength, the following minimum cross-sectional areas indicated in Table 55.2 shall be met. If there are special mechanical strength requirements for cables or conductors, then a larger cross-sectional area of conductor, selected in accordance with Chapter 52, may be required.

TABLE 55.2 – Minimum cross-sectional area of copper conductors in mm²

Application	Type of cable				
	Single-core		Two-core		Multicore
	Single-wire	Stranded	Screened	Unscreened	Screened or unscreened
Control circuits[a]	0.5	0.5	0.5	0.5	0.1
Data transfer	–	–	–	–	0.1

[a] Other auxiliary circuits may need a larger cross-sectional area of copper conductor, e.g. for measuring.

NOTE: The cross-sectional area of copper conductors is derived from Section 524.

557.5 Requirements for auxiliary circuits used for measurement

557.5.1 General

Measuring circuits are auxiliary circuits with dedicated requirements which are given in the following subclauses.

557.5.2 Auxiliary circuits for direct measurement of electrical quantities

An auxiliary circuit for direct measurement of electrical quantities shall be protected against the effects of a fault by one of the following means:

(i) Provision of a device for protection against fault current in accordance with Section 434. Where operation of the device could cause danger, or lead to a hazardous situation, such operation shall also cause disconnection of the main circuit

(ii) Simultaneous fulfilment of conditions (a) and (b) of Regulation 434.3.

557.5.3 Auxiliary circuits for measurement of electrical quantities via a transformer

557.5.3.1 Current transformer

Where a measurement device is connected to the main circuit via a current transformer, the following requirements shall be taken into account:

(i) The secondary side of the transformer in a low voltage installation shall not be earthed, except where the measurement can only be carried out with a connection to earth

(ii) Protective devices interrupting the circuit shall not be used on the secondary side of the transformer

(iii) Conductors on the secondary side of the transformer shall be insulated for the highest voltage of any live parts or shall be installed such that their insulation cannot come into contact with other live parts, e.g. contact with busbars

(iv) Terminals for temporary measurements shall be provided.

The above requirements do not apply to summation current transformers where hazardous voltages do not occur, e.g. equipment for insulation fault location according to BS EN 61557-9.

557.5.3.2 Voltage transformer

The secondary side of a voltage transformer shall be protected by a short-circuit protective device.

557.6 Functional considerations

557.6.1 Voltage supply

Where loss of voltage, i.e. voltage fluctuation, overvoltage or undervoltage, could cause the auxiliary circuit to be unable to perform its intended function, means to secure the continued operation of the auxiliary circuit shall be provided.

557.6.2 Quality of signals depending on the cable characteristics

The operation of an auxiliary circuit shall not be adversely affected by the characteristics, including impedance and length, of the cable between operational components.

The capacitance of the cable shall not impair the proper operation of an actuator in the auxiliary circuit. The cable characteristics and length shall be taken into account for the selection of switchgear and controlgear or electronic circuits.

NOTE: For an extensive auxiliary circuit, the use of a DC power supply or bus-system is recommended.

557.6.3 Measures to avoid the loss of functionality

An auxiliary circuit serving a special function where reliability is a concern will require additional design considerations to minimize the likelihood of wiring faults. These wiring faults could result in loss of function and/or loss of signal. Among the design considerations are:

(i) selection of appropriate installation methods for cables

(ii) selection of equipment where a fault from line to exposed-conductive-parts is not possible, e.g. Class II equipment

(iii) use of installation methods and equipment that are inherently short-circuit and earth fault proof.

For the design of installations and equipment that are inherently short-circuit and earth fault proof, the following shall be considered:

(a) Arrangements of single conductors with basic insulation, together with measures to prevent mutual contact and contact with exposed-conductive-parts. This may be achieved by:

- installation in (insulated) cable trunking systems, or
- installation in (insulated) conduit

(b) Arrangements of:

- single-core cables, or
- single-core, non-metallic-sheathed cables, or
- rubber-insulated flexible cables

(c) Provision of protection against mechanical damage and of safe distance from flammable material for non-metallic sheathed cables

(d) Arrangements of non-metallic-sheathed cables with nominal voltage U_0/U at least 0.6/1 kV (U_0 = conductor to Earth voltage, U = conductor to conductor voltage)

(e) Use of cables with an insulation which is self-extinguishing and flame-retardant

(f) Use of cables that are afforded physical protection by being buried, e.g. installation of cables in soil or concrete.

557.6.4 Current-limiting signal outputs

In earthed or unearthed auxiliary circuits with current-limiting signal outputs or electronically controlled protection against short-circuit conditions, respectively, the signal circuit shall be disconnected within 5 s if the respective measure operates. In special cases, a shorter disconnection time may be required.

For current-limiting signal outputs or electronically controlled protection of the signal output, respectively, automatic disconnection of supply may be omitted if a hazardous situation is not likely to occur.

557.6.5 Connection to the main circuit

557.6.5.1 Auxiliary circuits without direct connection to the main circuit

Electrical actuators, e.g. actuating relays, contactors, signalling lights, electromagnetic locking devices, shall be connected to the common conductor (see Figure 55.4):

(i) in earthed auxiliary circuits, at the earthed (common) conductor

(ii) in unearthed auxiliary circuits, at the common conductor.

Fig 55.4 – Configuration of an auxiliary circuit

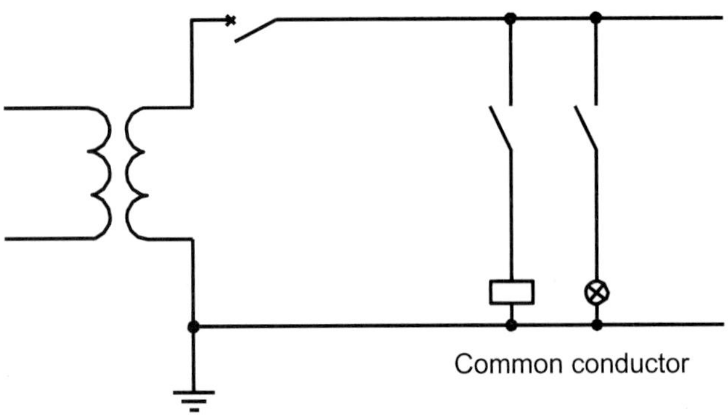

Exception: Switching elements of protective relays, e.g. overcurrent relays, may be installed between the earthed or the non-earthed conductor and a coil, provided that:

(a) this connection is contained inside a common enclosure, or

(b) it leads to a simplification of external control devices, e.g. conductor bars, cable drums, multiple connectors, and takes into account the requirements of Regulation 557.3.6.2.

557.6.5.2 Auxiliary circuits with direct connection to the main circuit

If the auxiliary circuit is:

(i) supplied between two line conductors (e.g. L1 and L2 of an IT system), two-pole switching contacts shall be used

(ii) connected to the earthed neutral of the main circuit, the requirements of Chapter 43 apply.

557.6.6 Plug-in connections

Interchangeability between multiple plug-in connections is permitted only where it will not result in mechanical damage or introduce a risk of fire, electric shock or injury to persons.

NOTE 1: These plug-in connections form a part of the auxiliary circuit(s) and may conduct different signals.

NOTE 2: Protection against interchangeability may be achieved by marking, polarization, design or electronic interlocking.

The connectors shall be secured by a means to prevent unintended disconnection.

558 NOT USED

559 LUMINAIRES AND LIGHTING INSTALLATIONS

559.1 Scope

This section applies to the selection and erection of luminaires and lighting installations intended to be part of the fixed installation.

NOTE 1: For outdoor lighting installations, extra-low voltage lighting installations and lighting installations in special locations, refer to Part 7.

The requirements of this section do not apply to:

(i) high voltage signs supplied at low voltage (such as neon tubes)

(ii) signs and luminous discharge tube installations operating from a no-load rated output voltage exceeding 1 kV but not exceeding 10 kV (BS EN 50107)

(iii) temporary festoon lighting.

NOTE 2: The requirements for high voltage signs are given in BS 559 and the BS EN 50107 series.

559.2 *Not used*

559.3 General requirements for installations

NOTE: See Table 55.3 for an explanation of the symbols used in luminaires, in controlgear for luminaires and in the installation of luminaires.

559.3.1 Every luminaire shall comply with the relevant standard for manufacture and test of that luminaire and shall be selected and erected to take account of the manufacturer's instructions.

559.3.2 For the purposes of this section, luminaires without transformers or convertors but which are fitted with extra-low voltage lamps connected in series to a low voltage supply shall be considered as low voltage equipment not extra-low voltage equipment. These luminaires shall be either Class I or Class II equipment.

559.3.3 Where a luminaire is installed in a pelmet, there shall be no adverse effects due to the presence or operation of curtains or blinds.

559.3.4 A track system for luminaires shall comply with the requirements of BS EN 60570.

559.4 Protection of the surroundings against thermal effects

559.4.1 General

In the selection and erection of a luminaire the thermal effects of radiant and convected energy on the surroundings shall be taken into account, including:

(i) the maximum permissible power dissipated by the lamps

(ii) the fire-resistance of adjacent material

- at the point of installation, and
- in the thermally affected areas

(iii) the minimum distance to combustible materials, including material in the path of a spotlight beam

(iv) the relevant markings on the luminaire.

559.5 Wiring systems

559.5.1 Connection to the fixed wiring

At each fixed lighting point one of the following shall be used for the termination of the wiring system:

(i) A ceiling rose complying with BS 67

(ii) A luminaire supporting coupler (LSC) complying with BS 6972 or BS 7001

(iii) A batten lampholder or a pendant set complying with BS EN 60598

(iv) A luminaire complying with BS EN 60598

(v) A suitable socket-outlet complying with BS 1363-2, BS 546 or BS EN 60309-2

(vi) A plug-in lighting distribution unit complying with BS 5733

(vii) A connection unit complying with BS 1363-4

(viii) Appropriate terminals enclosed in a box complying with the relevant part of BS EN 60670 series or BS 4662

(ix) A device for connecting a luminaire (DCL) outlet complying with BS EN 61995-1

(x) An installation coupler complying with BS EN 61535.

NOTE: In suspended ceilings one plug-in lighting distribution unit may be used for a number of luminaires.

559.5.1.201 A ceiling rose or lampholder shall not be installed in any circuit operating at a voltage normally exceeding 250 volts.

559.5.1.202 A ceiling rose shall not be used for the attachment of more than one outgoing flexible cable unless it is specially designed for multiple pendants.

559.5.1.203 Luminaire supporting couplers and devices for the connection of luminaires are designed specifically for the electrical connection of luminaires and shall not be used for the connection of any other equipment.

559.5.1.204 Lighting circuits incorporating B15, B22, E14, E27 or E40 lampholders shall be protected by an overcurrent protective device of maximum rating 16 A.

559.5.1.205 Bayonet lampholders B15 and B22 shall comply with BS EN 61184 and shall have the temperature rating T2 described in that standard.

559.5.1.206 In circuits of a TN or TT system, except for E14 and E27 lampholders complying with BS EN 60238, the outer contact of every Edison screw or single centre bayonet cap type lampholder shall be connected to the neutral conductor. This regulation also applies to track mounted systems.

559.5.1.207 A lighting installation shall be appropriately controlled.

NOTE: See Table 537.4 for guidance on the selection of suitable protective, isolation and switching devices.

559.5.1.208 Consideration shall be given to the provision of the neutral conductor, at each switch position, to facilitate the installation of electronic switching devices.

559.5.2 Fixing of luminaires

Adequate means to fix luminaires shall be provided.

The fixing means may be mechanical accessories (e.g. hooks or screws), boxes or enclosures which are able to support luminaires or supporting devices for connecting a luminaire.

In places where the fixing means is intended to support a luminaire, the fixing means shall be capable of carrying a mass of not less than 5 kg. If the mass of the luminaire is greater than 5 kg, a fixing means capable of supporting the mass of the luminaire shall be installed.

The installation of the fixing means shall be in accordance with the manufacturer's instructions.

The weight of luminaires and their eventual accessories, e.g. shades, shall be compatible with the mechanical capability of the ceiling or suspended ceiling or supporting structure where installed.

Any flexible cable between the fixing means and the luminaire shall be installed so that any expected stresses in the conductors, terminals and terminations will not impair the safety of the installation. (See also Table 4F3A of Appendix 4.)

559.5.3 Through wiring

559.5.3.1 The installation of through wiring in a luminaire is only permitted if the luminaire is designed for such wiring.

559.5.3.2 A cable for through wiring shall be selected in accordance with the temperature information on the luminaire or on the manufacturer's instruction sheet, if any, as follows:

(i) For a luminaire complying with BS EN 60598 but with temperature marking, cables suitable for the marked temperature shall be used

(ii) Unless specified in the manufacturer's instructions, for a luminaire complying with BS EN 60598 but with no temperature marking, heat-resistant cables are not required

(iii) In the absence of information, heat-resistant cables and/or insulated conductors of type H05S-U, H05S-K, H05SJ-K, H05SS-K (BS EN 50525 series) or equivalent shall be used.

559.5.4 Devices for connection of luminaires to the supply

If the luminaire does not provide a connecting device for connection to the supply, the connecting device shall be:

(i) terminals according to BS EN 60998, or

(ii) a device for connecting a luminaire (DCL) plug according to BS EN 61995-1, or

(iii) an installation coupler according to BS EN 61535, or

(iv) supporting coupler (LSC) plug according to BS 6972 or BS 7001, or

(v) a male connector (plug) of a plug-in lighting distribution unit according to BS 5733, or

(vi) another suitable and appropriate connecting device.

NOTE: For the installation of supply cables, see also Regulation 522.2.201.

559.5.5 Groups of luminaires

Groups of luminaires divided between the line conductors of a polyphase circuit with only one common neutral conductor shall be provided with at least one device disconnecting simultaneously all line conductors.

NOTE: See also Section 537.

559.5.6 Protection against heat and UV radiation effects within luminaires

External cables and cores of cables connected within a luminaire or passing through shall be so selected and erected that they will not suffer damage or deterioration due to heat and UV radiation generated by the luminaire or its lamps (e.g. shielding of the cable from heat and/or UV by means of heat/UV-resistant sleeving).

559.6 Independent lamp controlgear, e.g. ballasts

Only independent lamp controlgear marked as suitable for independent use, according to the relevant standard, shall be used external to a luminaire.

Only the following are permitted to be mounted on a flammable surface:

(i) A 'class P' thermally protected ballast/transformer.

(ii) A temperature declared thermally protected ballast/transformer.

NOTE: For an explanation of symbols used see Table 55.3.

559.7 Compensation capacitors

Compensation capacitors having a total capacitance exceeding 0.5 μF shall only be used in conjunction with discharge resistors. Capacitors and their marking shall be in accordance with BS EN 61048.

This requirement does not apply to capacitors forming part of the equipment.

559.8 Protection against electric shock for display stands for luminaires

Protection against electric shock for circuits supplying display stands for luminaires shall be provided by either:

(i) a SELV or PELV supply, or

(ii) a residual current device having a rated residual operating current not exceeding 30 mA which provides both automatic disconnection of supply according to Section 411 and additional protection according to Regulation 415.1.

559.9 Stroboscopic effect

In the case of lighting for premises where machines with moving parts are in operation, consideration shall be given to stroboscopic effects which can give a misleading impression of moving parts being stationary. Such effects may be avoided by selecting luminaires with suitable lamp controlgear, such as high frequency controlgear, or by distributing lighting loads across all the phases of a polyphase supply.

559.10 Ground-recessed luminaires

For ground-recessed luminaires, the selection and erection shall take account of the guidance given in Table A.1 of BS EN 60598-2-13.

TABLE 55.3 – Explanation of symbols used in luminaires, in controlgear for luminaires and in the installation of luminaires

BS EN 60598-1			
		Recessed luminaire not suitable for direct mounting on normally flammable surfaces	
		Surface mounted luminaire not suitable for direct mounting on normally flammable surfaces	
		Luminaire not suitable for covering with thermally insulating material	
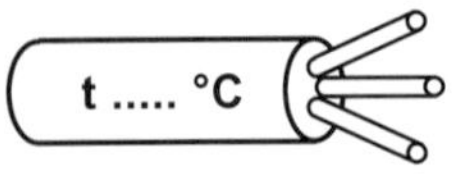	Use of heat-resistant supply cables, interconnecting cables, or external wiring (number of conductors of cable is optional) (BS EN 60598 series)		Luminaire for use with high pressure sodium lamps that require an external ignitor (BS EN 60598 series)
	Luminaire designed for use with bowl mirror lamps (BS EN 60598 series)		Luminaire for use with high pressure sodium lamps having an internal starting device (BS EN 60598 series)
t_a .... °C	Rated maximum ambient temperature (BS EN 60598 series)		Luminaire with limited surface temperature (BS EN 60598-2-24)
	Warning against the use of cool-beam lamps (BS EN 60598 series)		Short-circuit proof (inherently or non-inherently) safety isolating transformer (BS EN 61558-2-6)
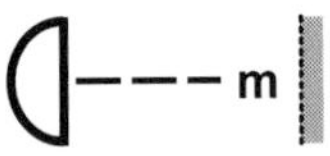	Minimum distance from lighted objects (metres) (BS EN 60598 series)		Temperature declared thermally protected lamp controlgear (… replaced by temperature) (BS EN 61347-1)
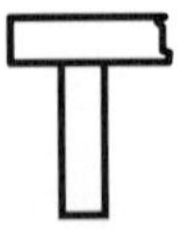	Rough service luminaire (BS EN 60598 series)		Electronic convertor for an extra-low voltage lighting installation
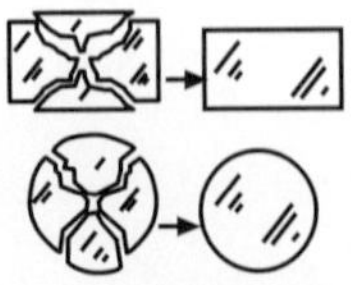	Replace any cracked protective screen (BS EN 60598 series)		Thermally protected lamp controlgear (class P) (BS EN 61347-1)
	Luminaire designed for use with self-shielded tungsten halogen lamps or self-shielded metal halide lamps only (BS EN 60598 series)		Independent ballast BS EN 60417 sheet No. 5138

CHAPTER 56

SAFETY SERVICES

CONTENTS

CHAPTER 56

SAFETY SERVICES

560.1 SCOPE

This chapter covers general requirements for safety services, selection and erection of electrical supply systems for safety services and electrical safety sources. Standby electrical supply systems are outside the scope.

This chapter does not apply to installations in hazardous areas (BE3), for which requirements are given in BS EN 60079-14.

NOTE: Examples of safety services include (this list is not exhaustive):

- Emergency lighting
- Fire pumps
- Fire rescue service lifts
- Fire detection and alarm systems
- CO detection and alarm systems
- Fire evacuation systems
- Smoke ventilation systems
- Fire services communication systems
- Essential medical systems
- Industrial safety systems.

560.2 Not used

560.3 Not used

560.4 CLASSIFICATION

560.4.1 An electrical safety service supply is either:

(i) a non-automatic supply, the starting of which is initiated by an operator, or

(ii) an automatic supply, the starting of which is independent of an operator.

An automatic supply is classified as follows, according to the maximum changeover time:

(a) No-break: an automatic supply which produces a continuous supply within specified conditions during the period of transition, for example as regards variations in voltage and frequency

(b) Very short break: an automatic supply available within 0.15 s

(c) Short break: an automatic supply available within 0.5 s

(d) Normal break: an automatic supply available within 5 s

(e) Medium break: an automatic supply available within 15 s

(f) Long break: an automatic supply available in more than 15 s.

560.4.2 The essential equipment for safety services shall be compatible with the changeover time in order to maintain the specified operation.

560.5 GENERAL

560.5.1 Safety services may be required to operate at all relevant times including during mains and local supply failure and through fire conditions. To meet this requirement specific sources, equipment, circuits and wiring are necessary. Some applications also have particular requirements, as in Regulations 560.5.2 and 560.5.3.

560.5.2 For safety services required to operate in fire conditions, the following two conditions shall be fulfilled:

(i) An electrical source for safety supply shall be so selected as to maintain a supply of adequate duration, and

(ii) all equipment of safety services shall be so provided, either by construction or by erection, with fire-resisting protection of adequate duration.

NOTE: The safety source is generally additional to the normal source. The normal source is, for example, the public supply network.

560.5.3 Where automatic disconnection of supply is used as a protective measure against electric shock, non-disconnection on the first fault is preferred. In IT systems, continuous insulation monitoring devices shall be provided which give an audible and visual indication in the event of a first fault.

560.5.4 A failure in the control or bus system of a normal installation shall not adversely affect the function of safety services.

560.6 ELECTRICAL SOURCES FOR SAFETY SERVICES

560.6.1 The following electrical sources for safety services are recognized:

(i) Storage batteries

(ii) Primary cells

(iii) Generator sets independent of the normal supply

(iv) A separate feeder of the supply network that is effectively independent of the normal feeder.

560.6.2 Safety sources for safety services shall be installed as fixed equipment and in such a manner that they cannot be adversely affected by failure of the normal source.

560.6.3 Safety sources shall be installed in a suitable location and be accessible only to skilled or instructed persons (BA5 or BA4).

560.6.4 The location of the safety source shall be properly and adequately ventilated so that exhaust gases, smoke or fumes from the safety source cannot penetrate areas occupied by one or more persons.

560.6.5 Separated independent feeders from a distributor's network shall not serve as electrical sources for safety services unless assurance can be obtained that the two supplies are unlikely to fail concurrently.

NOTE: Guidance on the specification of wiring systems for safety services may be found in BS 8519.

560.6.6 The safety source shall have sufficient capability to supply its related safety service.

560.6.7 A safety source may, in addition, be used for purposes other than safety services, provided that the availability for safety services is not thereby impaired. A fault occurring in a circuit for purposes other than safety services shall not cause the interruption of any circuit for safety services.

560.6.8 Special requirements for safety services having sources not capable of operation in parallel

560.6.8.1 Adequate precautions shall be taken to avoid the paralleling of sources.

NOTE: This may be achieved by mechanical interlocking.

560.6.8.2 Short-circuit protection and fault protection shall be provided for each source.

560.6.9 Special requirements for safety services having sources capable of operation in parallel

Short-circuit protection and fault protection shall be effective irrespective of whether the installation is supplied separately by either of the two sources or by both in parallel.

NOTE 1: The parallel operation of a private source with the public supply network is subject to authorization by the distribution network operator (DNO). This may require special devices, for example to prevent reverse power. Refer also to Section 551, the DCode (Distribution Code of licensed distribution operators) and ENA Engineering Recommendations G98 (up to and including 16 A per phase) and G99 (greater than 16 A per phase).

NOTE 2: Precautions may be necessary to limit current circulation in the connection between the neutral points of the sources, in particular the effect of triplen harmonics.

NOTE 3: Guidance on the specification of wiring systems for safety services may be found in BS 8519.

560.6.10 Central power supply sources

Batteries shall be of vented or valve-regulated maintenance-free type and shall be of heavy duty industrial design, for example cells complying with BS EN 60623 or the appropriate part of the BS EN 60896 series.

NOTE: The minimum design life of the batteries at 20 °C should be 10 years.

560.6.11 Low power supply sources

The power output of a low power supply system is limited to 500 W for 3-hour duration and 1500 W for a 1-hour duration. Batteries shall be of heavy duty industrial design, for example cells complying with BS EN 60623 or the appropriate part of the BS EN 60896 series are suitable.

NOTE: The minimum design life of the batteries at 20 °C should be 5 years.

560.6.12 Uninterruptible power supply sources (UPS)

Where an uninterruptible power supply is used, it shall:

(i) be able to operate distribution circuit protective devices, and

(ii) be able to start the safety devices when it is operating in the emergency condition from the inverter supplied by the battery, and

(iii) comply with the requirements of Regulation 560.6.10, and

(iv) comply with BS EN 62040-1 and BS EN 62040-3, as applicable.

560.6.13 Generator sets for safety services

Where a generating set is used as a safety source, it shall comply with BS 7698-12.

560.6.14 Monitoring of safety sources

The condition of the source for safety services (ready for operation, under fault conditions, feeding from the source for safety services) shall be monitored.

560.7 CIRCUITS OF SAFETY SERVICES

560.7.1 Except where the recommendations of other safety standards apply, circuits of safety services shall be independent of other circuits.

NOTE: This means that any electrical fault or any intervention or modification in one system must not affect the correct functioning of the other. This may necessitate separation by fire-resistant materials or different routes or enclosures.

560.7.2 Circuits of safety services shall not pass through locations exposed to fire risk (BE2) unless they are fire-resistant. The circuits shall not, in any case, pass through zones exposed to explosion risk (BE3).

NOTE: Where practicable, the passage of circuits through locations presenting a fire risk should be avoided.

560.7.3 In accordance with Regulation 433.3.3, protection against overload may be omitted where the loss of supply may cause a greater hazard. Where protection against overload is omitted, the occurrence of an overload shall be monitored.

560.7.4 Overcurrent protective devices shall be selected and erected so as to avoid an overcurrent in one circuit impairing the correct operation of circuits of safety services.

560.7.5 Switchgear and controlgear shall be clearly identified and grouped in locations accessible only to skilled or instructed persons (BA5 or BA4).

560.7.6 In equipment supplied by two different circuits, a fault occurring in one circuit shall not impair the protection against electric shock or the correct operation of the other circuit. Such equipment shall be connected to the protective conductors of both circuits, if necessary.

560.7.7 Safety circuit cables, other than metallic screened, fire-resistant cables, shall be adequately and reliably separated by distance or by barriers from other circuit cables, including other safety circuit cables.

NOTE: For battery cables, special requirements may apply.

560.7.8 Circuits for safety services, with the exception of wiring for fire and rescue service lift supply cables and wiring for lifts with special requirements, shall not be installed in lift wells or other flue-like openings.

NOTE: While fire-resistant cables will survive most fires, if they are located in an unstopped vertical well the upward air draught in a fire can generate excessive temperatures above the capabilities of the cable so they should be avoided as a route for safety systems.

560.7.9 In addition to a general schematic diagram, full details of all electrical safety sources shall be given. The information shall be maintained adjacent to the distribution board. A single-line diagram is sufficient.

560.7.10 A drawing or drawings of the electrical safety installations shall be available showing the exact location of:

(i) all electrical equipment and distribution boards, with equipment designations

(ii) safety equipment with final circuit designation and particulars and purpose of the equipment

(iii) special switching and monitoring equipment for the safety power supply (e.g. area switches, visual or acoustic warning equipment).

560.7.11 A list of all the current-using equipment permanently connected to the safety power supply, indicating the nominal electrical power, rated nominal voltage, current and starting current, together with its duration, shall be available.

NOTE: This information may be included in the circuit diagrams.

560.7.12 Operating instructions for safety equipment and electrical safety services shall be available. They shall take into account all the particulars of the installation.

560.8 WIRING SYSTEMS

560.8.1 One or more of the following wiring systems shall be utilised for safety services required to operate in fire conditions:

(i) Mineral insulated cable systems complying with BS EN 60702-1 and BS EN 60702-2 and BS EN 60332-1-2

(ii) Fire-resistant cables complying with IEC 60331-1, IEC 60331-2 or IEC 60331-3 and with BS EN 60332-1-2

(iii) Fire-resistant cables complying with test requirements of BS EN 50200, BS 8434 or BS 8491, appropriate for the cable size and with BS EN 60332-1-2

(iv) A wiring system maintaining the necessary fire and mechanical protection.

The wiring system selected shall meet the requirements of the relevant code of practice appropriate to the application and shall be mounted and installed in such a way that the circuit integrity will not be impaired during a fire.

NOTE 1: BS 5266, BS 5839 and BS 8519 specify cables to BS EN 60702-1, BS 7629-1 and BS 7846 as being suitable when appropriately selected for the application.

NOTE 2: Examples of a system maintaining the necessary fire and mechanical protection could be:

(i) constructional enclosures to maintain fire and mechanical protection, or

(ii) wiring systems in separate fire compartments.

560.8.2 Wiring for control and bus systems of safety services shall be in accordance with the same requirements as the wiring which is to be used for the safety services. This does not apply to circuits that do not adversely affect the operation of the safety equipment.

560.8.3 *Reserved for future use*

560.8.4 Circuits for safety services which can be supplied by direct current shall be provided with two-pole overcurrent protection mechanisms.

560.8.5 Switchgear and controlgear used for both AC and DC supply sources shall be suitable for both AC and DC operation.

560.9 EMERGENCY LIGHTING SYSTEMS

Emergency lighting systems shall comply with the relevant parts of BS 5266 series and BS EN 1838.

560.10 FIRE DETECTION AND FIRE ALARM SYSTEMS

Fire detection and fire alarm systems shall comply with the relevant parts of BS 5839 series.

560.11 LIFE SAFETY AND FIREFIGHTING APPLICATIONS

The selection and installation of power and control cable systems which are required to maintain their circuit integrity for life safety and firefighting applications shall comply with the relevant parts of BS 8519.

PART 6

INSPECTION AND TESTING

CONTENTS

CHAPTER 61 *NOT USED*

CHAPTER 62 *NOT USED*

CHAPTER 63 *NOT USED*

CHAPTER 64

INITIAL VERIFICATION

641 GENERAL

641.1 Every installation shall, during erection and on completion before being put into service, be inspected and tested to verify, so far as is reasonably practicable, that the requirements of BS 7671 have been met.

641.2 The result of the assessment of the fundamental principles, Section 131, the general characteristics required by Sections 311 to 313, together with the information required by Regulation 514.9.1, shall be made available to the person or persons carrying out the inspection and testing.

641.3 The verification shall include comparison of the results with relevant criteria to confirm that the requirements of BS 7671 have been met.

641.4 Precautions shall be taken to avoid danger to persons and livestock, and to avoid damage to property and installed equipment, during inspection and testing.

641.5 For an addition or alteration to an existing installation, it shall be verified that the addition or alteration complies with BS 7671 and does not impair the safety of the existing installation.

641.6 The verification shall be made by one or more skilled persons competent in such work.

641.7 On completion of the verification, according to Regulations 641.1 to 641.6, a certificate shall be prepared.

642 INSPECTION

642.1 Inspection shall precede testing and shall normally be done with that part of the installation under inspection disconnected from the supply.

642.2 The inspection shall be made to verify that the installed electrical equipment is:

(i) in compliance with the requirements of Section 511 (this may be ascertained by mark or by certification furnished by the installer or the manufacturer), and

(ii) correctly selected and erected in accordance with BS 7671, taking into account manufacturers' instructions, and

(iii) not visibly damaged or defective so as to impair safety.

642.3 The inspection shall include at least the checking of the following items where relevant:

(i) Connection of conductors

(ii) Identification of conductors

(iii) Routing of cables in prescribed zones, or protection against mechanical damage, in compliance with Section 522

(iv) Selection of conductors for current-carrying capacity and voltage drop, in accordance with the design

(v) Connection of single-pole devices for protection or switching in line conductors only

(vi) Correct connection of accessories and equipment

(vii) Presence of fire barriers, suitable seals and protection against thermal effects

(viii) Methods of protection against electric shock

(a) both basic protection and fault protection, i.e.:

- SELV
- PELV
- Double insulation
- Reinforced insulation

(b) basic protection (including measurement of distances, where appropriate), i.e.:

- protection by insulation of live parts
- protection by a barrier or an enclosure
- protection by obstacles
- protection by placing out of reach

(c) fault protection:

- (i) presence of adequate arrangements for alternative sources(s), where applicable
- (ii) choice and setting of protective and monitoring devices
- (iii) automatic disconnection of supply (including FELV)
- (iv) selection and installation of earthing arrangements, protective conductors and their connections confirmed for presence and sized in accordance with the design:
  - earth electrode(s), where applicable
  - earthing conductor
  - circuit protective conductors
  - protective bonding conductors
  - supplementary bonding conductors
  - earthing arrangements for combined protective and functional purposes
  - exposed-conductive-parts are connected to the earthing arrangement
- (v) non-conducting location (including measurement of distances, where appropriate)

  absence of protective conductors
- (vi) earth-free local equipotential bonding

  presence of earth-free protective bonding conductors
- (vii) electrical separation

(d) additional protection

- (ix) Prevention of mutual detrimental influence
- (x) Presence of appropriate devices for isolation and switching correctly located
- (xi) Presence of undervoltage protective devices
- (xii) Labelling of protective devices, switches and terminals
- (xiii) Selection of equipment and protective measures appropriate to external influences
- (xiv) Adequacy of access to switchgear and equipment
- (xv) Presence of danger notices and other warning signs
- (xvi) Presence of diagrams, instructions and similar information
- (xvii) Selection and erection of wiring systems
- (xviii) Erection methods
- (xix) Selection and installation of suitable SPDs where required
- (xx) Measures against electromagnetic disturbances.

The inspection shall include all particular requirements for special installations or locations (Part 7).

643 TESTING

643.1 General

The tests of Regulations 643.2 to 643.11, where relevant, shall be carried out and the results compared with relevant criteria.

Measuring instruments and monitoring equipment and methods shall be chosen in accordance with the relevant parts of BS EN 61557. If other measuring equipment is used, it shall provide no lesser degree of performance and safety.

When undertaking testing in a potentially explosive atmosphere, appropriate safety precautions in accordance with BS EN 60079-17 are necessary.

The tests of Regulations 643.2 to 643.6, where relevant, shall be carried out in that order before the installation is energized. Where the installation incorporates an earth electrode, the test of Regulation 643.7.2 shall also be carried out before the installation is energized.

If any test indicates a failure to comply, that test and any preceding test, the results of which may have been influenced by the fault indicated, shall be repeated after the fault has been rectified.

643.2 Continuity of conductors

643.2.1 The continuity of conductors and connections to exposed-conductive-parts and extraneous-conductive-parts, if any, shall be verified by a measurement of resistance of:

(i) protective conductors, including protective bonding conductors, and

(ii) in the case of ring final circuits, live conductors.

643.3 Insulation resistance

643.3.1 The insulation resistance shall be measured between:

(i) live conductors, and

(ii) live conductors and the protective conductor connected to the earthing arrangement. During this measurement, line and neutral conductors may be connected together.

643.3.2 The insulation resistance measured with the test voltages indicated in Table 64 shall be considered satisfactory if the main switchboard and each distribution circuit tested separately, with all its final circuits connected but with current-using equipment disconnected, has an insulation resistance not less than the appropriate value given in Table 64.

NOTE: More specific requirements are applicable for the wiring of fire detection and fire alarm systems in buildings, see BS 5839-1.

TABLE 64 –
Minimum values of insulation resistance

Circuit nominal voltage (V)	Test voltage DC (V)	Minimum insulation resistance (MΩ)
SELV and PELV	250	0.5
Up to and including 500 V with the exception of the above systems	500	1.0
Above 500 V	1000	1.0

Table 64 shall be applied when verifying insulation resistance between non-earthed protective conductors and Earth.

FELV circuits shall be tested at the same test voltage as that applied to the primary side of the source and shall meet all the test requirements for low voltage circuits.

643.3.3 Where connected equipment is likely to influence the measurement or result of the test, or be damaged, the test shall be applied prior to the connection of such equipment, in accordance with Table 64.

Following connection of the equipment, a test at 250 V DC shall be applied between live conductors and the protective conductor connected to the earthing arrangement. The insulation resistance shall have a value of at least 1 MΩ.

NOTE: Manufacturer's instructions may recommend some equipment to be disconnected during 250 V DC insulation resistance tests as it may influence the results of the test.

643.4 Protection by SELV, PELV or by electrical separation

It shall be verified that the separation of circuits is in accordance with Regulation 643.4.1 in the case of protection by SELV, Regulation 643.4.2 in the case of protection by PELV and Regulation 643.4.3 in the case of protection by electrical separation. The resistance values obtained in the tests of Regulations 643.4.1 to 643.4.3 shall be at least that of the circuit with the highest voltage present in accordance with Table 64.

643.4.1 Protection by SELV

The separation of the live parts from those of other circuits and from Earth, according to Section 414, shall be confirmed by a measurement of the insulation resistance. The resistance values obtained shall be in accordance with Table 64.

643.4.2 Protection by PELV

The separation of the live parts from other circuits, according to Section 414, shall be confirmed by a measurement of the insulation resistance. The resistance values obtained shall be in accordance with Table 64.

643.4.3 Protection by electrical separation

The separation of the live parts from those of other circuits and from Earth, according to Section 413, shall be confirmed by a measurement of the insulation resistance. The resistance values obtained shall be in accordance with Table 64. In the case of electrical separation with more than one item of current-using equipment, either by measurement or by calculation, it shall be verified that in case of two coincidental faults with negligible impedance between different line conductors and either the protective bonding conductor or exposed-conductive-parts connected to it, at least one of the faulty circuits shall be disconnected. The disconnection time shall be in accordance with that for the protective measure automatic disconnection of supply in a TN system.

643.5 Insulation resistance/impedance of floors and walls

643.5.1 Where it is necessary to comply with the requirements of Regulation 418.1, at least three measurements shall be made in the same location, one of these measurements being approximately 1 m from any accessible extraneous-conductive-part in the location. The other two measurements shall be made at greater distances. The measurement of resistance/impedance of insulating floors and walls is carried out with the system voltage to Earth at nominal frequency. The above series of measurements shall be repeated for each relevant surface of the location.

643.5.201 Any insulation or insulating arrangement of extraneous-conductive-parts intended to satisfy Regulation 418.1.4(iii) shall:

(i) when tested at 500 V DC, have an insulation resistance of not less than 1 MΩ, and

(ii) be able to withstand a test voltage of at least 2 kV AC rms, and

(iii) not pass a leakage current exceeding 1 mA in normal conditions of use.

643.6 Polarity

Where relevant, the polarity of the supply at the origin of the installation shall be verified before the installation is energized. Where single-pole switching devices are not permitted in the neutral conductor, a test shall be made to verify that all such devices are connected in the line conductor(s) only.

During the polarity test, it shall be verified that:

(i) every fuse and single-pole control and protective device is connected in the line conductor only, and

(ii) except for E14 and E27 lampholders to BS EN 60238, in circuits having an earthed neutral conductor, centre contact bayonet and Edison screw lampholders have the outer or screwed contacts connected to the neutral conductor, and

(iii) wiring has been correctly connected throughout the installation.

643.7 Protection by automatic disconnection of the supply

Where RCDs are applied also for protection against fire, the verification of the conditions for protection by automatic disconnection of the supply may be considered as satisfying the relevant requirements of Chapter 42.

643.7.1 General

The verification of the effectiveness of the measures for fault protection by automatic disconnection of supply is effected as follows:

a) TN system

Compliance with Regulation 411.4 shall be verified by:

(1) measurement of the earth fault loop impedance (see Regulation 643.7.3)

(2) verification of the characteristics and/or the effectiveness of the associated protective device. This verification shall be made:

- for overcurrent protective devices, by visual inspection or other appropriate methods (i.e. short-time or instantaneous tripping setting for circuit-breakers, current rating and type for fuses)
- for RCDs, by visual inspection and testing.

The effectiveness of automatic disconnection of supply by RCDs shall be verified using suitable test equipment according to BS EN 61557-6 (see Regulation 643.1) to confirm that the relevant requirements of Chapter 41 are met, taking into account the operating characteristic of the device.

NOTE: Regardless of RCD Type, effectiveness is deemed to have been verified where an RCD disconnects within the time stated below with an alternating current test at rated residual operating current ($I_{\Delta n}$):

- For general non-delay type, 300 ms maximum
- For delay 'S' type RCD, between 130 ms minimum and 500 ms maximum.

Where the effectiveness of the protective measure has been confirmed at a point located downstream of an RCD, the protection of the installation downstream from this point may be proved by confirmation of the continuity of the protective conductors.

b) TT system

Compliance with Regulation 411.5 shall be verified by:

(1) measurement of the resistance of the earth electrode for exposed-conductive-parts of the installation (see Regulation 643.7.2)

NOTE: Where a measurement of R_A is not practicable, the measured value of external earth fault loop impedance may be used.

(2) verification of the characteristics and/or effectiveness of the associated protective device. This verification shall be made:

- for overcurrent protective devices, by visual inspection or other appropriate methods (i.e. short-time or instantaneous tripping setting for circuit-breakers, current rating and type for fuses)
- for RCDs, by visual inspection and testing.

The effectiveness of automatic disconnection of supply by RCDs shall be verified using suitable test equipment according to BS EN 61557-6 (see Regulation 643.1) confirming that the relevant requirements of Chapter 41 are met, taking into account the operating characteristic of the device.

NOTE: Regardless of RCD Type, effectiveness is deemed to have been verified where an RCD disconnects within the time stated below with an alternating current test at rated residual operating current ($I_{\Delta n}$):

- For general non-delay type, 300 ms maximum
- For delay 'S' type RCD, between 130 ms minimum and 500 ms maximum.

Where the effectiveness of the protective measure has been confirmed at a point located downstream of an RCD, the protection of the installation downstream from this point may be proved by confirmation of the continuity of the protective conductors.

c) IT system

Compliance with the requirements of Regulation 411.6 shall be verified by calculation or measurement of the current I_d in case of a first fault on a live conductor.

A measurement is made only if calculation is not possible because all the parameters are not known. Precautions shall be taken while making the measurement in order to avoid the danger due to a double fault.

In the case of a double earth fault, the fault loop impedance shall be verified by calculation or measurement. Where the condition is similar to that of a TT system (see Chapter 41), verification shall be made as for a TT system (see Regulation 643.7.1, item b). Where conditions are similar to that of a TN system (see Chapter 41), verification by measurement shall be made as follows:

(i) for IT installations supplied from a local transformer, the earth fault loop impedance is measured by inserting a connection with negligible impedance between a live conductor and earth at the origin of the installation. The earth fault loop impedance measurement is made between a second live conductor and protective earth at the end of the circuit. Verification is achieved if the measured value is ≤ 50 % of the maximum allowed loop impedance

(ii) for IT systems connected to a public grid, the earth fault loop impedance is determined by verification of the continuity of the protective conductor and by measuring the loop impedance between two live conductors at the end of the circuit. Verification is achieved if the measured value is ≤ 50 % of the maximum permitted loop impedance. If verification is not achieved, more detailed measurements are necessary.

643.7.2 Earth electrode resistance

Where the earthing system incorporates an earth electrode as part of the installation, the electrode resistance to Earth shall be measured.

NOTE: Where a measurement of R_A is not practicable, the measured value of external earth fault loop impedance may be used.

643.7.3 Earth fault loop impedance and prospective fault current

643.7.3.1 Where protective measures are used which require a knowledge of earth fault loop impedance, the relevant impedances shall be measured, or determined by an alternative method.

An electrical continuity test shall be carried out according to Regulation 643.2 before carrying out the earth fault loop impedance measurement.

The measured earth fault loop impedance shall comply with Chapter 41.

Where the requirements of this regulation are not satisfied and supplementary protective equipotential bonding according to Chapter 41 is provided, the effectiveness of that bonding shall be verified.

NOTE 1: The validity of test readings taken with a fault loop impedance test instrument may be adversely affected by power converting equipment, such as inverters.

NOTE 2: Further information on measurement of earth fault loop impedance can be found in Appendix 3.

643.7.3.201 Prospective fault current

The prospective short-circuit current and prospective earth fault current shall be measured, calculated or determined by another method, at the origin and at other relevant points in the installation.

NOTE: Further information on the determination of prospective fault current can be found in Appendix 14.

643.8 Additional protection

The verification of the effectiveness of the measures applied for additional protection is fulfilled by visual inspection and testing. Where RCDs are required for additional protection, the effectiveness of automatic disconnection of supply by RCDs shall be verified using suitable test equipment according to BS EN 61557-6 (see Regulation 643.1) to confirm that the relevant requirements of Chapter 41 are met.

NOTE: Regardless of RCD Type, effectiveness is deemed to have been verified where an RCD disconnects within the time stated below with an alternating current test at rated residual operating current ($I_{\Delta n}$):

- For general non-delay type, 300 ms maximum.

Where additional protection is provided by supplementary protective equipotential bonding, the effectiveness of that bonding shall be checked against the requirements of Chapter 41.

643.9 Check of phase sequence

For polyphase circuits, it shall be verified that the phase sequence is maintained at all relevant points throughout the installation.

643.10 Functional testing

Equipment shall be subjected to functional testing, as appropriate, to verify that it is properly mounted, adjusted and installed and operates correctly in accordance with the relevant requirements of BS 7671. Examples of such equipment are:

- switchgear and controlgear assemblies, drives, controls and interlocks
- systems for emergency switching off and emergency stopping
- insulation monitoring.

NOTE 1: This list is not exhaustive.

Protective devices shall be submitted to a test of their function, as necessary, to check that they are properly installed and adjusted. Where fault protection and/or additional protection is provided by an RCD, the effectiveness of any test facility incorporated in the device shall be verified.

Where an AFDD is installed the effectiveness of any manually operated test facility shall be verified in accordance with the manufacturers' recommendations.

NOTE 2: This functional test does not replace the functional test indicated by the relevant standards.

643.11 Verification of voltage drop

Where required to verify compliance with Chapter 52, the voltage drop shall be evaluated by measurement or by calculation.

NOTE: Verification of voltage drop is not normally required during initial verification.

644 CERTIFICATION FOR INITIAL VERIFICATION

644.1 Except where Regulation 644.4.201 applies, upon completion of the verification of a new installation or an addition or alteration to an existing installation, including the replacement of a distribution board or consumer unit, an Electrical Installation Certificate based on the model given in Appendix 6 shall be issued to the person ordering the work.

644.1.1 For a new installation, any defect or omission revealed during the inspection and testing shall be corrected before the Certificate is issued.

644.1.2 For an addition and/or alteration to an existing installation, any defect or omission that will affect the safety of the addition or alteration that is revealed during inspection and testing shall be corrected before the Certificate is issued.

NOTE: See Regulation 132.16.

The person responsible for the new work, or a person authorized to act on their behalf, shall record on the Electrical Installation Certificate or the Minor Electrical Installation Works Certificate, any defects found, so far as is reasonably practicable, in the existing installation.

644.2 Not used

644.3 The Certificate shall include details of the extent of the work covered, and:

- schedule(s) of inspections, and
- schedule(s) of circuit details and test results.

The schedules shall be based on the models in Appendix 6.

644.4 The person or persons responsible for the design, construction and verification of the installation shall issue the Certificate, which takes account of their respective responsibilities, to the person ordering the work, together with the records mentioned in Regulation 644.3.

The recommendation for the interval between initial verification and the first periodic inspection shall be recorded on the Certificate.

644.4.201 Where electrical installation work does not include the provision of a new circuit or replacement of a distribution board or consumer unit, a Minor Electrical Installation Works Certificate, based on the model given in Appendix 6, may be provided for each circuit that has been added to or altered as an alternative to an Electrical Installation Certificate.

644.4.202 Electrical Installation Certificates and Minor Electrical Installation Works Certificates may be produced in any written or electronic form. Regardless of the media used for original certificates or their copies, their authenticity and integrity shall be verified by a reliable process or method. The process or method shall also verify that any copy is a true copy of the original.

644.5 Electrical Installation Certificates and Minor Electrical Installation Works Certificates shall be compiled and signed or otherwise authenticated by one or more skilled persons, competent to verify that the requirements of BS 7671 have been met.

CHAPTER 65

PERIODIC INSPECTION AND TESTING

651 GENERAL

651.1 Where required, periodic inspection and testing of every electrical installation shall be carried out in accordance with Regulations 651.2 to 651.5 in order to determine, so far as is reasonably practicable, whether the installation is in a satisfactory condition for continued service. Wherever possible, the documentation arising from the initial certification and any previous periodic inspection and testing shall be taken into account. Where no previous documentation is available, investigation of the electrical installation shall be undertaken prior to carrying out the periodic inspection and testing.

651.2 Periodic inspection shall be carried out without dismantling, or with partial dismantling, as required, supplemented by appropriate tests and measurements from Chapter 64, to provide for:

(i) the safety of persons and livestock against the effects of electric shock and burns

(ii) protection against damage to property by fire and heat arising from an electrical installation defect

(iii) confirmation of correct rating and setting of protective devices required by Chapter 41

(iv) confirmation of correct rating and setting of monitoring devices

(v) confirmation that the installation is not damaged or deteriorated so as to impair safety

(vi) the identification of installation defects and non-compliances with the requirements of the relevant parts of BS 7671, that may give rise to danger.

NOTE 1: A generic list of examples of items requiring inspection is given in Appendix 6.

NOTE 2: Existing installations may have been designed and installed to conform to previous editions of BS 7671, applicable at the time of their design and erection. This does not necessarily mean that they are unsafe.

Where a circuit is permanently monitored by an RCM or an IMD it is not necessary to measure the insulation resistance if the functioning of the RCM or IMD is correct.

The functioning of the RCM or IMD shall be verified.

651.3 Periodic inspection and testing shall not cause danger to persons or livestock and shall not cause damage to property or equipment even if the circuit is defective.

Measuring instruments and monitoring equipment and methods shall be chosen in accordance with the relevant parts of BS EN 61557. If other measuring equipment is used, it shall provide no less a degree of performance and safety.

651.4 Details of any damage, deterioration, defects or dangerous conditions shall be recorded in a report.

651.5 The periodic inspection and testing shall be carried out by one or more skilled persons competent in such work.

652 FREQUENCY OF PERIODIC INSPECTION AND TESTING

652.1 The frequency of periodic inspection and testing of an installation shall be determined having regard to the type of installation and equipment, its use and operation, the frequency and quality of maintenance and the external influences to which it may be subjected. The results and recommendations of previous certificates and condition reports shall also be taken into account.

652.2 In the case of an installation under an effective management system for preventative maintenance in normal use, periodic inspection and testing may be replaced by an adequate regime of continuous monitoring and maintenance of the installation and all its constituent equipment by one or more skilled persons competent in such work. Appropriate records shall be kept.

653 REPORTING FOR PERIODIC INSPECTION AND TESTING

653.1 Upon completion of the periodic inspection and testing of an existing installation, an Electrical Installation Condition Report based on the model given in Appendix 6 shall be produced.

653.2 The Report shall include the following:

- details of those parts of the installation that have been inspected and tested
- any limitations of the inspection and testing
- any damage, deterioration, defects or dangerous conditions
- any non-compliance with the requirements of BS 7671 which may give rise to danger
- inspection schedule(s) as appropriate to those detailed in Section 642
- schedules of circuit details and test results of the appropriate tests detailed in Section 643.

653.3 Not used

653.4 The Report shall indicate a recommended interval until the next inspection, supported by an explanation for the recommendation.

653.5 The Report shall be compiled and signed or otherwise authenticated by one or more skilled persons competent in such work.

653.6 The Report shall be issued to the person ordering the inspection and testing.

PART 7

SPECIAL INSTALLATIONS OR LOCATIONS

CONTENTS

SECTION 700

GENERAL

The particular requirements for each section (special installation or location) in Part 7 supplement or modify the general requirements contained in other parts of BS 7671.

The absence of reference to the exclusion of a part, a chapter, a section or a regulation means that the corresponding general regulations are applicable.

The number appearing after a section number generally refers to the corresponding chapter, section or regulation within Parts 1 to 6. The numbering does not, therefore, necessarily follow sequentially and new numbers have been added as required. Numbering of figures and tables takes the number of the section followed by a sequential number.

SECTION 701

LOCATIONS CONTAINING A BATH OR SHOWER

701.1 Scope

The particular requirements of this section apply to the electrical installations in locations containing a fixed bath (bath tub, birthing pool) or shower and to the surrounding zones as described in these regulations.

This section does not apply to emergency facilities such as emergency showers used in industrial areas or laboratories.

701.3 Assessment of general characteristics

701.32 Classification of external influences

701.32.1 General

When applying this section, the zones specified in Regulations 701.32.2 to 4 shall be taken into account. For fixed prefabricated bath or shower units, the zones are applied to the situation when the bath or shower basin is in its usable configuration(s).

Horizontal or inclined ceilings, walls with or without windows, doors, floors and fixed partitions may be taken into account where these effectively limit the extent of locations containing a bath or shower as well as their zones. Where the dimensions of fixed partitions are smaller than the dimensions of the relevant zones, e.g. partitions having a height lower than 2.25 m, the minimum distance in the horizontal and vertical directions shall be taken into account (see Figures 701.1 and 701.2).

For electrical equipment in parts of walls or ceilings limiting the zones specified in Regulations 701.32.2 to 701.32.4, but being part of the surface of that wall or ceiling, the requirements for the respective zone apply.

701.32.2 Description of zone 0

Zone 0 is the interior of the bath tub or shower basin (see Figures 701.1 and 701.2).

For showers without a basin, the height of zone 0 is 0.10 m and its surface extent has the same horizontal extent as zone 1 (see Figure 701.2(f)).

701.32.3 Description of zone 1

Zone 1 is limited by:

(i) the finished floor level and the horizontal plane corresponding to the highest fixed shower head or water outlet or the horizontal plane lying 2.25 m above the finished floor level, whichever is higher

(ii) the vertical surface:

  (a) circumscribing the bath tub or shower basin (see Figure 701.1)

  (b) at a distance of 1.20 m from the centre point of the fixed water outlet on the wall or ceiling for showers without a basin (see Figure 701.1(e) and (f)).

Zone 1 does not include zone 0.

The space under the bath tub or shower basin is considered to be zone 1. However, if the space under the bath tub or shower basin is only accessible with a tool, it is considered to be outside the zones.

701.32.4 Description of zone 2

Zone 2 is limited by:

(i) the finished floor level and the horizontal plane corresponding to the highest fixed shower head or water outlet or the horizontal plane lying 2.25 m above the finished floor level, whichever is higher

(ii) the vertical surface at the boundary of zone 1 and the parallel vertical surface at a distance of 0.60 m from the zone 1 border (see Figure 701.1).

For showers without a basin, there is no zone 2 but an increased zone 1 is provided by the horizontal dimension of 1.20 m mentioned in Regulation 701.32.3(ii)(b) (see Figure 701.1(e) and (f)).

701.41 Protection for safety: protection against electric shock

701.410.3 General requirements

701.410.3.5 The protective measures of obstacles and placing out of reach (Section 417) shall not be used.

701.410.3.6 The protective measures of non-conducting location (Regulation 418.1) and earth-free local equipotential bonding (Regulation 418.2) shall not be used.

701.411.3.3 Additional protection by RCDs

Additional protection by the use of one or more RCDs having the characteristics specified in Regulation 415.1.1 shall be provided for low voltage circuits:

(i) serving the location

(ii) passing through zones 1 and/or 2 not serving the location.

NOTE: See also Regulations 314.1(iv) and 531.3.2 concerning the avoidance of unwanted tripping.

701.413 Protective measure: Electrical separation

Protection by electrical separation shall only be used for:

(i) circuits supplying one item of current-using equipment, or

(ii) one single socket-outlet.

For electric floor heating systems, see Regulation 701.753.

701.414 Protective measure: Extra-low voltage provided by SELV or PELV

701.414.3 Sources for SELV and PELV

Where SELV or PELV is used in zones 0, 1 and/or 2, a source described in Regulation 414.3 (iv) shall not be used.

701.414.4.5 Requirements for SELV and PELV circuits

Where SELV or PELV is used, whatever the nominal voltage, basic protection for equipment in zones 0, 1 and 2 shall be provided by:

(i) basic insulation complying with Regulation 416.1, or

(ii) barriers or enclosures complying with Regulation 416.2.

701.415 Additional protection

701.415.2 Supplementary protective equipotential bonding

Local supplementary protective equipotential bonding according to Regulation 415.2 shall be established connecting together the terminals of the protective conductor of each circuit supplying Class I and Class II equipment to the accessible extraneous-conductive-parts, within a room containing a bath or shower, including the following:

(i) metallic pipes supplying services and metallic waste pipes (e.g. water, gas)

(ii) metallic central heating pipes and air conditioning systems

(iii) accessible metallic structural parts of the building (metallic door architraves, window frames and similar parts are not considered to be extraneous-conductive-parts unless they are connected to metallic structural parts of the building).

Supplementary protective equipotential bonding may be installed outside or inside rooms containing a bath or shower, preferably close to the point of entry of extraneous-conductive-parts into such rooms.

Where the location containing a bath or shower is in a building with a protective equipotential bonding system in accordance with Regulation 411.3.1.2, supplementary protective equipotential bonding may be omitted where all of the following conditions are met:

(iv) All final circuits of the location comply with the requirements for automatic disconnection according to Regulation 411.3.2

(v) All final circuits of the location have additional protection by means of an RCD in accordance with Regulation 415.1.1

(vi) All extraneous-conductive-parts of the location are effectively connected to the protective equipotential bonding according to Regulation 411.3.1.2.

NOTE: The effectiveness of the connection of extraneous-conductive-parts in the location to the main earthing terminal may be assessed, where necessary, by the application of Regulation 415.2.2.

701.5 Selection and erection of equipment

701.512.2 External influences

Installed electrical equipment shall have at least the following degrees of protection:

(i) In zone 0: IPX7

(ii) In zones 1 and 2: IPX4.

This requirement does not apply to shaver supply units complying with BS EN 61558-2-5 installed in zone 2 and located where direct spray from showers is unlikely.

Electrical equipment exposed to water jets, e.g. for cleaning purposes, shall have a degree of protection of at least IPX5.

701.512.3 Erection of switchgear, controlgear and accessories according to external influences

The following requirements do not apply to switches and controls which are incorporated in fixed current-using equipment suitable for use in that zone or to insulating pull cords of cord operated switches.

In zone 0:

switchgear or accessories shall not be installed.

In zone 1:

only switches of SELV circuits supplied at a nominal voltage not exceeding 12 V AC rms or 30 V ripple-free DC shall be installed, the safety source being installed outside zones 0, 1 and 2.

In zone 2:

switchgear, accessories incorporating switches or socket-outlets shall not be installed with the exception of:

(i) switches and socket-outlets of SELV circuits, the safety source being installed outside zones 0, 1 and 2, and

(ii) shaver supply units complying with BS EN 61558-2-5.

Except for SELV socket-outlets complying with Section 414 and shaver supply units complying with BS EN 61558-2-5, socket-outlets are prohibited within a distance of 2.5 m horizontally from the boundary of zone 1.

701.55 Current-using equipment

In zone 0, current-using equipment shall only be installed provided that all the following requirements are met:

(i) The equipment complies with the relevant standard and is suitable for use in that zone according to the manufacturer's instructions for use and mounting

(ii) The equipment is fixed and permanently connected

(iii) The equipment is protected by SELV at a nominal voltage not exceeding 12 V AC rms or 30 V ripple-free DC, the safety source being installed outside zones 0, 1 and 2.

In zone 1, only the following fixed and permanently connected current-using equipment shall be installed, provided it is suitable for installation in zone 1 according to the manufacturer's instructions:

(iv) Whirlpool units

(v) Electric showers

(vi) Shower pumps

(vii) Equipment protected by SELV or PELV at a nominal voltage not exceeding 25 V AC rms or 60 V ripple-free DC, the safety source being installed outside zones 0, 1 and 2

(viii) Ventilation equipment

(ix) Towel rails

(x) Water heating appliances

(xi) Luminaires.

701.753 Electric floor heating systems

For electric floor heating systems, only heating cables according to relevant product standards or thin sheet flexible heating elements according to the relevant equipment standard shall be erected provided that they have either a metal sheath or a metal enclosure or a fine mesh metallic grid. The fine mesh metallic grid, metal sheath or metal enclosure shall be connected to the protective conductor of the supply circuit. Compliance with the latter requirement is not required if the protective measure SELV is provided for the floor heating system.

For electric floor heating systems the protective measure 'protection by electrical separation' shall not be used.

Fig 701.1 – Examples of zone dimensions (plan)

NOT TO SCALE (See Regulation 701.32 for definitions of zones)

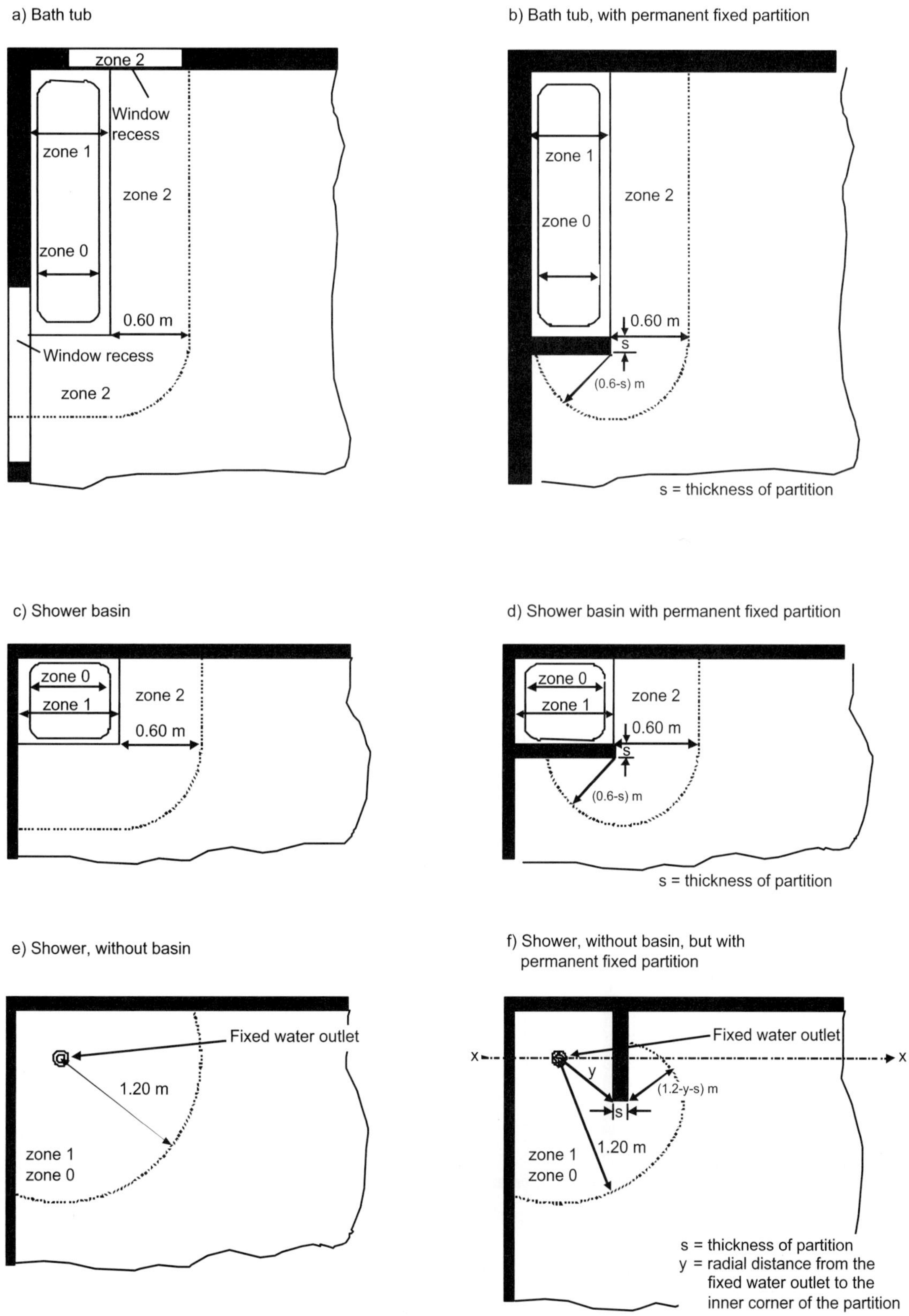

Fig 701.2 – Examples of zone dimensions (elevation)

NOT TO SCALE (See Regulation 701.32 for definitions of zones)

a) Bath tub

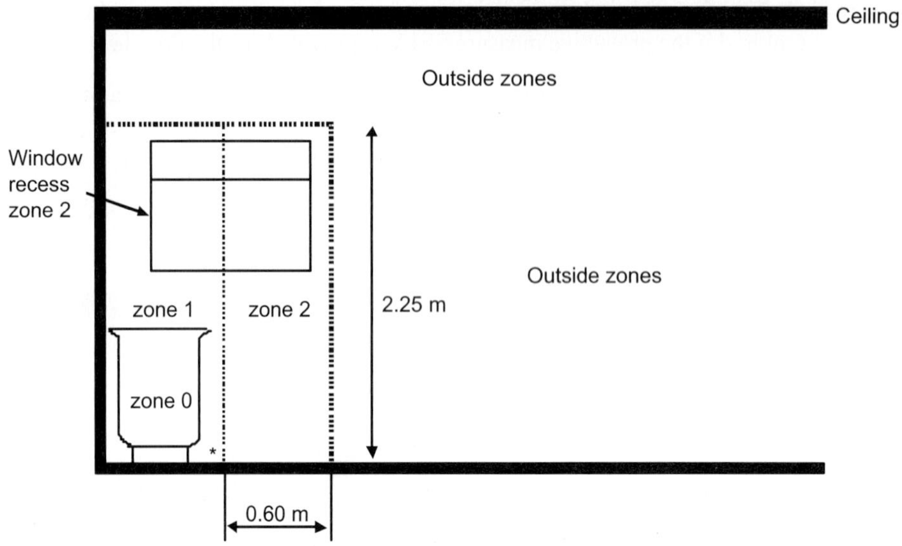

c) Shower basin

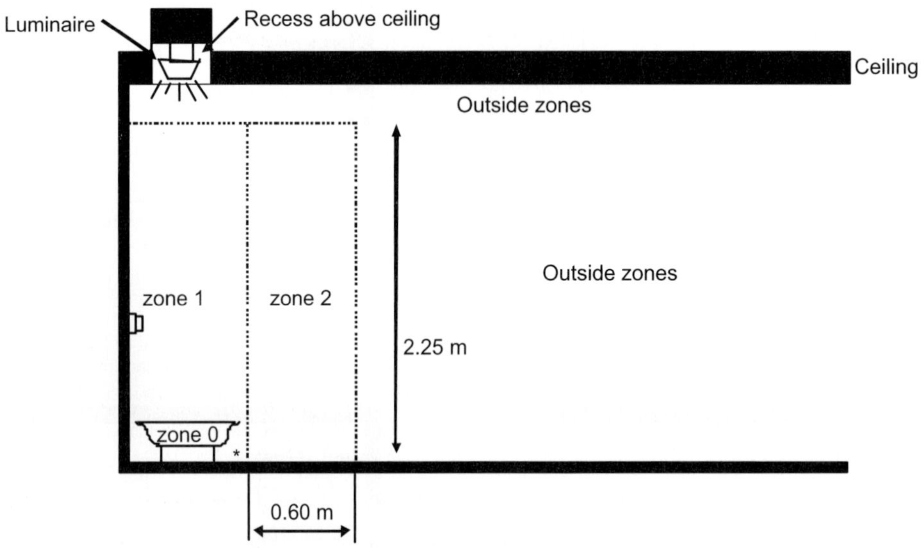

f) Shower without basin, but with permanent fixed partition

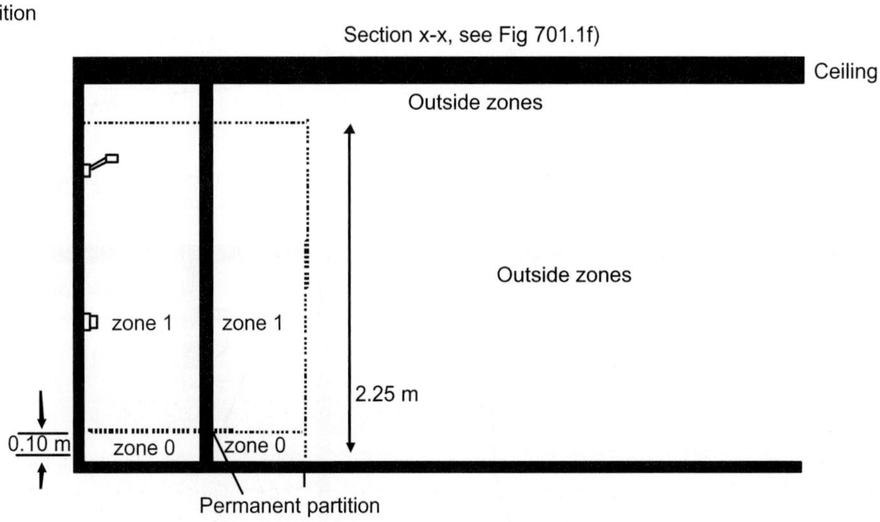

* Zone 1 if the space is accessible without the use of a tool.
Spaces under the bath accessible only with the use of a tool are outside the zones.

SECTION 702

SWIMMING POOLS AND OTHER BASINS

702.1 Scope, object and fundamental principles

702.11 Scope

The particular requirements of this section apply to the basins of swimming pools, the basins of fountains and the basins of paddling pools. The particular requirements also apply to the surrounding zones of these basins.

In these areas, in normal use, the risk of electric shock is increased by a reduction in body resistance and contact of the body with Earth potential. Swimming pools within the scope of an equipment standard are outside the scope of these regulations.

Except for areas especially designated as swimming pools, the requirements of this section do not apply to natural waters, lakes in gravel pits, coastal areas and the like.

702.3 Assessment of general characteristics

702.32 Classification of external influences

These requirements are based on the dimensions of three zones (examples are given in Figures 702.1 to 702.4).

Zones 1 and 2 may be limited by fixed partitions having a minimum height of 2.5 m.

(i) Zone 0

This zone is the interior of the basin of the swimming pool or fountain including any recesses in its walls or floors, basins for foot cleaning and waterjets or waterfalls and the space below them.

(ii) Zone 1

This zone is limited by:

- zone 0
- a vertical plane 2 m from the rim of the basin
- the floor or surface expected to be occupied by persons
- the horizontal plane 2.5 m above the floor or the surface expected to be occupied by persons.

Where the swimming pool or fountain contains diving boards, springboards, starting blocks, chutes or other components expected to be occupied by persons, zone 1 comprises the zone limited by:

- a vertical plane situated 1.5 m from the periphery of the diving boards, springboards, starting blocks, chutes and other components such as accessible sculptures, viewing bays and decorative basins
- the horizontal plane 2.5 m above the highest surface expected to be occupied by persons.

(iii) Zone 2

This zone is limited by:

- the vertical plane external to zone 1 and a parallel plane 1.5 m from the former
- the floor or surface expected to be occupied by persons
- the horizontal plane 2.5 m above the floor or surface expected to be occupied by persons.

There is no zone 2 for fountains.

702.4 Protection for safety

702.410.3 General requirements

702.410.3.4 Application of protective measures against electric shock

NOTE: Where a PME earthing facility is used as the means of earthing for the electrical installation of a swimming pool or other basin, it is recommended that an earth mat or earth electrode of suitably low resistance, e.g. 20 ohms or less, be installed and connected to the supplementary protective equipotential bonding.

702.410.3.4.1 Zones 0 and 1

Except for fountains as stated in Regulation 702.410.3.4.2, in zone 0 only protection by SELV at a nominal voltage not exceeding 12 V AC rms or 30 V ripple-free DC is permitted, the source for SELV being installed outside zones 0, 1 and 2.

Except for fountains as stated in Regulation 702.410.3.4.2, in zone 1 only protection by SELV at a nominal voltage not exceeding 25 V AC rms or 60 V ripple-free DC is permitted, the source for SELV being installed outside zones 0, 1 and 2.

Equipment for use in the interior of basins which is only intended to be in operation when people are not inside zone 0 shall be supplied by a circuit protected by:

(i) SELV (Section 414), the source for SELV being installed outside zones 0, 1 and 2. However, it is permitted to install the source for SELV in zone 2 if its supply circuit is protected by an RCD having the characteristics specified in Regulation 415.1.1, or

(ii) Automatic disconnection of supply (Section 411), using an RCD having the characteristics specified in Regulation 415.1.1, or

(iii) Electrical separation (Section 413), the source for electrical separation supplying only one item of current-using equipment and being installed outside zones 0, 1 and 2. However, it is permitted to install the source in zone 2 if its supply circuit is protected by an RCD having the characteristics specified in Regulation 415.1.1.

The socket-outlet of a circuit supplying such equipment and the control device of such equipment shall have a notice in order to warn the user that this equipment shall be used only when the swimming pool is not occupied by persons.

702.410.3.4.2 Zones 0 and 1 of fountains

In zones 0 and 1, one or more of the following protective measures shall be employed:

(i) SELV (Section 414), the source for SELV being installed outside zones 0 and 1

(ii) Automatic disconnection of supply (Section 411), using an RCD having the characteristics specified in Regulation 415.1.1

(iii) Electrical separation (Section 413), the source for electrical separation supplying only one item of current-using equipment and being installed outside zones 0 and 1.

702.410.3.4.3 Zone 2

One or more of the following protective measures shall be employed:

(i) SELV (Section 414), the source for SELV being installed outside zones 0, 1 and 2. However, it is permitted to install the source for SELV in zone 2 if its supply circuit is protected by an RCD having the characteristics specified in Regulation 415.1.1

(ii) Automatic disconnection of supply (Section 411), using an RCD having the characteristics specified in Regulation 415.1.1

(iii) Electrical separation (Section 413), the source for electrical separation supplying only one item of current-using equipment and being installed outside zones 0, 1 and 2. However, it is permitted to install the source in zone 2 if its supply circuit is protected by an RCD having the characteristics specified in Regulation 415.1.1.

There is no zone 2 for fountains.

702.410.3.5 The protective measures of obstacles and placing out of reach (Section 417) shall not be used.

702.410.3.6 The protective measures of non-conducting location (Regulation 418.1) and earth-free local equipotential bonding (Regulation 418.2) shall not be used.

702.414 Protective measure: Extra-low voltage provided by SELV or PELV

702.414.3 Sources for SELV and PELV

Where SELV or PELV is used in zones 0, 1 and/or 2, a source described in Regulation 414.3 (iv) shall not be used.

702.414.4 Requirements for SELV and PELV circuits

702.414.4.5 Where SELV is used, whatever the nominal voltage, basic protection shall be provided by:

(i) basic insulation complying with Regulation 416.1, or

(ii) barriers or enclosures complying with Regulation 416.2.

702.415 Additional protection

702.415.2 Additional protection: Supplementary protective equipotential bonding

All extraneous-conductive-parts in zones 0, 1 and 2 shall be connected by supplementary protective bonding conductors to the protective conductors of exposed-conductive-parts of equipment situated in these zones, in accordance with Regulation 415.2.

NOTE: The connection with the protective conductor may be provided in the proximity of the location, e.g. in an accessory or in a local distribution board.

702.5 Selection and erection of equipment

702.51 Common rules

702.512 Operational conditions and external influences

702.512.2 External influences

Electrical equipment shall have at least the following degree of protection according to BS EN 60529:

(i) zone 0: IPX8

(ii) zone 1: IPX4, IPX5 where water jets are likely to occur for cleaning purposes

(iii) zone 2: IPX2 for indoor locations, IPX4 for outdoor locations, IPX5 where water jets are likely to occur for cleaning purposes.

702.52 Wiring systems

702.520 General

The following regulations apply to surface wiring systems and to wiring systems embedded in the walls, ceilings or floors at a depth not exceeding 50 mm.

702.522 Selection and erection in relation to external influences

702.522.21 Erection according to the zones

In zones 0, 1 and 2, any metallic sheath or metallic covering of a wiring system shall be connected to the supplementary protective equipotential bonding.

NOTE: Cables should preferably be installed in conduits made of insulating material.

702.522.22 Limitation of wiring systems according to the zones

In zones 0 and 1, a wiring system shall be limited to that necessary to supply equipment situated in these zones.

702.522.23 Additional requirements for the wiring of fountains

For a fountain, the following additional requirements shall be met:

(i) A cable for electrical equipment in zone 0 shall be installed as far outside the basin rim as is reasonably practicable and run to the electrical equipment inside zone 0 by the shortest practicable route

(ii) In zone 1, a cable shall be selected, installed and provided with mechanical protection to medium severity (AG2) and the relevant submersion in water depth (AD8). The cable type H07RN8-F (BS EN 50525-2-21) is suitable up to a depth of 10 m of water. For depths of water greater than 10 m the cable manufacturer shall be consulted.

702.522.24 Junction boxes

A junction box shall not be installed in zones 0 or 1, except that for SELV circuits it is permitted to install junction boxes in zone 1.

702.53 Protection, isolation, switching, control and monitoring

In zones 0 or 1, switchgear or controlgear shall not be installed.

In zones 0 or 1, a socket-outlet shall not be installed.

In zone 2, a socket-outlet or a switch is permitted only where the supply circuit is protected by one of the following protective measures:

(i) SELV (Section 414), the source of SELV being installed outside zones 0, 1 and 2. However, it is permitted to install the source of SELV in zone 2 if its supply circuit is protected by an RCD having the characteristics specified in Regulation 415.1.1

(ii) Automatic disconnection of supply (Section 411), using an RCD having the characteristics specified in Regulation 415.1.1

(iii) Electrical separation (Section 413), the source for electrical separation supplying only one item of current-using equipment, or one socket-outlet, and being installed outside zones 0, 1 and 2. However, it is permitted to install the source in zone 2 if its supply circuit is protected by an RCD having the characteristics specified in Regulation 415.1.1.

702.55 Other equipment

702.55.1 Current-using equipment of swimming pools

In zones 0 and 1, it is only permitted to install fixed current-using equipment specifically designed for use in a swimming pool, in accordance with the requirements of Regulations 702.55.2 and 702.55.4.

Equipment which is intended to be in operation only when people are outside zone 0 may be used in all zones provided that it is supplied by a circuit protected according to Regulation 702.410.3.4.

It is permitted to install an electric heating unit embedded in the floor, provided that it:

(i) is protected by SELV (Section 414), the source of SELV being installed outside zones 0, 1 and 2. However, it is permitted to install the source of SELV in zone 2 if its supply circuit is protected by an RCD having the characteristics specified in Regulation 415.1.1, or

(ii) incorporates an earthed metallic sheath connected to the supplementary protective equipotential bonding specified in Regulation 702.415.2 and its supply circuit is additionally protected by an RCD having the characteristics specified in Regulation 415.1.1, or

(iii) is covered by an embedded earthed metallic grid connected to the supplementary protective equipotential bonding specified in Regulation 702.415.2 and its supply circuit is additionally protected by an RCD having the characteristics specified in Regulation 415.1.1.

702.55.2 Underwater luminaires for swimming pools

A luminaire for use in the water or in contact with the water shall be fixed and shall comply with BS EN 60598-2-18.

Underwater lighting located behind watertight portholes, and serviced from behind, shall comply with the appropriate part of BS EN 60598 and be installed in such a way that no intentional or unintentional conductive connection between any exposed-conductive-part of the underwater luminaires and any conductive parts of the portholes can occur.

702.55.3 Electrical equipment of fountains

Electrical equipment in zones 0 or 1 shall be provided with mechanical protection to medium severity (AG2), e.g. by use of mesh glass or by grids which can only be removed by the use of a tool.

A luminaire installed in zones 0 or 1 shall be fixed and shall comply with BS EN 60598-2-18.

An electric pump shall comply with the requirements of BS EN 60335-2-41.

702.55.4 Special requirements for the installation of electrical equipment in zone 1 of swimming pools and other basins

Fixed equipment designed for use in swimming pools and other basins (e.g. filtration systems, jet stream pumps) and supplied at low voltage is permitted in zone 1, subject to all the following requirements being met:

(i) The equipment shall be located inside an insulating enclosure providing at least Class II or equivalent insulation and providing protection against mechanical impact of medium severity (AG2)

This regulation applies irrespective of the classification of the equipment.

(ii) The equipment shall only be accessible via a hatch (or a door) by means of a key or a tool. The opening of the hatch (or door) shall disconnect all live conductors. The supply cable and the main disconnecting means shall be installed in a way which provides protection of Class II or equivalent insulation

(iii) The supply circuit of the equipment shall be protected by:

– SELV at a nominal voltage not exceeding 25 V AC rms or 60 V ripple-free DC, the source of SELV being installed outside zones 0, 1 and 2, or

– an RCD having the characteristics specified in Regulation 415.1.1, or

– electrical separation (Section 413), the source for electrical separation supplying a single fixed item of current-using equipment and being installed outside zones 0, 1 and 2.

For swimming pools where there is no zone 2, lighting equipment supplied by other than a SELV source at 12 V AC rms or 30 V ripple-free DC may be installed in zone 1 on a wall or on a ceiling, provided that the following requirements are fulfilled:

- The circuit is protected by automatic disconnection of the supply and additional protection is provided by an RCD having the characteristics specified in Regulation 415.1.1
- The height from the floor or surface expected to be occupied by persons is at least 2 m above the lower limit of zone 1.

In addition, every luminaire shall have an enclosure providing Class II or equivalent insulation and providing protection against mechanical impact of medium severity.

Fig 702.1 – zone dimensions for swimming pools and paddling pools

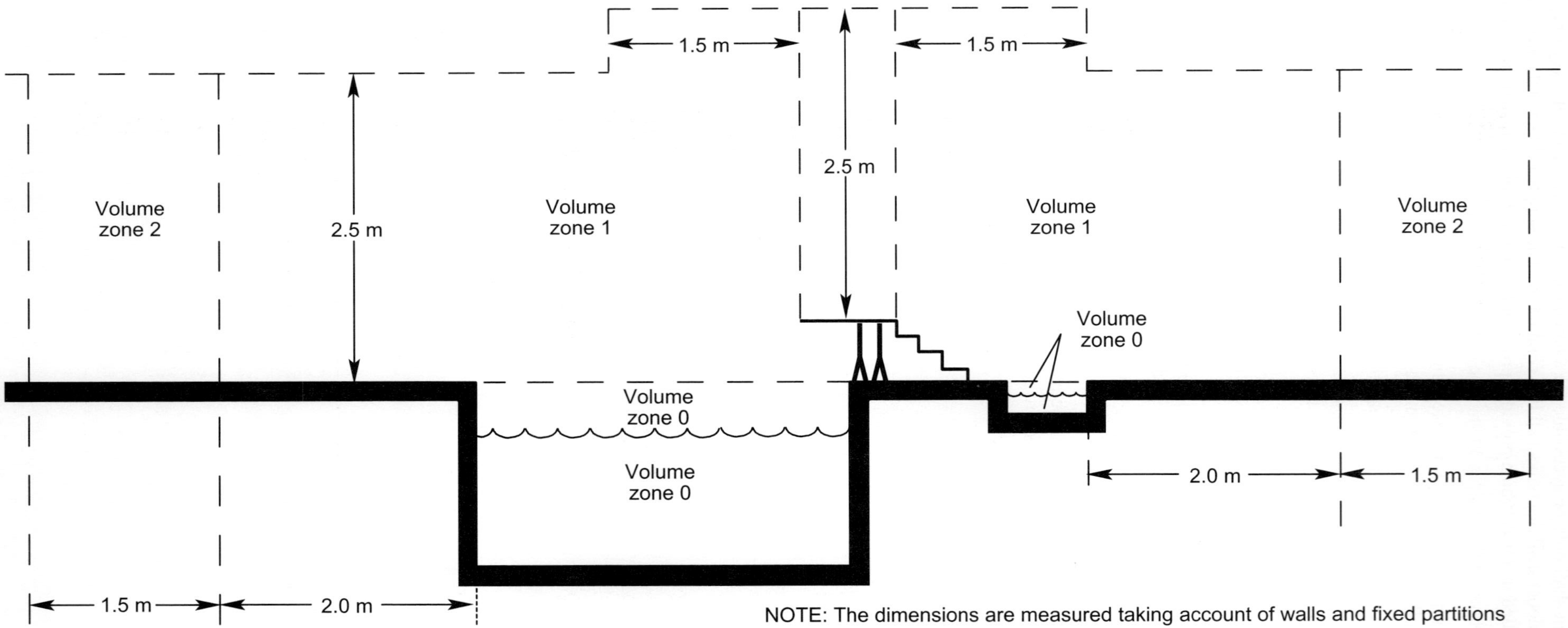

NOTE: The dimensions are measured taking account of walls and fixed partitions

Fig 702.2 – zone dimensions for basin above ground level

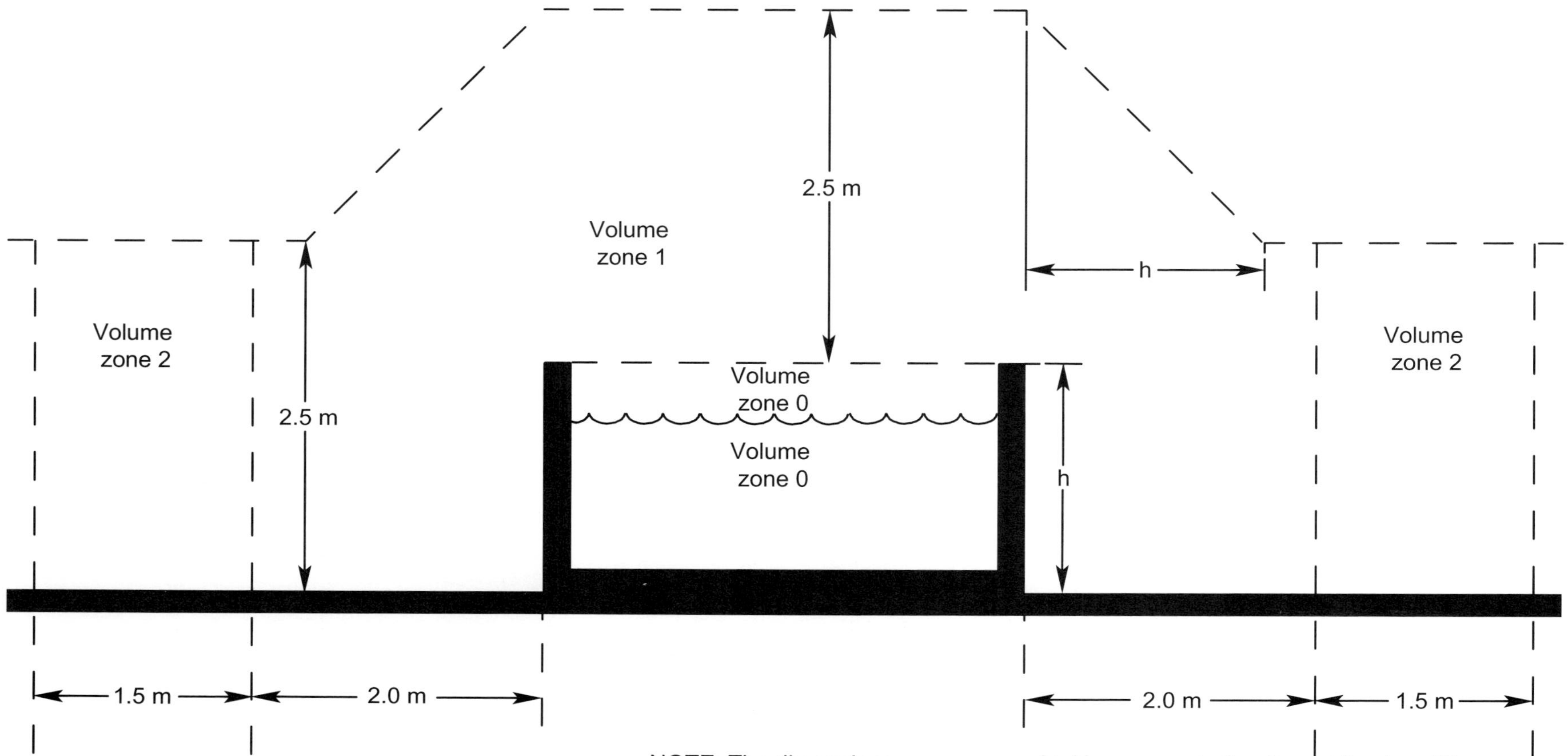

NOTE: The dimensions are measured taking account of walls and fixed partitions

Fig 702.3 – Example of zone dimensions (plan) with fixed partitions of height at least 2.5 m

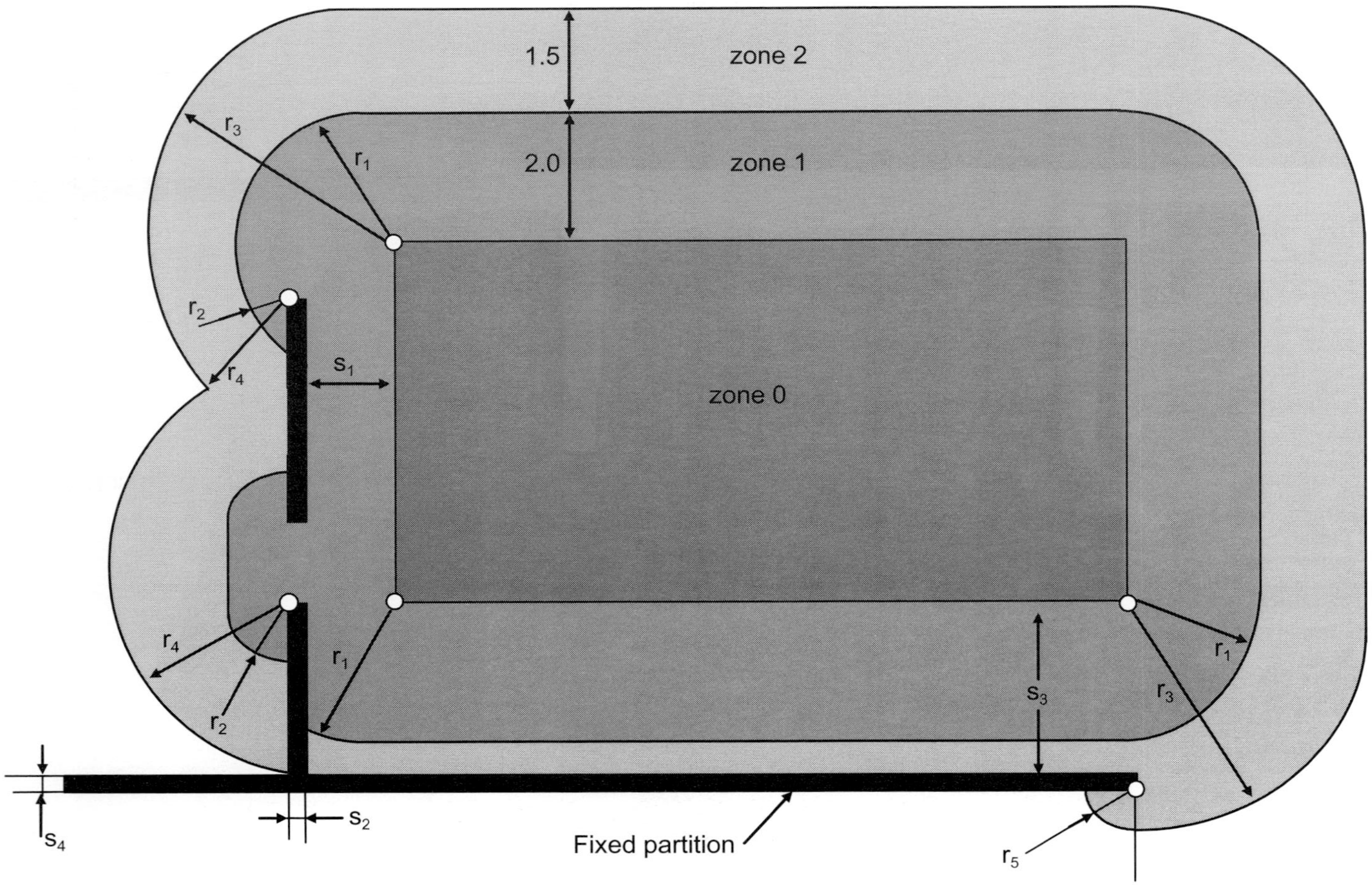

Dimensions in metres

$r_1 = 2$

$r_2 = r_1 - (s_1 + s_2)$

$r_3 = 3.5$

$r_4 = r_3 - (s_1 + s_2)$

$r_5 = r_3 - (s_3 + s_4)$

Fig 702.4 – Example of determination of the zones of a fountain

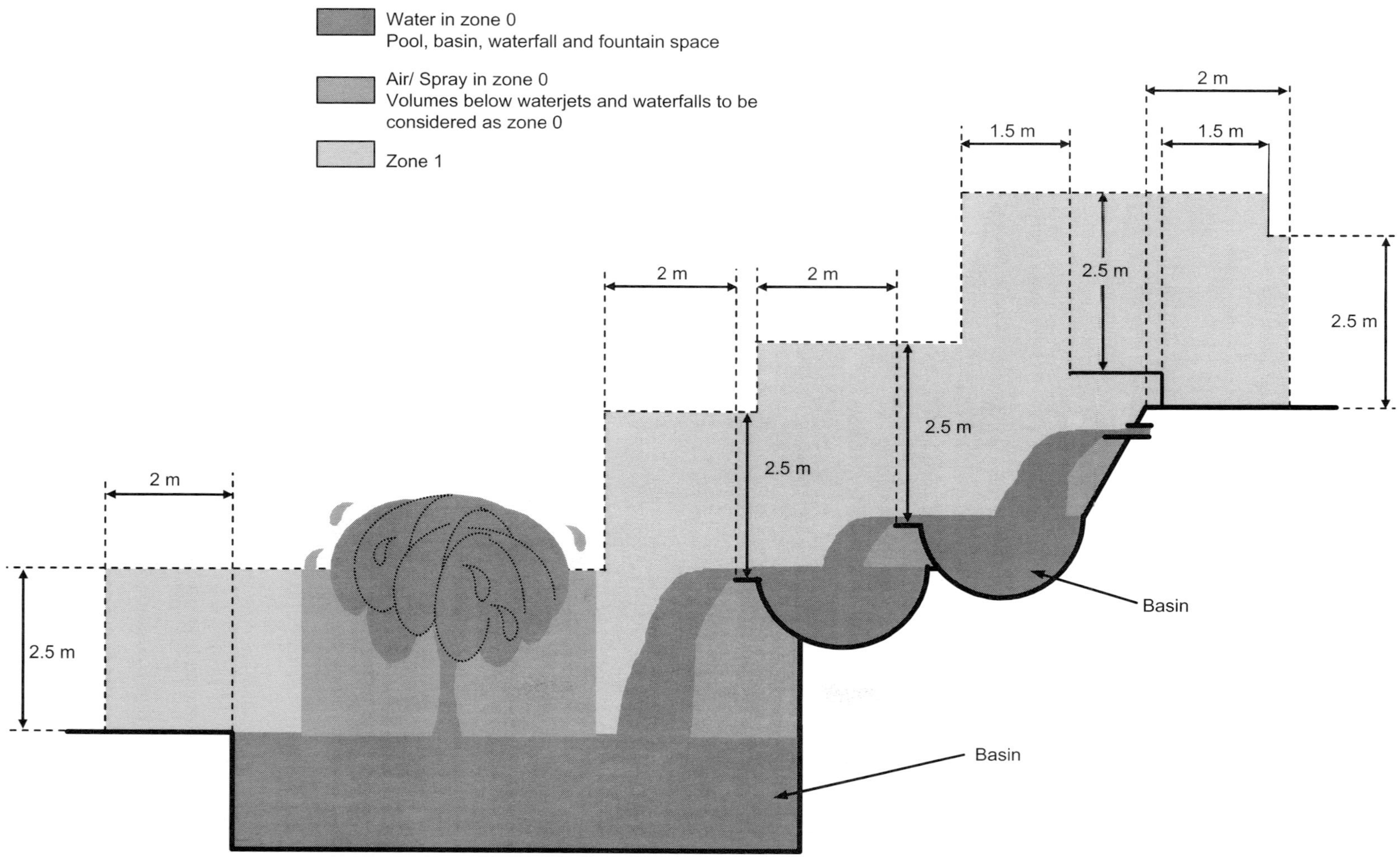

SECTION 703

ROOMS AND CABINS CONTAINING SAUNA HEATERS

703.1 Scope

The particular requirements of this section apply to:

(i) sauna cabins erected on site, e.g. in a location or in a room

(ii) the room where the sauna heater is, or the sauna heating appliances are installed. In this case the whole room is considered as the sauna.

The requirements of this section do not apply to prefabricated sauna cabins complying with a relevant equipment standard.

Where facilities such as showers etc. are installed, the requirements of Section 701 also apply.

703.3 Assessment of general characteristics

703.32 Classification of external influences

When applying these regulations, the zones specified in Regulations 703.32.1 to 703.32.3 shall be taken into account (see also Figure 703).

703.32.1 Description of zone 1

Zone 1 is the volume containing the sauna heater, limited by the floor, the cold side of the thermal insulation of the ceiling and a vertical surface circumscribing the sauna heater at a distance 0.5 m from the surface of the heater. If the sauna heater is located closer than 0.5 m to a wall, then zone 1 is limited by the cold side of the thermal insulation of that wall.

703.32.2 Description of zone 2

Zone 2 is the volume outside zone 1, limited by the floor, the cold side of the thermal insulation of the walls and a horizontal surface located 1.0 m above the floor.

703.32.3 Description of zone 3

Zone 3 is the volume outside zone 1, limited by the cold side of the thermal insulation of the ceiling and walls and a horizontal surface located 1.0 m above the floor.

703.41 Protection against electric shock

703.410.3 General requirements

703.410.3.5 The protective measures of obstacles and placing out of reach (Section 417) shall not be used.

703.410.3.6 The protective measures of non-conducting location (Regulation 418.1) and earth-free local equipotential bonding (Regulation 418.2) shall not be used.

703.411.3.3 Additional protection by RCDs

Additional protection shall be provided for all circuits of the sauna, by the use of one or more RCDs having the characteristics specified in Regulation 415.1.1. RCD protection need not be provided for the sauna heater unless such protection is recommended by the manufacturer.

703.414 Protective measure: Extra-low voltage provided by SELV or PELV

703.414.3 Sources for SELV and PELV

Where SELV or PELV is used in zones 1, 2 and/or 3, a source described in Regulation 414.3 (iv) shall not be used.

703.414.4.5 Where SELV or PELV is used, whatever the nominal voltage, basic protection shall be provided by:

(i) basic insulation complying with Regulation 416.1, or

(ii) barriers or enclosures complying with Regulation 416.2.

703.51 Selection and erection of equipment: Common rules

703.512.2 External influences

The equipment shall have a degree of protection of at least IPX4.

Where cleaning by use of water jets may be reasonably expected, electrical equipment shall have a degree of protection of at least IPX5.

Three zones are defined as shown in Figure 703:

(i) In zone 1: only the sauna heater and equipment belonging to the sauna heater shall be installed

(ii) In zone 2: there is no special requirement concerning heat-resistance of equipment

(iii) In zone 3: the equipment shall withstand a minimum temperature of 125 °C and the insulation and sheaths of cables shall withstand a minimum temperature of 170 °C (see also Regulation 703.52 for wiring).

703.52 Selection and erection of equipment: Wiring systems

The wiring system should be preferably installed outside the zones, i.e. on the cold side of the thermal insulation. Where the wiring system is installed on the warm side of the thermal insulation in zones 1 or 3, it shall be heat-resisting. Metallic sheaths and metallic conduits shall not be accessible in normal use.

703.53 Selection and erection of equipment: Isolation, switching, control and accessories

703.537.5 Switchgear and controlgear which forms part of the sauna heater equipment or of other fixed equipment installed in zone 2, may be installed within the sauna room or cabin in accordance with the manufacturer's instructions. Other switchgear and controlgear, e.g. for lighting, shall be placed outside the sauna room or cabin. Socket-outlets shall not be installed within the location containing the sauna heater.

703.55 Other equipment

Sauna heating appliances shall comply with BS EN 60335-2-53 and be installed in accordance with the manufacturer's instructions.

Fig 703 – zone dimensions for a sauna

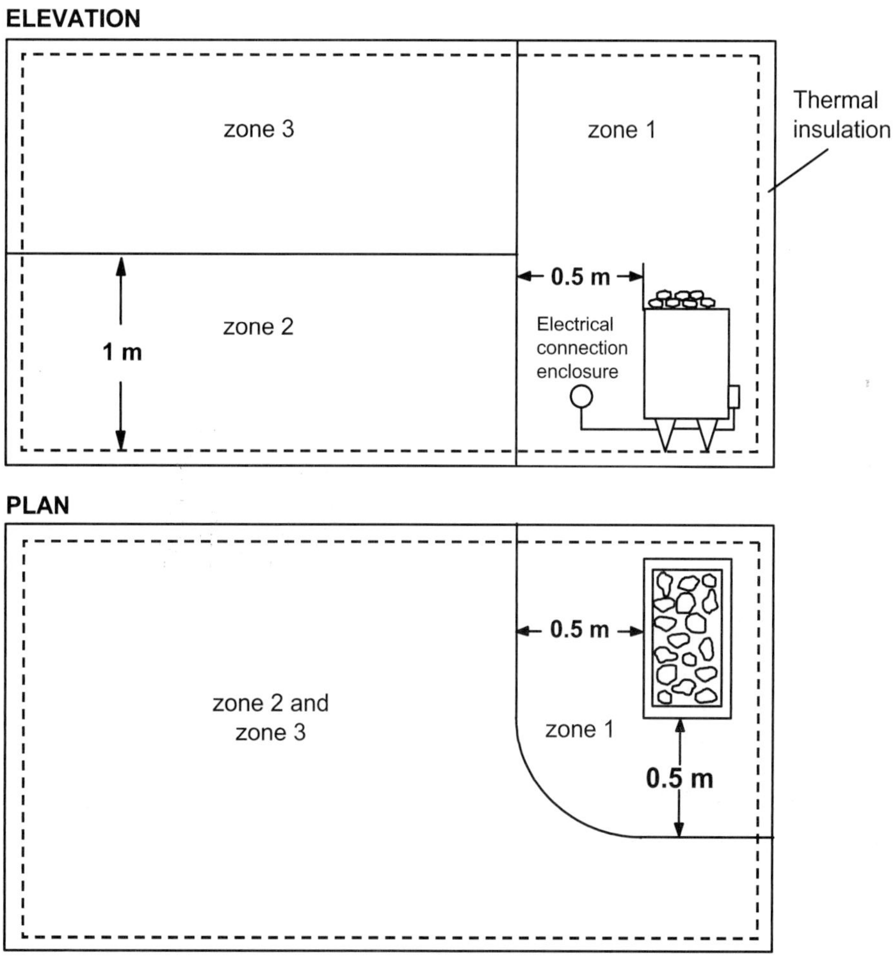

SECTION 704

CONSTRUCTION AND DEMOLITION SITE INSTALLATIONS

704.1 Scope

704.1.1 The particular requirements of this section apply to temporary installations for construction and demolition sites during the period of the construction or demolition work, including, for example, the following:

(i) construction work of new buildings

(ii) repair, alteration, extension or demolition of existing buildings or parts of existing buildings

(iii) engineering works

(iv) earthworks

(v) work of similar nature.

The requirements apply to fixed and movable installations.

This section does not apply to installations in administrative locations of construction sites (e.g. offices, cloakrooms, meeting rooms, canteens, restaurants, dormitories, toilets), where the general requirements of Parts 1 to 6 apply.

704.1.2 For special situations, further particular requirements apply, e.g. Section 706 for conducting locations with restricted movement.

704.31 Purposes, supplies and structure

704.313 Supplies

704.313.3 Equipment shall be identified with and be compatible with the particular supply from which it is energized and shall contain only components connected to one and the same installation, except for control or signalling circuits and inputs from standby supplies.

NOTE: A single construction site may be served by several sources of supply, e.g. public supply and generating set.

704.4 Protection for safety

704.41 Protection against electric shock

704.410 Introduction

704.410.3 General requirements

704.410.3.5 The protective measures of obstacles and placing out of reach (Section 417) shall not be used.

704.410.3.6 The protective measures of:

- non-conducting location,
- earth-free equipotential bonding, and
- electrical separation for the supply of more than one item of current-using equipment as specified in Regulation 418.3

shall not be used.

704.410.3.10 A circuit supplying a socket-outlet with a rated current up to and including 32 A and any other circuit supplying hand-held electrical equipment with rated current up to and including 32 A shall be protected by:

(i) reduced low voltage (Regulation 411.8), or

(ii) automatic disconnection of supply (Section 411) with additional protection provided by an RCD having the characteristics specified in Regulation 415.1.1, or

(iii) electrical separation of circuits (Section 413), each socket-outlet and item of hand-held electrical equipment being supplied by an individual transformer or by a separate winding of a transformer, or

(iv) SELV or PELV (Section 414).

Where electrical separation is used, special attention should be paid to the requirements of Regulation 413.3.4.

NOTE 1: The reduced low voltage system is strongly preferred for the supply to portable handlamps for general use and portable hand tools and local lighting up to 2 kW.

NOTE 2: The SELV system is strongly preferred for portable handlamps in confined or damp locations.

704.411 Protective measure: Automatic disconnection of supply

704.411.3 Requirements for fault protection

704.411.3.1 A PME earthing facility shall not be used for the means of earthing for an installation falling within the scope of this section unless all extraneous-conductive-parts are reliably connected to the main earthing terminal in accordance with Regulation 411.3.1.2.

NOTE: If the PME earthing facility is considered for use, see BS 7375. Protective bonding of all extraneous-conductive-parts is very difficult to achieve and maintain throughout the life of the installation.

704.411.3.2 Automatic disconnection in case of a fault

704.411.3.2.1 For any circuit supplying one or more socket-outlets with a rated current exceeding 32 A, Regulation 411.3.2.5 is not applicable. For any circuit supplying one or more socket-outlets with a rated current exceeding 32 A, an RCD having a rated residual operating current not exceeding 500 mA shall be provided to automatically interrupt the supply to the line conductors of a circuit or equipment in the event of a fault of negligible impedance between a line conductor and an exposed-conductive-part or a protective conductor in the circuit or equipment within the disconnection time required in Regulation 411.3.2.3 or 411.3.2.4 as appropriate.

704.414 Protective measure: Extra-low voltage provided by SELV or PELV

704.414.4 Requirements for SELV and PELV circuits

704.414.4.5 Irrespective of the nominal voltage, in AC and DC circuits, the requirement for basic protection shall be provided by the following:

(i) basic insulation complying with Regulation 416.1, or

(ii) barriers or enclosures complying with Regulation 416.2.

704.5 Selection and erection of equipment

704.51 Common rules

704.511 Compliance with standards

704.511.1 All assemblies on construction and demolition sites for the distribution of electricity shall be in compliance with the requirements of BS EN 61439-4.

Plugs and socket-outlets with a rated current of 16 A up to 125 A shall comply with the requirements of BS EN 60309-2.

Plugs and socket-outlets with a rated current exceeding 125 A up to 800 A and where interchangeability is not required shall comply with BS EN 60309-1.

704.512 Operational conditions and external influences

704.512.2 External influences

Consideration shall be given to the risk of damage to electrical equipment by corrosive substances, movement of structures and vehicles, wear and tear, tension, flexing, impact, abrasion, severing and ingress of liquids or solids.

704.52 Wiring systems

704.522.8 Other mechanical stresses (AJ)

704.522.8.11 For reduced low voltage systems, low temperature 3182/3/4/5A thermoplastic cable (BS 6004) or equivalent flexible cable shall be used. For applications exceeding reduced low voltage, flexible cable shall be H07RN-F (BS EN 50525-2-21) type or equivalent heavy duty flexible cable.

704.522.8.101 To avoid damage, cables should not be run across site roads or walkways. Where this is unavoidable, adequate protection against mechanical damage and contact with construction plant machinery shall be provided.

Particular attention shall be given to the protection of surface run and overhead cables against mechanical damage, given the environment and activities at the construction site.

Flexible cables that are subject to movement shall be H07RN-F (BS EN 50525-2-21) or equivalent, resistant to abrasion and to water.

704.53 Protection, isolation, switching, control and monitoring

704.537 Devices for isolation and switching

704.537.2 Devices for isolation

Current-using equipment shall be supplied via an Assembly for Construction Sites (ACS) incorporating overcurrent protective devices, devices affording fault protection, and socket-outlets if required.

Each ACS shall also incorporate suitable devices for the switching and isolation of the incoming supply.

A device for isolating the incoming supply shall be suitable for securing in the off position (see Regulation 537.2.4), for example, by providing a padlocking facility on the device or by locating the device inside a lockable enclosure.

Safety and standby supplies shall be connected by means of devices arranged to prevent interconnection of the different supplies.

SECTION 705

AGRICULTURAL AND HORTICULTURAL PREMISES

705.1 Scope

The particular requirements of this section apply to fixed electrical installations indoors and outdoors in agricultural and horticultural premises. Some of the requirements are also applicable to other locations that are in common buildings belonging to the agricultural and horticultural premises. Where special requirements also apply to residences and other locations in such common buildings this is stated in the text of the relevant regulations.

Rooms, locations and areas for residential premises and similar are not covered by this section.

NOTE: Section 705 does not cover electric fence installations. Refer to BS EN 60335-2-76.

705.41 Protection against electric shock

705.410.3 General requirements

705.410.3.5 The protective measures of obstacles and placing out of reach (Section 417) shall not be used.

705.410.3.6 The protective measures of non-conducting location (Regulation 418.1) and earth-free local equipotential bonding (Regulation 418.2) shall not be used.

705.411 Protective measure: Automatic disconnection of supply

705.411.1 General

In circuits, whatever the type of earthing system, the following disconnection devices shall be provided:

(i) In final circuits supplying socket-outlets with rated current not exceeding 32 A, an RCD having the characteristics specified in Regulation 415.1.1

(ii) In final circuits supplying socket-outlets with rated current more than 32 A, an RCD with a rated residual operating current not exceeding 100 mA

(iii) In all other circuits, RCDs with a rated residual operating current not exceeding 300 mA.

705.411.4 TN system

A TN-C system shall not be used. This requirement applies also to residences and other locations belonging to agricultural or horticultural premises according to the definition of 'Residences and other... premises' (see Part 2).

NOTE: This does not preclude the use of TN-C-S.

705.414 Protective measure: Extra-low voltage provided by SELV or PELV

705.414.4 Requirements for SELV and PELV circuits

705.414.4.5 Where SELV or PELV is used, whatever the nominal voltage, basic protection shall be provided by:

(i) basic insulation complying with Regulation 416.1, or

(ii) barriers or enclosures complying with Regulation 416.2.

705.415.2.1 Additional protection: Supplementary equipotential bonding

In locations intended for livestock, supplementary bonding shall connect all exposed-conductive-parts and extraneous-conductive-parts that can be touched by livestock. Where a metal grid is laid in the floor, it shall be included within the supplementary bonding of the location (Figure 705 shows an example of this, other suitable arrangements of a metal grid are not precluded).

Extraneous-conductive-parts in, or on, the floor, e.g. concrete reinforcement in general or reinforcement of cellars for liquid manure, shall be connected to the supplementary equipotential bonding.

It is recommended that spaced floors made of prefabricated concrete elements be part of the supplementary equipotential bonding. The supplementary equipotential bonding and the metal grid, if any, shall be erected so that it is durably protected against mechanical stresses and corrosion.

NOTE 1: This does not preclude the use of TN-C-S.

NOTE 2: Unless a metal grid is laid in the floor, the use of a PME earthing facility as the means of earthing for the electrical installation is not recommended.

705.42 Protection against thermal effects

NOTE: See Chapter 42 for the selection and erection of electrical equipment regarding locations with stored materials and animal feeds, etc.

705.422 Precautions where particular risks of fire exist

705.422.6 Electrical heating appliances used for the breeding and rearing of livestock shall comply with BS EN 60335-2-71 and shall be fixed so as to maintain an appropriate distance from livestock and combustible material, to minimize any risks of burns to livestock and of fire. For radiant heaters the clearance shall be not less than 0.5 m or such other clearance as recommended by the manufacturer.

705.422.7 For additional fire protection purposes in some circumstances, RCDs shall be installed with a rated residual operating current not exceeding 300 mA. RCDs shall disconnect all live conductors. Where improved continuity of service is required, RCDs not protecting socket-outlets shall be of the type S or have a time delay.

705.422.8 In locations where a fire risk exists, conductors of circuits supplied at extra-low voltage shall be protected either by barriers or enclosures affording a degree of protection of IPXXD or IP4X or, in addition to their basic insulation, by an enclosure of insulating material.

NOTE: For example, cables of the type H07RN-F (BS EN 50525-2-21) for outdoor use are in compliance with this requirement.

705.51 Selection and erection of equipment: Common rules

705.512 Operational conditions and external influences

705.512.2 External influences

In agricultural or horticultural premises, electrical equipment shall have a minimum degree of protection of IP44, when used under normal conditions. Where equipment of IP44 rating is not available, it shall be placed in an enclosure complying with IP44.

Where there are conditions of external influences >AD4, >AE3 and/or >AG1, socket-outlets shall be provided with the appropriate protection.

Protection may also be provided by the use of additional enclosures or by installation in building recesses.

These requirements do not apply to residential locations, offices, shops and locations with similar external influences belonging to agricultural and horticultural premises where, for socket-outlets, BS 1363-2 or BS 546 applies.

Where corrosive substances are present, e.g. in dairies or cattle sheds, the electrical equipment shall be adequately protected.

705.513 Accessibility

705.513.2 Accessibility by livestock

Electrical equipment generally shall be inaccessible to livestock. Equipment that is unavoidably accessible to livestock such as equipment for feeding and basins for watering, shall be adequately constructed and installed to avoid damage by, and to minimize the risk of injury to, livestock.

705.514 Identification and notices

705.514.9 Diagrams and documentation

705.514.9.3 The following documentation shall be provided to the user of the installation:

(i) A plan indicating the location of all electrical equipment
(ii) The routing of all concealed cables
(iii) A single-line distribution diagram
(iv) An equipotential bonding diagram indicating locations of bonding connections.

705.52 Selection and erection of equipment: Wiring systems

705.522 Selection and erection of wiring systems in relation to external influences

In locations accessible to, and enclosing, livestock, wiring systems shall be erected so that they are inaccessible to livestock or suitably protected against mechanical damage.

Overhead lines shall be insulated.

In areas of agricultural premises where vehicles and mobile agricultural machines are operated, the following methods of installation shall be applied:

(i) Cables shall be buried in the ground at a depth of at least 0.6 m with added mechanical protection
(ii) Cables in arable or cultivated ground shall be buried at a depth of at least 1 m
(iii) Self-supporting suspension cables shall be installed at a height of at least 6 m.

705.522.10 Special attention shall be given to the presence of different kinds of fauna, e.g. rodents.

705.522.16 Conduit systems, cable trunking systems and cable ducting systems

For locations where livestock is kept, external influences shall be classified AF4, and conduits shall have protection against corrosion of at least Class 2 (medium) for indoor use and Class 4 (high protection) outdoors according to BS EN 61386-21.

For locations where the wiring system may be exposed to impact and mechanical shock due to vehicles and mobile agricultural machines, etc, the external influences shall be classified AG3 and:

(i) conduits shall provide a degree of protection against impact of 5 J according to BS EN 61386-21
(ii) cable trunking and ducting systems shall provide a degree of protection against impact of 5 J according to BS EN 50085-2-1.

705.53 Selection and erection of equipment: Protection, isolation, switching, control and monitoring

Only electrical heating appliances with visual indication of the operating position shall be used.

705.537 Isolation and switching

705.537.2 Isolation

The electrical installation of each building or part of a building shall be isolated by a single isolation device according to Chapter 46.

Means of isolation of all live conductors, including the neutral conductor, shall be provided for circuits used occasionally, e.g. during harvest time.

The isolation devices shall be clearly marked according to the part of the installation to which they belong.

Devices for isolation and switching and devices for emergency stopping or emergency switching shall not be erected where they are accessible to livestock or in any position where access may be impeded by livestock.

705.54 Selection and erection of equipment: Earthing arrangements and protective conductors

705.544 Protective bonding conductors

705.544.2 Supplementary bonding conductors

Protective bonding conductors shall be protected against mechanical damage and corrosion, and shall be selected to avoid electrolytic effects.

For example, the following may be used:

(i) Hot-dip galvanized steel strip with dimensions of at least 30 mm × 3 mm

(ii) Hot-dip galvanized round steel of at least 8 mm diameter

(iii) Copper conductor having a minimum cross-sectional area of 4 mm^2.

Other suitable materials may be used.

705.55 Selection and erection of equipment: Other equipment

705.553.1 Socket-outlets

Socket-outlets of agricultural and horticultural premises shall comply with:

(i) BS EN 60309-1, or

(ii) BS EN 60309-2 where interchangeability is required, or

(iii) BS 1363 or BS 546 provided the rated current does not exceed 20 A.

705.56 Safety services

705.560.6 Automatic life support for high density livestock rearing

For high density livestock rearing, systems operating for the life support of livestock shall be taken into account as follows:

(i) Where the supply of food, water, air and/or lighting to livestock is not provided in the event of power supply failure, a secure source of supply shall be provided, such as an alternative or back-up supply (see also Section 551). For the supply of ventilation and lighting units separate final circuits shall be provided. Such circuits shall only supply electrical equipment necessary for the operation of the ventilation and lighting

(ii) Selectivity of the main circuits supplying the ventilation shall be provided in case of any overcurrent and/or short-circuit to Earth

(iii) Where electrically powered ventilation is necessary in an installation one of the following shall be provided:

(a) A standby electrical source ensuring sufficient supply for ventilation equipment, or

NOTE: A notice should be placed adjacent to the standby electrical source, indicating that it should be tested periodically according to the manufacturer's instructions.

(b) temperature and supply voltage monitoring. This can be achieved by one or more monitoring devices. The device(s) shall provide a visual or audible signal that can be readily observed by the user and shall operate independently from the normal supply.

Fig 705 – Example of supplementary equipotential bonding within a cattle shed

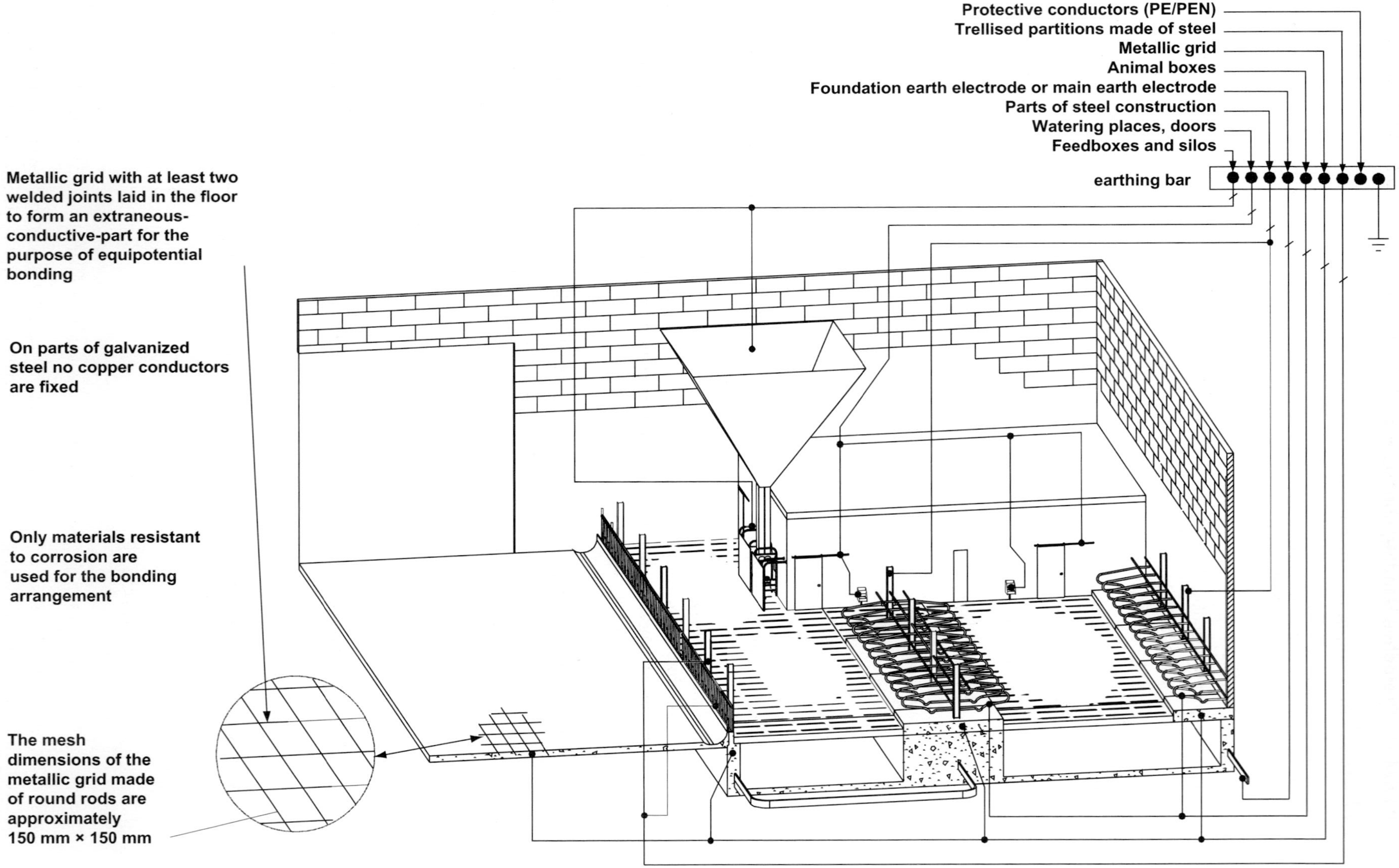

SECTION 706

CONDUCTING LOCATIONS WITH RESTRICTED MOVEMENT

706.1 Scope

The particular requirements of this section apply to:

(i) fixed equipment in conducting locations where movement of persons is restricted by the location, and

(ii) to supplies for mobile equipment for use in such locations.

A conducting location with restricted movement is comprised mainly of metallic or other conductive surrounding parts, within which it is likely that a person will come into contact through a substantial portion of the body with the metallic or other conductive surrounding parts and where the possibility of interrupting this contact is limited.

The particular requirements of this section do not apply to locations which allow a person freedom of bodily movement to work, enter and leave the location without physical constraint.

This section does not apply to electrical systems as defined in BS 7909 used in structures, sets, mobile units etc as used for public or private events, touring shows, theatrical, radio, TV or film productions and similar activities of the entertainment industry.

706.41 Protection against electric shock

706.410.3.3 For the supply to mobile equipment, one of the following protective measures shall be applied:

- SELV in accordance with Section 414, or
- Electrical separation for the supply of only one item of current-using equipment in accordance Section 413.

NOTE: An isolating transformer can have several secondary windings.

For the supply to fixed equipment, one of the following protective measures shall be applied:

- Automatic disconnection of the supply in accordance with Section 411 together with additional protection by supplementary equipotential bonding in accordance with Regulation 415.2, or
- SELV in accordance with Section 414, or
- PELV in accordance with Section 414, together with equipotential bonding between all exposed-conductive-parts and extraneous-conductive-parts and the connection of the PELV system to earth, or
- Electrical separation for the supply of only one item of current-using equipment in accordance with Section 413, or
- Double or reinforced insulation in accordance with Section 412 together with additional protection by the use of an RCD having a rated residual operating current not exceeding 30 mA.

706.410.3.5 The protective measures of obstacles and placing out of reach (Section 417) shall not be used.

706.411 Protective measure: Automatic disconnection of supply

706.411.1 General

706.411.1.1 Only circuits and the protective measures for supplying equipment indicated in Regulation 706.410.3.3 are permitted.

706.413 Protective measure: Electrical separation

706.413.3.2 The source shall be situated outside the conducting location with restricted movement, unless the source is part of the fixed installation within the location.

706.414 Protective measure: Extra-low voltage provided by SELV or PELV

706.414.3 Sources for SELV and PELV

Where SELV or PELV is used, a source described in Regulation 414.3 (iv) shall not be used.

706.414.3.101 A source for SELV or PELV shall be situated outside the conducting location with restricted movement, unless it is part of the fixed installation within the conducting location with restricted movement.

706.414.4 Requirements for SELV and PELV circuits

706.414.4.5 Where SELV or PELV is used, whatever the nominal voltage, basic protection shall be provided by:

(i) basic insulation complying with Regulation 416.1, or

(ii) barriers or enclosures complying with Regulation 416.2.

706.415 Additional protection

706.415.2 Additional protection: supplementary protective equipotential bonding

706.415.2.1 Where functional earthing is required, supplementary protective equipotential bonding shall be provided between all exposed-conductive-parts, extraneous-conductive-parts and the terminals for functional earthing.

SECTION 708

ELECTRICAL INSTALLATIONS IN CARAVAN / CAMPING PARKS AND SIMILAR LOCATIONS

NOTE: In order not to mix requirements on different subjects, such as those for electrical installations of caravan parks with those for electrical installations inside caravans, refer to:
- Section 708, which concerns electrical installations in caravan parks, camping parks and similar locations and
- Section 721, which concerns electrical installations in caravans and motor caravans.

708.1 Scope

The particular requirements contained in this section apply only to circuits intended to supply leisure accommodation vehicles, tents or residential park homes in caravan parks, camping parks and similar locations.

NOTE 1: For installations in caravans and motor caravans which are operated at 12 V DC, BS EN 1648-1 and BS EN 1648- 2 apply.

NOTE 2: For installations in caravans and motor caravans which are operated at voltages other than 12 V DC, Section 721 applies.

NOTE 3: The electrical installations of residential park homes are covered by the general requirements of BS 7671, together with the relevant particular requirements of Part 7.

708.3 Assessment of general characteristics

708.31 Purposes, supplies and structure

708.312 Conductor arrangement and system earthing

708.312.2 Types of system earthing

708.313 Supplies

The nominal supply voltage shall not exceed 230 V AC single-phase or 400 V AC three-phase or 48 V DC.

708.4 Protection for safety

708.41 Protection against electric shock

708.410.3 General requirements

708.410.3.5 The protective measures of obstacles and placing out of reach (Section 417) shall not be used.

708.410.3.6 The protective measures of non-conducting location (Regulation 418.1) and earth-free local equipotential bonding (Regulation 418.2) shall not be used.

708.411.4 TN system

The Electricity Safety, Quality and Continuity Regulations (ESQCR) prohibit the connection of a PME earthing facility to any metalwork in a leisure accommodation vehicle (including a caravan).

This does not preclude the use of a PME earthing facility as the means of earthing for other purposes, such as to the installations of permanent buildings.

NOTE: The requirements of other sections of Part 7 may also apply.

708.415 Additional protection

708.415.1 Residual current protective devices (RCDs)

Every socket-outlet shall be individually protected by an RCD having a rated residual operating current not exceeding 30 mA. Devices selected shall disconnect all live conductors.

A final circuit intended for the fixed connection of a supply to a mobile home or a residential park home shall be individually protected by an RCD having a rated residual operating current not exceeding 30 mA accessible to the consumer. Devices selected shall disconnect all live conductors.

NOTE: This final circuit will be from the connection point to the consumer and not the distribution circuit to the connection point.

708.5 Selection and erection of equipment

708.512 Operational conditions and external influences

708.512.1.1 Presence of water (AD)

Equipment shall be selected with a degree of protection of at least IPX4 in order to protect against water splashes (AD4).

708.512.1.2 Presence of solid foreign bodies (AE)

Equipment shall be selected or provided with a degree of protection of at least IP4X in order to protect against the ingress of very small objects (AE3).

708.512.2 External influences

NOTE: In a caravan park or camping park, special consideration is given to the protection of people, due to the fact that the human body may be in contact with earth potential, to the protection of wiring due to tent pegs or ground anchors and to the movement of heavy or high vehicles.

708.512.2.1.3 Impact (AG)

Equipment installed in a campsite shall be protected against mechanical damage (impact of high severity AG3). Protection of the equipment shall be afforded by one or more of the following:

(i) the position or location shall be selected to avoid damage by any reasonably foreseeable impact

(ii) local or general mechanical protection shall be provided

(iii) equipment shall be installed that complies with a minimum degree of protection against external mechanical impact of IK08 (see BS EN 62262).

708.521 Types of wiring system

708.521.7 Wiring systems in caravan parks

708.521.7.1 The preferred method of supply for feeding the caravan pitch or tent pitch electrical supply equipment is by means of underground distribution circuits.

708.521.7.2 Underground cables

An underground distribution circuit shall, unless provided with additional mechanical protection, be buried at a sufficient depth to avoid being damaged, e.g. by tent pegs or ground anchors or by the movement of vehicles.

NOTE 1: A depth of 0.6 m is generally considered as a minimum depth to fulfil this requirement. Alternatively, the cable may be installed outside the pitch or other area where tent pegs or ground anchors may be driven.

NOTE 2: For conduit systems buried underground, see BS EN 61386-24.

708.521.7.3 Overhead cables and overhead insulated conductors

Every overhead conductor shall be insulated.

Poles and other supports for overhead wiring shall be located or protected so that they are unlikely to be damaged by any foreseeable movement of vehicles.

Every overhead conductor shall be at a height above ground of not less than 6 m in all areas subject to the movement of vehicles and 3.5 m in all other areas.

708.53 Protection, isolation, switching, control and monitoring

708.533 Devices for protection against overcurrent

Every socket-outlet shall be individually protected by an overcurrent protective device, in accordance with the requirements of Chapter 43.

A fixed connection for a supply to a mobile home or residential park home shall be individually protected by an overcurrent protective device, in accordance with the requirements of Chapter 43.

708.537 Isolation and switching

708.537.2 Isolation

708.537.2.1 General

708.537.2.1.1 At least one means of isolation shall be installed in each distribution enclosure. This device shall disconnect all live conductors.

708.55 Other equipment

708.55.1 Socket-outlets

708.55.1.1 Every socket-outlet or connector shall either comply with:

(i) BS EN 60309-2 and shall be interlocked and classified to clause 6.1.5 of BS EN 60309-1 to prevent the socket contacts being live when accessible, or

(ii) be part of an interlocked self-contained product complying with BS EN 60309-4 and classified to clauses 6.1.101 and 6.1.102 of BS EN 60309-4 to prevent the socket contacts being live when accessible.

708.55.1.2 Caravan pitch electrical supply equipment shall be located adjacent to the pitch and not more than 20 m from the connection facility on the leisure accommodation vehicle or tent when on its pitch.

708.55.1.3 In order to avoid any hazard due to long connection cables, no more than 4 socket-outlets shall be grouped together in any one enclosure.

708.55.1.4 Every caravan pitch or tent pitch shall be supplied by at least one socket-outlet.

708.55.1.5 The current rating of socket-outlets shall be not less than 16 A.

708.55.1.6 The lowest part of any socket-outlet shall be placed at a height between 0.5 m and 1.5 m from the ground. In special cases of extreme environmental conditions, it is permitted to exceed the stated maximum height of 1.5 m. In such cases, special measures shall be taken to allow the safe insertion and withdrawal of plugs.

NOTE: This can be necessary if the caravan park or camping park is at risk of being flooded or if either location is used during winter after heavy snow falls.

708.55.1.7 Switchgear and controlgear assemblies used in caravan/tent pitch supplies shall comply with the requirements of BS EN IEC 61439-7.

708.553.1 Plugs and socket-outlets

708.553.1.8 Each socket-outlet and its enclosure forming part of the caravan pitch electrical supply equipment shall comply with BS EN 60309-2 and meet the degree of protection of at least IP44 in accordance with BS EN 60529.

708.553.1.14 Socket-outlet protective conductors shall not be connected to a PME earthing facility.

Fig 708 – Example of a 2-pole and protective conductor supply system between the caravan pitch supply equipment and the caravan or motor caravan

NOTE: See also Regulation 721.55.2.6

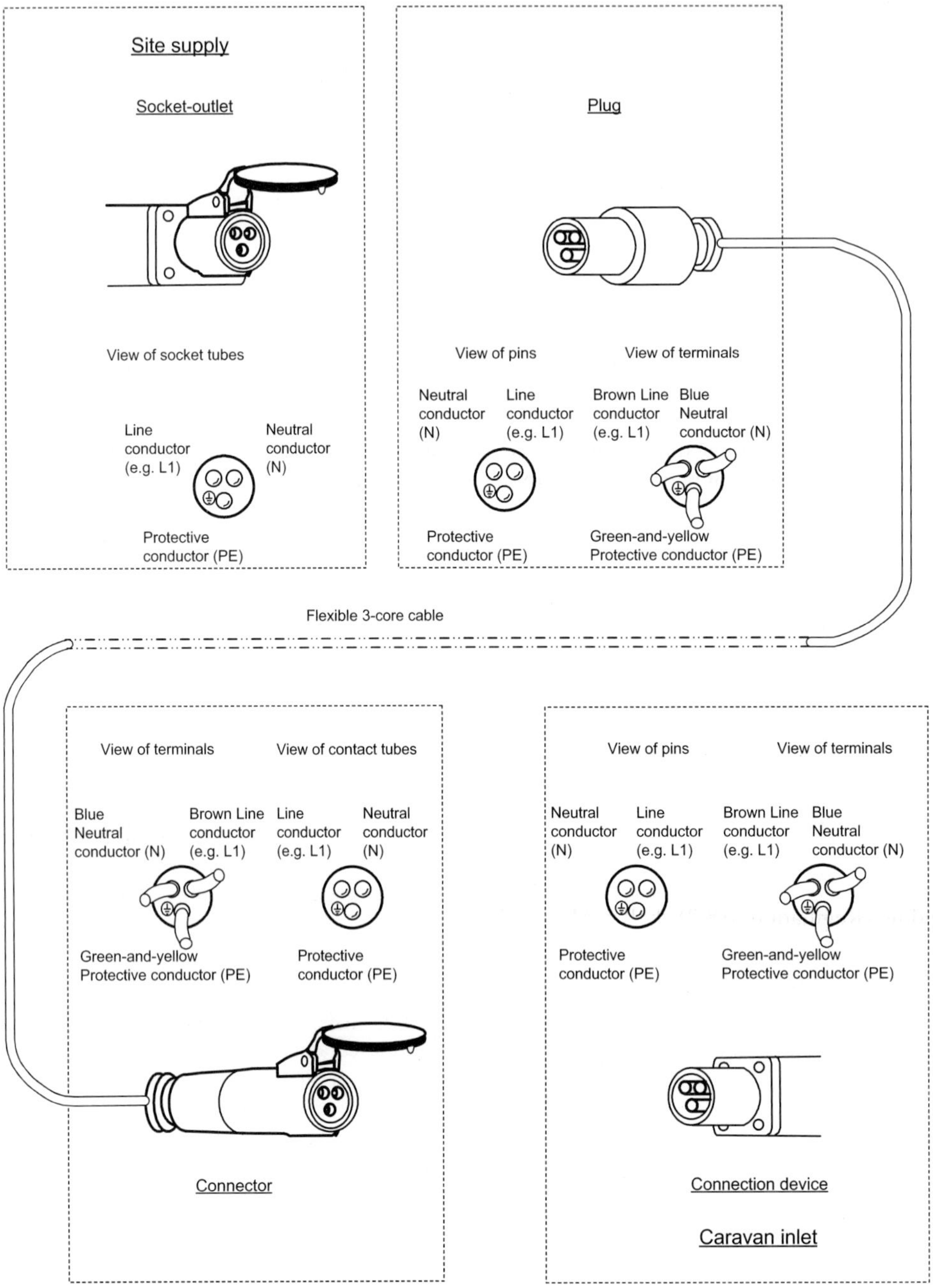

NOTE 1: See Regulation 708.55.1.1 for interlocking requirements.

NOTE 2: Typical requirements for cable extension sets.

The means of connection between the caravan pitch socket-outlet and the leisure accommodation vehicle should be an assembly of the following:

- a plug complying with BS EN 60309-2
- a flexible cable type to H05RN-F or H07RN-F (BS EN 50525-2-21) or equivalent, with a protective conductor and having the following characteristics:
  - continuous length 25 m (±2 m)
  - for current rating 16 A, minimum cross-sectional area: 2.5 mm^2. For a higher current rating, the cross-sectional area must be chosen so that secure tripping of the overcurrent protective device is achieved at the lowest fault current calculated at the end of the cable extension set
  - conductors to be identified in accordance with Table 51
- a connector complying with BS EN 60309-2.

SECTION 709

MARINAS AND SIMILAR LOCATIONS

709.1 Scope

The particular requirements of this section are applicable only to circuits intended to supply pleasure craft or houseboats in marinas and similar locations.

NOTE 1: In this section 'marina' means 'marina and similar locations'.

The particular requirements do not apply to the supply to houseboats if they are supplied directly from the public network.

The particular requirements do not apply to the internal electrical installations of pleasure craft or houseboats.

NOTE 2: For electrical installations of pleasure craft see BS EN 60092-507.

The electrical installations of houseboats shall comply with the general requirements of these Regulations together with the relevant particular requirements of Part 7.

For the remainder of the electrical installation of marinas and similar locations the general requirements of these Regulations together with the relevant particular requirements of Part 7 apply.

709.3 Assessment of general characteristics

709.313 Supplies

709.313.1.2 The nominal supply voltage of the installation for the supply to pleasure craft or houseboats shall be 230 V AC single-phase or 400 V AC three-phase.

709.41 Protection against electric shock

709.410.3 General requirements

709.410.3.5 The protective measures of obstacles and placing out of reach (Section 417) shall not be used.

709.410.3.6 The protective measures of non-conducting location (Regulation 418.1) and earth-free local equipotential bonding (Regulation 418.2) shall not be used.

709.411.4 TN system

The Electricity Safety, Quality and Continuity Regulations (ESQCR) prohibit the connection of a PME earthing facility to any metalwork in a boat.

This does not preclude the use of a PME earthing facility as the means of earthing for other purposes, such as to the installations of permanent buildings.

NOTE: The requirements of other sections of Part 7 may also apply.

709.5 Selection and erection of equipment

709.512 Operational conditions and external influences

709.512.2 External influences

For marinas, particular attention is given in this section to the likelihood of corrosive elements, movement of structures, mechanical damage, presence of flammable fuel and the increased risk of electric shock due to:

(i) presence of water

(ii) reduction in body resistance

(iii) contact of the body with Earth potential.

709.512.2.1.1 Presence of water (AD)

In marinas, equipment installed on or above a jetty, wharf, pier or pontoon shall be selected as follows, according to the external influences which may be present:

(i) Water splashes (AD4): IPX4

(ii) Water jets (AD5): IPX5

(iii) Water waves (AD6): IPX6.

709.512.2.1.2 Presence of solid foreign bodies (AE)

Equipment installed on or above a jetty, wharf, pier or pontoon shall be selected with a degree of protection of at least IP3X in order to protect against the ingress of small objects (AE2).

709.512.2.1.3 Presence of corrosive or polluting substances (AF)

Equipment installed on or above a jetty, wharf, pier or pontoon shall be suitable for use in the presence of atmospheric corrosive or polluting substances (AF2). If hydrocarbons are present, AF3 is applicable.

709.512.2.1.4 Impact (AG)

Equipment installed on or above a jetty, wharf, pier or pontoon shall be protected against mechanical damage (impact of medium severity AG2). Protection shall be afforded by one or more of the following:

(i) The position or location selected to avoid being damaged by any reasonably foreseeable impact
(ii) The provision of local or general mechanical protection
(iii) Installing equipment complying with a minimum degree of protection for external mechanical impact IK08 (see BS EN 62262).

709.521 Types of wiring system

709.521.1 Wiring systems of marinas

709.521.1.4 The following wiring systems are suitable for distribution circuits of marinas:

(i) Underground cables
(ii) Overhead cables or overhead insulated conductors
(iii) Cables with copper conductors and thermoplastic or elastomeric insulation and sheath installed within an appropriate cable management system taking into account external influences such as movement, impact, corrosion and ambient temperature
(iv) Mineral insulated cables with a PVC protective covering
(v) Cables with armouring and serving of thermoplastic or elastomeric material
(vi) Other cables and materials that are no less suitable than those listed above.

709.521.1.5 The following wiring systems shall not be used on or above a jetty, wharf, pier or pontoon:

(i) Cables in free air suspended from or incorporating a support wire, e.g. as installation method No. 35 in Table 4A2
(ii) Non-sheathed cables in cable management systems
(iii) Cables with aluminium conductors
(iv) Mineral insulated cables.

709.521.1.6 Cables shall be selected and installed so that mechanical damage due to tidal and other movement of floating structures is prevented.

Cable management systems shall be installed to allow the drainage of water by drainage holes and/or installation of the equipment on an incline.

709.521.1.7 Underground cables

Underground distribution cables shall, unless provided with additional mechanical protection, be buried at a sufficient depth to avoid being damaged, e.g. by heavy vehicle movement.

NOTE: A depth of 0.5 m is generally considered as a minimum depth to fulfil this requirement.

709.521.1.8 Overhead cables or overhead insulated conductors

All overhead conductors shall be insulated.

Poles and other supports for overhead wiring shall be located or protected so that they are unlikely to be damaged by any foreseeable vehicle movement.

Overhead conductors shall be at a height above ground of not less than 6 m in all areas subjected to vehicle movement and 3.5 m in all other areas.

709.531 Devices for fault protection by automatic disconnection of supply

709.531.2 RCDs

Socket-outlets shall be protected individually by an RCD having the characteristics specified in Regulation 415.1.1. Devices selected shall disconnect all poles, including the neutral.

Final circuits intended for fixed connection for the supply to houseboats shall be protected individually by an RCD having the characteristics specified in Regulation 415.1.1. The device selected shall disconnect all poles, including the neutral.

709.533 Devices for protection against overcurrent

Each socket-outlet shall be protected by an individual overcurrent protective device, in accordance with the requirements of Chapter 43.

A fixed connection for supply to each houseboat shall be protected individually by an overcurrent protective device, in accordance with the requirements of Chapter 43.

709.537 Isolation and switching

709.537.2 Isolation

709.537.2.1 General

709.537.2.1.1 At least one means of isolation shall be installed in each distribution cabinet. This switching device shall disconnect all live conductors including the neutral conductor. One isolating switching device for a maximum of four socket-outlets shall be installed.

709.55 Other equipment

709.553.1 Plugs and socket-outlets

709.553.1.8 Socket-outlets shall comply with BS EN 60309-1 above 63 A and BS EN 60309-2 up to 63 A. Every socket-outlet shall meet the degree of protection of at least IP44 or such protection shall be provided by an enclosure.

Where the codes AD5 or AD6 are applicable the degree of protection shall be at least either IPX5 or IPX6 respectively.

709.553.1.9 Every socket-outlet shall be located as close as practicable to the berth to be supplied.

Socket-outlets shall be installed in the distribution board or in separate enclosures.

709.553.1.10 In order to avoid any hazard due to long connection cords, a maximum of four socket-outlets shall be grouped together in one enclosure.

NOTE: See Figure 709.3 regarding the recommended instruction notice to be placed in marinas adjacent to each group of socket-outlets.

709.553.1.11 One socket-outlet shall supply only one pleasure craft or houseboat.

709.553.1.12 In general, single-phase socket-outlets with rated voltage 200 V – 250 V and rated current 16 A shall be provided.

Where greater demands are envisaged socket-outlets with higher current ratings shall be provided.

709.553.1.13 Socket-outlets shall be placed at a height of not less than 1 m above the highest water level. In the case of floating pontoons or walkways only, this height may be reduced to 300 mm above the highest water level provided that appropriate additional measures are taken to protect against the effects of splashing.

709.553.1.14 Socket-outlet protective conductors shall not be connected to a PME earthing facility.

Figs 709.1 & 709.2 – Examples of methods of obtaining supply in marina

Fig 709.1 – Connection to a single-phase mains supply with RCD

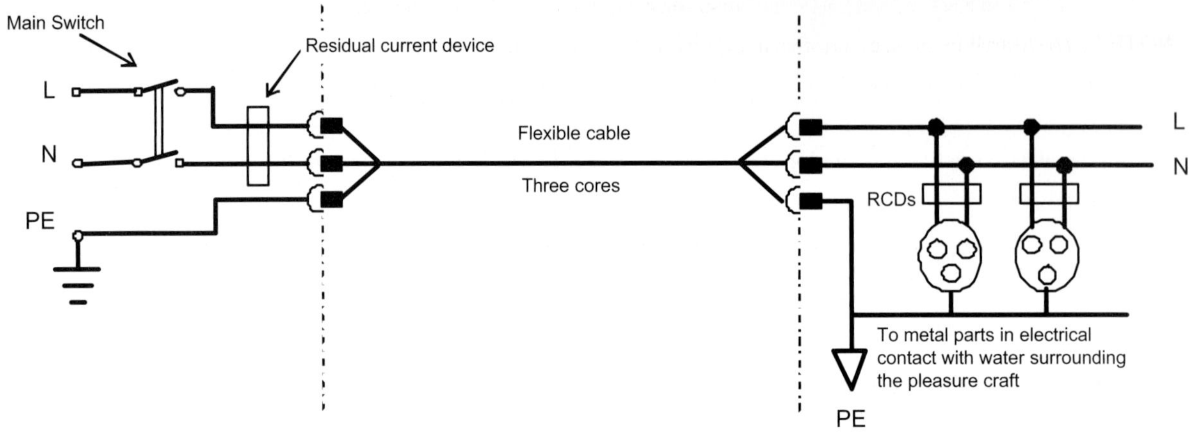

Fig 709.2 – Connection to a three-phase mains supply with RCD

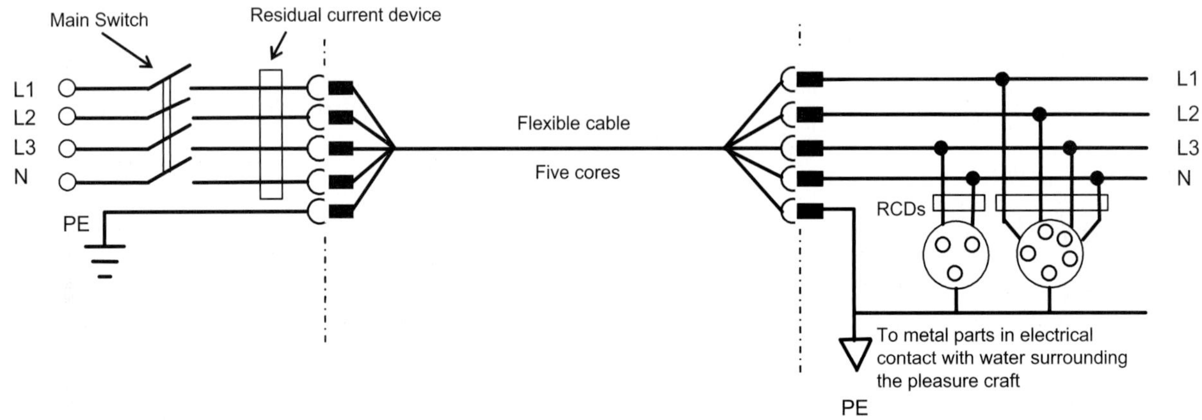

Fig 709.3 – Example of an instruction notice to be placed in marinas

NOTE 1: It is recommended that the marina operator provides every pleasure craft operator who wishes to connect a pleasure craft to an electrical supply with an up-to-date copy of this instruction notice.

NOTE 2: The instruction notice should contain, at least, the following:

INSTRUCTIONS FOR ELECTRICITY SUPPLY

BERTHING INSTRUCTIONS FOR CONNECTION TO SHORE SUPPLY

This marina provides power for use on your pleasure craft with a direct connection to the shore supply which is connected to Earth. Unless you have an isolating transformer fitted on board to isolate the electrical system on your craft from the shore supply system, corrosion through electrolysis could damage your craft or surrounding craft.

ON ARRIVAL

(i) Ensure the supply is switched off and disconnect all current-using equipment on the craft, before inserting the craft plug. Connect the flexible cable **firstly** at the pleasure-craft inlet socket and **then** at the marina socket-outlet.

(ii) The supply at this berth is * V, * Hz. The socket-outlet will accommodate a standard marina plug colour * (technically described as BS EN 60309-2, position 6 h).

(iii) For safety reasons, your craft must not be connected to any other socket-outlet than that allocated to you and the internal wiring on your craft must comply with the appropriate standards.

(iv) Every effort must be made to prevent the connecting flexible cable from falling into the water if it should become disengaged. For this purpose, securing hooks are provided alongside socket-outlets for anchorage at a loop of tie cord.

(v) For safety reasons, only one pleasure-craft connecting cable supplying one pleasure craft may be connected to any one socket-outlet.

(vi) The connecting flexible cable must be in one length, without signs of damage, and not contain joints or other means to increase its length.

(vii) The entry of moisture and salt into the pleasure-craft inlet socket may cause a hazard. Examine carefully and clean the plug and socket before connecting the supply.

(viii) It is dangerous to attempt repairs or alterations. If any difficulty arises, contact the marina management.

BEFORE LEAVING

(i) Ensure that the supply is switched off and disconnect all current-using equipment on the craft, before the connecting cable is disconnected and any tie cord loops are unhooked.

(ii) The connecting flexible cable should be disconnected **firstly** from the marina socket-outlet and **then** from the pleasure-craft inlet socket. Any cover that may be provided to protect the inlet from weather should be securely replaced. The connecting flexible cable should be coiled up and stored in a dry location where it will not be damaged.

* appropriate figures and colours to be inserted:

nominally 230 V 50 Hz blue - single-phase, and

nominally 400 V 50 Hz red - three-phase.

SECTION 710

MEDICAL LOCATIONS

710.1 Scope

The particular requirements of this section apply to patient healthcare facilities, such as, hospitals, private clinics, medical and dental practices, healthcare centres and dedicated medical rooms in the workplace to provide for the safety of patients and medical staff.

In these areas the risk to patients is increased due to:

- the reduction in body resistance, since the skin is often cut or broken, and
- the threat from failure of the supply, especially to life supporting equipment.

This section also applies to electrical installations in locations designed for medical research on patients.

The requirements of this section do not apply to medical electrical (ME) equipment.

NOTE 1: The requirements of other sections of Part 7 may also apply.

NOTE 2: It may be necessary to modify the existing electrical installation, in accordance with BS 7671, when a change of utilization of the location occurs. Special care should be taken where intracardiac and/or life supporting procedures are performed in existing installations.

NOTE 3: Where applicable, the requirements of this section can also be used in veterinary clinics.

NOTE 4: For ME equipment and ME systems, refer to the BS EN 60601 series.

NOTE 5: Care should be taken so that other installations do not compromise the level of safety provided by installations meeting the requirements of this section.

NOTE 6: Supporting information about electrical services supply and distribution in healthcare premises, is given in Health Technical Memorandum (HTM) 06-01 'Electrical Services Supply and Distribution' and HTM 06-02 'Electrical Safety Guidance for Low Voltage Systems', published by the Department of Health. There are equivalent guidance documents in the other devolved administrations (Scotland (SHTM), Wales (WHTM) and Northern Ireland (HTM)).

710.3 Assessment of general characteristics

In order to determine the classification and Group number of a medical location, it is necessary that the relevant medical staff indicate which medical procedures will take place within the location. Based on the intended use, the appropriate classification for the location shall be determined.

NOTE 1: Classification of a medical location is related to the type of contact between applied parts and the patient, the threat to the safety of the patient owing to a discontinuity (failure) of the electrical supply, as well as the purpose for which the location is used. (Guidance on the allocation of a Group number and classification of safety services for medical locations is shown in Annex A710.)

NOTE 2: To provide protection of patients from possible electrical hazards, additional protective measures are applied in medical locations. The type and description of these hazards can vary according to the treatment being administered. The purpose for which a location is to be used may justify areas with different classifications (Group 0, 1 or 2) for different medical procedures.

NOTE 3: Applied parts are defined by the particular standards for ME equipment.

NOTE 4: Where a medical location may be used for different medical procedures the requirements of the higher Group classification should be applied; refer to Annex A710.

710.31 Purposes, supplies and structure

710.312.2 Types of system earthing

PEN conductors shall not be used in medical locations and medical buildings downstream of the main distribution board.

NOTE: Regulation 8(4) of the Electricity Safety, Quality and Continuity Regulations (ESQCR) prohibits the use of PEN conductors in consumers' installations.

710.313 Supplies

710.313.1 General

In medical locations, the distribution system shall be designed and installed to facilitate the automatic changeover from the main distribution network to the electrical safety source feeding essential loads, as required by Regulation 560.5.

710.4 Protection for safety

710.41 Protection against electric shock

710.410.3 General requirements

710.410.3.5 The protective measures of obstacles and placing out of reach (Section 417) shall not be used.

710.410.3.6 The protective measures of:

- non-conducting location (Regulation 418.1),
- earth-free local equipotential bonding (Regulation 418.2), and
- electrical separation for the supply of more than one item of current-using equipment (Regulation 418.3) shall not be used.

NOTE: A medical IT system (Regulation 710.411.6) does not use electrical separation as the sole means of protection against electric shock.

710.411.3 Requirements for fault protection

710.411.3.2 Automatic disconnection in case of a fault

710.411.3.2.5 In medical locations of Group 1 and Group 2, the voltage presented between simultaneously accessible exposed-conductive-parts and/or extraneous-conductive-parts shall not exceed 25 V AC or 60 V DC.

710.411.4 TN system

Additional protection by RCDs having the characteristics specified in Regulation 415.1.1 shall be used:

(i) in final circuits of Group 1 with rated current not exceeding 32 A and below, and

(ii) in final circuits of Group 2, except those of the medical IT system specified in Regulation 710.411.6.

NOTE: It is also permissible to use RCDs above 32 A in Group 1 locations.

710.411.5 TT system

In medical locations of Group 1 and Group 2, RCDs shall be used as protective devices, except for circuits of an IT system specified in Regulation 710.411.6.

710.411.6 IT system

In Group 2 medical locations, an IT system complying with Regulations 710.411.6.3.1, 710.411.6.3.2 and 710.512.1.1 shall be used for final circuits of ME equipment and ME systems intended for life support and surgical applications within the patient environment, excluding:

(i) equipment with a rated power greater than 5 kVA

(ii) X-ray equipment

(iii) the supply of movements of fixed operating tables.

NOTE 1: The list of circuits in (i) to (iii) above is not exhaustive.

NOTE 2: For illustration of a typical theatre layout refer to Figure 710.2.

NOTE 3: A system constructed to the requirements of Regulation 710.411.6 is known as a medical IT system.

The total leakage current without load, including the medical IT transformer, of all circuits connected to a medical IT system shall not exceed 10 mA.

NOTE 4: The medical IT transformer output winding leakage current is limited to 0.5 mA without load. The leakage current due to cable capacitances of all final circuits will add to this value depending on the total cable length of all circuits connected to the transformer. The designer can calculate the total leakage current using capacitances determined from the cable manufacturer's data.

NOTE 5: It is recommended that the maximum length of a final circuit connected to the medical IT distribution board is limited to 25 m.

710.411.6.3.1 For each group of rooms serving the same function, at least one medical IT system is necessary. The IT system shall be equipped with an insulation monitoring device (MED-IMD) in accordance with Annex A and Annex B of BS EN 61557-8.

For each medical IT system, an audible and visual alarm system, incorporating the following components, shall be provided so that it can be permanently monitored by the medical staff and any alarm reported to technical staff:

(i) A green signal lamp to indicate normal operation

(ii) A yellow signal lamp which lights when the minimum value set for the insulation resistance is reached. It shall not be possible for this light to be cancelled or disconnected

(iii) An audible alarm which sounds when the minimum value set for the insulation resistance is reached. This audible alarm may be silenced

(iv) The yellow signal shall extinguish on removal of the fault and when the normal condition is restored.

Documentation shall be easily readable in the medical location and it shall include:

(v) the meaning of each type of signal, and

(vi) the procedure to be followed in case of an alarm at first fault.

710.411.6.3.2 The IT transformer shall be provided with monitoring of overload and over temperature.

710.411.6.3.3 In addition to an insulation monitoring device where a medical IT system serves more than one room or more than one patient place of treatment then an insulation fault location system in accordance with BS EN 61557-9 shall be installed.

710.411.7 Functional extra-low voltage (FELV)

In medical locations, functional extra-low voltage (FELV) shall not be used as a method of protection against electric shock.

710.414 Protective measure: Extra-low voltage provided by SELV or PELV

710.414.1 General

When using SELV and/or PELV circuits in medical locations of Group 1 and Group 2, the nominal voltage applied to current-using equipment shall not exceed 25 V AC rms or 60 V ripple-free DC. Protection by basic insulation of live parts as required by Regulation 416.1 or by barriers or enclosures as required by Regulation 416.2, shall be provided.

710.414.3 Where SELV or PELV is used in Groups 1 and/or 2 of a medical location, a source described in Regulation 414.3(iv) shall not be used.

710.414.4.1 In medical locations of Group 2, where PELV is used, exposed-conductive-parts of equipment, e.g. operating theatre luminaires, shall be connected to the circuit protective conductor.

710.415.1 Additional protection: RCDs

Where a medical IT system is used, additional protection by means of an RCD is not required.

710.415.2 Additional protection: Supplementary protective equipotential bonding

710.415.2.1 In each medical location of Group 1 and Group 2, supplementary protective equipotential bonding shall be installed and the supplementary bonding conductors shall be connected to an equipotential bonding busbar (EBB) for the purpose of equalizing potential differences between the following parts, which are located or that may be moved into the 'patient environment':

(i) Protective conductors

(ii) Extraneous-conductive-parts

(iii) Screening against electrical interference fields, if installed

(iv) Connection to conductive floor grids, if installed

(v) Metal screens of isolating transformers, via the shortest route to the earthing conductor.

Supplementary equipotential bonding connection points for the connection of ME equipment shall be available in Group 2 and should also be considered in Group 1 medical locations.

The designer in consultation with the end user shall determine the appropriate number of supplementary equipotential bonding connections.

NOTE 1: In a Group 1 medical location, a minimum of one supplementary equipotential bonding connection point per patient location is recommended. In a Group 2 medical location, a minimum number of four supplementary equipotential bonding connection points but not less than 25 % of the total number of individual medical IT socket-outlets provided per patient location is recommended.

NOTE 2: Manufacturers of fixed conductive non-electrical patient supports such as operating theatre tables, physiotherapy couches and dental chairs may require the equipment to be connected to the equipotential bonding conductor.

710.415.2.2 In medical locations of Group 1 and Group 2, the resistance of the protective conductors between the earth terminal of any socket-outlet (or fixed equipment) and any exposed-conductive-part and/or extraneous-conductive-part shall be such that the voltages given in 710.411.3.2.5 are not exceeded, and the measured resistance between the earth terminal of any socket-outlet (or fixed equipment) and any extraneous-conductive-part shall not exceed 0.2 Ω.

NOTE: In TN and TT systems, a value of 25 V AC or 60 V DC may be obtained by the provision of protective equipotential bonding in conjunction with circuit protective conductors for the particular circuit. In the case of TT systems a satisfactory value of R_A will also be required.

710.415.2.3 An equipotential bonding busbar (EBB) shall be located in or near each medical location it serves and be readily accessible with consideration to meeting local infection prevention and control requirements. The EBB shall not be located in a ceiling void due to infection risks associated with ceiling tile removal for inspection and testing purposes.

Infection prevention and control requirements shall include the use of suitable EBB enclosure.

The EBB shall be connected to the system earthing using a protective conductor having a cross-sectional area greater than or equal to the largest cross-sectional area of any conductor connected to the EBB.

Connections shall be so arranged that they are accessible, labelled, clearly visible and can be individually disconnected.

NOTE 1: Where an EBB is to be located 'near' to the medical location, this is generally considered to be immediately outside the room containing the respective patient environment.

NOTE 2: It is recommended that radial wiring patterns are used to avoid 'earth loops' that may exacerbate electromagnetic disturbances.

710.421.1.7 In medical locations of Group 1 and 2 Arc Fault Detection Devices (AFDDs) are not required to be installed. In medical locations of Group 0 Arc Fault Detection Devices (AFDDs) shall be used subject to a risk assessment.

710.444 Measures against electromagnetic disturbances

Special considerations have to be made concerning electromagnetic interference (EMI) and electromagnetic compatibility (EMC).

710.5 Selection and erection of equipment

710.51 Common rules

710.511.1 Distribution boards

Distribution boards shall meet the requirements of BS EN 61439 series.

Distribution boards for Group 2 locations shall be installed on the same floor level, immediately adjacent to or, where permitted, in the locations that they supply and shall be clearly labelled.

710.512 Operational conditions and external influences

710.512.1 Operational conditions

710.512.1.1 Transformers for medical IT systems

Transformers shall be in accordance with BS EN 61558-2-15. They shall be on the same floor level, within the same fire compartment as the medical location they supply, and the following additional requirements apply:

(i) The leakage current of the output winding to earth and the leakage current of the enclosure, when measured in no-load condition and with the transformer supplied at rated voltage and rated frequency, shall not exceed 0.5 mA

(ii) At least one single-phase transformer per room or functional group of rooms shall be used to form the medical IT system. The rated output of each transformer shall be not less than 0.5 kVA and shall not exceed 10 kVA. Where several transformers are needed to supply equipment in one room, they shall not be connected in parallel

(iii) If the supply of three-phase loads via a medical IT system is also required, a separate three-phase transformer shall be provided for this purpose.

For monitoring see Regulation 710.411.6.3.1.

Capacitors shall not be used in transformers for medical IT systems.

NOTE: Where the medical IT system cannot be located in the same fire compartment then additional measures may be provided to achieve compliance.

710.512.1.2 Power supply for medical locations of Group 2

In case of a fault or a failure of supply, a total loss of power in a Group 2 medical location shall be prevented.

NOTE: Supporting information is given in HTM 06-01. See also note 6 of Regulation 710.1.

710.512.2.1 Explosion risk

Electrical devices, e.g. socket-outlets and switches, installed below any medical-gas outlets for oxidizing or flammable gases shall be located at a distance of at least 0.2 m from the outlet (centre to centre), so as to minimize the risk of ignition of flammable gases.

NOTE: Requirements for ME equipment for use in conjunction with flammable gases and vapours are contained in BS EN 60601.

710.514.9 Diagrams and documentation

710.514.9.1 Plans of the electrical installation together with records, drawings, wiring diagrams and modifications relating to the medical location, shall be provided.

Information provided shall include but not be limited to:

(i) single-line overview diagrams showing the distribution system of the normal power supply and power supply for safety services in a single-line representation

(ii) distribution board block diagrams showing switchgear and controlgear and distribution boards in a single-line representation

(iii) schematic diagrams of controls

(iv) the verification of compliance with the requirements of standards

(v) functional description for the operation of the safety power supply services and of the safety power supply system

(vi) schematic drawing showing local supplementary equipotential bonding layout, location of the EBB and internal connections of the EBB

(vii) record schedule and test data for the supplementary protective equipotential bonding system, including the connected item, size of conductor and resistance measurement.

See also Regulation 710.65 Periodic inspection and testing.

710.52 Selection and erection of wiring systems

Any wiring system within Group 2 medical locations shall be exclusively for the use of equipment and accessories within that location.

NOTE: This exclusivity does not apply to Group 2 medical locations that are physically connected and are functionally dependent on each other such as a theatre and its associated anaesthetic preparation room.

710.53 Protection, isolation, switching, control and monitoring

710.531.2 Overcurrent protective devices - protection of wiring systems in medical locations of Group 2

Overload current protection shall not be used in either the primary or secondary circuit of the transformer of a medical IT system.

Overcurrent protection against overload and short-circuit currents is required for each final circuit.

NOTE 1: Overcurrent protective devices (e.g. fuses) may be used in the primary circuit of the transformer for short-circuit protection only.

NOTE 2: Figure 710.3 shows a typical medical IT system arrangement.

710.531.3 Types of RCDs

In medical locations of Group 1 and Group 2, where RCDs are required, only Type A according to BS EN 61008 or BS EN 61009 or Type B or Type F according to BS EN 62423 shall be selected, depending on the possible fault current arising. Type AC RCDs shall not be used.

710.537 Isolation and switching

710.537.1 General

Automatic changeover devices shall comply with BS EN 60947-6-1 and shall be arranged so that safe separation between supply lines is maintained.

710.55 Other equipment

710.553.1 Socket-outlet circuits in the medical IT system for medical locations of Group 2

At each patient's place of treatment, e.g. bedheads, the configuration of socket-outlets shall be as follows:

(i) Each socket-outlet supplied by an individually protected circuit, or

(ii) several socket-outlets separately supplied by a minimum of two circuits.

Socket-outlets intended to supply ME equipment shall be:

(iii) unswitched, and

(iv) identified with the colour blue, and

(v) clearly and permanently marked 'Medical Equipment Only'.

NOTE: Supporting information is given in HTM 06-01. See also note 6 of Regulation 710.1.

710.559 Luminaires and lighting installations

In medical locations of Group 1 and Group 2, a minimum of two lighting circuits shall be provided from separate sources of supply. At least one of the circuits shall be supplied from a system for safety services.

The luminaires of the escape route(s) shall be arranged on alternate circuits, one of which shall be supplied from the power supply source for safety services.

710.56 Safety services

A power supply for safety services is required which will maintain the supply for continuous operation for a defined period within a pre-set changeover time.

The safety power supply system shall automatically take over if the voltage of one or more incoming live conductors, at the main distribution board of the building, has dropped for more than 0.5 s and by more than 10 % in regard to the nominal voltage.

NOTE: A list of examples with suggested reinstatement times is given in Table A710 of Annex A710.

710.560.4 Classification

Classification of safety services is given in Regulation 560.4.1.

NOTE: Safety services provided for locations having differing classifications should meet that classification which gives the highest security of supply.

710.560.5.5 General requirements for safety power supply sources of Group 1 and Group 2

Primary cells are not allowed as safety power sources.

An additional main incoming power supply, from the general power supply, is not regarded as a source of the safety power supply.

The availability (readiness for service) of safety power sources shall be monitored and indicated at a suitable location.

710.560.5.6 In case of a failure of the general power supply source, the power supply for safety services shall be energized to feed the equipment stated in Regulations 710.560.6.1.1 to 710.560.6.1.3 with electrical energy for a defined period of time and within a predetermined changeover period.

710.560.5.7 Where socket-outlets are supplied from the safety power supply source they shall be readily identifiable according to their safety services classification.

710.560.6 Electrical sources for safety services

710.560.6.1 Detailed requirements for safety power supply services

NOTE: Also refer to Regulation 710.560.5.5.

710.560.6.1.1 Power supply sources with a changeover period less than or equal to 0.5 s

In the event of a voltage failure on one or more line conductors at the distribution board, a safety power supply source shall be used and be capable of providing power for a period of at least 3 h for the following:

(i) Luminaires of operating theatre tables

(ii) ME equipment containing light sources being essential for the application of the equipment, e.g. endoscopes, including associated essential equipment, e.g. monitors

(iii) Life-supporting ME equipment.

The duration of 3 h may be reduced to 1 h for items (ii) and (iii) if a power source meeting the requirements of Regulation 710.560.6.1.2 is installed.

The normal power supply shall be restored within a changeover period not exceeding 0.5 s.

NOTE: Supporting information relating to the autonomy of battery inverter units for theatre luminaires is given in HTM 06-01. See also note 6 of Regulation 710.1.

710.560.6.1.2 Power supply sources with a changeover period less than or equal to 15 s

Equipment meeting the requirements of Regulations 710.560.9.1 and 710.560.11 shall be connected within 15 s to a safety power supply source capable of maintaining it for a minimum period of 24 h, when the voltage of one or more live conductors at the main distribution board for the safety services has decreased by more than 10 % of the nominal value of supply voltage and for a duration greater than 3 s.

710.560.6.1.3 Power supply sources with a changeover period greater than 15 s

Equipment, other than that covered by Regulations 710.560.6.1.1 and 710.560.6.1.2, which is required for the maintenance of healthcare installations, shall be connected either automatically or manually to a safety power supply source capable of maintaining it for a minimum period of 24 h. This equipment may include, for example:

(i) Sterilization equipment

(ii) Technical building installations, in particular air conditioning, heating and ventilation systems, building services and waste disposal systems

(iii) Cooling equipment

(iv) Catering equipment

(v) Storage battery chargers.

710.560.7 Circuits of safety services

The circuit which connects the power supply source for safety services to the main distribution board shall be considered a safety circuit.

The circuit which connects the loads identified in Regulations 710.560.9.1 and 710.559 shall be considered a safety circuit.

710.560.9 Emergency lighting systems

710.560.9.1 Safety lighting

In the event of mains power failure, the changeover period to the safety services source shall not exceed 15 s. The necessary minimum illuminance shall be provided for the following:

(i) Emergency lighting and exit signs

(ii) Locations for switchgear and controlgear for emergency generating sets, for main distribution boards of the normal power supply and for power supply for safety services

(iii) Rooms in which essential services are intended. In each such room at least one luminaire shall be supplied from the power source for safety services

(iv) Locations of central fire alarm and monitoring systems

(v) Rooms of Group 1 medical locations; in each such room at least one luminaire shall be supplied from the power supply source for safety services

(vi) Rooms of Group 2 medical locations; a minimum of 90 % of the lighting shall be supplied from the power source for safety services.

The luminaires of the escape routes shall be arranged on alternate circuits.

710.560.11 Other services

Other services which may require a safety service supply with a changeover period not exceeding 15 s include, for example, the following:

(i) Firefighters lifts

(ii) Ventilation systems for smoke extraction

(iii) Paging/communication systems

(iv) ME equipment used in Group 2 medical locations which serves for surgical or other procedures of vital importance. Such equipment will be defined by responsible staff

(v) Electrical equipment of medical gas supply including compressed air, vacuum supply and narcosis (anaesthetics) exhaustion as well as their monitoring devices

(vi) Fire detection and fire alarms

(vii) Fire extinguishing systems.

710.6 Inspection and testing

NOTE: The testing of equipment connected to the electrical installation is outside the scope of this document. For ME equipment refer to BS EN 62353.

710.64 Initial verification

710.641 General

The dates and results of each verification shall be recorded.

The tests specified below under items (i) to (iii), in addition to the requirements of Chapter 64, shall be carried out, both prior to commissioning and after alteration or repairs and before re-commissioning:

(i) Complete functional tests of the insulation monitoring devices (IMDs) associated with the medical IT system including insulation failure, transformer high temperature, overload, discontinuity and the audible and/or visual alarms linked to them

(ii) Measurements of leakage current as specified by Regulations 710.411.6 and 710.512.1.1(i)

(iii) Measurements to verify that the resistance of the supplementary equipotential bonding is within the limits stipulated by Regulation 710.415.2.2.

NOTE: The tests specified within HTM 06-01 may also be required as part of client requirements. See also note 6 of Regulation 710.1.

710.65 Periodic inspection and testing

710.651 General

NOTE 1: In addition to the requirements of Chapter 65, the following procedures are recommended at the given intervals:

(i) Annually – Complete functional tests of the insulation monitoring devices (IMDs) associated with the medical IT system including insulation failure, transformer high temperature, overload, discontinuity and the audible/visual alarms linked to them.

(ii) Annually – Measurements to verify that the resistance of the supplementary protective equipotential bonding is within the limits stipulated by Regulation 710.415.2.2 and recorded in the same format required by Regulation 710.514.9.1(vii).

(iii) Every 3 years – Measurements of leakage current of the output circuit and of the enclosure of the medical IT transformers in no-load condition, as specified by Regulation 710.512.1.1(i).

NOTE 2: Supporting information on periodic inspection and testing is given in HTM 06-01. Client or local Health Authority requirements, if any, may apply. See also note 6 of Regulation 710.1.

Fig 710.1 – Example of patient environment (BS EN 60601)

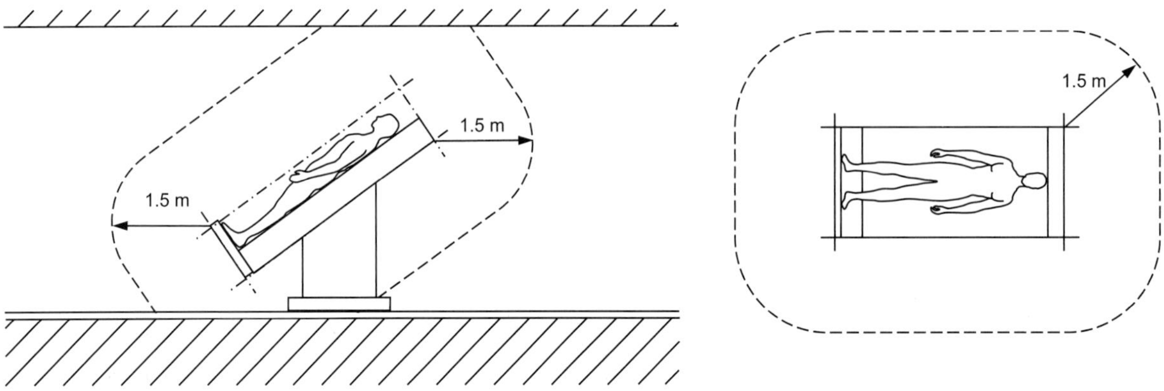

NOTE: The dimensions in the figure show the minimum extent of the patient environment in a free surrounding. This applies where the patient's position is predetermined; if not, all possible patient positions should be considered (see Patient environment, Part 2, Definitions).

Fig 710.2 – Typical theatre layout

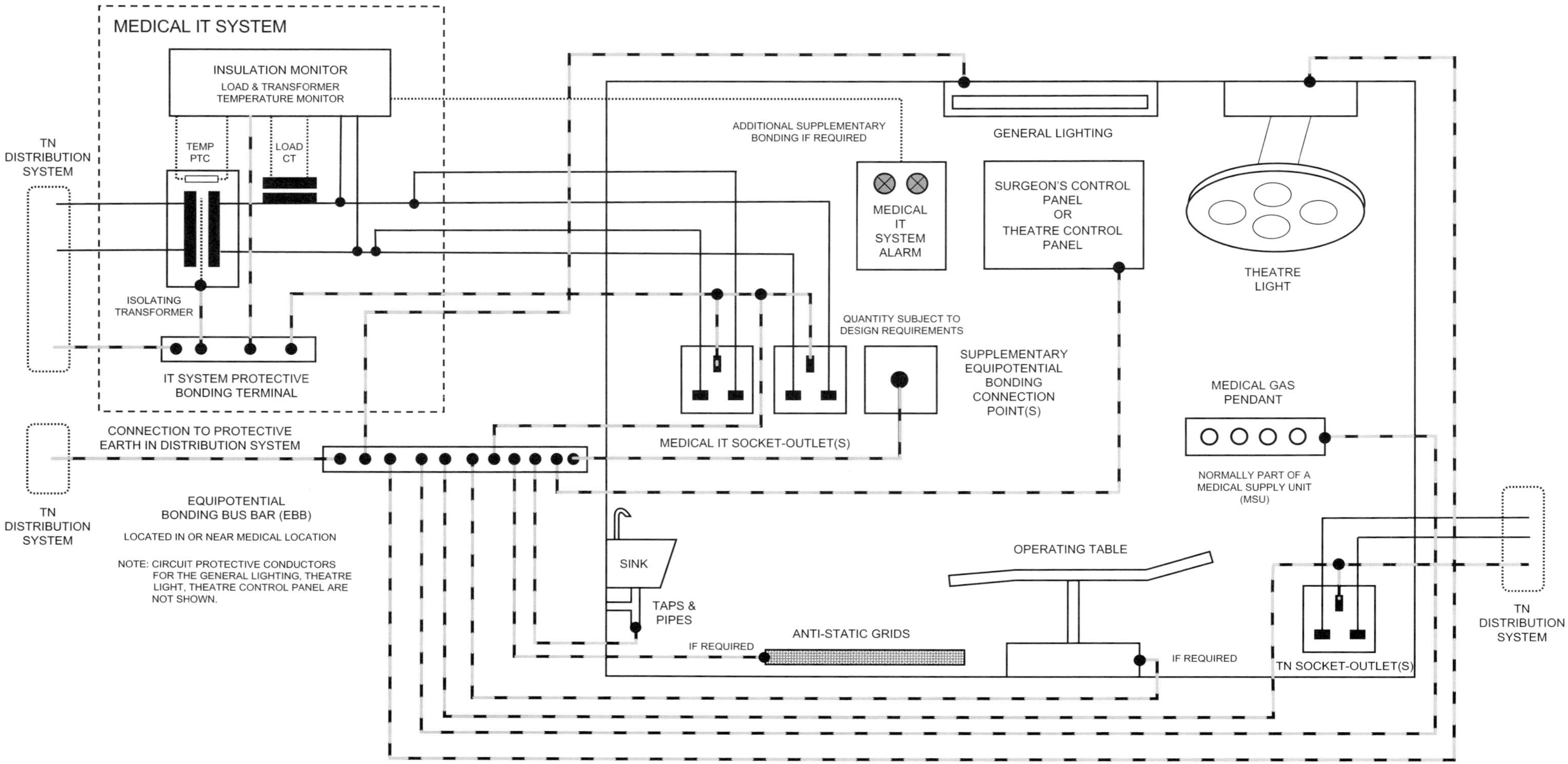

Fig 710.3 – Typical medical IT system arrangement

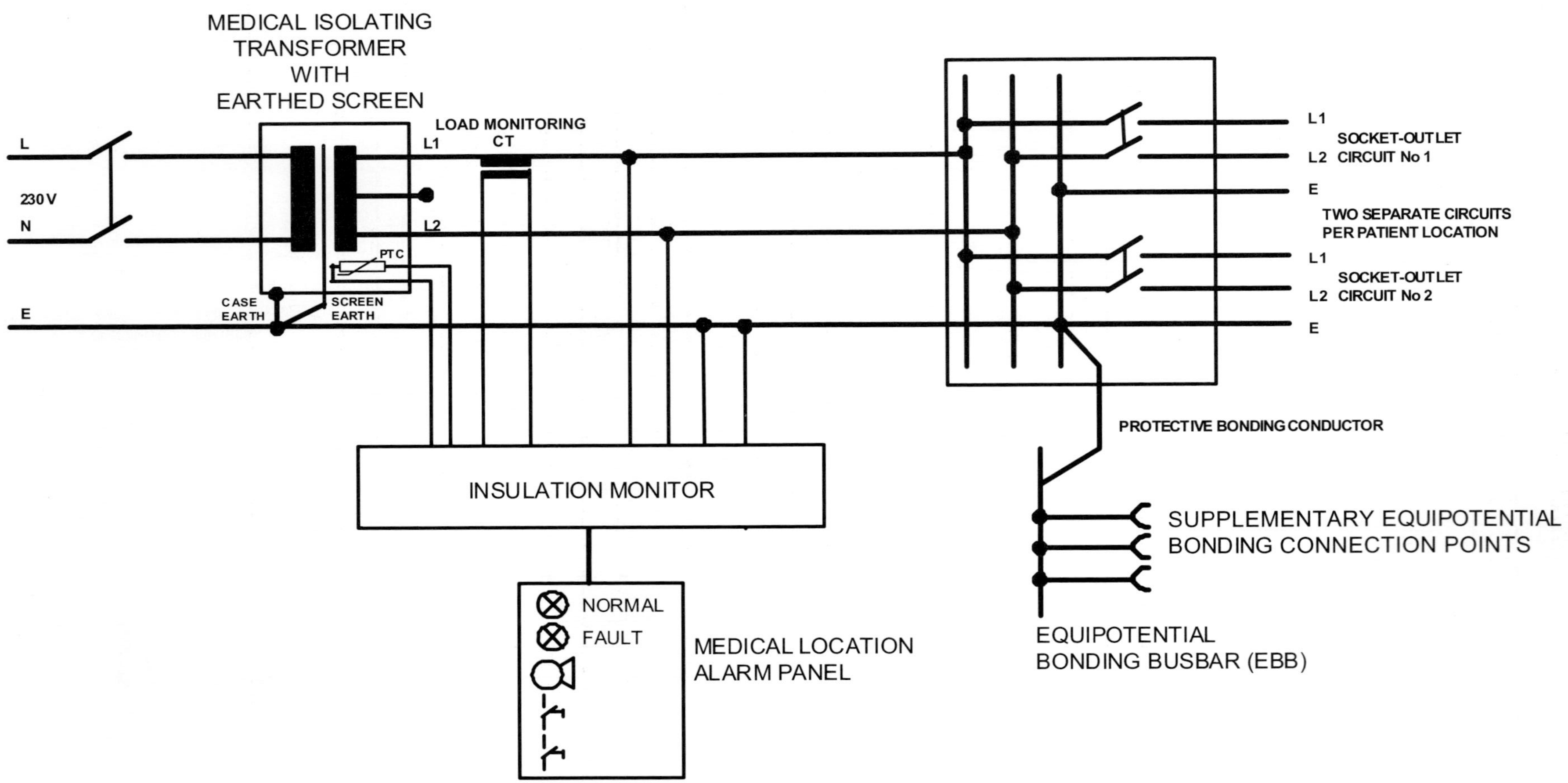

ANNEX A710

(Informative)

MEDICAL LOCATIONS

Examples for allocation of Group numbers and classification for safety services of medical locations

A definitive list of medical locations showing their assigned Groups is impracticable, as is the use to which locations (rooms) might be put. The accompanying list of examples is provided as a guide only and should be read in conjunction with Regulation 710.3. The requirements of Regulation 710.3 cannot be satisfied by using this table alone.

TABLE A710 – List of examples

Medical location	Group			Classification	
	0	1	2	≤ 0.5 s	> 0.5 s ≤ 15 s
1 Massage room	X	X			X
2 Bedrooms		X			X
3 Delivery room		X		X[a]	X
4 ECG, EEG, EHG room		X			X
5 Endoscopic room		X[b]		X	X[b]
6 Examination or treatment room		X		X	X
7 Urology room		X[b]		X	X[b]
8 Radiological diagnostic and therapy room		X	X	X	X
9 Hydrotherapy room		X			X
10 Physiotherapy room		X			X
11 Anaesthetic area			X	X[a]	X
12 Operating theatre			X	X[a]	X
13 Operating preparation room			X	X[a]	X
14 Operating plaster room			X	X[a]	X
15 Operating recovery room			X	X[a]	X
16 Heart catheterization room			X	X[a]	X
17 Intensive care room			X	X[a]	X
18 Angiographic examination room			X	X[a]	X
19 Haemodialysis room		X			X
20 Magnetic resonance imaging (MRI) room		X	X	X	X
21 Nuclear medicine		X			X
22 Premature baby room			X	X[a]	X
23 Intermediate Care Unit (IMCU)			X	X	X

a Specific luminaires, such as operating or procedure lights, that require a power supply within 0.5 s and life-support medical electrical equipment that requires a power supply within 0.5 s.
b Not being an operating theatre.

SECTION 711

EXHIBITIONS, SHOWS AND STANDS

711.1 Scope

The particular requirements of this section apply to the temporary electrical installations in exhibitions, shows and stands (including mobile and portable displays and equipment) to protect users. Unless specifically stated, this section does not apply to exhibits for which requirements are given in the relevant standards.

This section does not apply to the fixed electrical installation of the building, if any, in which the exhibition, show or stand may take place.

This section does not apply to electrical systems as defined in BS 7909 used in structures, sets, mobile units etc as used for public or private events, touring shows, theatrical, radio, TV or film productions and similar activities of the entertainment industry.

The requirements of other sections of Part 7 may also apply.

711.3 Assessment of general characteristics

711.313 Supplies

The nominal supply voltage of a temporary electrical installation in anexhibition, show or stand shall not exceed 230/400 V AC or 500 V DC.

711.41 Protection against electric shock

711.410.3 General requirements

711.410.3.4 A cable intended to supply temporary structures shall be protected at its origin by an RCD having a rated residual operating current not exceeding 300 mA. This device shall provide a delay by using a device in accordance with BS EN 60947-2, or be of the type S in accordance with BS EN 61008-1 or BS EN 61009-1 for selectivity with RCDs protecting final circuits.

NOTE: The requirement for cable protection relates to the increased risk of damage to cables in temporary locations.

711.410.3.5 The protective measures of obstacles and placing out of reach (Section 417) shall not be used.

711.410.3.6 The protective measures of:

- non-conducting location (Regulation 418.1),
- earth-free local equipotential bonding (Regulation 418.2), and
- electrical separation for the supply of more than one item of current-using equipment (Regulation 418.3) shall not be used.

711.410.3.101 Except for final circuits of safety services, all final circuits with a rated current not exceeding 32 A supplying socket-outlets or handheld electrical equipment, and all final circuits for lighting, shall use one of the following protective measures:

(i) automatic disconnection of supply (Section 411) together with additional protection by means of an RCD with a rated residual operating current not exceeding 30 mA, or

(ii) SELV or PELV (Section 414), or

(iii) electrical separation (Section 413), where each socket-outlet and item of hand-held electrical equipment is supplied by an individual isolating transformer or by separate windings of an isolating transformer.

711.411 Protective measure: Automatic disconnection of supply

711.411.3.1.2 Protective equipotential bonding

Structural metallic parts which are accessible from within the stand, vehicle, wagon, caravan or container shall be connected through the main protective bonding conductors to the main earthing terminal within the unit.

711.411.4 TN system

Except for a part of an installation within a building, a PME earthing facility shall not be used as the means of earthing for an installation falling within the scope of this section except where:

(i) the installation is continuously under the supervision of a skilled or instructed person(s), and

(ii) the suitability and effectiveness of the means of earthing has been confirmed before the connection is made.

711.414 Protective measure: Extra-low voltage provided by SELV or PELV

711.414.4.5 Where SELV or PELV is used, whatever the nominal voltage, basic protection shall be provided by:

(i) basic insulation complying with Regulation 416.1, or

(ii) by barriers or enclosures complying with Regulation 416.2 and affording a degree of protection of at least IPXXD or IP4X.

711.5 Selection and erection of equipment

711.51 Common rules

Switchgear and controlgear shall be placed in closed cabinets which can only be opened by the use of a key or a tool, except for those parts designed and intended to be operated by ordinary persons.

711.52 Wiring systems

Armoured cables or cables protected against mechanical damage shall be used wherever there is a risk of mechanical damage.

Wiring cables shall be copper, have a minimum cross-sectional area of 1.5 mm^2, and shall comply with an appropriate British or Harmonized Standard for either thermoplastic or thermosetting insulated electric cables.

Flexible cables shall not be laid in areas accessible to the public unless they are protected against mechanical damage.

711.521 Types of wiring system

711.521.101 Where no fire alarm system is installed in a building used for exhibitions etc. cable systems shall be either:

(i) flame retardant to BS EN 60332-1-2 or to a relevant part of the BS EN 60332-3 series, and low smoke to BS EN 61034-2, or

(ii) single-core or multicore unarmoured cables enclosed in metallic or non-metallic conduit or trunking, providing fire protection in accordance with BS EN 61386 series or BS EN 50085 series and providing a degree of protection of at least IP4X or IPXXD.

711.526 Electrical connections

711.526.1 Joints shall not be made in cables except where necessary as a connection into a circuit. Where joints are made, these shall either use connectors in accordance with relevant standards or be in enclosures with a degree of protection of at least IPXXD or IP4X.

Where strain can be transmitted to terminals the connection shall incorporate suitable cable anchorage(s).

711.53 Protection, isolation, switching, control and monitoring

711.536 Coordination of electrical equipment for protection, isolation, switching and control

711.536.4.1.4 Selectivity between RCDs

Selectivity between RCDs installed in series shall be provided.

711.537 Isolation and switching

711.537.2 Devices for isolation

711.537.2.1.1 Every separate temporary structure, such as a vehicle, a stand or a unit, intended to be occupied by one specific user and each distribution circuit supplying outdoor installations shall be provided with its own readily accessible and properly identifiable means of isolation.

The means of isolation shall be selected and erected in accordance with Section 462 and Regulation 537.2.

711.55 Other equipment

711.55.4 Electric motors

711.55.4.1 Isolation

Where an electric motor might give rise to a hazard, the motor shall be provided with an effective means of isolation on all poles and such means shall be adjacent to the motor which it controls (see BS EN 60204-1).

711.55.7 Socket-outlets

Where a floor mounted socket-outlet is installed, it shall be adequately protected from accidental ingress of water and have sufficient strength to be able to withstand the expected traffic load.

711.55.101 ELV transformers and electronic converters

Multiple connection extra-low voltage (ELV) transformers shall conform with BS EN 61558 (all parts) or provide an equivalent degree of safety.

ELV transformers and electronic converters shall be mounted out of arm's reach of ordinary persons.

Electronic converters shall conform with BS EN 61347 (all parts).

711.559 Luminaires and lighting installations

711.559.3 General requirements for installations

711.559.3.101 Luminaires

Luminaires mounted below 2.5 m (arm's reach) from floor level or otherwise accessible to accidental contact shall be firmly and adequately fixed, and so sited or guarded as to prevent risk of injury to persons or ignition of materials.

NOTE: In the case of outdoor lighting installations, Section 714 also applies.

711.559.3.102 Lampholders

Insulation-piercing lampholders shall not be used unless the cables and lampholders are compatible and the lampholders are non-removable once fitted to the cable.

711.559.3.103 Electrical discharge lamp installations

711.559.3.103.1 General

Circuits, other than the internal wiring of an exhibit or luminaire covered by a relevant product standard, operating at voltages exceeding 230/400 V AC derived from an installation of a lower nominal voltage, shall meet the requirements of Regulations 711.559.3.103.2 to 711.559.3.103.3.

711.559.3.103.2 Location

The sign or lamp shall be installed out of arm's reach or shall be adequately protected to reduce the risk of injury to persons.

711.559.3.103.3 Installation

The facia or stand fitting material behind luminous tube signs or lamps shall be non-ignitable.

Controlgear with output voltages higher than 230/400 V AC shall be mounted on non-ignitable material.

711.559.4.2 ELV lighting systems for filament lamps

Extra-low voltage systems for filament lamps shall comply with BS EN 60598-2-23.

711.6 Inspection and testing

The temporary electrical installations of exhibitions, shows and stands shall be inspected and tested on site in accordance with Chapter 64 after each assembly on site.

SECTION 712

SOLAR PHOTOVOLTAIC (PV) POWER SUPPLY SYSTEMS

712.1 Scope

This section applies to the electrical installation of PV generators intended to supply all or part of an installation and to feed electricity into the public grid or local distribution. In this section, the electrical equipment of a PV generator, like any other item of electrical equipment, is dealt with only so far as its selection and application in the installation is concerned. The electrical installation of a PV generator starts from a PV module or a set of PV modules connected in series with their cables, provided by the PV module manufacturer, up to the user's installation or the utility supply point. The requirements of this section apply to:

(i) PV generators for supply to an installation which is not connected to a system for distribution of electricity to the public,

(ii) PV generators for supply to an installation in parallel with a system for distribution of electricity to the public,

(iii) PV generators for supply to an installation as an alternative to a system for distribution of electricity to the public.

NOTE: Requirements for PV power supply systems which are intended for stand-alone operation are under consideration.

712.3 Assessment of general characteristics

712.31 Purposes, supplies and structure

712.312 System earthing

712.312.2 Type of earthing arrangement

Earthing of one of the live conductors of the DC side is permitted, if there is at least simple separation between the AC side and the DC side.

NOTE: Any connections with Earth on the DC side should be connected so as to avoid corrosion (see BS EN 13636 and BS EN 15112).

712.4 Protection for safety

712.41 Protection against electric shock

712.410.3 General requirements

712.410.3.5 The protective measure of obstacles and placing out of reach (Section 417) shall not be used.

712.410.3.6 The protective measures of:

- non-conducting location (Regulation 418.1),
- earth-free local equipotential bonding (Regulation 418.2), and
- electrical separation for the supply of more than one item of current-using equipment (Regulation 418.3) shall not be used.

712.410.101 Electrical equipment on the DC side shall be considered to be energized, even when the AC side is disconnected from the grid or when the inverter is disconnected from the DC side.

712.410.102 On the DC side, one of the following protective measures shall be applied:

(i) double or reinforced insulation (Section 412)

(ii) extra-low voltage (SELV or PELV) (Section 414).

712.412 Protective measure: Double or reinforced insulation

712.412.101 The electrical equipment, for example PV modules, wiring system (e.g. combiner box, cables) used on the DC side (up to the DC connection means of the PV inverter) shall be Class II or equivalent insulation.

NOTE: For wiring systems, see Regulation 412.2.4.

712.414 Protective measure: Extra-low voltage provided by SELV or PELV

712.414.1 General

712.414.1.1 For SELV and PELV systems, protective measures on the DC side, $U_{OC\ MAX}$ shall not exceed 120 V DC.

712.414.4.5 If the nominal voltage exceeds 30 V DC, basic protection shall be provided for SELV and PELV circuits.

712.42 Protection against thermal effects

712.421 Protection against fire caused by electrical equipment

712.421.101 Protection against the effects of insulation faults on the DC side in case of simple separation inside the inverter or on the AC side

712.421.101.1 An insulation monitoring device (IMD) shall be installed except where Regulation 712.421.101.2 applies, to verify the insulation status on the DC side throughout the life cycle of the PV array.

NOTE: Insulation monitoring devices (IMDs) complying with BS EN 61557-8 provide this function.

The monitoring function may be provided by an inverter with integrated insulation monitoring also capable of detecting insulation faults.

712.421.101.2 Where functional earthing is applied to a live conductor inside the inverter on the DC side, means shall be provided to ensure the interruption of the fault current in case of an insulation fault to earth in accordance with Table 712.2.

712.43 Protection against overcurrent

712.431 Requirements according to the nature of the circuits

712.431.101 In a PV array with N_S strings (above 2 strings) in parallel, protective devices shall be provided to protect each PV string where the following condition is met:

$$1.35\ I_{MOD\_MAX\_OCPR} < (N_S - 1)\ I_{SC\ MAX}$$

In a PV array with one PV string or two PV strings in parallel, no overcurrent protective device is required.

NOTE 1: If the inverter has several independent Maximum Power Point Trackers (MPPT) or devices with equivalent characteristics and no reverse current can flow from one input to another input by inverter design, then N_S is the number of strings connected to one individual DC input.

All strings connected in parallel shall have the same rated voltage, unless the same level of protection is provided by alternative means.

NOTE 2: In practice this means that unless individual optimizers or similar devices are used, each string shall have the same number of modules connected in series using equivalent modules.

712.431.102 For the protection of the PV strings, the use of protective devices as specified in 712.432 complies with the thermal withstand capability of the PV modules (as referred to in BS EN IEC 61730-1).

Where protective devices are required their rated current I_n shall comply with the following conditions:

$$1.1\ I_{SC\ MAX} \text{ of the string} \leq I_n \leq I_{MOD\_MAX\_OCPR}$$

NOTE: The coefficient 1.1 is a safety margin for untimely operation of the protection devices taking into account normal stress conditions.

The coefficient of 1.1 shall be adapted in case of special conditions, e.g. in case of reflections or special PV module technologies.

Each string shall be protected individually by a protective device.

If several parallel strings are protected by one overcurrent protective device, the rated current I_n shall comply with the following formula:

$$N_p \times 1.1\ I_{SC\ MAX} \leq I_n \leq I_{MOD\_MAX\_OCPR} - (N_p - 1) \times I_{SC\ MAX}$$

Where:

N_p is the number of parallel strings connected to the same overcurrent protective device.

712.432 Nature of protective devices

712.432.101 Where overcurrent protective devices are required on the DC side according to Regulation 712.431.101, both polarities shall be protected, irrespective of the installation's configuration.

712.432.102 Blocking diodes used to connect the PV strings in parallel shall not be relied upon as a means of protection against overcurrents.

712.432.103 The overcurrent protective devices of the DC side shall be either:

(i) gPV fuses in accordance with BS EN 60269-6 or fuse-combination units in accordance with BS EN 60947-3; or

(ii) circuit-breakers in accordance with BS EN 60947-2, BS EN 60898-2 or BS IEC 60898-3.

712.433 Protection against overload on the DC side

712.433.1 Overload protection may be omitted to PV string and PV array cables when the continuous current-carrying capacity of the cable is equal to or greater than 1.25 times $I_{SC\ STC}$ at any location.

712.433.101 Protection of PV string cable

Where overcurrent protection is required by Regulation 712.431.101, the continuous current-carrying capacity, I_Z, of the PV string cable shall be greater than or equal to the short-circuit maximum current of the string:

$I_{SC\ MAX}$ of the string $\leq I_Z$

In a PV array with N_S strings in parallel with N_S greater than 2, the maximum reverse current flowing in the PV string cable is $(N_S - 1)\ I_{SC\ MAX}$. One of the following measures shall be used:

(i) where PV string overcurrent protective devices are not required in accordance with Regulation 712.431.101, the continuous current-carrying capacity I_Z of the PV string cables shall be greater than or equal to the maximum reverse current:

$(N_S - 1)\ I_{SC\ MAX}$ of the string $\leq I_Z$

(ii) where PV string overcurrent protective devices are required in accordance with Regulation 712.431.101, the continuous current-carrying capacity I_Z of the PV string cables shall be greater than or equal to the rated current of the string protective device I_n:

$I_n \leq I_Z$

$I_{SC\ MAX}$ shall be determined in accordance with Regulation 712.433.101.1.

712.433.101.1 Calculation of $U_{OC\ MAX}$ and $I_{SC\ MAX}$

Without information of the expected minimum temperature of the site or without information about the temperature coefficient of the PV module, $U_{OC\ MAX}$ shall be chosen equal to 1.2 $U_{OC\ STC}$.

$I_{SC\ MAX}$ shall be chosen to be at least 1.25 $I_{SC\ STC}$.

NOTE 1: Under certain conditions, $I_{SC\ MAX}$ is increased to take into account environmental situations, for example increased reflection or solar intensity.

NOTE 2: For amorphous silicon modules, electrical characteristics during the first weeks of operation are higher than the specified characteristics. This phenomenon is indicated by the module manufacturer and has to be considered in the calculation of U_{OC} and $I_{SC\ MAX}$.

712.433.102 Protection of PV sub-array cable

For the protection of the PV sub-array cables, the following shall be considered:

(i) in a PV array with one or two sub-arrays, no overcurrent protective device of the PV sub-array cables is required. The continuous current-carrying capacity I_Z of the sub-array cable shall be greater than or equal to the short-circuit maximum current of the sub-array:

$I_{SC\ MAX}$ of the sub-array $\leq I_Z$

(ii) in a PV array with N_a sub-arrays in parallel with N_a greater than two, the maximum reverse current flowing in the PV sub-array cable is $(N_a - 1)\ I_{SC\ MAX}$. One of the following measures shall be used:

(a) where overcurrent protective device of the PV sub-array cables is not used, the continuous current-carrying capacity I_Z of the sub-array cables shall be greater than or equal to the maximum reverse current of the array:

$$(N_a - 1)\ I_{SC\ MAX} \text{ of the sub-array} \leq I_Z$$

(b) where overcurrent protective device of the PV sub-array cables is used, its rated current I_n and the continuous current-carrying capacity I_Z of the sub-array cables shall comply with the following conditions:

$$1.1\ I_{SC\ MAX} \text{ of the sub-array} \leq I_n \leq I_Z$$

NOTE: The coefficient 1.1 is a safety margin for untimely operation of the protection devices taking into account normal stress conditions.

The coefficient of 1.1 shall be adapted in case of special conditions, e.g. in case of reflections or special PV module technologies.

NOTE: Requirements for PV sub-arrays cable sizing are done in the same way as PV array cable.

712.433.103 Protection of PV array cable

The continuous current-carrying capacity I_Z of the PV array cable shall be greater than or equal to the maximum direct current of the PV array:

$$I_{SC\ MAX} \text{ of the array} \leq I_Z$$

712.433.104 Protection of PV AC supply cable

When defining the rated current of the overcurrent protective device for the AC supply cable, the design current of the inverter shall be taken into account.

The inverter design current is either the maximum AC current given by the inverter manufacturer or, if not available, 1.1 times its rated AC current.

712.44 Protection against voltage disturbances and electromagnetic disturbances

712.443 Protection against overvoltages of atmospheric origin or due to switching

712.443.101 Protection against transient overvoltage

Where protection against transient overvoltage is required by Section 443, such protection shall also be applied to the DC side of the PV installation.

712.444 Protection against electromagnetic disturbances

712.444.4.2.14 To minimize induced voltages, the area of all wiring loops shall be as small as possible.

712.5 Selection and erection of equipment

712.51 Common rules

712.511 Compliance with standards

712.511.101 PV modules shall comply with the requirements of the relevant electrical equipment standard, e.g. BS EN 61215-1-2, BS EN 61215-1-4 and BS EN 61215-1-3:2017.

712.511.102 Inverters shall comply with the requirements of the relevant electrical equipment standard, e.g. BS EN 62109-1 and BS EN 62109-2.

712.511.103 Assemblies and enclosures

Combiner box, distribution boards and switchgear assemblies shall comply with BS EN 61439 series. For residential premises and similar locations, enclosures may alternatively comply with BS EN 60670-24.

712.512 Operational conditions and external influences

712.512.1 Operational conditions

712.512.1.1 Voltage

For the selection of equipment for PV arrays, $U_{OC\ MAX}$ shall be considered as nominal voltage. $U_{OC\ MAX}$ shall be determined in accordance Regulation 712.433.101.1.

712.512.1.2 Current

For the selection of equipment for PV arrays, $I_{SC\ MAX}$ shall be considered as design current.

$I_{SC\ MAX}$ shall be determined in accordance with Regulation 712.433.101.1.

712.512.101 If blocking diodes are used, their reverse voltage shall be rated for $2U_{OC\ MAX}$ of the PV string and their rated current shall be not less than 1.1 $I_{SC\ MAX}$. The blocking diodes shall be connected in series with the PV strings.

712.512.102 Enclosures for electrical equipment installed outdoors shall have a degree of protection not less than IP44 in accordance with BS EN 60529 and a degree of protection against external mechanical impact not less than IK07 in accordance with BS EN 62262.

712.512.2.1 As specified by the manufacturer, the PV modules shall be installed in such a way that there is adequate heat dissipation under conditions of maximum solar radiation for the site.

712.514 Identification and notices

712.514.101 An instruction notice indicating the presence of a photovoltaic system on a building shall be fixed:

(i) at the origin of the electrical installation

(ii) at the metering position, if remote from the origin

(iii) at the consumer unit or distribution board to which the supply from the inverter is connected.

An example is shown in Figure 712.514.101.

Fig 712.514.101 – Indication showing the presence of a photovoltaic installation on a building

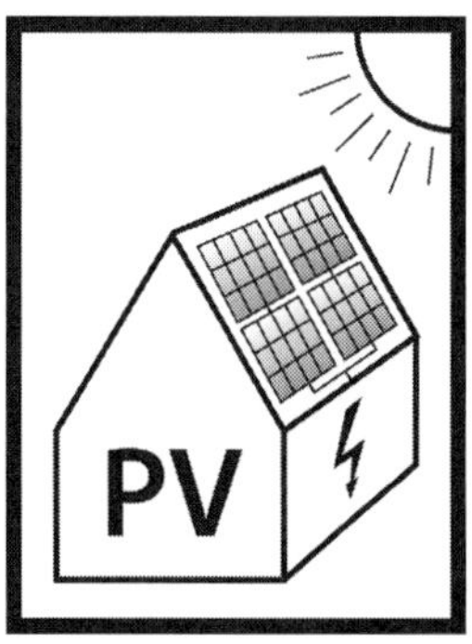

712.514.102 Each point of access to live parts on the DC side, such as distribution boards and combiner boxes, shall have a permanent warning notice indicating that live parts may be still energized after isolation, e.g. by the text, 'SOLAR DC - Live parts can remain energized after isolation'.

712.514.103 All inverters should bear a warning notice indicating that, before any servicing operation, the inverter shall be isolated on both DC and AC sides.

712.52 Wiring systems

712.521.101 Cables on the DC side shall be selected and erected so as to minimize the risk of earth faults and short-circuits. This shall be achieved by using:

(i) single-core cables having a non-metallic sheath, or

(ii) insulated (single-core) conductors installed in individually insulated conduit or trunking. Cable(s) shall not be placed directly on the surface of the roof.

Other types of wiring system providing an equivalent degree of safety are not precluded.

NOTE: BS EN 50618 describes cables intended for use at the direct current (DC) side of photovoltaic systems.

712.521.102 To minimize induced voltages due to lightning, the area of all loops should be as small as possible, in particular for the cabling of PV strings. The DC cables and any equipotential bonding conductor should run side by side.

712.523 Current-carrying capacities

712.523.101 For the design and sizing of cables subjected to direct heating of the underside of the PV module the ambient temperature shall be considered to be at least equal to 70 °C.

712.526 Electrical connections

712.526.1 Each pair of connectors shall be electrically and mechanically compatible and suitable for the environment. It is recommended to check with each manufacturer that each pair of connectors be compatible.

712.526.101 Connectors for the DC side

These connectors shall be selected in accordance with BS EN 62852:2015+A1:2020.

Connectors in a location accessible to persons other than skilled or instructed persons shall either be of a type which can only be disconnected by means of a key or a tool or be installed within an enclosure which can only be opened by means of a key or a tool.

712.53 Protection, isolation, switching, control and monitoring

712.531.3.5.1 Residual Current Devices

Where an RCD is used for protection of the PV AC supply circuit, the RCD shall be of Type B according to BS EN 62423 or BS EN 60947-2, unless:

(i) the inverter provides at least simple separation between the AC side and the DC side, or

(ii) the installation provides at least simple separation between the inverter and the RCD by means of separate windings of a transformer, or

(iii) the inverter does not require a Type B RCD as stated by the manufacturer of the inverter.

NOTE: In some circumstances other types of RCD may need to be selected based on inverter manufacturer's instructions. See also Regulation 531.3.3.

712.533 Devices for protection against overcurrent

712.533.101 Overcurrent protective devices of the DC side

The overcurrent protective devices of the DC side shall be:

(i) gPV fuses in accordance with BS EN 60269-6 or fuse-combination units according to BS EN 60947-3, or

(ii) circuit-breakers in accordance with BS EN 60947-2, BS EN 60898-2 or BS IEC 60898-3.

These devices shall comply with the following specific measures:

(a) the rated operating voltage (U_e) shall be greater than or equal to the voltage $U_{OC\ MAX}$ of the PV array

(b) the rated current I_n as defined in Regulation 712.431.102

(c) the rated breaking capacity shall be at least equal to $I_{SC\ MAX}$ of the PV array

(d) overcurrent protective devices shall be bidirectional.

712.534 Devices for protection against overvoltage

712.534.101 General

Where the PV system is installed within the protected volume of the lightning protection system (LPS), all power and signal cables or lines of the PV system shall be separated from all parts of the LPS in accordance with BS EN 62305-3.

NOTE 1: BS EN 62305-3 describes the calculation of the separation distance.

In addition to shielded lines, close and parallel wiring of DC positive and negative conductors is recommended in order to reduce electromagnetic injection into the PV array.

NOTE 2: BS EN 62305-4 provides detailed information on the design, installation rules and calculation of magnetic fields and induced voltages or currents of shielding, conductor shielding and wiring.

712.534.102 Selection of SPDs on the DC side

SPDs installed on the DC side of the PV installation shall comply with BS EN 61643-31.

When the inverter incorporates SPDs on the DC side, they are considered as fulfilling the surge protective function only if the manufacturer of the inverter specifies their suitability for use for the DC side of PV installation. Otherwise, protection shall be provided by external SPDs.

NOTE: Varistors included in the inverter are not considered as an SPD.

The voltage protection level U_p of external SPDs shall be determined in relation to the characteristics of the devices incorporated in the inverters. The inverter manufacturer shall in that case provide the voltage level needed for the selection of external SPDs.

712.534.102.1 Selection of SPD Types

Generally, SPDs shall be Type 2. If protection against effects of direct lightning strokes is specified and separation distance, s, is not kept in accordance with BS EN 62305-3, Type 1 SPDs shall be used (generally in conjunction with Type 2 SPDs).

712.534.102.2 Selection of SPD voltage protection level U_p

Where no information is provided by the manufacturer, the rated impulse withstand voltage U_W shall be considered equal to that in Table 712.1 for the modules and conversion electrical equipment.

TABLE 712.1 – Rated impulse withstand voltage U_W where no information is provided

<table>
<tr><th rowspan="2">$U_{OC\ MAX}$
V</th><th colspan="2">U_W
V</th></tr>
<tr><th>PV generator and other electrical equipment</th><th>Inverter</th></tr>
<tr><td>100</td><td>800</td><td rowspan="4">2 500 (minimum requirement)</td></tr>
<tr><td>150</td><td>1 500</td></tr>
<tr><td>300</td><td>2 500</td></tr>
<tr><td>424</td><td>4 000</td></tr>
<tr><td>600</td><td>4 000</td><td rowspan="3">4 000</td></tr>
<tr><td>800</td><td>5 000</td></tr>
<tr><td>849</td><td>6 000</td></tr>
<tr><td>1 000</td><td>6 000</td><td>6 000</td></tr>
<tr><td>1 500</td><td>8 000</td><td>8 000</td></tr>
<tr><td colspan="3">NOTE 1: U_{imp} according BS EN IEC 60664-1 overvoltage category II.
NOTE 2: This table does not apply to the AC side of the inverter.</td></tr>
</table>

712.534.102.3 Selection of SPD maximum continuous operating voltage U_{cpv}

The value of the maximum voltage acceptable by the surge protective device U_{cpv} shall be selected according to the maximum no-load voltage of the PV array $U_{OC\ MAX}$. The voltage U_{cpv} shall be greater than or equal to the maximum voltage $U_{OC\ MAX}$ of the PV array.

An SPD shall be selected and arranged with regard to the maximum voltage $U_{OC\ MAX}$ between:

- its live terminals (+ and - terminals), and
- its live terminals (+ and - terminals) and earth.

712.534.102.4 Selection of SPD nominal discharge current I_{nspd}

The minimum value of nominal discharge current I_{nspd} of Type 2 SPDs shall be 5 kA.

NOTE: A nominal discharge current exceeding the minimum value will give the surge protective device a longer service life.

712.534.102.5 Selection of SPD short-circuit current rating I_{SCPV} and of protective device associated with the surge protective device

The surge protective device shall be provided with an external disconnecting device, if required by the manufacturer; these devices shall be designed to automatically disconnect the SPD when a failure of the SPD occurs and to be able to operate whatever is the current produced by the PV modules.

NOTE: An external disconnecting device is required as an SPD could fail to a short-circuit condition.

The short-circuit current rating, I_{SCPV}, shall be selected according to the maximum current that can be delivered by the photovoltaic generator $I_{SC\ MAX}$. The current I_{SCPV} shall be greater than or equal to $I_{SC\ MAX}$ of the PV array. Surge protective devices for which this parameter is not given shall not be used.

712.534.102.6 Type 1 SPD impulse current I_{imp}

The selection of impulse current I_{imp} for SPD Type 1 shall be done according to PD CLC/TS 51643-32. If the impulse current I_{imp} cannot be calculated, I_{imp} shall not be less than 12.5 kA. This value is referring to lightning protection level (LPL) III.

712.534.104 Erection of SPDs in the DC side

SPDs in the DC side shall be located as close as possible to the inverter.

To provide protection, additional SPDs may be required, further from the inverter.

NOTE 1: An example is where the distance between the entrance of the DC cable into a building and the inverter is longer than 10 m.

NOTE 2: The surge voltage level on the electrical equipment depends on its distance away from the SPD. Beyond 10 m, the value of this voltage can be doubled due to the effect of resonance (amplification phenomena due to the high frequencies of the lightning surges).

712.534.105 Connection of SPDs

When SPDs are installed on both AC and DC sides of the inverter in separate switchboard assemblies, it is recommended to minimize the distance between those switchboards.

For DC and AC SPD connections to the main earthing terminal, the conductors shall have a minimum cross-section of 6 mm^2 copper for Type 2 SPDs and 16 mm^2 copper for Type 1 SPDs.

712.537 Isolation and switching

712.537.2 Isolation

712.537.2.101 To allow maintenance and replacement of the inverter, means of isolating the inverter from the DC side and the AC side shall be provided.

NOTE: Further requirements with regard to the isolation of a PV installation operating in parallel with the public supply system are given in Regulation 551.7.

712.537.2.2 Devices for isolation

712.537.2.2.101 A switch disconnector or a circuit-breaker suitable for isolation shall be provided on the DC side of the inverter.

712.537.2.2.103 Disconnection of the functional bonding conductor

An automatic disconnecting device shall be connected in series with the functional bonding conductor and shall be rated for the following conditions:

(i) the short-circuit maximum current of the PV array $I_{SC\ MAX}$

(ii) the maximum voltage of the PV array $U_{OC\ MAX}$

(iii) the maximum rated current given in Table 712.2.

TABLE 712.2 – Rated current of automatic disconnecting device in the functional bonding conductor

Total PV array power rating (peak value) kW	Maximum rated current I_n of the automatic disconnecting device A
≤ 25	1
> 25 - 50	2
> 50 - 100	3
> 100 - 250	4
> 250	5

712.537.2.2.104 Measures to prevent DC on-load interruption

In order to prevent arcing, every device without breaking capacity that could be used to open a DC circuit shall be secured against inadvertent or unauthorized operation. This may be achieved by locating the device in a lockable space or enclosure or by padlocking.

NOTE: Examples of devices to which this requirement applies are SPD carriages and fuse carriers.

712.537.2.2.105 Remote operation of devices for opening operation located in the combiner box

Remotely operated devices within a combiner box to provide isolation under specific fault conditions may be installed.

712.538.101 Monitoring devices

An IMD shall be selected in accordance with BS EN 61557-8.

Where the IMD is an integral part of the inverter, the IMD function has to be in accordance with BS EN 62109-2 or BS EN 61557-8.

712.54 Earthing arrangements and protective conductors

712.542 Earthing arrangements

712.542.3.101 Functional bonding conductor

The bonding conductors (insulated or bare) shall have a minimum cross-sectional area of not less than of 4 mm^2 copper equivalent.

712.542.101 Equipotential bonding of PV metal structures

Where equipotential bonding is necessary to prevent the accumulation of electrostatic charge, the metallic structures supporting the PV modules including the metallic cable management systems shall be bonded.

The bonding conductor shall be connected to any suitable earthing terminal. Where metal structures are of aluminium, appropriate terminations are necessary to prevent electrolytic corrosion.

712.542.102 Functional bonding arrangements on the DC side

A live conductor of the DC side of the inverter may be bonded only if there is galvanic isolation between the AC installation and the DC side provided by means of a transformer having electrically separate windings.

The transformer may be either internal or external to the inverter. The transformer winding connected to the inverter shall not be earthed, and the inverter shall be suitable for this.

The bonding shall be made at a single point of the DC side of the inverter.

The bonding shall be located between the disconnection device and the DC connection means of the inverter.

712.551 Additional requirements for installations where the generating set may operate in parallel with other sources including systems for distribution of electricity to the public

712.551.7.2 On the AC side, the PV supply cable shall be connected to the supply side of the overcurrent protective device for automatic disconnection of circuits supplying current-using equipment.

712.6 Verification

712.6.101 Inspection and testing

NOTE: Having complied with the relevant requirements of Part 6, additional requirements for system documentation, commissioning tests and inspection are given in BS EN 62446 series.

Fig 712.1 – PV installation - General schematic - One array

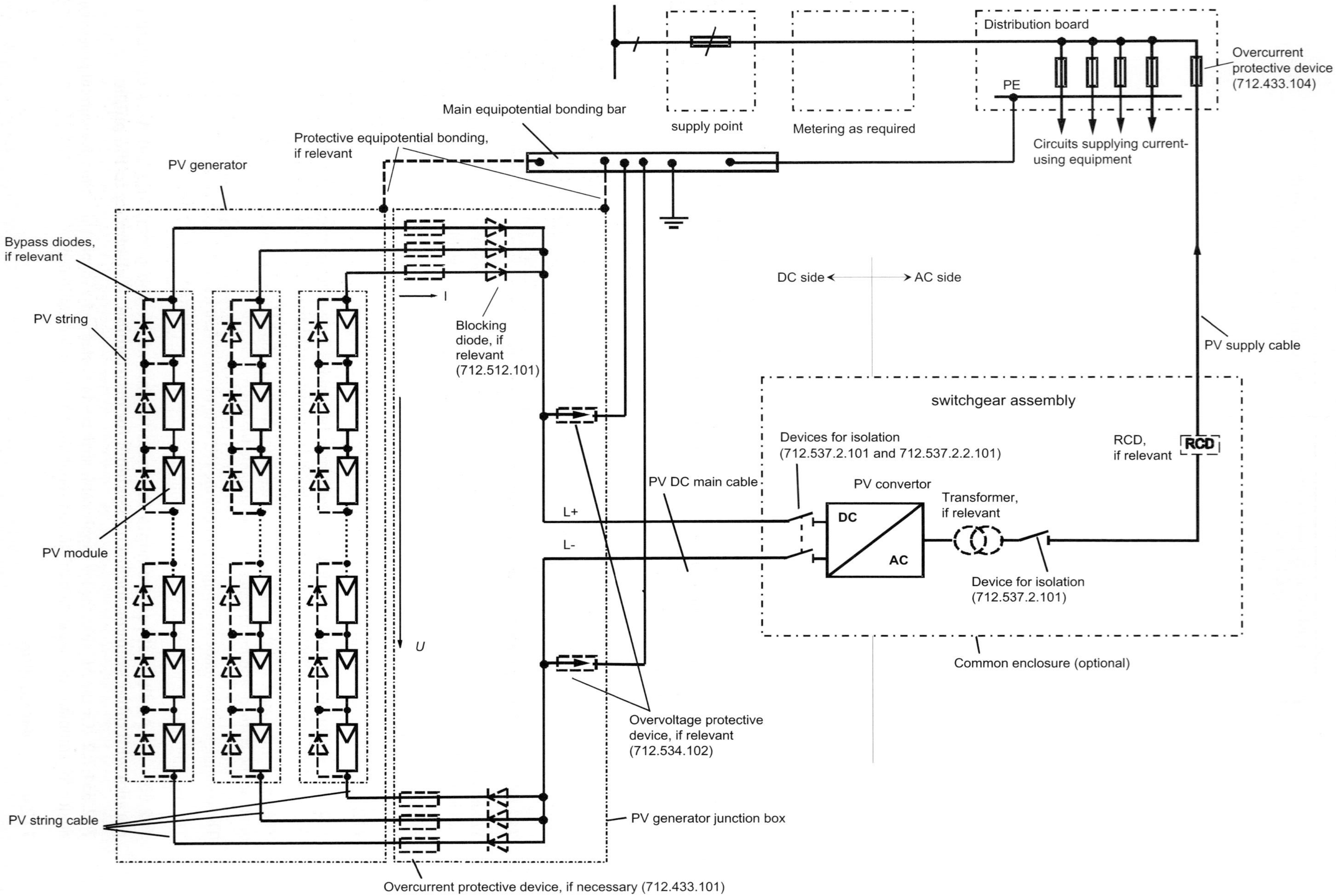

Fig 712.2 – PV installation - Example with two or more arrays

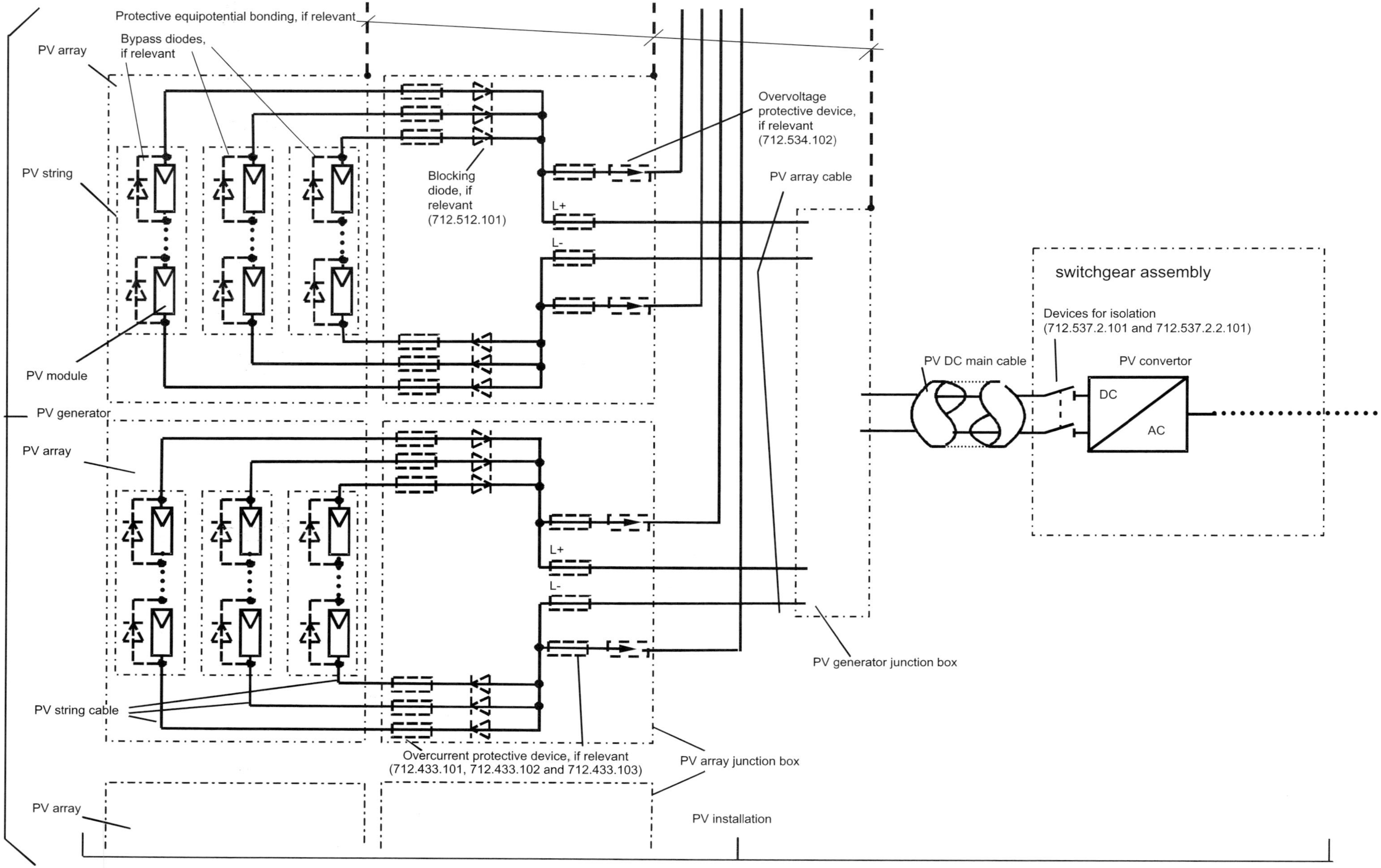

SECTION 714

OUTDOOR LIGHTING INSTALLATIONS

714.1 Scope

This section applies to outdoor lighting installations comprising one or more luminaires, a wiring system and accessories, and to highway power supplies and street furniture.

The following are included in outdoor lighting installations:

(i) Lighting installations such as those for roads, parks, car parks, gardens, places open to the public, sporting areas, illumination of monuments and floodlighting

(ii) Other lighting arrangements in places such as telephone kiosks, bus shelters, advertising panels and town plans

(iii) Road signs.

The following are excluded:

(iv) Temporary festoon lighting

(v) Luminaires fixed to the outside of a building and supplied directly from the internal wiring of that building

(vi) Road traffic signal systems.

714.4 Protection for safety

714.41 Protection against electric shock

714.410.3 General requirements

714.410.3.6 The protective measures of non-conducting location (Regulation 418.1) and earth-free local equipotential bonding (Regulation 418.2) shall not be used.

714.411 Protective measure: Automatic disconnection of supply

714.411.201 Where the protective measure automatic disconnection of supply is used, all live parts of electrical equipment shall be protected by insulation or by barriers or enclosures providing basic protection. A door in street furniture, used for access to electrical equipment, shall not be used as a barrier or an enclosure.

714.411.202 A maximum disconnection time of 5 s shall apply to all circuits feeding fixed equipment used in highway power supplies for compliance with Regulation 411.3.2.3 (TN system) or 411.3.2.4 (TT system).

714.411.203 Where street furniture is connected directly to a distributor's PME network, the earthing conductor and any main protective bonding conductors to any extraneous-conductive parts shall have a minimum copper equivalent cross-sectional area of 6 mm^2 for PEN conductors of the supply with copper equivalent cross-sectional areas up to 10 mm^2. For larger sized neutral conductors, the main bonding shall comply with Table 54.8.

714.411.2.201 Provisions for basic protection

For every accessible enclosure live parts shall only be accessible with a key or a tool, unless the enclosure is in a location where only skilled or instructed person(s) have access.

A door giving access to electrical equipment and located less than 2.50 m above ground level shall be locked with a key or shall require the use of a tool for access. In addition, basic protection shall be provided when the door is open either by the use of equipment having at least a degree of protection IPXXB or IP2X by construction or by installation, or by installing a barrier or an enclosure giving the same degree of protection.

For a luminaire at a height of less than 2.80 m above ground level, access to the light source shall only be possible after removing a barrier or an enclosure requiring the use of a tool.

714.411.3.1 Protective earthing and protective equipotential bonding

714.411.3.1.2 Protective equipotential bonding

A metallic structure (such as a fence, grid etc.), which is in the proximity of but is not part of the outdoor lighting installation need not be connected to the main earthing terminal.

714.411.3.4 Additional protection

Lighting that is accessible to the public shall have additional protection by an RCD having the characteristics specified in Regulation 415.1.1.

Examples include:

(i) gardens, spaces open to the public
(ii) telephone kiosks
(iii) bus shelters
(iv) advertising panels and town plans.

This list is not exhaustive.

The following are excluded:

(v) Street lighting & traffic signage
(vi) Illumination of monuments
(vii) Platform lighting at rail / bus stations.

714.5 Selection and erection of equipment

714.51 Common rules

714.512 Operational conditions and external influences

714.512.2 External influences

714.512.2.1 The following classes are generally recommended:

(i) Ambient temperature: AA2 and AA4 (from −40 °C to +40 °C)
(ii) Climatic conditions: AB2 and AB4 (relative humidity between 5 % and 100 %).

714.512.2.105 Electrical equipment shall have, by construction or by installation, a degree of protection of at least IP33.

714.514.12 Identification and notices

714.514.12.201 The requirements for notices for:

(i) periodic inspection and testing (Regulation 514.12.1) and
(ii) the testing of RCDs (Regulation 514.12.2)

need not be applied where the installation is subject to a programmed inspection and testing procedure.

714.514.12.202 On every temporary supply unit there shall be an externally mounted durable label stating the maximum sustained current to be supplied from that unit.

714.537 Isolation and switching

714.537.2 Devices for isolation

714.537.2.1 General

714.537.2.1.1 Every circuit shall be capable of being isolated individually from each of the live supply conductors, except as detailed in Regulation 461.2.

714.537.2.1.201 Where it is intended that isolation and switching is carried out only by skilled or instructed persons and subject to suitable provisions being made so that precautions can be taken to prevent any equipment from being inadvertently or unintentionally energized, for TN systems, the means of switching the supply on load and the means of isolation is permitted to be provided by a suitably rated fuse carrier.

714.537.2.1.202 Where the distributor's cut-out is used as the means of isolation of a highway power supply the approval of the distributor shall be obtained.

SECTION 715

EXTRA-LOW VOLTAGE LIGHTING INSTALLATIONS

715.1 Scope

This section applies to extra-low voltage lighting installations supplied from a source with a maximum rated voltage of 50 V AC or 120 V DC.

715.4 Protection for safety

715.41 Protection against electric shock

715.411.7.201 Functional extra-low voltage (FELV)

FELV shall not be used.

715.414 Protective measure: Extra-low voltage provided by SELV or PELV

An extra-low voltage luminaire without provision for the connection of a protective conductor shall be installed only as part of a SELV system. Where bare conductors are used, the nominal voltage shall not exceed 25 V AC or 60 V DC according to Regulation 414.4.5.

A safety isolating transformer for an extra-low voltage lighting installation shall comply with BS EN 61558-2-6 and shall meet at least one of the requirements of Regulation 715.422.106.

NOTE 1: For an explanation of symbols used see Table 55.3.

Parallel operation of transformers in the secondary circuit is allowed only if they are also paralleled in the primary circuit and the transformers have identical electrical characteristics.

An electronic convertor for an extra-low voltage lighting installation shall comply with BS EN 61347-2-2, Annex 1 for incandescent lamps or BS EN 61347-2-13, Annex 1 for LEDs.

NOTE 2: For an explanation of symbols used see Table 55.3.

Parallel operation of convertors to BS EN 61347-2-2 or BS EN 61347-2-13 is not permitted.

715.42 Protection against thermal effects

715.422.106 Fire risk of transformers

Transformers shall be either:

(i) protected on the primary side by a protective device complying with the requirements of Regulation 715.422.107.2, or

(ii) short-circuit proof (both inherently and non-inherently).

NOTE: For an explanation of symbols used see Table 55.3.

715.422.107 Fire risk due to short-circuit

715.422.107.1 Where both the live circuit conductors are uninsulated, either:

(i) they shall be provided with a protective device complying with the requirements of Regulation 715.422.107.2, or

(ii) they shall be supplied from a transformer or convertor, the power of which does not exceed 200 VA, or

(iii) the system shall comply with BS EN 60598-2-23.

715.422.107.2 A device providing protection against the risk of fire shall meet all the following requirements:

(i) The device shall continuously monitor the power demand of the luminaires

(ii) The device shall automatically disconnect the supply circuit within 0.3 s in the case of a short-circuit or failure which causes a power increase of more than 60 W

(iii) The device shall provide automatic disconnection while the supply circuit is operating with reduced power (for example, by gating control or a regulating process or a lamp failure) if there is a failure which causes a power increase of more than 60 W

(iv) The device shall provide automatic disconnection upon connection of the supply circuit if there is a failure which causes a power increase of more than 60 W

(v) The device shall be fail-safe.

NOTE: Account needs to be taken of starting currents.

715.43 Protection against overcurrent

715.430.104 Self-resetting overcurrent protective devices

The use of self-resetting overcurrent protective devices is permitted only for transformers up to 50 VA.

715.5 Selection and erection of equipment

715.52 Wiring systems

715.521 Types of wiring system

715.521.1 The following wiring systems shall be used:

(i) Insulated conductors in conduit systems according to BS EN 6186 series or cable trunking/ducting systems according to BS EN 50085 series
(ii) Rigid cables
(iii) Flexible cables
(iv) Systems for ELV lighting according to BS EN 60598-2-23
(v) Track systems according to BS EN 60570
(vi) Bare conductors (see Regulation 715.521.106).

Where parts of the ELV lighting installation are accessible, the requirements of Section 423 also apply.

Metallic structural parts of buildings, for example pipe systems, or parts of furniture, shall not be used as live conductors.

715.521.106 Bare conductors

If the nominal voltage does not exceed 25 V AC or 60 V DC, bare conductors may be used provided that the extra-low voltage lighting installation complies with all the following requirements:

(i) The lighting installation shall be designed, and installed or enclosed in such a way that the risk of a short-circuit is reduced to a minimum
(ii) The conductors used shall have a cross-sectional area according to Regulation 715.524
(iii) The conductors shall not be placed directly on combustible material.

For suspended bare conductors, at least one conductor and its terminals shall be insulated for that part of the circuit between the transformer and the short-circuit protective device to prevent a short-circuit.

715.521.107 Suspended systems

Suspension devices for extra-low voltage luminaires, including supporting conductors, shall be capable of carrying five times the mass of the luminaires (including their lamps) intended to be supported, but not less than 5 kg.

Terminations and connections of conductors shall be made by screw terminals or screwless clamping devices complying with BS EN 60998-2-1 or BS EN 60998-2-2.

Safety of the installation due to expected stresses in the conductors shall be in accordance with Regulation 559.5.2.

Insulation piercing connectors and termination wires which rely on counterweights hung over suspended conductors to maintain the electrical connection shall not be used.

The suspended system shall be fixed to walls or ceilings by insulated distance cleats and shall be continuously accessible throughout the route.

715.524 Cross-sectional area of conductors

715.524.201 The minimum cross-sectional area of the ELV conductors for connection to the output terminals or terminations of transformers/convertors shall be chosen according to the load current.

In the case of systems with luminaires suspended from the conductors, the minimum cross-sectional area of the ELV conductors for connection to the output terminals or terminations of transformers/convertors shall be 4 mm^2, for mechanical reasons.

715.525 Voltage drop in consumers' installations

In ELV lighting installations, if the voltage drop between the transformer and the furthest luminaire does not exceed 5 % of the nominal voltage of the ELV installation it shall be deemed to comply with Section 525.

715.53 Protection, isolation, switching, control and monitoring

715.530.3 General and common requirements

715.530.3.104 Protective devices and SELV sources

Protective devices may be located above false ceilings, which are removable or easily accessible, provided that information is given about the presence and location of such devices.

SELV sources, protective devices or similar equipment mounted above false ceilings or in a similar place shall be permanently connected.

SELV sources and their protective devices shall be installed so as to:

(i) avoid mechanical stress on their electrical connections, and

(ii) be adequately supported, and

(iii) avoid overheating of the equipment due to thermal insulation.

715.537 Isolation and switching

715.537.1.1 Where transformers are operated in parallel, the primary circuits shall be permanently connected to a common isolating device.

SECTION 717

MOBILE OR TRANSPORTABLE UNITS

717.1 Scope

The particular requirements of this section apply to AC and DC installations for mobile or transportable units.

For the purposes of this section, the term 'unit' is intended to mean a vehicle and/or mobile or transportable structure in which all or part of an electrical installation is contained.

Units are either:

(i) of the mobile type, e.g. vehicles (self-propelled or towed), or

(ii) of the transportable type, e.g. containers or cabins.

Examples of the units include technical and facilities vehicles for the entertainment industry, medical or health screening services, welfare units, promotion & demonstration, firefighting, workshops, offices, transportable catering units etc.

The requirements are not applicable to:

(iii) generating sets

(iv) marinas and pleasure craft

(v) mobile machinery in accordance with BS EN 60204-1

(vi) caravans to Section 721

(vii) traction equipment of electric vehicles

(viii) electrical equipment required by a vehicle to allow it to be driven safely or used on the highway.

Additional requirements shall be applied where necessary for units including showers, or for medical locations, etc.

NOTE: Guidance on temporary electrical systems for events, entertainment and related purposes is given in BS 7909.

717.132 Design

The design of the installation within a mobile or transportable unit shall take into account the characteristics of the variety of supplies to which the unit might be connected and any limitation shall be clearly stated in the user documentation. See also Regulation 717.514.

717.313 Supplies

One or more of the following methods shall be used to supply a unit:

(i) Connection to a low voltage generating set, located inside the unit, in accordance with Section 551 (see Figure 717.1)

(ii) Connection to a low voltage electrical supply external to the unit, in which the protective measures are effective (see Figure 717.3), the supply derived from either a fixed electrical installation or a generating set in accordance with Section 551

(iii) Connection to a low voltage electrical supply external to the unit, and where internal protective measures are provided by the use of simple separation, in accordance with Section 413 (see Figures 717.4, 717.5, 717.6 and 717.7 showing alternative forms of fault protection within the unit).

NOTE 1: In cases (i), (ii) and (iii), an earth electrode may be provided where supplies are used external to the vehicle (see Regulation 717.411.4).

NOTE 2: In the case of Figure 717.4, an earth electrode may be necessary for protective purposes (see Regulation 717.411.6.2(ii)).

NOTE 3: Simple separation or electrical separation is appropriate, for example, where information technology equipment is used in the unit or where a reduction of electromagnetic disturbances is necessary, or if high protective conductor currents are to be expected (use of frequency convertors), and/or if the supply to the unit comes from alternative supply systems (as is the case in disaster management).

The sources, means of connection or separation may be within the unit.

NOTE 4: Where there is a potential hazard due to moving the unit whilst connected to an external installation, it is recommended that the unit is equipped with an electrical interlock, warning, alarm or other appropriate means to reduce the risk.

NOTE 5: For the purpose of this section, power inverters or frequency convertors supplied from the unit's electrical system or an auxiliary system driven by the unit's prime mover are also considered as generating sets.

Power inverters or frequency convertors shall include electrical separation where both the DC supply and the AC neutral point are earthed.

717.4 Protection for safety

717.41 Protection against electric shock

717.410.3 General requirements

717.410.3.5 The protective measures of obstacles and placing out of reach (Section 417) shall not be used.

717.410.3.6 The protective measure of non-conducting location (Regulation 418.1) shall not be used.

The protective measure of earth-free local equipotential bonding (Regulation 418.2) is not recommended.

717.411 Protective measure: Automatic disconnection of supply

717.411.3.1.2 Protective equipotential bonding

Structural metallic parts of the unit shall be connected through the main protective bonding conductors to the main earthing terminal within the unit. The main protective bonding conductors shall be finely stranded.

NOTE: Cable types H05V-K and H07V-K to BS EN 50525-2-31 are considered appropriate.

717.411.4 TN system

A PME earthing facility shall not be used as the means of earthing for mobile and transportable units falling within the scope of this section except where one of the following applies:

(i) the installation is continuously under the supervision of a skilled person (electrically) or instructed person (electrically), competent in such work and the suitability and effectiveness of the means of earthing has been confirmed before the connection is made, or

(ii) the unit is located within, or outdoors on an upper storey or roof of, a building or structure containing the electrical installation which supplies the unit.

717.411.6 IT system

717.411.6.2 An IT system can be provided by:

(i) an isolating transformer or a low voltage generating set, with an insulation monitoring device or an insulation fault location system, both without automatic disconnection of the supply in case of the first fault and without a need of connection to an earthing installation (see Figure 717.7); the second fault shall be automatically disconnected by overcurrent protective devices according to Regulation 411.6.5, or

(ii) a transformer providing simple separation, e.g. in accordance with BS EN 61558-1, with an RCD and an earth electrode installed to provide automatic disconnection in the case of failure in the transformer providing the simple separation (see Figure 717.4)

717.413 Protective measure: Electrical separation

Electrical separation can be provided by the use of a transformer providing simple separation, e.g. meeting the requirements of BS EN 61558-1, in accordance with Regulation 413.1.3. It shall be used only where:

(i) an insulation monitoring device is installed to provide automatic disconnection of supply in the case of a first fault between live parts and the conductive structure of the unit (see Figure 717.5), or

(ii) an RCD and an earth electrode are installed to provide automatic disconnection in the case of failure in the transformer providing the electrical separation (see Figure 717.4). Each socket-outlet intended to supply current-using equipment outside the unit shall be protected individually by an RCD having the characteristics specified in Regulation 415.1.1.

717.415 Additional protection

717.415.1 Additional protection by an RCD having the characteristics specified in Regulation 415.1.1, shall be provided for every socket-outlet intended to supply current-using equipment outside the unit, with the exception of socket-outlets which are supplied from circuits with protection by:

(i) SELV, or

(ii) PELV, or

(iii) electrical separation, with an insulation monitoring device; see Regulation 717.413(i).

717.5 Selection and erection of equipment

717.51 Common requirements

717.514 Identification and notices

A permanent notice of such durable material as to be likely to remain easily legible throughout the life of the installation, shall be fixed to the unit in a prominent position, preferably adjacent to each supply inlet connector. The notice should state in clear and unambiguous terms the following:

(i) The types of supply which may be connected to the unit and any limitations on use imposed by the designer

(ii) The voltage rating of the unit

(iii) The number of supplies, phases and their configuration

(iv) The on-board earthing arrangement

(v) The maximum power requirement of the unit.

717.52 Wiring systems

717.52.1 Where the supply to the mobile or transportable unit is provided by means of a plug and socket-outlet, flexible cables in accordance with H07RN-F (BS EN 50525-2-21), or cables of equivalent design, having a minimum cross-sectional area of 2.5 mm^2 copper, shall be used for connecting the unit to the supply. The flexible cable shall enter the unit by an insulating inlet in such a way as to minimize the possibility of any insulation damage or fault which might energize the exposed-conductive-parts of the unit.

717.52.2 The wiring system shall be installed using one or more of the following:

(i) Unsheathed flexible or stranded (minimum of 7 strands) cable with thermoplastic or thermosetting insulation to BS EN 50525-2-31, -3-31 or BS EN 50525-3-41 installed in conduit in accordance with the appropriate part of BS EN 61386 series or in trunking or ducting in accordance with the appropriate part of BS EN 50085 series

(ii) Sheathed flexible cable with thermoplastic or thermosetting insulation to BS EN 50525-2-11, -2-21, -3-11 or -3-21, if precautionary measures are taken such that no mechanical damage is likely to occur due to any sharp-edged parts or abrasion.

All cables shall, as a minimum, meet the requirements of BS EN 60332-1-2.

Conduits shall comply with BS EN 61386-21, BS EN 61386-22 or BS EN 61386-23.

717.528.3 Proximity to non-electrical services

717.528.3.4 No electrical equipment, including wiring systems, except ELV equipment for gas supply control, shall be installed in any gas cylinder storage compartment.

ELV cables and electrical equipment may only be installed within the LPG cylinder compartment if the installation serves the operation of the gas cylinder (e.g. indication of empty gas cylinder) or is for use within the compartment. Such electrical installations and components shall be constructed and installed so that they are not a potential source of ignition.

Where cables have to run through such a compartment, they shall be protected against mechanical damage by installation within a conduit system complying with the appropriate part of the BS EN 61386 series or within a ducting system complying with the appropriate part of the BS EN 50085 series.

Where installed, this conduit or ducting system shall be able to withstand an impact equivalent to AG3 without visible physical damage.

717.55 Other equipment

717.55.1 Where the means of connection is a plug and socket-outlet, mounted, accessed or used outside the unit and used to connect the unit to the supply, or supply other equipment, it shall comply with the appropriate parts of BS EN 60309-2 series and shall meet with the following requirements:

(i) Plugs shall have an enclosure of insulating material

(ii) Connecting devices, plugs and socket-outlets, with an enclosure as necessary, shall afford a degree of protection of at least IP44 when in use or connected and protection of at least IP55 when not connected, e.g. when the unit is in transit

(iii) The inlet (with 'male' contacts) shall be situated on the unit.

717.55.2 *Not used*

717.55.3 Generating sets able to produce voltages other than SELV or PELV, mounted in a mobile unit, shall automatically be switched off in case of an accident to the unit (e.g. event causing the release of airbags). If this requirement is difficult to implement an emergency switch, easily accessible, shall be installed.

717.551.6 Additional requirements for installations where the generating set provides a supply as a switched alternative to the normal supply to the installation

Live conductors from different power supplies shall not be interconnected.

Protective conductors, including functional earthing conductors, from different earthing systems shall only be interconnected where suitable precautions have been taken into account; see also Regulation 542.1.3.3.

Plugs and socket-outlets shall comply with the appropriate parts of BS EN 60309 series, except those intended for special equipment, such as broadcasting equipment where combined connectors for information signals and power supply are used.

717.551.7.2 Additional requirements for installations where the generating set may operate in parallel with other sources including systems for distribution of electricity to the public

A generating set used as an additional source of supply in parallel with another source shall only be connected on the supply side of all the protective devices for the final circuits of the installation.

Protective conductors, including functional earthing conductors, from different earthing systems shall only be interconnected where suitable precautions have been taken into account; see also Regulation 542.1.3.3.

Fig 717.1 – An example of a connection to a low voltage generating set located inside the unit, with or without an earth electrode

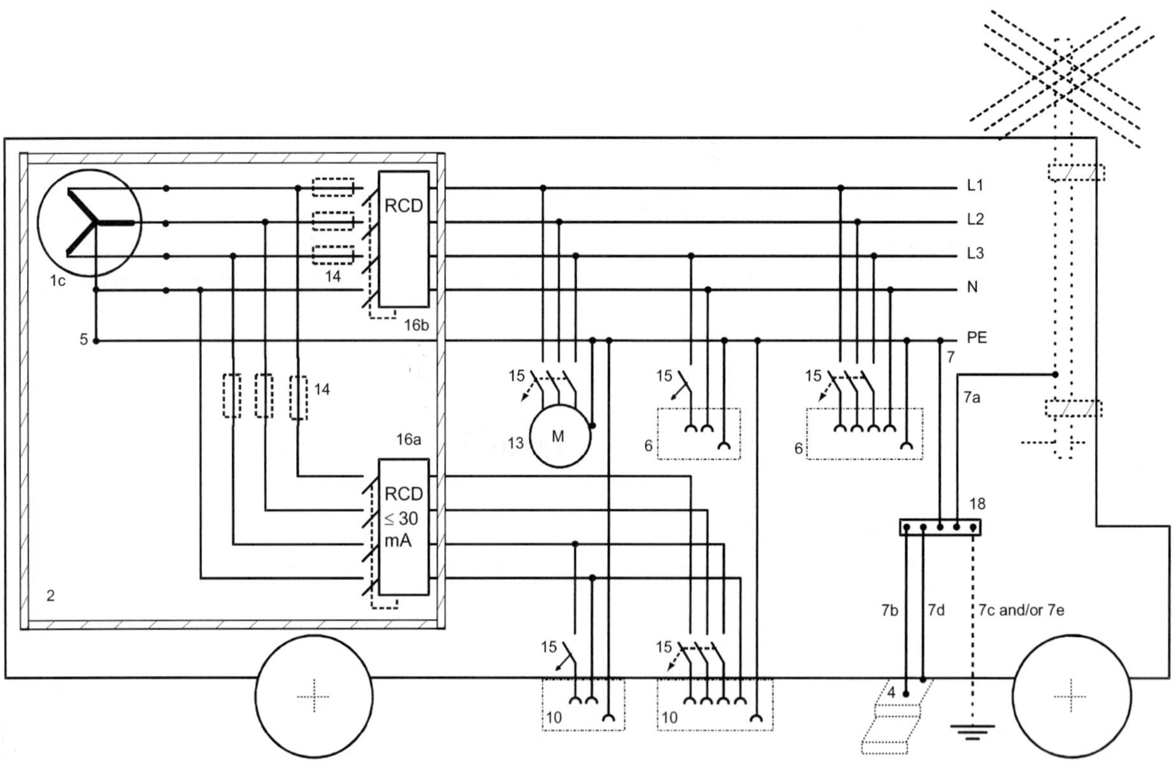

Fig 717.2 – Reserved for future use

Fig 717.3 – An example of a connection to a low voltage electrical supply external to the unit in which the protective measures are effective, the supply derived from either a fixed electrical installation or a generating set, with or without an earth electrode at the unit

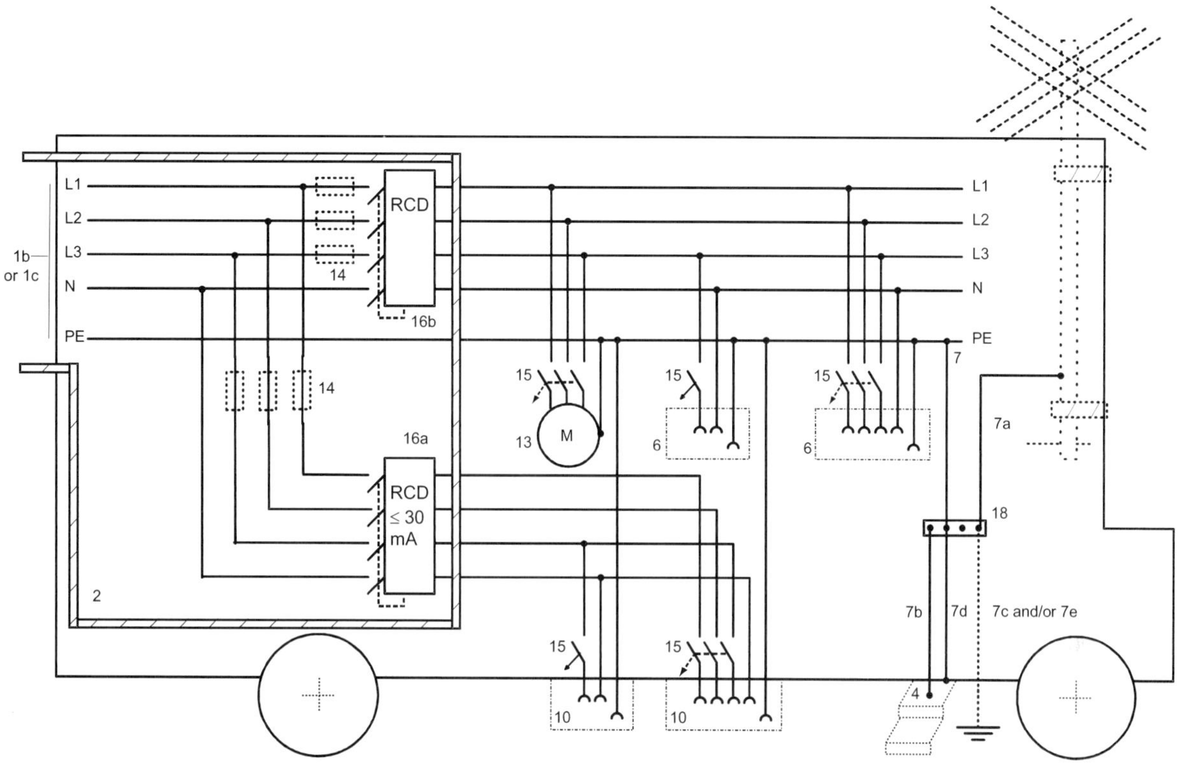

NOTE: Where a PME earthing facility is used, see Regulation 717.411.4.

Fig 717.4 – An example of a connection to a low voltage electrical supply external to the unit, derived from either a fixed electrical installation or a generating set with any type of earthing arrangement using simple separation and an internal IT system, with an earth electrode

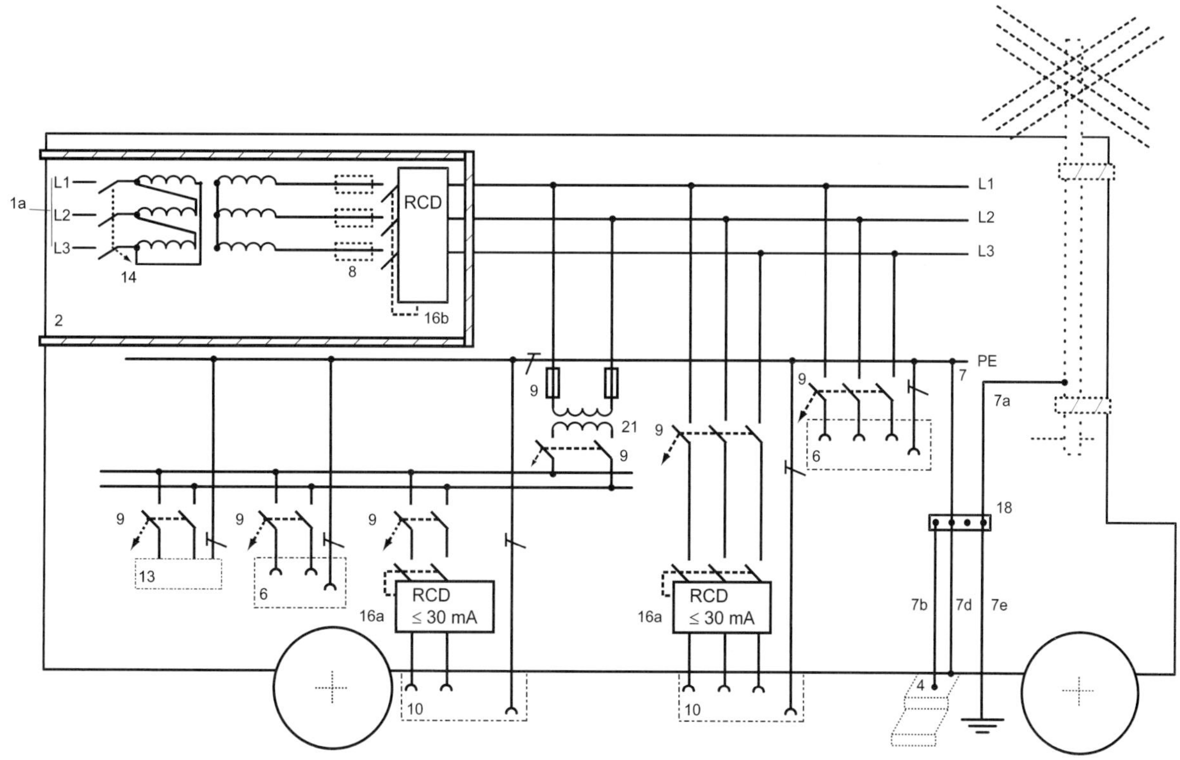

Fig 717.5 – An example of a connection to a low voltage electrical supply external to the unit, using simple separation and an internal IT system with an insulation monitoring device and automatic disconnection of supply on the occurrence of a first fault, with earth electrode

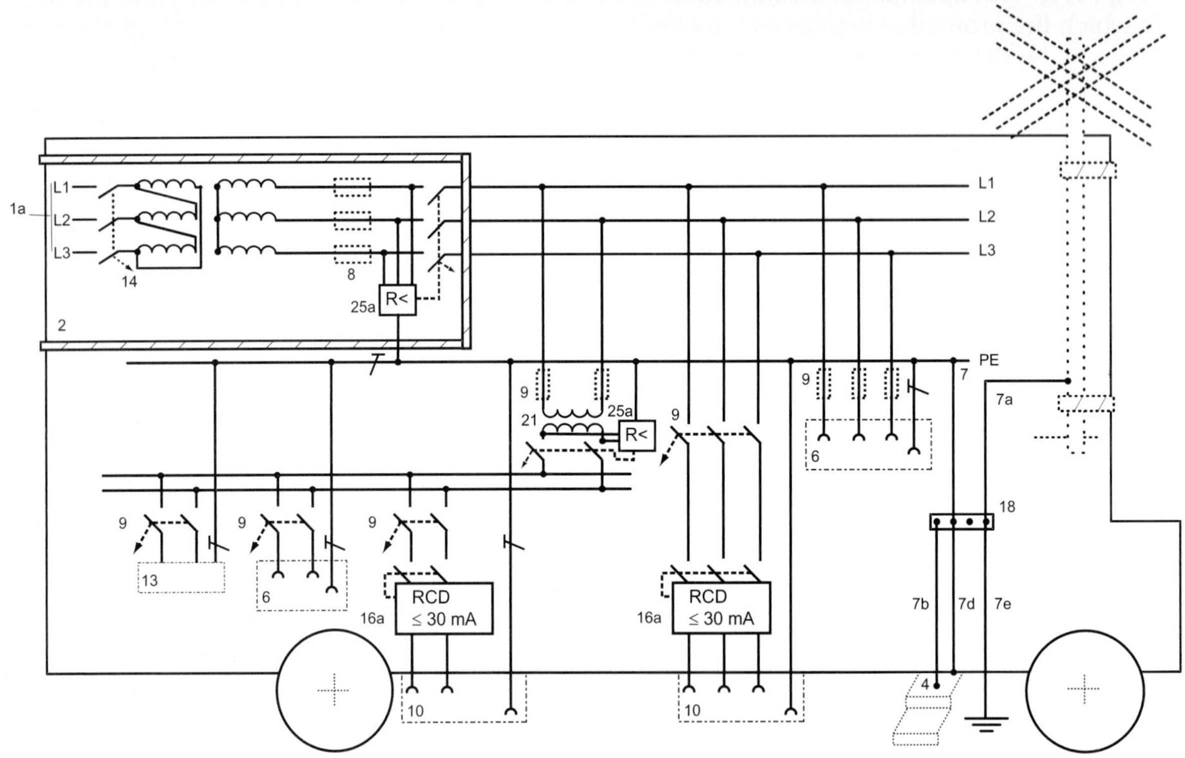

Fig 717.6 – An example of a connection to a low voltage electrical supply external to the unit, with any type of earthing arrangement using an internal TN system with simple separation

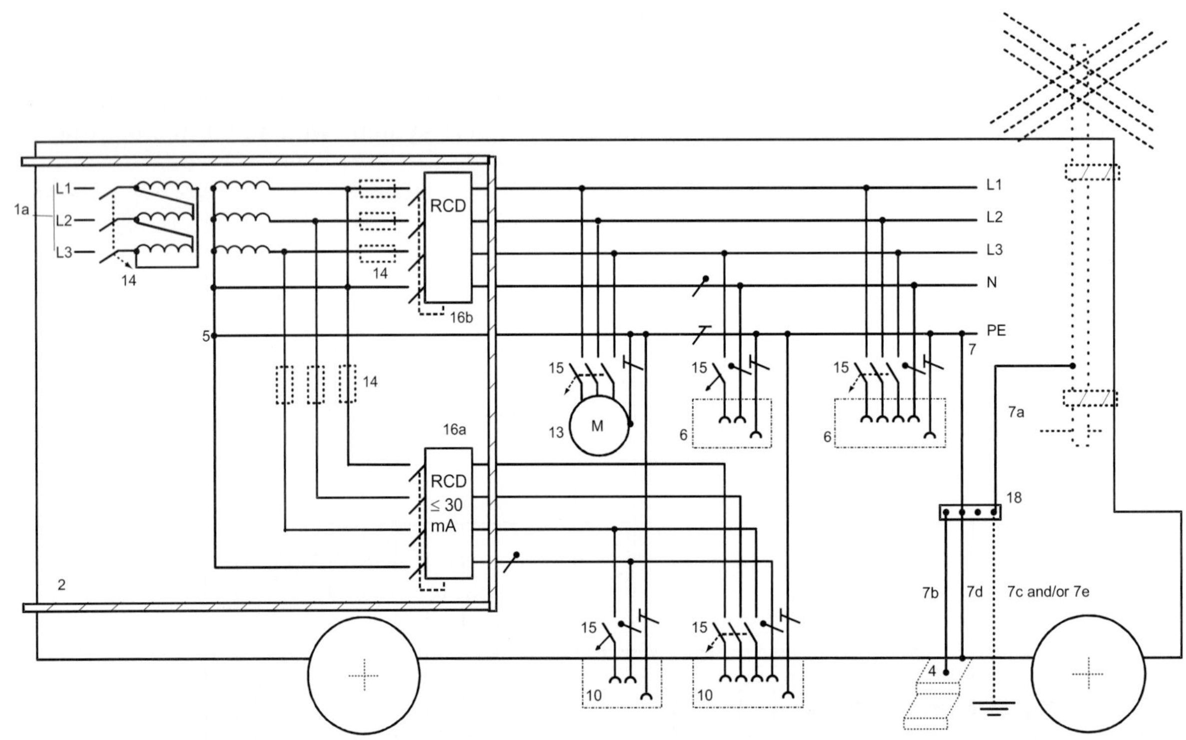

Fig 717.7 – An example of a connection to a low voltage electrical supply external to the unit, with any type of earthing arrangement using simple separation with an internal IT system with automatic disconnection on the occurrence of a second fault

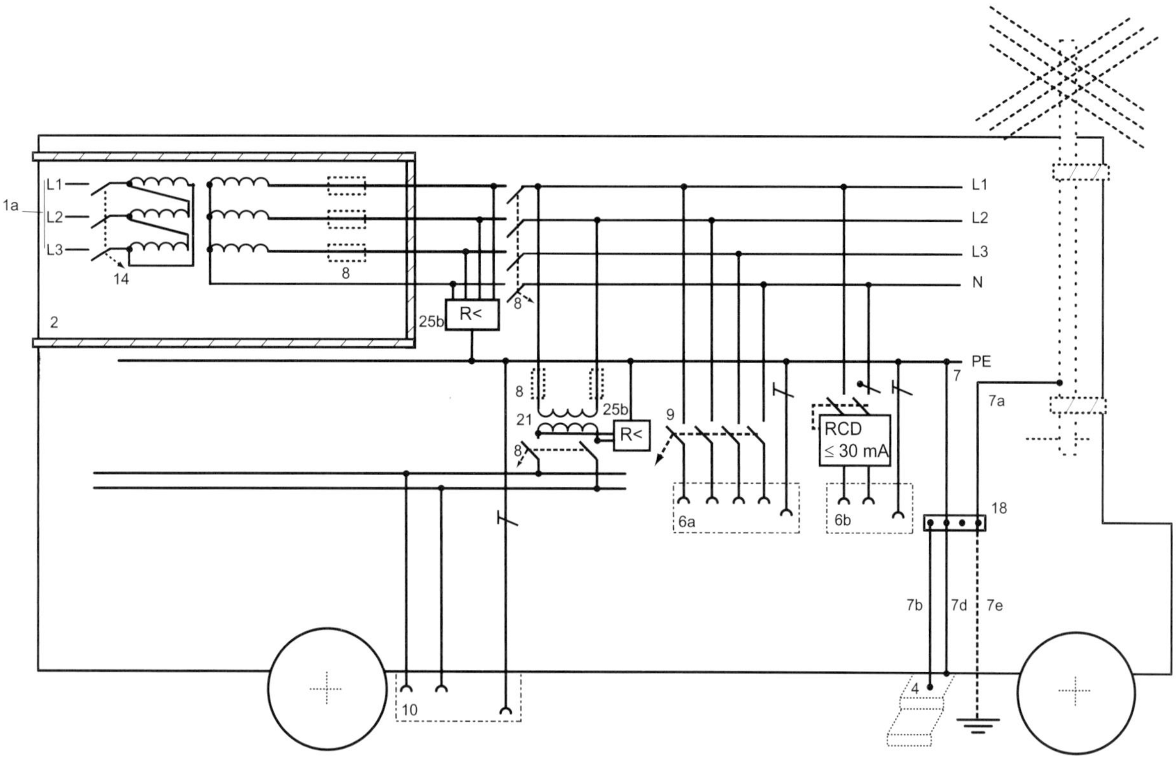

Key to Figures 717.1 to 717.7

1a Connection of the unit to an external supply through an on board transformer with simple separation

1b Connection of the unit to a supply in which the protective measures are effective

1c Connection to an LV generator set in accordance with Section 551

2 Class II or equivalent enclosure up to the first protective device providing automatic disconnection of supply

4 Conductive external staircase, if any

5 Connection of the neutral point (or, if not available, a line conductor) to the conductive structure of the unit

6 Socket-outlets for use exclusively within the unit

6a Socket-outlets for use exclusively within the unit for reasons of continuity of supply in the event of first fault

6b Socket-outlets for general use if explicitly required (operation of the RCD in the event of first fault cannot be excluded)

7 Protective equipotential bonding in accordance with Regulation 717.411.3.1.2

7a to an antenna pole, if any

7b to the conductive external stairs, if any, in contact with the ground

7c to a functional earth electrode, if required

7d to the conductive structure of the unit

7e to an earth electrode for protective purposes, if required

8 Protective devices, if required, for overcurrent and/or for protection by disconnection of supply in case of a second fault

9 Protective devices for overcurrent and for automatic disconnection of supply in case of a second fault

10 Socket-outlets for current-using equipment for use outside the unit

13 Current-using equipment for use exclusively within the unit

14 Overcurrent protective device, if required

15 Overcurrent protective device

16a RCD having the characteristics specified in Regulation 415.1.1 for protection by automatic disconnection of supply for circuits of equipment for use outside the unit

16b RCD for protection by automatic disconnection of supply for circuits of equipment for use inside the unit: see Regulations 411.4.4 and 411.5.3. Where an internal IT system is installed, see also Regulation 411.6.4

18 Main earthing terminal or bar

21 Transformer with at least simple separation, e.g. 230 V current-using equipment

25a Insulation monitoring device providing disconnection on the first fault, see Figure 717.5

25b Insulation monitoring device or insulation fault location system including monitoring of the N conductor if distributed (disconnection only in the event of second fault), see Figure 717.7

SECTION 721
ELECTRICAL INSTALLATIONS IN CARAVANS AND MOTOR CARAVANS

NOTE: In order not to mix requirements on different subjects, such as those for electrical installations of caravan parks with those for electrical installations inside caravans, refer to:
- Section 708, which concerns electrical installations in caravan parks, camping parks and similar locations and
- Section 721, which concerns electrical installations in caravans and motor caravans.

721.1 Scope

The particular requirements of this section apply to the electrical installations of caravans and motor caravans at nominal voltages not exceeding 230/400 V AC or 48 V DC.

They do apply to those electrical circuits and equipment intended for the use of the caravan for habitation purposes.

They do not apply to those electrical circuits and equipment for automotive purposes, nor to installations covered by BS EN 1648-1 and BS EN 1648-2.

They do not apply to the electrical installations of mobile homes, residential park homes or transportable units.

NOTE 1: For mobile homes and residential park homes the general requirements apply.

NOTE 2: For transportable units see Section 717.

For the purposes of this section, caravans and motor caravans are referred to as 'caravans'.

The particular requirements of some other sections of Part 7 may also apply to such installations in caravans, e.g. Section 701.

721.31 Purposes, supplies and structure

721.313 Supplies

721.313.1.2 The nominal supply system voltage shall be chosen from BS EN 60038.

The nominal AC supply voltage of the installation of the caravan shall not exceed 230 V single-phase or 400 V three-phase.

The nominal DC supply voltage of the installation of the caravan shall not exceed 48 V.

721.4 Protection for safety

721.41 Protection against electric shock

721.410.3 General requirements

721.410.3.5 The protective measures of obstacles and placing out of reach (Section 417) shall not be used.

721.410.3.6 The protective measures of non-conducting location (Regulation 418.1) and earth-free local equipotential bonding (Regulation 418.2) shall not be used.

721.411 Protective measure: Automatic disconnection of supply

721.411.3.1 Protective earthing and protective equipotential bonding

721.411.3.1.2 Protective equipotential bonding

Structural metallic parts which are accessible from within the caravan shall be connected through main protective bonding conductors to the main earthing terminal within the caravan.

721.413 Protective measure: Electrical separation

The protective measure of electrical separation shall not be used, except for a shaver supply unit complying with BS EN 61558-2-5.

721.414 Protective measure: Extra-low voltage provided by SELV and PELV

Any part of a caravan installation operating at extra-low voltage shall comply with the requirements of Section 414.

For extra-low voltage DC power sources, a maximum of 48 V is allowed. In exceptional cases, when AC extra-low voltage is required, the voltage (rms) is not allowed to exceed 48 V.

NOTE: The requirements of Section 721 are also applicable to extra low voltage DC installations. See Annex A721 for recommendations that may be applied in addition.

721.415.1 Additional protection: RCDs

Where protection by automatic disconnection of supply is used, an RCD with a rated residual operating current not exceeding 30 mA, complying with BS EN 60947-2 (Annex B), BS EN 61008-1, BS EN 61009-1 or BS EN 62423 breaking all live conductors, shall be provided having the characteristics specified in 415.1.1.

Each supply inlet shall be directly connected to its associated RCD.

NOTE: This implies that there may not be any junctions in this connection.

721.43 Protection against overcurrent

721.43.1 Final circuits

Each final circuit shall be protected by an overcurrent protective device which disconnects all live conductors of that circuit.

721.5 Selection and erection of equipment

721.51 Common rules

721.510 Introduction

721.510.3 General

Where there is more than one electrically independent installation, each independent installation shall be supplied by a separate connecting device and shall be segregated in accordance with the relevant requirements of BS 7671.

721.514 Identification and notices

721.514.1 General

Instructions for use shall be provided with the caravan so that the caravan can be used safely.

The instructions shall comprise:

(i) a description of the installation
(ii) a description of the function of the RCD(s) and the use of the test button(s)
(iii) a description of the function of the main isolating switch
(iv) the text of the instructions of Figure 721.

If it is necessary to take precautions during user maintenance, appropriate details shall be given.

Fig 721 – Instructions for electricity supply

INSTRUCTIONS FOR ELECTRICITY SUPPLY

TO CONNECT

1. Before connecting the caravan installation to the mains supply, check that:
   (a) the supply available at the caravan pitch supply point is suitable for the caravan electrical installation and appliances, and
   (b) the voltage, frequency and current ratings are suitable, and
   (c) the caravan main switch is in the OFF position.

   Also, prior to use, examine the supply flexible cable to ensure there is no visible damage or deterioration.
2. Open the cover to the appliance inlet provided at the caravan supply point, if any, and insert the connector of the supply flexible cable.
3. Raise the cover of the electricity outlet provided on the pitch supply point and insert the plug of the supply cable.

THE CARAVAN SUPPLY FLEXIBLE CABLE MUST BE FULLY UNCOILED TO AVOID DAMAGE BY OVERHEATING

4. Switch on at the caravan main isolating switch.
5. Check the operation of residual current devices (RCDs) fitted in the caravan by pressing the test button(s) and reset.

IN CASE OF DOUBT OR, IF AFTER CARRYING OUT THE ABOVE PROCEDURE THE SUPPLY DOES NOT BECOME AVAILABLE, OR IF THE SUPPLY FAILS, CONSULT THE CARAVAN PARK OPERATOR OR THE OPERATOR'S AGENT OR A QUALIFIED ELECTRICIAN.

TO DISCONNECT

6. Switch off at the caravan main isolating switch, unplug the cable first from the caravan pitch supply point and then from the caravan inlet connector.

PERIODIC INSPECTION

Preferably not less than once every three years and annually if the caravan is used frequently, the caravan electrical installation and supply cable should be inspected and tested and a report on their condition obtained as prescribed in BS 7671 Requirements for Electrical Installations published by the Institution of Engineering and Technology & BSI.

721.521 Types of wiring system

721.521.2 The wiring systems shall be installed using one or more of the following:

(i) Insulated single-core cables, with flexible class 5 conductors, in conduits, conduit system, trunking or trunking system

(ii) Insulated single-core cables, with stranded class 2 conductors (minimum of 7 strands), in conduit, conduit system, trunking or trunking system

(iii) Sheathed flexible cables.

All cables shall, as a minimum, meet the requirements of BS EN 60332-1-2.

Conduits and conduit systems shall comply with the relevant part of BS EN 61386.

Cable trunking, cable trunking systems and cable ducting systems shall comply with the relevant part of BS EN 50085.

721.522 Selection and erection of wiring systems in relation to external influences

721.522.7 Vibration (AH)

721.522.7.1 As the wiring will be subjected to vibration, all wiring shall be protected against mechanical damage either by location or by enhanced mechanical protection. Wiring passing through metalwork shall be protected by means of suitable bushes or grommets, securely fixed in position. Precautions shall be taken to avoid mechanical damage due to sharp edges or abrasive parts.

721.522.8 Other mechanical stresses (AJ)

721.522.8.1.3 All cables, unless enclosed in rigid conduit, and all flexible conduit shall be supported at intervals not exceeding 0.4 m for vertical runs and 0.25 m for horizontal runs.

721.524 Cross-sectional areas of conductors

721.524.1 The cross-sectional area of every conductor shall be not less than 1.5 mm^2.

721.528 Proximity of wiring systems to other services

721.528.1 Proximity to electrical services

Cables of low voltage systems shall be run separately from the cables of extra-low voltage systems, in such a way, so far as is reasonably practicable, that there is no risk of physical contact between the two wiring systems.

721.528.2 Proximity to non-electrical services

721.528.2.1 No electrical equipment, including wiring systems, except ELV equipment for gas supply control, shall be installed in any gas cylinder storage compartment.

ELV cables and electrical equipment may only be installed within the LPG cylinder compartment if the installation serves the operation of the gas cylinder (e.g. indication of empty gas cylinder) or is for use within the compartment. Such electrical installations and components shall be constructed and installed so that they are not a potential source of ignition.

Where cables have to run through such a compartment, they shall pass through the compartment at a height of not less than 500 mm above the base of the cylinders and shall be protected against mechanical damage by installation within a conduit system complying with the appropriate part of the BS EN 61386 series or within a ducting system complying with the appropriate part of the BS EN 50085 series.

Where installed, this conduit or ducting system shall be able to withstand an impact equivalent to AG3 without visible physical damage.

721.53 Protection, isolation, switching, control and monitoring

721.537 Isolation and switching

721.537.2 Devices for Isolation

721.537.2.1.1 Each installation shall be provided with a main disconnector which shall disconnect all live conductors and which shall be suitably placed for ready operation within the caravan. In an installation consisting of only one final circuit, the isolating switch may be the overcurrent protective device fulfilling the requirements for isolation.

721.537.2.1.1.1 A notice of such durable material as to be likely to remain easily legible throughout the life of the installation, shall be permanently fixed near the main isolating switch inside the caravan, bearing the text shown in Figure 721 in the appropriate language(s) in indelible and easily legible characters.

721.543 Protective conductors

721.543.2 Types of protective conductor

721.543.2.1 Circuit protective conductors shall be incorporated in a multicore cable or in a conduit together with the live conductors.

721.544.1 Protective bonding conductors

721.544.1.1 The terminations of protective bonding conductors connecting the conductive structure of the unit shall be accessible and protected against corrosion.

721.55 Other equipment

721.55.1 Inlets

721.55.1.1 Any AC electrical inlet on the caravan shall be an appliance inlet complying with BS EN 60309-1. If interchangeability is required the inlet shall comply with BS EN 60309-2.

721.55.1.2 The inlet shall be installed:

(i) not more than 1.8 m above ground level, and

(ii) in a readily accessible position, and

(iii) such that it shall have a minimum degree of protection of IP44 with or without a connector engaged, and

(iv) such that it shall not protrude significantly beyond the body of the caravan.

721.55.2 Accessories

721.55.2.2 Every socket-outlet supplied at extra-low voltage shall have its voltage visibly marked.

721.55.2.3 Where an accessory is located in a position in which it is exposed to the effects of moisture it shall be constructed or enclosed so as to provide a degree of protection not less than IP44.

721.55.2.4 Each luminaire in a caravan shall preferably be fixed directly to the structure or lining of the caravan. Where a pendant luminaire is installed in a caravan, provision shall be made for securing the luminaire to prevent damage when the caravan is in motion.

Accessories for the suspension of pendant luminaires shall be suitable for the mass suspended and the forces associated with vehicle movement.

721.55.2.5 A luminaire intended for dual voltage operation shall comply with the appropriate standard.

721.55.2.6 The means of connection to the caravan pitch socket-outlet shall be supplied with the caravan and shall comprise the following (see Figure 708):

(i) A plug complying with BS EN 60309-2, and

(ii) a flexible cable of continuous length 25 m (±2 m), having a harmonized code designation of H05RN-F or H07RN-F (BS EN 50525-2-21) or equivalent, incorporating a protective conductor, with conductors to be identified according to Table 51 and of a cross-sectional area in accordance with Table 721, and

(iii) a connector, if any, compatible with the appliance inlet installed under Regulation 721.55.1.

TABLE 721 – Minimum cross-sectional areas of flexible cables for caravan connection

Rated current A	Minimum cross-sectional area mm^2
16	2.5
25	4
32	6
63	16
100	35

ANNEX A721 (Informative)

GUIDANCE FOR EXTRA-LOW VOLTAGE DC INSTALLATIONS

NOTE: In general, the requirements of Section 721 are also applicable to an extra-low voltage DC installation. The following requirements should be applied in addition.

A721.31 Purposes, supplies and structure

A721.313 Supplies

A721.313.4 Sources of supply

The supply should be obtained from one or more of the following sources:

(i) The electrical installation of the towing vehicle
(ii) An auxiliary battery mounted in the caravan
(iii) A low voltage DC supply via a transformer/rectifier unit complying with BS EN 60335-1 and BS EN 61558-2-6
(iv) A DC generator that is driven by any form of energy
(v) Solar photovoltaic (PV) power supply systems.

A721.514 Identification and notices

A721.514.1 General

The following information should be provided in the instructions for use and should be in the official language/s of the country in which the caravan is to be sold:

(i) A warning worded as follows: 'Any replacement of anauxiliary battery should be of the same type and specification as that originally fitted'
(ii) Instructions on the maintenance and recharging of anauxiliary battery where it is fitted. Where a battery charger is provided, instructions on its safe use should be included
(iii) Instructions on selecting and installing an auxiliary battery, in a compartment, if the caravan installation is designed for the installation of an auxiliary battery
(iv) Details of the warning notice specified in A721.55.3.7 and its importance for safety
(v) In order to provide for safe operation of the electrical installation, a simplified diagram of the wiring of the ELV and LV installation, with details of the cable colours and/or marking and the nominal values of the overcurrent protective devices
(vi) Type of appliances that can be used and from what source of supply
(vii) Instructions for the correct operation and maintenance of fitted appliances, as supplied by the appliance manufacturer
(viii) A warning worded as follows: 'Always disconnect the electrical connector between the towing vehicle and the caravan before connecting an LV supply to the caravan and before charging the caravan battery by any other means.'

A721.515 Prevention of mutual detrimental influence

A721.515.2 The ELV installation should be so installed that the protective measures of the LV installation for basic protection or for fault protection are not impaired.

It should be verified that the protective conductors of the LV installationare not loaded by the operating currents of the ELV installation.

A721.521 Types of wiring system

A721.521.2 Cables should be of stranded construction and should comply with BS 6004, BS 7211 or an appropriate part of BS EN 50525.

A721.523 Current-carrying capacities of cables

A721.523.1 The cross-sectional areas of the fixed wiring should be such that the permissible voltage drop is not exceeded.

A721.525 Voltage drop in consumers' installations

Under normal service conditions the voltage at the terminals of any fixedcurrent-using equipment should be greater than the lower limit corresponding to the British or Harmonized Standard relevant to the equipment. Where the equipment is not the subject of a British or Harmonized Standard, the voltage at the terminals should be such as not to impair the safe functioning of that equipment. In the absence of precise data a voltage drop of 0.8 V from the power supply to the equipment may be allowed.

The voltage drop between the plug of the connector to the towing vehicle or LV battery charger and the auxiliary battery should not exceed 0.3 V.

The charging current I_c (A) to determine the voltage drop is established by the following formula:

$$I_c = \frac{c \times 0.1}{t}$$

where:

I_c is the charging current in A

c is the battery capacity in Ah

t is the charging period in h.

NOTE: Some battery manufacturers now ratebatteries in Watt/hours (Wh).

A721.528 Proximity of wiring systems to other services

A721.528.2 Proximity to non-electrical services

A721.528.2.5 Cable runs and LPG installations

Cables including those used for automotive purposes should not be run through a compartment or housing intended for liquefied petroleum gas storage cylinders. Where cables have to run through such a compartment or housing, they should be run at a height of not less than 500 mm above the base of the cylinders, and such cables should be protected against mechanical damage by installation within a continuous gas tight conduit or duct passing through the compartment.

Where installed, this conduit or duct should be able to withstand an impact equivalent to AG3 without visible physical damage.

ELV cables and electrical equipment are only to be installed within the LPG cylinder compartment or housing if the installation serves the operation of the gas cylinders (e.g.indication of empty gas cylinders) or is for use within the compartment or housing. Such electrical installations and components should be constructed and installed so that they are not a source of ignition and are in accordance with the relevant standards for any hazardous area classification of the compartment or housing.

A721.53 Protection, isolation, switching, control and monitoring

A721.533 Devices for protection against overcurrent

A721.533.1 General requirements

A721.533.1.5 The overcurrent protective device for the power supply from the towing vehicle should be fitted as near as possible to the auxiliary battery, but in no case more than 1 m away. The overcurrent protective device for the auxiliary battery should be fitted at the end of the battery cable and before the fixed installation. The ELV output of the transformer/rectifier unit and of the DC generator should be provided with an overcurrent protective device installed as near as possible to the unit or generator and, in all cases, upstream of the distribution circuits.

A721.533.1.6 Overcurrent protective devices should be either fuse links according to ISO 8820 or suitable circuit-breakers complying with BS EN 60947-2, BS EN 60898-2 or BS IEC 60898-3.

A721.533.1.7 Fuses should be protected to prevent accidental damage.

A721.533.1.8 Overcurrent protective devices should not be fitted in a fuel storage compartment or fuel storage housing intended for the storage of liquefied petroleum gas (LPG) cylinders or in the compartment for housing an auxiliary battery.

A721.55 Other equipment

A721.55.1 Inlets

The inlet, when the plug is disconnected, should be protected against the ingress of water, foreign bodies and accidental damage.

A721.55.2 Accessories

A721.55.2.6 The means of connection to the towing vehicle should be supplied with the caravan and comprise the following:

(i) a plug complying with BS AU 149a and BS AU 177a or BS ISO 11446, and

(ii) a flexible cable with the number of cores with the minimum cross-sectional area and the allocation according to Table A721 and a length not exceeding 5 m, and

(iii) a connector complying with BS AU 149a and BS AU 177a or BS ISO 11446.

TABLE A721 – Functional allocation and cross-sectional areas of cores for caravan connectors

Core No.	Function	Contact numbers		Minimum-cross-sectional area mm^2
		BS ISO 11446	**BS AU 149a**	
1	Left-hand direction – indicator light	1	1	1.5
2	Rear fog light	2	2	1.5
3	Common return for core Nos. 1,2 and 4 to 8	3*	3*	2.5
4	Right-hand direction – indicator light	4	4	1.5
5	Right-hand rear position and marker lights, and rear registration-plate illumination device	5	5	1.5
6	Stop lights	6	6	1.5
7	Left-hand rear position and marker lights, and rear registration-plate illumination device	7	7	1.5
			BS AU 177a	
8	Reversing light	8	1	1.5
9	Continuous power supply	9	4	2.5
10	Power supply controlled by ignition switch	10	6	2.5
11	Return for core No. 10	11*	7*	2.5
12	Coding for coupled trailer	12	2	-
13	Return for core No. 9	13*	3*	2.5
14	No allocation	-	5	1.5

* These return circuits should not be connected electrically in the trailer.

A721.55.3 Auxiliary batteries

A721.55.3.1 Type of battery

An auxiliary battery should be of the rechargeable type.

NOTE: Non-rechargeable batteries are not auxiliary batteries. They may be used in caravans, provided that they are used in circuits separated from other sources of electrical supply.

A721.55.3.2 Capacity

An auxiliary battery should have a minimum capacity of 40 Ah at 20 h discharge rate.

NOTE: It is recommended to use a battery designed to be discharged over long periods at a relatively low current.

A721.55.3.3 Terminals

Auxiliary battery terminals should be clearly and durably marked '+' and '–'. Connections to auxiliary battery terminals should be securely clamped or bolted to provide continuous contact and should be insulated unless the auxiliary battery is provided with an insulating device.

A721.55.3.4 Location

An auxiliary battery should be placed in a separate compartment, with easy access for maintenance or removal, and secured to prevent movement of the battery, e.g. when the caravan is in motion.

A721.55.3.5 Auxiliary battery compartment

A tray should be installed under an auxiliary battery if the electrolyte of this battery is liquid.

The tray should be sufficient to hold 20 % of the volume of the electrolyte.

The interior of an auxiliary battery compartment should be ventilated and protected against the corrosive effect of acid-laden gases, either by:

(i) installing a sealed auxiliary battery that incorporates an external ventilating kit that is taken to the exterior of the caravan, or

(ii) installing an auxiliary battery in an enclosed battery compartment that is protected internally against corrosion and is ventilated to the exterior of the caravan by means of a suitable tube with a minimum inside diameter of 10 mm at the top of the auxiliary battery compartment, in accordance with the battery manufacturer's instructions or as supplied by the manufacturer of the auxiliary battery, or

(iii) ventilating the compartment at low level and high level to the exterior of the caravan and constructing the interior of the compartment, including the sides of the ventilator openings, of acid-resistant material or providing it with an anticorrosive finish. If the compartment opens into the interior of the caravan, the lid should provide an air seal. The minimum free area of ventilation should be not less than 80 mm^2 at low level and not less than 80 mm^2 at high level.

If an auxiliary battery is not provided, then the position and instructions for the installation of the battery and compartment, in accordance with (i), (ii) or (iii), should be included in the instructions for use and a notice should be fixed in or near the proposed location stating: 'For instructions on auxiliary battery installation, see the instructions for use'.

The requirements concerning the protection against corrosion and ventilation are not applicable if batteries with bound electrolytes are used.

Where the manufacturer makes no provision for the installation of an auxiliary battery, the following statement should be made in the instructions for use: 'This caravan has not been designed to accommodate an auxiliary battery. Do not fit one.'

A721.55.3.6 Auxiliary battery cables

Cables from an auxiliary battery should be protected by additional sheathing or taping from the battery terminal up to the overcurrent protective device.

A721.55.3.7 Warning notice

A warning notice should be fixed in a prominent position near the auxiliary battery or displayed on the lid of the auxiliary battery compartment. This warning should be in the official language(s) of the country in which the caravan is to be sold and should state: 'Switch off all appliances and lamps before disconnecting the auxiliary battery.'

The auxiliary battery compartment should be additionally marked 'Smoking prohibited' in accordance with BS EN ISO 7010 (symbols ISO P002 and P003) and in the language(s) of the country in which the caravan is to be sold.

A721.55.4 Other sources of supply

A721.55.4.1 Generators and transformer/rectifier unit

If a supply is obtained from a generator or from a low voltage supply via a transformer/rectifier unit, the extra-low voltage at the output terminals of the supply unit should be maintained between 11 V minimum and 14 V maximum with applied loads varying from 0.5 A minimum up to the maximum rated load of the supply unit. Over the same load range, alternating voltage ripple should not exceed 1.2 V peak-to-peak.

A721.55.4.2 Regenerative sources

Regenerative energy sources, such as wind energy, solar energy etc., should be installed only for charging batteries.

Regenerative energy sources should only be operated with a device which prevents overcharging of the battery(ies).

A721.55.5 Charging of auxiliary battery and operation of refrigerator

A721.55.5.1 The circuit to charge an auxiliary battery should be separate from a circuit to operate a refrigerator.

A721.55.5.2 The charging circuit for an auxiliary battery should be completed only when the ignition of the towing vehicle is switched on.

A721.55.5.3 The 12 V heating facility of a refrigerator should be completed only when the ignition of the towing vehicle is switched on. This may be performed by a device built into the refrigerator.

A721.55.6 Terminal block

If the connection between the connecting cable(s) and the caravan's fixed wiring is by means of a terminal block, it should have a protective cover. If the terminal block is positioned externally it should have a cover with a degree of protection of at least IP34 according to BS EN 60529.

A721.55.7 Appliances

A721.55.7.1 General

The caravan manufacturer's technical specification should state whether an ELV appliance is suitable for use with a supply obtained from a DC generator or a transformer/rectifier unit.

Appliances suitable for operation on both 12 V AC and 12 V DC systems are allowed provided that AC and DC systems are segregated and interconnection is prevented.

A721.55.7.2 Selection and connection of appliances

All appliances should be fitted and connected in accordance with the appliance manufacturer's instructions. Where polarity-sensitive appliances are fitted and connected, only those should be used that have terminals clearly marked '–' and '+', or that have two conductors, indicating polarity by colour or by identification tags or sleeves marked '–' or '+'.

A721.55.8 Socket-outlets

ELV socket-outlets should be two-pole non-reversible and should be of a different type from those provided for any low voltage installation. The voltage and maximum power rating of the circuit should be stated on or adjacent to the socket-outlets.

A721.55.9 Battery charger

If a battery charger is connected to a low voltage AC supply, it should comply with the relevant clauses of BS EN 60335-2-29. The DC output should either be electronically regulated or the maximum DC output of the charger in amperes should be limited to 10 % of the capacity of the auxiliary battery in A h at 20 h discharge rate.

A721.55.10 External lights

Lights, such as door lamps, fixed outside on a caravan should be constructed or enclosed to provide protection against the ingress of water with a degree of protection of at least IP34 according to BS EN 60529.

SECTION 722

ELECTRIC VEHICLE CHARGING INSTALLATIONS

NOTE: Definitions relating to Section 722 can be found in Part 2 under electric vehicle (EV).

722.1 Scope

The particular requirements of this section apply to circuits intended to supply electric vehicles for charging purposes.

The requirements of this section do not apply to electric vehicle charging points that:

(i) employ inductive charging

(ii) charge mobility scooters and similar vehicles of 10 A and less.

722.3 Assessment of general characteristics

722.31 Purposes, supplies and structure

722.311 Maximum demand and diversity

722.311.201 Load curtailment, including load reduction or disconnection, either automatically or manually, may be taken into account when determining maximum demand of the installation or part thereof.

722.312 Conductor arrangement and system earthing

722.312.2.1 TN systems

A circuit supplying charging equipment for electric vehicles shall not include a PEN conductor.

722.4 Protection for safety

722.41 Protection against electric shock

722.410.3 General requirements

722.410.3.5 The protective measures of obstacles and placing out of reach (Section 417) shall not be used.

722.410.3.6 The protective measures of non-conducting location (Regulation 418.1) and earth-free local equipotential bonding (Regulation 418.2) shall not be used.

722.411.4 TN system

722.411.4.1 A PME earthing facility shall not be used as the means of earthing for the protective conductor contact of a charging point located outdoors or that might reasonably be expected to be used to charge a vehicle located outdoors unless one of the following methods is used:

(i) *Deleted by BS 7671:2018+A2:2022*

(ii) The main earthing terminal of the installation is connected to an installation earth electrode by a protective conductor complying with Regulation 544.1.1. The resistance of the earth electrode to Earth shall be such that the maximum voltage between the main earthing terminal of the installation and Earth in the event of an open-circuit fault in the PEN conductor of the low voltage network supplying the installation does not exceed 70 V rms.

NOTE: Annex 722, item A722.3 gives guidance on determining the maximum resistance required for the earth electrode in (ii). See Note 1.

(iii) Protection against electric shock is provided by a device which electrically disconnects the vehicle from the live conductors of the supply and from protective earth in accordance with Regulation 543.3.3.101(ii) within 5 s in the event of the voltage between the circuit protective conductor and Earth exceeding 70 V rms due to an open-circuit fault in the PEN conductor of the low voltage network. The device need not operate if the voltage exceeds 70 V rms for less than 4 s. The device shall provide isolation and be selected in accordance with Table 537.4. Closing or resetting of the device shall be possible only if the voltage between the circuit protective conductor and Earth does not exceed 70 V rms. Equivalent means of functionality could be included within the charging equipment. See Note 1.

NOTE: Annex 722, item A722.4 gives guidance on (iii).

(iv) Protection against electric shock is provided by a device which electrically disconnects the vehicle from the live conductors of the supply and from protective earth in accordance with Regulation 543.3.3.101(ii) within 5 s in the event of the utilisation voltage at the charging point, between the line and neutral conductors, being greater than 253 V rms or less than 207 V rms. The device shall provide isolation and be selected in accordance with Table 537.4. Equivalent means of functionality could be included within the charging equipment. Closing or resetting of the device shall be possible only if the voltage between line and neutral conductors is in the range 207 to 253 V rms. See Note 1.

(v) Protection against electric shock is provided by the use of an alternative device to those in (iii) or (iv) which does not result in a lesser degree of safety than using (iii) or (iv). Equivalent means of functionality could be included within the charging equipment. The device (or means of functionality) shall operate by electrically disconnecting the vehicle from the live conductors of the supply and from protective earth in accordance with Regulation 543.3.3.101(ii). It shall provide isolation and be selected in accordance with Table 537.4. See Note 1.

NOTE 1: See Section 511. BS 7671 does not deal with the safety requirements for the construction of electrical equipment.

Where equipment to be used is not covered by a British or Harmonized Standard, or where there is no British or Harmonized Standard for the functionality of a piece of equipment used, it is the responsibility of the electrical installation designer or other person responsible for specifying the installation to establish that:

(i) the equipment meets the requirements of the Electrical Equipment (Safety) Regulations 2016 (as amended), the Electromagnetic Compatibility Regulations 2016 (as amended) and other relevant legislation, and

(ii) the equipment has either a CE, UKCA or UKNI mark and a Declaration of Conformity; where third-party approval is required, the equipment is marked appropriately.

The Declaration of Conformity is to be appended to the certification for initial verification.

Where the above is satisfied, it is not considered to be a departure from BS 7671:2018+A2:2022.

NOTE 2: See also Regulation 641.5 when undertaking an addition or alteration to an existing installation.

Where buried in the ground, a protective conductor connected to an earth electrode for the purposes of (ii) or (iii) shall have a cross-sectional area not less than that stated in Table 54.1.

Protective conductors and exposed-conductive-parts downstream of a protective device provided for the purposes of (iii), (iv) and (v) shall have no connection to:

(a) any protective conductors or exposed-conductive-parts of any circuit not protected by the same protective device, or

(b) any extraneous-conductive-part.

NOTE 3: Creating a TT earthing system for charging equipment or the whole installation as an alternative to using a PME earthing facility with one of methods (ii) to (v) above may not be an appropriate solution due to the inability to provide sufficient separation from buried metalwork connected to the supply PEN conductor.

722.413 Protective measure: Electrical separation

722.413.1.2 This protective measure shall be limited to the supply of one electric vehicle supplied from one unearthed source. The circuit shall be supplied through a fixed isolating transformer complying with BS EN 61558-2-4.

NOTE: An example of an arrangement for the supply of a Class I electric vehicle charging point from a separated source is shown in Annex A722, Item A722.5.

722.5 Selection and erection of equipment

722.51 Common rules

722.511 Compliance with standards

722.511.1 Where an EV charging point is built into a low voltage switchgear or controlgear assembly the requirements of the relevant part of BS EN 61439 series shall apply.

722.511.101 EV charging equipment shall comply with the appropriate parts of the BS EN 61851 series.

722.512 Operational conditions and external influences

722.512.2 External influences

722.512.2.201 Presence of water (AD)

Where installed outdoors, the equipment shall provide a degree of protection of at least IPX4.

722.512.2.202 Presence of solid foreign bodies (AE)

Where installed outdoors, the equipment shall provide a degree of protection of at least IP4X.

722.512.2.203 Impact (AG)

Equipment installed in public areas and car park sites shall be protected against mechanical damage (impact of high severity AG3). Protection of the equipment shall be afforded by one or more of the following:

- the position or location shall be selected to avoid damage by any reasonably foreseeable impact
- local or general mechanical protection shall be provided
- equipment shall be installed that complies with a minimum degree of protection against external mechanical impact of IK08 in accordance with the requirements of BS EN 62262.

722.53 Protection, isolation, switching, control and monitoring

722.531 Devices for protection against electric shock by automatic disconnection of supply

722.531.3 Residual current devices (RCDs)

722.531.3.1 RCDs shall disconnect all live conductors.

722.531.3.101 Unless supplied by a circuit using the protective measure of electrical separation, each charging point incorporating a socket-outlet or vehicle connector complying with the BS EN 62196 series shall be protected individually by an RCD of Type A, Type F or Type B and having a rated residual operating current not exceeding 30 mA.

Except where provided by the EV charging equipment, protection against DC fault currents shall be provided by:

(i) an RCD Type B, or

(ii) an RCD Type A or Type F in conjunction with a residual direct current detecting device (RDC-DD) complying with BS IEC 62955 as appropriate to the nature of the residual and superimposed currents and recommendation of the manufacturer of the charging equipment.

RCDs shall comply with one of the following standards: BS EN 61008-1, BS EN 61009-1, BS EN 60947-2 or BS EN 62423.

NOTE 1: Types of RCD are described in Regulation 531.3.3 in respect of their behaviour when exposed to DC components and frequencies.

NOTE 2: Requirements for the selection and erection of RCDs in the case of supplies using DC vehicle connectors according to the BS EN 62196 series are under consideration.

NOTE 3: An RCD Type A or Type F in conjunction with an RDC-DD can be arranged with the RDC-DD inside the EV charging equipment and the Type A or Type F RCD upstream in either the charging equipment or the installation.

722.533 Devices for protection against overcurrent

722.533.101 Each charging point shall be supplied individually by a final circuit protected by an overcurrent protective device complying with BS EN 60947-2, BS EN 60947-6-2 or BS EN 61009-1 or with the relevant parts of the BS EN 60898 series or the BS EN 60269 series.

NOTE: The electric vehicle charging equipment may have multiple charging points.

722.55 Other equipment

722.55.101 Socket-outlets and connectors

722.55.101.0.201.1 Each AC charging point shall incorporate:

(i) one socket-outlet complying with BS 1363-2 marked 'EV' on its rear and, except where there is no possibility of confusion, a label shall be provided on the front face or adjacent to the socket-outlet or its enclosure stating: 'suitable for electric vehicle charging', or

(ii) one socket-outlet or connector complying with BS EN 60309-2 which is interlocked and classified to clause 6.1.5 of BS EN 60309-1 to prevent the socket contacts being live when accessible, or

(iii) one socket-outlet or connector complying with BS EN 60309-2 which is part of an interlocked self-contained product complying with BS EN 60309-4 and classified to clauses 6.1.101 and 6.1.102 to prevent the socket contacts being live when accessible, or

(iv) one Type 1 vehicle connector complying with BS EN 62196-2 for use with mode 3 charging only, or

(v) one Type 2 socket-outlet or vehicle connector complying with BS EN 62196-2 for use with mode 3 charging only, or

(vi) one Type 3 socket-outlet or vehicle connector complying with BS EN 62196-2 for use with mode 3 charging only.

NOTE: Vehicle manufacturers' instructions should be taken into account when determining the type of socket-outlet to be installed.

722.55.101.0.201.2 Each socket-outlet shall be installed in a distribution board in accordance with Regulation 722.51 or in its appropriate enclosure (e.g. flush or surface mounted socket-outlet box) and mounted in a fixed position.

Portable socket-outlets shall not be used but tethered vehicle connectors are permitted.

722.55.101.3 One socket-outlet or vehicle connector shall supply only one electric vehicle.

722.55.101.4 In EV charging modes 3 and 4, an electrical or mechanical system shall be provided to prevent the plugging/unplugging of the plug unless the socket-outlet or the vehicle connector has been switched off from the supply.

722.55.101.6 Precautions on supply of the fixed installation by the EV

NOTE: Requirements for precautions on supply of the fixed installation by the EV are under consideration.

722.551.7.2 The socket-outlet or vehicle connector shall comply with BS EN 62196 series.

722.826.3.201 Where a device complying with 722.411.4.1 (iii), (iv) or (v) is used to protect EV charging equipment in a PEI, the vehicle shall be electrically disconnected from the live conductors of the supply and from protective earth in accordance with Regulation 543.3.3.101(ii) in island mode unless the manufacturer of the device confirms that the device will continue to provide protection in island mode.

ANNEX A722 (Informative)

GUIDANCE FOR TN SYSTEMS WHERE PME CONDITIONS APPLY

A722.1 Neutral current of a three-phase installation

Where the power factors of the currents in all three phases are similar and triple harmonics can be neglected, it may be assumed that the neutral current of a three-phase installation is given by:

$$I_m = \sqrt{I_{L1}^2 + I_{L2}^2 + I_{L3}^2 - I_{L1}I_{L2} - I_{L1}I_{L3} - I_{L2}I_{L3}}$$

NOTE: The maximum neutral current (I_m) occurs under conditions of maximum imbalance, not necessarily maximum overall demand.

A722.2 *Deleted by BS 7671:2018+A2:2022*

A722.3 Earth electrode resistance

For the purposes of condition (ii) of Regulation 722.411.4.1, the sum of the resistances of the earth electrode and the protective conductor connecting it to the main earthing terminal must meet the following condition, as applicable.

For a single-phase installation:

$$R_{A\ ev} \leq \frac{70\ U_0 C_{max}}{I_{inst}(U_0 C_{max} - 70)}$$

For a three-phase installation:

$$R_{A\ ev} \leq \frac{70 U_0 C_{max}}{I_m U_0 C_{max} - 70(I_{L1} + I_{L2} + I_{L3})}$$

NOTE 1: Earth electrodes with a resistance above 200 Ω may be unstable. Where the above three-phase formula leads to an earth electrode resistance exceeding 200 Ω, or

$$I_m > \frac{70(I_{L1}+I_{L2}+I_{L3})}{U_0 C_{max}}$$

then the maximum value of $R_{A\ ev}$ should be taken as 200 Ω.

NOTE 2: The above currents and voltage are magnitudes only; they are not phasors.

NOTE 3: In determining I_m, I_{L1}, I_{L2} and I_{L3}, allowance must be made for single-phase vehicles being charged from three-phase charging points.

A722.4 Guidance on voltage monitoring device described in Regulation 722.411.4.1(iii)

Regulation 722.411.4.1(iii) describes a device that measures the voltage between the circuit protective conductor of the electric vehicle charging equipment and Earth. During a PEN failure in the supply network, the neutral of a TN-C-S supply is no longer considered to be reliably connected to Earth, and a device that measures the voltages between the following points will fail to provide equivalent safety to the device described in Regulation 722.411.4.1(iii):

(i) the circuit protective conductor and neutral

(ii) the circuit protective conductor and the consumer's main earthing terminal.

Suitable arrangements include measurement of the voltage between either:

(a) the circuit protective conductor and a suitable measurement earth electrode, or

(b) the circuit protective conductor and a reference point derived from the line conductors of a three-phase system, provided that suitable precautions are also taken to disconnect the device when the supply to one or more line conductors is interrupted.

Where used, measurement earth electrodes should be located at a sufficient distance from other earth electrodes and/or other buried exposed-conductive-parts and/or extraneous-conductive-parts connected to the PME earthing terminal to reduce transfer of earth potential rise on the PME system during a PEN failure.

Applying Clauses 9.6.1 and 9.6.2 of BS 7430:2011+A1:2015, in conditions where only buried exposed- and/or extraneous-conductive-parts equivalent to vertical and horizontal earth electrodes are being considered, a separation distance of at least 8 m reduces the voltage transfer from the PME earthing system to less than 20 % and, in these conditions, the voltage operated device should be arranged to operate at 56 V.

It is recommended that the minimum separation distance is 2 m, and at this separation distance the device should be arranged to operate at 40 V.

Where other types of earth electrodes are used, separation distances should be increased, or the operating voltage decreased, accordingly, using the guidance in BS 7430.

To protect a person touching the body of an electric vehicle, connected to the charging point, when an open-circuit fault occurs in the PEN conductor of the supply, an enhanced safety provision may be offered. This may be achieved by a device that provides shorter disconnection times, depending on the detected voltage between the circuit protective conductor and Earth but also taking into account the need to limit the risk of unwanted tripping. Examples of shorter disconnection times are shown in Table A722 below. These disconnection times are based on IEC Technical Report IEC TR 60479-5 and also on comparable requirements already included in Chapter 41 of BS 7671.

TABLE A722 – Maximum disconnection times in relation to the voltage between the circuit protective conductor and Earth*

70 V	100 V	200 V	400 V
1 s	0.7 s	0.2 s	0.04 s

* Total voltage drop across the sensing element of the device and the earth electrode (if any) in series.

A722.5 Example arrangement of a separated system described in Regulation 722.413

An example arrangement of a separated system as described in Regulation 722.413 is shown in Figure A722.

Isolating transformers may have high inrush currents, and overcurrent protective devices in the primary circuit should be selected accordingly to prevent unwanted-tripping.

Fig A722 – Example of a separated system where protective conductor monitoring is required

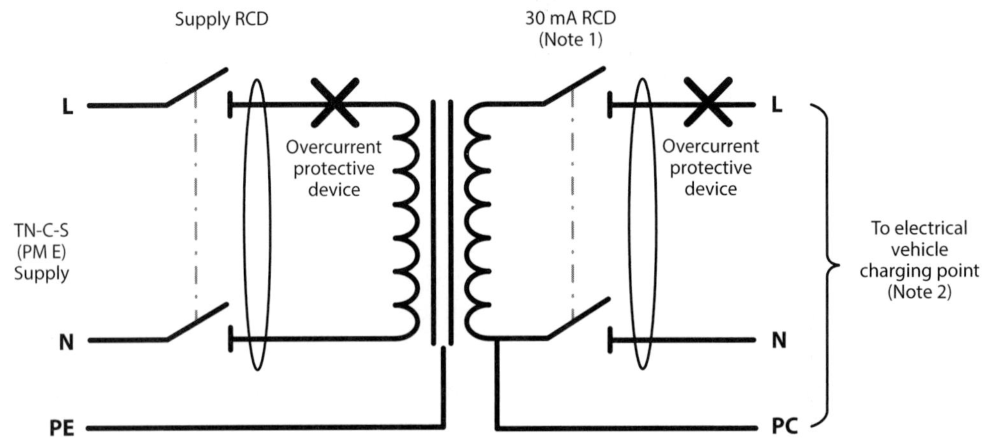

Key

L	line conductor
N	neutral conductor
PE	protective earthing conductor
PC	protective conductor

NOTE 1: The RCD on the secondary, complying with Regulation 722.531.3.101, should be placed as close as possible to the transformer.

NOTE 2: Only one charging point may be supplied by each transformer, see Regulation 722.413.1.2.

SECTION 729

OPERATING AND MAINTENANCE GANGWAYS

729.1 Scope

The particular requirements of this section apply to basic protection and other aspects relating to the operation or maintenance of switchgear and controlgear within areas including gangways, where access is restricted to skilled or instructed person(s).

729.3 Assessment of general characteristics

For restricted access areas the following apply:

(i) They shall be clearly and visibly marked by appropriate signs

(ii) They shall not provide access to unauthorised persons

(iii) Doors provided for closed restricted access areas shall allow easy evacuation by opening without the use of a key, tool or any other device not being part of the opening mechanism.

729.513 Accessibility

729.513.2 Requirements for operating and maintenance gangways

The width of gangways and access areas shall be adequate for work, operational access, emergency access, emergency evacuation and for transport of equipment.

Gangways shall permit at least a 90 degree opening of equipment doors or hinged panels (see also Annex A729).

729.513.2.1 Restricted access areas where basic protection is provided by barriers or enclosures

Where basic protection is provided by barriers or enclosures in accordance with Chapter 41, the following minimum dimensions apply (see Figure 729.1):

(i)	Gangway width including between: barriers or enclosures and switch handles or circuit-breakers in the most onerous position, and barriers or enclosures or switch handles or circuit-breakers in the most onerous position and the wall	700 mm
(ii)	Gangway width between barriers or enclosures or other barriers or enclosures and the wall	700 mm
(iii)	Height of gangway to barrier or enclosure above floor	2000 mm
(iv)	Live parts placed out of reach, see Regulation 417.3	2500 mm

NOTE: Where additional workspace is needed e.g. for special switchgear and controlgear assemblies, larger dimensions may be required.

Fig 729.1 – Gangways in installations with protection by barriers or enclosures

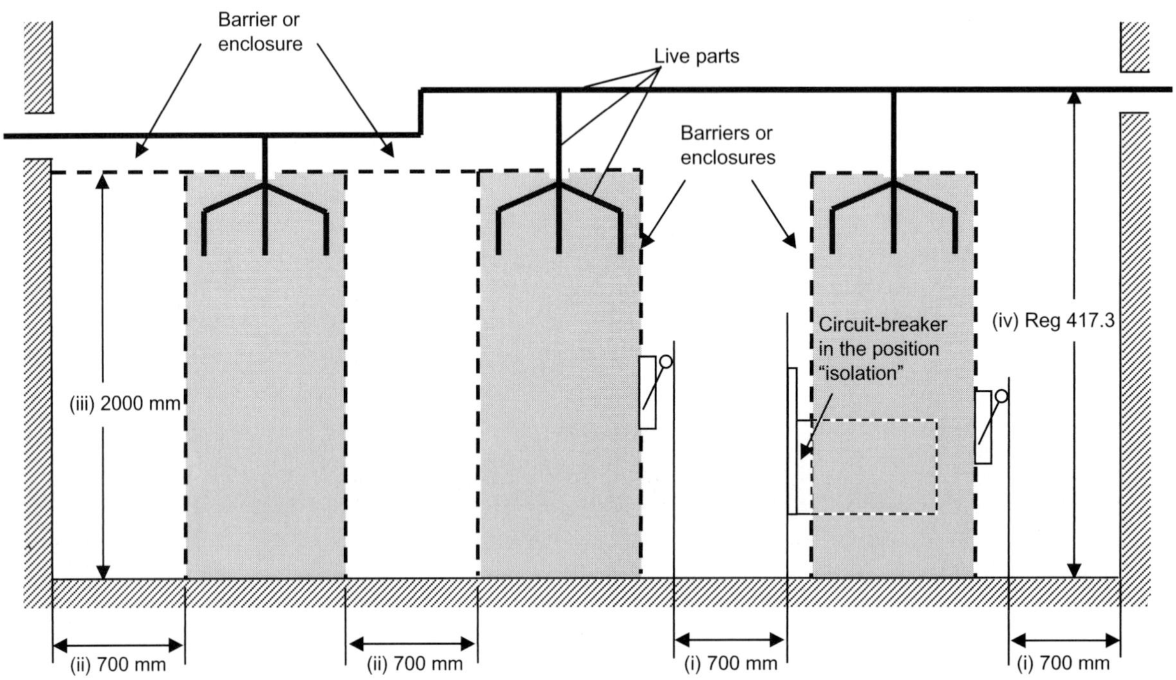

NOTE: The above dimensions apply after barriers and enclosures have been fixed and with circuit-breakers and switch handles in the most onerous position, including "isolation".

729.513.2.2 Restricted access areas where the protective measure of obstacles is applied

Where the protective measure of obstacles is used, the requirements of Section 417, Obstacles and placing out of reach, apply. The measure is for application in those parts of installations controlled or supervised by skilled persons.

The following minimum dimensions apply (see Figure 729.2):

(i)	Gangway width including between: obstacles and switch handles or circuit-breakers in the most onerous position, and obstacles or switch handles or circuit-breakers in the most onerous position and the wall.	700 mm
(ii)	Gangway width between obstacles or other obstacles and the wall	700 mm
(iii)	Height of gangway to obstacles above floor	2000 mm
(iv)	Live parts placed out of reach, see Regulation 417.3	2500 mm

Fig 729.2 – Gangways in installations with protection by obstacles

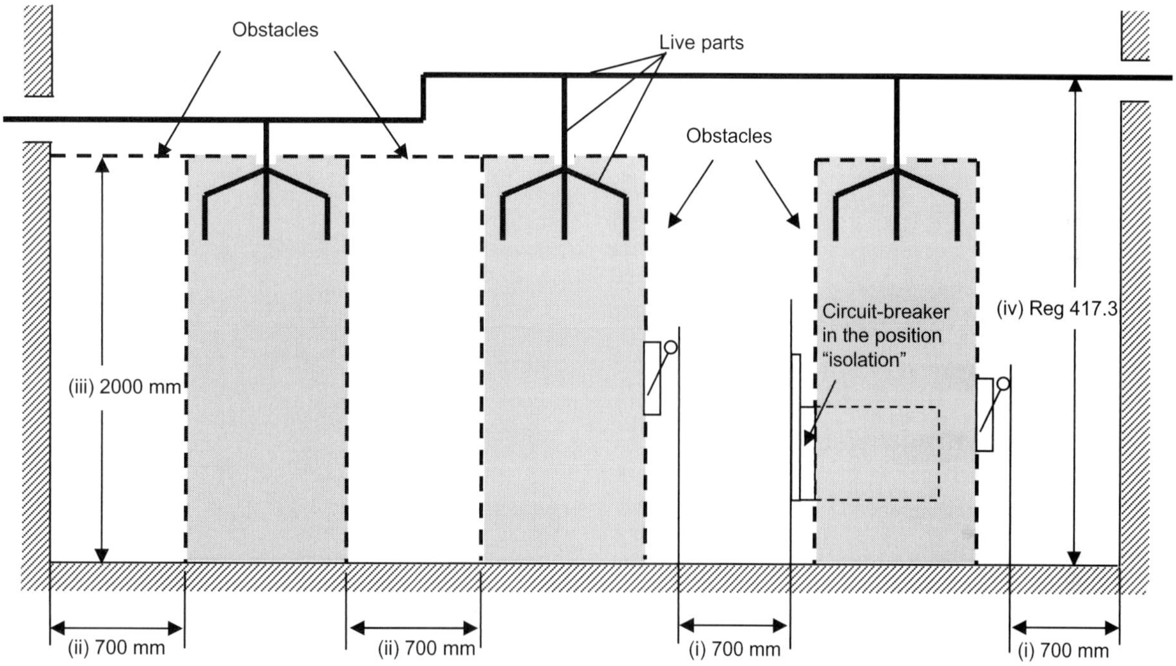

NOTE: The above dimensions apply after all obstacles have been fixed and with circuit-breakers and switch handles in the most onerous position, including "isolation".

729.513.2.3 Access to gangways

Gangways longer than 10 m shall be accessible from both ends.

NOTE 1: This may be accomplished by placement of the equipment a minimum of 700 mm from all walls (see Figure 729.3) or by providing an access door, if needed, on the wall against which the equipment is positioned.

Closed restricted access areas with a length exceeding 20 m shall be accessible by doors from both ends.

NOTE 2: For closed restricted access areas with a length exceeding 6 m, accessibility from both ends is recommended.

Fig 729.3 – Examples of positioning of doors in closed restricted access areas

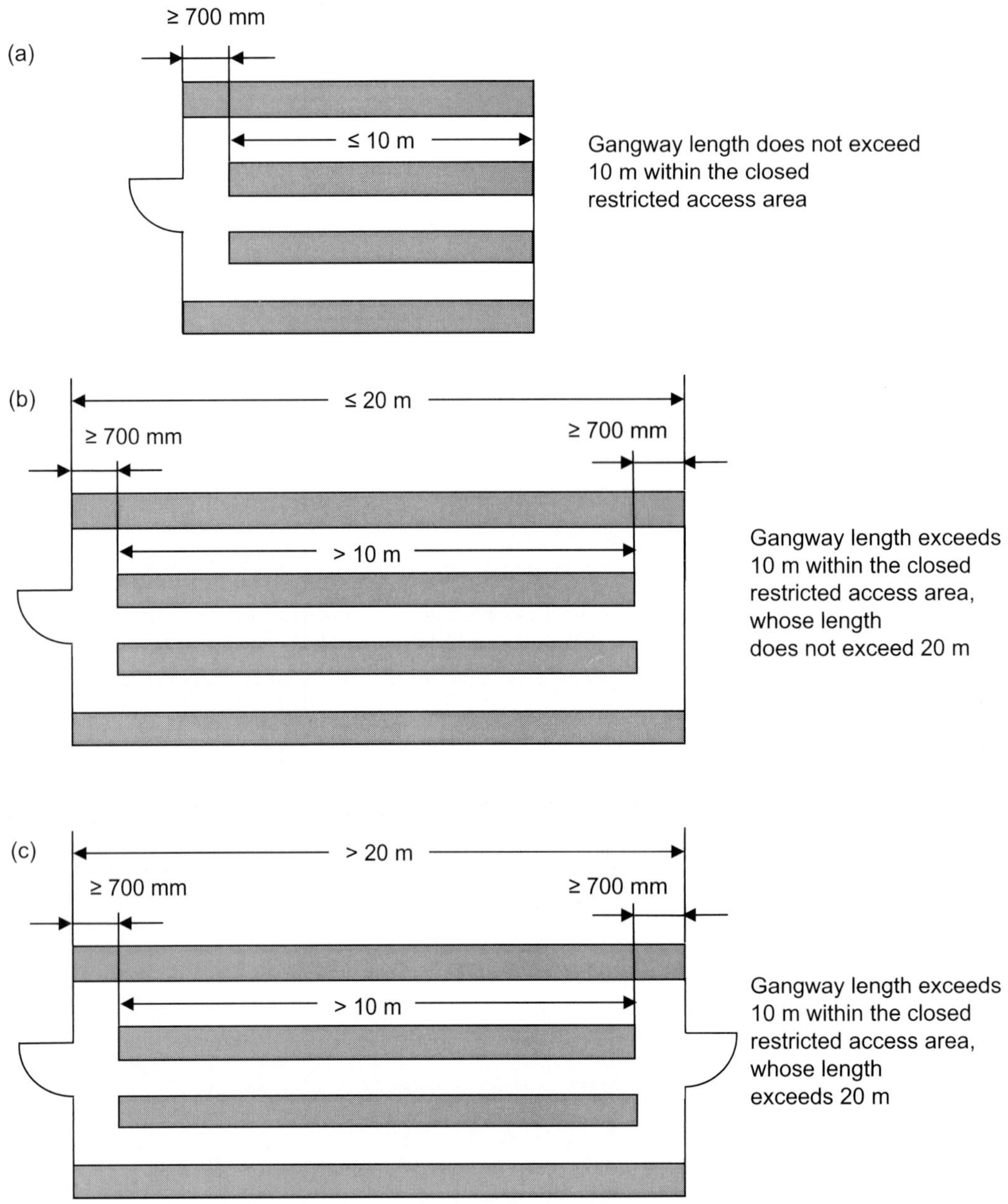

Doors giving access to gangways shall open outwards (see Figure 729.3) and they shall have the following minimum dimensions:

(i) width 700 mm

(ii) height 2000 mm.

ANNEX A729

(Normative)

ADDITIONAL REQUIREMENTS FOR CLOSED RESTRICTED ACCESS AREAS

A729.1 Evacuation

For reason of easy evacuation the doors of any equipment inside the location shall close in the direction of the evacuation route (see Figure A729.1). Gangways shall permit at least a 90 degree opening of equipment doors or hinged panels (see Figure A729.2).

Fig A729.1 – Minimum passing width in case of evacuation – Case 1

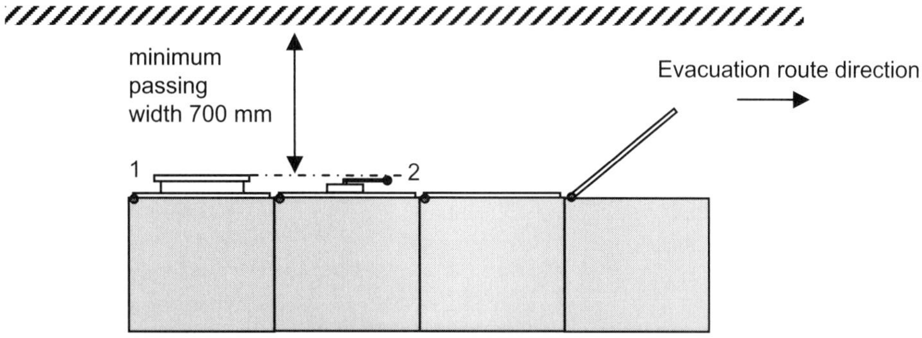

1 Circuit-breaker in the isolated position

2 Handles (e.g. for controls or equipment)

In the case of doors which can be fixed in the open position and circuit-breakers which are withdrawn fully for maintenance (completely extracted) a minimum distance of 500 mm shall be complied with between the door edge or circuit-breaker/ equipment edge and the opposite limitation of the gangway (see Figure A729.2).

Fig A729.2 – Minimum passing width in case of evacuation – Case 2

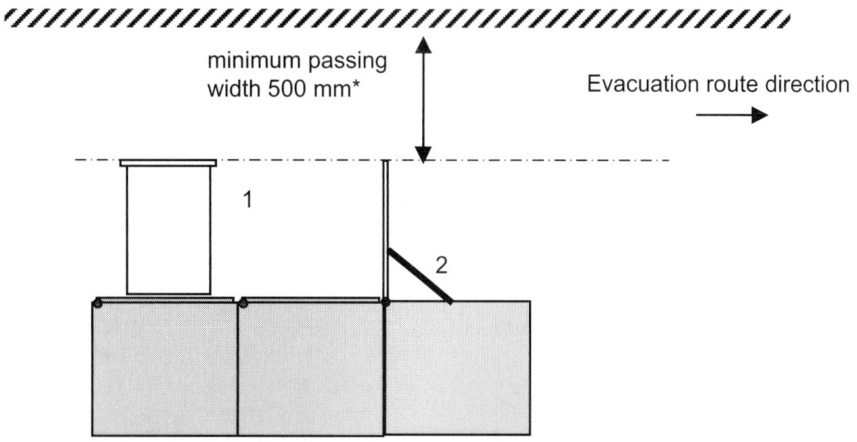

1 Circuit-breaker in the position "completely extracted"

2 Fixing device of a door

* The minimum width gangway of 500 mm shall be taken into consideration between the wall and the circuit-breaker in the position "completely extracted" and equipment door in the 90 degree position.

SECTION 730

ONSHORE UNITS OF ELECTRICAL SHORE CONNECTIONS FOR INLAND NAVIGATION VESSELS

730.1 Scope

The particular requirements of this section apply to onshore installations dedicated to the supply of inland navigation vessels for commercial and administrative purposes, berthed in ports and berths.

NOTE 1: For supplies to pleasure craft or houseboats in marinas and similar locations, see Section 709.

NOTE 2: Additional requirements that do not relate to electrical installation are given in BS EN 15869-1 and BS EN 15869-2.

The particular requirements do not apply to the onboard installations of inland navigation vessels including their connection cables. Additional requirements on the onboard installation are given in BS EN 15869-3.

730.31 Purposes, supplies and structure

730.312 Conductor arrangement and system earthing

730.313 Supplies

730.313.1.101 The nominal supply voltage of the installation for the supply to connections for inland navigation vessels shall be 230 V AC single-phase and 400 V AC three-phase.

NOTE: An arrangement diagram of an electrical shore connection is shown in BS EN 15869-1, and an overview diagram of an electrical power-supply station with two connector units is shown in BS EN 15869-2.

730.313.1.102 Galvanic separation

Where a fixed onshore isolating transformer is used to prevent galvanic currents circulating between the hull of the vessel and metallic parts on the shore side, equipment complying with BS EN 61558-2-4 shall be used.

The protective conductor (PE) of the supply to the isolating transformer shall not be connected to the earth terminal in the socket-outlet supplying the inland navigation vessel.

730.4 Protection for safety

730.41 Protection against electric shock

730.410.3.5 The protective measures of obstacles and placing out of reach (Section 417) shall not be used.

730.410.3.6 The protective measures of non-conducting location (Regulation 418.1) and earth-free local equipotential bonding (Regulation 418.2) shall not be used.

730.411.4 TN system

Regulation 9(4) of the Electricity Safety, Quality and Continuity Regulations (ESQCR) prohibits the connection of a PME earthing facility to any metalwork in a boat.

This does not preclude the use of a PME earthing facility as the means of earthing for other purposes, such as to the installations of permanent buildings.

NOTE: The requirements of other sections of Part 7 may also apply.

730.5 Selection and erection of equipment

730.512 Operational conditions and external influences

730.512.2 External influences

730.512.2.101 Degree of protection

Equipment shall be selected with a minimum degree of protection of IP44.

730.521 Types of wiring system

730.521.101 Wiring systems of berths, ports and floating landing stages

730.521.101.1 Berths and ports

The following wiring systems and cables are suitable for distribution circuits in berths and ports:

(i) underground cables

(ii) overhead cables

(iii) cables with copper conductors and thermoplastic or elastomeric insulation and installed within an appropriate cable management system taking into account external influences such as movement, impact, corrosion and ambient temperature

(iv) mineral-insulated cables with thermoplastic protective covering

(v) armoured cables with a thermoplastic or elastomeric covering.

Other cables and materials that are at least as suitable as those listed above may be used.

730.521.101.2 Floating landing stages

Wiring systems and cables shall be suitable for the movement of floating landing stages. The following wiring systems and cables are suitable for distribution circuits on floating landing stages:

(i) cables with copper conductors and thermoplastic or elastomeric insulation and installed within an appropriate cable management system taking into account external influences such as movement, impact, corrosion and ambient temperature

(ii) armoured cables with a thermoplastic or elastomeric covering.

Other cables and materials that are at least as suitable as those listed in (i) or (ii) may be used.

730.521.101.3 Cables and cable management systems

730.521.101.3.1 General

Cables and cable management systems shall be selected and installed so that mechanical damage due to tidal and other movement of floating structures is prevented.

Cable management systems shall be installed to allow the drainage of water/condensate, e.g. by sloping away and/or drainage holes.

730.521.101.3.2 Underground cables

Underground distribution circuits shall, unless provided with additional mechanical protection, be buried at a sufficient depth to avoid being damaged, e.g. by movement of vehicles.

NOTE 1: A depth of 0.6 m is generally considered as a minimum depth to fulfil this requirement.

NOTE 2: For conduit systems buried underground, see BS EN 61386-24.

730.521.101.3.3 Overhead cables

Overhead cables shall not be used over waterways.

Poles and other supports for overhead wiring shall be located or protected so that they are unlikely to be damaged by any foreseeable movement of vehicles.

Overhead cables shall be at a height above ground of not less than 6 m in all areas subjected to movement of vehicles and 3.5 m in all other areas.

Any overhead conductors shall be insulated.

730.53 Protection, isolation, switching, control and monitoring

730.531 Devices for protection against electric shock by automatic disconnection of supply

730.531.3 Residual current protective devices (RCDs)

Socket-outlets with a rated current not exceeding 63 A shall be individually protected by an RCD providing additional protection in accordance with Regulation 415.1 having a rated residual operating current not exceeding 30 mA.

The RCD selected shall disconnect all live conductors, i.e. line and neutral.

Socket-outlets with a rated current exceeding 63 A shall be individually protected by an RCD having a rated residual operating current not exceeding 300 mA. The RCD selected shall disconnect all live conductors, i.e. line and neutral.

NOTE: The purpose of these RCDs is to protect the shore supply and the flexible cable. It is not intended to provide protection for onboard circuits, which are outside the scope of this section.

730.533 Devices for protection against overcurrent

Socket-outlets shall be individually protected by an overcurrent protective device.

730.537 Isolation and switching

730.537.2 Devices for isolation

730.537.2.1 At least one means of isolation shall be installed for each distribution board. This device shall disconnect all live conductors.

730.55 Other equipment

730.55.1 Socket-outlets

730.55.1.1 Socket-outlets shall comply with BS EN 60309-1 and BS EN 60309-4 and socket-outlets with a current rating not exceeding 125 A shall comply with BS EN 60309-2.

Where interchangeability is not required, socket-outlets shall comply with BS EN 60309-1 and BS EN 60309-4 and need not comply with BS EN 60309-2.

730.55.1.2 Socket-outlets shall be located as close as practicable to the berth to be supplied.

730.55.1.3 No more than four socket-outlets shall be grouped together in any one enclosure.

730.55.1.4 Each socket-outlet shall supply only one vessel.

730.55.1.6 Socket-outlets shall be placed in an enclosure in accordance with BS EN 15869-2.

730.553.13 Socket-outlets shall be placed at a height of not less than 1 m above the highest water level. In the case of floating pontoons or walkways only, this height may be reduced to 0.3 m above the highest water level provided that appropriate additional measures are taken to protect against the effects of splashing.

ANNEX A730

(Informative)

Examples of methods of obtaining supply

NOTE: In Figures A730.1 to A730.4, functional switches are not shown.

Fig A730.1 – Direct connection to a single-phase mains supply

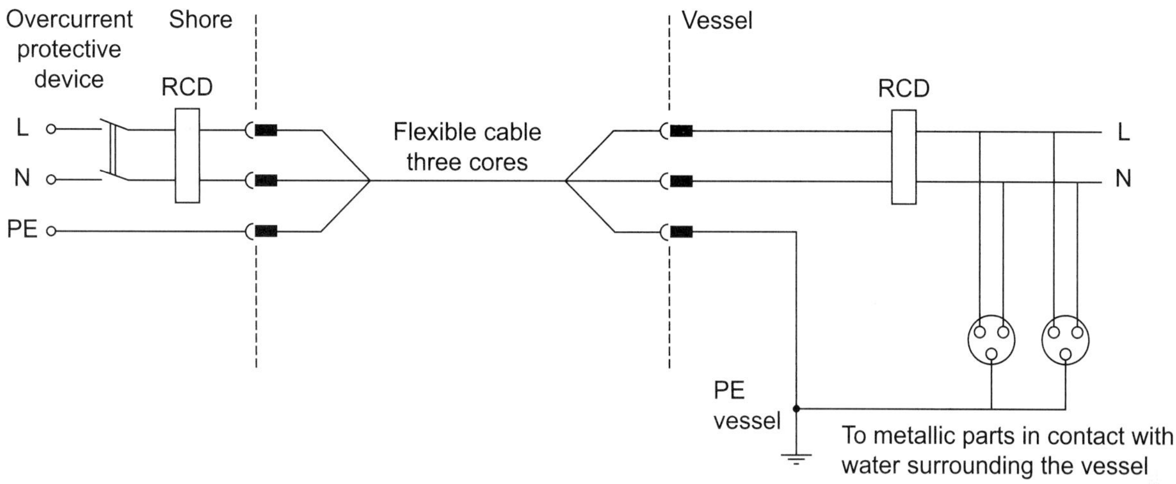

There is a risk of electrolytic corrosion resulting from circulating galvanic currents in the protective conductor to shore.

Fig A730.2 – Direct connection to a single-phase mains supply with an isolating transformer on the vessel

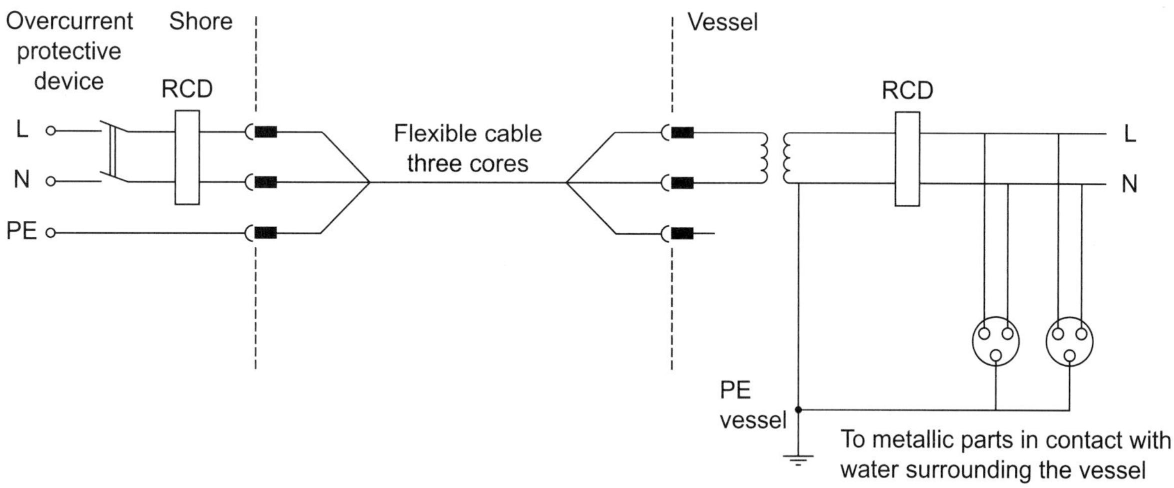

No connection shall be made between the PE conductor of the vessel and the PE conductor of the shore supply (see Regulation 730.313.1.102). This is to prevent galvanic currents circulating between the hull of the vessel and metallic parts on the shore side.

Fig A730.3 – Direct connection to a three-phase mains supply

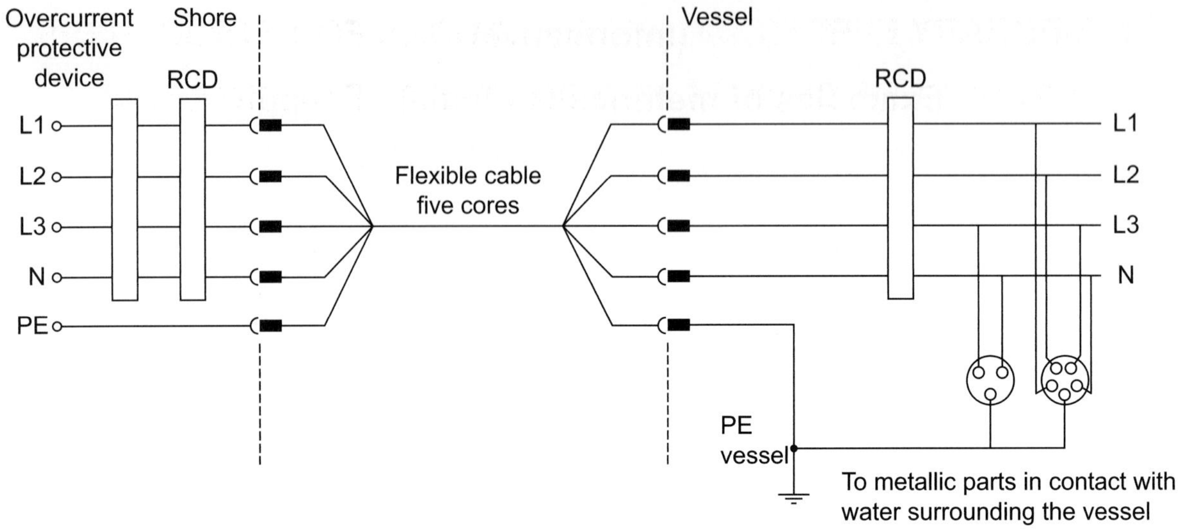

There is a risk of electrolytic corrosion resulting from circulating galvanic currents in the protective conductor to shore.

Fig A730.4 – Direct connection to a three-phase mains supply with an isolating transformer on the vessel

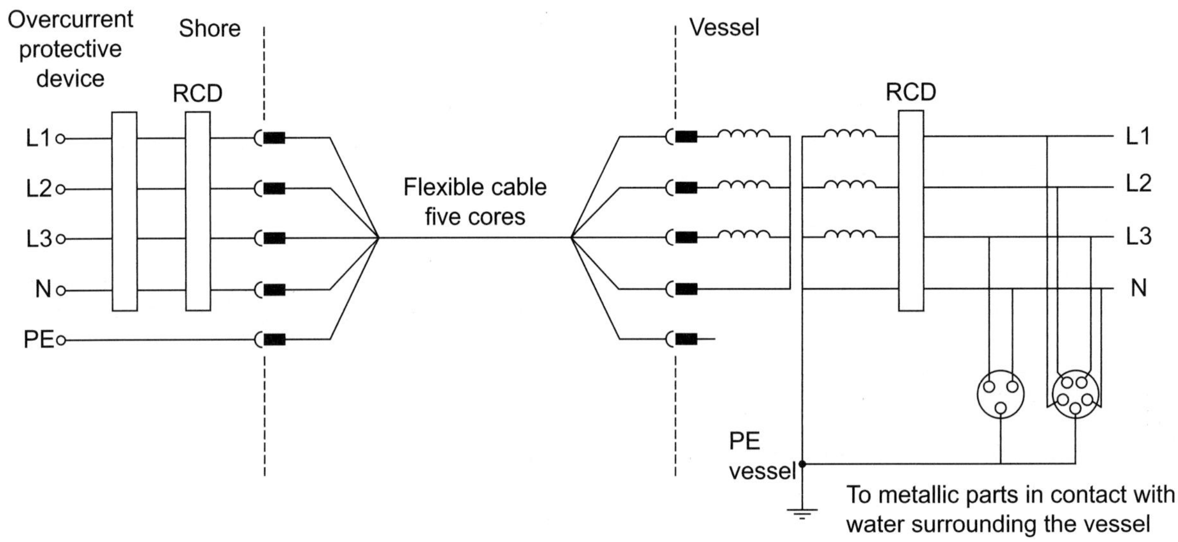

No connection shall be made between the PE conductor of the vessel and the PE conductor of the shore supply (see Regulation 730.313.1.102). This is to prevent galvanic currents circulating between the hull of the vessel and metallic parts on the shore side.

SECTION 740

TEMPORARY ELECTRICAL INSTALLATIONS FOR STRUCTURES, AMUSEMENT DEVICES AND BOOTHS AT FAIRGROUNDS, AMUSEMENT PARKS AND CIRCUSES

740.1 Scope, object and fundamental principles

740.1.1 Scope

This section specifies the minimum electrical installation requirements to facilitate the safe design, installation and operation of temporarily erected mobile or transportable electrical machines and structures which incorporate electrical equipment. The machines and structures are intended to be installed repeatedly, without loss of safety, temporarily, at fairgrounds, amusement parks, circuses or similar places.

The object of this section is to define the electrical installation requirements for such structures and machines, being either integral parts or constituting the total amusement device.

This section does not apply to the internal electrical wiring of machines (see BS EN 60204-1).

NOTE 1: Guidance on temporary electrical systems for entertainment and related purposes is given in BS 7909.

NOTE 2: The permanent electrical installation is excluded from the scope.

740.3 Assessment of general characteristics

740.31 Purposes, supplies and structure

740.313 Supplies

740.313.1.1 Voltage

The nominal supply voltage of temporary electrical installations in booths,stands and amusement devices shall not exceed 230/400 V AC or 440 V DC.

740.313.3 Supply from the public network

Irrespective of the number of sources of supply, the line and neutral conductors from different sources shall not be interconnected downstream of the origin of the temporary electrical installation. The instructions of the operator for the supply of the system to the public shall be followed.

740.4 Protection for safety

740.41 Protection against electric shock

740.410.3 General requirements

Automatic disconnection of supply to the temporary electrical installation shall be provided at the origin of the installation by one or more RCDs with a rated residual operating current not exceeding 300 mA. The RCD shall incorporate a time delay in accordance with BS EN 60947-2 or be of the type S in accordance with BS EN 61008-1 or BS EN 61009-1 where necessary to provide selectivity with RCDs protecting final circuits.

740.410.3.5 The protective measure of obstacles (Regulation 417.2) shall not be used.

Placing out of arm's reach is acceptable for electric dodgems (see Regulation 740.55.9).

740.410.3.6 The protective measures of non-conducting location (Regulation 418.1) and earth-free local equipotential bonding (Regulation 418.2) shall not be used.

740.411 Protective measure: Automatic disconnection of supply

NOTE: For supplies to AC motors, RCDs, where used, should be of the time-delayed type in accordance with BS EN 60947-2 or be of the type S in accordance with BS EN 61008-1 or BS EN 61009-1 where necessary to prevent unwanted tripping.

740.411.4 TN system

740.411.4.1 A PME earthing facility shall not be used as the means of earthing for an installation falling within the scope of this section.

NOTE: The Electricity Safety, Quality and Continuity Regulations (ESQCR) prohibit the use of a PME earthing facility as the means of earthing for the installation of a caravan or similar construction.

740.411.4.3 Where the type of system earthing is TN, a PEN conductor shall not be used downstream of the origin of the temporary electrical installation.

740.411.6 IT system

Where an alternative system is available, an IT system shall not be used. IT systems, however, may be used for DC applications where continuity of service is needed.

740.415 Additional protection

740.415.1 Additional protection: RCDs

All final circuits for:

(i) lighting,
(ii) socket-outlets rated up to 32 A, and
(iii) mobile equipment connected by means of a flexible cable with a current-carrying capacity up to 32 A

shall be protected by RCDs having the characteristics specified in Regulation 415.1.1.

NOTE: The requirement for additional protection relates to the increased risk of damage to cables.

The supply to a battery-operated emergency lighting circuit shall be connected to the same RCD protecting the lighting circuit.

This requirement does not apply to:

(iv) circuits protected by SELV or PELV, or
(v) circuits protected by electrical separation, or
(vi) lighting circuits placed out of arm's reach, provided they are not supplied by socket-outlets complying with BS 1363-2 or BS EN 60309-1.

740.415.2 Additional protection: Supplementary protective equipotential bonding

740.415.2.1 In locations intended for livestock, supplementary bonding shall connect all exposed-conductive-parts and extraneous-conductive-parts that can be touched by livestock. Where a metal grid is laid in the floor, it shall be included within the supplementary bonding of the location (see Figure 705).

Extraneous-conductive-parts in, or on, the floor, e.g. concrete reinforcement in general or reinforcement of cellars for liquid manure, shall be connected to the supplementary protective equipotential bonding.

It is recommended that spaced floors made of prefabricated concrete elements be part of the equipotential bonding (see Figure 705).The supplementary protective equipotential bonding and the metal grid, if any, shall be erected so that it is durably protected against mechanical stresses and corrosion.

740.42 Protection against thermal effects

740.422.3 Nature of processed or stored materials

740.422.3.7 A motor which is automatically or remotely controlled and which is not continuously supervised shall be fitted with a manually reset protective device against excess temperature.

740.5 Selection and erection of equipment

740.51 Common rules

Switchgear and controlgear shall be placed in cabinets which can be opened only by the use of a key or a tool, except for those parts designed and intended to be operated by ordinary persons (BA1) as defined in Part 2 (see also Appendix 5).

740.512 Operational conditions and external influences

740.512.2 External influences

Electrical equipment shall have a degree of protection of at least IP44.

740.52 Wiring systems

740.521 Types of wiring system

740.521.1 Cables and cable management systems

Conduit systems shall comply with the relevant part of the BS EN 61386 series, cable trunking systems and cable ducting systems shall comply with the relevant part 2 of BS EN 50085, and tray and ladder systems shall comply with BS EN 61537.

All cables shall meet the requirements of BS EN 60332-1-2.

Cables shall have a minimum rated voltage of 450/750 V, except that, within amusement devices, cables having a minimum rated voltage of 300/500 V may be used.

The routes of cables buried in the ground shall be marked at suitable intervals. Buried cables shall be protected against mechanical damage.

NOTE 1: Conduit classified as 450 N regarding protection against compression and classified as normal regarding protection against impact, according to BS EN 61386-24, is considered to fulfil the above requirement.

Armoured cables or cables protected against mechanical damage shall be used wherever there is a risk of mechanical damage due to external influence, e.g. > AG2. Mechanical protection shall be used in public areas and in areas where wiring systems are crossing roads or walkways.

The following methods are considered to meet the above requirements:

(i) conduit systems complying with BS EN 61386-21 with a classification of heavy regarding protection against compression, a classification of heavy regarding protection against impact, and, for metallic and composite conduit systems, class 3 protection against corrosion (i.e. medium protection inside and high protection outside)

(ii) cable trunking systems and cable ducting systems complying with BS EN 50085 series with a classification 5 J regarding protection against impact.

Where subjected to movement, wiring systems shall be of flexible construction. Where flexible conduit systems are provided they shall comply with BS EN 61386-23.

NOTE 2: Cables of type H07RN-F or H07BN4-F (BS EN 50525-2-21) together with conduit complying with BS EN 61386-23 are deemed to satisfy this requirement.

740.526 Electrical connections

Joints shall not be made in cables except where necessary as a connection into a circuit. Where joints are made, these shall either use connectors in accordance with the relevant British or Harmonized Standard or the connection shall be made in an enclosure with a degree of protection of at least IPXXD or IP4X.

Where strain can be transmitted to terminals the connection shall incorporate cable anchorage(s).

740.53 Switchgear and controlgear

740.537 Isolation and switching

740.537.1 General

Every electrical installation of a booth, stand or amusement device shall have its own means of isolation, switching and overcurrent protection, which shall be readily accessible.

740.537.2.1.1 Every separate temporary electrical installation for amusement devices and each distribution circuit supplying outdoor installations shall be provided with its own readily accessible and properly identified means of isolation.

740.537.2.2 Devices for isolation

A device for isolation shall disconnect all live conductors (line and neutral conductors).

740.55 Other equipment

740.55.1 Lighting installation

740.55.1.1 Luminaires

Every luminaire and decorative lighting chain shall be installed so as not to impair its ingress protection, and be securely attached to the structure or support intended to carry it. Its weight shall not be carried by the supply cable, unless it has been selected and erected for this purpose.

Luminaires and decorative lighting chains mounted less than 2.5 m (arm's reach) above floor level or otherwise accessible to accidental contact, shall be firmly fixed and so sited or guarded as to prevent risk of injury to persons or ignition of materials. Access to the fixed light source shall only be possible after removing a barrier or an enclosure which shall require the use of a tool.

Lighting chains shall use H05RN-F or HO7RN-F (BS EN 50525-2-21) cable or equivalent.

NOTE: Lighting chains may be used in any length provided the overcurrent protective device in the circuit is properly rated.

740.55.1.2 Lampholders

Insulation-piercing lampholders shall not be used unless the cables and lampholders are compatible and the lampholders are non-removable once fitted to the cable.

740.55.1.3 Lamps in shooting galleries

All lamps in shooting galleries and other sideshows where projectiles are used shall be suitably protected against accidental damage.

740.55.1.4 Floodlights

Where transportable floodlights are used, they shall be mounted so that the luminaire is inaccessible. Supply cables shall be flexible and have adequate protection against mechanical damage.

740.55.1.5 Fire risks from luminaires and floodlights

Luminaires and floodlights shall be so fixed and protected that a focusing or concentration of heat is not likely to cause ignition of any material.

740.55.3 Electric discharge lamp installations

Installations of any luminous tube, sign or lamp on a booth, stand or amusement device with an operating voltage higher than 230/400 V AC shall comply with Regulations 740.55.3.1 and 740.55.3.2.

740.55.3.1 Location

The luminous tube, sign or lamp shall be installed out of arm's reach or be adequately protected to reduce the risk of injury to persons.

740.55.3.2 Emergency switching device

A separate circuit shall be used to supply luminous tubes, signs or lamps, which shall be controlled by an emergency switch. The switch shall be easily visible, accessible and marked in accordance with the requirements of the local authority.

740.55.5 Safety isolating transformers and electronic convertors

Safety isolating transformers shall comply with BS EN 61558-2-6 or provide an equivalent degree of safety.

A manually reset protective device shall protect the secondary circuit of each transformer or electronic convertor.

Safety isolating transformers and electronic convertors shall be mounted out of arm's reach or be mounted in a location that provides equal protection, e.g. in a panel or room with adequate ventilation that can only be accessed by skilled or instructed persons. Such access shall be provided only to facilitate inspection, testing and maintenance.

Electronic convertors shall conform to BS EN 61347-2-2.

Enclosures containing rectifiers and transformers shall be adequately ventilated and the vents shall not be obstructed when in use.

740.55.7 Plugs and socket-outlets

An adequate number of socket-outlets shall be installed to allow the user requirements to be met safely.

NOTE 1: In booths, stands and for fixed installations, one socket-outlet for each square metre or linear metre of wall is generally considered adequate.

Socket-outlets dedicated to lighting circuits placed out of arm's reach (in accordance with Regulation 740.415.1) shall be encoded or marked according to their purpose.

When used outdoors, plugs, socket-outlets and couplers shall comply with:

(i) BS EN 60309-2, or

(ii) where interchangeability is not required, BS EN 60309-1.

However, socket-outlets according to the relevant National Standard may also be installed if they have suitable mechanical protection (equivalent to the requirements of BS EN 60309-1) and a rated current not exceeding 16 A.

NOTE 2: Suitable mechanical protection may be provided by the socket-outlet or by an enclosure.

740.55.8 Electrical supply

At each amusement device, there shall be a connection point readily accessible and permanently marked to indicate the following essential characteristics:

(i) Rated voltage

(ii) Rated current

(iii) Rated frequency.

740.55.9 Electric dodgems

Electric dodgems shall only be operated at voltages not exceeding 50 V AC or 120 V DC. The circuit shall be electrically separated from the supply mains by means of a transformer in accordance with BS EN 61558-2-4 or a motor-generator set.

740.551 Low voltage generating sets

740.551.8 Generators

All generators shall be so located or protected as to prevent danger and injury to people through inadvertent contact with hot surfaces and dangerous parts.

Electrical equipment associated with the generator shall be mounted securely and, if necessary, on anti-vibration mountings.

Where a generator supplies a temporary installation, forming part of a TN, TT or IT system, care shall be taken to verify that the earthing arrangements are in accordance with Regulation 542.1 and, where earth electrodes are used, with Regulation 542.2.

The neutral conductor of the star-point of the generator shall, except for an IT system, be connected to the exposed-conductive-parts of the generator.

740.6 Inspection and testing

The electrical installation between its origin and any electrical equipment shall be inspected and tested in accordance with the requirements of Part 6 after each assembly onsite.

NOTE 1: Internal electrical wiring of roller coasters, electric dodgems and similar equipment are not considered as part of the verification.

NOTE 2: In special cases the number of the tests may be modified according to the type of temporary electrical installation.

SECTION 753

HEATING CABLES AND EMBEDDED HEATING SYSTEMS

753.1 Scope

This section applies to embedded electric heating systems for surface heating. It also applies to electric heating systems for de-icing, frost prevention and similar applications. Both indoor and outdoor systems are covered.

Heating systems for industrial and commercial applications complying with relevant parts of BS EN 60519, BS EN 62395 and BS EN 60079 are not covered.

NOTE: Examples of heating systems covered by this section are heating systems for walls, ceilings, floors, roofs, drainpipes, gutters, pipes, stairs, roadways, and non-hardened compacted areas (e.g. football fields, lawns).

753.4 Protection for safety

753.41 Protection against electric shock

753.410.3 General requirements

753.410.3.5 The protective measures of obstacles and placing out of reach (Section 417) shall not be used.

753.410.3.6 The protective measures of non-conducting location (Regulation 418.1) and earth-free local equipotential bonding (Regulation 418.2) shall not be used.

753.411 Protective measure: Automatic disconnection of supply

753.411.3.2 RCDs having the characteristics specified in Regulation 415.1.1 shall be used as disconnecting devices. In the case of heating units which are delivered from the manufacturer without exposed-conductive-parts, a suitable conductive covering, for example, a metal grid with a spacing of not more than 30 mm, shall be provided on site as an exposed-conductive-part above the floor heating elements or under the ceiling heating elements, and connected to the protective conductor of the electrical installation.

NOTE 1: Where Class II floor or ceiling heating units complying with the requirements of Regulations 412.1.1 and 412.2.1.1 are used, the mesh metal grid detailed in Regulation 753.411.3.2 is not required.

NOTE 2: Limitation of the rated heating power to 7.5 kW/230 V or 13 kW/400 V downstream of a 30 mA RCD may avoid unwanted tripping due to leakage current.

753.412 Protective measure: Double or reinforced insulation

753.412.1.201 The use of this protective measure as the sole protective measure for a wall heating system is not permitted.

Where this protective measure is used as the sole protective measure for floor or ceiling heating units, complying with the requirements of Regulation 412.2.1.1, the mesh metal grid detailed in Regulation 753.411.3.2 is not required.

753.413 Protective measure: Electrical separation

Electrical separation shall not be used for wall heating systems (Section 413).

753.415 Additional protection: RCDs

753.415.1 Circuits supplying heating units shall have additional protection by the use of RCDs having the characteristics specified in Regulation 415.1.1. Time delayed type RCDs shall not be used.

753.42 Protection against thermal effects

753.423 Protection against burns

In floor areas where contact with skin or footwear is possible, the surface temperature of the floor shall be limited.

NOTE: For example, to no more than 35 °C for floor heating systems. For additional information reference can be made to CENELEC Guide 29.

753.424 Protection against overheating

753.424.101 To meet the requirements of Chapter 42, special care shall be taken to prevent the heating elements creating high temperatures to adjacent material. This may be achieved by using heating units with temperature self-limiting functions or by separation with heat-resistant materials. The latter may be accomplished by placing on a metal sheet, in metal conduit or at a distance of at least 10 mm in air from the ignitable structure.

NOTE: Dependent on adjacent material it may be prudent to consider a larger separation distance.

753.424.201 For floor or ceiling heating systems in buildings, one or more of the following measures shall be applied within the zone where heating units are installed to limit the temperature to a maximum of 80 °C:

(i) Appropriate design of the heating system
(ii) Appropriate installation of the heating system
(iii) Use of protective devices.

Heating units shall be connected to the electrical installation via cold tails or suitable terminals.

Heating units shall be inseparably connected to cold tails, for example, by a crimped connection.

753.424.102 For wall heating systems, the heating units shall be provided with a metal sheath or metal enclosure or fine mesh metallic grid. The metal sheath or metal enclosure or fine mesh metallic grid shall be connected to the protective conductor of the supply circuit.

NOTE: This requirement is intended to protect against the effects of overheating caused by a short-circuit between live conductors due to penetration of an embedded heating unit.

753.5 Selection and erection of electrical equipment

753.51 Common rules

753.511 Compliance with standards

Flexible sheet heating elements shall comply with the requirements of BS EN 60335-2-96. Heating cables shall comply with IEC 60800.

753.514 Identification

753.514.1 General

The designer of the installation/heating system or installer shall provide documentation for each heating system, containing the following details:

(i) Manufacturer and type of heating units
(ii) Number of heating units installed
(iii) Length/area of heating units
(iv) Rated power
(v) Surface power density
(vi) Layout of the heating units in the form of a sketch, drawing or picture
(vii) Position/depth of heating units
(viii) Position of junction boxes
(ix) Cables, earthed conductive shields and the like
(x) Rated voltage
(xi) Rated resistance (cold) of heating units
(xii) Rated current of overcurrent protective device
(xiii) Rated residual operating current of RCD
(xiv) The insulation resistance of the heating installation and the test voltage used
(xv) Product information containing provisions about approved materials in contact with the heating units, with necessary instructions for installation.

This documentation shall be fixed to, or adjacent to, the distribution board of the heating system.

Furthermore, the requirements of Figure 753 apply.

753.515 Prevention of mutual detrimental influence

753.515.1 Electric heating systems shall be selected and erected so as to avoid any harmful influence between the heating system and any electrical or non-electrical installations envisaged.

If materials other than those recommended are used in the surroundings of the heating units the manufacturer shall be consulted.

NOTE: For example, an electric heating system can affect other parts of the electrical installation such as reducing the current-carrying capacity of cables of other circuits due to high local ambient temperature caused by the heating system.

753.515.101 Heating units shall not cross expansion joints of the building or structure.

753.52 Wiring systems

753.520 Introduction

753.520.4 Heating-free areas

For the necessary attachment of room fittings, heating-free areas shall be provided in such a way that the heat emission is not prevented by such fittings.

753.522 Selection and erection of wiring systems in relation to external influences

753.522.1 Ambient temperature (AA)

753.522.1.3 For cold leads (circuit wiring) and control leads installed in the zone of heated surfaces, the increase of ambient temperature shall be taken into account.

753.522.4 Presence of solid foreign bodies (AE)

753.522.4.3 Where heating units are installed there shall be heating-free areas where drilling and fixing by screws and the like may be carried out without risk of damage to the units.

753.522.6 Impact (AG)

753.522.6.201 The heating-free areas for floor or ceiling heating units shall be readily identifiable.

753.554.4 Heating conductors and cables

753.554.4.201 A heating cable intended for laying directly in soil, concrete, cement screed or other material used for road and building construction shall be:

(i) capable of withstanding mechanical damage under the conditions that can reasonably be expected to prevail during its installation, and

(ii) constructed of material that will be resistant to damage from dampness and/or corrosion under normal conditions of service.

753.554.4.202 A heating cable laid directly in soil, a road or the structure of a building shall be installed so that it:

(i) is completely embedded in the substance it is intended to heat, and

(ii) does not suffer damage in the event of movement normally to be expected in it or the substance in which it is embedded, and

(iii) complies in all respects with the manufacturer's instructions and recommendations.

Fig 753 – Information for the user of the installation

A description of the heating system shall be provided to the person ordering the work.

The description shall contain at least the following information:

a) Description of the construction of the heating system, especially the installation depth of the heating units;

b) Location diagram with information concerning

- the distribution of the heating circuits and their rated power;
- the position of the heating units in each room;
- particularities which have been taken into account when installing the heating units, for example, heating-free areas, complementary heating zones, heating-free areas for fixing means penetrating into the covering material;

c) Data on the control equipment used, with relevant circuit diagrams and the dimensioned position of floor temperature and weather conditions sensors, if any;

d) Data on the type of heating units and their maximum operating temperature.

The installer shall inform the owner that the description of the heating system includes all necessary information, for example for repair work.

Instructions for use shall be provided to the person ordering the work upon completion. One copy of the instructions for use shall be permanently fixed in or near each relevant distribution board.

The instructions for use shall include at least the following data:

a) Description of the heating system and its function;

b) Operation of the heating installation in the first heating period in the case of a new building, for example, regarding drying out;

c) Operation of the control equipment for the heating system in the dwelling area and the complementary heating zones, if any;

d) Information on restrictions on placing of furniture or similar.

- additional floor coverings for example, carpets with a thickness of >10 mm may lead to higher floor temperatures which can adversely affect the performance of the heating system
- pieces of furniture solidly covering the floor and/or built-in cupboards shall only be placed on heating-free areas
- furniture, such as carpets, seating and rest furniture with pelmets, which in part do not solidly cover the floor, may not be placed in complementary heating zones, if any;

e) In the case of ceiling heating systems, restrictions regarding the height of furniture. Cupboards of room height may be placed only below the area of ceiling where no heating elements are installed;

f) Dimensioned position of complementary heating zones and placing areas;

g) Statement that, in the case of thermal floor, wall and ceiling heating systems, no fixing shall be made into the floor, wall or ceiling respectively. Excluded from this requirement are heating-free areas. Alternatives shall be given, where applicable.

PART 8

FUNCTIONAL REQUIREMENTS

CONTENTS

CHAPTER 82

PROSUMER'S LOW VOLTAGE ELECTRICAL INSTALLATIONS

CONTENTS

CHAPTER 82

PROSUMER'S LOW VOLTAGE ELECTRICAL INSTALLATIONS

821 SCOPE

This chapter provides additional requirements, measures and recommendations for the design, erection and verification of all types of low voltage electrical installation according to Chapter 11, including local production and/or storage of energy in order to achieve compatibility with the existing and future ways to deliver electrical energy to current-using equipment or to the public network by means of local sources. Such electrical installations are designated as prosumer's electrical installations (PEIs).

This chapter provides requirements for PEIs to achieve safe operation, sustainability and efficient use of energy when integrated into smart grids.

This chapter does not cover electrical sources for safety services including associated electrical installations and standby electrical supply systems for a secure continuity of supply, which are operated only occasionally and for short periods in parallel with the distribution grid for testing purposes.

822 INTERACTION OF SMART GRID AND PEI

822.1 Main objectives

Smart grids and prosumers' electrical installations interact with each other and a dynamic power demand/response concept should be implemented.

The smart grid has an impact on prosumers' electrical installations on the:

(i) consideration of the prosumer's needs, taking into account the constraint of the public network

(ii) design and configuration of the installation to allow load shedding (according to BS IEC 60364-8-1) and source selection by the Electrical Energy Management System (EEMS).

The user should be able to give different inputs to the EEMS, depending on the contract with the Distribution System Operator (DSO).

Both the production of energy by renewable sources such as PV or wind turbines and the consumption patterns of loads within the installation are intermittent. Therefore, it is suggested that storage capacity is installed within the PEI if availability in island mode or stand-alone mode is needed, or to maximize the self-consumption in connected mode.

822.2 Safety

The implementation of the requirements provided shall not impair the safety of the PEI. In case of change of any energy supply configuration (e.g. from network supply to local power supplies) all protective measures shall continue to be operational or shall be automatically replaced by other protective measures providing an equivalent level of safety.

822.3 Proper functioning

It is essential for the operation of the smart grid that the electrical installation remains reliable and available for the maximum possible time while the power quality parameters are also maximized by using appropriate protection measures and other good installation practices.

These requirements are important for the use of the island mode. It is essential for the PEI to comply with similar requirements on stability, availability and quality in island mode as for the connected mode.

822.4 Implementation of PEI

The PEI shall be designed, constructed and commissioned taking account of the constraints from the DNO/electricity supplier and the needs expressed by the end user.

An EEMS shall be incorporated which combines information and/or data, from/to the DNO/electricity supplier, the availability of energy by the local sources and the user's needs.

823 PEI CONCEPT

A PEI is a set of electrical equipment that has the following functions (see Figure 82.1):

(i) supply (e.g. connection to public power supply, local generator, photovoltaic systems, wind turbines, and batteries)

(ii) distribution

(iii) consumption (e.g. motors, heating systems, lighting, lifts)

(iv) energy management (e.g. load shedding equipment, monitoring device).

NOTE: A battery can be considered as a generator and as a load.

An uninterruptible power supply (UPS) is not considered to form part of a PEI as the purpose of a UPS is only to supply downstream critical loads and is not capable of reverse feeding the public network and/or current-using equipment in the upstream part of the electrical installation.

The general principles of the PEI are described in Annex A82.

Fig 82.1 – Example of a PEI

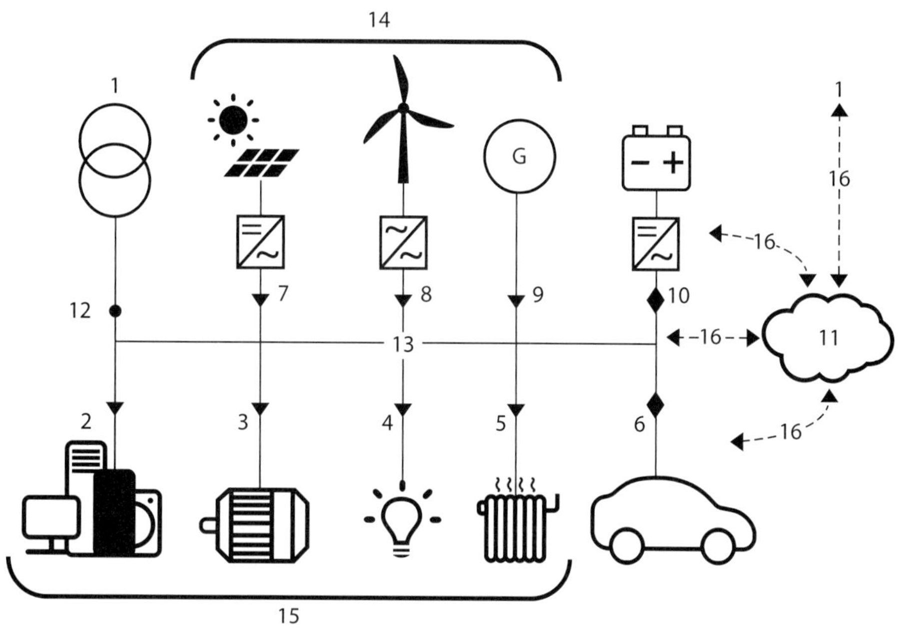

Key

1	Public network	9	Other generators
2	Home appliances and electronic devices	10	Electric storage
3	Motors	11	EEMS
4	Lighting	12	Origin of installation
5	Heaters	13	Local distribution
6	Electric vehicles	14	Local generation
7	Solar inverter	15	Local consumption
8	Wind inverter	16	Management signals

In a PEI, an installation owner may consider independently the supervision and the control of different power supplies connected to the electrical installation in order to supply efficiently and cost-effectively all the electrical loads connected to this low voltage electrical installation.

824 TYPES OF PEI

824.1 General

There are different types of PEI:

(i) individual (see Regulation 824.3)
(ii) collective (see Regulation 824.4)
(iii) shared (see Regulation 824.5).

Each type of PEI can be arranged in the different operating modes defined in Regulation 824.2.

824.2 Operating modes

The main operating modes may be adopted for each type of PEI (individual, collective or shared). They are:

(i) direct feeding mode (see Annex B82.1.1, B82.2.1, B82.3.1)
(ii) reverse feeding mode (see Annex B82.1.3, B82.2.3, B82.3.3)
(iii) island mode (see Annex B82.1.2, B82.2.2, B82.3.2).

Storage units can supply current-using equipment or be charged by local power supplies or by the public network, except in island mode.

Local power supplies can supply current-using equipment or local storage units or the public network, except in island mode.

Transfer from/to the direct feeding mode to island mode and vice versa can be achieved by operating the switching device for islanding. This can be either directly controlled (manually or remotely) or automatically controlled.

Switching from one mode to another can be done if the generators and/or converters are synchronized with the network (provided that the requirements of Section 551 are met).

See Annex B82 for examples of operating modes.

Selection of the possible operating modes may depend on the contract with the DNO.

Technical requirements for the design of the PEI according to the selected operating mode are provided in Section 826.

824.3 Individual PEI

An individual PEI is characterized by one electrical installation having the capability to both consume and produce electrical energy, and having a management system for its operation.

An example of individual PEI is provided in Figure 82.2.

Fig 82.2 – Example of electrical design of individual PEI

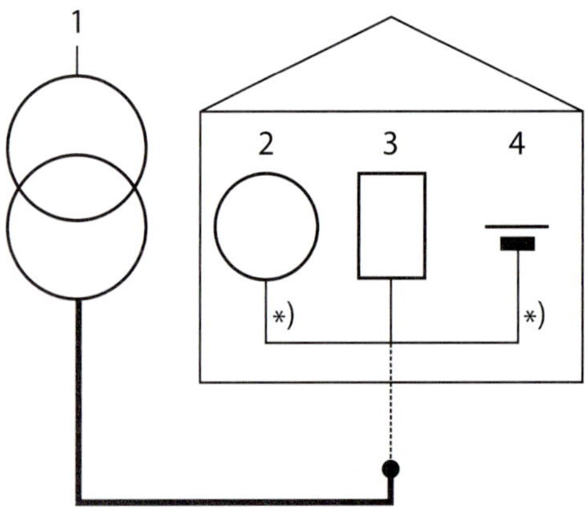

Key

1 Public network
2 Power supplies
3 Loads
4 Storage units
*) Optional (at least, one of them shall be present)

824.4 Collective PEI

The different power supplies may supply all involved prosumers through either the distribution system within the PEI or the DNO distribution system, if agreed by the DNO.

A group of prosumers (e.g. a group of single private houses, flats in a building, shops in a mall) may co-operate and coordinate their resources to erect common electrical power supplies such as in the example provided in Figure 82.3. In this case, all private electrical installations are considered as consumers. Only one separate unit producing electrical energy is managed for the group of consumers.

Fig 82.3 – Example of electrical design of collective PEI using DNO distribution system

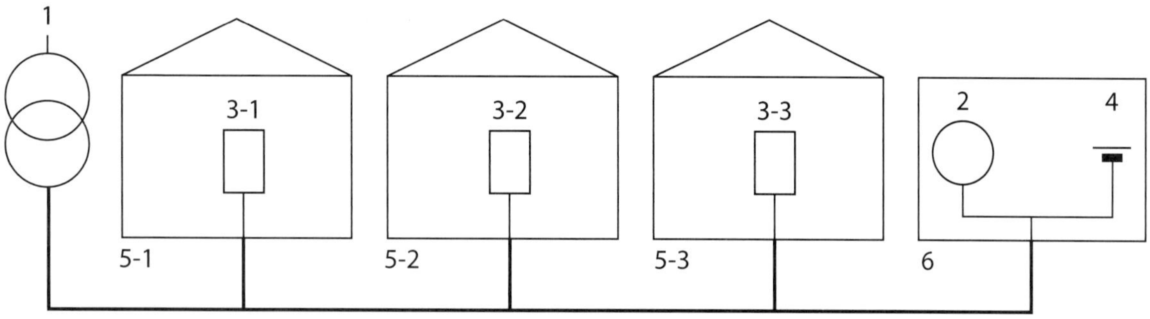

Key

1	Public network
2	Power supplies
3-1	Load 1
3-2	Load 2
3-3	Load 3
4	Storage units
5-1	Consumer 1
5-2	Consumer 2
5-3	Consumer 3
6	Producer

Where connection between all involved prosumers is using a distribution system within the PEI, then the aggregation of all prosumers installations corresponds to one single PEI seen by the DNO (see Figure 82.4).

Fig 82.4 – Example of electrical design of collective PEI with distribution system within PEI

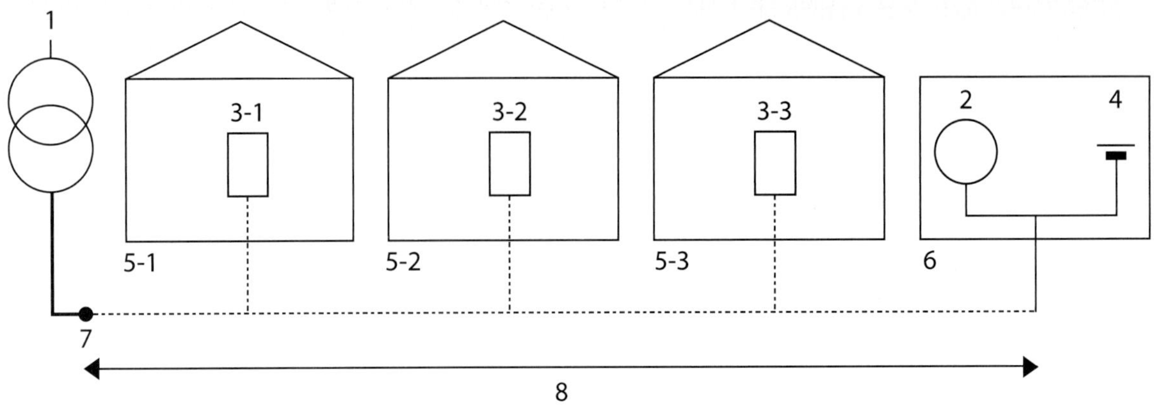

Key

1	Public network
2	Power supplies
3-1	Load 1
3-2	Load 2
3-3	Load 3
4	Storage units
5-1	Consumer 1
5-2	Consumer 2
5-3	Consumer 3
6	Producer
7	Origin of the distribution system within PEI
8	Distribution system within PEI

Where the connection between all involved consumers uses the public distribution network in combination with a distribution system within the PEI, the origin of the PEI for each consumer corresponds to the incoming service of each individual prosumer (see Figure 82.5).

Fig 82.5 – Example of electrical design of collective PEI with distribution system within PEI in parallel with DNO distribution system

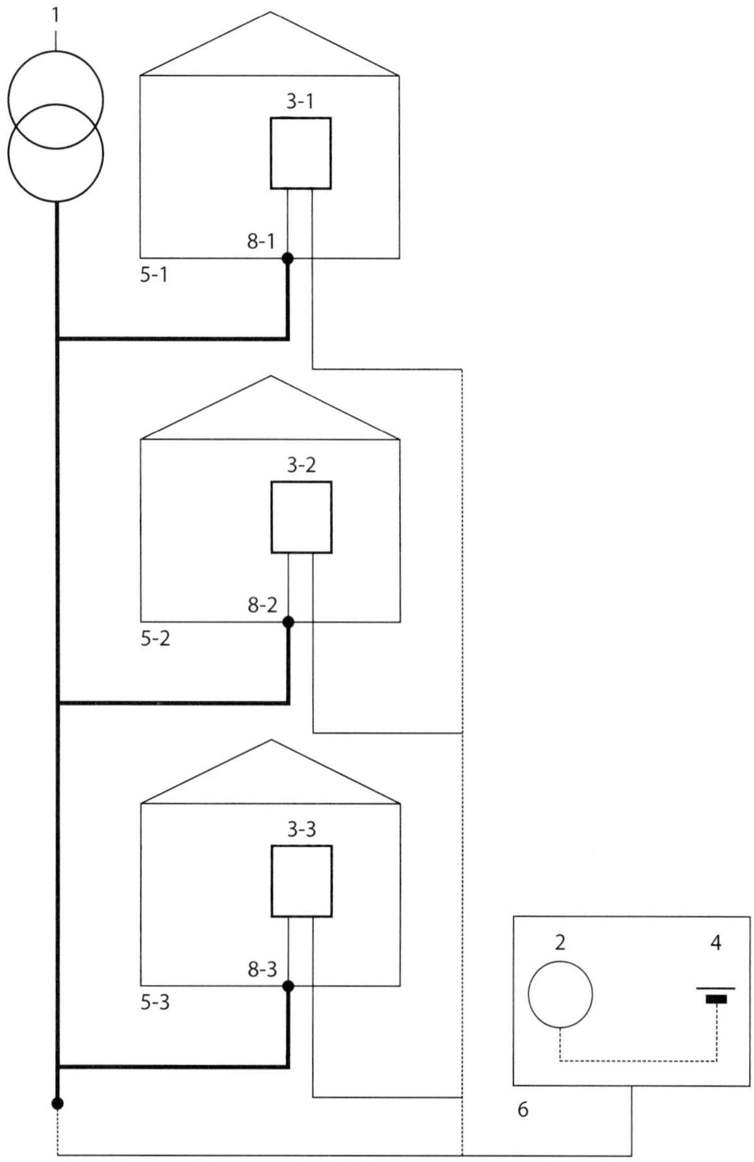

Key

1	Public network
2	Power supplies
3-1	Load 1
3-2	Load 2
3-3	Load 3
4	Storage units
5-1	Consumer 1
5-2	Consumer 2
5-3	Consumer 3
6	Producer
8-1	Origin of installation 1
8-2	Origin of installation 2
8-3	Origin of installation 3

For the collective PEI, each electrical installation is considered as a consumer installation, while a common source of local production is considered as a producer installation. Consumer and producer installations shall be considered as separate.

824.5 Shared PEI

The different power supplies may supply all the involved prosumers through a distribution system within the PEI or through the DNO distribution system, if agreed by the DNO.

Individual premises, such as a residential estate or business park may group their interests in accepting to share their supply with their neighbours from their own local production. Each building owner may have installed private renewable energy power sources which can either supply the private electrical installation or supply the group of private electrical installations. Such a system is named shared PEI. Examples of shared PEI are provided in Figure 82.6 and in Figure 82.7.

Fig 82.6 – Example of electrical design of shared PEI using DNO distribution system

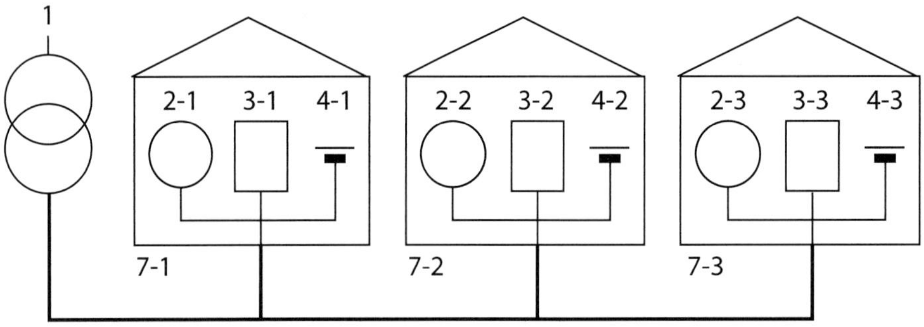

Key

1	Public network
2-1	Power supply 1
2-2	Power supply 2
2-3	Power supply 3
3-1	Load 1
3-2	Load 2
3-3	Load 3
4-1	Storage unit 1
4-2	Storage unit 2
4-3	Storage unit 3
7-1	Prosumer 1
7-2	Prosumer 2
7-3	Prosumer 3

Fig 82.7 – Example of electrical design of shared PEI with distribution system within PEI

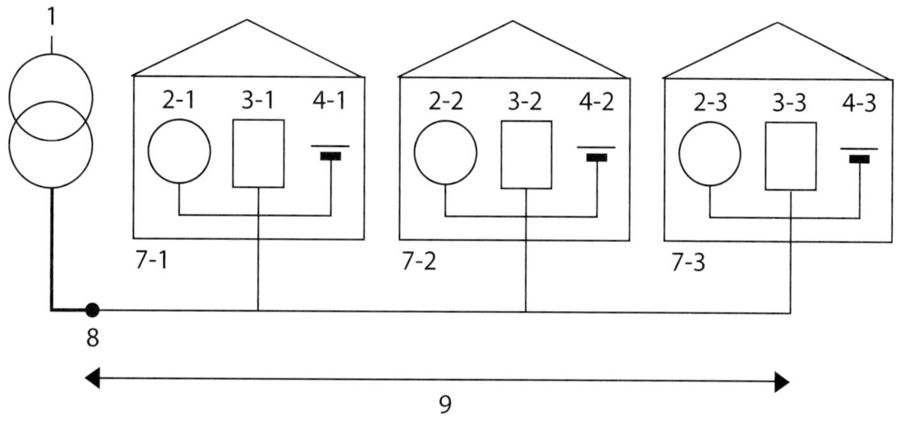

Key

1	Public network
2-1	Power supply 1
2-2	Power supply 2
2-3	Power supply 3
3-1	Load 1
3-2	Load 2
3-3	Load 3
4-1	Storage unit 1
4-2	Storage unit 2
4-3	Storage unit 3
7-1	Prosumer 1
7-2	Prosumer 2
7-3	Prosumer 3
8	Origin of shared PEI
9	Shared electrical installation within PEI

Where connection between all involved prosumers is using the DNO distribution system network in combination with the distribution system within the PEI, the origin of the PEI for each prosumer corresponds to the incoming service of each individual prosumer (see Figure 82.8).

Fig 82.8 – Example of electrical design of shared PEI with distribution system within PEI in parallel with DNO distribution system

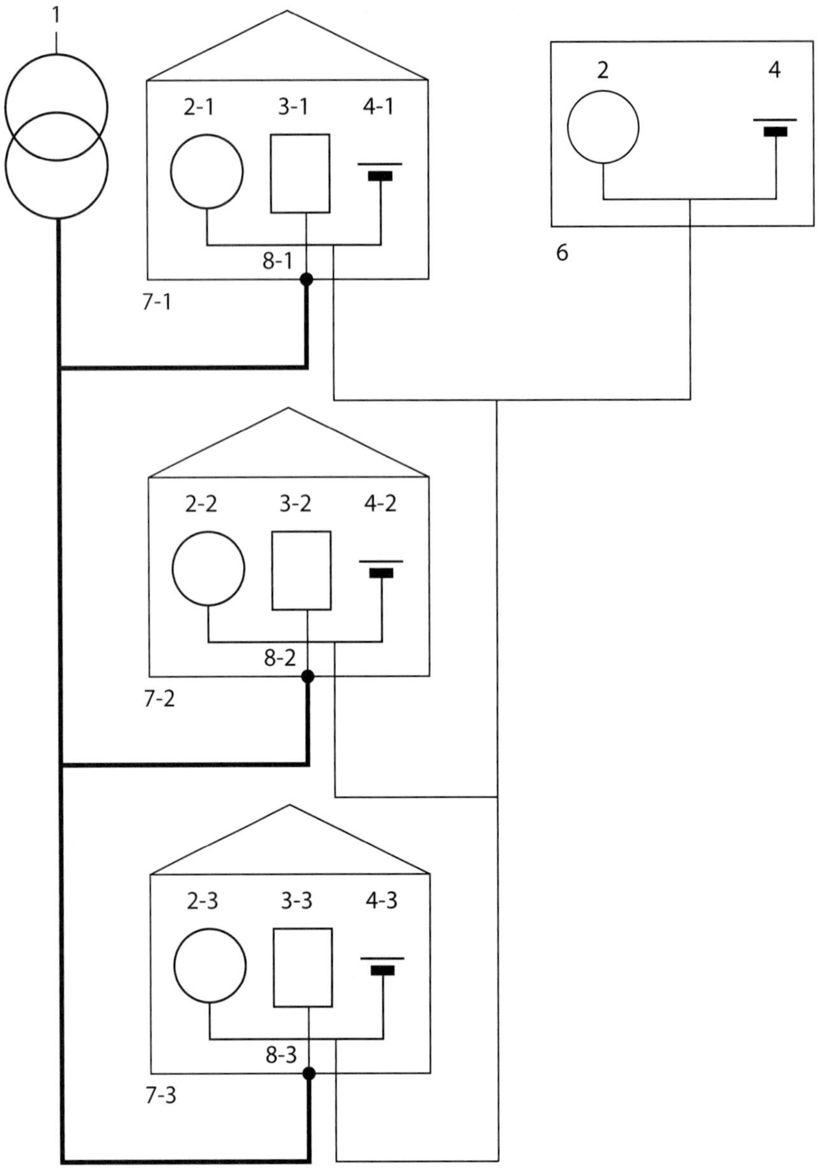

Key

1 Public network
2 Power supplies
2-1 Power supply 1
2-2 Power supply 2
2-3 Power supply 3
3-1 Load 1
3-2 Load 2
3-3 Load 3
4-1 Storage unit 1
4-2 Storage unit 2
4-3 Storage unit 3
6 Producer
7-1 Prosumer 1
7-2 Prosumer 2
7-3 Prosumer 3
8-1 Origin of installation 1
8-2 Origin of installation 2
8-3 Origin of installation 3

For the shared PEI, each electrical installation can either produce or consume electrical energy. Therefore, each of them shall be considered as a prosumer. The electricity produced by the power supply of one prosumer may be used locally by this prosumer or may be used by other prosumers of the group.

The different power supplies may supply all involved prosumers through a distribution system within the PEI or through the DNO distribution system, if agreed by the DNO.

825 ELECTRICAL ENERGY MANAGEMENT SYSTEM (EEMS)

825.1 General

An EEMS shall monitor and control the operation of all power supplies, the load of the storage units and the operation of loads.

The concept and the design architecture of the PEI mainly depends on the concept of electrical energy management system (see BS IEC 60364-8-1). The objectives of the EEMS are to:

(i) control the connection of the PEI to the smart grid
(ii) manage locally the electrical energy production
(iii) manage locally the electrical energy consumption
(iv) manage the electrical energy procurement from the DSO.

The following are examples of functions that may be handled by the EEMS:

(v) power sources and loads management
(vi) multi source connection management
(vii) propose load control (load shedding and load shifting)
(viii) bidirectional exchange of information with the DSO
(ix) backup system management by means of electrical energy storage units and power sources
(x) control the electrical energy flow from and to electrical energy storage units
(xi) monitor the voltage quality
(xii) provide an interface for the end user.

The EEMS may be installed as a separate item of equipment or be integrated within other equipment. Annex D82 shows examples of implementation of an EEMS.

825.2 Architecture of EEMS

In an individual, collective or shared PEI, one or several EEMSs shall monitor and control the operation of loads, all power supplies and storage units.

For collective or shared PEI, this EEMS can be either a central system or the association of the different EEMSs sharing together all the relevant information.

Consumption of each electrical installation forming part of the individual, collective or shared PEI shall be monitored, as well as all local electricity production.

To do so, energy meters, performance measuring and monitoring devices (PMD) or other measuring equipment shall be installed in appropriate/required locations.

See Annex D82.

826 TECHNICAL ISSUES

826.1 Safety issues

826.1.1 Protection against electric shock

826.1.1.1 General

The PEI shall be able to operate in any intended operating mode as defined in Regulation 824.2. According to needs, a PEI can change the operating mode at any time and may come back to the initial operating mode also at any time (for instance, from direct feeding mode to island mode and then back to direct feeding mode).

Protection of persons and properties shall be provided in all operating modes.

This is particularly the case for the protection of persons against electric shock when the automatic disconnection of supply is used as a protective measure for all intended operating modes. In this case, the system earthing used for all intended operating modes may differ and may depend on the operating mode.

(i) In connected mode, without galvanic separation, the PEI remains connected to the public distribution network. Therefore, the system earthing of the PEI shall be the same as that of the public distribution network.

(ii) In island mode, the PEI is disconnected from the public distribution network. Therefore, the system earthing of the PEI may differ from the one used within the public distribution network.

In island mode, where ADS is the protective measure, the PEI shall meet the requirements of Regulation 551.4.3.2.1.

A switching device suitable for isolation according to Chapter 53 shall disconnect all live conductors from the public distribution network.

826.1.1.2 System earthing

826.1.1.2.1 Earthing arrangement

The earthing arrangement needed for operating the PEI in island mode shall be connected to a suitable earth electrode complying with Regulation 542.2.2. Any connection to the means of earthing derived from the public distribution network need not be disconnected.

Any change of type of system earthing for island mode shall be reversible as operation in this mode can be temporary, and the PEI can then be operated again in connected mode.

When the installation is connected to different sources in parallel, simultaneously accessible exposed-conductive-parts and/or extraneous-conductive-parts shall be connected to the same earthing arrangement individually, in groups or collectively.

NOTE: Two parts are deemed to be simultaneously accessible if they are not more than 2.50 m apart (see Figure 417).

826.1.1.2.2 Neutral conductor

When in island mode, all live conductors shall be disconnected from the DNO supply.

To prevent incorrect operation of RCDs the use of a neutral switch device shall connect the neutral and the earth of the PEI without overlapping with switching of the DNO neutral.

826.1.1.2.3 Connection to the local earthing arrangement

A transfer switching device meeting the requirements of Regulation 537.1.5 shall be inserted in the conductor used for the connection to the earthing arrangement of the local star/mid-point or the local exposed-conductive-parts.

826.1.1.3 Selection of protective device

Operation of the protective device shall be in accordance with the maximum disconnection times as required by Chapter 41 and the requirements of Regulation 551.2.

The selection of protective devices against electric shock shall consider the minimum earth fault current value (between one of the line conductors and PE conductor). The minimum earth fault current value may depend on the operating mode (see Figure 82.9).

Fig 82.9 – Estimation of the minimum earth fault current according to the operating mode (connected and island)

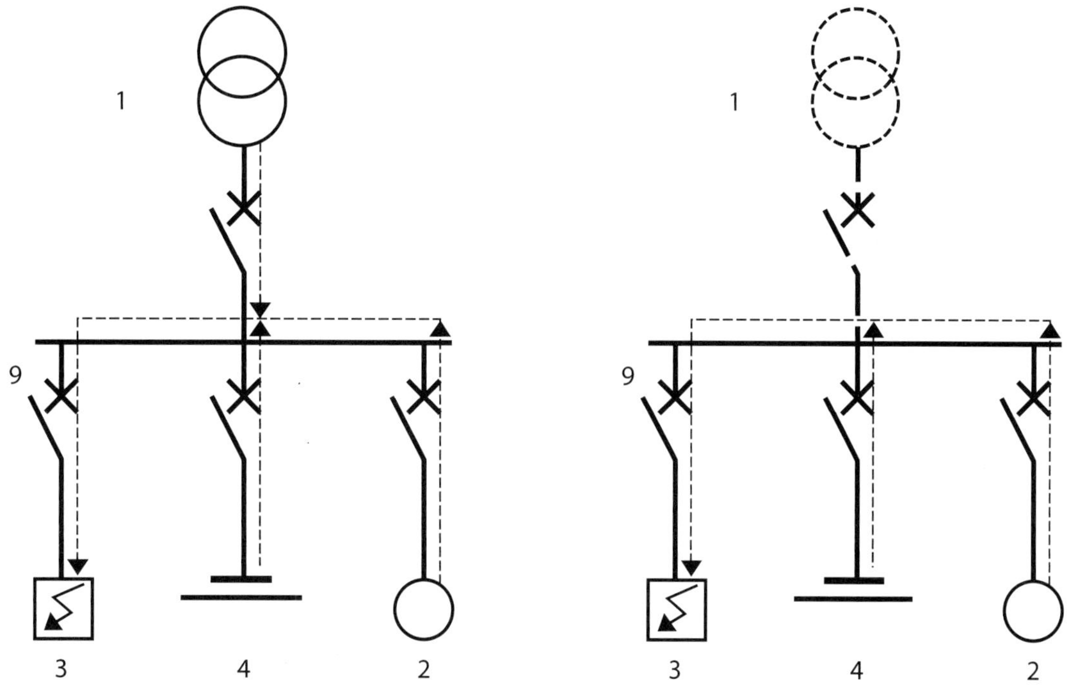

Key

1	Public network
2	Power supplies
3	Loads
4	Storage units
9	Overcurrent protection device (OCPD)

(i) In TN and IT systems, an overcurrent protective device (OCPD) may be used as a protective device giving automatic disconnection in the event of a fault. The system earthing may change depending on the operating mode, the selection of the overcurrent device shall consider each operating mode.

(ii) In connected mode, the fault current is supplied by the public network, the local power supply and by the local storage units. The main contribution will certainly come from the public network.

(iii) In island mode, the earth fault current is only supplied by the local power supplies and by the local storage units. The local power supplies may be current sources (e.g. solar photovoltaic cells) with a very low earth fault current value. The local energy storage units may have their contribution to the earth-fault current value limited by the rectifiers used for their connection to an AC installation.

Therefore, in TN and IT systems, overcurrent protective devices used for fault protection shall be selected for each source, considering the minimum contribution of each one (island mode).

NOTE: It could be reached e.g. using a double setting on the same device, two devices coordinated or taking into account the minimum setting of the worst condition.

In a TT system, as the earth fault current is limited by the earth electrode resistance, there is no practical interaction of the operating mode as far as the level of fault current is concerned. See Regulation 411.5.

826.1.1.4 Isolation of the installation

Where an installation is supplied from more than one source, a main switch suitable for isolation (e.g. switch disconnector) shall be provided for each source of supply. A durable warning notice shall be permanently fixed in the vicinity of these main switches in such a position that any person seeking to operate any of these main switches will be warned of the need to operate all such switches to achieve isolation of the installation. Alternatively, a suitable interlock system shall be provided. See Section 462 and Regulation 537.1.2.

826.1.1.5 Switching device for island mode

Switching devices for island mode, introduced in Regulation 824.2, shall comply with Regulation 512.1.2 and their relevant product standard, and be suitable for isolation.

826.1.2 Protection against overcurrent

826.1.2.1 Magnitude of overcurrent

Overload and short-circuit currents shall be determined at every point of the PEI where a protective device is installed:

(i) for all possible configurations of each type of PEI, and

(ii) for situations corresponding to the minimum and maximum current magnitudes.

In all cases, compliance with Chapter 43 shall be fulfilled.

NOTE 1: The operating mode influences greatly the overcurrent magnitude. In particular, in island mode, the short-circuit current will have a different magnitude from the direct feeding mode and reverse feeding mode.

NOTE 2: The solar photovoltaic cells are considered to be current mode power sources with a very low short-circuit current. In the case where batteries are connected in parallel with the solar photovoltaic cells, the short-circuit current will be higher.

The selection of overcurrent protective devices shall consider:

(iii) the maximum short-circuit current (e.g. connected mode) for the selection of breaking capacity, and

(iv) the minimum short-circuit current (e.g. island mode) for the setting of the tripping characteristics of the short-circuit protective device.

826.1.2.2 Location of overcurrent protective devices

Selection and erection of overcurrent protective devices shall take account of all possible directions of current flow and polarity.

Connection of a source of supply via switchgear or controlgear assemblies shall comply with Regulation 551.7.2.

Overload and short-circuit protection shall be placed at the origin of the circuit. Exceptions are permitted (see Regulations 433.2 and 433.3).

The origin of the circuit may differ depending on the arrangements used for the PEI. Where the PEI is supplied from the public network the origin of the circuit is located at one end of the circuit. Where the PEI is supplied from local power supplies (including local storage units) the origins may be at the other physical locations on the circuit.

In this case a short-circuit protective device is installed at each possible origin of the relevant distribution circuit, each short-circuit protective device shall be selected and, where applicable, adjusted for each type of power supply (see Figure 82.10). To reduce multiple short-circuit protective devices on the same circuit, it is recommended to connect local power supplies and local storage units directly to the main distribution board.

Fig 82.10 – Example of double short-circuit protection for the same circuit

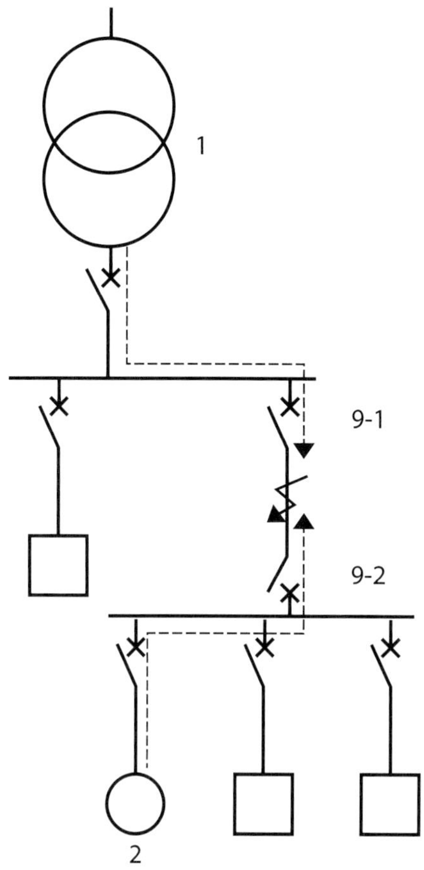

Key

1	Public network
2	Power supplies
9-1	Overcurrent protection device (OCPD 1)
9-2	Overcurrent protection device (OCPD 2)

826.1.2.3 Combined short-circuit protection

Where combined short-circuit protection is used in the electrical installation in accordance with Regulation 434.5.1, coordination between two or more short-circuit protective devices shall consider all possible configurations of power supplies. This requires that the installation designer considers all possible short-circuit currents through all short-circuit protective devices and to verify if coordination between two or more short-circuit protective devices is achieved.

NOTE: In some configurations, the short-circuit current through coordinated protective devices may differ depending on the connected power supplies.

As a consequence of the above, the designer, when considering combined short-circuit protection, shall take account of all possible fault current magnitudes depending on:

(i) location of fault

(ii) various possible combinations of power supplies connected to the PEI and

(iii) various operating modes.

826.1.3 Outage of the public network

When the public network is not energized, prosumers shall operate their private individual PEI in island mode or automatically disconnect all local power supplies.

As control devices and protective devices may be operated more frequently than in non PEI, selection of components is recommended to be made according to the increased duty requirements (in terms of electrical and mechanical operations).

826.1.4 Protection against transient overvoltages

Switching overvoltages in a PEI may be more frequent and perhaps greater than in a non PEI installation (e.g. due to the switching between sources, load shedding, load shifting). Consideration is to be given to the installation of surge protective devices for the protection of the PEI installation and equipment against switching overvoltages.

826.2 Interaction with the public network

The PEI shall comply with all the supply requirements (e.g. voltage, frequency). See also Annex C82.

826.3 Energy storage

Consideration shall be given to the inrush current and other capabilities of local energy storage in the design of the system, especially when switching between operating modes.

826.4 Design for flexibility of load and generators (demand/response)

Where applicable the electrical installation shall be designed for the operability of load shedding. See BS HD 60364-8-1.

826.5 Electric vehicle charging

An electric vehicle (EV) capable of operating as a prosumer represents a particular case of a load and local storage unit. Such a vehicle will not be permanently connected to the PEI.

Electric vehicle charging equipment (see Section 722) for a vehicle operating as a prosumer should be managed by the EEMS as defined in Regulation 825.1.

826.6 Selectivity between protective devices

Selectivity between protective devices consists of the coordination between two or more protective devices such that on the incidence of an overcurrent or a fault current only the protective device on the load side (downstream) operates while the other protective device or devices do not. See Section 536.

Therefore, when considering selectivity issues, it is important to know where the power supply is connected to the installation (see Figure 82.11).

As a consequence, selectivity shall be considered according to all possible fault currents depending on:

(i) location of fault
(ii) various possible combination of power supplies connected to the PEI and
(iii) various operating modes.

Selectivity can concern the following types of protective device:

(iv) overload protective device
(v) short-circuit protective device
(vi) residual current device.

Fig 82.11 – Example of selectivity with various power supplies

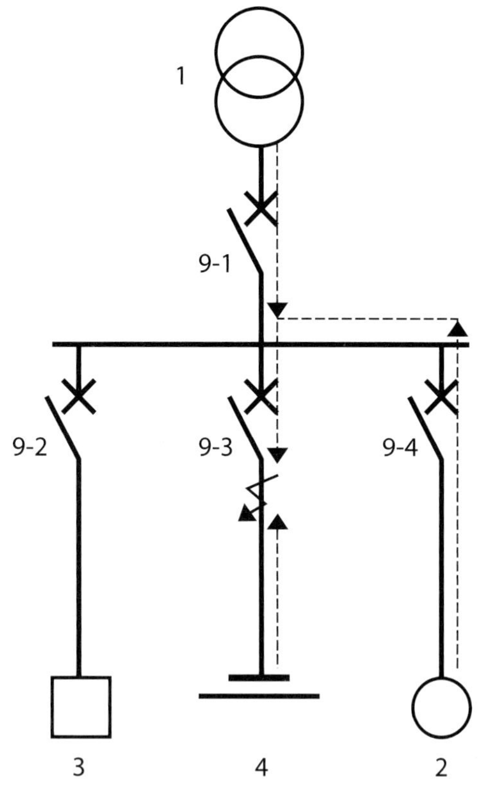

Key

1	Public network
2	Power supplies
3	Loads
4	Storage units
9-1	Overcurrent protection device (OCPD 1)
9-2	Overcurrent protection device (OCPD 2)
9-3	Overcurrent protection device (OCPD 3)
9-4	Overcurrent protection device (OCPD 4)

In this example, selectivity is usually proposed between OCPD 1 and OCPD 2 in the case of normal supply from the DSO transformer. OCPD 3 and OCPD 4 are considered at the same level of selectivity as OCPD 1. In island mode, for instance, power supply can be local.
In this case, selectivity needs to be considered between OCPD 2 and OCPD 3 and/or OCPD 4.

826.7 Testing and verification

The validity of test readings taken with a fault loop impedance test instrument may be adversely affected by power converting equipment (such as inverters) within the prosumer's installation. In such cases an alternative method of determining prospective fault current and earth fault loop impedance shall be used. See Note 1 of Regulation 643.7.3.1.

Annex A82
(Informative)
OBJECTIVES AND CONCEPT OF PEI

The concept of PEI has been developed to provide an appropriate answer to the following issues.

(i) The role of the end user: to control the usage of electrical energy taking into account their needs and the effects on the DNO's supply. The PEI concept keeps the role of the end user as central. Where the PEI has a storage capability, the end user can take advantage of low demand to store the energy when its price may be lower.

(ii) Active energy management: the end user should be able to permanently monitor and control their own electricity consumption and their own electricity production as a result of an active energy management system. These systems are intended to balance the local consumption with the local production and the supply from/to the DNO. The active energy management system should also communicate with the DSO for exchanging or receiving of information for controlling purposes, for example receiving signals from the DSO when urgent electricity consumption reduction is needed.

(iii) Renewable sources: deployment of renewable energy at the DSO side and/or at the end user side is of prime importance for the reduction of CO_2 emissions. In addition, local renewable power supplies such as photovoltaic systems and wind turbines can play an important role by supplying local current-using equipment when necessary (during daily peak electricity consumption).

(iv) Storage: electrical energy may be stored locally (e.g. batteries) to be used when required. Storage may also be used to reduce reliance on the traditional utility supply. Storage may also be used to hold a reserve of excess energy generated by renewable sources that may not function when their ability to function becomes impaired (e.g. solar at night, wind turbines in calm weather).

Annex B82
(Informative)
OPERATING MODES

B82.1 Operating modes for individual PEI

B82.1.1 Direct feeding mode

In this mode, the PEI is supplied from the public network. The PEI acts as a consumer. See Figure B82.1.

Current-using equipment inside the PEI is supplied either from the public network and/or from the local power supplies and/or from local energy storage unit(s), if any.

Local energy storage units are either charged from public network and/or from local power supply, or supplying the local current-using equipment.

NOTE: For clarity, the diagrams in this Annex are shown omitting protective and isolation devices and the earthing arrangement of the installations.

Fig B82.1 – Example of electrical design of individual PEI operating in direct feeding mode

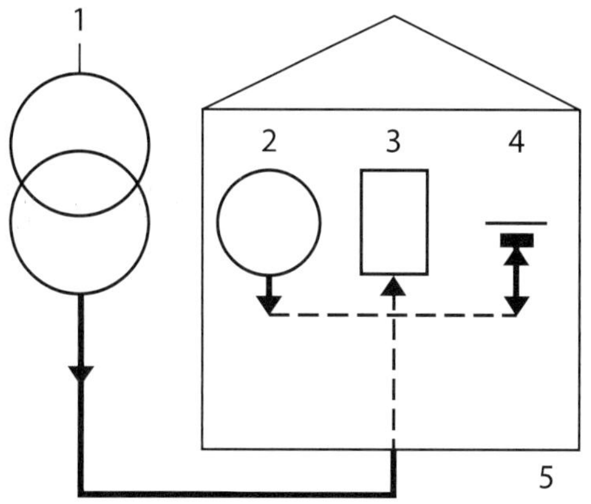

Key

1 Public network
2 Power supplies
3 Loads
4 Storage units
5 Consumer

B82.1.2 Island mode

In this mode, the PEI is disconnected from the public network. See Figure B82.2.

Current-using equipment inside the PEI is supplied either from the local power supplies and/or from local energy storage unit(s), if any.

Local energy storage units are charged from the local power supply, or supplying the local current-using equipment.

Load shedding may be recommended if this operating mode is considered to be economic, and consequently should remain in place for as long as is necessary. In this case charging the storage units should be avoided.

Fig B82.2 – Example of electrical design of individual PEI operating in island mode

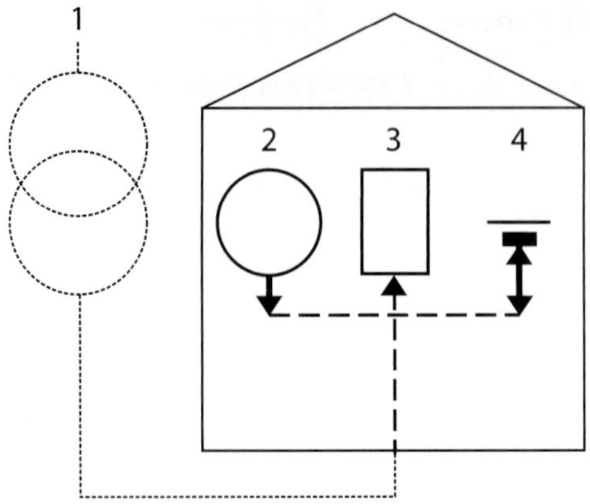

Key

1	Public network
2	Power supplies
3	Loads
4	Storage units

B82.1.3 Reverse feeding mode

In this mode, the PEI is supplying the public network. The PEI acts as a producer. See Figure B82.3.

Current-using equipment inside the PEI is supplied either from the local power supplies and/or from local energy storage unit(s), if any.

Local energy storage units are charged from the local power supply or may be used to supply the local current-using equipment or the public network.

Fig B82.3 – Example of electrical design of individual PEI operating in reverse feeding mode

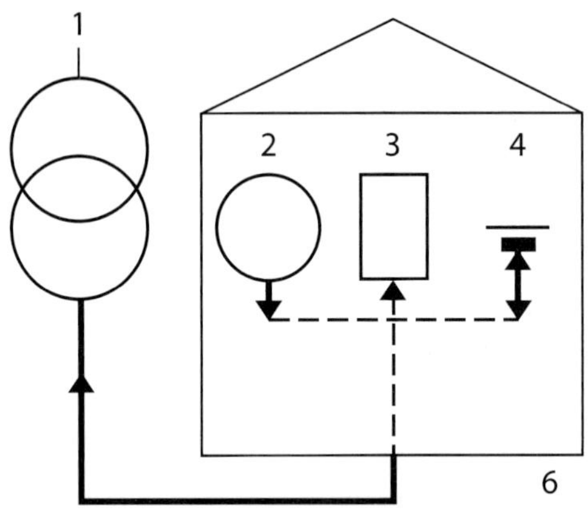

Key

1	Public network
2	Power supplies
3	Loads
4	Storage units
6	Producer

B82.2 Operating modes for collective PEI

B82.2.1 Direct feeding mode

The public network supplies all connected current-using equipment of all private electrical installations, and may also charge the private collective storage units.

Collective power supplies may also be used to charge the storage units and to supply current-using equipment or can be switched off.

Collective storage units may also be used to supply current-using equipment, charged by the local power supply, by the public network or can be switched off.

See Figure B82.4 and Figure B82.5.

Fig B82.4 – Example of electrical design of collective PEI operating in direct feeding mode with one single electrical installation

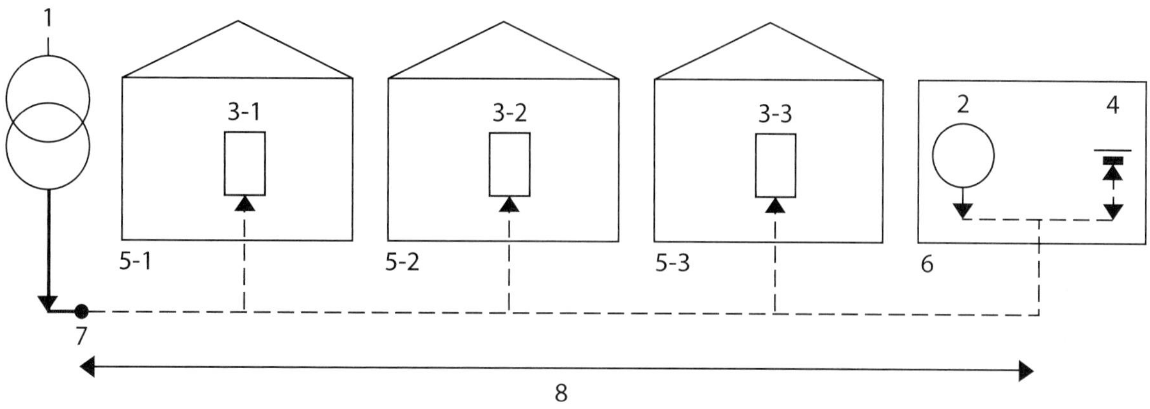

Key

1	Public network
2	Power supplies
3-1	Load 1
3-2	Load 2
3-3	Load 3
4	Storage units
5-1	Consumer 1
5-2	Consumer 2
5-3	Consumer 3
6	Producer
7	Origin of the distribution system within PEI
8	Distribution system within PEI

Fig B82.5 – Example of electrical design of collective PEI operating in direct feeding mode with several electrical installations

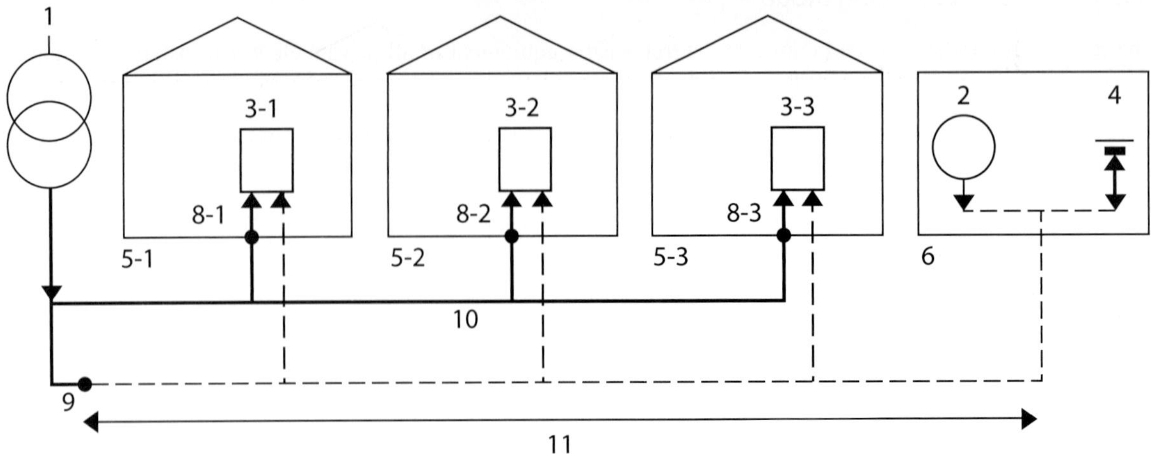

Key

1	Public network
2	Power supplies
3-1	Load 1
3-2	Load 2
3-3	Load 3
4	Storage units
5-1	Consumer 1
5-2	Consumer 2
5-3	Consumer 3
6	Producer
8-1	Origin of installation 1
8-2	Origin of installation 2
8-3	Origin of installation 3
9	Origin of the distribution system within PEI
10	DSO distribution system
11	Distribution system within PEI

B82.2.2 Island mode

Similarly to the individual PEI, all electrical installations should be disconnected from the public network when electrical energy is not supplied by the DNO.

Electricity consumed by connected current-using equipment of all private electrical installations is produced by the private resources of local power supplies or by the local storage units.

Local storage units may also be charged from the local power supplies.

See Figure B82.6 and Figure B82.7.

Fig B82.6 – Example of electrical design of collective PEI operating in island mode with one single electrical installation

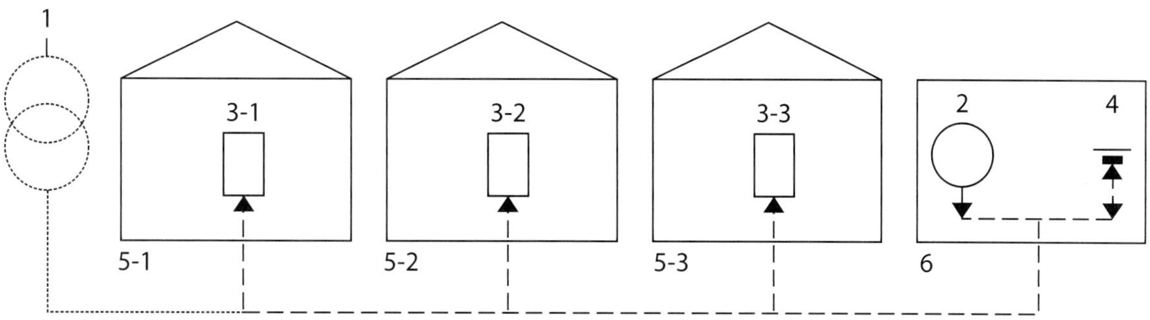

Key

1	Public network
2	Power supplies
3-1	Load 1
3-2	Load 2
3-3	Load 3
4	Storage units
5-1	Consumer 1
5-2	Consumer 2
5-3	Consumer 3
6	Producer

Fig B82.7 – Example of electrical design of collective PEI operating in island mode with several electrical installations

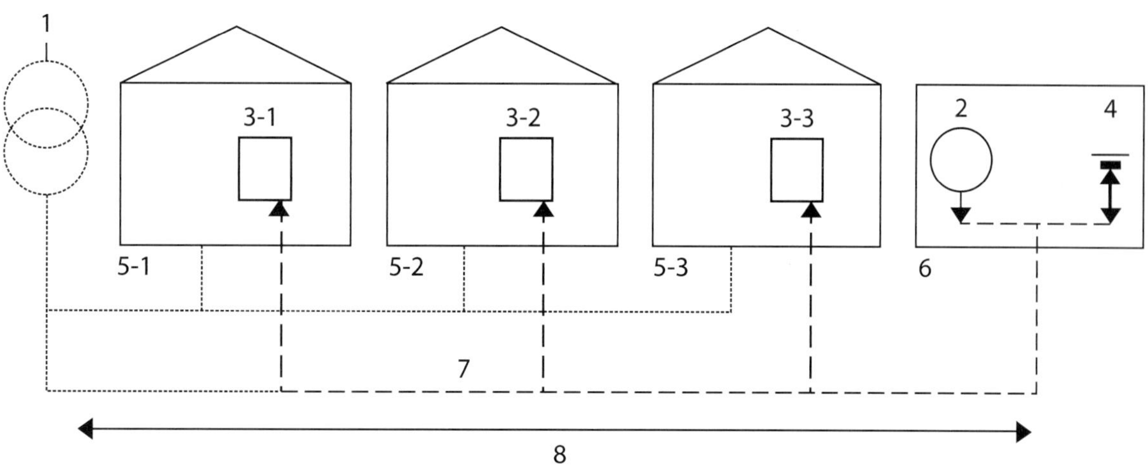

Key

1	Public network
2	Power supplies
3-1	Load 1
3-2	Load 2
3-3	Load 3
4	Storage units
5-1	Consumer 1
5-2	Consumer 2
5-3	Consumer 3
6	Producer
7	DSO distribution system
8	Distribution system within PEI

Any measure against the occurrence of voltage disturbances or interruptions should be considered when designing the installation working in island mode.

B82.2.3 Reverse feeding mode

In this operating mode only one prosumer is supplying electrical energy to the public network. See Figure B82.8 and Figure B82.9.

The electrical energy produced locally is:

(i) used by the group of end users through a local installation connecting all local prosumers or for charging the local storage units and

(ii) the surplus electrical energy produced locally or from local storage units is supplied to the DNO.

Fig B82.8 – Example of electrical design of collective PEI operating in reverse feeding mode with one single electrical installation

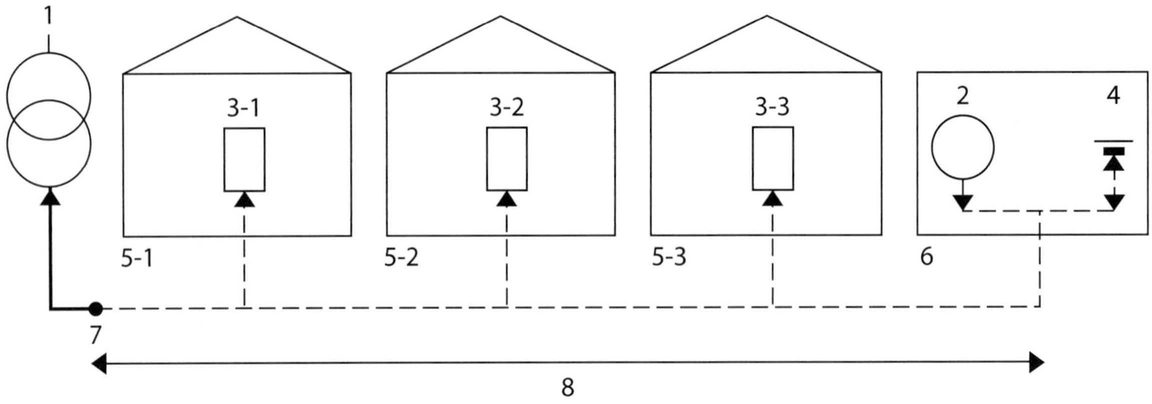

Key

1	Public network
2	Power supplies
3-1	Load 1
3-2	Load 2
3-3	Load 3
4	Storage units
5-1	Consumer 1
5-2	Consumer 2
5-3	Consumer 3
6	Producer
7	Origin of the distribution system within PEI
8	Distribution system within PEI

Fig B82.9 – Example of electrical design of collective PEI operating in reverse feeding mode with several electrical installations

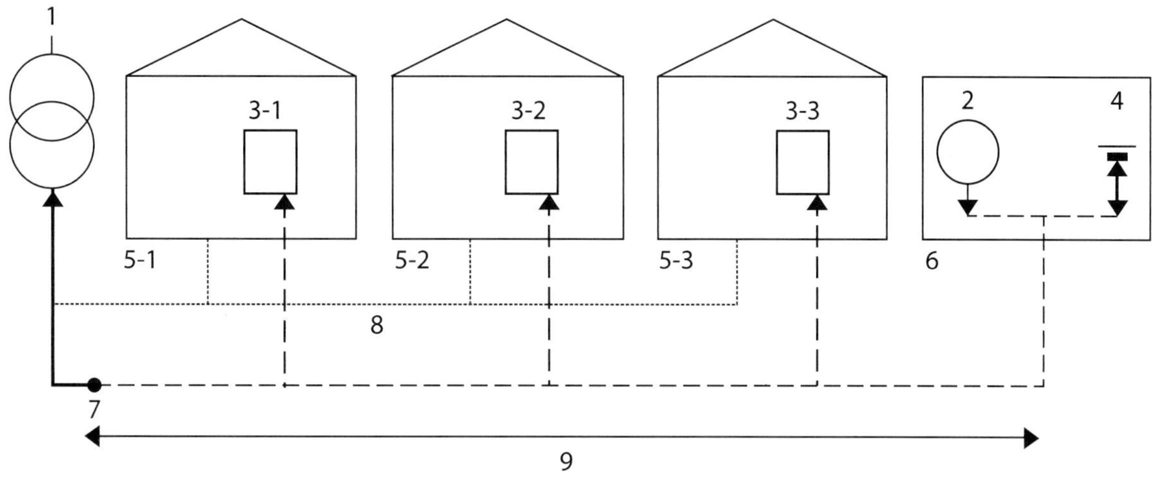

Key

1	Public network
2	Power supplies
3-1	Load 1
3-2	Load 2
3-3	Load 3
4	Storage units
5-1	Consumer 1
5-2	Consumer 2
5-3	Consumer 3
6	Producer
7	Origin of the distribution system within PEI
8	DSO distribution system
9	Distribution system within PEI

B82.3 Operating modes for shared PEI

B82.3.1 Direct feeding mode

All storage units may be charged either from:

(i) the common public network, or
(ii) the local power supplies, or
(iii) power supplies part of the group of prosumers.

In this operating mode, current-using equipment may be supplied either from:

(iv) the common public network, or from
(v) any local power supply part of the group of prosumers, or from
(vi) any local storage unit part of the group of prosumers.

See Figure B82.10 and Figure B82.11.

Fig B82.10 – Example of electrical design of shared PEI operating in direct feeding mode with one single electrical installation

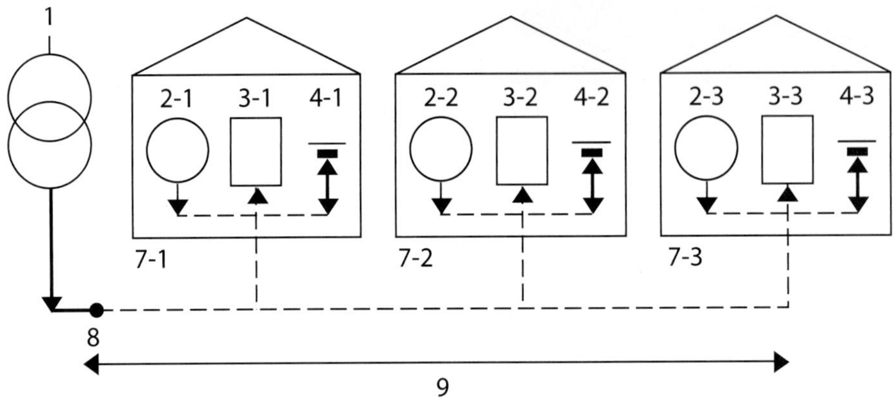

Key

1	Public network
2-1	Power supply 1
2-2	Power supply 2
2-3	Power supply 3
3-1	Load 1
3-2	Load 2
3-3	Load 3
4-1	Storage unit 1
4-2	Storage unit 2
4-3	Storage unit 3
7-1	Prosumer 1
7-2	Prosumer 2
7-3	Prosumer 3
8	Origin of the distribution system within PEI
9	Distribution system within PEI

Fig B82.11 – Example of electrical design of shared PEI operating in direct feeding mode with several electrical installations

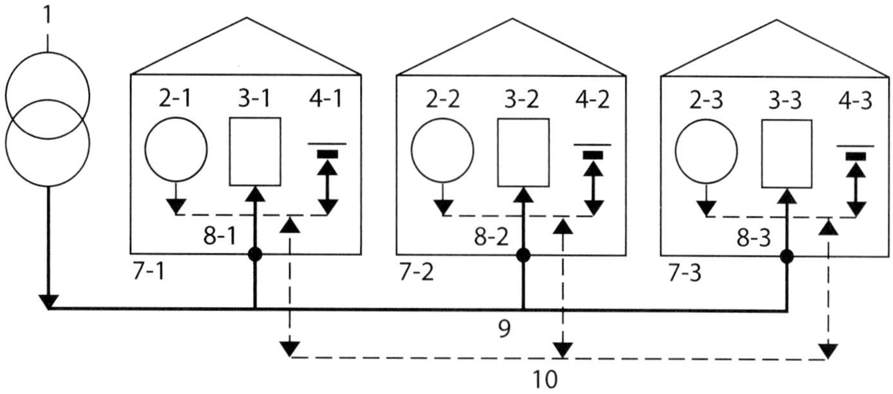

Key

1	Public network
2-1	Power supply 1
2-2	Power supply 2
2-3	Power supply 3
3-1	Load 1
3-2	Load 2
3-3	Load 3
4-1	Storage unit 1
4-2	Storage unit 2
4-3	Storage unit 3
7-1	Prosumer 1
7-2	Prosumer 2
7-3	Prosumer 3
8-1	Origin of installation 1
8-2	Origin of installation 2
8-3	Origin of installation 3
9	DSO distribution system
10	Distribution system within PEI

Local power supply may supply local loads, charge local storage units, run in reverse feeding mode or may be switched off.

B82.3.2 Island mode

Similar to the collective PEI, all private electrical installations should be disconnected from the public network during this operating mode.

All current-using equipment should be supplied by any local power supply part of the group of prosumers.

All local storage units should be charged from any local power supply part of the group of prosumers.

See Figure B82.12 and Figure B82.13

Fig B82.12 – Example of electrical design of shared PEI operating in island mode with one single electrical installation

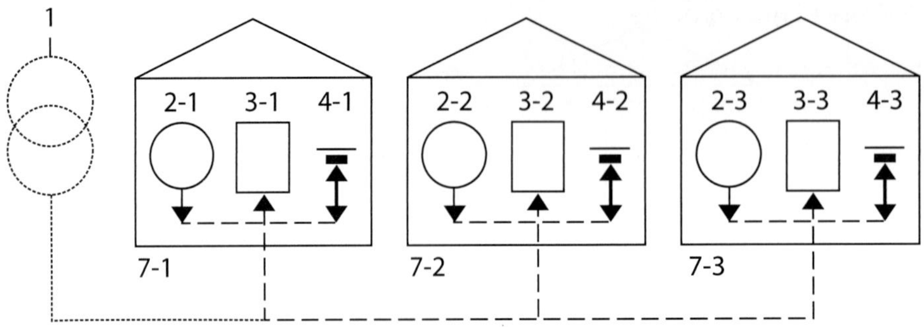

Key

1	Public network
2-1	Power supply 1
2-2	Power supply 2
2-3	Power supply 3
3-1	Load 1
3-2	Load 2
3-3	Load 3
4-1	Storage unit 1
4-2	Storage unit 2
4-3	Storage unit 3
7-1	Prosumer 1
7-2	Prosumer 2
7-3	Prosumer 3

Fig B82.13 – Example of electrical design of shared PEI operating in island mode with several electrical installations

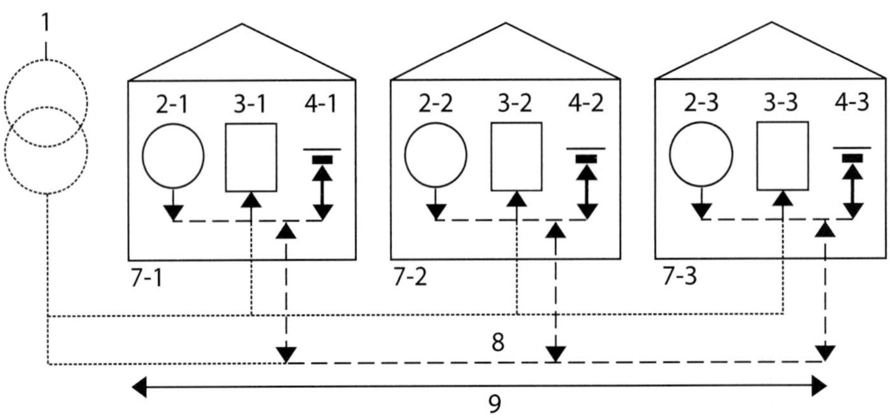

Key

1	Public network
2-1	Power supply 1
2-2	Power supply 2
2-3	Power supply 3
3-1	Load 1
3-2	Load 2
3-3	Load 3
4-1	Storage unit 1
4-2	Storage unit 2
4-3	Storage unit 3
7-1	Prosumer 1
7-2	Prosumer 2
7-3	Prosumer 3
8	DSO distribution system
9	Distribution system within PEI

Any measure against the occurrence of voltage disturbances or interruptions should be considered when designing the installation working in island mode.

B82.3.3 Reverse feeding mode

In this operating mode, part of or the entire electrical energy produced locally by all local power supplies or from local storage units may be supplied to the public network.

All current-using equipment in operation should also be supplied by all local power supplies and/or local storage unit(s).

See Figure B82.14.

Fig B82.14 – Example of electrical design of shared PEI operating in reverse feeding mode

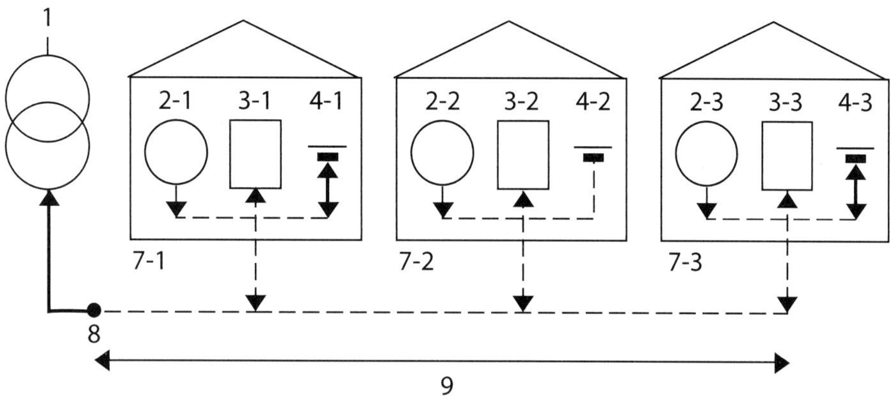

Key

1	Public network
2-1	Power supply 1
2-2	Power supply 2
2-3	Power supply 3
3-1	Load 1
3-2	Load 2
3-3	Load 3
4-1	Storage unit 1
4-2	Storage unit 2
4-3	Storage unit 3
7-1	Prosumer 1
7-2	Prosumer 2
7-3	Prosumer 3
8	Origin of the distribution system within PEI
9	Distribution system within PEI

Annex C82
(Informative)
INTERACTION WITH THE PUBLIC NETWORK

C82.1 General

Usually the nominal power of each PEI is sufficiently low compared to the nominal power of the low voltage public power network and even lower when compared to the HV public network, so that there is no noticeable influence of the PEI on the stability of public networks. However, high numbers of PEIs connected to the same public network could cause instability of these public networks resulting in the disconnection of some lines (distribution or transmission). PEIs should be designed in such a way that their dynamical influence on the stability of public networks is reduced or, even better, that they improve the dynamic stability of these public networks.

C82.2 National grid codes compliance with active and reactive power control

Local energy supplies used in the PEI are controlled by electronic devices such as inverters. Consequently, it is possible to monitor and control the active power and the reactive power injected into the public network from the PEI. These electronic devices may adjust the magnitude and the phase angle of the output voltage allowing the control of the active and reactive power consumed by or injected to the public network.

NOTE: In the UK these requirements are covered by the Distribution Code (www.dcode.org.uk).

C82.3 Voltage control

Correct voltage regulation avoids circulation of reactive current between power supplies. In some architectures, electrical power sources of several prosumers may not be far from each other. Line inductance between each source is not high enough to limit currents circulating between prosumers' sources.

NOTE: In the UK these requirements are covered by the Distribution Code (www.dcode.org.uk).

C82.4 Frequency control

Frequency control for each local power supply should be installed in order to adapt the frequency to the nominal value when disconnecting from the public network for operating in island mode.

This frequency control should also consider the case where several local power supplies operate in parallel.

NOTE: In the UK these requirements are covered by the Distribution Code (www.dcode.org.uk).

C82.5 Load shedding strategy

The electrical energy management system (EEMS) of a PEI should control the local loads and the local generation of energy. In response to the public network needs through, for instance, communication facilities, the local PEI may shed some local loads. In accordance with the contract agreed between the DSO and the prosumer, the EEMS of the PEI may adapt the consumption of energy from the public network and/or the generation of local energy according to the request from the DSO.

NOTE: In the UK these requirements are covered by the Distribution Code (www.dcode.org.uk).

Annex D82

(Informative)

ARCHITECTURE OF PEI

D82.1 Architecture of individual PEI

An example of architecture of an individual PEI is given in Figure D82.1.

Fig D82.1 – Example of type of architecture of individual PEI

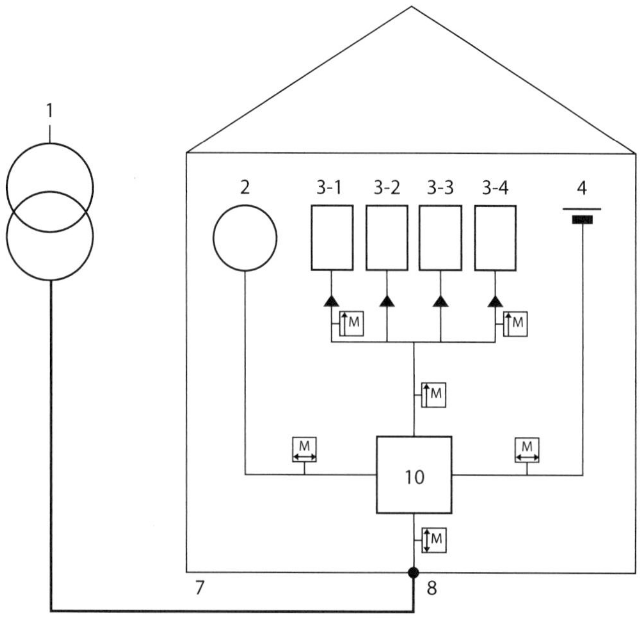

Key

1	Public network
2	Power supplies
3-1	Load 1
3-2	Load 2
3-3	Load 3
3-4	Load 4
4	Storage units
7	Prosumer
8	Origin of installation
10	Electrical energy management system (EEMS)
M	Energy meter/Measuring equipment (M)
↔	Direction of energy flow

D82.2 Architecture of collective PEI

In the collective PEI configuration, common local power supplies (e.g. renewable power sources) and common local energy storage units (e.g. batteries) are shared by a community of end users. Each prosumer may also have the possibility to connect its own private power supply for individual purposes or for common purposes. See Figure D82.2.

Fig D82.2 – Example of type of architecture of collective PEI

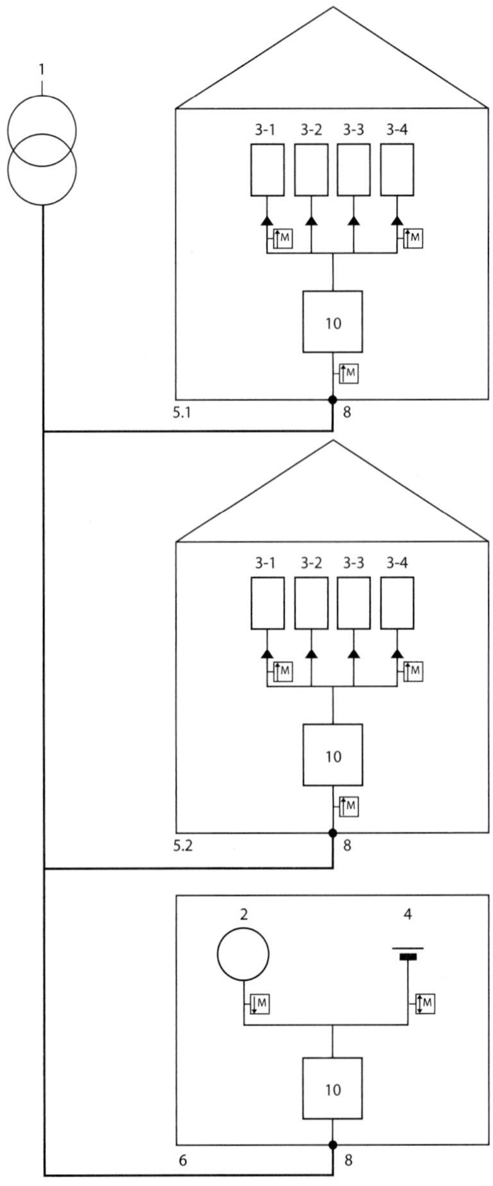

Key

1	Public network
2	Power supplies
3-1	Load 1
3-2	Load 2
3-3	Load 3
4	Storage units
5-1	Consumer 1
5-2	Consumer 2
6	Producer
8	Origin of installation
10	Electrical energy management system (EEMS)
M	Energy meter/Measuring equipment (M)
↔	Direction of energy flow

D82.3 Architecture of shared PEI

In a shared PEI configuration, a group of individual PEIs with their own local power supplies and local energy storage units are connected to the same low-voltage public network. See Figure D82.3.

Fig D82.3 – Example of type of architecture of shared PEI

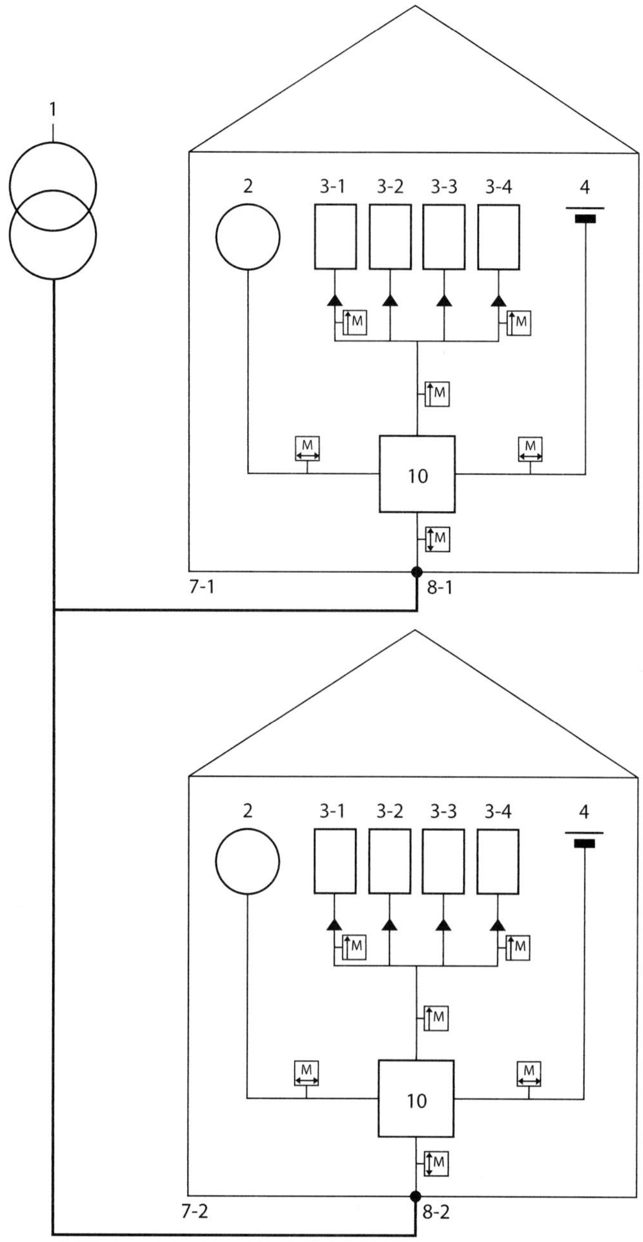

Key

1	Public network
2	Power supplies
3-1	Load 1
3-2	Load 2
3-3	Load 3
3-4	Load 4
4	Storage units
7-1	Producer 1
7-2	Producer 2
8-1	Origin of installation 1
8-2	Origin of installation 2
10	Electrical energy management system (EEMS)
M	Energy meter/Measuring equipment (M)
⟷	Direction of energy flow

APPENDICES

APPENDIX 1 (Normative)

BRITISH STANDARDS TO WHICH REFERENCE IS MADE IN BS 7671

NOTE: Certain British Standards have been withdrawn since the issue of the previous Edition. From the date of withdrawal, certificates and marks already awarded may continue to apply to production until a date specified in the superseding standard. During the period between these dates, the withdrawn standard may be specified in contracts. However, it should be noted that this appendix may not list such standards, as only current British Standards are listed with some references to superseded standards. Where standards are not dated they are a multiple standard.

Standard Number	Title	References
BS 67:1987 (1999)	Specification for ceiling roses	416.2.4 note 559.5.1
BS 88	The term "BS 88 series", when used in these Regulations, means BS 88-1, -2 and -3.	432.4 433.1.204 533.1.1 Table 537.4 Appx 4 sec 4 Appx 8 sec 4
BS 88-1:1975	Low-voltage fuses - Part 1: General requirements. This standard has been withdrawn.	
BS 88-2:2010	Low-voltage fuses - Part 2: Supplementary requirements for fuses for use by authorized persons (fuses mainly for industrial application) Examples of standardized systems of fuses A to K. BS 88-2:2010 replaced BS 88-2.2:1988 and BS 88-6:1988 which have been withdrawn.	Part 2 gG Part 2 gM Table 41.2 Table 41.4 Table 41.6 433.1.201 Appx 3 Figs 3A3(a)-(c)
BS 88-2.2:1988	Specification for fuses for use by authorized persons (mainly for industrial application). Additional requirements for fuses with fuse-links for bolted connections Replaced by BS 88-2:2010 and withdrawn 1/3/2010.	
BS 88-3:2010	Low-voltage fuses - Part 3: Supplementary requirements for fuses for use by unskilled persons (fuses mainly for household and similar applications) Examples of standardized systems of fuses A to F BS 88-3:2010 replaced BS 1361:1971 which has been withdrawn.	Table 41.2 Table 41.4 433.1.201 533.1.2.2 Appx 3 Fig 3A1
BS 88-6:1988	Specification of supplementary requirements for fuses of compact dimensions for use in 240/415 V AC industrial and commercial electrical installations Replaced by BS 88-2:2010 and withdrawn 1/3/2010.	
BS 196:1961 Withdrawn	Specification for protected-type non-reversible plugs, socket-outlets cable-couplers and appliance-couplers with earthing contacts for single-phase AC circuits up to 250 volts This standard has been withdrawn as the products have almost been replaced by those manufactured to BS EN 60309-1:1999 and BS EN 60309-2:1999.	
BS 476	Fire tests on building materials and structures	
BS 476-4:1970	Non-combustible test for materials	526.5(iii)
BS 476-12:1991	Method of test for ignitability of products by direct flame impingement	554.4.1
BS 546:1950 (2017)	Specification. Two-pole and earthing-pin plugs, socket-outlets and socket-outlet adaptors	Table 537.4 Table 55.1 553.1.5 559.5.1(v) 705.512.2 705.553.1(iii)
BS 559:2009 (2019)	Specification for the design and construction of signs for publicity, decorative and general purposes	110.1.3(i) 559.1 note 2
BS 646:1958 +A2:2013 (2018)	Specification. Cartridge fuse-links (rated up to 5 amperes) for AC and DC service BS 646 remains current but the requirements for type B fuse-links have been replaced by BS 2950:1958.	533.1.1 Table 55.1 553.1.5(i)

Standard Number	Title	References
BS 951:2009 (2019)	Electrical earthing. Clamps for earthing and bonding. Specification	514.13.1
BS 1361:1971 (1986)	Specification for cartridge fuses for AC circuits in domestic and similar premises Replaced by BS 88-3:2010 and withdrawn 30/3/2010.	
BS 1362:1973 (1991)	Specification for general purpose fuse links for domestic and similar purposes (primarily for use in plugs)	Table 41.2 Table 41.4 533.1.1 533.1.2.2 Table 55.1 553.1.5(i)
BS 1363	13 A plugs, socket-outlets, adaptors and connection units	433.1.204 Table 55.1 553.1.201 553.1.5 705.553.1(iii) Appx 15
BS 1363-1:2016 +A1:2018	Specification for rewirable and non-rewirable 13 A fused plugs	Table 537.4
BS 1363-2:2016 +A1:2018	Specification for 13 A switched and unswitched socket-outlets	Table 537.4 559.5.1(v) 705.512.2 722.55.101.0.201.1(i) 740.415.1(vi)
BS 1363-3:2016 +A1:2018	Specification for adaptors	
BS 1363-4:2016 +A1:2018	Specification for 13A fused connection units switched and unswitched	Table 537.4 559.5.1(vii) Appx 15 Figs 15A & 15B
BS 3036:1958 (2017)	Specification. Semi-enclosed electric fuses (ratings up to 100 amperes and 240 volts to earth)	Part 2 C_f Table 41.2 Table 41.4 432.4 433.1.1 note 1 433.1.202 433.1.204 533.1.1 533.1.2.2 Table 537.4 Appx 3 Fig 3A2(a) 3A2(b) Appx 4 sec 3 sec 4 sec 5.1 sec 5.1.1(iii) sec 6.1
BS 3535	Replaced by BS EN 61558-2-5:2010 Safety of transformers, reactors, power supply units and combinations thereof. Particular requirements and tests for transformer for shavers, power supply units for shavers and shaver supply units.	Appx 6 EIC Checklist item 12.3 Appx 6 Cond. Report item 6.3
BS 3676	Switches for household and similar fixed electrical installations. Specification for general requirements. Replaced by BS EN 60669-1:1999+A2:2008	
BS 3858:1992 (2019)	Specification for binding and identification sleeves for use on electric cables and wires	514.3.2
BS 4177:1992 (2015)	Specification for cooker control units	Table 537.4
BS 4444:1989 (2017)	Guide to electrical earth monitoring and protective conductor proving	543.3.4 543.7.1.202(iii) 543.7.1.203(iv)
BS 4573:1970+A5:2016	Specification for 2-pin reversible plugs and shaver socket-outlets	553.1.5(ii)
BS 4662:2006 +A1:2009 (2019)	Boxes for flush mounting of electrical accessories. Requirements, test methods and dimensions	530.4.2 559.5.1(viii)

Standard Number	Title	References
BS 5266	Emergency lighting	110.1.3(ii) 528.1 note 2 560.8.1 note 1 560.9
BS 5266-1:2016	Emergency lighting. Code of practice for the emergency lighting of premises	See BS 5266
BS 5467:2016	Electric cables. Thermosetting insulated, armoured cables of rated voltages of 600/1000 V and 1900/3300 V for fixed installations.	522.6.204(i) Appx 4 Table 4A3
BS 5655	Lifts and service lifts	110.2(x)
BS 5655-1:1986	Safety rules for the construction and installation of electric lifts (Applicable only to the modernization of existing lift installations) BS EN 81-1:1998+A3:2009 available but still current.	
BS 5655-2:1988	Safety rules for the construction and installation of hydraulic lifts (Applicable only to the modernization of existing lift installations) BS EN 81-2:1998+A3:2009 available but still current.	
BS 5655-11:2005 (2017)	Code of practice for the undertaking of modifications to existing electric lifts (Applicable only to the modernization of existing lift installations.)	
BS 5655-12:2005 (2017)	Code of practice for the undertaking of modifications to existing hydraulic lifts (Applicable only to the modernization of existing lift installations.)	
BS 5656	Safety rules for the construction and installation of escalators and passenger conveyors	110.2(x)
BS 5733:2010 +A1:2014 (2019)	Specification for general requirements for electrical accessories	411.3.3 note 4 526.3(vi) 530.4.2 Table 537.4 559.5.1(vi) 559.5.4(v)
BS 5803-5:1985	Thermal insulation for use in pitched roof spaces in dwellings Specification for installation of man-made mineral fibre thermal insulation mats.	Appx 4 Table 4A2 items 100 to 103 Table 4D5
BS 5839	Fire detection and fire alarm systems for buildings	110.1.3(v) 528.1 note 2 560.8.1 note 1 560.10
BS 5839-1:2017	Code of practice for design, installation, commissioning and maintenance of systems in non-domestic premises	643.3.2 note 1
BS 6004:2012	Electric cables. PVC insulated and PVC sheathed cables for voltages up to and including 300/500 V, for electric power and lighting	704.522.8.11 A721.521.2 Appx 4 Table 4A3 Appx 15 Figs 15A & 15B note
BS 6007:2006 superseded, withdrawn	Electric cables. Single core unsheathed heat resisting cables for voltages up to and including 450/750 V, for internal wiring see BS EN 50525.	
BS 6217 Withdrawn	Guide to graphical symbols - see note at end of this table	Appx 1 note
BS 6220:1983 (1999)	*Deleted by BS 7671:2008, Corrigendum (July 2008)*	
BS 6231:2006	*Deleted by BS 7671:2018*	
BS 6346:1997 (2005)	Electric cables. PVC insulated, armoured cables for voltages of 600/1000 V and 1900/3300 V **NOTE 1:** This was withdrawn in 2011 but is retained within BS 7671 for historical purposes.	Appx 4 Table 4A3
BS 6351	*Deleted by BS 7671:2018*	

Standard Number	Title	References
BS 6500:2000	Electric cables. Flexible cords rated up to 300/500 V, for use with appliances and equipment intended for domestic, office and similar environments. **NOTE 2:** This was withdrawn in 2012 but is retained within BS 7671 for historical purposes. Replaced by some parts of BS EN 50525	
BS 6701:2016 +A1:2017	Telecommunications equipment and telecommunications cabling. Specification for installation, operation and maintenance	110.1.3(vi) 444.1(iii) 528.2 note 2
BS 6724:2016	Electric cables. Thermosetting insulated, armoured cables for voltages of 600/1000 V and 1900/3300 V, for fixed installations, having low emission of smoke and corrosive gases when affected by fire. Specification.	522.6.204(i) Appx 4 Table 4A3
BS 6891:2015 +A1:2019	Specification for the installation and maintenance of low pressure gas installation pipework of up to 35 mm (R1$\frac{1}{4}$) on premises	528.3.4 note
BS 6907	Electrical installations for open-cast mines and quarries	110.1.3(viii)
BS 6972:1988 (2017)	Specification for general requirements for luminaire supporting couplers for domestic, light industrial and commercial use	Table 537.4 559.5.1(ii) 559.5.4(iv)
BS 6991:1990 (2017)	Specification for 6/10 A, two-pole weather-resistant couplers for household, commercial and light industrial equipment	553.2.1
BS 7001:1988 (2017)	Specification for interchangeability and safety of a standardized luminaire supporting coupler	559.5.1(ii) 559.5.4(iv)
BS 7211:2012	Electric cables. Thermosetting insulated and thermoplastic sheathed cables for voltages up to and including 450/750 V for electric power and lighting and having low emission of smoke and corrosive gases when affected by fire	A721.521.2 Appx 4 Table 4A3
BS 7288:2016 +A1:2021	Specification for residual current devices with or without overcurrent protection for socket-outlets for household and similar uses	531.3.4.1 531.3.4.1 note 2 531.3.4.2 531.3.4.2 note 531.3.6 Table 537.4
BS 7375:2010	Code of practice for distribution of electricity on construction and building sites	704.411.3.1 note
BS 7430:2011 +A1:2015	Code of practice for protective earthing of electrical installations	542.2.2 note 542.3.1 note A722.4
BS 7454:1991 +A1:2008 (2010)	Method for calculation of thermally permissible short-circuit currents, taking into account non-adiabatic heating effects	Table 43.1 note 2 543.1.3
BS 76291:2015 +A1:2019	Electric cables. Specification for 300/500 V fire resistant, screened, fixed installation cables having low emission of smoke and corrosive gases when affected by fire. Multicore cables	422.6(ii) 560.8.1 note 1 Appx 4 Table 4A3
BS 7697:1993 (2010)	Nominal voltages for low voltage public electricity supply systems Withdrawn and replaced by BS EN 60038:2011.	Appx 2 sec 15
BS 7698-12:1998	Reciprocating internal combustion engine driven alternating current generating sets. Emergency power supply to safety devices	560.6.13
BS 7769	Electric cables. Calculation of the current rating. (Some parts of the BS 7769 series are now numbered BS IEC 60287 series, eventually all parts will be renumbered.)	523.3 Appx 4 sec 1 sec 2.1 sec 2.2 Table 4B3 note 2 Table 4C2 notes 1 & 2 Table 4C3 notes 1 & 2 Appx 10 sec 1 note, sec 2 para 8
BS 7769-1.1:1997	Has been superseded/withdrawn and replaced by BS IEC 60287-1-1:2006+A1:2014	

Standard Number	Title	References
BS 7769-1.2:1994 (2005)	Current rating equations (100% load factor) and calculation of losses. Sheath eddy current loss factors for two circuits in flat formation	
BS 7769-2.2:1997 (2005)	Thermal resistance. A method for calculating reduction factors for groups of cables in free air, protected from solar radiation	
BS 7769-2-2.1:1997 (2006)	Thermal resistance. Calculation of thermal resistance. Section 2.1: Calculation of thermal resistance	
BS 7846:2015	Electric cables. Thermosetting insulated, armoured, fire-resistant cables of rated voltage 600/1000 V for fixed installations, having low emission of smoke and corrosive gases when affected by fire. Specification	422.6(ii) 522.6.204(i) 560.8.1 note 1 Appx 4 Table 4A3
BS 7889:2012 (2017)	Electric cables. Thermosetting insulated, unarmoured cables with a voltage of 600/1000 V, for fixed installations	Appx 4 Table 4A3
BS 7909:2011 (2016)	Code of practice for temporary electrical systems for entertainment and related purposes	110.1.3(ix) 706.1 711.1 717.1 note 740.1.1 note 1
BS 7919:2001 (2006)	Electric cables. Flexible cables rated up to 450/750V, for use with appliances and equipment intended for industrial and similar environments Withdrawn in 2012 and replaced by some parts of BS EN 50525.	
BS 8434	Methods of test for assessment of the fire integrity of electric cables	560.8.1(iii)
BS 8436:2011	Electric cables. Specification for 300/500 V screened electric cables having low emission of smoke and corrosive gases when affected by fire, for use in walls, partitions and building voids. Multicore cables	522.6.204(i) Appx 4 Table 4A3
BS 8450:2006 (2017)	Code of practice for installation of electrical and electronic equipment in ships	110.2(iv)
BS 8488:2009 +A1:2010	Specification for prefabricated wiring systems intended for permanent connection in fixed installations	521.201
BS 8491:2008	Method for assessment of fire integrity of large diameter power cables for use as components for smoke and heat control systems and certain other active fire safety systems	560.8.1(iii)
BS 8519:2010 (2015)	Selection and installation of fire-resistant power and control cable systems for life safety and fire-fighting applications. Code of practice	110.1.3(x) 528.1 note 2 560.6.5 note 560.6.9 note 3 560.8.1 note 1 560.11
BS 8573:2012 (2017)	Electric cables. Thermosetting insulated, non-armoured cables with a voltage of 600/1000V, for fixed installations, having low emission of smoke and corrosive gases when affected by fire	422.6(ii) Appx 4 Table 4A3
BS 8592:2016	Electric cables. Thermosetting insulated, non-armoured, fire-resistant, single core non-sheathed cables of rated voltage 450/750 V, having low emission of smoke and corrosive gases when affected by fire	422.6(ii) Table 4A3
BS 9999:2017	Code of practice for fire safety in the design, management and use of buildings	110.1.3(x) 422.1
BS AU 149a:1980 (1987)	Specification for electrical connections between towing vehicles and trailers with 6 V or 12 V electrical equipment: type 12 N (normal). Withdrawn in 2014.	A721.55.2.6(i) A721.55.2.6(iii) Table A721
BS AU 177a:1980 (1987)	Specification for electrical connections between towing vehicles and trailers with 6 V or 12 V electrical equipment: type 12 S (supplementary). Withdrawn in 2014.	A721.55.2.6(i) A721.55.2.6(iii) Table A721
BS EN 81	Safety rules for the construction and installation of lifts	110.2(x) 528.3.5
BS EN 81-1:1998 +A3:2009	Electric lifts (also known as BS 5655-1:1986 Lifts and service lifts ... etc.) Withdrawn in 2017 and replaced by BS EN 81-20 and BS EN 81-50.	
BS EN 81-20	Passenger and goods passenger lifts	

Standard Number	Title	References
BS EN 81-50	Examinations and tests. Design rules, calculations, examinations and tests of lift components	
BS EN 115 BS EN 115-1 BS EN 115-2	Safety of escalators and moving walks. Construction and installation Rules for the improvement of safety of existing escalators and moving walks	110.2(x)
BS EN 1648	Leisure accommodation vehicles	
BS EN 1648-1:2018	12 V direct current extra low voltage electrical installations. Caravans	708.1 note 1 721.1
BS EN 1648-2:2018	12 V direct current extra low voltage electrical installations. Motor caravans	708.1 note 1 721.1
BS EN 1838:2013	Lighting applications - Emergency lighting	110.1.3(ii) 560.9
BS EN 12954:2019	Cathodic protection of buried or immersed metallic structures. General principles and application for pipelines	110.2(xiii)
BS EN 13636:2004 (2018)	Cathodic protection of buried metallic tanks and related piping	110.2(xiii) 712.312.2 note
BS EN 14505:2005 (2019)	Cathodic protection of complex structures	110.2(xiii)
BS EN 15112:2006 (2011)	External cathodic protection of well casing BS 7361-1 was withdrawn and has been replaced by this.	712.312.2 note
BS EN 15869	Inland navigation vessels	
BS EN 15869-1:2019	Electrical shore connection, three phase current 400 V, up to 63 A, 50 Hz. General requirements	730.1 note 2 730.313.1.101 note
BS EN 15869-2:2019	Electrical shore connection, three phase current 400 V, up to 63 A, 50 Hz. Onshore unit, safety requirements	730.1 note 2 730.313.1.101 note 730.55.1.6
BS EN 15869-3:2019	Electrical shore connection, three phase current 400 V, up to 63 A, 50 Hz. On-board unit, safety requirements	730.1
BS EN 50085	Cable trunking and cable ducting systems for electrical installations	412.2.4.1(ii)(b) 422.2.1(iv) 422.3.4 521.6 527.1.5 527.1.6 530.4.3 711.521.101(ii) 715.521.1(i) 717.52.2(i) 717.528.3.4 721.521.2 721.528.2.1 740.521.1 740.521.1(ii)
BS EN 50085-1:2005 +A1:2013	General requirements.	422.4.204
BS EN 50085-2-1:2006 +A1:2011	Cable trunking systems and cable ducting systems intended for mounting on walls and ceilings	522.6.204(iii) 705.522.16(ii)
BS EN 50086	Specification for conduit systems for cable management. Withdrawn and replaced by BS EN 61386.	
BS EN 50107	Signs and luminous-discharge-tube installations operating from a no-load rated output voltage exceeding 1 kV but not exceeding 10 kV	110.1.3(i) 559.1(ii) 559.1 note 2
BS EN 50107-1:2002 (2019)	General requirements	
BS EN 50107-2:2005 (2019)	Requirements for earth-leakage and open-circuit protective devices	

Standard Number	Title	References
BS EN 50117-4-1: 2008 +A1:2013	Coaxial cables – Sectional specification for cables for BCT cabling in accordance with EN 50173 – Indoor drop cables for systems operating at 5 MHz – 3 000 MHz	Table A444.1(v)
BS EN 50174	Information technology – Cabling installation	443.1.1 444.1(v) 444.4.1 note A444.4 A444.5 528.2 note 2
BS EN 50174-1:2018	Installation specification and quality assurance	444.4.10(i)
BS EN 50174-2:2018	Installation planning and practices inside buildings	444.4.10(ii) A444.4
BS EN 50174-3:2013 +A1:2017	Information technology. Cabling installation. Installation planning and practices outside buildings	A444.4
BS EN 50200:2015	Method of test for resistance to fire of unprotected small cables for use in emergency circuits	560.8.1(iii)
BS EN 50266	Common test methods for cables under fire conditions. Test for vertical flame spread of vertically-mounted bunched wires or cables. Standard (all parts) withdrawn and replaced by BS EN 60332.	
BS EN 50281	Electrical apparatus for use in the presence of combustible dust	110.1.3(iv)
BS EN 50288	Multi-element metallic cables used in analogue and digital communication and control	Table A444.1(iv)
BS EN 50310:2016	Telecommunications bonding networks for buildings and other structures	444.1(iv) 444.4.10(iii) A444.1 note A444.1.3
BS EN 50428:2005 +A2:2009	Switches for household and similar fixed installations. Collateral standard Switches and related accessories for use in home and building electronic systems (HBES). Replaced by BS EN 60669-2-5:2016	Table 537.4
BS EN 50522:2010	Earthing of power installations exceeding 1 kV AC	442.2 442.2 note 3 442.2.1 note 1
BS EN 50525	Electric cables – Low voltage energy cables of rated voltages up to and including 450/750 V (U_0/U)	559.5.3.2(iii) A721.521.2 Appx7 Table 7D
BS EN 50525-2-11, 2-12, 2-21, 2-23, 2-51, 2-82, 3-11, 3-21	Electric cables. Low voltage energy cables of rated voltages up to and including 450/750 V (U_0/U). *(flexible cables)*	422.3.201 note 2 702.522.23(ii) 704.522.8.11 704.522.8.101 705.422.8 note Fig 708 note 2 717.52.1 717.52.2(i) 717.52.2(ii) 721.55.2.6(ii) 740.521.1 note 2 740.55.1.1 Appx 4 Table 4A3
BS EN 50525-2-31, 3-41	Electric cables. Low voltage energy cables of rated voltages up to and including 450/750 V (U_0/U). Cables for general applications. Single core non-sheathed cables with thermoplastic PVC insulation	717.411.3.1.2 note 717.52.2(i) Appx 4 Table 4A3
BS EN 50525-3-31:2011	Electric cables. Low voltage energy cables of rated voltages up to and including 450/750 V (U_0/U). Cables with special fire performance. Single core non-sheathed cables with halogen-free thermoplastic insulation, and low emission of smoke	717.52.2(i)
BS EN 50549-1:2019	Requirements for generating plants to be connected in parallel with distribution networks. Connection to a LV distribution network. Generating plants up to and including Type B	551.1 note 551.7.4 551.7.5 note 551.7.6

Standard Number	Title	References
BS EN 50565-1:2014	Electric cables. Guide to use for cables with a rated voltage not exceeding 450/750 V (U_0/U). General guidance	422.3.201 note 1 521.9.1 note
BS EN 50575	Power, control and communication cables - Cables for general applications in construction works subject to reaction to fire requirements.	Appx 2, item 17
BS EN 50618:2014	Electric cables for photovoltaic systems	712.521.101 note
BS EN 60034-30-1:2014	Rotating electrical machines. Efficiency classes of line operated AC motors (IE code)	App 17 sec 17.13 note 1 App 17 sec 17.13
BS EN 60038:2011	CENELEC standard voltages	Table 443.2 note a 557.3.5.3 557.3.5.4.2 721.313.1.2
BS EN 60068-2-11:1999	Environmental testing. Test methods. Tests. Test KA. Salt mist	Appx 5 AF2
BS EN 60073:2002	Basic and safety principles for man-machine interface, marking and identification. Coding principles for indicators and actuators	514.1.1 Appx 11 sec 3
BS EN 60079	Electrical apparatus for explosive gas atmospheres	110.1.3(iii) & (iv) 753.1 App 5 BE3
BS EN 60079-10-1:2015 *(split into 10-1 and 10-2)*	Explosive atmospheres. Classification of areas. Explosive gas atmospheres	
BS EN 60079-14:2014	Explosive atmospheres. Electrical installations design, selection and erection	422.3 560.1
BS EN 60079-17:2014	Explosive atmospheres. Electrical installations inspection and maintenance	643.1
BS EN 60092-507:2015	Electrical installations in ships – small vessels	110.2(iv) 709.1 note 2
BS EN 60146-2:2000	Semiconductor convertors. General requirements and line commutated convertors. Self-commutated semiconductor converters including direct DC converters	414.2 note 2
BS EN 60204	Safety of machinery. Electrical equipment of machines	110.2(xi) 460 Scope 464.1 note 3 537.2.3 note 537.3.2.1 note 537.3.3.1 note
BS EN 60204-1:2018	General requirements	711.55.4.1 717.1(v) 740.1.1
BS EN 60228:2005	Conductors of insulated cables	Appx 4 sec 1
BS EN 60238:2004 +A2:2011	Edison screw lampholders Withdrawn in 2018 and replaced by BS EN IEC 60238:2018+A1:2018	416.2.4 note 559.5.1.206 643.6(ii)
BS EN 60255-26:2013	Measuring relays and protection equipment. electromagnetic compatibility requirements	Appx 5 AM-24-2
BS EN 60269 BS EN 60269-1:2007 +A2:2014	Low-voltage fuses General requirements	722.533.101 A53.8
BS HD 60269-2:2010, BS 88-2:2010	Low-voltage fuses. Supplementary requirements for fuses for use by authorized persons (fuses mainly for industrial application). Examples of standardized systems of fuses A to J	
BS HD 60269-2:2013	Low-voltage fuses - Part 2: Supplementary requirements for fuses for use by authorized persons (fuses mainly for industrial application) Examples of standardized systems of fuses A to K' (also numbered BS 88-2:2013).	

Standard Number	Title	References
BS EN 60269-3:1995	Supplementary requirements for fuses for use by unskilled persons (fuses mainly for household and similar applications) Replaced by BS 88-3:2007 and BS EN 60269-1:2007+A2:2014 and withdrawn 1/3/2010.	
BS HD 60269-3:2010	Low-voltage fuses - Part 3: Supplementary requirements for fuses for use by unskilled persons (fuses mainly for household and similar applications) Examples of standardized systems of fuses A to F (also numbered BS 88-3:2010).	
BS EN 60269-6:2011	Low-voltage fuses. Supplementary requirements for fuse-links for the protection of solar photovoltaic energy systems	712.533.101
BS HD 60364-8-1:2019	Low-voltage electrical installations. Functional aspects. Energy efficiency	822.1 825.1 826.4
BS EN 60309	Plugs, socket-outlets and couplers for industrial purposes	Table 537.4 551.4.4 note 717.551.6
BS EN 60309-1:1999 +A2:2012	General requirements	704.511.1 705.553.1 708.55.1.1 709.553.1.8 721.55.1.1 722.55.101.0.201.1(ii) 730.55.1.1 740.415.1(vi) 740.55.7(ii)
BS EN 60309-2:1999 +A2:2012	Dimensional interchangeability requirements for pin and contact-tube accessories	543.7.1.202(ii) Table 55.1 553.1.5 553.2.1 559.5.1(v) 704.511.1 705.553.1 708.55.1.1 708.553.1.8 Fig 708 note 2 709.553.1.8 Fig 709.3 717.55.1 721.55.1.1 721.55.2.6(ii) 722.55.101.0.201.1(ii) 722.55.101.0.201.1(iii) 730.55.1.1 740.55.7(i)
BS EN 60309-4:2007 +A1:2012	Switched socket-outlets and connectors, with or without interlock	708.55.1.1 722.55.101.0.201.1(iii) 730.55.1.1
BS EN 60320-1:2015	Appliance couplers for household and similar general purposes. General requirements	553.2.1
BS EN 60332-1-2:2004 +A11:2016	Tests on electric and optical fibre cables under fire conditions. Test for vertical flame propagation for a single insulated wire or cable. Procedure for 1 kW pre-mixed flame	422.1(ii) 422.2.1(i) 422.3.4 422.4.203 527.1.3 560.8.1(i)(ii) & (iii) 711.521(i) 717.52.2 721.521.2 740.521.1

Standard Number	Title	References
BS EN 60332-3	Tests on electric and optical fibre cables under fire conditions. Test for vertical flame spread of vertically-mounted bunched wires or cables	422.1(i) 422.2.1(i) 422.3.4 422.3.4 note 3 422.5 527.1.3 711.521.101
BS EN 60335-1:2012 +A13:2017	Household and similar electrical appliances. Safety. General requirements	A721.313.4(iii)
BS EN 60335-2-29:2004 +A11:2018	Particular requirements for battery chargers	A721.55.9
BS EN 60335-2-41:2003 +A2:2010	Particular requirements for pumps	702.55.3
BS EN 60335-2-53:2011	Particular requirements for sauna heating appliances and infrared cabins	703.55
BS EN 60335-2-71:2003	Particular requirements for electrical heating appliances for breeding and rearing animals	705.422.6
BS EN 60335-2-76:2005 +A2:2015	Particular requirements for electric fence energizers	110.2(xii) 705.1 note
BS EN 60335-2-96:2002 +A2:2009	Particular requirements for flexible sheet heating elements for room heating	110.1.3(vii) 753.511
BS EN 60417	Graphical symbols for use on equipment. Now withdrawn. See Note at end of this Table.	412.2.1.1 note Table 55.3
BS EN 60447:2004	Basic and safety principles for man-machine interface, marking and identification. Actuating principles	514.1.1 Appx 11 sec 3
BS EN 60519	Safety in installations electroheating and electromagnetic processing	753 Scope
BS EN 60529:1992 +A2:2013	Specification for degrees of protection provided by enclosures (IP code)	522.3.1 522.4.1 527.2.3(i) & (ii) 702.512.2 708.553.1.8 712.512.102 A721.55.6 A721.55.10 Appx 5 AD2, AE
BS EN 60570:2003+A1:2018	Electrical supply track systems for luminaires Replaces BS EN 60570:1997 and BS EN 60570-2-1:1995 which have both been withdrawn.	527.1.5 559.3.4 715.521.1(v)
BS EN 60598	Luminaires	559.5.1(iii) & (iv) 559.5.3.2(i) & (ii) Table 55.3 702.55.2
BS EN 60598-1:2015 +A1:2018	Luminaires. General requirements and tests	Table 55.3
BS EN 60598-2-13:2006 +A2:2016	Luminaires. Particular requirements. Ground recessed luminaires	559.10
BS EN 60598-2-18:1994 +A1:2012	Luminaires. Particular requirements. Luminaires for swimming pools and similar applications	702.55.2 702.55.3
BS EN 60598-2-23:1997	Particular requirements. Extra-low voltage lighting systems for filament lamps	711.559.4.2 715.422.107.1(iii) 715.521.1(iv)
BS EN 60598-2-24:2013	Particular requirements. Luminaires with limited surface temperatures	422.3.2 note 422.3.8(iii) Table 55.3
BS EN 60601	Medical electrical equipment. General requirements for basic safety and essential performance	710.1 note 4 710.512.2.1 note Fig 710.1
BS EN 60617	Graphical symbols for diagrams. Now withdrawn and replaced by IEC 60617. See Note at end of this table.	Appx 1 note

Standard Number	Title	References
BS EN 60623:2017	Secondary cells and batteries containing alkaline or other non-acid electrolytes. Vented nickel-cadmium prismatic rechargeable single cells	560.6.10 560.6.11
BS EN 60669	Switches for household and similar fixed electrical installations	
BS EN 60669-1:2018	General requirements. This replaces BS 3676.	416.2.4 note Table 537.4
BS EN 60669-2-1:2004 +A12:2010	Particular requirements. Electronic switches	Table 537.4
BS EN 60669-2-2:2006	Particular requirements. Electromagnetic remote-control switches (RCS)	536.4.2.3 536.4.3.2 Table 537.4 Table A53.1
BS EN 60669-2-3:2006	Particular requirements. Time delay switches (TDS)	Table 537.4
BS EN 60669-2-4:2005	Particular requirements. Isolating switches	536.4.2.3 536.4.3.2 Table 537.4 Table A53.1
BS EN 60669-2-6:2012	Switches for household and similar fixed electrical installations. Particular requirements. Fireman's switches for exterior and interior signs and luminaires	537.4.1
BS EN 60670	Boxes and enclosures for electrical accessories for household and similar fixed electrical installations	530.4.2 559.5.1(viii)
BS EN 60670-24:2013	Particular requirements for enclosures for housing protective devices and other power dissipating electrical equipment	712.511.103
BS EN 60670-1:2005 +A1:2013	General requirements	422.5 note 3
BS EN 60670-22:2006	Particular requirements for connecting boxes and enclosures	521.8.3 Appx 15 Figs 15A & 15B
BS EN 60684	Flexible insulating sleeving	543.3.201
BS EN 60702-1:2002 +A1:2015	Mineral insulated cables and their terminations with a rated voltage not exceeding 750 V. Cables	422.6(i) 433.1.204 522.6.204(i) Table 52.1 note a 543.4.8 560.8.1(i) 560.8.1 note 1 Appx 4 Table 4A3
BS EN 60702-2:2002 +A1:2015	Mineral insulated cables and their terminations with a rated voltage not exceeding 750 V	560.8.1(i)
BS EN 60896-11:2003	Stationary lead-acid batteries. General requirements and methods of test. Vented types. General requirements and methods of test	560.6.10 560.6.11
BS EN 60898 series	Specification for circuit-breakers for overcurrent protection for household and similar installations	Table 41.3 Table 41.6 432.4 433.1.201 433.1.204 533.1.1 Table 537.4 Table A53.1 722.533.101 Appx 3 Fig 3A4 Fig 3A5 Fig 3A6 Appx 4 sec 4 Appx 8 sec 4
BS EN 60898-1:2019	Circuit breakers for AC operation	434.5.2 536.4.2.1 Table A53.1 Appx 4 sec 5.5.2

Standard Number	Title	References
BS EN 60898-2:2006	Circuit-breakers for AC and DC operation BS EN 60898-2:2001 remained current up until 1 June 2010.	434.5.2 533.1.1 Table A53.1 712.533.101(ii) A721.533.1.6
BS EN 60904-3:2016	Photovoltaic devices. Measurement principles for terrestrial photovoltaic (PV) solar devices with reference spectral irradiance data Replaced by BS EN IEC 60904-3:2019	Part 2 STC
BS EN 60947	Low-voltage switchgear and control gear	533.1.1
BS EN 60947-2:2017 +A1:2020	Circuit-breakers	411.4.4 note 2 432.4 433.1.201 433.1.204 531.1.1 531.3.2(v) 531.3.4.2 533.1.1 533.1.3 536.4.1.4 note 3 536.4.2.1 536.4.3.1 Table 537.4 Table A53.1 711.410.3.4 712.531.3.5.1 712.533.101(ii) 721.415.1 A721.533.1.6 722.531.3.101 722.533.101 740.410.3 740.411 note Appx 4 sec 4 Appx 8 sec 4
BS EN 60947-3:2009 +A2:2015	Switches, disconnectors, switch-disconnectors and fuse-combination units	536.4.2.3 536.4.2.3 note 536.4.3.2 Table 537.4 537.4.1 Table A53.1 712.533.101(i)
BS EN 60947-4-1:2010 +A1:2012	Contactors and motor starters – Electromechanical contactor and motor starters. Replaced by BS EN IEC 60947-4-1:2019	435.2 536.4.2.2 536.4.3.1 Table 537.4
BS EN 60947-4-2:2012	Contactors and motor-starters. AC semiconductor motor controllers and starters	Table 537.4
BS EN 60947-4-3:2014	Contactors and motor-starters. AC semiconductor controllers and contactors for non-motor loads	Table 537.4
BS EN 60947-5-1:2017	Control circuit devices and switching elements – Electromechanical control circuit devices	Table 537.4
BS EN 60947-6-1:2005 +A1:2014	Multiple function equipment – Transfer switching equipment	536.4.3.2 Table 537.4 Table A53.1 710.537.1
BS EN 60947-6-2:2003	Multiple function equipment – Control and protective switching devices (or equipment) (CPS)	Table 537.4 Table A53.1 722.533.101
BS EN 60947-7	Specification for low-voltage switchgear and controlgear	526.2 note 1
BS EN 60947-7-1:2009	Ancillary equipment. Terminal blocks for copper conductors	
BS EN 60947-7-2:2009	Ancillary equipment. Protective conductor terminal blocks for copper conductors	

Standard Number	Title	References
BS EN 60950-1:2006 +A2:2013	Information technology equipment. Safety. General requirements Replaced by BS EN 62368-1:2-14+A11:2017	444.4.9
BS EN 60998	Connecting devices for low-voltage circuits for household and similar purposes	526.2 note 1 559.5.4(i)
BS EN 60998-2-1:2004	Particular requirements for connecting devices as separate entities with screw-type clamping units	715.521.107
BS EN 60998-2-2:2004	Particular requirements for connecting devices as separate entities with screwless-type clamping units	715.521.107
BS EN 61000	Electromagnetic compatibility (EMC) BS EN 61000 is a multiple part standard	515.2 Appx 4 sec 5.5.1
BS EN 61000-2	Electromagnetic compatibility (EMC). Environment	Appx 5 AM
BS EN 61000-4	Electromagnetic compatibility (EMC). Testing and measurement techniques	Appx 5 AM
BS EN 61000-4-5:2014 +A1:2017	Electromagnetic compatibility (EMC). Testing and measurement techniques - Surge immunity test.	443.6.1
BS EN 61000-6	Electromagnetic compatibility (EMC). Generic standards.	Table A444.1(i) & (ii)
BS EN 61008	Residual current operated circuit-breakers without integral overcurrent protection for household and similar uses (RCCBs)	531.3.4.1 531.3.4.2 531.3.6 536.4.1.4 note 2 536.4.2.4 536.4.3.2 Table 537.4 Table A53.1
BS EN 61008-1:2012 +A12:2017	Residual current operated circuit-breakers without integral overcurrent protection for household and similar uses (RCCBs) General rules	411.4.204 Table 41.5 534.4.7 note 1 711.410.3.4 721.415.1 722.531.3.101 740.410.3 740.411 note
BS EN 61009	Residual current operated circuit-breakers with integral overcurrent protection for household and similar uses (RCBOs)	531.3.4.1 531.3.4.2 531.3.6 533.1.1 536.4.1.4 note 2 Table 537.4 Table A53.1
BS EN 61009-1:2012 +A12:2016	Electrical accessories. Residual current operated circuit-breakers with integral overcurrent protection for household and similar uses (RCBOs) General rules	Table 41.3 411.4.204 Table 41.5 Table 41.6 432.4 433.1.201 433.1.204 434.5.2 534.4.7 note 1 Table 537.4 711.410.3.4 721.415.1 722.531.3.101 722.533.101 740.410.3 740.411 note Fig 3A4 Fig 3A5 Fig 3A6 Appx 4 sec 4
BS EN 61034-2:2005 +A1:2013	Measurement of smoke density of cables burning under defined conditions Test procedure and requirements	422.1 422.2.1(ii) 711.521.101(i)

Standard Number	Title	References
BS EN 61048:2006 +A1:2016	Auxiliaries for lamps. Capacitors for use in tubular fluorescent and other discharge lamp circuits. General and safety requirements	559.7
BS EN 61082-1:2015	Preparation of documents used in electrotechnology. Rules	514.9.2
BS EN 61095:2009	Specification for electromechanical contactors for household and similar purposes	536.4.2.2 536.4.3.1 Table 537.4 Table A53.1
BS EN 61140:2016	Protection against electric shock. Common aspects for installation and equipment	Part 2 Class I ... Class II ... Class III ... 410 412.2.4.1 note 1 414.1.1(iii) Appx 5 BC
BS EN 61184:2017	Bayonet lampholders	416.2.4 note 559.5.1.205
BS EN 61215	Crystalline silicon terrestrial photovoltaic (PV) modules. Design qualification and type approval Replaced by BS EN 61215-1:2016, BS EN 61215-1-1:2016, BS EN 61215:2017	712.511.101
BS EN 61215-1-2:2017	Terrestrial photovoltaic (PV) modules. Design qualification and type approval. Special requirements for testing of thin-film Cadmium Telluride (CdTe) based photovoltaic (PV) modules	712.511.101
BS EN 61215-1-3:2017	Terrestrial photovoltaic (PV) modules. Design qualification and type approval. Special requirements for testing of thin-film amorphous silicon based photovoltaic (PV) modules	712.511.101
BS EN 61215-1-4:2017	Terrestrial photovoltaic (PV) modules. Design qualification and type approval. Special requirements for testing of thin-film Cu(In,Ga)(S,Se)2 based photovoltaic (PV) modules	712.511.101
BS EN 61241-10:2004	Replaced by BS EN 60079-10-2:2015.	712.511.101
BS EN 61241-14:2004	Selection and installation Withdrawn in 2012 and replaced by BS EN 60079-14:2014.	422.3
BS EN 61347	Lamp controlgear	
BS EN 61347-1:2015	General and safety requirements. This remains current but has been superseded.	Table 55.3
BS EN 61347-2-2:2012	Particular requirements for DC or AC supplied electronic step-down convertors for filament lamps	715.414 715.414 note 2 740.55.5
BS EN 61347-2-13:2014 +A1:2017	Lamp controlgear. Particular requirements for DC or AC supplied electronic controlgear for LED modules	715.414 715.414 note 2
BS EN 61386	Conduit systems for cable management	412.2.4.1(ii)(b) 422.2.1(iii) A444.1.4 521.6 527.1.5 527.1.6 711.521.101(ii) 715.521.1(i) 717.52.2(i) 717.528.3.4 721.521.2 721.528.2.1 740.521.1
BS EN 61386-1:2008 +A1:2019	General requirements. Replaced by BS EN 61386-17:2008 but remains current.	422.3.4(i) 422.4.204
BS EN 61386-21:2004 +A11:2010	Particular requirements. Rigid conduit systems	522.6.204(ii) 705.522.16 705.522.16(i) 717.52.2 740.521.1(i)

Standard Number	Title	References
BS EN 61386-22:2004 +A11:2010	Particular requirements. Pliable conduit systems	717.52.2
BS EN 61386-23:2004 +A11:2010	Particular requirements. Flexible conduit systems	717.52.2 740.521.1 740.521.1 note 2
BS EN 61386-24:2010	Particular requirements. Conduit systems buried underground	522.8.10 note 708.521.7.2 note 2 730.521.101.3.2 note 2 740.521.1 note 1 Appx 4 Table 4B3
BS EN 61439 series	Low-voltage switchgear and controlgear assemblies	412.2.1.1 422.5 note 3 530.4.2 536.4.5 536.4.201 536.4.202 536.4.203 543.2.2(ii) 557.1 710.511.1 712.511.103 722.511.1
BS EN 61439-1	General rules	Part 2 LV switchge...
BS EN 61439-3:2012 (2019)	Low-voltage switchgear and controlgear assemblies. Distribution boards intended to be operated by ordinary persons (DBO)	421.1.201 536.4.201 536.4.201 note Appx 14
BS EN 61439-4:2013	Particular requirements for assemblies for construction sites (ACS)	704.511.1
BS EN 61439-6:2012	Particular requirements for busbar trunking systems (busways)	421.1.7 note 434.5.3 521.4 527.1.5 527.1.6 Appx 8 sec 1 Appx 8 sec 2 Appx 10 sec 2 note
BS EN 61534	Powertrack systems	421.1.7 note 422.2.1(vi) 422.3.4 434.5.3 521.4 527.1.5 527.1.6 Appx 8 sec 1 Appx 8 sec 2 Appx 10 sec 2 note
BS EN 61535:2009 +A1:2013	Installation couplers intended for permanent connection to fixed installations	Part 2: Prefab wiring system 521.201 526.2 note 1 553.2.1 559.5.1(x) 559.5.4(iii)
BS EN 61537:2007	Cable tray systems and cable ladder systems for cable management	422.2.1(v) 422.3.4 521.6 527.1.5 527.1.6 740.521.1
BS EN 61557	Electrical safety in low voltage distribution systems up to 1000 V AC and 1500 V DC. Equipment for testing, measuring or monitoring of protective measures. General requirements	643.1 651.3

Standard Number	Title	References
BS EN 61557-6:2007	Effectiveness of residual current devices (RCD) in TT, TN and IT systems.	643.7.1(a) 643.7.1(b) 643.8
BS EN 61557-8:2015	Insulation monitoring devices for IT systems	Part 2 Symbols (Ra) 538.1.1 557.3.4.3 710.411.6.3.1 712.421.101.1 712.538.101
BS EN 61557-9:2015	Equipment for insulation fault location in IT systems	411.6.4 538.2 557.5.3.1 710.411.6.3.3
BS EN 61557-12:2008	Electrical safety in low voltage distribution systems up to 1000 V AC and 1500 V DC Equipment for testing, measuring or monitoring of protective measures. Performance measuring and monitoring devices (PMD)	App 17 sec 17.9
BS EN 61558-1:2005 +A1:2009	Safety of power transformers, power supplies, reactors and similar products. General requirements and tests Replaced by BS EN IEC 61558-1:2019	411.8.4.1(i) 717.411.6.2(ii) 717.413
BS EN 61558-2-1:2007	Particular requirements and tests for separating transformers and power supplies incorporating separating transformers for general applications	444.4.9
BS EN 61558-2-4:2009	Particular requirements and tests for isolating transformers and power supply units incorporating isolating transformers	444.4.9 722.413.1.2 730.313.1.102 740.55.9
BS EN 61558-2-5:2010	Particular requirements and tests for transformer for shavers, power supply units for shavers and shaver supply units	553.1.5(ii) 701.512.2 701.512.3(ii) 701.512.3 721.413 Appx 6 Init. Insp. sched. item 12.3 Appx 6 Cond. Report item 6.3
BS EN 61558-2-6:2009	Particular requirements and test for safety isolating transformers and power supply units incorporating safety isolating transformers	414.3(i) 444.4.9 Table 55.3 715.414 A721.313.4(iii) 740.55.5
BS EN 61558-2-8:2010	Particular requirements and tests for transformers and power supply units for bells and chimes	414.3(i)
BS EN 61558-2-15:2012	Particular requirements and tests for isolating transformers for the supply of medical locations	444.4.9 710.512.1.1
BS EN 61558-2-23:2010	Particular requirements and tests for transformers for construction sites.	411.8.4.1(i)
BS EN 61643	Low-voltage surge protective devices	534.1
BS EN 61643-11:2012 +A11:2018	Surge protective devices connected to low-voltage power systems. Requirements and test methods	534.4.4.1 534.4.4.6 note 1 & 2 Appx 16 Table 16A note 1, 2 & 3
BS EN 61643-21:2001 +A2:2013	Surge protective devices connected to telecommunications and signalling networks. Performance requirements and testing methods	
BS EN 61643-31	Low-voltage surge protective devices. Requirements and test methods for SPDs for photovoltaic installations	712.534.102
BS EN 61851-1	Electric vehicle conductive charging system. General requirements Withdrawn and replaced by BS EN 62752:2016, BS EN IEC 61851-1:2019	Part 2 EV Mode 1 722.511.101

Standard Number	Title	References
BS EN 61936-1:2010 +A1:2014	Power installations exceeding 1 kV AC - Part 1: Common rules	442.2.3
BS EN 61995-1:2008 +A1:2017	Devices for the connection of luminaires for household and similar purposes. General requirements	Table 537.4 559.5.1(ix) 559.5.4(ii)
BS EN 62020:1999	Electrical accessories. Residual current monitors for household and similar uses (RCMs)	532.3 538.4.1
BS EN 62040-1:2008 +A1:2013	Uninterruptible power systems (UPS). General and safety for UPS	560.6.12(iv)
BS EN 62040-3:2011	Uninterruptible power systems (UPS). Method of specifying the performance and test requirements	560.6.12(iv)
BS EN 62109-1:2010	Safety of power converters for use in photovoltaic power systems. General requirements	712.511.102
BS EN 62109-2:2011	Safety of power converters for use in photovoltaic power systems. Particular requirements for inverters	712.511.102 712.538.101
BS EN 62196	Plugs, socket-outlets, vehicle connectors and vehicle inlets. Conductive charging of electric vehicles.	722.531.3.101 note 2 722.551.7.2
BS EN 62196-2:2017	Plugs, socket-outlets, vehicle connectors and vehicle inlets. Conductive charging of electric vehicles. Dimensional compatibility and interchangeability requirements for AC pin and contact-tube accessories	722.55.101.0.201.1(iv) 722.55.101.0.201.1(v) 722.55.101.0.201.1(vi)
BS EN 62208:2011	Empty enclosures for low-voltage switchgear and controlgear assemblies. General requirements	530.4.2
BS EN 62262:2002	Degrees of protection provided by enclosures for electrical equipment against external mechanical impacts (IK code)	708.512.2.1.3 709.512.2.1.4(iii) 709.512.2.1.4 712.512.102 722.512.2.203
BS EN 62305	Protection against lightning	110.2(ix) 131.6.2 note 411.3.1.2 421.1.1 note 2 444.5.2(iv) 528.1 note 1 534.1 534.4.4.4.2 (ii) 541.3 Appx 16 Table 16A
BS EN 62305-1:2011	General principles	542.2.3 note Appx 5 AQ3
BS EN 62305-2:2012	Risk management	443.1.1 note 1 & 8 534.4.4.4.2(i) & (ii)
BS EN 62305-3:2011	Physical damage to structures and life hazard	712.534.101 712.534.101 note 1 712.534.102.1
BS EN 62305-4:2011	Electrical and electronic systems within structures	443.1.1 534.1 534.4.1.1 note 534.4.1.6 note 712.534.101 note 2 Fig 16A5 note 4 Appx 16 Table 16A
BS EN 62353:2014	Medical electrical equipment. Recurrent test and after repair of medical electrical equipment	710.6 note
BS EN 62368-1:2014 +A1:2017	Audio/video, information and communication technology equipment. Safety requirements	444.4.9
BS EN 62395-1:2013	Electrical resistance trace heating systems for industrial and commercial applications. General and testing requirements	
BS EN 62395-2:2013	Electrical resistance trace heating systems for industrial and commercial applications	753.1

Standard Number	Title	References
BS EN 62423:2012	Type F and type B residual current operated circuit-breakers with and without integral overcurrent protection for household and similar uses	411.4.4 note 2 531.3.4.1 531.3.4.2 531.3.6 533.1.1 Table A53.1 710.531.3 712.531.3.5.1 721.415.1 722.531.3.101
BS EN 62446-1:2016 +A1:2018	Photovoltaic (PV) systems. Requirements for testing, documentation and maintenance. Grid connected systems. Documentation, commissioning tests and inspection	712.6.101
BS EN 62477-1:2012 +A1:2017	Safety requirements for power electronic converter systems and equipment. General	419.2
BS EN 62606:2013 +A1:2017	General requirements for arc fault detection devices	421.1.7 532.6 Table 537.4
BS EN 62852:2015 +A1:2020	Connectors for DC-application in photovoltaic systems. Safety requirements and tests	712.526.101
BS EN 81346-1:2009	Industrial systems, installations and equipment and industrial product. Structuring principles and reference designations. Basic rules	514.9.2
BS EN IEC 606641: 2020	Insulation coordination for equipment within low-voltage systems Principles, requirements and tests	Table 44.2 note 1 443.6.1 534.1 Table 712.1
BS EN IEC 60721	Classification of environmental conditions	
BS EN IEC 60721-3-3: 2019	Classification of groups of environmental parameters and their severities. Stationary use at weather protected locations	Appx 5 AA, AB, AD ...
BS EN IEC 60721-3-4: 2019	Classification of groups of environmental parameters and their severities. Stationary use at non-weather protected locations	Appx 5 AA, AB, AD ...
BS EN IEC 61439-7: 2020	Low-voltage switchgear and controlgear assemblies. Assemblies for specific applications such as marinas, camping sites, market squares, electric vehicle charging stations	708.55.1.7
BS EN IEC 61730-1: 2018	Photovoltaic (PV) module safety qualification. Requirements for construction	712.431.102 712.511.101
BS EN IEC/IEEE 82079-1: 2020	Preparation of instructions for use. Structuring, content and presentation. General principles and detailed requirements	514.9.2 Appx 11
BS EN ISO 7010:2020 +A1:2020	Graphical symbols. Safety colours and safety signs. Registered safety signs	514.9.2 A721.55.3.7 Appx 11
BS EN ISO 9241-300: 2008 (2018)	Ergonomics of human-system interaction. Introduction to electronic visual display requirements	Appx 11
BS EN ISO 10133:2017	Small craft. Electrical systems. Extra-low-voltage d.c. installations.	110.2(iv)
BS EN ISO 11925-2: 2010	Reaction to fire tests. Ignitability of products subjected to direct impingement of flame. Single-flame source test	421.1.3
BS EN ISO 12696:2016	Cathodic protection of steel in concrete	110.2(xiii)
BS EN ISO 13174:2012 (2018)	Cathodic protection of harbour installations	110.2(xiii)
BS EN ISO 13297:2018	Small craft. Electrical systems. Alternating current installations	110.2(iv)
BS ISO 11446:2012	Road vehicles. Connectors for the electrical connection of towing and towed vehicles. 13-pole connectors for vehicles with 12 V nominal supply voltage not intended to cross water fords	A721.55.2.6(i) & (iii) Table A721

NOTE on graphical symbols IEC 60617 is the central standards database for electrotechnical symbols. They used to be covered by BS 6217, BS EN 60417 and BS EN 60617. In 2002 IEC launched an 'on-line' database format for the symbol library, available on subscription from the IEC website. Following this decision, in 2004 CENELEC decided to cease publication of EN 60417 in 'paper' form, to withdraw the then-existing standards and formally to adopt the IEC database without any changes for use in Europe. Consequently, the British Standard versions have now also been withdrawn.

Other standards to which reference is made in BS 7671		
BS IEC 60287	Electric cables. Calculation of the current rating (Some parts of the BS 7769 series are now numbered BS IEC 60287 series, eventually all parts will be renumbered.)	523.3 Appx 4 sec 1 sec 2.1 sec 2.2 Table 4B3 note 2 Table 4C2 note 1 & 2 Table 4C3 note 1 & 2 Appx 10 sec 1 note, sec 2 para 8
BS IEC 60287-1-1:2006 +A1:2014	Cable rating equations (100% load factor) and calculation of losses (general)	Appx 4 sec 5.6
BS IEC 60898-3:2019	Electrical accessories. Circuit-breakers for overcurrent protection for household and similar installations. Circuit-breakers for DC operation	712.432.103 712.533.101 A721.533.1.6
BS IEC 61000-5-2:1997	Electromagnetic compatibility (EMC). Installation and mitigation guidelines. Earthing and cabling	444.1(vi)
BS IEC 62955:2018	Residual direct current detecting device (RDC-DD) to be used for mode 3 charging of electric vehicles	722.531.3.101(ii)
BS ISO 3864-1:2011	Graphical symbols. Safety colours and safety signs. Design principles for safety signs and safety markings	514.9.2
BS ISO 7001:2007 +A4:2017	Graphical symbols. Public information symbols	Appx 11 sec 2

Standard Number	Title	References
HD 472 S1+A1	BS 7697:1993 (2004) - Nominal voltages for low voltage public electricity supply systems	Appx 2 sec 15
HD 60364 series	Low-voltage electrical installations	Notes on the plan ...
HD 60364-1:2008 /A11:2017	Fundamental principles, assessment of general characteristics..., definitions	Preface
HD 60364-4-41:2017 +A12:2019	Protection for safety. Protection against electric shock	Preface
HD 60364-4-42:2011 /A11:2021	Protection for safety - Protection against thermal effects	Preface Appx 5 CA2 Appx 5 CB2
HD 60364-4-43:2010	Protection against overcurrent	Preface
HD 60364-4-442:2012	Protection for safety. Protection of low-voltage installations against temporary overvoltages due to earth faults in the high-voltage system and due to faults in the low voltage system	Preface
HD 60364-4-443:2016	Protection against transient overvoltages of atmospheric origin or due to switching	Preface
HD 60364-4-444:2010 (2012)	Measures against electromagnetic disturbances	Preface
HD 60364-5-51:2009 /A12:2017	Selection and erection of electrical equipment – Common rules	Preface Appx 5 Appx 5 BE2
HD 60364-5-52:2011 /A11:2017	Selection and erection of electrical equipment – Wiring systems	Preface Chap 52 Appx 5 CB2, CB3 & CB4
HD 60364-5-53:2015	Protection, isolation, switching, control and monitoring	Preface
HD 60364-5-534:2016	Selection and erection of electrical equipment - Devices for protection against transient overvoltages	Preface
HD 60364-5-54:2011 /A11:2017	Earthing arrangements and protective conductors	Preface

HD 60364-5-551:2010 /A11:2016	Low voltage generating sets	Preface
HD 60364-5-557:2013 /A11:2016	Auxiliary circuits	Preface
HD 60364-5-559:2012 /A11:2017	Selection and erection of electrical equipment - Luminaires and lighting installations	Preface
HD 60364-6:2016 /A12:2017	Initial Verification	Preface
HD 60364-7-701:2007 /A12:2017	Locations containing a bath or shower	Preface
HD 60364-7-702:2010	Swimming pools and other basins	Preface
HD 60364-7-703:2005	Rooms and cabins containing sauna heaters	Preface
HD 60364-7-704:2018	Construction and demolition site installations	Preface
HD 60364-7-705:2007 /A12:2017	Agricultural and horticultural premises	Preface
HD 60364-7-706:2007 /A1:2020	Conducting locations with restricted movement	Preface
HD 60364-7-708:2017	Caravan parks, camping parks and similar locations	Preface
HD 60364-7-709:2009 /A12:2019	Marinas and similar locations	Preface
HD 60364-7-710:2012 /AC:2013	Medical locations	Preface
HD 60364-7-711:2019	Exhibitions, shows and stands	Preface
HD 60364-7-712:2016	Photovoltaic (PV) systems	Preface
HD 60364-7-714:2012	External lighting installations	Preface
HD 60364-7-715:2012 /A11:2017	Extra-low voltage lighting installations	Preface
HD 60364-7-717:2010 /AC:2014	Mobile or transportable units	Preface
HD 60364-7-721:2019	Electrical installations in caravans and motor caravans	Preface
HD 60364-7-722:2018	Supplies for electric vehicles	Preface
HD 60364-7-729:2009 /A11:2017	Operating or maintenance gangways	Preface
HD 60364-7-730:2015	Onshore units of electrical shore connections for inland navigation vessels	Preface
HD 60364-7-740:2006 /A11:2017	Temporary electrical installations for structures, amusement devices at fairgrounds, amusement parks and circuses	Preface
HD 60364-7-753:2014 /AC:2014	Heating cables and embedded heating systems	Preface
HD 60364-8-1:2019 /AC:2019-06	Functional aspects - energy efficiency	Preface
HD 60364-8-2:2018 /A12:2021	Prosumer's low-voltage electrical installations	Preface
IEC 60038:2009	IEC standard voltages	TABLE 443.2
IEC 60050	International Electrotechnical Vocabulary	Part 2
IEC 60331 IEC 60331-1:2018 IEC 60331-2:2018 IEC 60331-3:2018	Tests for electric cables under fire conditions - Circuit integrity. Test method for fire with shock at a temperature of at least 830°C for cables of rated voltage up to and including 0,6/1,0 kV, and with an overall diameter exceeding 20 mm and with an overall diameter not exceeding 20 mm tested in a metal enclosure	 560.8.1(ii) 560.8.1(ii) 560.8.1(ii)
IEC 60364-4-44:2007 /AMD2:2018	Protection against voltage disturbances and electromagnetic disturbances	Preface
IEC 60364-5-52:2009 (2011)	Selection and erection of wiring systems	Preface
IEC 60417	Graphical symbols for use on equipment	Appx 11 sec 2

IEC 60502-1:2021	Power cables with extruded insulation and their accessories for rated voltages from 1 kV (U_m = 1,2 kV) up to 30 kV (U_m = 36 kV) - Part 1: Cables for rated voltages of 1 kV (U_m = 1,2 kV) and 3 kV (U_m = 3,6 kV)	Table 52.1 Appx 4 sec 1
IEC 60617	Central standards database of graphical symbols for use in electrotechnical diagram	514.9.1 Table 537.4 note 1 Appx 11 sec 2
IEC 60800:2021	Heating cables with a rated voltage of 300/500 V for comfort heating and prevention of ice formation	753.511
IEC 61140:2016	Protection against electric shock - Common aspects for installation and equipment	414.1.1
IEC TS 61201:2007	Extra-low voltage (ELV). Limit values. Also known as PD 6536.	414.2 note 3
IEC TR 60479-5:2007 /COR1:2013	Effects of current on human beings and livestock –. Part 5: Touch voltage threshold values for physiological effects	A722.4
ISO 8820	Road vehicles. Fuse-links	A721.533.1.6
PD 6536:1992	Extra-low voltage (ELV). Limit values. Also known as IEC 61201.	414.2 note 3
PD CLC/TR 50480:2011	Technical Report - Determination of cross-sectional area of conductors and selection of protective devices	533.3 note Appx 4 sec 2.5
PD CLC/TS 51643-32: 2020	Low-voltage surge protective devices. Surge protective devices connected to the DC side of photovoltaic installations. Selection and application principles	712.534.102.6
PD IEC/TR 62350:2006	Guidance for the correct use of residual current-operated protective devices (RCDs) for household and similar use	531.3.3 note 2
DD CLC/TS 61643-12: 2009	Low-voltage surge protective devices. Surge protective devices connected to low-voltage power distribution systems. Selection and application principles	534.1 534.1 note 2 534.4.1.6 534.4.2 534.4.4.1 note 534.4.4.2 note 1 & 3 534.4.4.4 Fig 534.9 note Fig 16A5 note 1
PD CLC/TS 61643-22: 2016	Low-voltage surge protective devices. Surge protective devices connected to telecommunications and signalling networks. Selection and application principles	443.1.1

APPENDIX 2 (Informative)

STATUTORY REGULATIONS AND ASSOCIATED MEMORANDA

1. In the United Kingdom the following classes of electrical installations are required to comply with the Statutory Regulations indicated below. The regulations listed represent the principal legal requirements. Information concerning these regulations may be obtained from the appropriate authority also indicated below.

Provisions relating to electrical installations are also to be found in other legislation relating to particular activities.

(i)	Distributors' installations generally, subject to certain exemptions	Electricity Safety, Quality and Continuity Regulations 2002 as amended	Health and Safety Executive
			Secretary of State for Scotland
			Secretary of State for Business, Energy and Industrial Strategy
		Electricity Safety, Quality and Continuity Regulations (Northern Ireland) 2012	Department of Enterprise, Trade and Investment
(ii)	Buildings generally subject to certain exemptions	The Building Regulations 2010 (as amended) (for England and Wales)	Department for Levelling Up, Housing and Communities
		The Building (Scotland) Regulations 2004 (as amended)	The Scottish Government
		Building Regulations (Northern Ireland) 2012	Department of Finance and Personnel
(iii)	Work activity. Places of work. Non-domestic installations	The Electricity at Work Regulations 1989 as amended	Health and Safety Executive
		The Electricity at Work Regulations (Northern Ireland) 1991	Health and Safety Executive for Northern Ireland
(iv)	Cinematograph installations	Cinematograph (Safety) Regulations 1955, as amended made under the Cinematograph Act, 1909, and/or Cinematograph Act, 1952	The Secretary of State for the Home Office, and The Scottish Government
(v)	Machinery	The Supply of Machinery (Safety) Regulations 2008	Department for Business, Energy and Industrial Strategy
(vi)	Theatres and other places licensed for public entertainment, music, dancing, etc.	Conditions of licence under: (a) in England and Wales, The Local Government Licensing Act 2003	(a) Department for Culture, Media and Sport
		(b) in Scotland, The Civic Government (Scotland) Act 1982	(b) The Scottish Government
(vii)	High voltage luminous tube signs	As (a) and (b) above	As (a) and (b) above

2. Failure to comply in a consumer's installation in the United Kingdom with the requirements of Chapter 13 of BS 7671:2018 +A2:2022 places the distributor in the position of not being compelled to commence or, in certain circumstances, to continue to give, a supply of energy to that installation.

Under Regulation 26 of the Electricity Safety, Quality and Continuity Regulations (ESQCR), any dispute which may arise between a consumer and the distributor having reference to the consumer's installation shall be determined by a person nominated by the Secretary of State (or the Scottish Government in relation to disputes arising in Scotland) on the application of the consumer or consumer's authorized agent or the distributor.

Under Regulation 26 of the Electricity Safety, Quality and Continuity Regulations (Northern Ireland) 2012 (ESQCR (NI) 2012), any dispute which may arise between a consumer and the distributor having reference to the consumer's installation shall be determined by a person nominated by the Department of Enterprise, Trade and Investment on the application of the consumer or consumer's authorized agent or the distributor.

Regulation 28 of the ESQCR 2002 (or ESQCR(NI)2012) requires distributors to provide the following information to relevant persons free of charge:

The maximum prospective short-circuit current at the supply terminals

The maximum earth loop impedance of the earth fault path outside the installation (Z_e)

The type and rating of the distributor's protective device or devices nearest to the supply terminals

The type of earthing system applicable to the connection

The number of phases of the supply

The frequency of the supply and the extent of the permitted variations

The voltage of the supply and the extent of the permitted variations.

3. Where it is intended to use protective multiple earthing the distributor and the consumer must comply with the ESQCR 2002 (or ESQCR(NI)2012).

4. For further guidance on the application of the Electricity at Work Regulations (including the Northern Ireland version), reference may be made to the following publication:

(i) Electricity at Work Regulations 1989 Guidance on Regulations (HSR25).

5. For installations in potentially explosive atmospheres reference should be made to:

(i) the Electricity at Work Regulations 1989

(ii) the Dangerous Substances and Explosive Atmospheres Regulations (DSEAR) 2002

(iii) the Petroleum (Consolidation) Regulations 2014

(iv) the Equipment and Protective Systems Intended for Use in Potentially Explosive Atmospheres Regulations 2016

(v) relevant British or Harmonized Standards.

Under the Petroleum (Consolidation) Regulations 2014 Petroleum Enforcing Authorities are empowered to grant licences in respect of premises where petroleum spirit is stored and the authorities may attach such conditions as they think fit, the requirements may vary from one local authority to another. Guidance may be obtained from the Energy Institute (APEA/IP) publication Design, Construction, Modification, Maintenance and Decommissioning of Filling Stations.

6. For installations on caravan parks, the requirements of the licensing authority should be ascertained. Model Standards were issued by the Department of the Environment in 1977 under the Caravan Sites and Control of Development Act 1960, as amended, as guidance for local authorities.

7. For installations in theatres and other places of entertainment, the requirements of the licensing authority should be ascertained.

8. The Electrical Equipment (Safety) Regulations 2016, administered by the Department for Business, Energy & Industrial Strategy (BEIS), contain requirements for safety of equipment designed or suitable for general use. Information on the application of the Regulations is given in guidance issued by BEIS.

9. The Plugs and Sockets etc. (Safety) Regulations 1994 made under the Consumer Safety Act 1978, administered by the Department for Business, Energy and Industrial Strategy, contains requirements for the safety of plugs, sockets, adaptors and fuse links etc. designed for use at a voltage of not less than 200 volts.

10. The Health and Safety (Safety Signs and Signals) Regulations 1996 require employers to provide safety signs. Guidance from the Health and Safety Executive L64, Safety Signs and Signals, specifies signs including emergency escape, first aid and fire safety signs.

11. The Management of Health and Safety at Work Regulations 1999 require employers and self-employed persons to assess risks to workers and others who may be affected by their work or business. This is intended to enable them to identify measures they need to take to comply with the law. Guidance from the Health and Safety Executive INDG163 gives advice on these regulations. For the purposes of BS 7671, a risk assessment should involve an appropriate electrically skilled person.

12. The Provision and Use of Work Equipment Regulations 1998 require employers to ensure that all work equipment is suitable for the purpose for which it is used, is properly maintained and that appropriate training is given. The Health and Safety Executive has published an Approved Code of Practice L22, Safe use of work equipment, which gives advice that has special legal status.

13. The Electromagnetic Compatibility Regulations 2016 provide requirements for electrical and electronic products for electromagnetic compatibility.

14. Other Regulations, as amended, relevant to electrical installation include:

The Personal Protective Equipment at Work Regulations 2002 (HSE Publication L25)

The Personal Protective Equipment (Enforcement) Regulations 2018

The Workplace (Health, Safety and Welfare) Regulations 1992 (HSE Publication L24)

The Manual Handling Operations Regulations 1992 (HSE Publication L23)

The Work at Height Regulations 2005

The Construction (Design and Management) Regulations 2015 (HSE Publication L153).

15. In November 1988 the European electrical standards body CENELEC agreed on harmonization of low voltage electricity supplies within Europe (CENELEC document HD 472 S1), implemented by BS 7697 Nominal voltages for low voltage public electricity supply systems. The measure is intended to harmonize mains electricity supplies at 230 V within Europe. CENELEC proposed three stages of harmonization. Two stages of harmonization have taken place, these being shown below.

Effective date	Nominal voltage	Permitted tolerance	Permitted voltage range
Pre-1995	240 V	+6 % / −6 %	225.6 – 254.4 V
1 January 1995	230 V	+10 % / −6 %	216.2 – 253.0 V

16. Distribution Code [of licensed Distribution Operators of Great Britain]

The Electricity Act 1989 (as amended by the Utilities Act 2000 and the Energy Act 2004) requires the DNOs under the terms of their licences to produce and comply with the Distribution Code.

The Distribution Code has to be complied with by the DNO and by potential and existing Generators, Suppliers and Customers connected to or seeking connection to the DNO's distribution system.

17. The Construction Products Regulation, administered by the Department for Levelling Up, Housing and Communities, lays down requirements relating to obligations of manufacturers in respect of testing, certifying and placing upon the market, products intended for permanent installation in buildings and construction works. The electrical products covered are 'Power, control and communication cables' in respect of their reaction to fire performance. The Regulations empower the UK to determine levels of performance required for particular installations. In the UK, DLUHC has not exercised these powers in respect of reaction to fire for cables. Guidance may be given elsewhere, including in European and British Standards. For example, guidance for telecommunication cabling is in BS 6701 *Telecommunications equipment and telecommunications cabling - Specification for installation, operation and maintenance.*

The generic reaction to fire requirements for all cables are given in BS EN 50575 *Power, control and communication cables - Cables for general applications in construction works subject to reaction to fire requirements.* (All other requirements – electrical, mechanical, constructional, transmission – remain in existing product standards.)

APPENDIX 3 (Informative)

TIME/CURRENT CHARACTERISTICS OF OVERCURRENT PROTECTIVE DEVICES

This appendix gives the time/current characteristics of the following overcurrent protective devices:

FUSES:

Figure 3A1	Fuses to BS 88-3 – fuse system C
Figures 3A2(a) & 3A2(b)	Semi-enclosed fuses to BS 3036
Figures 3A3(a) & 3A3(b)	Fuses to BS 88-2 – fuse systems E (bolted) and G (clip-in)
Figure 3A3(c)	Fuses to BS 88-2 – fuse system E (bolted)

CIRCUIT-BREAKERS:

Figure 3A4	Type B to BS EN 60898 and the overcurrent characteristics of RCBOs to BS EN 61009-1
Figure 3A5	Type C to BS EN 60898 and the overcurrent characteristics of RCBOs to BS EN 61009-1
Figure 3A6	Type D to BS EN 60898 and the overcurrent characteristics of RCBOs to BS EN 61009-1

In all of these cases time/current characteristics are based on the slowest operating times for compliance with the relevant requirements of BS 7671 and have been used as the basis for determining the limiting values of earth fault loop impedance prescribed in Chapter 41.

Maximum earth fault loop impedance

Regulation 411.3.2 specifies maximum disconnection times for circuits.

The maximum values of earth fault loop impedance to achieve the disconnection time vary with the different types of protective devices and also between manufacturers. Wherever possible designers should use the manufacturer's specific data.

Alternatively, the impedance values given in Tables 41.3 and 41.6 can be used for BS EN 60898 circuit-breakers. These values are far more onerous and in some cases may be difficult to achieve without installing larger sized cpcs.

Regulations 411.4.4 to 411.4.204 state that the maximum earth fault loop impedance for a protective device is:

$$Z_s = \frac{U_0 \times C_{min}}{I_a}$$

where:

U_0 is the nominal AC rms line voltage to Earth.

I_a is the current in amperes (A) causing operation of the protective device within the specified time.

C_{min} is the minimum voltage factor to take account of voltage variations depending on time and place, changing of transformer taps and other considerations.

NOTE: For a low voltage supply given in accordance with the Electricity Safety, Quality and Continuity Regulations (ESQCR) as amended, C_{min} is given the value 0.95.

The tabulated values are applicable for supplies from distribution network operators. For other supplies the designer will need to determine the nominal voltage and calculate Z_s accordingly.

When impedance measurements are made at ambient temperature the procedure hereinafter described may be followed to take into account the increase of resistance of the conductors with the increase of temperature due to load current. For TN and TT systems in which protection by automatic disconnection is provided by overcurrent protective devices, such measurements are made to verify the maximum impedance value stated by the device manufacturer is not exceeded. In the absence of manufacturers' data, the values stated in tables may be used.

The requirements of Regulation 411.4.4 or 411.5.4, as appropriate, are considered to be met when the measured value of earth fault loop impedance satisfies the following equation:

$$Z_s(m) \leq 0.8 \frac{U_0 \times C_{min}}{I_a}$$

where:

$Z_s(m)$	is the measured impedance of the earth fault current loop up to the most distant point of the relevant circuit from the origin of the installation (Ω)
0.8	is a factor to take into account the increase of resistance of the conductors with the increase of temperature due to load current
U_0	is the nominal AC rms line voltage to Earth (V)
I_a	is the current in amperes causing operation of the protective device within the time stated in Table 41.1 or within 5 s according to the conditions stated in Regulation 411.3.2.3
C_{min}	is the minimum voltage factor to take account of voltage variations depending on time and place, changing of transformer taps and other considerations.

NOTE: For a low voltage supply given in accordance with the Electricity Safety, Quality and Continuity Regulations (ESQCR) as amended, C_{min} is given the value 0.95.

NOTE: The above is one method of correcting for the effects of temperature difference; other methods are not precluded.

TABLE 3A – *Deleted by BS 7671:2018+A2:2022*

Fig 3A1 – Fuses to BS 88-3 fuse system C

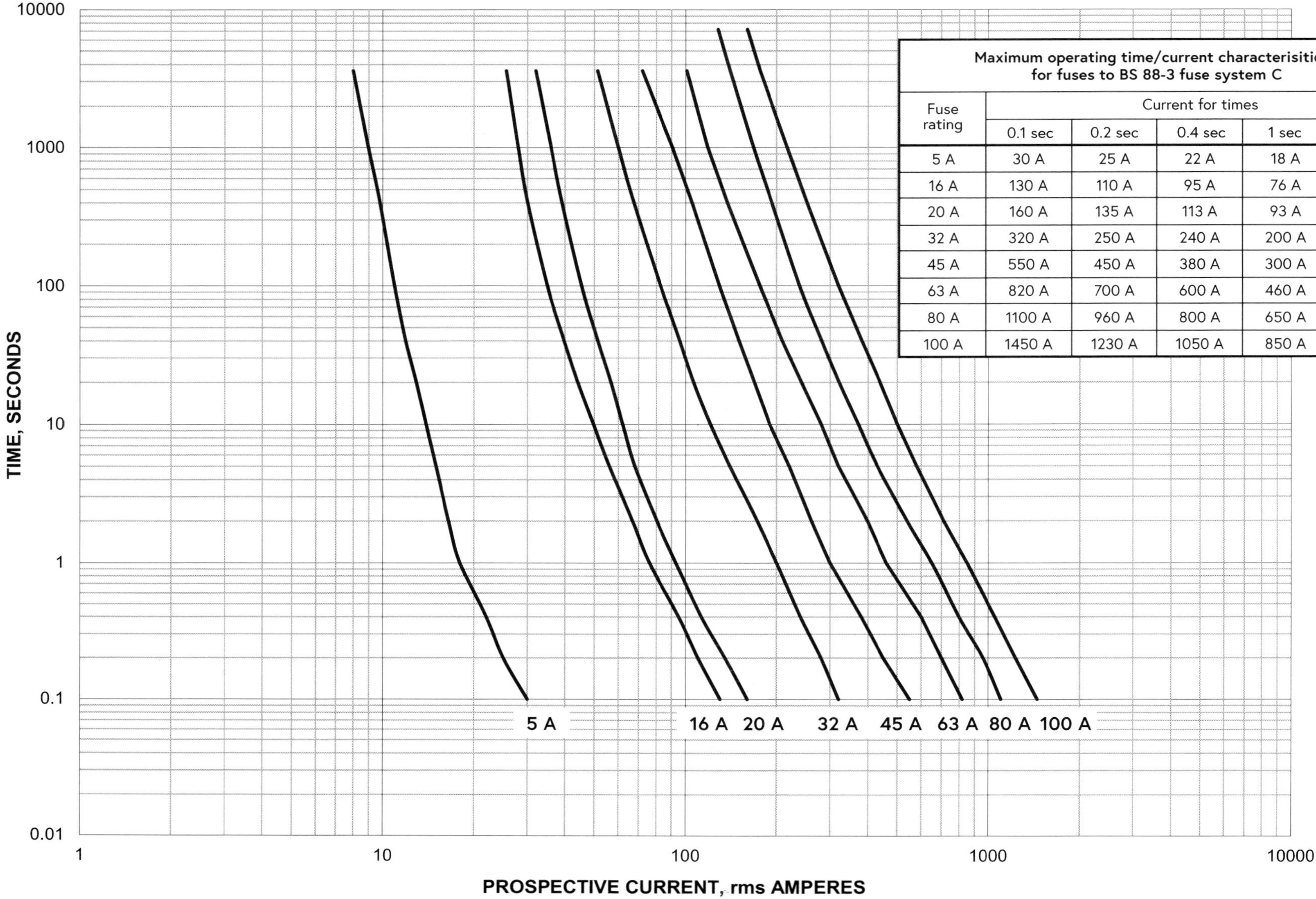

Maximum operating time/current characterisitics for fuses to BS 88-3 fuse system C

Fuse rating	Current for times				
	0.1 sec	0.2 sec	0.4 sec	1 sec	5 secs
5 A	30 A	25 A	22 A	18 A	15 A
16 A	130 A	110 A	95 A	76 A	56 A
20 A	160 A	135 A	113 A	93 A	68 A
32 A	320 A	250 A	240 A	200 A	140 A
45 A	550 A	450 A	380 A	300 A	220 A
63 A	820 A	700 A	600 A	460 A	320 A
80 A	1100 A	960 A	800 A	650 A	430 A
100 A	1450 A	1230 A	1050 A	850 A	580 A

Fig 3A2(a) – Semi-enclosed fuses to BS 3036

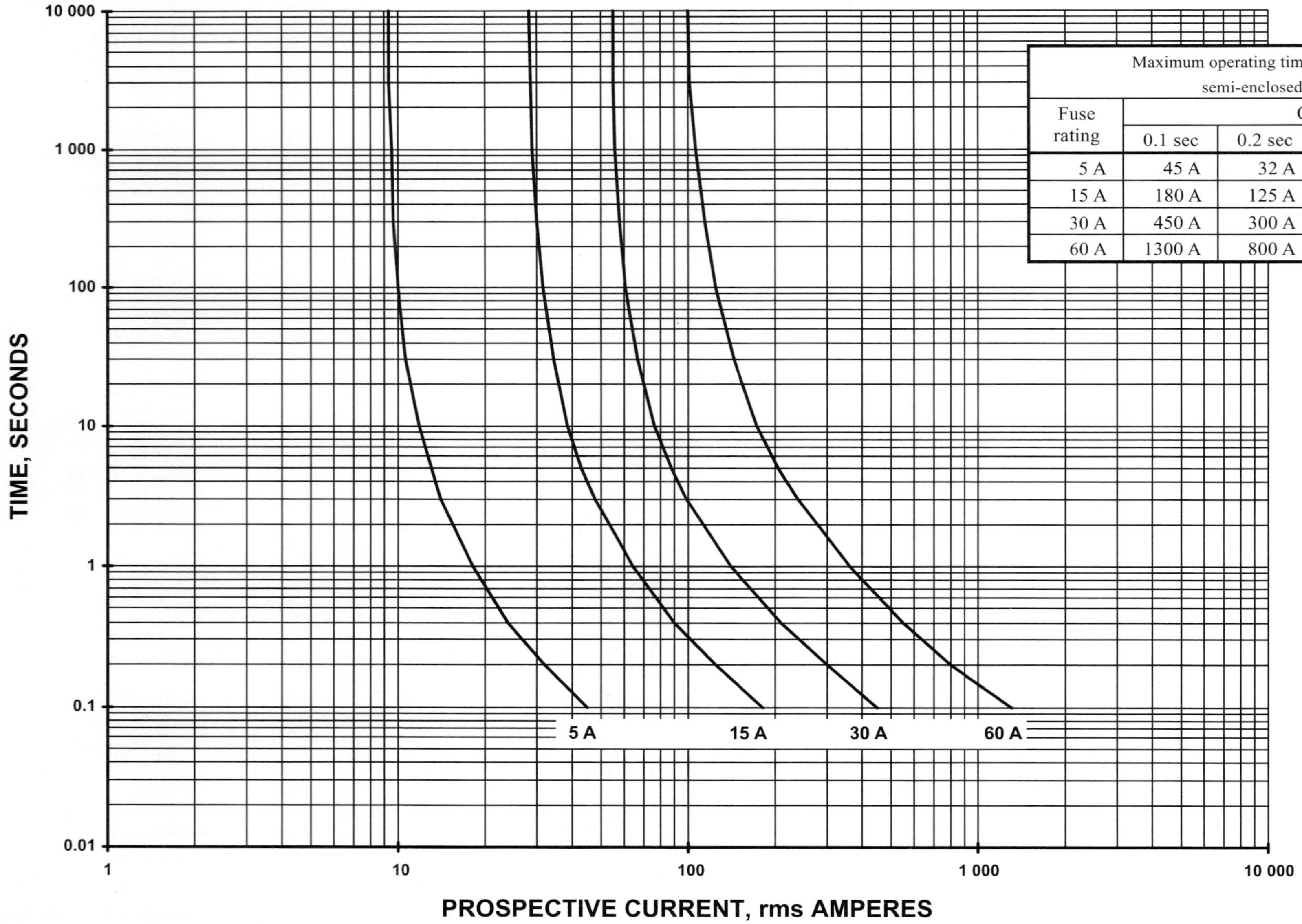

Maximum operating time/current characteristics for semi-enclosed fuses to BS 3036					
Fuse rating	Current for time				
	0.1 sec	0.2 sec	0.4 sec	1 sec	5 secs
5 A	45 A	32 A	24 A	18 A	13 A
15 A	180 A	125 A	90 A	65 A	43 A
30 A	450 A	300 A	210 A	140 A	87 A
60 A	1300 A	800 A	550 A	360 A	205 A

Fig 3A2(b) – Semi-enclosed fuses to BS 3036

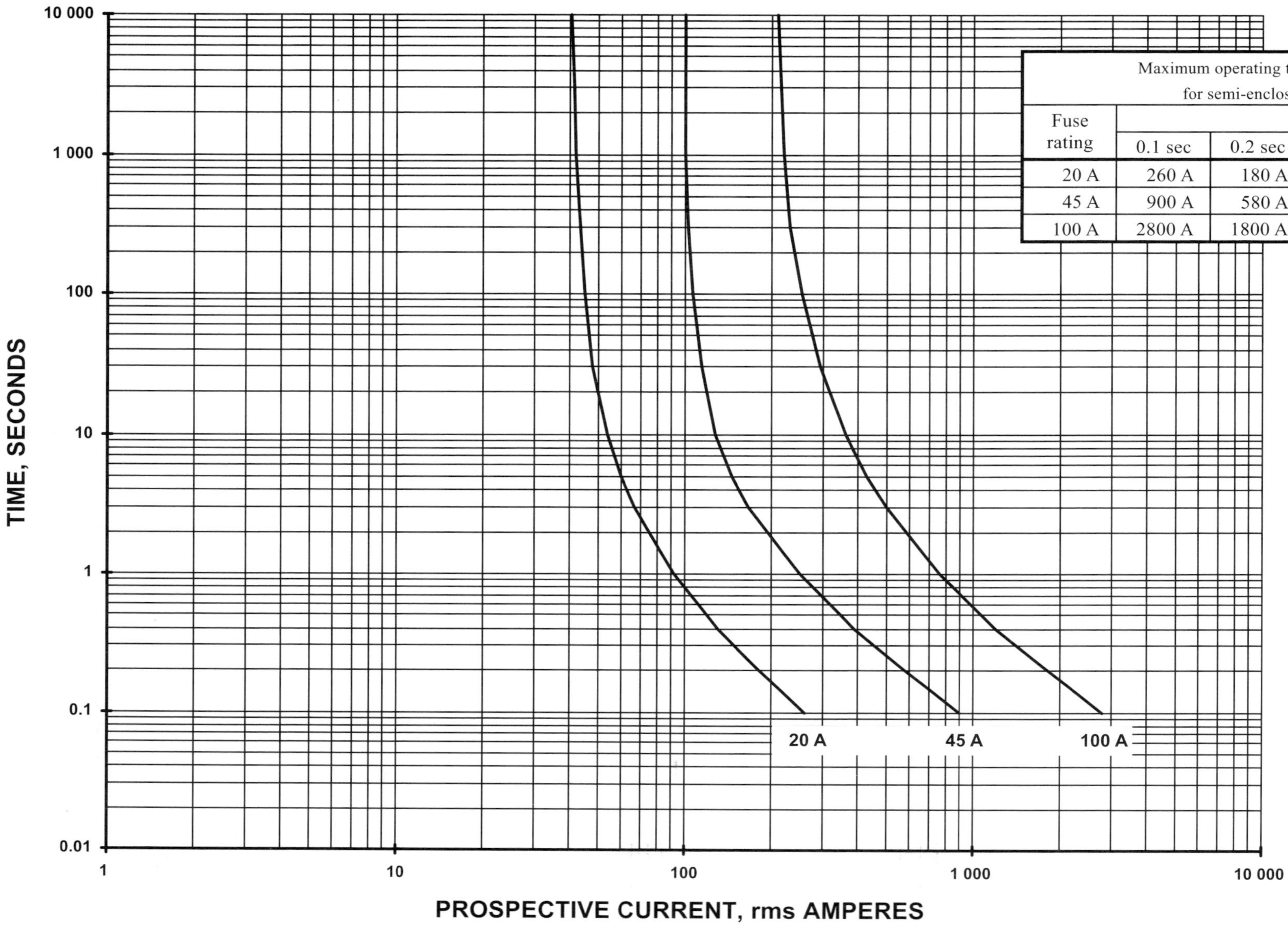

Maximum operating time/current characteristics for semi-enclosed fuses to BS 3036					
Fuse rating	Current for time				
	0.1 sec	0.2 sec	0.4 sec	1 sec	5 secs
20 A	260 A	180 A	130 A	92 A	60 A
45 A	900 A	580 A	390 A	250 A	145 A
100 A	2800 A	1800 A	1200 A	770 A	430 A

Fig 3A3(a) – Fuses to BS 88-2 fuse systems E and G

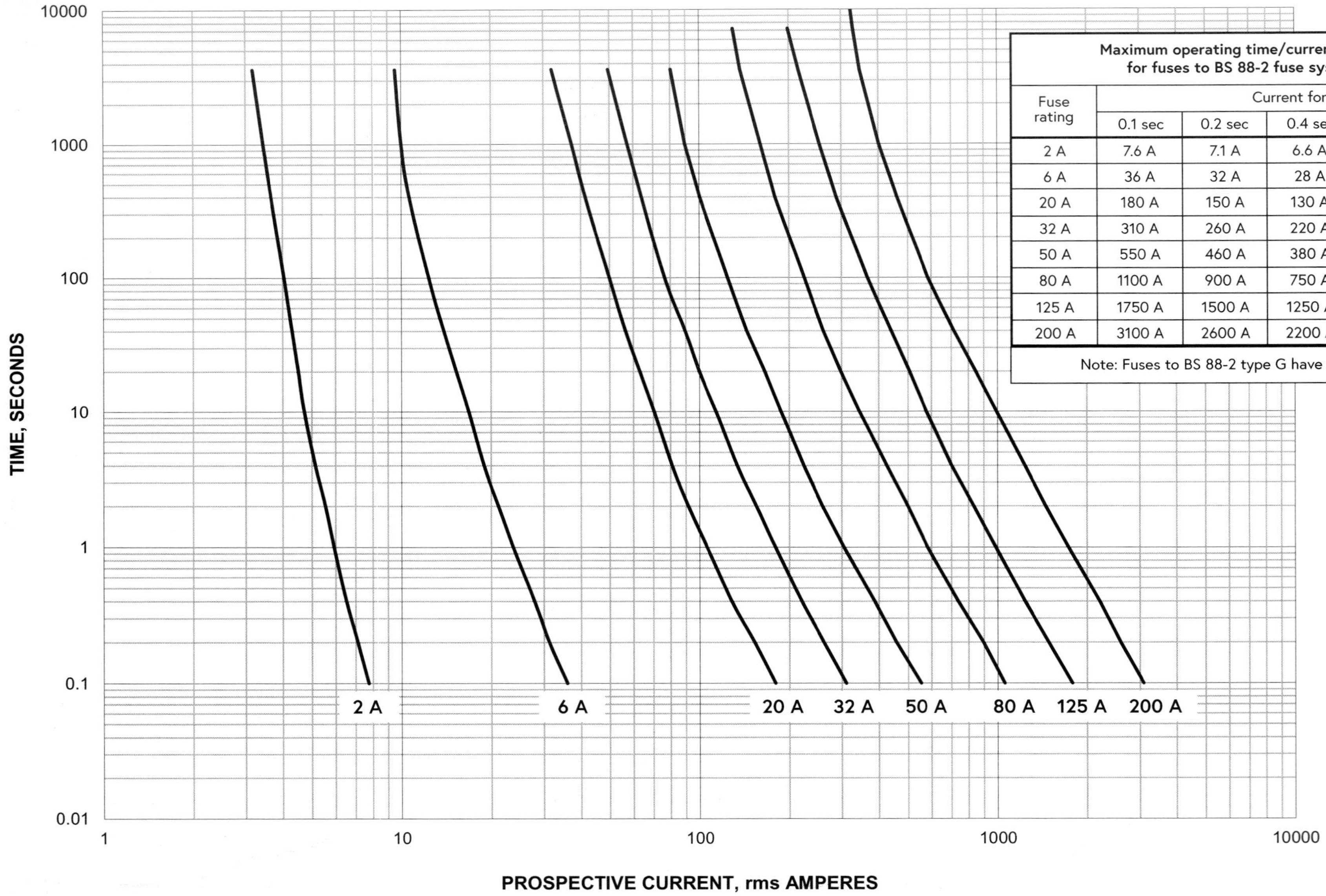

Maximum operating time/current characterisitics for fuses to BS 88-2 fuse systems E and G

Fuse rating	Current for times				
	0.1 sec	0.2 sec	0.4 sec	1 sec	5 secs
2 A	7.6 A	7.1 A	6.6 A	5.9 A	5.0 A
6 A	36 A	32 A	28 A	24 A	18 A
20 A	180 A	150 A	130 A	110 A	78 A
32 A	310 A	260 A	220 A	180 A	125 A
50 A	550 A	460 A	380 A	300 A	220 A
80 A	1100 A	900 A	750 A	580 A	400 A
125 A	1750 A	1500 A	1250 A	1000 A	680 A
200 A	3100 A	2600 A	2200 A	1700 A	1200 A

Note: Fuses to BS 88-2 type G have ratings up to 125 A

Fig 3A3(b) – Fuses to BS 88-2 fuse systems E and G

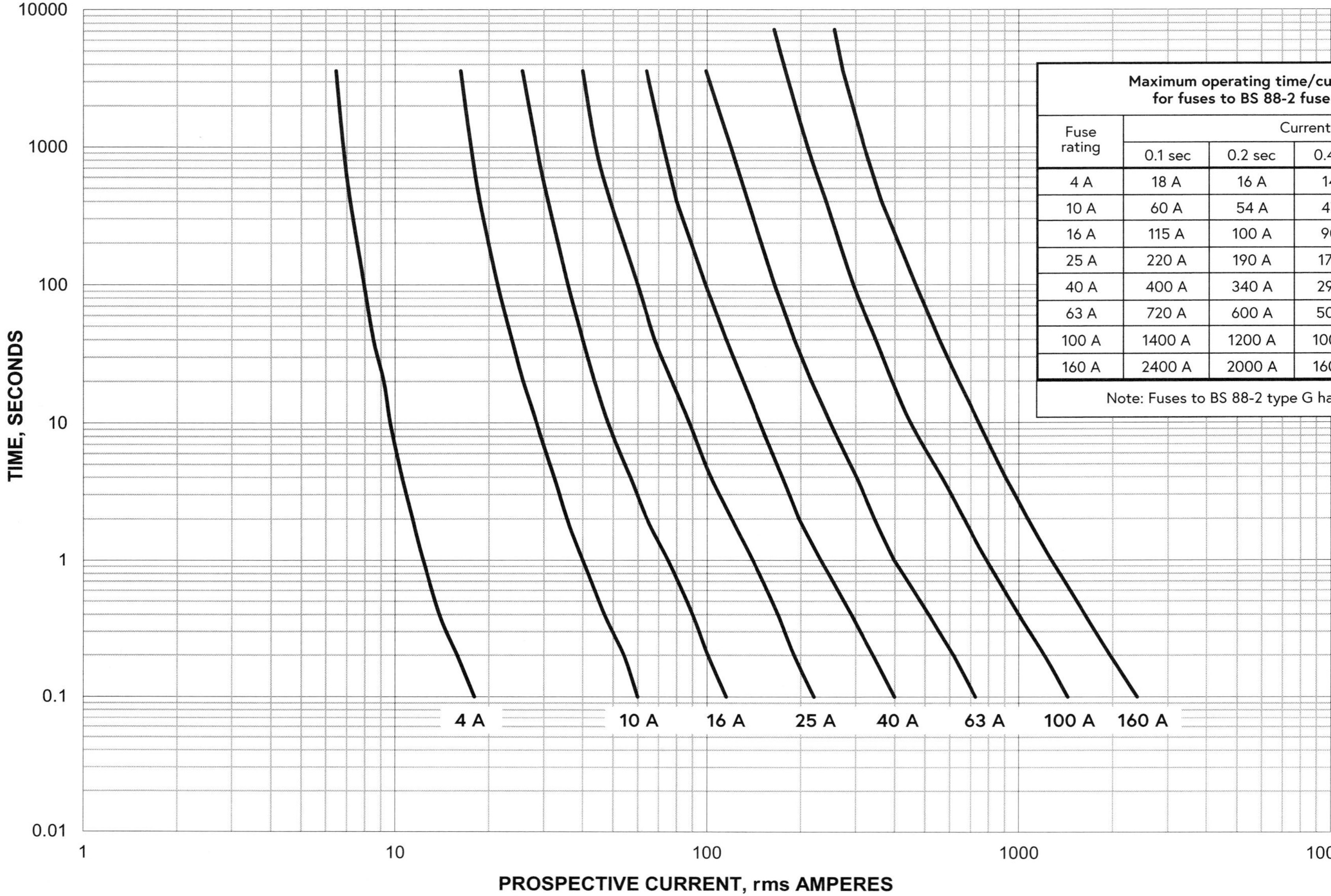

Maximum operating time/current characterisitics for fuses to BS 88-2 fuse systems E and G

Fuse rating	Current for times				
	0.1 sec	0.2 sec	0.4 sec	1 sec	5 secs
4 A	18 A	16 A	14 A	12 A	10.5 A
10 A	60 A	54 A	47 A	40 A	32 A
16 A	115 A	100 A	90 A	75 A	55 A
25 A	220 A	190 A	170 A	140 A	100 A
40 A	400 A	340 A	290 A	230 A	170 A
63 A	720 A	600 A	500 A	400 A	280 A
100 A	1400 A	1200 A	1000 A	780 A	520 A
160 A	2400 A	2000 A	1600 A	1300 A	820 A

Note: Fuses to BS 88-2 type G have ratings up to 125 A

Fig 3A3(c) – Fuses to BS 88-2 fuse system E

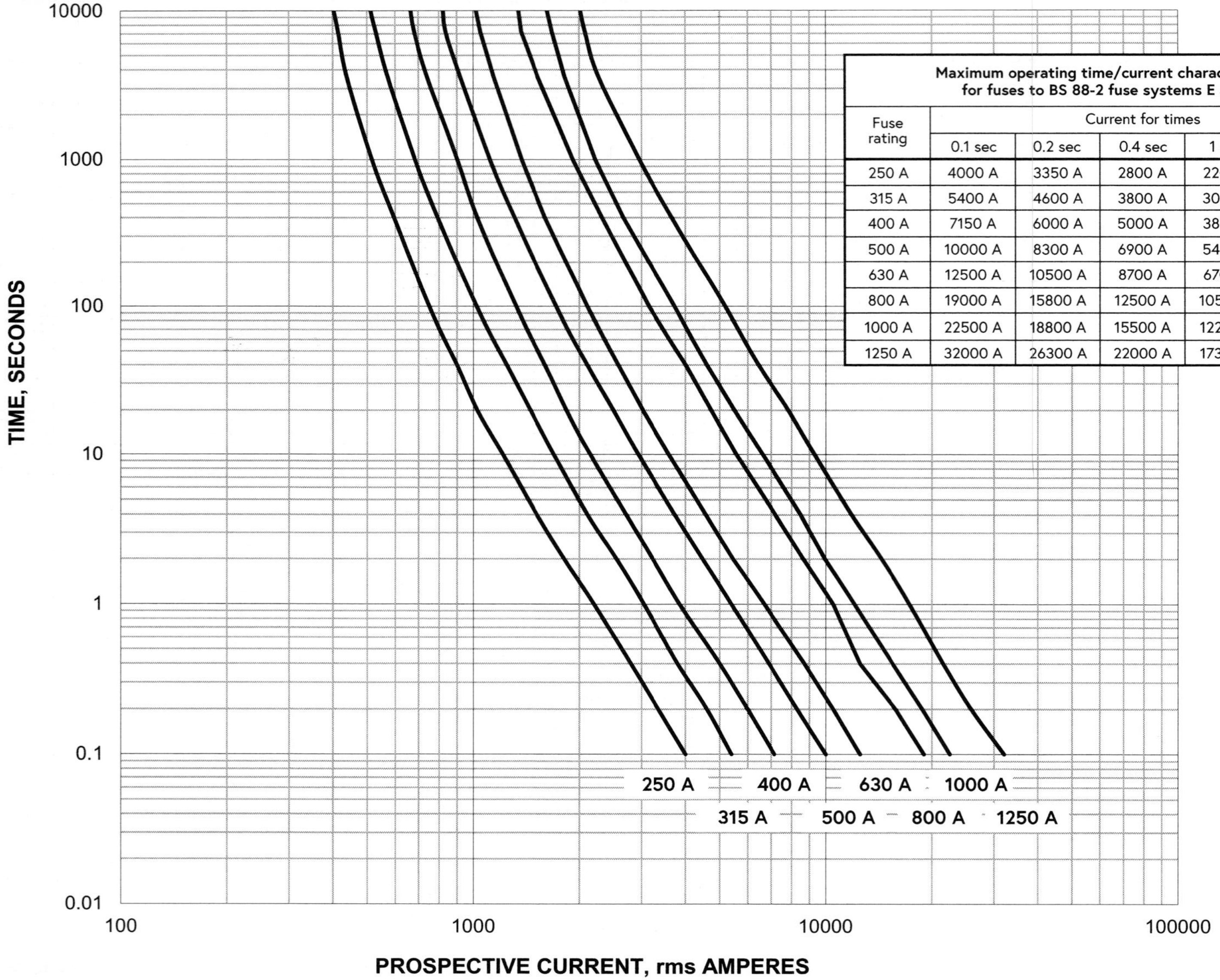

Maximum operating time/current characterisitics for fuses to BS 88-2 fuse systems E and G

Fuse rating	Current for times				
	0.1 sec	0.2 sec	0.4 sec	1 sec	5 secs
250 A	4000 A	3350 A	2800 A	2200 A	1440 A
315 A	5400 A	4600 A	3800 A	3050 A	2000 A
400 A	7150 A	6000 A	5000 A	3850 A	2580 A
500 A	10000 A	8300 A	6900 A	5400 A	3590 A
630 A	12500 A	10500 A	8700 A	6700 A	4420 A
800 A	19000 A	15800 A	12500 A	10500 A	6870 A
1000 A	22500 A	18800 A	15500 A	12200 A	8000 A
1250 A	32000 A	26300 A	22000 A	17300 A	10950 A

Fig 3A4 – Type B circuit-breakers to BS EN 60898 and RCBOs to BS EN 61009-1

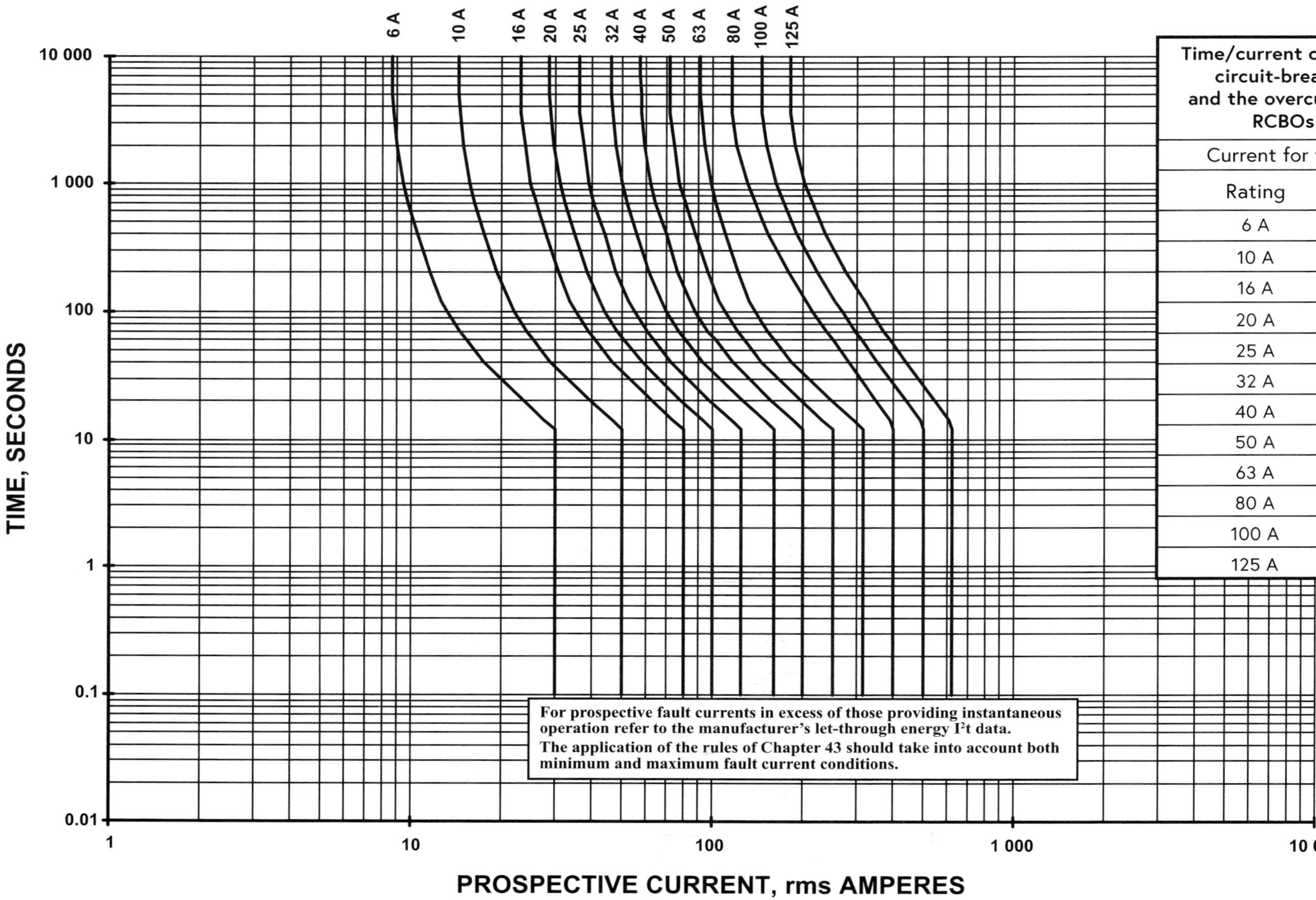

Time/current characteristics for Type B circuit-breakers to BS EN 60898 and the overcurrent characteristics of RCBOs to BS EN 61009-1	
Current for time, 0.1 sec to 5 secs	
Rating	Current
6 A	30 A
10 A	50 A
16 A	80 A
20 A	100 A
25 A	125 A
32 A	160 A
40 A	200 A
50 A	250 A
63 A	315 A
80 A	400 A
100 A	500 A
125 A	625 A

Fig 3A5 – Type C circuit-breakers to BS EN 60898 and RCBOs to BS EN 61009-1

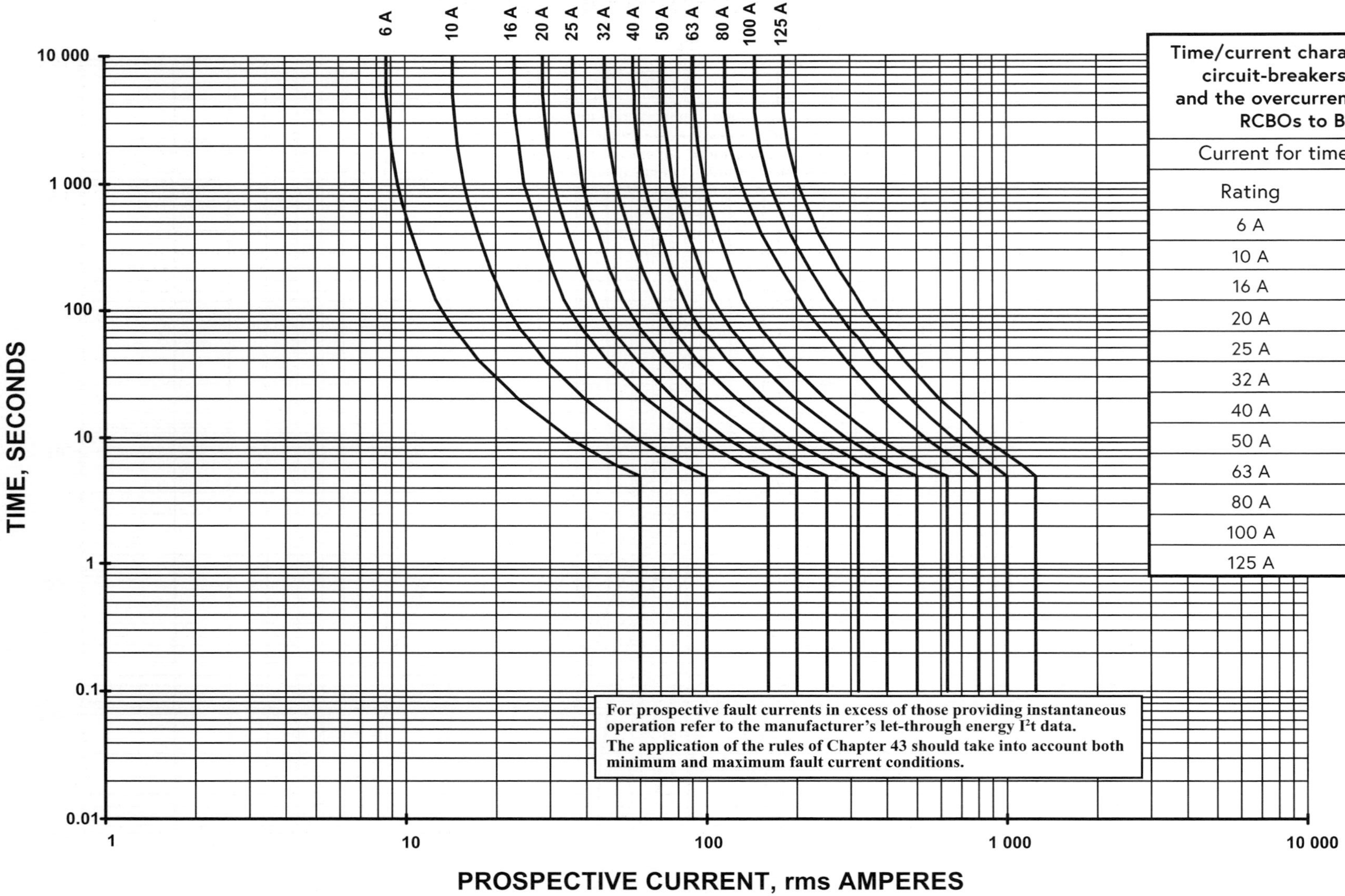

Time/current characteristics for Type C circuit-breakers to BS EN 60898 and the overcurrent characteristics of RCBOs to BS EN 61009-1	
Current for time, 0.1 sec to 5 secs	
Rating	Current
6 A	60 A
10 A	100 A
16 A	160 A
20 A	200 A
25 A	250 A
32 A	320 A
40 A	400 A
50 A	500 A
63 A	630 A
80 A	800 A
100 A	1000 A
125 A	1250 A

Fig 3A6 – Type D circuit-breakers to BS EN 60898 and RCBOs to BS EN 61009-1

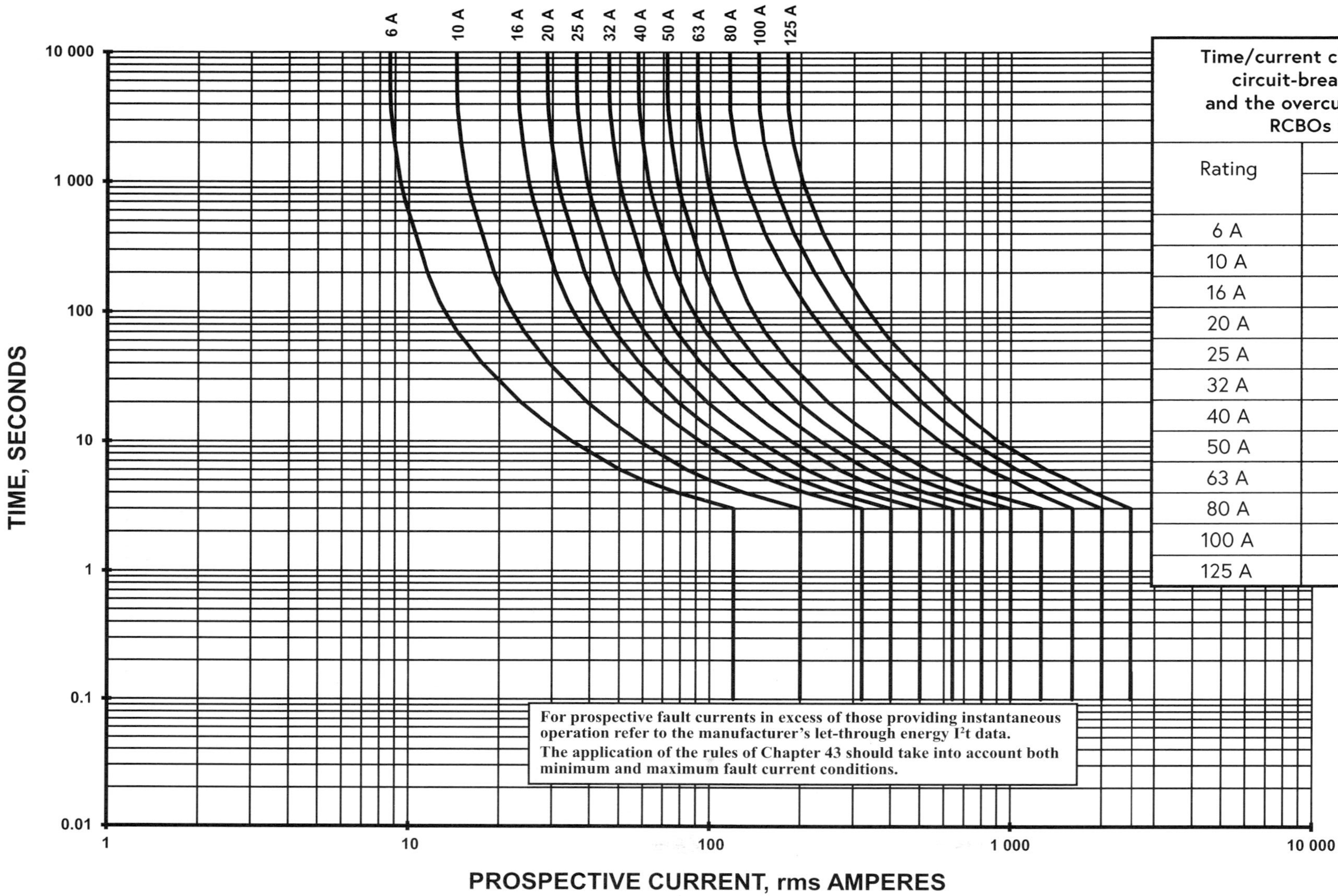

Time/current characteristics for Type D circuit-breakers to BS EN 60898 and the overcurrent characteristics of RCBOs to BS EN 61009-1

Rating	Current for time	
	0.4 sec	5 secs
6 A	120 A	60 A
10 A	200 A	100 A
16 A	320 A	160 A
20 A	400 A	200 A
25 A	500 A	250 A
32 A	640 A	320 A
40 A	800 A	400 A
50 A	1000 A	500 A
63 A	1260 A	630 A
80 A	1600 A	800 A
100 A	2000 A	1000 A
125 A	2500 A	1250 A

APPENDIX 4 (Informative)

CURRENT-CARRYING CAPACITY AND VOLTAGE DROP FOR CABLES

CONTENTS

Tables:

APPENDIX 4 (Informative)

CURRENT-CARRYING CAPACITY AND VOLTAGE DROP FOR CABLES

1 INTRODUCTION

The recommendations of this appendix are intended to provide for a satisfactory life of conductors and insulation subjected to the thermal effects of carrying current for prolonged periods of time in normal service. Other considerations affect the choice of cross-sectional area of conductors, such as the requirements for protection against electric shock (Chapter 41), protection against thermal effects (Chapter 42), overcurrent protection (Chapter 43), voltage drop (Section 525), and limiting temperatures for terminals of equipment to which the conductors are connected (Section 526).

This appendix applies to non-sheathed and sheathed cables having a nominal voltage rating not exceeding 1 kV AC or 1.5 kV DC.

The values in Tables 4D1A to 4J4A have been derived in accordance with the methods given in BS 7769 (BS IEC 60287) using such dimensions as specified in the international standard IEC 60502-1 and conductor resistances given in BS EN 60228. Known practical variations in cable construction (e.g. form of conductor) and manufacturing tolerances result in a spread of possible dimensions and hence current-carrying capacities for each conductor size. Tabulated current-carrying capacities have been selected in such a way as to take account of this spread of values with safety and to lie on a smooth curve when plotted against conductor cross-sectional area.

For multicore cables having conductors with a cross-sectional area of 25 mm^2 or larger, either circular or shaped conductors are permissible. Tabulated values have been derived from dimensions appropriate to shaped conductors.

All the current-carrying capacities given are based on the ambient temperature and conductor/sheath operating temperature stated in Tables 4D1A to 4F2A and 4G1A to 4J4A.

2 CIRCUIT PARAMETERS

2.1 Ambient Temperature

The current-carrying capacities in this appendix are based upon the following reference ambient temperatures:

(i) For non-sheathed and sheathed cables in air, irrespective of the Installation Method: 30 °C

(ii) For buried cables, either directly in the soil or in ducts in the ground: 20 °C.

Where the ambient temperature in the intended location of the non-sheathed or sheathed cables differs from the reference ambient temperature, the appropriate rating factors given in Tables 4B1 and 4B2 are to be applied to the values of current-carrying capacity set out in Tables 4D1A to 4J4A. For buried cables, further correction is not needed if the soil temperature exceeds the selected ambient temperature by an amount up to 5 °C for only a few weeks a year.

The rating factors in Tables 4B1 and 4B2 do not take account of the increase, if any, due to solar or other infrared radiation. Where non-sheathed or sheathed cables are subject to such radiation, the current-carrying capacity may be derived by the methods specified in BS 7769 (BS IEC 60287).

2.2 Soil Thermal Resistivity

The current-carrying capacities tabulated in this appendix for cables in the ground are based upon a soil thermal resistivity of 2.5 K.m/W and are intended to be applied to cables laid in and around buildings. For other installations, where investigations establish more accurate values of soil thermal resistivity appropriate for the load to be carried, the values of current-carrying capacity may be derived by the methods of calculation given in BS 7769 (BS IEC 60287) or obtained from the cable manufacturer.

In locations where the effective soil thermal resistivity is higher than 2.5 K.m/W, an appropriate reduction in current-carrying capacity should be made or the soil immediately around the cables should be replaced by a more suitable material. Such cases can usually be recognized by very dry ground conditions. Rating factors for soil thermal resistivities other than 2.5 K.m/W are given in Table 4B3.

2.3 Groups of cables containing more than one circuit

2.3.1 Methods of Installation A to D in Table 4A2

Current-carrying capacities given in Tables 4D1A to 4J4A apply to single circuits consisting of:

(i) two non-sheathed cables or two single-core cables, or one two-core cable

(ii) three non-sheathed cables or three single-core cables, or one three-core cable.

Where more non-sheathed cables, other than bare mineral insulated cables not exposed to touch, are installed in the same group, the group rating factors specified in Tables 4C1 to 4C3 need to be applied.

NOTE: The group rating factors have been calculated on the basis of prolonged steady-state operation at a 100 % load factor for all live conductors. Where the loading is less than 100 % as a result of the conditions of operation of the installation, the group rating factors may be higher.

2.3.2 Methods of Installation E and F in Table 4A2

The current-carrying capacities of Tables 4D1A to 4J4A apply to these Reference Methods.

For installations on perforated trays, cleats and similar, current-carrying capacities for both single circuits and groups are obtained by multiplying the capacities given for the relevant arrangements of non-sheathed or sheathed cables in free air, as indicated in Tables 4D1A to 4J4A, by the applicable group rating factors given in Tables 4C4 and 4C5. No group rating factors are required for bare mineral insulated cables not exposed to touch, Tables 4G1A and 4G2A refer.

NOTE 1: Group rating factors have been calculated as averages for the range of conductor sizes, cable types and installation condition considered. Attention is drawn to the notes under each table. In some instances, a more precise calculation may be required.

NOTE 2: Group rating factors have been calculated on the basis that the group consists of similar, equally loaded non-sheathed or sheathed cables. Where a group contains various sizes of non-sheathed or sheathed cables, caution should be exercised over the current loading of the smaller cables (see 2.3.3 below).

NOTE 3: A group of similar cables is taken to be a group where the current-carrying capacity of all the cables is based on the same maximum permissible conductor temperature and where the range of conductor sizes in the group spans not more than three adjacent standard sizes.

2.3.3 Groups of cables containing different sizes

Tabulated group rating factors are applicable to groups consisting of similar equally loaded cables. The calculation of rating factors for groups containing different sizes of equally loaded sheathed or non-sheathed cables is dependent on the total number in the group and the mix of sizes. Such factors cannot be tabulated but must be calculated for each group. The method of calculation of such factors is outside the scope of this appendix. Two specific examples of where such calculations may be advisable are given below.

2.3.3.1 Groups in conduit systems, cable trunking systems or cable ducting systems

For a group containing different sizes of non-sheathed or sheathed cables in conduit systems, cable trunking systems or cable ducting systems, a simple formula for calculation of the group rating factor is:

$$C_g = \frac{1}{\sqrt{n}}$$

where

C_g is the group rating factor

n is the number of circuits in the group.

The group rating factor obtained by this equation will reduce the danger of overloading the smaller sizes but may lead to under-utilization of the larger sizes. Such under-utilization can be avoided if large and small sizes of non-sheathed or sheathed cable are not mixed in the same group.

The use of a method of calculation specifically intended for groups containing different sizes of non-sheathed or sheathed cable in conduit will produce a more precise group rating factor.

2.3.3.2 Groups of cables on trays

Where a group contains different sizes of non-sheathed or sheathed cable, caution must be exercised over the current loading of the smaller sizes. It is preferable to use a method of calculation specifically intended for groups containing different sizes of non-sheathed or sheathed cables.

The group rating factor obtained in accordance with the formula in 2.3.3.1 will provide a value which may be safely applied, but which may result in under-utilisation.

2.4 Conductors

The current-carrying capacities and voltage drops tabulated in this appendix are based on cables having solid conductors (Class 1), or stranded conductors (Class 2), except for Tables 4F1A to 4F3B. To obtain the correct current-carrying capacity or voltage drop for cable types similar to those covered by Tables 4D1, 4D2,4E1 and 4E2 but with flexible conductors (Class 5), the tabulated values are multiplied by the following factors:

Cable size	Current-carrying capacity	Voltage drop
≤ 16 mm^2	0.95	1.10
≥ 25 mm^2	0.97	1.06

2.5 Other calculations

In addition to calculations related to current-carrying capacity, overload protection and voltage drop described in this appendix, other calculations are also required for the design of an electrical installation. These include calculations of fault current under various conditions. The equations given in Technical Report PD CLC/TR 50480 are recommended for calculating circuit impedances, fault currents and other parameters. National Annex NA of PD CLC/TR 50480 provides details of additional and alternative calculation methods that are intended for use in the UK.

The UK National Annex includes calculation methods for the following:

(i) two non-sheathed cables or two single-core cables, or one two-core cable

(ii) three non-sheathed cables or three single-core cables, or one three-core cable

- (a) Cables in steel conduit
- (b) In steel trunking
- (c) Steel wire armoured cables
- (d) External cpc in parallel with armour
- (e) Aluminium wire armoured single-core cables.

2.6 Cable surrounded by thermal insulation (C_i)

Where a cable is to be totally surrounded by thermal insulation for less than 0.5 m the current-carrying capacity of the cable shall be reduced appropriately depending on the size of cable, length in insulation and thermal properties of the insulation. The derating factors shown below are appropriate to conductor sizes up to 10 mm^2 in thermal insulation having a thermal conductivity (λ) greater than 0.04 $Wm^{-1}K^{-1}$.

Length in insulation (mm)	**Derating factor**
50	0.88
100	0.78
200	0.63
400	0.51
≥ 500	0.5

3 RELATIONSHIP OF CURRENT-CARRYING CAPACITY TO OTHER CIRCUIT PARAMETERS

The relevant symbols used in the BS 7671 are as follows:

I_z the current-carrying capacity of a cable for continuous service, under the particular installation conditions concerned.

I_t the value of current tabulated in this appendix for the type of cable and installation method concerned, for a single circuit in the ambient temperature stated in the current-carrying capacity tables.

I_b the design current of the circuit, i.e. the current intended to be carried by the circuit in normal service.

I_n the rated current or current setting of the protective device.

I_2 the operating current (i.e. the fusing current or tripping current for the conventional operating time) of the device protecting the circuit against overload.

C a rating factor to be applied where the installation conditions differ from those for which values of current-carrying capacity are tabulated in this appendix. The various rating factors are identified as follows:

- C_a for ambient temperature
- C_c for circuits buried in the ground
- C_d for depth of burial
- C_f for semi-enclosed fuse to BS 3036
- C_g for grouping
- C_i for thermal insulation
- C_s for thermal resistivity of soil.

The rated current or current setting of the protective device (I_n) must not be less than the design current (I_b) of the circuit, and the rated current or current setting of the protective device (I_n) must not exceed the lowest of the current-carrying capacities (I_z) of any of the conductors of the circuit.

Where the overcurrent device is intended to afford protection against overload, I_2 must not exceed 1.45 I_z and I_n must not exceed I_z (see paragraph 4 below).

Where the overcurrent device is intended to afford fault current protection only, I_n can be greater than I_z and I_2 can be greater than 1.45 I_z. The protective device must be selected for compliance with Regulation 434.5.2.

4 OVERLOAD PROTECTION

Where overload protection is required, the type of protection does not affect the current-carrying capacity of a cable for continuous service (I_z) but it may affect the choice of conductor size. The operating conditions of a cable are influenced not only by the limiting conductor temperature for continuous service, but also by the conductor temperature which might be attained during the conventional operating time of the overload protective device, in the event of an overload.

This means that the operating current of the protective device must not exceed 1.45 I_z. Where the protective device is a fuse to BS 88 series, a circuit-breaker to BS EN 60898 or BS EN 60947-2 or a residual current circuit-breaker with integral overcurrent protection to BS EN 61009-1 (RCBO), this requirement is satisfied by selecting a value of I_z not less than I_n.

In practice, because of the standard steps in ratings of fuses and circuit-breakers, it is often necessary to select a value of I_n exceeding I_b. In that case, because it is also necessary for I_z in turn to be not less than the selected value of I_n, the choice of conductor cross-sectional area may be dictated by the overload conditions and the current-carrying capacity (I_z) of the conductors will not always be fully utilised.

The size needed for a conductor protected against overload by a BS 3036 semi-enclosed fuse can be obtained by the use of a rating factor, 1.45/2 = 0.725, which results in the same degree of protection as that afforded by other overload protective devices. This factor is to be applied to the nominal rating of the fuse as a divisor, thus indicating the minimum value of I_t required of the conductor to be protected. In this case also, the choice of conductor size is dictated by the overload conditions and the current-carrying capacity (I_z) of the conductors cannot be fully utilised.

The tabulated current-carrying capacities for cables direct in ground or in ducts in the ground, given in this appendix, are based an ambient temperature of 20 °C. The factor of 1.45 that is applied in Regulation 433.1.1 when considering overload protection assumes that the tabulated current-carrying capacities are based on an ambient temperature of 30 °C. To achieve the same degree of overload protection where a cable is "in a duct in the ground" or "buried direct" as compared with other installation methods a rating factor of 0.9 is applied as a multiplier to the tabulated current-carrying capacity.

5 DETERMINATION OF THE SIZE OF CABLE TO BE USED

Having established the design current (I_b) of the circuit under consideration, the appropriate procedure described in paragraphs 5.1 and 5.2 below will enable the designer to determine the size of the cable it will be necessary to use.

As a preliminary step it is useful to identify the length of the cable run and the permissible voltage drop for the equipment being supplied, as this may be an overriding consideration (see Section 525 and paragraph 6 of this appendix). The permissible voltage drop in mV, divided by I_b and by the length of run, will give the value of voltage drop in mV/A/m which can be tolerated. A voltage drop not exceeding that value is identified in the appropriate table and the corresponding cross-sectional area of conductor needed on this account can be read off directly before any other calculations are made.

The conductor size necessary from consideration of the conditions of normal load and overload is then determined. All rating factors affecting I_z (i.e. for factors for ambient temperature, grouping and thermal insulation) can, if desired, be applied to the values of I_t as multipliers. This involves a process of trial and error until a cross-sectional area is reached so that I_z is not less than I_b and not less than I_n of any protective device it is intended to select. In any event, if a rating factor for protection by a semi-enclosed fuse is necessary, this has to be applied to I_n as a divisor. It is therefore more convenient to apply all the rating factors to I_n as divisors.

This method is used in items 5.1 and 5.2 and produces a value of current and that value (or the next larger value) can be readily located in the appropriate table of current-carrying capacity and the corresponding cross-sectional area of conductor can be identified directly. It should be noted that the value of I_t appearing against the chosen cross-sectional area is not I_z. It is not necessary to know I_z where the size of conductor is chosen by this method.

5.1 Where overload protection is afforded by a device listed in Regulation 433.1.201 or a semi-enclosed fuse to BS 3036

5.1.1 For single circuits

(i) Divide the rated current of the protective device (I_n) by any applicable rating factors for ambient temperature (C_a) , soil thermal resistivity (C_s) and depth of burial (C_d) given in Tables 4B1 to 4B4.

For cables installed above ground C_s and $C_d = 1$.

(ii) Then further divide by any applicable rating factor for thermal insulation (C_i).

(iii) Then further divide by the applicable rating factor for the type of protective device or installation condition (C_f, C_c):

$$I_t \geq \frac{I_n}{C_a\, C_s\, C_d\, C_i\, C_f\, C_c} \qquad \text{Equation 1}$$

(a) Where the protective device is a semi-enclosed fuse to BS 3036, $C_f = 0.725$. Otherwise $C_f = 1$

(b) Where the cable installation method is 'in a duct in the ground' or 'buried direct', $C_c = 0.9$. For cables installed above ground $C_c = 1$.

The size of cable to be used is to be such that its tabulated current-carrying capacity (I_t) is not less than the value of rated current of the protective device adjusted as above.

5.1.2 For groups

(i) In addition to the factors given in 5.1.1, divide the rated current of the protective device (I_n) by the applicable rating factor for grouping (C_g) given in Tables 4C1 to 4C6:

$$I_t \geq \frac{I_n}{C_g\, C_a\, C_s\, C_d\, C_i\, C_f\, C_c} \qquad \text{Equation 2}$$

Alternatively, I_t may be obtained from the following formulae, provided that the circuits of the group are not liable to simultaneous overload:

$$I_t \geq \frac{I_b}{C_g\, C_a\, C_s\, C_d\, C_i\, C_f\, C_c} \qquad \text{Equation 3}$$

$$I_t \geq \frac{1}{C_a\, C_s\, C_d\, C_i}\sqrt{\left(\frac{I_n}{C_f\, C_c}\right)^2 + 0.48\, I_b^2\left(\frac{1 - C_g^2}{C_g^2}\right)} \qquad \text{Equation 4}$$

The size of cable to be used is to be such that its tabulated single-circuit current-carrying capacity (I_t) is not less than the value of I_t calculated in accordance with equation 2 above or, where equations 3 and 4 are used, not less than the larger of the resulting two values of I_t.

5.2 Where overload protection is not required

Where Regulation 433.3.1 applies, and the cable under consideration is not required to be protected against overload, the design current of the circuit (I_b) is to be divided by any applicable rating factors, and the size of the cable to be used is to be such that its tabulated current-carrying capacity (I_t) for the installation method concerned is not less than the value of I_b adjusted as above, i.e.:

$$I_t \geq \frac{I_b}{C_g \, C_a \, C_s \, C_d \, C_i \, C_c}$$

Equation 5

NOTE: Where overload protection is not required $C_c = 1$.

5.3 Other frequencies

Current ratings stated in the tables are for DC and 50/60 Hz AC. The current-carrying capacity of cables carrying, for example, balanced 400 Hz AC compared with the current-carrying capacity at 50 Hz, may be no more than 50 %. For small cables (e.g. as may be used to supply individual loads), the difference in the 50 Hz and the 400 Hz current-carrying capacities may be negligible. Current rating and voltage drop vary with frequency. Suitable ratings should be obtained from the manufacturer.

5.4 Effective current-carrying capacity

The current-carrying capacity of a cable corresponds to the maximum current that can be carried in specified conditions without the conductors exceeding the permissible limit of steady-state temperature for the type of insulation concerned.

The values of current tabulated represent the effective current-carrying capacity only where no rating factor is applicable. Otherwise, the current-carrying capacity corresponds to the tabulated value multiplied by the appropriate factor or factors for ambient temperature, grouping and thermal insulation as well as depth of burial and soil thermal resistivity, for buried cables, as applicable. Where harmonic currents are present further factors may need to be applied. See section 5.5 of this appendix.

Irrespective of the type of overcurrent protective device associated with the conductors concerned, the ambient temperature rating factors to be used when calculating current-carrying capacity (as opposed to those used when selecting cable sizes) are those given in Tables 4B1 and 4B2.

5.5 Rating factors for triple harmonic currents in four-core and five-core cables with four cores carrying current

5.5.1 Rating factors

Regulation 523.6.3 states that, where the neutral conductor carries current without a corresponding reduction in load of the line conductors, the neutral conductor shall be taken into account in ascertaining the current-carrying capacity of the circuit.

This section is intended to cover the situation where there is current flowing in the neutral of a balanced three-phase system. Such neutral currents are due to the line currents having a harmonic content which does not cancel in the neutral. The most significant harmonic which does not cancel in the neutral is usually the third harmonic. The magnitude of the neutral current due to the third harmonic may exceed the magnitude of the power frequency line current. In such a case the neutral current will have a significant effect on the current-carrying capacity of the cables of the circuit.

The rating factors given in this appendix apply to balanced three-phase circuits; it is recognized that the situation is more onerous if only two of the three phases are loaded. In this situation, the neutral conductor will carry the harmonic currents in addition to the unbalanced current. Such a situation can lead to overloading of the neutral conductor.

Equipment likely to cause significant harmonic currents includes, for example, variable-speed motor drives, fluorescent lighting banks and DC power supplies such as those found in computers. Further information on harmonic disturbances can be found in BS EN 61000.

The rating factors given in the following table only apply to cables where the neutral conductor is within a four-core or five-core cable and is of the same material and cross-sectional area as the line conductors. These rating factors have been calculated on the basis of third harmonic currents measured with respect to the fundamental frequency of the line current. Where the total harmonic distortion is more than 15 %, due to the third harmonic or multiples thereof, e.g. 9th, 15th, etc. then lower rating factors are applicable. Where there is an imbalance between phases of more than 50 % then lower rating factors may be applicable.

The tabulated rating factors, when applied to the current-carrying capacity of a cable with three loaded conductors, will give the current-carrying capacity of a cable with four loaded conductors where the current in the fourth conductor is due to harmonics. The rating factors also take the heating effect of the harmonic current in the line conductors into account.

Where the neutral current is expected to be higher than the line current then the cable size should be selected on the basis of the neutral current.

Where the cable size selection is based on a neutral current which is not significantly higher than the line current it is necessary to reduce the tabulated current-carrying capacity for three loaded conductors.

If the neutral current is more than 135 % of the line current and the cable size is selected on the basis of the neutral current then the three line conductors will not be fully loaded. The reduction in heat generated by the line conductors offsets the heat generated by the neutral conductor to the extent that it is not necessary to apply any rating factor to the current-carrying capacity for three loaded conductors, to take account of the effect of four loaded conductors.

TABLE 4Aa – Rating factors for triple harmonic currents in four-core and five-core cables

Third harmonic content of line current* %	Rating factor	
	Size selection is based on line current	Size selection is based on neutral current
0 – 15	1.0	–
>15 – 33	0.86	–
>33 – 45	–	0.86
> 45	–	1.0

* **NOTE:** The third harmonic content expressed as total harmonic distortion.

5.5.2 Example of the application of rating factor for third harmonic currents

Consider a three-phase circuit with a design load (fundamental current) of 58 A is to be installed using a four-core 90 °C thermosetting insulated cable. The cable will be installed in a group with 3 other circuits on a perforated cable tray (method E or F) in an expected maximum ambient temperature of 35 °C. The cable will be protected at its origin using a circuit-breaker to BS EN 60898-1.

Case 1: Load does not produce third harmonic currents

The design current, I_b, of the three-phase load is 58 A.

To satisfy Regulation 433.1.1, $I_n \geq I_b$, so the rated current of the circuit-breaker, I_n, is selected to be 63 A. The required tabulated current-carrying capacity, I_t, under the above operational conditions is to satisfy:

$I_t \geq \frac{I_n}{C_a C_g}$ (where circuits of the group are assumed to be liable to simultaneous overload)

From Table 4B1, $C_a = 0.96$ and from Table 4C1, $C_g = 0.77$

$$I_t = \frac{63}{0.96 \times 0.77} = 85.2 \text{ A}$$

From Table 4E4A, a 16 mm² cable with copper conductors and steel wire armour has a tabulated current-carrying capacity of 99 A and hence it is suitable if third harmonic currents are not present in the circuit.

Case 2: Load produces an additional third harmonic content – THD-i = 20 %

For the second case it is assumed that the above load is expected to produce third harmonic distortion of 20 % in addition to the fundamental line current.

The fundamental line current of the above load is 58 A. Since the third harmonic content is between 15-33 %, the cable sizing is based upon the line current. Because the load has 20 % third harmonic the design current used for the selection of the protective device is given by:

$$I_{bh} = 58 \times \sqrt{1^2 + 0.2^2} = 59.1 \text{ A}$$

where: I_{bh} = design current including the effect of third harmonic currents

To satisfy Regulation 433.1.1, $I_n \geq I_b$, so the rated current of the circuit-breaker, I_n, is selected to be 63 A. In addition, to comply with the Regulation 431.2.3, overcurrent detection must be provided for the neutral conductor. Therefore, a 4-pole protective device with overcurrent protection of the neutral should be provided. The required tabulated current-carrying capacity, I_t, under the above operational conditions is to satisfy:

$I_t \geq \frac{I_n}{C_a\, C_g \times 0.86}$ (where circuits of the group are assumed to be liable to simultaneous overload)

The factor of 0.86 is taken from the above table.

Applying the above grouping and temperature rating factors, the required tabulated current-carrying capacity is found to be 99.1 A.

From Table 4E4A, a 16 mm^2 cable has a tabulated current-carrying capacity of 99 A, thus a rule-based system may select a 25 mm^2 cable whereas a designer may exercise judgement and select a 16 mm^2 cable.

Case 3: Load produces third harmonic content – THD-i = 42 %

The load is expected to produce third harmonic distortion of 42 % of the fundamental line current.

The fundamental line current of the above load is 58 A. Since the third harmonic content is between 33-45 %, the cable sizing is based upon the neutral current with a rating factor of 0.86 applied to the current-carrying capacity of the cable. In addition, to comply with Regulation 431.2.3, overcurrent detection must be provided for the neutral conductor. Therefore, a 4-pole protective device with overcurrent protection of the neutral should be provided.

The neutral current arising from third harmonics is given by:

$$I_{bn} = \frac{3\,h}{100} I_{nL}$$

where: I_{bn} = neutral current due to third harmonic currents

I_{nL} = fundamental line current

h = third harmonic as a percentage of the fundamental line current.

Hence, the neutral current of the circuit is $I_{bn} = 3 \times 0.42 \times 58 = 73$ A.

Therefore, the design current of the circuit due to third harmonics is 73 A.

To satisfy Regulation 433.1.1, $I_n \geq I_b$, so the rated current of the circuit-breaker, I_n, is selected to be 80 A. The required tabulated current-carrying capacity, I_t, under the above operational conditions is to satisfy:

$I_t \geq \frac{I_n}{C_a\, C_g \times 0.86}$ (where circuits of the group are assumed to be liable to simultaneous overload)

Applying the above grouping and temperature rating factors, the required tabulated current-carrying capacity is found to be 125.8 A.

From Table 4E4A, a 25 mm^2 cable is necessary to compensate for the additional thermal effect due to third harmonic current.

All the above cable selections are based on the current-carrying capacity of the cable; voltage drop and other aspects of design have not been considered.

5.6 Harmonic currents in line conductors

Section 5.5 covers the effect of additive harmonic currents flowing in the neutral conductor. The rating factors given in section 5.5 take account of the heating effect of the third harmonic in the neutral as well as the heating effect of the third harmonic in each of the line conductors.

Where other harmonics are present, e.g. 5th, 7th etc, the heating effect of these harmonics in the line conductors has to be taken into account. For smaller sizes, less than 50 mm^2, the effect of harmonic currents can be taken into account by applying the following factor, C_h, to the fundamental design current.

$$C_h = \sqrt{\frac{I_f^2 + \ldots + I_{hn}^2}{I_f^2}}$$

where: I_f = 50 Hz current

I_{hn} = n^{th} harmonic current

For larger conductor sizes the increase in conductor resistance, due to skin and proximity effects, at higher frequencies has to be taken into account. The resistance at harmonic frequencies can be calculated using the equations given in BS IEC 60287-1-1.

6 TABLES OF VOLTAGE DROP

In the tables, values of voltage drop are given for a current of one ampere for a metre run, i.e. for a distance of 1 m along the route taken by the cables, and represent the result of the voltage drops in all the circuit conductors. The values of voltage drop assume that the conductors are at their maximum permitted normal operating temperature.

The values in the tables, for AC operation, apply to frequencies in the range 49 to 61 Hz and for single-core armoured cables the tabulated values apply where the armour is bonded to earth at both ends. The values of voltage drop for cables operating at higher frequencies may be substantially greater.

For a given run, to calculate the voltage drop (in mV) the tabulated value of voltage drop per ampere per metre for the cable concerned has to be multiplied by the length of the run in metres and by the current the cable is intended to carry, namely, the design current of the circuit (I_b) in amperes. For three-phase circuits the tabulated mV/A/m values relate to the line voltage and balanced conditions have been assumed.

For cables having conductors of 16 mm^2 or less cross-sectional area, their inductances can be ignored and $(mV/A/m)_r$ values only are tabulated. For cables having conductors greater than 16 mm^2 cross-sectional area the impedance values are given as $(mV/A/m)_z$, together with the resistive component $(mV/A/m)_r$ and the reactive component $(mV/A/m)_x$.

The direct use of the tabulated $(mV/A/m)_r$ or $(mV/A/m)_z$ values, as appropriate, may lead to pessimistically high calculated values of voltage drop or, in other words, to unnecessarily low values of permitted circuit lengths. For example, where the design current of a circuit is significantly less than the effective current-carrying capacity of the chosen cable, the actual voltage drop would be less than the calculated value because the conductor temperature (and hence their resistance) will be less than that on which the tabulated mV/A/m had been based.

As regards power factor in AC circuits, the use of the tabulated mV/A/m values (for the larger cable sizes, the tabulated $(mV/A/m)_z$ values) leads to a calculated value of the voltage drop higher than the actual value. In some cases it may be advantageous to take account of the load power factor when calculating voltage drop.

Where a more accurate assessment of the voltage drop is desirable the following methods may be used.

6.1 Correction for operating temperature

For cables having conductors of cross-sectional area 16 mm^2 or less, the design value of mV/A/m is obtained by multiplying the tabulated value by a factor C_t, given by:

$$C_t = \frac{230 + t_p - \left(C_a^2 C_g^2 C_s^2 C_d^2 - \frac{I_b^2}{I_t^2}\right)(t_p - 30)}{230 + t_p} \qquad \text{Equation 6}$$

where t_p is the maximum permitted normal operating temperature (°C).

This equation applies only where the overcurrent protective device is other than a BS 3036 fuse and where the actual ambient temperature is equal to or greater than 30 °C.

NOTE: For convenience, the above equation is based on the approximate resistance-temperature coefficient of 0.004 per °C at 20 °C for both copper and aluminium conductors.

For cables having conductors of cross-sectional area greater than 16 mm^2, only the resistive component of the voltage drop is affected by the temperature and the factor C_t is therefore applied only to the tabulated value of $(mV/A/m)_r$ and the design value of $(mV/A/m)_z$ is given by the vector sum of C_t $(mV/A/m)_r$ and $(mV/A/m)_x$.

For very large conductor sizes, where the resistive component of voltage drop is much less than the corresponding reactive part (i.e. when $x/r \geq 3$), this rating factor need not be considered.

6.2 Correction for load power factor

For cables having conductors of cross-sectional area 16 mm^2 or less, the design value of mV/A/m is obtained approximately by multiplying the tabulated value by the power factor of the load, cos Ø.

For cables having conductors of cross-sectional area greater than 16 mm^2, the design value of mV/A/m is given approximately by:

$$\cos Ø\,(\text{tabulated}(\text{mV/A/m})_r) + \sin Ø\,(\text{tabulated}(\text{mV/A/m})_x)$$

For single-core cables in flat formation the tabulated values apply to the outer cables and may underestimate for the voltage drop between an outer cable and the centre cable for cross-sectional areas above 240 mm^2, and power factors greater than 0.8.

6.3 Correction for both operating temperature and load power factor

For paragraphs 6.1 and 6.2 above, where it is considered appropriate to correct the tabulated mV/A/m values for both operating temperature and load power factor, the design figure for mV/A/m is given by:

(i) for cables having conductors of cross-sectional area 16 mm^2 or less

$C_t \cos Ø$ (tabulated mV/A/m)

(ii) for cables having conductors of cross-sectional area greater than 16 mm^2

$C_t \cos Ø\,(\text{tabulated } (\text{mV/A/m})_r) + \sin Ø\,(\text{tabulated } (\text{mV/A/m})_x)$.

6.4 Voltage drop in consumers' installations

The voltage drop between the origin of an installation and any load point should not be greater than the values in the table below expressed with respect to the value of the nominal voltage of the installation.

The calculated voltage drop should include any effects due to harmonic currents.

TABLE 4Ab – Voltage drop

	Lighting	Other uses
(i) Low voltage installations supplied directly from a public low voltage distribution system	3 %	5 %
(ii) Low voltage installation supplied from private LV supply (*)	6 %	8 %

(*) The voltage drop within each final circuit should not exceed the values given in (i).

Where the wiring systems of the installation are longer than 100 m, the voltage drops indicated above may be increased by 0.005 % per metre of the wiring system beyond 100 m, without this increase being greater than 0.5 %.

The voltage drop is determined from the demand of the current-using equipment, applying diversity factors where applicable, or from the value of the design current of the circuit.

NOTE 1: A greater voltage drop may be acceptable for a motor circuit during starting and for other equipment with a high inrush current, provided that in both cases the voltage variations remain within the limits specified in the relevant equipment standard.

NOTE 2: The following temporary conditions are excluded:
- voltage transients
- voltage variations due to abnormal operation.

7 METHODS OF INSTALLATION

Table 4A2 lists the methods of installation for which this appendix provides guidance for the selection of the appropriate cable size. Table 4A3 lists the appropriate tables for selection of current ratings for specific cable constructions. The Reference Methods are those methods of installation for which the current-carrying capacities given in Tables 4D1A to 4J4A have been determined (see 7.1 below).

The use of other methods is not precluded and in that case the evaluation of current-carrying capacity may need to be based on experimental work.

7.1 Reference Methods

The Reference Methods are those methods of installation for which the current-carrying capacity has been determined by test or calculation.

NOTE 1: It is impractical to calculate and publish current ratings for every installation method, since many would result in the same current rating. Therefore a suitable (limited) number of current ratings have been calculated which cover all of the installation methods stated in Table 4A2 and have been called Reference Methods.

Reference Method A, for example, Installation Methods 1 and 2 of Table 4A2 (non-sheathed cables and multicore cables in conduit in a thermally insulated wall).

The wall consists of an outer weatherproof skin, thermal insulation and an inner skin of wood or wood-like material having a thermal conductance of at least 10 W/m^2K. The conduit is fixed such that it is close to, but not necessarily touching, the inner skin. Heat from the cables is assumed to escape through the inner skin only. The conduit can be metal or plastic.

Reference Method B, for example, Installation Method 4 of Table 4A2 (non-sheathed cables in conduit mounted on a wooden or masonry wall) and Installation Method 5 of Table 4A2 (multicore cable in conduit on a wooden or masonry wall).

The conduit is mounted on a wooden wall such that the gap between the conduit and the surface is less than 0.3 times the conduit diameter. The conduit can be metal or plastic. Where the conduit is fixed to a masonry wall the current-carrying capacity of the non-sheathed or sheathed cable may be higher.

Reference Method C (clipped direct), for example, Installation Method 20 of Table 4A2 (single-core or multicore cables on a wooden or masonry wall).

Cable mounted on a wooden wall so that the gap between the cable and the surface is less than 0.3 times the cable diameter. Where the cable is fixed to or embedded in a masonry wall the current-carrying capacity may be higher.

NOTE 2: The term 'masonry' is taken to include brickwork, concrete, plaster and similar (but excluding thermally insulating materials).

Reference Method D, for example, Installation Method 70 of Table 4A2 (multicore armoured cable in conduit or in cable ducting in the ground).

The cable is drawn into a 100 mm diameter plastic, earthenware or metallic duct laid in direct contact with soil having a thermal resistivity of 2.5 K.m/W and at a depth of 0.7 m. The values given for this method are those stated in this appendix and are based on conservative installation parameters. If the specific installation parameters are known (thermal resistance of the ground, ground ambient temperature, cable depth), reference can be made to the cable manufacturer or the ERA 69-30 series of publications, which may result in a smaller cable size being selected.

NOTE 3: The current-carrying capacity for cables laid in direct contact with soil having a thermal resistivity of 2.5 Km/W and at a depth of 0.7 m is approximately 10 % higher than the values tabulated for Reference Method D.

Reference Methods E, F and G, for example, Installation Methods 31 to 35 of Table 4A (single-core or multicore cables in free air).

The cable is supported such that the total heat dissipation is not impeded. Heating due to solar radiation and other sources is to be taken into account. Care is to be taken that natural air convection is not impeded. In practice, a clearance between a cable and any adjacent surface of at least 0.3 times the cable external diameter for multicore cables or 1.0 times the cable diameter for single-core cables is sufficient to permit the use of current-carrying capacities appropriate to free air conditions.

7.2 Other Methods

Cable on a floor: Reference Method C applies for current rating purposes.

Cable under a ceiling: This installation may appear similar to Reference Method C but because of the reduction in natural air convection, Reference Method B is to be used for the current rating.

Cable tray systems: A perforated cable tray has a regular pattern of holes that occupy at least 30 % of the area of the base of the tray. The current-carrying capacity for cables attached to perforated cable trays should be taken as Reference Method E or F. The current-carrying capacity for cables attached to unperforated cable trays (no holes or holes that occupy less than 30 % of the area of the base of the tray) is to be taken as Reference Method C.

Cable ladder system: This is a construction which offers a minimum of impedance to the air flow around the cables, i.e. supporting metalwork under the cables occupies less than 10 % of the plan area. The current-carrying capacity for cables on ladder systems should be taken as Reference Method E or F.

Cable cleats, cable ties and cable hangers: Cable supports hold the cable at intervals along its length and permit substantially complete free air flow around the cable. The current-carrying capacity for cable cleats, cable ties and cable hangers should be taken as Reference Method E or F.

Cable installed in a ceiling: This is similar to Reference Method A. It may be necessary to apply the rating factors due to higher ambient temperatures that may arise in junction boxes and similar mounted in the ceiling.

NOTE: Where a junction box in the ceiling is used for the supply to a luminaire, the heat dissipation from the luminaire may provide higher ambient temperatures than permitted in Tables 4D1A to 4J4A (see also Regulation 522.2.1). The temperature may be between 40 °C and 50 °C, and a rating factor according to Table 4B1 must be applied.

General notes to all tables in this appendix

NOTE 1: Current-carrying capacities are tabulated for methods of installation which are commonly used for fixed electrical installations. The tabulated capacities are for continuous steady-state operation (100 % load factor) for DC or AC of nominal frequency 50 Hz and take no account of harmonic content.

NOTE 2: Table 4A2 itemizes the reference methods of installation to which the tabulated current-carrying capacities refer.

TABLE 4A1 – Schedule of Installation Methods in relation to conductors and cables

Conductors and cables		Installation Method							
		Without fixings	Clipped direct	Conduit systems	Cable trunking systems*	Cable ducting systems	Cable ladder, cable tray, cable brackets	On insulators	Support wire
Bare conductors		np	np	np	np	np	np	P	np
Non-sheathed cable		np	np	P[1]	P[1 2]	P[1]	np[1]	P	np
Sheathed cables (including armoured and mineral insulated)	Multicore	P	P	P	P	P	P	N/A	P
	Single-core	N/A	P	P	P	P	P	N/A	P

P Permitted.

np Not permitted.

N/A Not applicable, or not normally used in practice.

* including skirting trunking and flush floor trunking

[1] Non-sheathed cables which are used as protective conductors or protective bonding conductors need not be laid in conduits or ducts

[2] Non-sheathed cables are acceptable if the trunking system provides at least the degree of protection IPXXD or IP4X and if the cover can only be removed by means of a tool or a deliberate action.

TABLE 4A2 - Schedule of Installation methods of cables (including Reference Methods) for determining current-carrying capacity

NOTE 1: The illustrations are not intended to depict actual product or installation practices but are indicative of the method described.

NOTE 2: The installation and reference methods stated are in line with IEC. However, not all methods have a corresponding rating for all cable types.

Installation Method			Reference Method to be used to determine current-carrying capacity
Number	**Examples**	**Description**	
1	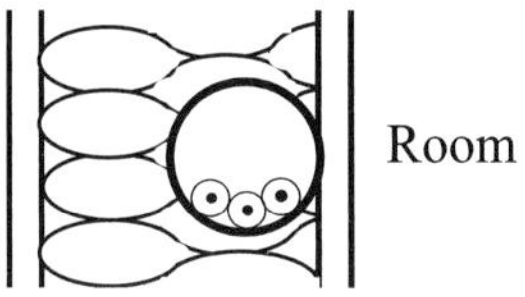	Non-sheathed cables in conduit in a thermally insulated wall with an inner skin having a thermal conductance of not less than 10 W/m²K [c]	**A**
2	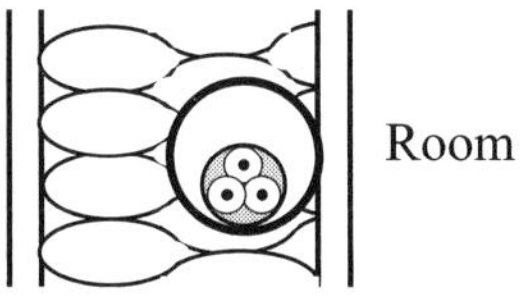	Multicore cable in conduit in a thermally insulated wall with an inner skin having a thermal conductance of not less than 10 W/m²K [c]	**A**
3	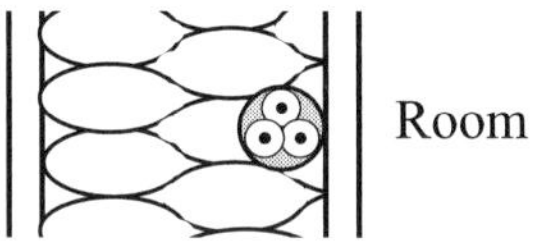	Multicore cable direct in a thermally insulated wall with an inner skin having a thermal conductance of not less than 10 W/m²K [c]	**A**
4	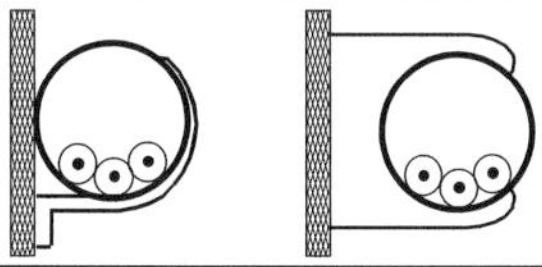	Non-sheathed cables in conduit on a wooden or masonry wall or spaced less than 0.3 × conduit diameter from it [c]	**B**
5	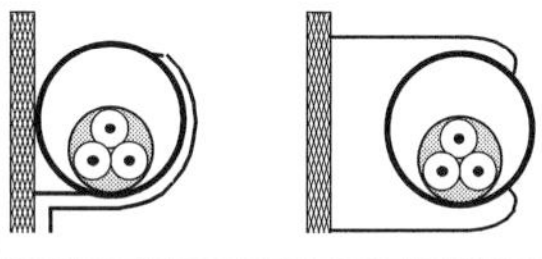	Multicore cable in conduit on a wooden or masonry wall or spaced less than 0.3 × conduit diameter from it [c]	**B**
6 7	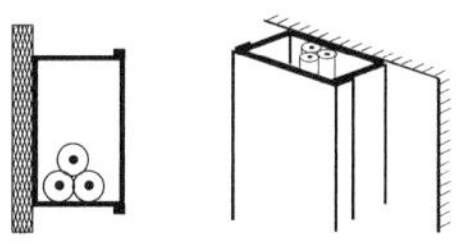 6 7	Non-sheathed cables in cable trunking on a wooden or masonry wall 6 - run horizontally [b] 7 - run vertically [b, c]	**B**
8 9	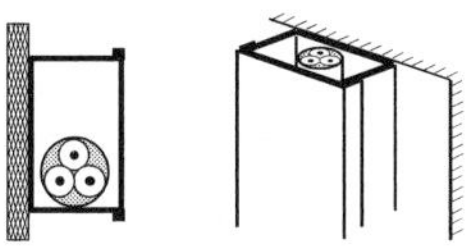 8 9	Multicore cable in cable trunking on a wooden or masonry wall 8 - run horizontally [b] 9 - run vertically [b, c]	**B***
10 11	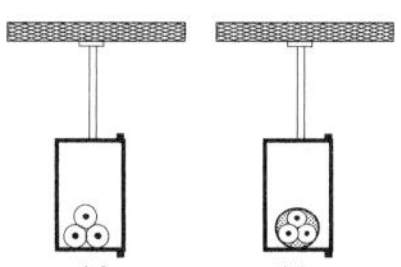 10 11	Non-sheathed cables in suspended cable trunking [b]	**B**
		Multicore cable in suspended cable trunking [b]	**B**
12	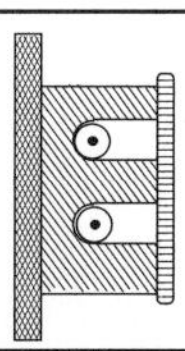	Non-sheathed cables run in mouldings [c, e]	**A**

b Values given for Reference Method B in Appendix 4 are for a single circuit. Where there is more than one circuit in the trunking the group rating factor given in Table 4C1 is applicable, irrespective of the presence of an internal barrier or partition.

c Care is needed where the cable runs vertically and ventilation is restricted. The ambient temperature at the top of the vertical section can be much higher.

e The thermal resistivity of the enclosure is assumed to be poor because of the material of construction and possible air spaces. Where the construction is thermally equivalent to Installation Methods 6 or 7, Reference Method B may be used.

* Still under consideration in IEC.

TABLE 4A2 *(continued)*

Installation Method			Reference Method to be used to determine current-carrying capacity
Number	**Examples**	**Description**	
13 *14*		*Not used*	
15	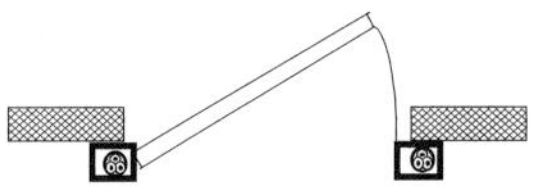	Non-sheathed cables in conduit or single-core or multicore cable in architrave [c, f]	**A**
16		Non-sheathed cables in conduit or single-core or multicore cable in window frames [c, f]	**A**
20	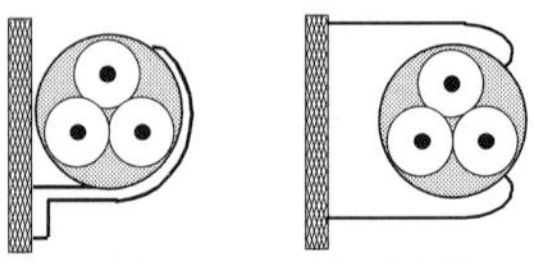	Single-core or multicore cables: - fixed on (clipped direct), or spaced less than 0.3 × cable diameter from a wooden or masonry wall [c]	**C**
21	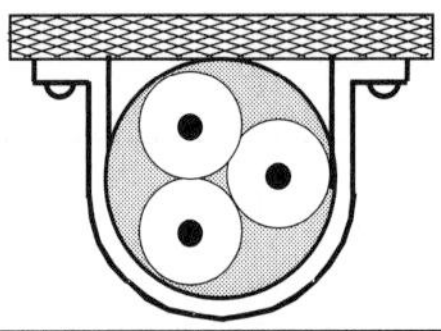	Single-core or multicore cables: - fixed directly under a wooden or masonry ceiling	**B** (Higher than standard ambient temperatures may occur with this installation method)
22	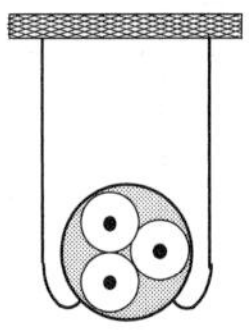	Single-core or multicore cables: - spaced from a ceiling	**E, F or G*** (Higher than standard ambient temperatures may occur with this installation method)
23		*Not used*	
30	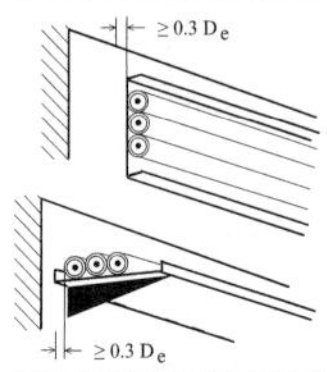	Single-core or multicore cables: - on unperforated tray run horizontally or vertically [c, h]	**C** with item 2 of Table 4C1
31	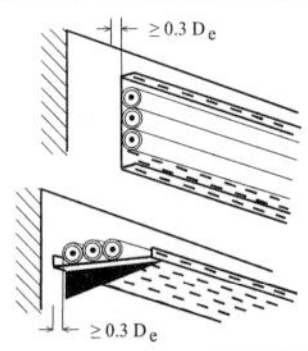	Single-core or multicore cables: - on perforated tray run horizontally or vertically [c, h]	**E or F**
32	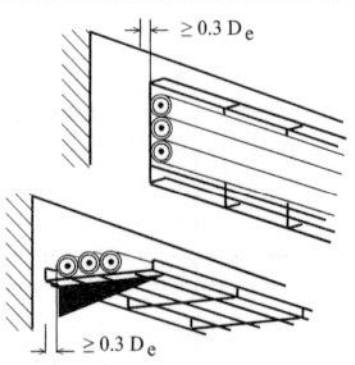	Single-core or multicore cables: - on brackets or on a wire mesh tray run horizontally or vertically [c, h]	**E or F**

c Care is needed where the cable runs vertically and ventilation is restricted. The ambient temperature at the top of the vertical section can be much higher.

f The thermal resistivity of the enclosure is assumed to be poor because of the material of construction and possible air spaces. Where the construction is thermally equivalent to Installation Methods 6, 7, 8 or 9, Reference Method B may be used.

h D_e = the external diameter of a multicore cable:
- 2.2 × the cable diameter when three single-core cables are bound in trefoil, or
- 3 × the cable diameter when three single-core cables are laid in flat formation.

* Still under consideration in IEC.

TABLE 4A2 *(continued)*

Installation Method			Reference Method to be used to determine current-carrying capacity
Number	**Examples**	**Description**	
33	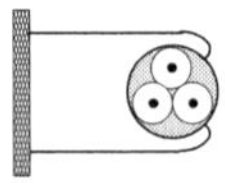	Single-core or multicore cables: - spaced more than 0.3 times the cable diameter from a wall	**E, F or G**[g]
34	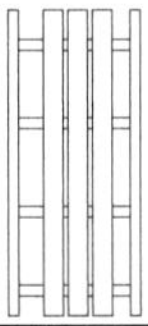	Single-core or multicore cables: - on a ladder[c]	**E or F**
35	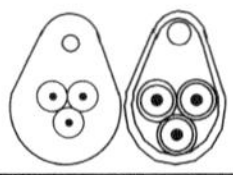	Single-core or multicore cable suspended from or incorporating a support wire or harness	**E or F**
36		Bare or non-sheathed cables on insulators	**G**
40	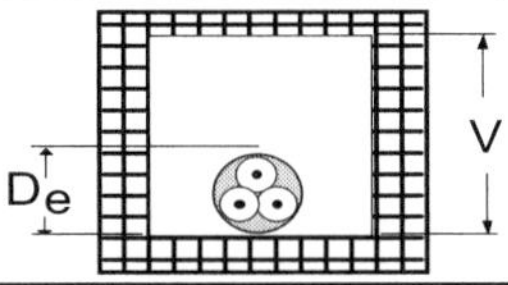	Single-core or multicore cable in a building void [c, h, i]	Where $1.5\ D_e \leq V < 20\ D_e$ use **B**
41		Non-sheathed cables in conduit in a building void in masonry having a thermal resistivity not greater than 2 K.m/W [c, i, j]	Where $1.5\ D_e \leq V$ use **B**
42	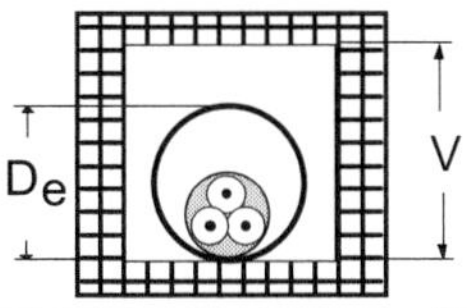	Single-core or multicore cable in conduit in a building void in masonry having a thermal resistivity not greater than 2 K.m/W [c, j]	Where $1.5\ D_e \leq V$ use **B**
43	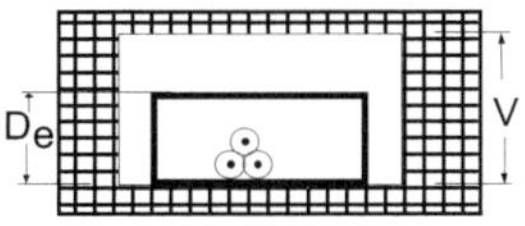	Non-sheathed cables in cable ducting in a building void in masonry having a thermal resistivity not greater than 2 K.m/W [c, i, j]	Where $1.5\ D_e \leq V$ use **B**
44	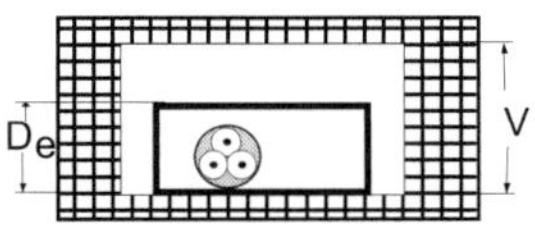	Single-core or multicore cable in cable ducting in a building void in masonry having a thermal resistivity not greater than 2 K.m/W [c, i, j]	Where $1.5\ D_e \leq V$ use **B**
45	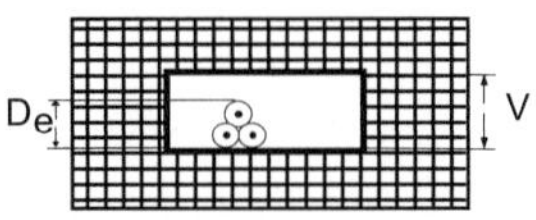	Non-sheathed cables in cable ducting in masonry having a thermal resistivity not greater than 2 K.m/W [c, h, i]	Where $1.5\ D_e \leq V < 50\ D_e$ use **B**
46	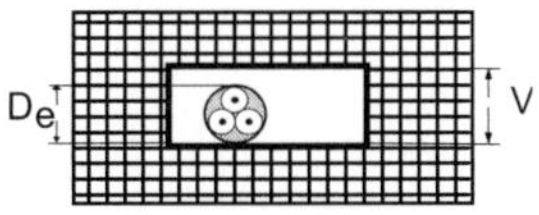	Single-core or multicore cable in cable ducting in masonry having a thermal resistivity not greater than 2 K.m/W [c, h, i]	Where $1.5\ D_e \leq V < 50\ D_e$ use **B**
47	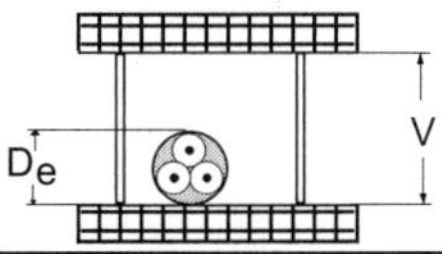	Single-core or multicore cable: - in a ceiling void - in a suspended floor [h, i]	Where $1.5\ D_e \leq V < 50\ D_e$ use **B**

c Care is needed where the cable runs vertically and ventilation is restricted. The ambient temperature at the top of the vertical section can be much higher.

g The factors in Table 4C1 may also be used.

h D_e = the external diameter of a multicore cable:
- 2.2 × the cable diameter when three single-core cables are bound in trefoil, or
- 3 × the cable diameter when three single-core cables are laid in flat formation.

i V = the smaller dimension or diameter of a masonry duct or void, or the vertical depth of a rectangular duct, floor or ceiling void or channel.

j D_e = external diameter of conduit or vertical depth of cable ducting.

TABLE 4A2 *(continued)*

Installation Method			Reference Method to be used to determine current-carrying capacity
Number	**Examples**	**Description**	
50	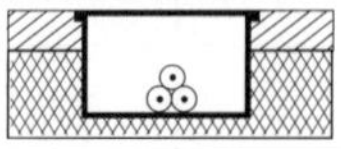	Non-sheathed cables in flush cable trunking in the floor	**B**
51	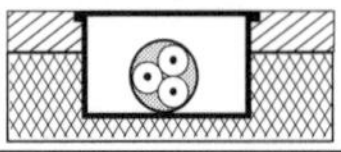	Multicore cable in flush cable trunking in the floor	**B**
52	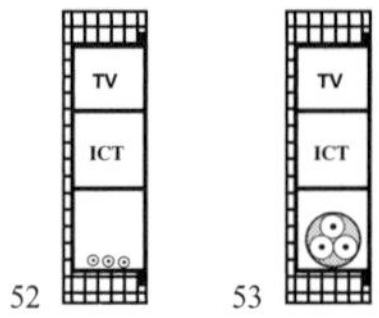	Non-sheathed cables in flush trunking[c]	**B**
53		Multicore cable in flush trunking [c]	**B**
54	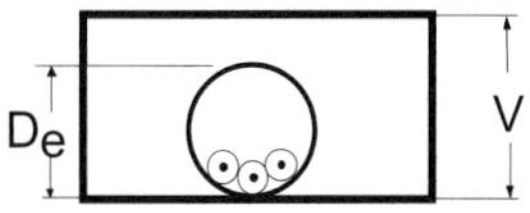	Non-sheathed cables or single-core cables in conduit in an unventilated cable channel run horizontally or vertically [c, i, k, m]	Where $1.5\ D_e \leq V$ use **B**
55	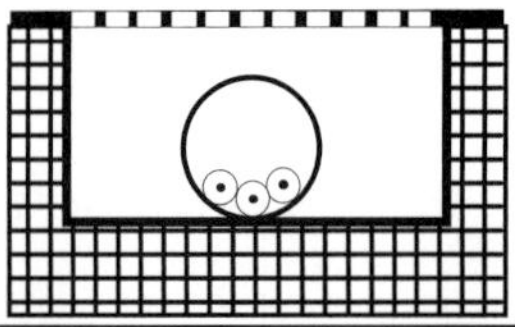	Non-sheathed cables in conduit in an open or ventilated cable channel in the floor [l, m]	**B**
56	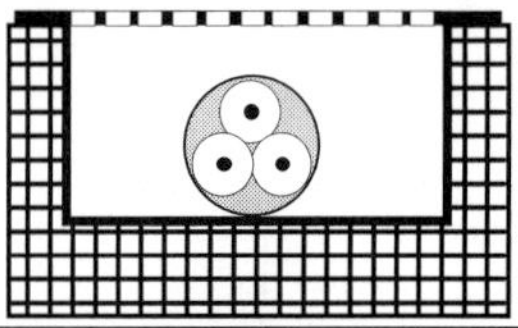	Sheathed single-core or multicore cable in an open or ventilated cable channel run horizontally or vertically [m]	**B**
57	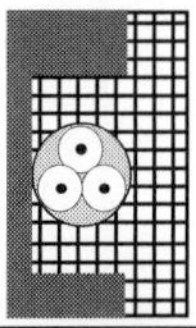	Single-core or multicore cable direct in masonry having a thermal resistivity not greater than 2 K.m/W - without added mechanical protection [n, o]	**C**
58	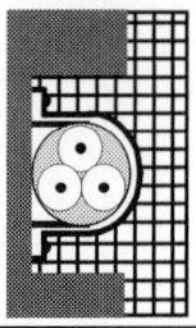	Single-core or multicore cable direct in masonry having a thermal resistivity not greater than 2 K.m/W - with added mechanical protection [n, o] (e.g. capping)	**C**
59	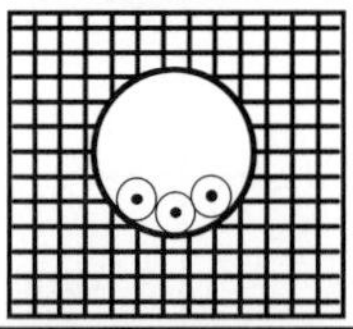	Non-sheathed cables or single-core cables in conduit in masonry having a thermal resistivity not greater than 2 K.m/W [o]	**B**
60	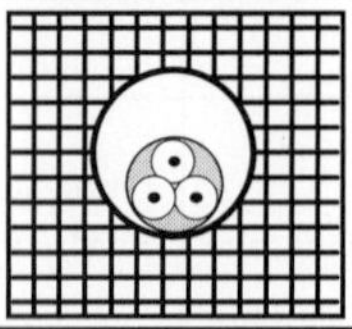	Multicore cables in conduit in masonry having a thermal resistivity not greater than 2 K.m/W [o]	**B**

c Care is needed where the cable runs vertically and ventilation is restricted. The ambient temperature at the top of the vertical section can be much higher.

k D_e = external diameter of conduit.

i V = the smaller dimension or diameter of a masonry duct or void, or the vertical depth of a rectangular duct, floor or ceiling void or channel. The depth of the channel is more important than the width.

l For multicore cable installed as Method 55, use current-carrying capacity for Reference Method B.

m It is recommended that these Installation Methods are used only in areas where access is restricted to authorized persons so that the reduction in current-carrying capacity and the fire hazard due to the accumulation of debris can be prevented.

n For cables having conductors not greater than 16 mm², the current-carrying capacity may be higher.

o Thermal resistivity of masonry is not greater than 2 K.m/W. The term masonry is taken to include brickwork, concrete, plaster and the like (excludes thermally insulating materials).

TABLE 4A2 *(continued)*

Installation Method			Reference Method to be used to determine current-carrying capacity
Number	**Examples**	**Description**	
70	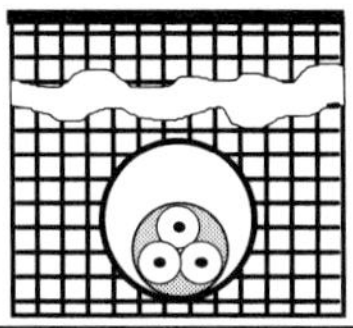	Multicore armoured cable in conduit or in cable ducting in the ground	**D** **For multicore armoured cable only**
71		*Not used*	
72	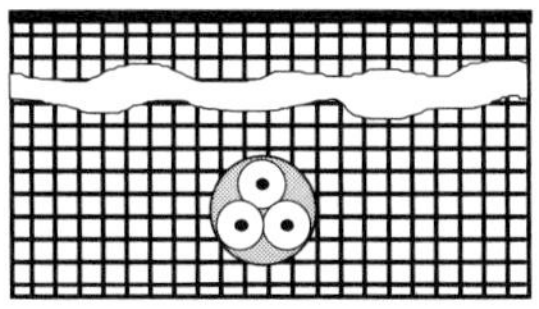	Sheathed, armoured or multicore cables direct in the ground: - without added mechanical protection (see note)	**D**
73	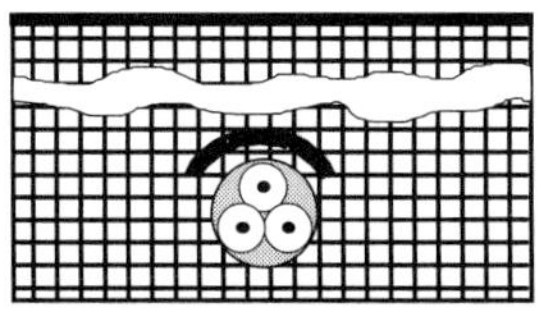	Sheathed, armoured or multicore cables direct in the ground: - with added mechanical protection (e.g. cable covers) (see note)	**D**

NOTE: The inclusion of directly buried cables is satisfactory where the soil thermal resistivity is of the order of 2.5 K.m/W. For lower soil resistivities, the current-carrying capacity for directly buried cables is appreciably higher than for cables in ducts.

TABLE 4A2 *(continued - Installation methods for flat twin and earth cables in thermal insulation)*

Installation Method			Reference Method to be used to determine current-carrying capacity
Number	**Examples**	**Description**	
100	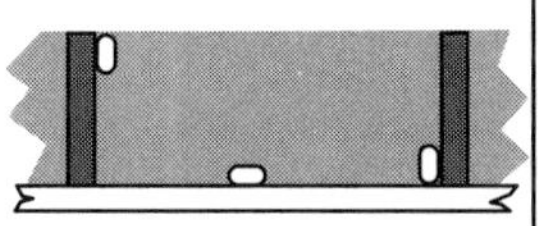	Installation methods for flat twin and earth cable clipped direct to a wooden joist, or touching the plasterboard ceiling surface, above a plasterboard ceiling with thermal insulation not exceeding 100 mm in thickness having a minimum U value of 0.1 W/m^2K	Table 4D5
101	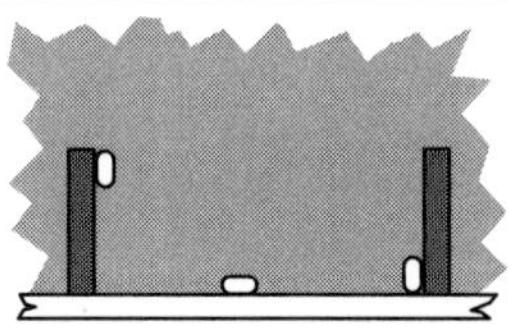	Installation methods for flat twin and earth cable clipped direct to a wooden joist, or touching the plasterboard ceiling surface, above a plasterboard ceiling with thermal insulation exceeding 100 mm in thickness having a minimum U value of 0.1 W/m^2K	Table 4D5
102	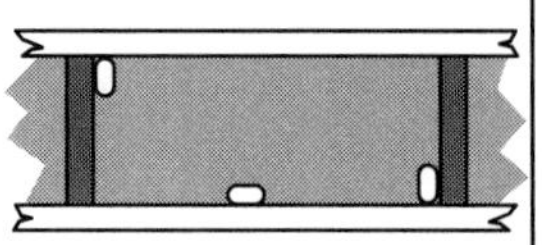	Installation methods for flat twin and earth cable in a stud wall with thermal insulation with a minimum U value of 0.1 W/m^2K with the cable touching the inner wall surface, or touching the plasterboard ceiling surface, and the inner skin having a minimum U value of 10 W/m^2K	Table 4D5
103	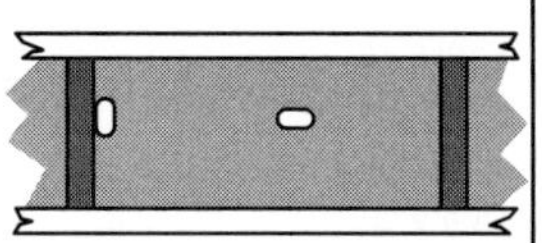	Installation methods for flat twin and earth cable in a stud wall with thermal insulation with a minimum U value of 0.1 W/m^2K with the cable not touching the inner wall surface	Table 4D5

Wherever practicable, a cable is to be fixed in a position such that it will not be covered with thermal insulation.

Regulation 523.9, BS 5803-5: Appendix C: Avoidance of overheating of electric cables, Building Regulations Approved Document B and Thermal insulation: avoiding risks, BR 262, BRE, 2001 refer.

TABLE 4A2 *(continued - Installation methods for cables enclosed in infloor concrete troughs)*

Installation Method			Reference Method to be used to determine current-carrying capacity
Number	**Examples**	**Description**	
117	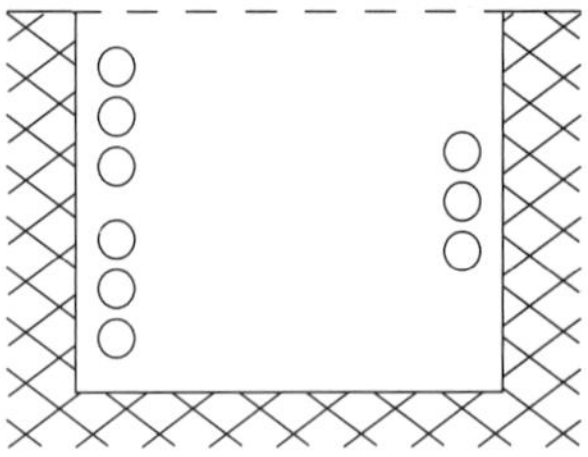	Cables supported on the wall of an open or ventilated infloor concrete trough with spacing as follows: - Sheathed single-core cables in free air (any supporting metalwork under the cables occupying less than 10 % of plan area). - Two or three cables vertically one above the other, minimum distance between cable surfaces equal to the overall cable diameters, distance from the wall not less than ½ the cable diameter. - Two or three cables horizontally with spacing as above.	**E or F**
118	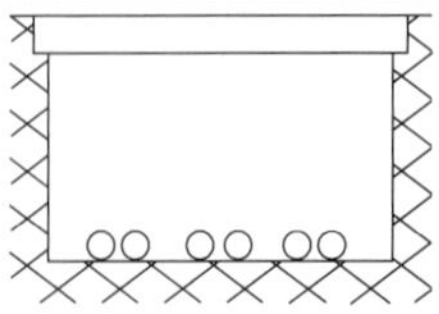	Cables in enclosed trench 450 mm wide by 300 mm deep (minimum dimensions) including 100 mm cover - Two to six single–core cables with surfaces separated by a minimum of one cable diameter	**E or F** **using rating factors in Table 4C6**
	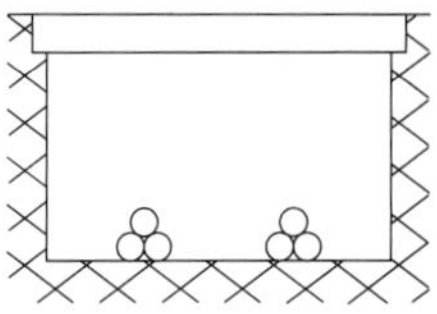	- One or two groups of three single-core cables in trefoil formation	
	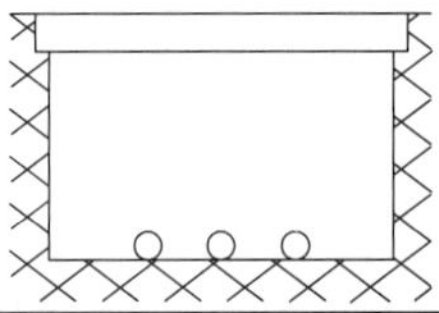	- One to four 2-core cables or one to three cables of 3 or 4 cores with all cables separated by a minimum of 50 mm	
119	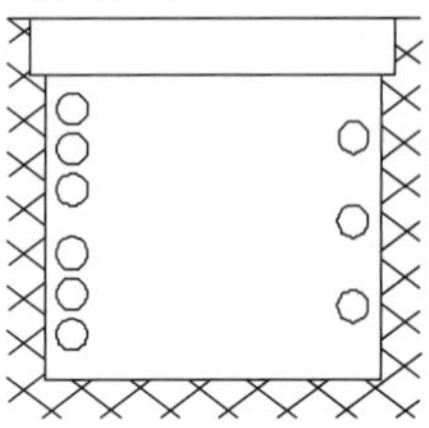	Cables enclosed in an infloor concrete trough 450 mm wide by 600 mm deep (minimum dimensions) including 100 mm cover. Six to twelve single-core cables arranged in flat groups of two or three on the vertical trench wall with cables separated by one cable diameter and a minimum of 50 mm between groups.	**E or F** **using rating factors in Table 4C6**
	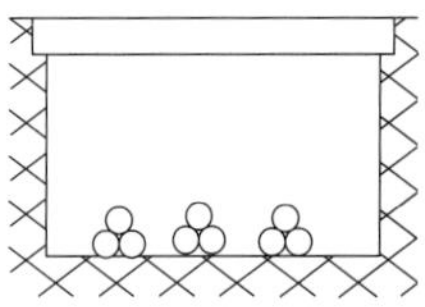	or two to four groups of three single-core cables in trefoil formation with a minimum of 50 mm between trefoil formations.	
	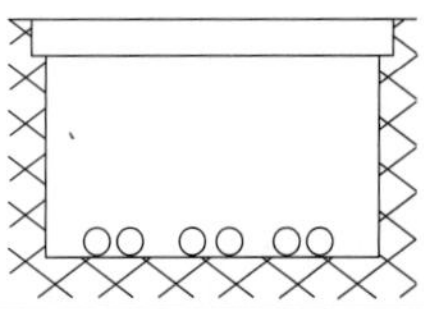	or four to eight 2-core cables or three to six cables of 3 or 4 cores with cables separated by a minimum of 75 mm. All cables spaced at least 25 mm from trench wall.	

TABLE 4A2 *(continued - Installation methods for cables enclosed in floor concrete troughs)*

Installation Method			Reference Method to be used to determine current-carrying capacity
Number	**Examples**	**Description**	
120	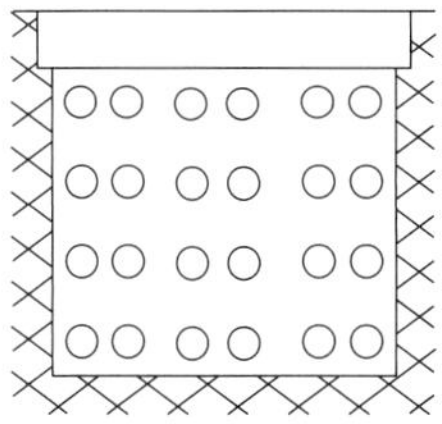 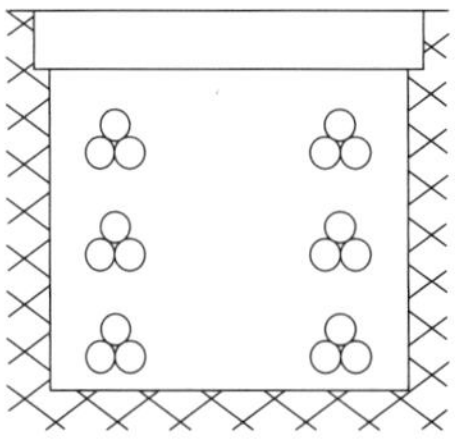 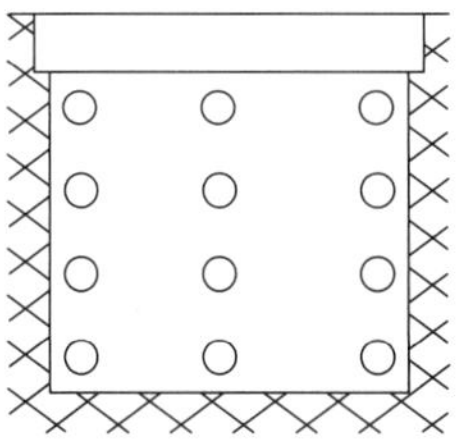	Cables enclosed in an infloor concrete trough 600 mm wide by 760 mm deep (minimum dimensions) including 100 mm cover. Twelve to twenty-four single-core cables arranged in either flat formation of two or three cables in a group with cables separated by one cable diameter and each cable group separated by a minimum of 50 mm either horizontally or vertically or single-core cables in trefoil formation with each group or trefoil formation separated by a minimum of 50 mm either horizontally or vertically or eight to sixteen 2-core cables or six to twelve cables of 3 or 4 cores with cables separated by a minimum of 75 mm either horizontally or vertically. All cables spaced at least 25 mm from trench wall.	**E or F** **using rating factors in Table 4C6**

TABLE 4A3 – Schedule of cable specifications and current rating tables

Specification number	Specification title	Applicable current rating Tables	Conductor operating temperature
BS 5467	Electric cables – Thermosetting insulated, armoured cables for voltages of 600/1000 V and 1900/3300 V.	4E3, 4E4	90 °C
BS 6004	Electric cables – PVC insulated, non-armoured cables for voltages up to and including 450/750 V, for electric power, lighting and internal wiring (fixed installation).	4D1, 4D2, 4D5	70 °C
	Low temperature PVC insulated and sheathed flexible cable (flexible cables).	4F3	60 °C
BS 6346 (withdrawn) Retained here for historical purposes	Electric cables – PVC insulated, armoured cables for voltages of 600/1000 V and 1900/3300 V.	4D3, 4D4, 4H3, 4H4, 4J3, 4J4	70 °C
BS 6724	Electric cables – Thermosetting insulated, armoured cables for voltages of 600/1000 V and 1900/3300 V, having low emission of smoke and corrosive gases when affected by fire.	4E3, 4E4	90 °C
BS 7211	Electric cables – Thermosetting insulated, non-armoured cables for voltages up to and including 450/750 V, for electric power, lighting and internal wiring, and having low emission of smoke and corrosive gases when affected by fire.	4E1, 4E2	90 °C
BS 7629-1	Electric cables – Specification for 300/500 V fire-resistant electric cables having low emission of smoke and corrosive gases when affected by fire - Part 1: Multicore cables.	4D2	70 °C
BS 7846	Electric cables – Thermosetting insulated, armoured, fire-resistant cables of rated voltage 600/1000 V having low emission of smoke and corrosive gases when affected by fire. Specification	4E3, 4E4	90 °C
BS 7889	Electric cables – Thermosetting insulated, PVC sheathed, unarmoured cables for a voltage of 600/1000 V.	4E1, 4E2	90 °C
BS 8436	Electric cables – 300/500 V screened electric cables having low emission of smoke and corrosive gases when affected by fire, for use in walls, partitions and building voids - multicore cables.	4D2	70 °C
BS 8573	Electric cables – Thermosetting insulated, non-armoured cables with a voltage of 600/1000 V, for fixed installations, having low emission of smoke and corrosive gases when affected by fire.	4E1, 4E2	90 °C
BS 8592	Electric Cables – Thermosetting insulated, non-armoured, fire-resistant, single core non-sheathed cables of rated voltage 450/750 V, having low emission of smoke and corrosive gases when affected by fire.	4E1	90 °C
BS EN 50525-2-11, 2-12, 2-21, 2-51, 2-82, 3-11, 3-21	Electric cables – Low voltage energy cables of rated voltages up to and including 450/750 V (flexible cables).	4F1, 4F2, 4F3	60 °C, 90 °C 110 °C, 180 °C
BS EN 50525-2-31, 3-41	Electric cables – Low voltage energy cables of rated voltages up to and including 450/750 V (fixed installation)	4D1, 4E1	70 °C, 90 °C
BS EN 60702-1	Mineral insulated cables and their terminations with a rated voltage not exceeding 750 V – Part 1: cables.	4G1, 4G2	70 °C**, 105 °C**

** Sheath operating temperature.

TABLE 4B1 – Rating factors (C_a) for ambient air temperatures other than 30 °C

Ambient temperature[a] °C	Insulation				
				Mineral[a]	
	60 °C thermosetting	70 °C thermoplastic	90 °C thermosetting	Thermoplastic covered or bare and exposed to touch 70 °C	Bare and not exposed to touch 105 °C
25	1.04	1.03	1.02	1.07	1.04
30	1.00	1.00	1.00	1.00	1.00
35	0.91	0.94	0.96	0.93	0.96
40	0.82	0.87	0.91	0.85	0.92
45	0.71	0.79	0.87	0.78	0.88
50	0.58	0.71	0.82	0.67	0.84
55	0.41	0.61	0.76	0.57	0.80
60	–	0.50	0.71	0.45	0.75
65	–	–	0.65	–	0.70
70	–	–	0.58	–	0.65
75	–	–	0.50	–	0.60
80	–	–	0.41	–	0.54
85	–	–	–	–	0.47
90	–	–	–	–	0.40
95	–	–	–	–	0.32

a For higher ambient temperatures, consult manufacturer.

TABLE 4B2 – Rating factors (C_a) for ambient ground temperatures other than 20 °C

Ground temperature °C	Insulation	
	70 °C thermoplastic	90 °C thermosetting
10	1.10	1.07
15	1.05	1.04
20	1.00	1.00
25	0.95	0.96
30	0.89	0.93
35	0.84	0.89
40	0.77	0.85
45	0.71	0.80
50	0.63	0.76
55	0.55	0.71
60	0.45	0.65
65	–	0.60
70	–	0.53
75	–	0.46
80	–	0.38

TABLE 4B3 – Rating factors (C_s) for cables buried direct in the ground or in an underground conduit system to BS EN 50086-2-4 for soil thermal resistivities other than 2.5 K.m/W to be applied to the current-carrying capacities for Reference Method D

Thermal resistivity, K.m/W	0.5	0.8	1	1.2	1.5	2	2.5	3
Rating factor for cables in buried ducts	1.28	1.20	1.18	1.13	1.1	1.05	1	0.96
Rating factor for direct buried cables	1.88	1.62	1.5	1.40	1.28	1.12	1	0.90

NOTE 1: The rating factors given have been averaged over the range of conductor sizes and types of installation included in the relevant tables in this appendix. The overall accuracy of rating factors is within ± 5 %.

NOTE 2: Where more precise values are required they may be calculated by methods given in BS 7769 (BS IEC 60287).

NOTE 3: The rating factors are applicable to ducts buried at depths of up to 0.8 m.

TABLE 4B4 – Rating factors (C_d) for depths of laying other than 0.7 m for direct buried cables and cables in buried ducts

Depth of laying, m	Buried direct	In buried ducts
0.5	1.03	1.02
0.7	1.00	1.00
1	0.97	0.98
1.25	0.95	0.96
1.5	0.94	0.95
1.75	0.93	0.94
2	0.92	0.93
2.5	0.90	0.92
3	0.89	0.91

TABLE 4B5 – Rating factors for cables having more than 4 loaded cores

Number of loaded cores	5	6	7	10	12	14	19
Rating factor	0.72	0.67	0.63	0.56	0.53	0.51	0.45
Number of loaded cores	24	27	30	37	44	46	48
Rating factor	0.42	0.40	0.39	0.36	0.34	0.33	0.33

NOTE 1: The current-carrying capacity for a cable in the size range 1.5 to 4 mm², having more than 4 loaded cores, is obtained by multiplying the current-carrying capacity of a 2-core, having the same insulation type, by the factor selected from this table. The current-carrying capacity for the 2-core cable is that for the installation condition to be used for the multicore cable.

NOTE 2: If, due to known operating conditions, a core is expected to carry not more than 30 % of its current-carrying capacity in the multicore cable it may be ignored for the purpose of determining the number of cores in the cable.

NOTE 3: If, due to known operating conditions, a core is expected to carry not more than 30 % of its rating, after applying the rating factor for the total number of current-carrying cores, it may be ignored for the purpose of obtaining the rating factor for the number of loaded cores.

For example, the current-carrying capacity of a cable having N loaded cores would normally be obtained by multiplying the current-carrying capacity of a 2-core, having the same insulation type, by the factor selected from this table for N cores. That is $I_{z1c} = I_{t2c} \times C_{gN}$

where:

I_{z1c} is the current-carrying capacity of the multicore cable after applying the rating factor for the total number of current-carrying cores

I_{t2c} is the tabulated current-carrying capacity of a 2-core cable, having the same insulation type as the multicore cable

C_{gN} is the rating factor from Table 4B5 for the total number of current-carrying cores

However, if M cores in the cable carry loads which are not greater than $0.3 \times I_{t2c} \times C_{gN}$, the current-carrying capacity can be obtained by using the rating factor corresponding to (N-M) cores.

The 'not greater than $0.3 \times I_{t2c} \times C_{gN}$' calculation should be applied before the adjacent multicore cable grouping factor, if applicable, from Table 4C1. The 30 % rule should not be further applied to any adjacent cable grouping factor calculations.

I_{z1c} should be greater than or equal to I_n or I_b as appropriate, divided by the relevant rating factor(s) C, that is $I_{z1c} \geq I_n$ or I_b / C

TABLE 4C1 – Rating factors for one circuit or one multicore cable or for a group of circuits, or a group of multicore cables, to be used with current-carrying capacities of Tables 4D1A to 4J4A

Item	Arrangement (cables touching)	Number of circuits or multicore cables												To be used with current-carrying capacities, Reference Method
		1	2	3	4	5	6	7	8	9	12	16	20	
1.	Bunched in air, on a surface, embedded or enclosed	1.00	0.80	0.70	0.65	0.60	0.57	0.54	0.52	0.50	0.45	0.41	0.38	A to F
2.	Single layer on wall or floor	1.00	0.85	0.79	0.75	0.73	0.72	0.72	0.71	0.70	0.70	0.70	0.70	C
3.	Single layer multicore on a perforated horizontal or vertical cable tray system	1.00	0.88	0.82	0.77	0.75	0.73	0.73	0.72	0.72	0.72	0.72	0.72	E
4.	Single layer multicore on cable ladder system or cleats etc.	1.00	0.87	0.82	0.80	0.80	0.79	0.79	0.78	0.78	0.78	0.78	0.78	

NOTE 1: These factors are applicable to uniform groups of cables, equally loaded.

NOTE 2: Where horizontal clearances between adjacent cables exceed twice their overall diameter, no rating factor need be applied.

NOTE 3: The same factors are applied to:
- groups of two or three single-core cables;
- multicore cables.

NOTE 4: If a group consists of both two- and three-core cables, the total number of cables is taken as the number of circuits, and the corresponding factor is applied to the tables for two loaded conductors for the two-core cables, and to the Tables for three loaded conductors for the three-core cables.

NOTE 5: If a group consists of *n* single-core cables it may either be considered as *n*/2 circuits of two loaded conductors or *n*/3 circuits of three loaded conductors.

NOTE 6: The rating factors given have been averaged over the range of conductor sizes and types of installation included in Tables 4D1A to 4J4A and the overall accuracy of tabulated values is within 5 %.

NOTE 7: For some installations and for other methods not provided for in the above table, it may be appropriate to use factors calculated for specific cases, see for example Tables 4C4 and 4C5.

NOTE 8: Where cables having differing conductor operating temperature are grouped together, the current rating is to be based upon the lowest operating temperature of any cable in the group.

NOTE 9: If, due to known operating conditions, a cable is expected to carry not more than 30 % of its *grouped* rating, it may be ignored for the purpose of obtaining the rating factor for the rest of the group.

For example, a group of N loaded cables would normally require a group rating factor of C_g applied to the tabulated I_t.

However, if M cables in the group carry loads which are not greater than 0.3 C_gI_t amperes the other cables can be sized by using the group rating factor corresponding to (N-M) cables.

TABLE 4C2 – Rating factors for more than one circuit, cables buried directly in the ground – Reference Method D in Tables 4D4A to 4J4A multicore cables

Number of circuits	Cable-to-cable clearance (α)				
	Nil (cables touching)	One cable diameter	0.125 m	0.25 m	0.5 m
2	0.75	0.80	0.85	0.90	0.90
3	0.65	0.70	0.75	0.80	0.85
4	0.60	0.60	0.70	0.75	0.80
5	0.55	0.55	0.65	0.70	0.80
6	0.50	0.55	0.60	0.70	0.80

Multicore cables

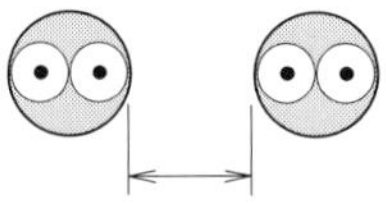

α

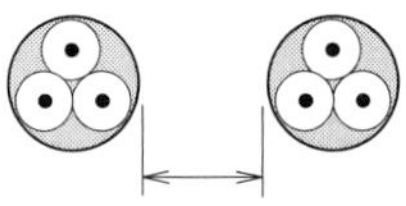

α

NOTE 1: Values given apply to an installation depth of 0.7 m and a soil thermal resistivity of 2.5 K.m/W. These are average values for the range of cable sizes and types quoted for Tables 4D4A to 4J4A. The process of averaging, together with rounding off, can result in some cases in errors of up to ±10 %. (Where more precise values are required they may be calculated by methods given in BS 7769 (BS IEC 60287).)

NOTE 2: In case of a thermal resistivity lower than 2.5 K.m/W the rating factors can, in general, be increased and can be calculated by the methods given in BS 7769 (BS IEC 60287).

TABLE 4C3 – Rating factors for more than one circuit, single cables in ducts buried in the ground – Reference Method D in Tables 4D4A to 4J4A (Multicore cables in single-way ducts)

Number of ducts	Duct-to-duct clearance (α)			
	Nil (ducts touching)	0.25 m	0.5 m	1.0 m
2	0.85	0.90	0.95	0.95
3	0.75	0.85	0.90	0.95
4	0.70	0.80	0.85	0.90
5	0.65	0.80	0.85	0.90
6	0.60	0.80	0.80	0.90

Multicore cables

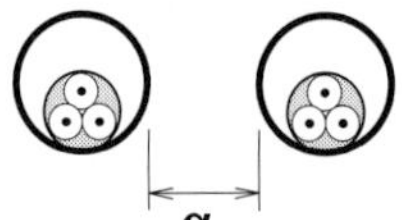

α

NOTE 1: Values given apply to an installation depth of 0.7 m and a soil thermal resistivity of 2.5 K.m/W. They are average values for the range of cable sizes and types quoted for Tables 4D4A to 4J4A. The process of averaging, together with rounding off, can result in some cases in errors of up to ±10 %. (Where more precise values are required they may be calculated by methods given in BS 7769 (BS IEC 60287).)

NOTE 2: In case of a thermal resistivity lower than 2.5 K.m/W the rating factors can, in general, be increased and can be calculated by the methods given in BS 7769 (BS IEC 60287).

TABLE 4C4 – Rating factors for groups of more than one multicore cable, to be applied to reference current-carrying capacities for multicore cables in free air – Reference Method E in Tables 4D2A to 4J4A

Installation Method in Table 4A2			Number of trays or ladders	1	2	3	4	6	9
Perforated cable tray systems (Note 3)	31	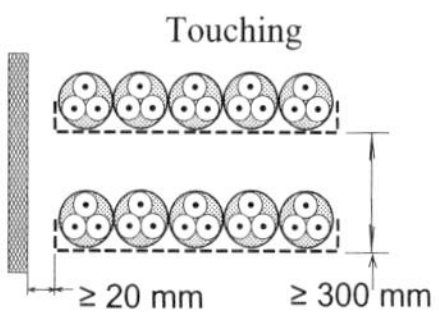	1	See item 3 of Table 4C1					
			2	1.00	0.87	0.80	0.77	0.73	0.68
			3	1.00	0.86	0.79	0.76	0.71	0.66
			6	1.00	0.84	0.77	0.73	0.68	0.64
		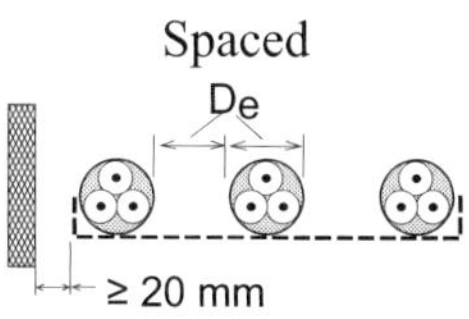	1	1.00	1.00	0.98	0.95	0.91	–
			2	1.00	0.99	0.96	0.92	0.87	–
			3	1.00	0.98	0.95	0.91	0.85	–
Vertical perforated cable tray systems (Note 4)	31	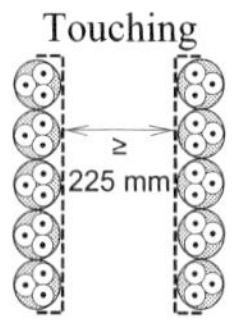	1	See item 3 of Table 4C1					
			2	1.00	0.88	0.81	0.76	0.71	0.70
		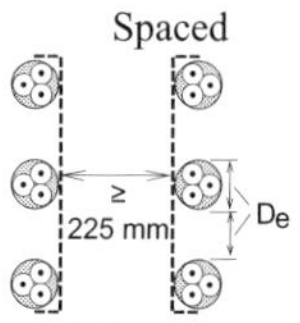	1	1.00	0.91	0.89	0.88	0.87	–
			2	1.00	0.91	0.88	0.87	0.85	–
Unperforated cable tray systems	30	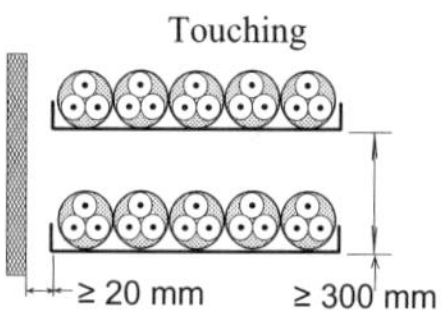	1	0.97	0.84	0.78	0.75	0.71	0.68
			2	0.97	0.83	0.76	0.72	0.68	0.63
			3	0.97	0.82	0.75	0.71	0.66	0.61
			6	0.97	0.81	0.73	0.69	0.63	0.58
Cable ladder systems, cleats, wire mesh tray, etc. (Note 3)	32 33 34	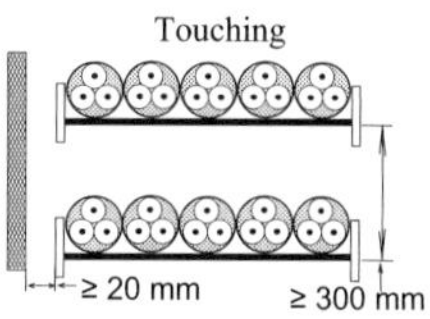	1	See item 4 of Table 4C1					
			2	1.00	0.86	0.80	0.78	0.76	0.73
			3	1.00	0.85	0.79	0.76	0.73	0.70
			6	1.00	0.84	0.77	0.73	0.68	0.64
		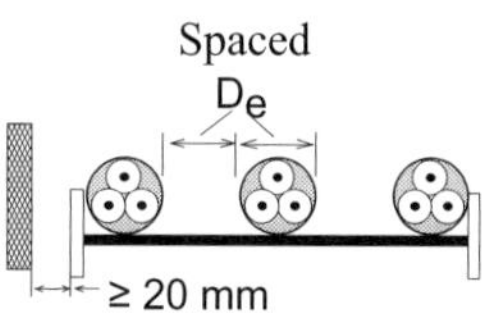	1	1.00	1.00	1.00	1.00	1.00	–
			2	1.00	0.99	0.98	0.97	0.96	–
			3	1.00	0.98	0.97	0.96	0.93	–

NOTE 1: Values given are averages for the cable types and range of conductor sizes considered in Tables 4D2A to 4J4A. The spread of values is generally less than 5 %.

NOTE 2: Factors apply to single layer groups of cables as shown above and do not apply when cables are installed in more than one layer touching each other. Values for such installations may be significantly lower and must be determined by an appropriate method.

NOTE 3: Values are given for vertical spacing between cable trays of 300 mm and at least 20 mm between cable trays and wall. For closer spacing the factors should be reduced.

NOTE 4: Values are given for horizontal spacing between cable trays of 225 mm with cable trays mounted back-to-back. For closer spacing the factors should be reduced.

TABLE 4C5 – Rating factors for groups of one or more circuits of single-core cables to be applied to reference current-carrying capacity for one circuit of single-core cables in free air – Reference Method F in Tables 4D1A to 4J3A

Installation Method in Table 4A2			Number of trays or ladders	Number of three-phase circuits per tray or ladder			Use as a multiplier to rating for
				1	2	3	
Perforated cable tray systems (Note 3)	31	Touching 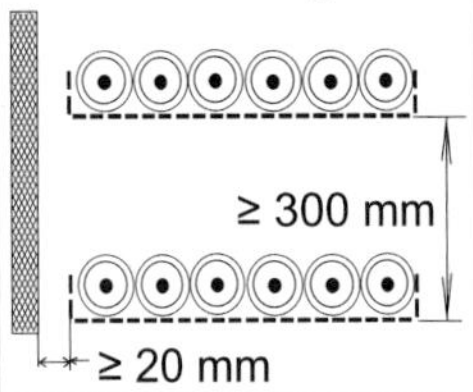	1	0.98	0.91	0.87	Three cables in horizontal formation
			2	0.96	0.87	0.81	
			3	0.95	0.85	0.78	
Vertical perforated cable tray systems (Note 4)	31	Touching 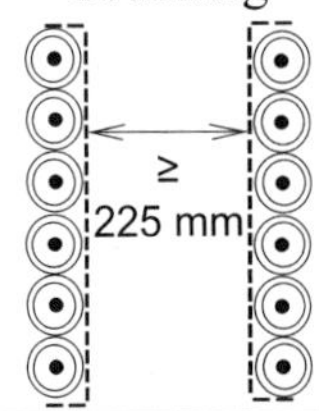	1	0.96	0.86	–	Three cables in vertical formation
			2	0.95	0.84	–	
Cable ladder systems, cleats, wire mesh tray, etc. (Note 3)	32	Touching 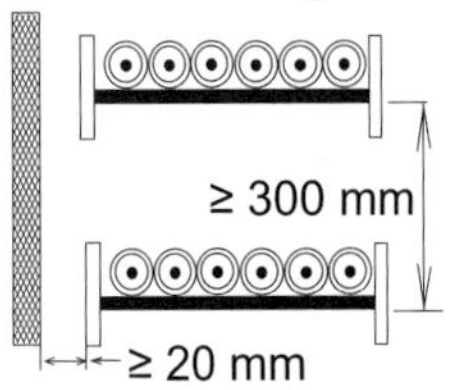	1	1.00	0.97	0.96	Three cables in horizontal formation
	33		2	0.98	0.93	0.89	
	34		3	0.97	0.90	0.86	
Perforated systems (Note 3)	31	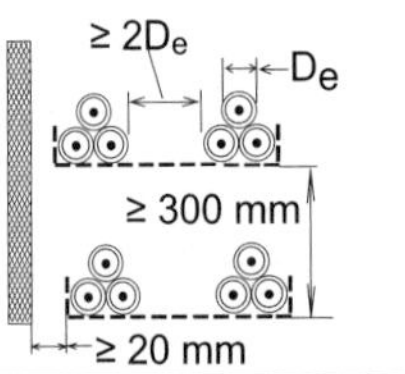	1	1.00	0.98	0.96	Three cables in trefoil formation
			2	0.97	0.93	0.89	
			3	0.96	0.92	0.86	
Vertical perforated cable tray systems (Note 4)	31	Spaced 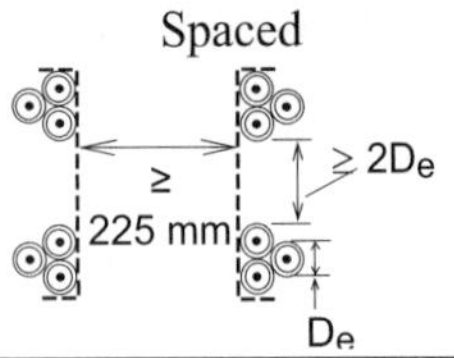	1	1.00	0.91	0.89	
			2	1.00	0.90	0.86	
Cable ladder systems, cleats, wire mesh tray, etc. (Note 3)	32	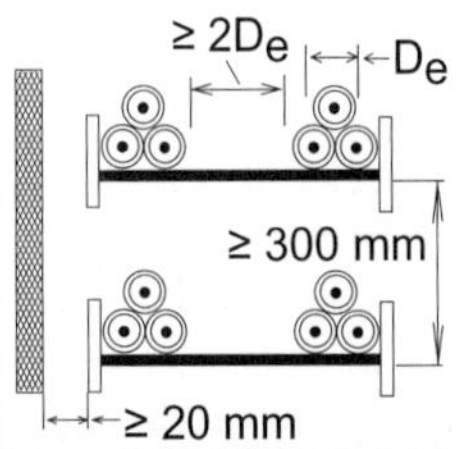	1	1.00	1.00	1.00	
	33		2	0.97	0.95	0.93	
	34		3	0.96	0.94	0.90	

NOTE 1: Values given are averages for the cable types and range of conductor sizes considered in Tables 4D1A to 4J3A. The spread of values is generally less than 5 %.

NOTE 2: Factors apply to single layer groups of cables (or trefoil groups) as shown above and do not apply when cables are installed in more than one layer touching each other. Values for such installations may be significantly lower and must be determined by an appropriate method.

NOTE 3: Values are given for vertical spacing between cable trays of 300 mm and at least 20 mm between cable trays and wall. For closer spacing the factors should be reduced.

NOTE 4: Values are given for horizontal spacing between cable trays of 225 mm with cable trays mounted back-to-back. For closer spacing the factors should be reduced.

NOTE 5: For circuits having more than one cable in parallel per phase, each three-phase set of conductors is to be considered as a circuit for the purpose of this table.

NOTE 6: Polyphase and neutral in a quad formation may also be acceptable. Reference should be made to manufacturers' specific data for current-carrying capacity and voltage drop values.

TABLE 4C6 – Rating factors for cables enclosed in infloor concrete troughs (Installation Methods 118 to 120 of Table 4A2)

The rating factors tabulated below relate to the disposition of cables illustrated in items 118 to120 of Table 4A2 and are applicable to the current-carrying capacities for Reference Methods E and F as given in the relevant tables of this appendix.

Conductor cross-sectional area	Rating factor									
	Installation method 118				Installation method 119			Installation method 120		
	2 single-core cables, or 1 three- or four-core cables	3 single-core cables, or 2 two-core cables	4 single-core cables, or 2 three- or four-core cables	6 single-core cables, 4 two-core cables, or 3 three- or four-core cables	6 single-core cables, 4 two-core cables, or 3 three- or four-core cables	8 single-core cables, or 4 three- or four-core cables	12 single-core cables, 8 two-core cables, or 6 three- or four-core cables	12 single-core cables, 8 two-core cables, or 6 three- or four-core cables	18 single-core cables, 12 two-core cables, or 9 three- or four-core cables	24 single-core cables, 16 two-core cables, or 12 three- or four-core cables
1	2	3	4	5	6	7	8	9	10	11
(mm^2)										
4	0.93	0.90	0.87	0.82	0.86	0.83	0.76	0.81	0.74	0.69
6	0.92	0.89	0.86	0.81	0.86	0.82	0.75	0.80	0.73	0.68
10	0.91	0.88	0.85	0.80	0.85	0.80	0.74	0.78	0.72	0.66
16	0.91	0.87	0.84	0.78	0.83	0.78	0.71	0.76	0.70	0.64
25	0.90	0.86	0.82	0.76	0.81	0.76	0.69	0.74	0.67	0.62
35	0.89	0.85	0.81	0.75	0.80	0.74	0.68	0.72	0.66	0.60
50	0.88	0.84	0.79	0.74	0.78	0.73	0.66	0.71	0.64	0.59
70	0.87	0.82	0.78	0.72	0.77	0.72	0.64	0.70	0.62	0.57
95	0.86	0.81	0.76	0.70	0.75	0.70	0.63	0.68	0.60	0.55
120	0.85	0.80	0.75	0.69	0.73	0.68	0.61	0.66	0.58	0.53
150	0.84	0.78	0.74	0.67	0.72	0.67	0.59	0.64	0.57	0.51
185	0.83	0.77	0.73	0.65	0.70	0.65	0.58	0.63	0.55	0.49
240	0.82	0.76	0.71	0.63	0.69	0.63	0.56	0.61	0.53	0.48
300	0.81	0.74	0.69	0.62	0.68	0.62	0.54	0.59	0.52	0.46
400	0.80	0.73	0.67	0.59	0.66	0.60	0.52	0.57	0.50	0.44
500	0.78	0.72	0.66	0.58	0.64	0.58	0.51	0.56	0.48	0.43
630	0.77	0.71	0.65	0.56	0.63	0.57	0.49	0.54	0.47	0.41

NOTES:

1. The factors in Table 4C6 are applicable to groups of cables all of one size. The value of current derived from application of the appropriate factors is the maximum current to be carried by any of the cables in the group.
2. If, due to known operating conditions, a cable is expected to carry not more than 30 % of its *grouped* rating, it may be ignored for the purpose of obtaining the rating factor for the rest of the group.
3. Where cables having different conductor operating temperatures are grouped together the current rating should be based on the lowest operating temperature of any cable in the group.
4. When the number of cables used differs from those stated in the table, the rating factor for the next higher stated number of cables should be used.

TABLE 4D1A – Single-core 70 °C thermoplastic insulated cables, non-armoured, with or without sheath (COPPER CONDUCTORS)

Ambient temperature: 30 °C
Conductor operating temperature: 70 °C

CURRENT-CARRYING CAPACITY (amperes):

Conductor cross-sectional area	Reference Method A (enclosed in conduit in thermally insulating wall etc.)		Reference Method B (enclosed in conduit on a wall or in trunking etc.)		Reference Method C (clipped direct)		Reference Method F (in free air or on a perforated cable tray horizontal or vertical)				
							Touching			Spaced by one diameter	
	2 cables, single-phase AC or DC	3 or 4 cables, three-phase AC	2 cables, single-phase AC or DC	3 or 4 cables, three-phase AC	2 cables, single-phase AC or DC flat and touching	3 or 4 cables, three-phase AC flat and touching or trefoil	2 cables, single-phase AC or DC flat	3 cables, three-phase AC flat	3 cables, three-phase AC trefoil	2 cables, single-phase AC or DC or 3 cables three-phase AC flat	
										Horizontal	Vertical
1	2	3	4	5	6	7	8	9	10	11	12
(mm²)	(A)	(A)	(A)	(A)	(A)	(A)	(A)	(A)	(A)	(A)	(A)
1	11	10.5	13.5	12	15.5	14	-	-	-	-	-
1.5	14.5	13.5	17.5	15.5	20	18	-	-	-	-	-
2.5	20	18	24	21	27	25	-	-	-	-	-
4	26	24	32	28	37	33	-	-	-	-	-
6	34	31	41	36	47	43	-	-	-	-	-
10	46	42	57	50	65	59	-	-	-	-	-
16	61	56	76	68	87	79	-	-	-	-	-
25	80	73	101	89	114	104	131	114	110	146	130
35	99	89	125	110	141	129	162	143	137	181	162
50	119	108	151	134	182	167	196	174	167	219	197
70	151	136	192	171	234	214	251	225	216	281	254
95	182	164	232	207	284	261	304	275	264	341	311
120	210	188	269	239	330	303	352	321	308	396	362
150	240	216	300	262	381	349	406	372	356	456	419
185	273	245	341	296	436	400	463	427	409	521	480
240	321	286	400	346	515	472	546	507	485	615	569
300	367	328	458	394	594	545	629	587	561	709	659
400	-	-	546	467	694	634	754	689	656	852	795
500	-	-	626	533	792	723	868	789	749	982	920
630	-	-	720	611	904	826	1005	905	855	1138	1070
800	-	-	-	-	1030	943	1086	1020	971	1265	1188

COPPER CONDUCTORS

NOTE:

For cables having flexible conductors, see section 2.4 of this Appendix for adjustment factors for current-carrying capacity and voltage drop.

VOLTAGE DROP (per ampere per metre): Conductor operating temperature: 70 °C

Conductor cross-sectional area	2 cables, DC	2 cables, single-phase AC									3 or 4 cables, three-phase AC											
		Reference Methods A & B (enclosed in conduit or trunking)			Reference Methods C & F (clipped direct, on tray or in free air)						Reference Methods A & B (enclosed in conduit or trunking)			Reference Methods C & F (clipped direct, on tray or in free air)								
					Cables touching			Cables spaced*						Cables touching, Trefoil			Cables touching, Flat			Cables spaced*, Flat		
1	2	3			4			5			6			7			8			9		
(mm²)	(mV/A/m)	(mV/A/m)			(mV/A/m)			(mV/A/m)			(mV/A/m)			(mV/A/m)			(mV/A/m)			(mV/A/m)		
1	44		44			44			44			38			38			38			38	
1.5	29		29			29			29			25			25			25			25	
2.5	18		18			18			18			15			15			15			15	
4	11		11			11			11			9.5			9.5			9.5			9.5	
6	7.3		7.3			7.3			7.3			6.4			6.4			6.4			6.4	
10	4.4		4.4			4.4			4.4			3.8			3.8			3.8			3.8	
16	2.8		2.8			2.8			2.8			2.4			2.4			2.4			2.4	
		r	x	z	r	x	z	r	x	z	r	x	z	r	x	z	r	x	z	r	x	z
25	1.75	1.80	0.33	1.80	1.75	0.20	1.75	1.75	0.29	1.80	1.50	0.29	1.55	1.50	0.175	1.50	1.50	0.25	1.55	1.50	0.32	1.55
35	1.25	1.30	0.31	1.30	1.25	0.195	1.25	1.25	0.28	1.30	1.10	0.27	1.10	1.10	0.170	1.10	1.10	0.24	1.10	1.10	0.32	1.15
50	0.93	0.95	0.30	1.00	0.93	0.190	0.95	0.93	0.28	0.97	0.81	0.26	0.85	0.80	0.165	0.82	0.80	0.24	0.84	0.80	0.32	0.86
70	0.63	0.65	0.29	0.72	0.63	0.185	0.66	0.63	0.27	0.69	0.56	0.25	0.61	0.55	0.160	0.57	0.55	0.24	0.60	0.55	0.31	0.63
95	0.46	0.49	0.28	0.56	0.47	0.180	0.50	0.47	0.27	0.54	0.42	0.24	0.48	0.41	0.155	0.43	0.41	0.23	0.47	0.40	0.31	0.51
120	0.36	0.39	0.27	0.47	0.37	0.175	0.41	0.37	0.26	0.45	0.33	0.23	0.41	0.32	0.150	0.36	0.32	0.23	0.40	0.32	0.30	0.44
150	0.29	0.31	0.27	0.41	0.30	0.175	0.34	0.29	0.26	0.39	0.27	0.23	0.36	0.26	0.150	0.30	0.26	0.23	0.34	0.26	0.30	0.40
185	0.23	0.25	0.27	0.37	0.24	0.170	0.29	0.24	0.26	0.35	0.22	0.23	0.32	0.21	0.145	0.26	0.21	0.22	0.31	0.21	0.30	0.36
240	0.180	0.195	0.26	0.33	0.185	0.165	0.25	0.185	0.25	0.31	0.17	0.23	0.29	0.160	0.145	0.22	0.160	0.22	0.27	0.160	0.29	0.34
300	0.145	0.160	0.26	0.31	0.150	0.165	0.22	0.150	0.25	0.29	0.14	0.23	0.27	0.130	0.140	0.190	0.130	0.22	0.25	0.130	0.29	0.32
400	0.105	0.130	0.26	0.29	0.120	0.160	0.20	0.115	0.25	0.27	0.12	0.22	0.25	0.105	0.140	0.175	0.105	0.21	0.24	0.100	0.29	0.31
500	0.086	0.110	0.26	0.28	0.098	0.155	0.185	0.093	0.24	0.26	0.10	0.22	0.25	0.086	0.135	0.160	0.086	0.21	0.23	0.081	0.29	0.30
630	0.068	0.094	0.25	0.27	0.081	0.155	0.175	0.076	0.24	0.25	0.08	0.22	0.24	0.072	0.135	0.150	0.072	0.21	0.22	0.066	0.28	0.29
800	0.053		-		0.068	0.150	0.165	0.061	0.24	0.25		-		0.060	0.130	0.145	0.060	0.21	0.22	0.053	0.28	0.29
1000	0.042		-		0.059	0.150	0.160	0.050	0.24	0.24		-		0.052	0.130	0.140	0.052	0.20	0.21	0.044	0.28	0.28

NOTE: * Spacings larger than one cable diameter will result in a larger voltage drop.

TABLE 4D2A – Multicore 70 °C thermoplastic insulated and thermoplastic sheathed cables, non-armoured (COPPER CONDUCTORS)

Ambient temperature: 30 °C

CURRENT-CARRYING CAPACITY (amperes): Conductor operating temperature: 70 °C

Conductor cross-sectional area	Reference Method A (enclosed in conduit in thermally insulating wall etc.)		Reference Method B (enclosed in conduit on a wall or in trunking etc.)		Reference Method C (clipped direct)		Reference Method E (in free air or on a perforated cable tray etc, horizontal or vertical)	
	1 two-core cable*, single-phase AC or DC	1 three-core cable* or 1 four-core cable, three-phase AC	1 two-core cable*, single-phase AC or DC	1 three-core cable* or 1 four-core cable, three-phase AC	1 two-core cable*, single-phase AC or DC	1 three-core cable* or 1 four-core cable, three-phase AC	1 two-core cable*, single-phase AC or DC	1 three-core cable* or 1 four-core cable, three-phase AC
1	2	3	4	5	6	7	8	9
(mm^2)	(A)	(A)	(A)	(A)	(A)	(A)	(A)	(A)
1	11	10	13	11.5	15	13.5	17	14.5
1.5	14	13	16.5	15	19.5	17.5	22	18.5
2.5	18.5	17.5	23	20	27	24	30	25
4	25	23	30	27	36	32	40	34
6	32	29	38	34	46	41	51	43
10	43	39	52	46	63	57	70	60
16	57	52	69	62	85	76	94	80
25	75	68	90	80	112	96	119	101
35	92	83	111	99	138	119	148	126
50	110	99	133	118	168	144	180	153
70	139	125	168	149	213	184	232	196
95	167	150	201	179	258	223	282	238
120	192	172	232	206	299	259	328	276
150	219	196	258	225	344	299	379	319
185	248	223	294	255	392	341	434	364
240	291	261	344	297	461	403	514	430
300	334	298	394	339	530	464	593	497
400	-	-	470	402	634	557	715	597

* with or without a protective conductor

NOTE:
For cables having flexible conductors, see section 2.4 of this Appendix for adjustment factors for current-carrying capacity and voltage drop.

TABLE 4D2B

VOLTAGE DROP (per ampere per metre): Conductor operating temperature: 70 °C

Conductor cross-sectional area	Two-core cable, DC	Two-core cable, single-phase AC			Three- or four-core cable, three-phase AC		
1	2	3			4		
(mm^2)	(mV/A/m)	(mV/A/m)			(mV/A/m)		
1	44	44			38		
1.5	29	29			25		
2.5	18	18			15		
4	11	11			9.5		
6	7.3	7.3			6.4		
10	4.4	4.4			3.8		
16	2.8	2.8			2.4		
		r	x	z	r	x	z
25	1.75	1.75	0.170	1.75	1.50	0.145	1.50
35	1.25	1.25	0.165	1.25	1.10	0.145	1.10
50	0.93	0.93	0.165	0.94	0.80	0.140	0.81
70	0.63	0.63	0.160	0.65	0.55	0.140	0.57
95	0.46	0.47	0.155	0.50	0.41	0.135	0.43
120	0.36	0.38	0.155	0.41	0.33	0.135	0.35
150	0.29	0.30	0.155	0.34	0.26	0.130	0.29
185	0.23	0.25	0.150	0.29	0.21	0.130	0.25
240	0.180	0.190	0.150	0.24	0.165	0.130	0.21
300	0.145	0.155	0.145	0.21	0.135	0.130	0.185
400	0.105	0.115	0.145	0.185	0.100	0.125	0.160

TABLE 4D3A – Single-core armoured 70 °C thermoplastic insulated cables (non-magnetic armour) (COPPER CONDUCTORS)

Ambient temperature: 30 °C

Conductor operating temperature: 70 °C

CURRENT-CARRYING CAPACITY (amperes):

Conductor cross-sectional area	Reference Method C (clipped direct)		Reference Method F (in free air or on a perforated cable tray, horizontal or vertical)								
	Touching		Touching			Spaced by one cable diameter					
	2 cables, single-phase AC or DC flat	3 or 4 cables, three-phase AC flat	2 cables, single-phase AC or DC flat	3 cables, three-phase AC flat	3 cables, three-phase AC trefoil	2 cables, DC		2 cables, single-phase AC		3 or 4 cables, three-phase AC	
						Horizontal	Vertical	Horizontal	Vertical	Horizontal	Vertical
1	2	3	4	5	6	7	8	9	10	11	12
(mm²)	(A)	(A)	(A)	(A)	(A)	(A)	(A)	(A)	(A)	(A)	(A)
50	193	179	205	189	181	229	216	229	217	230	212
70	245	225	259	238	231	294	279	287	272	286	263
95	296	269	313	285	280	357	340	349	332	338	313
120	342	309	360	327	324	415	396	401	383	385	357
150	393	352	413	373	373	479	458	449	429	436	405
185	447	399	469	422	425	548	525	511	489	490	456
240	525	465	550	492	501	648	622	593	568	566	528
300	594	515	624	547	567	748	719	668	640	616	578
400	687	575	723	618	657	885	851	737	707	674	632
500	763	622	805	673	731	1035	997	810	777	721	676
630	843	669	891	728	809	1218	1174	893	856	771	723
800	919	710	976	777	886	1441	1390	943	905	824	772
1000	975	737	1041	808	945	1685	1627	1008	967	872	816

TABLE 4D3B

VOLTAGE DROP (per ampere per metre): Conductor operating temperature: 70 °C

Conductor cross-sectional area	2 cables, DC	Reference Methods C & F (clipped direct, on tray or free air)														
		2 cables, single-phase AC						3 or 4 cables, three-phase AC								
		touching			spaced*			trefoil and touching			flat and touching			flat and spaced*		
1	2	3			4			5			6			7		
(mm²)	(mV/A/m)	(mV/A/m)			(mV/A/m)			(mV/A/m)			(mV/A/m)			(mV/A/m)		
		r	x	z	r	x	z	r	x	z	r	x	z	r	x	z
50	0.93	0.93	0.22	0.95	0.92	0.30	0.97	0.80	0.190	0.82	0.79	0.26	0.84	0.79	0.34	0.86
70	0.63	0.64	0.21	0.68	0.66	0.29	0.72	0.56	0.180	0.58	0.57	0.25	0.62	0.59	0.32	0.68
95	0.46	0.48	0.20	0.52	0.51	0.28	0.58	0.42	0.175	0.45	0.44	0.25	0.50	0.47	0.31	0.57
120	0.36	0.39	0.195	0.43	0.42	0.28	0.50	0.33	0.170	0.37	0.36	0.24	0.43	0.40	0.30	0.50
150	0.29	0.31	0.190	0.37	0.34	0.27	0.44	0.27	0.165	0.32	0.30	0.24	0.38	0.34	0.30	0.45
185	0.23	0.26	0.190	0.32	0.29	0.27	0.39	0.22	0.160	0.27	0.25	0.23	0.34	0.29	0.29	0.41
240	0.180	0.20	0.180	0.27	0.23	0.26	0.35	0.175	0.160	0.23	0.20	0.23	0.30	0.24	0.28	0.37
300	0.145	0.160	0.180	0.24	0.190	0.26	0.32	0.140	0.155	0.21	0.165	0.22	0.28	0.20	0.28	0.34
400	0.105	0.140	0.175	0.22	0.180	0.24	0.30	0.120	0.130	0.195	0.160	0.21	0.26	0.21	0.25	0.32
500	0.086	0.120	0.170	0.21	0.165	0.23	0.29	0.105	0.145	0.180	0.145	0.20	0.25	0.190	0.24	0.30
630	0.068	0.105	0.165	0.195	0.150	0.22	0.27	0.091	0.145	0.170	0.135	0.195	0.23	0.175	0.22	0.28
800	0.053	0.095	0.160	0.185	0.145	0.21	0.25	0.082	0.140	0.160	0.125	0.180	0.22	0.170	0.195	0.26
1000	0.042	0.091	0.155	0.180	0.140	0.190	0.24	0.079	0.135	0.155	0.125	0.165	0.21	0.165	0.170	0.24

NOTE: * Spacings larger than one cable diameter will result in a larger voltage drop.

TABLE 4D4A – Multicore armoured 70 °C thermoplastic insulated cables (COPPER CONDUCTORS)

Ambient temperature: 30 °C
Ground ambient temperature: 20 °C
Conductor operating temperature: 70 °C

CURRENT-CARRYING CAPACITY (amperes):

Conductor cross-sectional area	Reference Method C (clipped direct)		Reference Method E (in free air or on a perforated cable tray etc, horizontal or vertical)		Reference Method D (direct in ground or in ducting in ground, in or around buildings)	
	1 two-core cable, single-phase AC or DC	1 three- or four-core cable, three-phase AC	1 two-core cable, single-phase AC or DC	1 three- or four-core cable, three-phase AC	1 two-core cable, single-phase AC or DC	1 three- or four-core cable, three-phase AC
1	2	3	4	5	6	7
(mm^2)	(A)	(A)	(A)	(A)	(A)	(A)
1.5	21	18	22	19	22	18
2.5	28	25	31	26	29	24
4	38	33	41	35	37	30
6	49	42	53	45	46	38
10	67	58	72	62	60	50
16	89	77	97	83	78	64
25	118	102	128	110	99	82
35	145	125	157	135	119	98
50	175	151	190	163	140	116
70	222	192	241	207	173	143
95	269	231	291	251	204	169
120	310	267	336	290	231	192
150	356	306	386	332	261	217
185	405	348	439	378	292	243
240	476	409	516	445	336	280
300	547	469	592	510	379	316
400	621	540	683	590	-	-

TABLE 4D4B

VOLTAGE DROP (per ampere per metre): Conductor operating temperature: 70 °C

Conductor cross-sectional area	Two-core cable, DC	Two-core cable, single-phase AC			Three- or four-core cable, three-phase AC		
1	2	3			4		
(mm^2)	(mV/A/m)	(mV/A/m)			(mV/A/m)		
1.5	29	29			25		
2.5	18	18			15		
4	11	11			9.5		
6	7.3	7.3			6.4		
10	4.4	4.4			3.8		
16	2.8	2.8			2.4		
		r	x	z	r	x	z
25	1.75	1.75	0.170	1.75	1.50	0.145	1.50
35	1.25	1.25	0.165	1.25	1.10	0.145	1.10
50	0.93	0.93	0.165	0.94	0.80	0.140	0.81
70	0.63	0.63	0.160	0.65	0.55	0.140	0.57
95	0.46	0.47	0.155	0.50	0.41	0.135	0.43
120	0.36	0.38	0.155	0.41	0.33	0.135	0.35
150	0.29	0.30	0.155	0.34	0.26	0.130	0.29
185	0.23	0.25	0.150	0.29	0.21	0.130	0.25
240	0.180	0.190	0.150	0.24	0.165	0.130	0.21
300	0.145	0.155	0.145	0.21	0.135	0.130	0.185
400	0.105	0.115	0.145	0.185	0.100	0.125	0.160

TABLE 4D5 – 70 °C thermoplastic insulated and sheathed flat cable with protective conductor (COPPER CONDUCTORS)

Ambient temperature: 30 °C
Conductor operating temperature: 70 °C

CURRENT-CARRYING CAPACITY (amperes) and VOLTAGE DROP (per ampere per metre):

Conductor cross-sectional area	Method 100# (above a plasterboard ceiling covered by thermal insulation not exceeding 100 mm in thickness)	Method 101# (above a plasterboard ceiling covered by thermal insulation exceeding 100 mm in thickness)	Method 102# (in a stud wall with thermal insulation with cable touching the inner wall surface)	Method 103# (in a stud wall with thermal insulation with cable not touching the inner wall surface)	Reference Method A* (enclosed in conduit in an insulated wall)	Reference Method B* (enclosed in conduit on a wall or in trunking etc.)	Reference Method C* (clipped direct)	Voltage drop (per ampere per metre)
1	2	3	4	5	6	7	8	9
(mm^2)	(A)	(A)	(A)	(A)	(A)	(A)	(A)	(mV/A/m)
1	13	10.5	13	8	11.5	13	16	44
1.5	16	13	16	10	14.5	16.5	20	29
2.5	21	17	21	13.5	20	23	27	18
4	27	22	27	18.5	26	30	37	11
6	34	27	35	23.5	32	38	47	7.3
10	45	36	47	32	44	52	64	4.4
16	57	46	63	42.5	57	69	85	2.8

A* For full installation method refer to Table 4A2 Installation Method 2 but for flat twin and earth cable

B* For full installation method refer to Table 4A2 Installation Method 5, 8 or 9 but for flat twin and earth cable

C* For full installation method refer to Table 4A2 Installation Method 20 but for flat twin and earth cable

100# For full installation method refer to Table 4A2 Installation Method 100

101# For full installation method refer to Table 4A2 Installation Method 101

102# For full installation method refer to Table 4A2 Installation Method 102

103# For full installation method refer to Table 4A2 Installation Method 103

Wherever practicable, a cable is to be fixed in a position such that it will not be covered with thermal insulation.

Regulation 523.9, BS 5803-5: Appendix C: Avoidance of overheating of electric cables.

Building Regulations Approved Document B and Thermal insulation: avoiding risks, BR 262, BRE, 2001 refer.

THIS PAGE HAS BEEN INTENTIONALLY LEFT BLANK

TABLE 4E1A – Single-core 90 °C thermosetting insulated cables, non-armoured, with or without sheath (COPPER CONDUCTORS)

Ambient temperature: 30 °C
Conductor operating temperature: 90 °C

CURRENT-CARRYING CAPACITY (amperes):

Conductor cross-sectional area	Reference Method A (enclosed in conduit in thermally insulating wall etc.)		Reference Method B (enclosed in conduit on a wall or in trunking etc.)		Reference Method C (clipped direct)		Reference Method F (in free air or on a perforated cable tray etc horizontal or vertical etc) Touching			Reference Method G (in free air) Spaced by one cable diameter	
	2 cables, single-phase AC or DC	3 or 4 cables, three-phase AC	2 cables, single-phase AC or DC	3 or 4 cables, three-phase AC	2 cables, single-phase AC or DC flat and touching	3 or 4 cables, three-phase AC flat and touching or trefoil	2 cables, single-phase AC or DC flat	3 cables, three-phase AC flat	3 cables, three-phase AC trefoil	2 cables, single-phase AC or DC or 3 cables three-phase AC flat	
										Horizontal	Vertical
1	2	3	4	5	6	7	8	9	10	11	12
(mm²)	(A)	(A)	(A)	(A)	(A)	(A)	(A)	(A)	(A)	(A)	(A)
1	14	13	17	15	19	17.5	-	-	-	-	-
1.5	19	17	23	20	25	23	-	-	-	-	-
2.5	26	23	31	28	34	31	-	-	-	-	-
4	35	31	42	37	46	41	-	-	-	-	-
6	45	40	54	48	59	54	-	-	-	-	-
10	61	54	75	66	81	74	-	-	-	-	-
16	81	73	100	88	109	99	-	-	-	-	-
25	106	95	133	117	143	130	161	141	135	182	161
35	131	117	164	144	176	161	200	176	169	226	201
50	158	141	198	175	228	209	242	216	207	275	246
70	200	179	253	222	293	268	310	279	268	353	318
95	241	216	306	269	355	326	377	342	328	430	389
120	278	249	354	312	413	379	437	400	383	500	454
150	318	285	393	342	476	436	504	464	444	577	527
185	362	324	449	384	545	500	575	533	510	661	605
240	424	380	528	450	644	590	679	634	607	781	719
300	486	435	603	514	743	681	783	736	703	902	833
400	-	-	683	584	868	793	940	868	823	1085	1008
500	-	-	783	666	990	904	1083	998	946	1253	1169
630	-	-	900	764	1130	1033	1254	1151	1088	1454	1362
800	-	-	-	-	1288	1179	1358	1275	1214	1581	1485

NOTES:

1. Where it is intended to connect the cables in this table to equipment or accessories designed to operate at a temperature lower than the maximum operating temperature of the cable, the cables should be rated at the maximum operating temperature of the equipment or accessory (see Regulation 512.1.5).
2. Where it is intended to group a cable in this table with other cables, the cable should be rated at the lowest of the maximum operating temperatures of any of the cables in the group (see Regulation 512.1.5).
3. For cables having flexible conductors see section 2.4 of this appendix for adjustment factors for current-carrying capacity and voltage drop.

TABLE 4E1B

VOLTAGE DROP (per ampere per metre):

Conductor operating temperature: 90°C

Conductor cross-sectional area	2 cables, DC	2 cables, single-phase AC									3 or 4 cables, three-phase AC											
		Reference Methods A & B (enclosed in conduit or trunking)			References Methods C, F & G (clipped direct, on tray or in free air)						Reference Methods A & B (enclosed in conduit or trunking)			Reference Methods C, F & G (clipped direct, on tray or in free air)								
					Cables touching			Cables spaced*						Cables touching, Trefoil			Cables touching, Flat			Cables spaced*, Flat		
1	2	3			4			5			6			7			8			9		
(mm²)	(mV/A/m)	(mV/A/m)			(mV/A/m)			(mV/A/m)			(mV/A/m)			(mV/A/m)			(mV/A/m)			(mV/A/m)		
1	46	46			46			46			40			40			40			40		
1.5	31	31			31			31			27			27			27			27		
2.5	19	19			19			19			16			16			16			16		
4	12	12			12			12			10			10			10			10		
6	7.9	7.9			7.9			7.9			6.8			6.8			6.8			6.8		
10	4.7	4.7			4.7			4.7			4.0			4.0			4.0			4.0		
16	2.9	2.9			2.9			2.9			2.5			2.5			2.5			2.5		
		r	x	z	r	x	z	r	x	z	r	x	z	r	x	z	r	x	z	r	x	z
25	1.85	1.85	0.31	1.90	1.85	0.190	1.85	1.85	0.28	1.85	1.60	0.27	1.65	1.60	0.165	1.60	1.60	0.190	1.60	1.60	0.27	1.65
35	1.35	1.35	0.29	1.35	1.35	0.180	1.35	1.35	0.27	1.35	1.15	0.25	1.15	1.15	0.155	1.15	1.15	0.180	1.15	1.15	0.26	1.20
50	0.99	1.00	0.29	1.05	0.99	0.180	1.00	0.99	0.27	1.00	0.87	0.25	0.90	0.86	0.155	0.87	0.86	0.180	0.87	0.86	0.26	0.89
70	0.68	0.70	0.28	0.75	0.68	0.175	0.71	0.68	0.26	0.73	0.60	0.24	0.65	0.59	0.150	0.61	0.59	0.175	0.62	0.59	0.25	0.65
95	0.49	0.51	0.27	0.58	0.49	0.170	0.52	0.49	0.26	0.56	0.44	0.23	0.50	0.43	0.145	0.45	0.43	0.170	0.46	0.43	0.25	0.49
120	0.39	0.41	0.26	0.48	0.39	0.165	0.43	0.39	0.25	0.47	0.35	0.23	0.42	0.34	0.140	0.37	0.34	0.165	0.38	0.34	0.24	0.42
150	0.32	0.33	0.26	0.43	0.32	0.165	0.36	0.32	0.25	0.41	0.29	0.23	0.37	0.28	0.140	0.31	0.28	0.165	0.32	0.28	0.24	0.37
185	0.25	0.27	0.26	0.37	0.26	0.165	0.30	0.25	0.25	0.36	0.23	0.23	0.32	0.22	0.140	0.26	0.22	0.165	0.28	0.22	0.24	0.33
240	0.190	0.21	0.26	0.33	0.20	0.160	0.25	0.195	0.25	0.31	0.185	0.22	0.29	0.170	0.140	0.22	0.170	0.165	0.24	0.170	0.24	0.29
300	0.155	0.175	0.25	0.31	0.160	0.160	0.22	0.155	0.25	0.29	0.150	0.22	0.27	0.140	0.140	0.195	0.135	0.160	0.21	0.135	0.24	0.27
400	0.120	0.140	0.25	0.29	0.130	0.155	0.20	0.125	0.24	0.27	0.125	0.22	0.25	0.110	0.135	0.175	0.110	0.160	0.195	0.110	0.24	0.26
500	0.093	0.120	0.25	0.28	0.105	0.155	0.185	0.098	0.24	0.26	0.100	0.22	0.24	0.090	0.135	0.160	0.088	0.160	0.180	0.085	0.24	0.25
630	0.072	0.100	0.25	0.27	0.086	0.155	0.175	0.078	0.24	0.25	0.088	0.21	0.23	0.074	0.135	0.150	0.071	0.160	0.170	0.068	0.23	0.24
800	0.056		-		0.072	0.150	0.170	0.064	0.24	0.25		-		0.062	0.130	0.145	0.059	0.155	0.165	0.055	0.23	0.24
1000	0.045		-		0.063	0.150	0.165	0.054	0.24	0.24		-		0.055	0.130	0.140	0.050	0.155	0.165	0.047	0.23	0.24

NOTE: * Spacings larger than one cable diameter will result in a larger voltage drop.

TABLE 4E2A – Multicore 90 °C thermosetting insulated and thermoplastic sheathed cables, non-armoured (COPPER CONDUCTORS)

COPPER CONDUCTORS

CURRENT-CARRYING CAPACITY (amperes):

Ambient temperature: 30 °C
Conductor operating temperature: 90 °C

Conductor cross-sectional area	Reference Method A (enclosed in conduit in thermally insulating wall etc.)		Reference Method B (enclosed in conduit on a wall or in trunking etc.)		Reference Method C (clipped direct)		Reference Method E (free air or on a perforated cable tray etc, horizontal or vertical)	
	1 two-core cable*, single-phase AC or DC	1 three- or four-core cable*, three-phase AC	1 two-core cable*, single-phase AC or DC	1 three- or four-core cable*, three-phase AC	1 two-core cable*, single-phase AC or DC	1 three- or four-core cable*, three-phase AC	1 two-core cable*, single-phase AC or DC	1 three- or four-core cable*, three-phase AC
1	2	3	4	5	6	7	8	9
(mm²)	(A)	(A)	(A)	(A)	(A)	(A)	(A)	(A)
1	14.5	13	17	15	19	17	21	18
1.5	18.5	16.5	22	19.5	24	22	26	23
2.5	25	22	30	26	33	30	36	32
4	33	30	40	35	45	40	49	42
6	42	38	51	44	58	52	63	54
10	57	51	69	60	80	71	86	75
16	76	68	91	80	107	96	115	100
25	99	89	119	105	138	119	149	127
35	121	109	146	128	171	147	185	158
50	145	130	175	154	209	179	225	192
70	183	164	221	194	269	229	289	246
95	220	197	265	233	328	278	352	298
120	253	227	305	268	382	322	410	346
150	290	259	334	300	441	371	473	399
185	329	295	384	340	506	424	542	456
240	386	346	459	398	599	500	641	538
300	442	396	532	455	693	576	741	621
400	-	-	625	536	803	667	865	741

* with or without a protective conductor

NOTES:

1. Where it is intended to connect the cables in this table to equipment or accessories designed to operate at a temperature lower than the maximum operating temperature of the cable, the cables should be rated at the maximum operating temperature of the equipment or accessory (see Regulation 512.1.5).
2. Where it is intended to group a cable in this table with other cables, the cable should be rated at the lowest of the maximum operating temperatures of any of the cables in the group (see Regulation 512.1.5).
3. For cables having flexible conductors see section 2.4 of this appendix for adjustment factors for current-carrying capacity and voltage drop.

TABLE 4E2B

VOLTAGE DROP (per ampere per metre): Conductor operating temperature: 90 °C

Conductor cross-sectional area	Two-core cable, DC	Two-core cable, single-phase AC			Three- or four-core cable, three-phase AC		
1	2	3			4		
(mm^2)	(mV/A/m)	(mV/A/m)			(mV/A/m)		
1	46	46			40		
1.5	31	31			27		
2.5	19	19			16		
4	12	12			10		
6	7.9	7.9			6.8		
10	4.7	4.7			4.0		
16	2.9	2.9			2.5		
		r	x	z	r	x	z
25	1.85	1.85	0.160	1.90	1.60	0.140	1.65
35	1.35	1.35	0.155	1.35	1.15	0.135	1.15
50	0.98	0.99	0.155	1.00	0.86	0.135	0.87
70	0.67	0.67	0.150	0.69	0.59	0.130	0.60
95	0.49	0.50	0.150	0.52	0.43	0.130	0.45
120	0.39	0.40	0.145	0.42	0.34	0.130	0.37
150	0.31	0.32	0.145	0.35	0.28	0.125	0.30
185	0.25	0.26	0.145	0.29	0.22	0.125	0.26
240	0.195	0.200	0.140	0.24	0.175	0.125	0.21
300	0.155	0.160	0.140	0.21	0.140	0.120	0.185
400	0.120	0.130	0.140	0.190	0.115	0.120	0.165

TABLE 4E3A – Single-core armoured 90 °C thermosetting insulated cables (non-magnetic armour) (COPPER CONDUCTORS)

Ambient temperature: 30 °C
Conductor operating temperature: 90 °C

CURRENT-CARRYING CAPACITY (amperes):

Conductor cross-sectional area	Reference Method C (clipped direct)		Reference Method F (in free air or on a perforated cable tray, horizontal or vertical)									
	Touching		Touching			Spaced by one cable diameter						
	2 cables, single-phase AC or DC flat	3 or 4 cables, three-phase AC flat	2 cables, single-phase AC or DC flat	3 cables, three-phase AC flat	3 cables, three-phase AC trefoil	2 cables, DC		2 cables, single-phase AC		3 or 4 cables, three-phase AC		
						Horizontal	Vertical	Horizontal	Vertical	Horizontal	Vertical
1	2	3	4	5	6	7	8	9	10	11	12
(mm²)	(A)	(A)	(A)	(A)	(A)	(A)	(A)	(A)	(A)	(A)	(A)
50	237	220	253	232	222	284	270	282	266	288	266
70	303	277	322	293	285	356	349	357	337	358	331
95	367	333	389	352	346	446	426	436	412	425	393
120	425	383	449	405	402	519	497	504	477	485	449
150	488	437	516	462	463	600	575	566	539	549	510
185	557	496	587	524	529	688	660	643	614	618	574
240	656	579	689	612	625	815	782	749	714	715	666
300	755	662	792	700	720	943	906	842	805	810	755
400	853	717	899	767	815	1137	1094	929	889	848	797
500	962	791	1016	851	918	1314	1266	1032	989	923	871
630	1082	861	1146	935	1027	1528	1474	1139	1092	992	940
800	1170	904	1246	987	1119	1809	1744	1204	1155	1042	978
1000	1261	961	1345	1055	1214	2100	2026	1289	1238	1110	1041

NOTES:

1. Where it is intended to connect the cables in this table to equipment or accessories designed to operate at a temperature lower than the maximum operating temperature of the cable, the cables should be rated at the maximum operating temperature of the equipment or accessory (see Regulation 512.1.5).
2. Where it is intended to group a cable in this table with other cables, the cable should be rated at the lowest of the maximum operating temperatures of any of the cables in the group (see Regulation 512.1.5).

TABLE 4E3B

VOLTAGE DROP (per ampere per metre): Conductor operating temperature: 90 °C

Conductor cross-sectional area	2 cables, DC	Reference Methods C & F (clipped direct, on tray or in free air)														
		2 cables, single-phase AC						3 or 4 cables, three-phase AC								
		touching			spaced*			trefoil and touching			flat and touching			flat and spaced*		
1	2	3			4			5			6			7		
(mm^2)	(mV/A/m)	(mV/A/m)			(mV/A/m)			(mV/A/m)			(mV/A/m)			(mV/A/m)		
		r	x	z	r	x	z	r	x	z	r	x	z	r	x	z
50	0.98	0.99	0.21	1.00	0.98	0.29	1.00	0.86	0.180	0.87	0.84	0.25	0.88	0.84	0.33	0.90
70	0.67	0.68	0.200	0.71	0.69	0.29	0.75	0.59	0.170	0.62	0.60	0.25	0.65	0.62	0.32	0.70
95	0.49	0.51	0.195	0.55	0.53	0.28	0.60	0.44	0.170	0.47	0.46	0.24	0.52	0.49	0.31	0.58
120	0.39	0.41	0.190	0.45	0.43	0.27	0.51	0.35	0.165	0.39	0.38	0.24	0.44	0.41	0.30	0.51
150	0.31	0.33	0.185	0.38	0.36	0.27	0.45	0.29	0.160	0.33	0.31	0.23	0.39	0.34	0.29	0.45
185	0.25	0.27	0.185	0.33	0.30	0.26	0.40	0.23	0.160	0.28	0.26	0.23	0.34	0.29	0.29	0.41
240	0.195	0.21	0.180	0.28	0.24	0.26	0.35	0.180	0.155	0.24	0.21	0.22	0.30	0.24	0.28	0.37
300	0.155	0.170	0.175	0.25	0.195	0.25	0.32	0.145	0.150	0.21	0.170	0.22	0.28	0.20	0.27	0.34
400	0.115	0.145	0.170	0.22	0.180	0.24	0.30	0.125	0.150	0.195	0.160	0.21	0.27	0.20	0.27	0.33
500	0.093	0.125	0.170	0.21	0.165	0.24	0.29	0.105	0.145	0.180	0.145	0.20	0.25	0.190	0.24	0.31
630	0.073	0.105	0.165	0.195	0.150	0.23	0.27	0.092	0.145	0.170	0.135	0.195	0.24	0.175	0.23	0.29
800	0.056	0.090	0.160	0.190	0.145	0.23	0.27	0.086	0.140	0.165	0.130	0.180	0.23	0.175	0.195	0.26
1000	0.045	0.092	0.155	0.180	0.140	0.21	0.25	0.080	0.135	0.155	0.125	0.170	0.21	0.165	0.180	0.24

NOTE: * Spacings larger than one cable diameter will result in a larger voltage drop.

TABLE 4E4A – Multicore armoured 90 °C thermosetting insulated cables (COPPER CONDUCTORS)

Air ambient temperature: 30 °C
Ground ambient temperature: 20 °C
Conductor operating temperature: 90 °C

CURRENT-CARRYING CAPACITY (amperes):

Conductor cross-sectional area	Reference Method C (clipped direct)		Reference Method E (in free air or on a perforated cable tray etc, horizontal or vertical)		Reference Method D (direct in ground or in ducting in ground, in or around buildings)	
	1 two-core cable, single-phase AC or DC	1 three- or 1 four-core cable, three-phase AC	1 two-core cable, single-phase AC or DC	1 three- or 1 four-core cable, three-phase AC	1 two-core cable, single-phase AC or DC	1 three- or 1 four-core cable, three-phase AC
1	2	3	4	5	6	7
(mm^2)	(A)	(A)	(A)	(A)	(A)	(A)
1.5	27	23	29	25	25	21
2.5	36	31	39	33	33	28
4	49	42	52	44	43	36
6	62	53	66	56	53	44
10	85	73	90	78	71	58
16	110	94	115	99	91	75
25	146	124	152	131	116	96
35	180	154	188	162	139	115
50	219	187	228	197	164	135
70	279	238	291	251	203	167
95	338	289	354	304	239	197
120	392	335	410	353	271	223
150	451	386	472	406	306	251
185	515	441	539	463	343	281
240	607	520	636	546	395	324
300	698	599	732	628	446	365
400	787	673	847	728	-	-

NOTES:

1. Where it is intended to connect the cables in this table to equipment or accessories designed to operate at a temperature lower than the maximum operating temperature of the cable, the cables should be rated at the maximum operating temperature of the equipment or accessory (see Regulation 512.1.5).
2. Where it is intended to group a cable in this table with other cables, the cable should be rated at the lowest of the maximum operating temperatures of any of the cables in the group (see Regulation 512.1.5).

TABLE 4E4B

VOLTAGE DROP (per ampere per metre): Conductor operating temperature: 90 °C

Conductor cross-sectional area	Two-core cable, DC	Two-core cable, single-phase AC			Three- or four-core cable, three-phase AC		
1	2	3			4		
(mm^2)	(mV/A/m)	(mV/A/m)			(mV/A/m)		
1.5	31	31			27		
2.5	19	19			16		
4	12	12			10		
6	7.9	7.9			6.8		
10	4.7	4.7			4.0		
16	2.9	2.9			2.5		
		r	x	z	r	x	z
25	1.85	1.85	0.160	1.90	1.60	0.140	1.65
35	1.35	1.35	0.155	1.35	1.15	0.135	1.15
50	0.98	0.99	0.155	1.00	0.86	0.135	0.87
70	0.67	0.67	0.150	0.69	0.59	0.130	0.60
95	0.49	0.50	0.150	0.52	0.43	0.130	0.45
120	0.39	0.40	0.145	0.42	0.34	0.130	0.37
150	0.31	0.32	0.145	0.35	0.28	0.125	0.30
185	0.25	0.26	0.145	0.29	0.22	0.125	0.26
240	0.195	0.20	0.140	0.24	0.175	0.125	0.21
300	0.155	0.16	0.140	0.21	0.140	0.120	0.185
400	0.120	0.13	0.140	0.190	0.115	0.120	0.165

TABLE 4F1A – 60 °C thermosetting insulated flexible cables with sheath, non-armoured (COPPER CONDUCTORS)

CURRENT-CARRYING CAPACITY (amperes): Ambient temperature: 30 °C
Conductor operating temperature: 60 °C

Conductor cross-sectional area	Single-phase AC or DC	Three-phase AC	Single-phase AC or DC
	1 two-core cable, with or without protective conductor	1 three-core, four-core or five-core cable	2 single-core cables, touching
1	2	3	4
(mm^2)	(A)	(A)	(A)
4	30	26	-
6	39	34	-
10	51	47	-
16	73	63	-
25	97	83	-
35	-	102	140
50	-	124	175
70	-	158	216
95	-	192	258
120	-	222	302
150	-	255	347
185	-	291	394
240	-	343	471
300	-	394	541
400	-	-	644
500	-	-	738
630	-	-	861

NOTES:

1. The current ratings tabulated are for cables in free air but may also be used for cables resting on a surface. If the cable is to be wound on a drum on load the ratings should be reduced in accordance with NOTE 2 below and for cables which may be covered, NOTE 3 below.

2. *Flexible cables wound on reeling drums*

   The current ratings of cables used on reeling drums are to be reduced by the following factors:

   a) Radial type drum
      - ventilated: 85 %
      - unventilated: 75 %

   b) Ventilated cylindrical type drum
      - 1 layer of cable: 85 %
      - 2 layers of cable: 65 %
      - 3 layers of cable: 45 %
      - 4 layers of cable: 35 %

   A radial type drum is one where spiral layers of cable are accommodated between closely spaced flanges; if fitted with solid flanges the ratings given above should be reduced and the drum is described as non-ventilated. If the flanges have suitable apertures the drum is described as ventilated.

   A ventilated cylindrical cable drum is one where layers of cable are accommodated between widely spaced flanges and the drum and end flanges have suitable ventilating apertures.

3. Where cable may be covered over or coiled up whilst on load, or the air movement over the cable restricted, the current rating should be reduced.

   It is not possible to specify the amount of reduction but the table of rating factors for reeling drums can be used as a guide.

TABLE 4F1B

VOLTAGE DROP (per ampere per metre): Conductor operating temperature: 60 °C

Conductor cross-sectional area	Two-core cable, DC	Two-core cable, single-phase AC			1 three-core, four-core or five-core cable, three-phase AC			2 single-core cables, touching: DC	2 single-core cables, touching: Single-phase AC*		
1	2	3			4			5	6		
(mm^2)	(mV/A/m)	(mV/A/m)			(mV/A/m)			(mV/A/m)	(mV/A/m)		
4	12	12			10			-	-		
6	7.8	7.8			6.7			-	-		
10	4.6	4.6			4.0			-	-		
16	2.9	2.9			2.5			-	-		
		r	x	z	r	x	z		r	x	z
25	1.80	1.80	0.175	1.85	1.55	0.150	1.55	-	-	-	-
35	-	-	-	-	1.10	0.150	1.15	1.31	1.31	0.21	1.32
50	-	-	-	-	0.83	0.145	0.84	0.91	0.91	0.21	0.93
70	-	-	-	-	0.57	0.140	0.58	0.64	0.64	0.20	0.67
95	-	-	-	-	0.42	0.135	0.44	0.49	0.49	0.195	0.53
120	-	-	-	-	0.33	0.135	0.36	0.38	0.38	0.190	0.43
150	-	-	-	-	0.27	0.130	0.30	0.31	0.31	0.190	0.36
185	-	-	-	-	0.22	0.130	0.26	0.25	0.25	0.190	0.32
240	-	-	-	-	0.170	0.130	0.21	0.190	0.195	0.185	0.27
300	-	-	-	-	0.135	0.125	0.185	0.150	0.155	0.180	0.24
400	-	-	-	-	-	-	-	0.115	0.120	0.175	0.21
500	-	-	-	-	-	-	-	0.090	0.099	0.170	0.20
630	-	-	-	-	-	-	-	0.068	0.079	0.170	0.185

NOTE: * A larger voltage drop will result if the cables are spaced.

TABLE 4F2A – 90 °C and 180 °C thermosetting insulated flexible cables with sheath, non-armoured (COPPER CONDUCTORS)

CURRENT-CARRYING CAPACITY (amperes): Ambient temperature: 30 °C
Conductor operating temperature: 90 °C

Conductor cross-sectional area	Single-phase AC or DC	Three-phase AC	Single-phase AC or DC
	1 two-core cable, with or without protective conductor	1 three-core, four-core or five-core cable	2 single-core cables, touching
1	2	3	4
(mm²)	(A)	(A)	(A)
4	42	37	-
6	55	49	-
10	76	66	-
16	103	89	-
25	136	119	-
35	-	146	200
50	-	177	250
70	-	225	310
95	-	273	369
120	-	316	432
150	-	363	497
185	-	414	564
240	-	487	673
300	-	560	773
400	-	-	924
500	-	-	1062
630	-	-	1242

RATING FACTOR FOR AMBIENT TEMPERATURE

90 °C thermosetting insulated cables:

Ambient temperature	35 °C	40 °C	45 °C	50 °C	55 °C	60 °C	65 °C	70 °C	75 °C	80 °C	85 °C
Rating factor	0.95	0.91	0.86	0.82	0.76	0.70	0.64	0.57	0.50	0.40	0.28

180 °C thermosetting insulated cables:

Ambient temperature	35 to 90 °C	95 °C	100 °C	105 °C	110 °C	115 °C	120 °C	125 °C	130 °C	135 °C	140 °C	145 °C
Rating factor	1.0	0.96	0.91	0.86	0.81	0.76	0.70	0.64	0.57	0.50	0.40	0.28

NOTES:

1. The current ratings tabulated are for cables in free air but may also be used for cables resting on a surface. If the cable is to be wound on a drum on load the ratings should be reduced in accordance with NOTE 2 below and for cables which may be covered, NOTE 3 below.

2. *Flexible cables wound on reeling drums*
   The current ratings of cables used on reeling drums are to be reduced by the following factors:
   a) Radial type drum
      ventilated: 85 %
      unventilated: 75 %
   b) Ventilated cylindrical type drum
      1 layer of cable: 85 %
      2 layers of cable: 65 %
      3 layers of cable: 45 %
      4 layers of cable: 35 %

   A radial type drum is one where spiral layers of cable are accommodated between closely spaced flanges; if fitted with solid flanges the ratings given above should be reduced and the drum is described as non-ventilated. If the flanges have suitable apertures the drum is described as ventilated.

   A ventilated cylindrical cable drum is one where layers of cable are accommodated between widely spaced flanges and the drum and end flanges have suitable ventilating apertures.

3. Where cable may be covered over or coiled up whilst on load, or the air movement over the cable restricted, the current rating should be reduced. It is not possible to specify the amount of reduction but the table of rating factors for reeling drums can be used as a guide.

4. For 180 °C cables, the rating factors for ambient temperature allow a conductor operating temperature up to 150 °C. Consult the cable manufacturer for further information.

5. Where it is intended to connect the cables in this table to equipment or accessories designed to operate at a temperature lower than the maximum operating temperature of the cable, the cables should be rated at the maximum operating temperature of the equipment or accessory (see Regulation 512.1.5).

6. Where it is intended to group a cable in this table with other cables, the cable should be rated at the lowest of the maximum operating temperatures of any of the cables in the group (see Regulation 512.1.5).

TABLE 4F2B

VOLTAGE DROP (per ampere per metre): Conductor operating temperature: 90 °C

Conductor cross-sectional area	1 two-core or 2 single-core cables, DC	Two-core cable, single-phase AC			1 three-core, four-core or five-core cable, three-phase AC			2 single-core cables touching Single-phase AC*		
1	2	3			4			5		
(mm^2)	(mV/A/m)	(mV/A/m)			(mV/A/m)			(mV/A/m)		
4	13.2	13.2			11.1			-		
6	8.5	8.5			7.4			-		
10	5.1	5.1			4.4			-		
16	3.2	3.2			2.7			-		
		r	x	z	r	x	z	r	x	z
25	2.03	2.03	0.175	2.04	1.73	0.15	1.73	-	-	-
35	1.42	-	-	-	1.22	0.15	1.23	1.44	0.21	1.46
50	1.00	-	-	-	0.91	0.145	0.93	1.00	0.21	1.02
70	0.71	-	-	-	0.62	0.14	0.64	0.71	0.20	0.73
95	0.54	-	-	-	0.47	0.135	0.49	0.54	0.195	0.57
120	0.42	-	-	-	0.37	0.135	0.39	0.42	0.190	0.46
150	0.34	-	-	-	0.29	0.130	0.32	0.34	0.190	0.39
185	0.27	-	-	-	0.24	0.130	0.27	0.27	0.190	0.33
240	0.21	-	-	-	0.188	0.130	0.23	0.21	0.185	0.28
300	0.167	-	-	-	0.147	0.125	0.195	0.173	0.180	0.25
400	0.127	-	-	-	-	-	-	0.132	0.175	0.22
500	0.100	-	-	-	-	-	-	0.107	0.170	0.20
630	0.074	-	-	-	-	-	-	0.085	0.170	0.190

NOTES:

1. The voltage drop figures given above are based on a conductor operating temperature of 90 °C and are therefore not accurate when the operating temperature is in excess of 90 °C. In the case of the 180 °C cables with a conductor temperature of 150 °C the above resistive values should be increased by a factor of 1.2.
2. *A larger voltage drop will result if the cables are spaced.

TABLE 4F3A - Flexible cables, non-armoured (COPPER CONDUCTORS)

CURRENT-CARRYING CAPACITY (amperes): and MASS SUPPORTABLE (kg):

Conductor cross-sectional area	Current-carrying capacity		Maximum mass supportable by twin flexible cable (see Regulations 522.7.2 and 559.5.2)
	Single-phase AC	Three-phase AC	
1	2	3	4
(mm^2)	(A)	(A)	(kg)
0.5	3	3	2
0.75	6	6	3
1	10	10	5
1.25	13	-	5
1.5	16	16	5
2.5	25	20	5
4	32	25	5

Where cable is on a reel see the notes to Table 4F1A.

RATING FACTOR FOR AMBIENT TEMPERATURE

60 °C thermoplastic or thermosetting insulated cable:

Ambient temperature	35 °C	40 °C	45 °C	50 °C	55 °C
Rating factor	0.91	0.82	0.71	0.58	0.41

90 °C thermoplastic or thermosetting insulated cable:

Ambient temperature	35 to 50 °C	55 °C	60 °C	65 °C	70 °C
Rating factor	1.0	0.96	0.83	0.67	0.47

110 °C flexible cable:

Ambient temperature	35 to 80 °C	85 °C	90 °C	95 °C	100 °C	105 °C
Rating factor	1.0	0.96	0.85	0.74	0.60	0.42

150 °C flexible cable:

Ambient temperature	35 to 120 °C	125 °C	130 °C	135 °C	140 °C	145 °C
Rating factor	1.0	0.96	0.85	0.74	0.60	0.42

Glass fibre flexible cable:

Ambient temperature	35 to 50 °C	155 °C	160 °C	165 °C	170 °C	175 °C
Rating factor	1.0	0.92	0.82	0.71	0.57	0.40

TABLE 4F3B

VOLTAGE DROP (per ampere per metre):

Conductor operating temperature: 60 °C

Conductor cross-sectional area	DC or single-phase AC	Three-phase AC
1	2	3
(mm^2)	(mV/A/m)	(mV/A/m)
0.5	93	80
0.75	62	54
1	46	40
1.25	37	-
1.5	32	27
2.5	19	16
4	12	10

NOTE: * The tabulated values above are for 60 °C thermoplastic or thermosetting insulated flexible cables and for other types of flexible cable they are to be multiplied by the following factors:

For	90 °C thermoplastic or thermosetting insulated	1.09
	110 °C	1.17
	150 °C	1.31
	185 °C glass fibre	1.43

TABLE 4G1A – Mineral insulated cables thermoplastic covered or bare and exposed to touch (COPPER CONDUCTORS AND SHEATH)

CURRENT-CARRYING CAPACITY (amperes):

Ambient temperature: 30 °C
Sheath operating temperature: 70 °C

Conductor cross-sectional area	Reference Method C (clipped direct)			Reference Methods E, F and G (in free air or on a perforated cable tray etc, horizontal or vertical)				
	Single-phase AC or DC	Three-phase AC		Single-phase AC or DC	Three-phase AC			
	2 single-core cables touching or 1 two-core cable	3 single-core cables in trefoil or 1 three-core or four-core cable	3 single-core cables flat and touching, horizontal or vertical	2 single-core cables touching or 1 two-core cable	3 single-core cables in trefoil or 1 three-core or four-core cable	3 single-core cables flat and touching	3 single-core cables flat and spaced by one cable diameter: vertical	3 single-core cables flat and spaced by one cable diameter: horizontal
1	2	3	4	5	6	7	8	9
(mm²)	(A)	(A)	(A)	(A)	(A)	(A)	(A)	(A)
Light duty 500 V								
1	18.5	15	17	19.5	16.5	18	20	23
1.5	23	19	21	25	21	23	26	29
2.5	31	26	29	33	28	31	34	39
4	40	35	38	44	37	41	45	51
Heavy duty 750 V								
1	19.5	16	18	21	17.5	20	22	25
1.5	25	21	23	26	22	26	28	32
2.5	34	28	31	36	30	34	37	43
4	45	37	41	47	40	45	49	56
6	57	48	52	60	51	57	62	71
10	77	65	70	82	69	77	84	95
16	102	86	92	109	92	102	110	125
25	133	112	120	142	120	132	142	162
35	163	137	147	174	147	161	173	197
50	202	169	181	215	182	198	213	242
70	247	207	221	264	223	241	259	294
95	296	249	264	317	267	289	309	351
120	340	286	303	364	308	331	353	402
150	388	327	346	416	352	377	400	454
185	440	371	392	472	399	426	446	507
240	514	434	457	552	466	496	497	565

NOTES:

1. For single-core cables, the sheaths of the circuit are assumed to be connected together at both ends.
2. For bare cables exposed to touch, the tabulated values should be multiplied by 0.9.

TABLE 4G1B

VOLTAGE DROP (per ampere per metre): Sheath operating temperature 70 °C

Conductor cross-sectional area	Single-phase AC or DC						Three-phase AC											
	2 single-core cables touching			1 two-core cable			1 three- or four-core cable			3 single-core cables in trefoil formation			3 single-core cables flat and touching			3 single-core cables flat and spaced by one cable diameter*		
1	2			3			4			5			6			7		
(mm²)	(mV/A/m)			(mV/A/m)			(mV/A/m)			(mV/A/m)			(mV/A/m)			(mV/A/m)		
1	42			42			36			36			36			36		
1.5	28			28			24			24			24			24		
2.5	17			17			14			14			14			14		
4	10			10			9.1			9.1			9.1			9.1		
6	7			7			6.0			6.0			6.0			6.0		
10	4.2			4.2			3.6			3.6			3.6			3.6		
16	2.6			2.6			2.3			2.3			2.3			2.3		
	r	x	z	r	x	z	r	x	z	r	x	z	r	x	z	r	x	z
25	1.65	0.200	1.65	1.65	0.145	1.65	1.45	0.125	1.45	1.45	0.170	1.45	1.45	0.25	1.45	1.45	0.32	1.50
35	1.20	0.195	1.20	-	-	-	-	-	-	1.05	0.165	1.05	1.05	0.24	1.10	1.05	0.31	1.10
50	0.89	0.185	0.91	-	-	-	-	-	-	0.78	0.160	0.80	0.79	0.24	0.83	0.82	0.31	0.87
70	0.62	0.180	0.64	-	-	-	-	-	-	0.54	0.155	0.56	0.55	0.23	0.60	0.58	0.30	0.65
95	0.46	0.175	0.49	-	-	-	-	-	-	0.40	0.150	0.43	0.41	0.22	0.47	0.44	0.29	0.53
120	0.37	0.170	0.41	-	-	-	-	-	-	0.32	0.150	0.36	0.33	0.22	0.40	0.36	0.28	0.46
150	0.30	0.170	0.34	-	-	-	-	-	-	0.26	0.145	0.30	0.29	0.21	0.36	0.32	0.27	0.42
185	0.25	0.165	0.29	-	-	-	-	-	-	0.21	0.140	0.26	0.25	0.21	0.32	0.28	0.26	0.39
240	0.190	0.160	0.25	-	-	-	-	-	-	0.165	0.140	0.22	0.21	0.20	0.29	0.26	0.25	0.36

NOTE: * Spacings larger than one cable diameter will result in a larger voltage drop.

COPPER CONDUCTORS

TABLE 4G2A – Mineral insulated cables bare and neither exposed to touch nor in contact with combustible materials (COPPER CONDUCTORS AND SHEATH)

CURRENT-CARRYING CAPACITY (amperes):

Ambient temperature: 30 °C
Sheath operating temperature: 105 °C

Conductor cross-sectional area	Reference Method C (clipped direct)			Reference Methods E, F and G (in free air or on a perforated cable tray etc, horizontal or vertical)				
	Single-phase AC or DC	Three-phase AC		Single-phase AC or DC	Three-phase AC			
	2 single-core cables touching or 1 two-core cable	3 single-core cables in trefoil or 1 three-core or four-core cable	3 single-core cables flat and touching, horizontal or vertical	2 single-core cables touching or 1 two-core cable	3 single-core cables in trefoil or 1 three-core or four-core cable	3 single-core cables flat and touching	3 single-core cables flat a spaced by one cable diame vertical	horizont
1	2	3	4	5	6	7	8	9
(mm²)	(A)	(A)	(A)	(A)	(A)	(A)	(A)	(A)
Light duty 500 V								
1	22	19	21	24	21	23	26	29
1.5	28	24	27	31	26	29	33	37
2.5	38	33	36	41	35	39	43	49
4	51	44	47	54	46	51	56	64
Heavy duty 750 V								
1	24	20	24	26	22	25	28	32
1.5	31	26	30	33	28	32	35	40
2.5	42	35	41	45	38	43	47	54
4	55	47	53	60	50	56	61	70
6	70	59	67	76	64	71	78	89
10	96	81	91	104	87	96	105	120
16	127	107	119	137	115	127	137	157
25	166	140	154	179	150	164	178	204
35	203	171	187	220	184	200	216	248
50	251	212	230	272	228	247	266	304
70	307	260	280	333	279	300	323	370
95	369	312	334	400	335	359	385	441
120	424	359	383	460	385	411	441	505
150	485	410	435	526	441	469	498	565
185	550	465	492	596	500	530	557	629
240	643	544	572	697	584	617	624	704

NOTES:

1. For single-core cables, the sheaths of the circuit are assumed to be connected together at both ends.
2. No rating factor for grouping need be applied.
3. Where a conductor operates at a temperature exceeding 70 °C it should be ascertained that the equipment connected to the conductor is suitable for the conductor operating temperature (see Regulation 512.1.5).

TABLE 4G2B

VOLTAGE DROP (per ampere per metre): Sheath operating temperature 105 °C

Conductor cross-sectional area	Single-phase AC or DC						Three-phase AC											
	2 single-core cables touching			1 two-core cable			1 three- or four-core cable			3 single-core cables in trefoil formation			3 single-core cables flat and touching			3 single-core cables flat and spaced by one cable diameter*		
1		2			3			4			5			6			7	
(mm^2)		(mV/A/m)			(mV/A/m)			(mV/A/m)			(mV/A/m)			(mV/A/m)			(mV/A/m)	
1		47			47			40			40			40			40	
1.5		31			31			27			27			27			27	
2.5		19			19			16			16			16			16	
4		12			12			10			10			10			10	
6		7.8			7.8			6.8			6.8			6.8			6.8	
10		4.7			4.7			4.1			4.1			4.1			4.1	
16		3.0			3.0			2.6			2.6			2.6			2.6	
	r	x	z	r	x	z	r	x	z	r	x	z	r	x	z	r	x	z
25	1.85	0.180	1.85	1.85	0.145	1.85	1.60	0.125	1.60	1.60	0.160	1.65	1.60	0.23	1.65	1.60	0.31	1.65
35	1.35	0.175	1.35	-	-	-	-	-	-	1.15	0.155	1.20	1.15	0.23	1.20	1.20	0.30	1.25
50	1.00	0.170	1.00	-	-	-	-	-	-	0.87	0.150	0.88	0.88	0.22	0.91	0.90	0.29	0.95
70	0.69	0.165	0.71	-	-	-	-	-	-	0.60	0.145	0.62	0.61	0.22	0.65	0.63	0.29	0.70
95	0.51	0.160	0.54	-	-	-	-	-	-	0.45	0.140	0.47	0.46	0.21	0.50	0.48	0.28	0.56
120	0.41	0.160	0.44	-	-	-	-	-	-	0.36	0.135	0.38	0.37	0.21	0.42	0.39	0.28	0.48
150	0.33	0.155	0.36	-	-	-	-	-	-	0.29	0.135	0.32	0.31	0.20	0.37	0.34	0.27	0.43
185	0.27	0.150	0.31	-	-	-	-	-	-	0.23	0.130	0.27	0.26	0.20	0.33	0.29	0.26	0.39
240	0.21	0.150	0.26	-	-	-	-	-	-	0.180	0.130	0.22	0.22	0.195	0.29	0.26	0.25	0.36

NOTE: * Spacings larger than one cable diameter will result in a larger voltage drop.

TABLE 4H1A – Single-core 70 °C thermoplastic insulated cables, non-armoured, with or without sheath (ALUMINIUM CONDUCTORS)

CURRENT-CARRYING CAPACITY (amperes):

Ambient temperature: 30 °C
Conductor operating temperature: 70 °C

Conductor cross-sectional area	Reference Method A (enclosed in conduit in thermally insulating wall etc.)		Reference Method B (enclosed in conduit on a wall or in trunking etc.)		Reference Method C (clipped direct)		Reference Method F (in free air on a perforated cable tray, horizontal or vertical)				
							Touching			Spaced by one diameter	
	2 cables, single-phase AC or DC	3 or 4 cables, three-phase AC	2 cables, single-phase AC or DC	3 or 4 cables, three-phase AC	2 cables, single-phase AC or DC flat and touching	3 or 4 cables, three-phase AC flat and touching or trefoil	2 cables, single-phase AC or DC flat	3 cables, three-phase AC flat	3 cables, three-phase AC trefoil	2 cables, single-phase AC or DC or 3 cables three-phase AC flat	
										Horizontal	Vertical
1	2	3	4	5	6	7	8	9	10	11	12
(mm²)	(A)	(A)	(A)	(A)	(A)	(A)	(A)	(A)	(A)	(A)	(A)
50	93	84	118	104	125	110	149	133	128	169	152
70	118	107	150	133	160	140	192	173	166	217	196
95	142	129	181	161	195	170	235	212	203	265	241
120	164	149	210	186	226	197	273	247	237	308	282
150	189	170	234	204	261	227	316	287	274	356	327
185	215	194	266	230	298	259	363	330	316	407	376
240	252	227	312	269	352	305	430	392	375	482	447
300	289	261	358	306	406	351	497	455	434	557	519
380	-	-	413	352	511	472	543	502	507	625	584
480	-	-	477	405	591	546	629	582	590	726	680
600	-	-	545	462	679	626	722	669	680	837	787
740	-	-	-	-	771	709	820	761	776	956	902
960	-	-	-	-	900	823	953	886	907	1125	1066
1200	-	-	-	-	1022	926	1073	999	1026	1293	1229

TABLE 4H1B

VOLTAGE DROP (per ampere per metre): Conductor operating temperature: 70 °C

Conductor cross-sectional area	2 cables, DC	2 cables, single-phase AC									3 or 4 cables, three-phase AC														
		Reference Methods A & B (enclosed in conduit or trunking)			Reference Methods C & F (clipped direct, on tray or in free air)						Reference Methods A & B (enclosed in conduit or trunking)			Reference Methods C & F (clipped direct, on tray or in free air)								
					Cables touching			Cables spaced*						Cables touching, Trefoil			Cables touching, Flat			Cables spaced*, Flat		
1	2	3			4			5			6			7			8			9		
(mm²)	(mV/A/m)	(mV/A/m)			(mV/A/m)			(mV/A/m)			(mV/A/m)			(mV/A/m)			(mV/A/m)			(mV/A/m)		
		r	x	z	r	x	z	r	x	z	r	x	z	r	x	z	r	x	z	r	x	z
50	1.55	1.60	0.30	1.60	1.55	0.190	1.55	1.55	0.28	1.55	1.35	0.26	1.40	1.35	0.165	1.35	1.35	0.24	1.35	1.35	0.32	1.40
70	1.05	1.10	0.30	1.15	1.05	0.185	1.05	1.05	0.27	1.10	0.94	0.26	0.97	0.91	0.160	0.92	0.91	0.24	0.94	0.91	0.31	0.96
95	0.77	0.81	0.29	0.86	0.77	0.185	0.79	0.77	0.27	0.82	0.70	0.25	0.74	0.67	0.160	0.69	0.67	0.23	0.71	0.67	0.31	0.74
120	0.61	0.64	0.29	0.70	0.61	0.180	0.64	0.61	0.27	0.67	0.55	0.25	0.61	0.53	0.155	0.55	0.53	0.23	0.58	0.53	0.31	0.61
150	0.49	0.51	0.28	0.59	0.49	0.175	0.52	0.49	0.26	0.55	0.45	0.24	0.51	0.42	0.155	0.45	0.42	0.23	0.48	0.42	0.30	0.52
185	0.39	0.42	0.28	0.50	0.40	0.175	0.43	0.39	0.26	0.47	0.36	0.24	0.44	0.34	0.150	0.37	0.34	0.23	0.41	0.34	0.30	0.46
240	0.30	0.32	0.27	0.42	0.30	0.170	0.35	0.30	0.26	0.40	0.28	0.24	0.37	0.26	0.150	0.30	0.26	0.22	0.35	0.26	0.30	0.40
300	0.24	0.26	0.27	0.37	0.24	0.170	0.30	0.24	0.26	0.35	0.23	0.23	0.32	0.21	0.145	0.26	0.21	0.22	0.31	0.21	0.30	0.36
380	0.190	0.22	0.27	0.35	0.195	0.165	0.26	0.195	0.25	0.32	0.190	0.23	0.30	0.170	0.145	0.22	0.170	0.22	0.28	0.170	0.29	0.34
480	0.150	0.18	0.26	0.32	0.155	0.165	0.23	0.155	0.25	0.29	0.155	0.23	0.27	0.140	0.140	0.195	0.140	0.22	0.26	0.135	0.29	0.32
600	0.120	0.150	0.26	0.30	0.130	0.160	0.21	0.125	0.25	0.28	0.125	0.22	0.26	0.110	0.140	0.180	0.110	0.22	0.24	0.110	0.29	0.31
740	0.099	-	-	-	0.105	0.160	0.190	0.100	0.25	0.27	-	-	-	0.094	0.135	0.165	0.094	0.21	0.23	0.089	0.29	0.30
960	0.075	-	-	-	0.086	0.155	0.180	0.082	0.24	0.26	-	-	-	0.077	0.135	0.155	0.077	0.21	0.22	0.071	0.29	0.29
1200	0.060	-	-	-	0.074	0.155	0.170	0.068	0.24	0.25	-	-	-	0.066	0.135	0.150	0.066	0.21	0.22	0.059	0.28	0.29

NOTE: * Spacings larger than one cable diameter will result in a larger voltage drop.

TABLE 4H2A – Multicore 70 °C thermoplastic insulated and thermoplastic sheathed cables, non-armoured (ALUMINIUM CONDUCTORS)

Ambient temperature: 30 °C
Conductor operating temperature: 70 °C

CURRENT-CARRYING CAPACITY (amperes):

Conductor cross-sectional area	Reference Method A (enclosed in conduit in thermally insulating wall etc.)		Reference Method B (enclosed in conduit on a wall or in trunking etc.)		Reference Method C (clipped direct)		Reference Method E (in free air or on a perforated cable tray etc, horizontal or vertical)	
	1 two-core cable, single-phase AC or DC	1 three- or four-core cable, three-phase AC	1 two-core cable, single-phase AC or DC	1 three- or four-core cable, three-phase AC	1 two-core cable, single-phase AC or DC	1 three- or four-core cable, three-phase AC	1 two-core cable, single-phase AC or DC	1 three- or four-core cable, three-phase AC
1	2	3	4	5	6	7	8	9
(mm^2)	(A)	(A)	(A)	(A)	(A)	(A)	(A)	(A)
16	44	41	54	48	66	59	73	61
25	58	53	71	62	83	73	89	78
35	71	65	86	77	103	90	111	96
50	86	78	104	92	125	110	135	117
70	108	98	131	116	160	140	173	150
95	130	118	157	139	195	170	210	183
120	-	135	-	160	-	197	-	212
150	-	155	-	176	-	227	-	245
185	-	176	-	199	-	259	-	280
240	-	207	-	232	-	305	-	330
300	-	237	-	265	-	351	-	381

TABLE 4H2B

VOLTAGE DROP (per ampere per metre): Conductor operating temperature: 70 °C

Conductor cross-sectional area	Two-core cable, DC	Two-core cable, single-phase AC			Three- or four-core cable, three-phase AC		
1	2	3			4		
(mm^2)	(mV/A/m)	(mV/A/m)			(mV/A/m)		
16	4.5	4.5			3.9		
		r	x	z	r	x	z
25	2.9	2.9	0.175	2.9	2.5	0.150	2.5
35	2.1	2.1	0.170	2.1	1.80	0.150	1.80
50	1.55	1.55	0.170	1.55	1.35	0.145	1.35
70	1.05	1.05	0.165	1.05	0.90	0.140	0.92
95	0.77	0.77	0.160	0.79	0.67	0.140	0.68
120	-	-	-	-	0.53	0.135	0.55
150	-	-	-	-	0.42	0.135	0.44
185	-	-	-	-	0.34	0.135	0.37
240	-	-	-	-	0.26	0.130	0.30
300	-	-	-	-	0.21	0.130	0.25

TABLE 4H3A – Single-core armoured 70 °C thermoplastic insulated cables (non-magnetic armour) (ALUMINIUM CONDUCTORS)

CURRENT-CARRYING CAPACITY (amperes):

Ambient temperature: 30 °C
Conductor operating temperature: 70 °C

Conductor cross-sectional area	Reference Method C (clipped direct)		Reference Method F (in free air or on a perforated cable tray, horizontal or vertical)								
	Touching		Touching			Spaced by one cable diameter					
	2 cables, single-phase AC or DC flat	3 or 4 cables, three-phase AC flat	2 cables, single-phase AC or DC flat	3 cables, three-phase AC flat	3 cables, three-phase AC trefoil	2 cables, DC		2 cables, single-phase AC		3 or 4 cables, three-phase AC	
						Horizontal	Vertical	Horizontal	Vertical	Horizontal	Vertical
1	2	3	4	5	6	7	8	9	10	11	12
(mm^2)	(A)	(A)	(A)	(A)	(A)	(A)	(A)	(A)	(A)	(A)	(A)
50	143	133	152	141	131	167	157	168	159	169	155
70	183	168	194	178	168	214	202	212	200	213	196
95	221	202	234	214	205	261	247	259	245	255	236
120	255	233	270	246	238	303	288	299	285	293	272
150	294	267	310	282	275	349	333	340	323	335	312
185	334	303	352	319	315	400	382	389	371	379	354
240	393	354	413	374	372	472	452	457	437	443	415
300	452	405	474	427	430	545	523	520	498	505	475
380	518	452	543	479	497	638	613	583	559	551	518
480	586	501	616	534	568	742	715	655	629	604	568
600	658	550	692	589	642	859	828	724	696	656	618
740	728	596	769	642	715	986	952	802	770	707	666
960	819	651	868	706	808	1171	1133	866	832	770	726
1200	893	692	952	756	880	1360	1317	938	902	822	774

TABLE 4H3B

VOLTAGE DROP (per ampere per metre): Conductor operating temperature: 70 °C

Conductor cross-sectional area	2 cables, DC	Reference Methods C & F (clipped direct, on tray or in free air)														
		2 cables, single-phase AC						3 or 4 cables, three-phase AC								
		touching			spaced*			trefoil and touching			flat and touching			flat and spaced*		
1	2	3			4			5			6			7		
(mm^2)	(mV/A/m)	(mV/A/m)			(mV/A/m)			(mV/A/m)			(mV/A/m)			(mV/A/m)		
		r	x	z	r	x	z	r	x	z	r	x	z	r	x	z
50	1.55	1.55	0.23	1.55	1.55	0.31	1.55	1.35	0.195	1.35	1.35	0.27	1.35	1.30	0.34	1.35
70	1.05	1.05	0.22	1.10	1.05	0.30	1.10	0.92	0.190	0.93	0.93	0.26	0.96	0.95	0.33	1.00
95	0.77	0.78	0.21	0.81	0.81	0.29	0.86	0.68	0.185	0.70	0.70	0.25	0.75	0.73	0.32	0.80
120	0.61	0.62	0.21	0.66	0.65	0.29	0.71	0.54	0.180	0.57	0.57	0.25	0.62	0.60	0.32	0.68
150	0.49	0.50	0.20	0.54	0.53	0.28	0.60	0.44	0.175	0.47	0.46	0.24	0.52	0.50	0.31	0.58
185	0.39	0.41	0.195	0.45	0.44	0.28	0.52	0.35	0.170	0.39	0.38	0.24	0.45	0.42	0.30	0.51
240	0.30	0.32	0.190	0.37	0.34	0.27	0.44	0.28	0.165	0.32	0.30	0.23	0.38	0.33	0.29	0.44
300	0.24	0.26	0.185	0.32	0.28	0.26	0.39	0.22	0.160	0.27	0.24	0.23	0.34	0.28	0.29	0.40
380	0.190	0.22	0.185	0.28	0.26	0.25	0.36	0.185	0.155	0.24	0.22	0.22	0.32	0.27	0.26	0.38
480	0.150	0.180	0.180	0.25	0.22	0.25	0.33	0.155	0.155	0.22	0.195	0.22	0.29	0.24	0.25	0.35
600	0.120	0.150	0.175	0.23	0.195	0.24	0.31	0.130	0.150	0.200	0.170	0.21	0.27	0.21	0.24	0.32
740	0.097	0.135	0.170	0.22	0.180	0.23	0.29	0.115	0.145	0.185	0.160	0.20	0.26	0.200	0.22	0.30
960	0.075	0.115	0.160	0.200	0.165	0.21	0.27	0.100	0.140	0.175	0.150	0.185	0.24	0.190	0.195	0.27
1200	0.060	0.110	0.155	0.190	0.160	0.180	0.24	0.094	0.140	0.170	0.145	0.160	0.22	0.185	0.165	0.25

NOTE: * Spacings larger than one cable diameter will result in a larger voltage drop.

TABLE 4H4A – Multicore armoured 70 °C thermoplastic insulated cables (ALUMINIUM CONDUCTORS)

Air Ambient temperature: 30 °C
Ground Ambient temperature: 20 °C
Conductor operating temperature: 70 °C

CURRENT-CARRYING CAPACITY (amperes):

Conductor cross-sectional area	Reference Method C (clipped direct)		Reference Method E (in free air or on a perforated cable tray etc, horizontal or vertical)		Reference Method D (direct in ground or in ducting in ground, in or around buildings)	
	1 two-core cable, single-phase AC or DC	1 three- or 1 four-core cable, three-phase AC	1 two-core cable, single-phase AC or DC	1 three- or 1 four-core cable, three-phase AC	1 two-core cable, single-phase AC or DC	1 three- or 1 four-core cable, three-phase AC
1	2	3	4	5	6	7
(mm^2)	(A)	(A)	(A)	(A)	(A)	(A)
16	68	58	71	61		
25	89	76	94	80	77	64
35	109	94	115	99	93	77
50	131	113	139	119	109	91
70	165	143	175	151	135	112
95	199	174	211	186	159	132
120	-	202	-	216	-	150
150	-	232	-	250	-	169
185	-	265	-	287	-	190
240	-	312	-	342	-	218
300	-	360	-	399	-	247

TABLE 4H4B

VOLTAGE DROP (per ampere per metre): Conductor operating temperature: 70 °C

Conductor cross-sectional area	Two-core cable, DC	Two-core cable, single-phase AC			Three- or four-core cable, three-phase AC		
1	2	3			4		
(mm^2)	(mV/A/m)	(mV/A/m)			(mV/A/m)		
16	4.5	4.5			3.9		
		r	x	z	r	x	z
25	2.9	2.9	0.175	2.9	2.5	0.150	2.5
35	2.1	2.1	0.170	2.1	1.80	0.150	1.80
50	1.55	1.55	0.170	1.55	1.35	0.145	1.35
70	1.05	1.05	0.165	1.05	0.90	0.140	0.92
95	0.77	0.77	0.160	0.79	0.67	0.140	0.68
120	-	-	-	-	0.53	0.135	0.55
150	-	-	-	-	0.42	0.135	0.44
185	-	-	-	-	0.34	0.135	0.37
240	-	-	-	-	0.26	0.130	0.30
300	-	-	-	-	0.21	0.130	0.25

TABLE 4J1A – Single-core 90 °C thermosetting insulated cables, non-armoured, with or without sheath (ALUMINIUM CONDUCTORS)

Ambient temperature: 30 °C
Conductor operating temperature: 90 °C

CURRENT-CARRYING CAPACITY (amperes):

Conductor cross-sectional area	Reference Method A (enclosed in conduit in thermally insulating wall etc.)		Reference Method B (enclosed in conduit on a wall or in trunking etc.)		Reference Method C (clipped direct)		Reference Method F (in free air or on a perforated cable tray horizontal or vertical etc.) Touching			Reference Method G (in free air) Spaced by one cable diameter	
	2 cables, single-phase AC or DC	3 or 4 cables, three-phase AC	2 cables, single-phase AC or DC	3 or 4 cables, three-phase AC	2 cables, single-phase AC or DC flat and touching	3 or 4 cables, three-phase AC flat and touching or trefoil	2 cables, single-phase AC or DC flat	3 cables, three-phase AC flat	3 cables, three-phase AC trefoil	2 cables, single-phase AC or DC or 3 cables three-phase AC flat: Horizontal	Vertical
1	2	3	4	5	6	7	8	9	10	11	12
(mm²)	(A)	(A)	(A)	(A)	(A)	(A)	(A)	(A)	(A)	(A)	(A)
50	125	113	157	140	154	136	184	165	159	210	188
70	158	142	200	179	198	174	237	215	206	271	244
95	191	171	242	217	241	211	289	264	253	332	300
120	220	197	281	251	280	245	337	308	296	387	351
150	253	226	307	267	324	283	389	358	343	448	408
185	288	256	351	300	371	323	447	413	395	515	470
240	338	300	412	351	439	382	530	492	471	611	561
300	387	344	471	402	508	440	613	571	544	708	652
380	-	-	-	-	658	594	679	628	638	798	742
480	-	-	-	-	765	692	786	728	743	927	865
600	-	-	-	-	871	791	903	836	849	1058	990
740	-	-	-	-	1001	911	1025	951	979	1218	1143
960	-	-	-	-	1176	1072	1191	1108	1151	1440	1355
1200	-	-	-	-	1333	1217	1341	1249	1307	1643	1550

NOTES:

1. Where it is intended to connect the cables in this table to equipment or accessories designed to operate at a temperature lower than the maximum operating temperature of the cable, the cables should be rated at the maximum operating temperature of the equipment or accessory (see Regulation 512.1.5).
2. Where it is intended to group a cable in this table with other cables, the cable should be rated at the lowest of the maximum operating temperatures of any of the cables in the group (see Regulation 512.1.5).

TABLE 4J1B

VOLTAGE DROP (per ampere per metre): Conductor operating temperature: 90 °C

Conductor cross-sectional area	2 cables, DC	2 cables, single-phase AC									3 or 4 cables, three-phase AC											
		Reference Methods A & B (enclosed in conduit or trunking)			References Methods C, F & G (clipped direct, on tray or in free air)						Reference Methods A & B (enclosed in conduit or trunking)			Reference Methods C, F & G (clipped direct, on tray or in free air)								
					Cables touching			Cables spaced*						Cables touching, Trefoil			Cables touching, Flat			Cables spaced*, Flat		
1	2	3			4			5			6			7			8			9		
(mm²)	(mV/A/m)	(mV/A/m)			(mV/A/m)			(mV/A/m)			(mV/A/m)			(mV/A/m)			(mV/A/m)			(mV/A/m)		
		r	x	z	r	x	z	r	x	z	r	x	z	r	x	z	r	x	z	r	x	z
50	1.65	1.70	0.30	1.72	1.65	0.190	1.66	1.65	0.28	1.68	1.44	0.26	1.46	1.44	0.165	1.45	1.44	0.24	1.46	1.44	0.32	1.48
70	1.13	1.17	0.30	1.21	1.12	0.185	1.14	1.12	0.27	1.15	1.00	0.26	1.04	0.97	0.160	0.98	0.97	0.24	1.00	0.97	0.31	1.02
95	0.82	0.86	0.29	0.91	0.82	0.185	0.84	0.82	0.27	0.94	0.75	0.25	0.79	0.71	0.160	0.73	0.71	0.23	0.75	0.71	0.31	0.78
120	0.65	0.68	0.29	0.74	0.65	0.180	0.67	0.65	0.27	0.70	0.59	0.25	0.64	0.57	0.155	0.59	0.57	0.23	0.61	0.57	0.31	0.64
150	0.53	0.54	0.28	0.61	0.52	0.175	0.55	0.52	0.26	0.58	0.48	0.24	0.54	0.45	0.155	0.47	0.45	0.23	0.50	0.45	0.30	0.54
185	0.42	0.45	0.28	0.53	0.43	0.175	0.46	0.42	0.26	0.49	0.38	0.24	0.45	0.36	0.150	0.39	0.36	0.23	0.43	0.36	0.30	0.47
240	0.32	0.34	0.27	0.43	0.32	0.170	0.36	0.32	0.26	0.41	0.30	0.24	0.38	0.28	0.150	0.32	0.28	0.22	0.35	0.28	0.30	0.41
300	0.26	0.28	0.27	0.38	0.26	0.170	0.31	0.26	0.26	0.36	0.25	0.23	0.34	0.22	0.145	0.27	0.22	0.22	0.31	0.22	0.30	0.37
380	0.20	-	-	-	0.21	0.165	0.27	0.21	0.25	0.33	0.20	0.23	0.31	0.180	0.145	0.23	0.180	0.22	0.28	0.180	0.29	0.34
480	0.160	-	-	-	0.170	0.165	0.23	0.165	0.25	0.30	0.165	0.23	0.28	0.150	0.140	0.20	0.150	0.22	0.27	0.145	0.29	0.32
600	0.130	-	-	-	0.140	0.160	0.21	0.135	0.25	0.28	0.135	0.22	0.26	0.120	0.140	0.185	0.120	0.22	0.25	0.120	0.29	0.31
740	0.105	-	-	-	0.115	0.160	0.19	0.110	0.25	0.27	-	-	-	0.100	0.135	0.170	0.100	0.21	0.23	0.095	0.29	0.30
960	0.080	-	-	-	0.092	0.155	0.18	0.087	0.24	0.26	-	-	-	0.082	0.135	0.160	0.082	0.21	0.23	0.076	0.29	0.30
1200	0.064	-	-	-	0.079	0.155	0.17	0.073	0.24	0.25	-	-	-	0.070	0.135	0.150	0.070	0.21	0.22	0.063	0.28	0.29

NOTE: * Spacings larger than one cable diameter will result in a larger voltage drop.

TABLE 4J2A – Multicore 90 °C thermosetting insulated and thermoplastic sheathed cables, non-armoured (ALUMINIUM CONDUCTORS)

Ambient temperature: 30 °C
Conductor operating temperature: 90 °C

CURRENT-CARRYING CAPACITY (amperes)

Conductor cross-sectional area	Reference Method A (enclosed in conduit in thermally insulating wall etc.)		Reference Method B (enclosed in conduit on a wall or in trunking etc.)		Reference Method C (clipped direct)		Reference Method E (in free air or on a perforated cable tray etc, horizontal or vertical)	
	1 two-core cable, single-phase AC or DC	1 three- or four-core cable, three-phase AC	1 two-core cable, single-phase AC or DC	1 three- or four-core cable, three-phase AC	1 two-core cable, single-phase AC or DC	1 three- or four-core cable, three-phase AC	1 two-core cable, single-phase AC or DC	1 three- or four-core cable, three-phase AC
1	2	3	4	5	6	7	8	9
(mm²)	(A)	(A)	(A)	(A)	(A)	(A)	(A)	(A)
16	60	55	72	64	84	76	91	77
25	78	71	94	84	101	90	108	97
35	96	87	115	103	126	112	135	120
50	115	104	138	124	154	136	164	146
70	145	131	175	156	198	174	211	187
95	175	157	210	188	241	211	257	227
120	-	180	-	216	-	245	-	263
150	-	206	-	240	-	283	-	304
185	-	233	-	272	-	323	-	347
240	-	273	-	318	-	382	-	409
300	-	313	-	364	-	440	-	471

NOTES:

1. Where it is intended to connect the cables in this table to equipment or accessories designed to operate at a temperature lower than the maximum operating temperature of the cable, the cables should be rated at the maximum operating temperature of the equipment or accessory (see Regulation 512.1.5).
2. Where it is intended to group a cable in this table with other cables, the cable should be rated at the lowest of the maximum operating temperatures of any of the cables in the group (see Regulation 512.1.5).

TABLE 4J2B

VOLTAGE DROP (per ampere per metre): Conductor operating temperature: 90 °C

Conductor cross-sectional area	Two-core cable, DC	Two-core cable, single-phase AC			Three- or four-core cable, three-phase AC		
1	2	3			4		
(mm^2)	(mV/A/m)	(mV/A/m)			(mV/A/m)		
16	4.8	4.8			4.2		
		r	x	z	r	x	z
25	3.1	3.1	0.165	3.1	2.7	0.140	2.7
35	2.2	2.2	0.160	2.2	1.90	0.140	1.95
50	1.60	1.65	0.160	1.65	1.40	0.135	1.45
70	1.10	1.10	0.155	1.15	0.96	0.135	0.97
95	0.82	0.82	0.150	0.84	0.71	0.130	0.72
120	-	-	-	-	0.56	0.130	0.58
150	-	-	-	-	0.45	0.130	0.47
185	-	-	-	-	0.37	0.130	0.39
240	-	-	-	-	0.28	0.125	0.31
300	-	-	-	-	0.23	0.125	0.26

TABLE 4J3A – Single-core armoured 90 °C thermosetting insulated cables (non-magnetic armour) ALUMINIUM CONDUCTOR

Ambient temperature: 30 °C
Conductor operating temperature: 90 °C

CURRENT-CARRYING CAPACITY (amperes):

Conductor cross-sectional area	Reference Method C (clipped direct)		Reference Method F (in free air or on a perforated cable tray, horizontal or vertical)								
	Touching		Touching			Spaced by one cable diameter					
	2 cables, single-phase AC or DC flat	3 or 4 cables, three-phase AC flat	2 cables, single-phase AC or DC flat	3 cables, three-phase AC flat	3 cables, three-phase AC trefoil	2 cables, DC		2 cables, single-phase AC		3 or 4 cables, three-phase AC	
						Horizontal	Vertical	Horizontal	Vertical	Horizontal	Vertical
1	2	3	4	5	6	7	8	9	10	11	12
(mm^2)	(A)	(A)	(A)	(A)	(A)	(A)	(A)	(A)	(A)	(A)	(A)
50	179	165	192	176	162	216	197	212	199	215	192
70	228	209	244	222	207	275	253	269	254	270	244
95	276	252	294	267	252	332	307	328	310	324	296
120	320	291	340	308	292	384	357	378	358	372	343
150	368	333	390	352	337	441	411	429	409	424	394
185	419	378	444	400	391	511	480	490	467	477	447
240	494	443	521	468	465	605	572	576	549	554	523
300	568	508	597	536	540	701	666	654	624	626	595
380	655	573	688	608	625	812	780	735	704	693	649
480	747	642	786	685	714	942	906	825	790	765	717
600	836	706	880	757	801	1076	1036	909	872	832	780
740	934	764	988	824	897	1250	1205	989	950	890	835
960	1056	838	1121	911	1014	1488	1435	1094	1052	970	911
1200	1163	903	1236	990	1118	1715	1658	1187	1141	1043	980

NOTES:

1. Where it is intended to connect the cables in this table to equipment or accessories designed to operate at a temperature lower than the maximum operating temperature of the cable, the cables should be rated at the maximum operating temperature of the equipment or accessory (see Regulation 512.1.5).
2. Where it is intended to group a cable in this table with other cables, the cable should be rated at the lowest of the maximum operating temperatures of any of the cables in the group (see Regulation 512.1.5).

TABLE 4J3B

VOLTAGE DROP (per ampere per metre): Conductor operating temperature: 90 °C

Conductor cross-sectional area	2 cables, DC	Reference Method C & F (clipped direct, on tray or in free air)														
		2 cables, single-phase AC						3 or 4 cables, three-phase AC								
		touching			spaced*			trefoil and touching			flat and touching			flat and spaced*		
1	2	3			4			5			6			7		
(mm^2)	(mV/A/m)	(mV/A/m)			(mV/A/m)			4 (mV/A/m)			(mV/A/m)			(mV/A/m)		
		r	x	z	r	x	z	r	x	z	r	x	z	r	x	z
50	1.60	1.60	0.22	1.60	1.60	0.30	1.60	1.40	0.185	1.40	1.40	0.26	1.40	1.35	0.34	1.40
70	1.10	1.10	0.21	1.15	1.10	0.29	1.15	0.96	0.180	0.98	0.97	0.25	1.00	0.99	0.33	1.05
95	0.82	0.83	0.20	0.85	0.85	0.29	0.90	0.71	0.175	0.74	0.74	0.25	0.78	0.76	0.32	0.83
120	0.66	0.66	0.20	0.69	0.69	0.28	0.74	0.57	0.170	0.60	0.60	0.24	0.64	0.63	0.31	0.70
150	0.52	0.53	0.195	0.57	0.56	0.28	0.62	0.46	0.170	0.49	0.49	0.24	0.54	0.52	0.30	0.60
185	0.42	0.43	0.190	0.47	0.46	0.27	0.54	0.38	0.165	0.41	0.40	0.24	0.47	0.44	0.30	0.53
240	0.32	0.34	0.185	0.39	0.37	0.27	0.45	0.29	0.160	0.34	0.32	0.23	0.39	0.35	0.29	0.46
300	0.26	0.27	0.185	0.33	0.30	0.26	0.40	0.24	0.160	0.29	0.26	0.23	0.34	0.29	0.29	0.41
380	0.21	0.23	0.180	0.29	0.26	0.25	0.36	0.195	0.155	0.25	0.23	0.22	0.32	0.27	0.27	0.38
480	0.160	0.185	0.175	0.25	0.23	0.25	0.34	0.160	0.155	0.22	0.20	0.21	0.29	0.24	0.26	0.35
600	0.130	0.160	0.175	0.24	0.20	0.24	0.31	0.135	0.150	0.20	0.175	0.21	0.27	0.22	0.25	0.33
740	0.105	0.140	0.170	0.22	0.190	0.22	0.29	0.120	0.145	0.190	0.165	0.195	0.26	0.21	0.22	0.30
960	0.080	0.120	0.160	0.20	0.170	0.21	0.27	0.105	0.140	0.175	0.150	0.180	0.24	0.195	0.195	0.28
1200	0.064	0.105	0.160	0.190	0.155	0.20	0.25	0.093	0.135	0.165	0.140	0.175	0.22	0.180	0.185	0.26

NOTE: * Spacings larger than one cable diameter will result in a larger voltage drop.

TABLE 4J4A – Multicore armoured 90 °C thermosetting insulated cables (ALUMINIUM CONDUCTORS)

Air Ambient temperature: 30 °C
Ground Ambient temperate: 20 °C
Conductor operating temperature: 90 °C

CURRENT-CARRYING CAPACITY (amperes):

Conductor cross-sectional area	Reference Method C (clipped direct)		Reference Method E (in free air or on a perforated cable tray etc, horizontal or vertical)		Reference Method D (direct in ground or in ducting in ground, in or around buildings)	
	1 two-core cable, single-phase AC or DC	1 three- or 1 four-core cable, three-phase AC	1 two-core cable, single-phase AC or DC	1 three- or 1 four-core cable, three-phase AC	1 two-core cable, single-phase AC or DC	1 three- or 1 four-core cable, three-phase AC
1	2	3	4	5	6	7
(mm²)	(A)	(A)	(A)	(A)	(A)	(A)
16	82	71	85	74	71	59
25	108	92	112	98	90	75
35	132	113	138	120	108	90
50	159	137	166	145	128	106
70	201	174	211	185	158	130
95	242	214	254	224	186	154
120	-	249	-	264	-	174
150	-	284	-	305	-	197
185	-	328	-	350	-	220
240	-	386	-	418	-	253
300	-	441	-	488	-	286

NOTES:

1. Where it is intended to connect the cables in this table to equipment or accessories designed to operate at a temperature lower than the maximum operating temperature of the cable, the cables should be rated at the maximum operating temperature of the equipment or accessory (see Regulation 512.1.5).
2. Where it is intended to group a cable in this table with other cables, the cable should be rated at the lowest of the maximum operating temperatures of any of the cables in the group (see Regulation 512.1.5).

TABLE 4J4B

VOLTAGE DROP (per ampere per metre): Conductor operating temperature: 90 °C

Conductor cross-sectional area	Two-core cable, DC	Two-core cable, single-phase AC			Three- or four-core cable, three-phase AC		
1	2	3			4		
(mm^2)	(mV/A/m)	(mV/A/m)			(mV/A/m)		
16	4.8	4.8			4.2		
		r	x	z	r	x	z
25	3.1	3.1	0.165	3.1	2.7	0.140	2.7
35	2.2	2.2	0.160	2.2	1.90	0.140	1.95
50	1.60	1.65	0.160	1.65	1.40	0.135	1.45
70	1.10	1.10	0.155	1.15	0.96	0.135	0.97
95	0.82	0.82	0.150	0.84	0.71	0.130	0.72
120	-	-	-	-	0.56	0.130	0.58
150	-	-	-	-	0.45	0.130	0.47
185	-	-	-	-	0.37	0.130	0.39
240	-	-	-	-	0.28	0.125	0.31
300	-	-	-	-	0.23	0.125	0.26

APPENDIX 5 (Informative)

CLASSIFICATION OF EXTERNAL INFLUENCES

This appendix gives the classification and codification of external influences.

NOTE 1: The appendix is an extract from HD 60364-5-51.

Each condition of external influence is designated by a code comprising a group of two capital letters and a number, as follows:

The first letter relates to the general category of external influence:

A	Environment
B	Utilization
C	Construction of buildings

The second letter relates to the nature of the external influence:

. . .	**A**
. . .	**B**
. . .	**C**

The number relates to the class within each external influence:

. . .	. . .	**1**
. . .	. . .	**2**
. . .	. . .	**3**

For example, the code **AA4** signifies:

A	= Environment
AA	= Environment - Ambient temperature
AA4	= Environment - Ambient temperature in the range of –5 °C to +40 °C .

NOTE 2: The codification given in this appendix is not intended to be used for marking equipment.

The following key is for use when referencing the list of external influences:

[a]	May necessitate certain supplementary precautions (e.g. special lubrication).
[b]	This means that ordinary equipment will operate safely under the described external influences.
[c]	This means that special arrangements should be made, for example, between the designer of the installation and the equipment manufacturer, e.g. for specially designed equipment.

CONCISE LIST OF EXTERNAL INFLUENCES

Environment	**A**	**AA**	***Ambient (°C)***		**AF**	***Corrosion***	AM8	Radiated magnetic fields
		AA1	-60 °C	+5 °C	AF1	Negligible	AM9	Electric fields
		AA2	-40 °C	+5 °C	AF2	Atmospheric	AM21	High-frequency etc. . .
		AA3	-25 °C	+5 °C	AF3	Intermittent	AM22	Conducted. . .nano. . .
		AA4	-5 °C	+40 °C	AF4	Continuous	AM23	Conducted. . .micro. . .
		AA5	+5 °C	+40 °C			AM24	Conducted oscillatory
		AA6	+5 °C	+60 °C	**AG**	***Impact***	AM25	Radiated HF
		AA7	-25 °C	+55 °C	AG1	Low	AM31	Electrostatic discharges
		AA8	-50 °C	+40 °C	AG2	Medium	AM41	Ionization
					AG3	High		
		AB	***Temperature and Humidity***				**AN**	***Solar***
					AH	***Vibration***	AN1	Low
		AC	***Altitude (metres)***		AH1	Low	AN2	Medium
		AC1	≤ 2000 metres		AH2	Medium	AN3	High
		AC2	> 2000 metres		AH3	High		
							AP	***Seismic***
		AD	***Water***		**AJ**	***Other mechanical stresses***	AP1	Negligible
		AD1	Negligible				AP2	Low
		AD2	Drops		**AK**	***Flora***	AP3	Medium
		AD3	Sprays		AK1	No hazard	AP4	High
		AD4	Splashes		AK2	Hazard		
		AD5	Jets				**AQ**	***Lightning***
		AD6	Waves		**AL**	***Fauna***	AQ1	Negligible
		AD7	Immersion		AL1	No hazard	AQ2	Indirect
		AD8	Submersion		AL2	Hazard	AQ3	Direct
		AE	***Foreign Bodies***		**AM**	***Electromagnetic. . .***	**AR**	***Movement of air***
		AE1	Negligible		AM1	Level	AR1	Low
		AE2	Small		AM2	Signalling voltages	AR2	Medium
		AE3	Very small		AM3	Voltage amplitude variations	AR3	High
		AE4	Light dust		AM4	Voltage unbalance		
		AE5	Moderate dust		AM5	Power frequency variations	**AS**	**Wind**
		AE6	Heavy dust		AM6	Induced low-frequency voltage	AS1	Low
					AM7	DC current in AC voltage	AS2	Medium
							AS3	High
Utilization	**B**	**BA**	***Capability***		**BC**	***Contact with Earth***	**BE**	***Materials***
		BA1	Ordinary		BC1	None	BE1	No risk
		BA2	Children		BC2	Low	BE2	Fire risk
		BA3	Disabled		BC3	Frequent	BE3	Explosion risk
		BA4	Instructed		BC4	Continuous	BE4	Contamination risk
		BA5	Skilled					
					BD	***Evacuation***		
		BB	***Resistance***		BD1	Normal		
					BD2	Difficult		
					BD3	Crowded		
					BD4	Difficult and crowded		
Buildings	**C**	**CA**	***Materials***		**CB**	***Structure***		
		CA1	Non-combustible		CB1	Negligible		
		CA2	Combustible		CB2	Fire propagation		
					CB3	Structural movement		
					CB4	Flexible		

A ENVIRONMENT:

Code	External influences	Characteristics required for selection and erection of equipment	Reference for information only
A	*Environmental conditions*		
AA	*Ambient temperature* The ambient temperature is that of the ambient air where the equipment is to be installed It is assumed that the ambient temperature includes the effects of other equipment installed in the same location The ambient temperature to be considered for the equipment is the temperature at the place where the equipment is to be installed resulting from the influence of all other equipment in the same location, when operating, not taking into account the thermal contribution of the equipment to be installed Lower and upper limits of ranges of ambient temperature:		
AA1	–60 °C +5 °C	Specially designed equipment or appropriate arrangements[a]	Includes temperature range of BS EN 60721-3-3, class 3K8, with high air temperature restricted to +5 °C. Part of temperature range of BS EN 60721-3-4, class 4K4, with low air temperature restricted to –60 °C and high air temperature restricted to +5 °C
AA2	–40 °C +5 °C		Part of temperature range of BS EN 60721-3-3, class 3K7, with high air temperature restricted to +5 °C. Includes part of temperature range of BS EN 60721-3-4, class 4K3, with high air temperature restricted to +5 °C
AA3	–25 °C +5 °C		Part of temperature range of BS EN 60721-3-3, class 3K6, with high air temperature restricted to +5 °C. Includes temperature range of BS EN 60721-3-4, class 4K1, with high air temperature restricted to +5 °C
AA4	–5 °C +40 °C	Normal (in certain cases special precautions may be necessary)	Part of temperature range of BS EN 60721-3-3, class 3K5, with high air temperature restricted to +40 °C
AA5	+5 °C +40 °C	Normal[b]	Identical to temperature range of BS EN 60721-3-3, class 3K3

A ENVIRONMENT *(cont.)*

Code	External influences						Characteristics required for selection and erection of equipment	Reference for information only
AA6	+5 °C +60 °C						Specially designed equipment or appropriate arrangements[a]	Part of temperature range of BS EN 60721-3-3, class 3K7, with low air temperature restricted to +5 °C and high air temperature restricted to +60 °C. Includes temperature range of BS EN 60721-3-4, class 4K4 with low air temperature restricted to +5 °C
AA7	–25 °C +55 °C						Specially designed equipment or appropriate arrangements[a]	– Identical to temperature range of BS EN 60721-3-3, class 3K6
AA8	–50 °C +40 °C Ambient temperature classes are applicable only where humidity has no influence The average temperature over a 24 h period must not exceed 5 °C below the upper limits Combination of two ranges to define some environments may be necessary. Installations subject to temperatures outside the ranges require special consideration							– Identical to temperature range of BS EN 60721-3-4, class 4K3
AB	*Atmospheric humidity*							
	Air temperature °C **a) low** **b) high**		**Relative humidity** % **c) low** **d) high**		**Absolute humidity** g/m³ **e) low** **f) high**			
AB1	–60	+5	3	100	0.003	7	Indoor and outdoor locations with extremely low ambient temperatures Appropriate arrangements should be made[c]	Includes temperature range of BS EN 60721-3-3, class 3K8, with high air temperature restricted to +5 °C. Part of temperature range of BS EN 60721-3-4, class 4K4, with low air temperature restricted to – 60 °C and high air temperature restricted to +5 °C
AB2	–40	+5	10	100	0.1	7	Indoor and outdoor locations with low ambient temperatures Appropriate arrangements should be made[c]	Part of temperature range of BS EN 60721-3-3, class 3K7, with high temperature restricted to +5 °C. Part of temperature range of BS EN 60721-3-4, class 4K4, with low air temperature restricted to – 40 °C and high air temperature restricted to +5 °C

A ENVIRONMENT *(cont.)*

Code	External influences						Characteristics required for selection and erection of equipment	Reference for information only
	Air temperature °C **a) low**	**b) high**	**Relative humidity** % **c) low**	**d) high**	**Absolute humidity** g/m^3 **e) low**	**f) high**		
AB3	–25	+5	10	100	0.5	7	Indoor and outdoor locations with low ambient temperatures Appropriate arrangements should be made[c]	Part of temperature range of BS EN 60721-3-3, class 3K6, with high air temperature restricted to +5 °C. Includes temperature range of BS EN 60721-3-4, class 4K1, with high air temperature range restricted to +5 °C
AB4	–5	+40	5	95	1	29	Weather protected locations having neither temperature nor humidity control. Heating may be used to raise low ambient temperatures Normal[b]	Identical with temperature range of BS EN 60721-3-3, class 3K5. The high air temperature restricted to +40 °C
AB5	+5	+40	5	85	1	25	Weather protected locations with temperature control Normal[b]	Identical with temperature range of BS EN 60721-3-3, class 3K3
AB6	+5	+60	10	100	1	35	Indoor and outdoor locations with extremely high ambient temperatures, influence of cold ambient temperatures is prevented. Occurrence of solar and heat radiation Appropriate arrangements should be made[c]	Part of temperature range of BS EN 60721-3-3, class 3K7, with low air temperature restricted to +5 °C and high air temperature restricted to +60 °C. Includes temperature range of BS EN 60721-3-4, class 4K4, with low air temperature restricted to +5 °C
AB7	–25	+55	10	100	0.5	29	Indoor weather protected locations having neither temperature nor humidity control, the locations may have openings directly to the open air and be subjected to solar radiation Appropriate arrangements must be made[c]	Identical to temperature range of BS EN 60721-3-3, class 3K6
AB8	–50	+40	15	100	0.04	36	Outdoor and non-weather protected locations, with low and high temperatures Appropriate arrangements should be made[c]	Identical to temperature range of BS EN 60721-3-4, class 4K3

A ENVIRONMENT *(cont.)*

Code	External influences	Characteristics required for selection and erection of equipment	Reference for information only
AC	*Altitude*		
AC1	≤2 000 m	Normal[b]	
AC2	>2 000 m	May necessitate special precautions such as the application of derating factors **NOTE:** For some equipment special arrangements may be necessary at altitudes of 1 000 m and above	
AD	*Presence of water*		
AD1	Negligible	IPX0 Probability of presence of water is negligible. Location in which the walls do not generally show traces of water but may do so for short periods, for example in the form of vapour which good ventilation dries rapidly	BS EN 60529 BS EN 60721-3-4, class 4Z6
AD2	Free-falling drops	IPX1 or IPX2 Possibility of vertically falling drops Location in which water vapour occasionally condenses as drops or where steam may occasionally be present	BS EN 60529 BS EN 60721-3-3, class 3Z7
AD3	Sprays	IPX3 Possibility of water falling as a spray at an angle up to 60° from the vertical Locations in which sprayed water forms a continuous film on floors and/or walls	BS EN 60529 BS EN 60721-3-3, class 3Z8 BS EN 60721-3-4, class 4Z7
AD4	Splashes	IPX4 Possibility of splashes from any direction Locations where equipment may be subjected to splashed water; this applies, for example, to certain external luminaires, construction site equipment	BS EN 60529 BS EN 60721-3-3, class 3Z9 BS EN 60721-3-4, class 4Z7
AD5	Jets	IPX5 Possibility of jets of water from any direction Locations where hose water is used regularly (yards, car-washing bays)	BS EN 60529 BS EN 60721-3-3, class 3Z10 BS EN 60721-3-4, class 4Z8
AD6	Waves	IPX6 Possibility of water waves Seashore locations such as piers, beaches, quays, etc	BS EN 60529 BS EN 60721-3-4, class 4Z9
AD7	Immersion	IPX7 Locations which may be flooded and/or where the equipment is immersed as follows: • Equipment with a height of less than 850 mm is located in such a way that its lowest point is not more than 1 000 mm below the surface of the water • Equipment with a height equal to or greater than 850 mm is located in such a way that its highest point is not more than 150 mm below the surface of the water	
AD8	Submersion	IPX8 Possibility of permanent and total covering by water Locations such as swimming pools where electrical equipment is permanently and totally covered with water	

A ENVIRONMENT *(cont.)*

Code	External influences	Characteristics required for selection and erection of equipment	Reference for information only
AE	*Presence of foreign solid bodies*	IPXX see also Section 416	
AE1	Negligible	IP0X	BS EN 60529
		The quantity or size of dust or foreign solid bodies is not significant	BS EN 60721-3-3, class 3S1
			BS EN 60721-3-4, class 4S1
AE2	Small objects (2.5 mm)	IP3X	BS EN 60529
		Presence of foreign solid bodies where the smallest dimension is not less than 2.5 mm	BS EN 60721-3-3, class 3S2
		Tools and small objects are examples of foreign solid bodies of which the smallest dimension is at least 2.5 mm	BS EN 60721-3-4, class 4S2
AE3	Very small objects (1 mm)	IP4X	BS EN 60529
		Presence of foreign solid bodies where the smallest dimension is not less than 1 mm	BS EN 60721-3-3, class 3S3
		Wires are examples of foreign solid bodies of which the smallest dimension is not less than 1 mm	BS EN 60721-3-4, class 4S3
AE4	Light dust }	IP5X	BS EN 60529
		Presence of dust	BS EN 60721-3-3, class 3S2
		if dust penetration is not harmful to the functioning of the equipment	BS EN 60721-3-4, class 4S2
AE5	Moderate dust }	IP6X	BS EN 60529
		Presence of dust	BS EN 60721-3-4, class 3S3
		if dust penetration is harmful to the functioning of the equipment	BS EN 60721-3-3, class 4S3
AE6	Heavy dust }	IP6X	BS EN 60721-3-3, class 3S4
		Presence of dust	BS EN 60721-3-4, class 4S4
		Dust must not penetrate equipment	
AF	*Presence of corrosive of polluting substances*		
AF1	Negligible	Normal[b]	BS EN 60721-3-3, class 3C1
			BS EN 60721-3-4, class 4C1
AF2	Atmospheric	The presence of corrosive or polluting substances of atmospheric origin is significant. Installations situated by the sea or near industrial zones producing serious atmospheric pollution, such as chemical works, cement works; this type of pollution arises especially in the production of abrasive, insulating or conductive dusts	BS EN 60721-3-3, class 3C2
			BS EN 60721-3-4, class 4C2
		According to the nature of substances (for example, satisfaction of salt mist test according to BS EN 60068-2-11)	
AF3	Intermittent or accidental	Intermittent or accidental subjection to corrosive or polluting chemical substances being used or produced	BS EN 60721-3-3, class 3C3
			BS EN 60721-3-4, class 4C3
		Locations where some chemical products are handled in small quantities and where these products may come only accidentally into contact with electrical equipment; such conditions are found in factory laboratories, other laboratories or in locations where hydrocarbons are used (boiler-rooms, garages, etc.)	
		Protection against corrosion according to equipment specification	

A ENVIRONMENT *(cont.)*

Code	External influences	Characteristics required for selection and erection of equipment	Reference for information only
AF4	Continuous	Continuously subject to corrosive or polluting chemical substances in substantial quantity, e.g. chemical works Equipment specially designed according to the nature of substances	BS EN 60721-3-3, class 3C4 BS EN 60721-3-4, class 4C4
AG	*Mechanical stress:* *Impact*		
AG1	Low severity	Normal, e.g. household and similar equipment	BS EN 60721-3-3, classes 3M1/3M2/3M3 BS EN 60721-3-4, classes 4M1/4M2/4M3
AG2	Medium severity	Standard industrial equipment, where applicable, or reinforced protection	BS EN 60721-3-3, classes 3M4/3M5/3M6 BS EN 60721-3-4, classes 4M4/4M5/4M6
AG3	High severity	Reinforced protection	BS EN 60721-3-3, classes 3M7/3M8 BS EN 60721-3-4, classes 4M7/4M8

A ENVIRONMENT *(cont.)*

Code	External influences	Characteristics required for selection and erection of equipment	Reference for information only
AH	*Vibration*		
AH1	Low severity	Residential premises and similar conditions where the effects of vibration are generally negligible Normal	BS EN 60721-3-3, classes 3M1/3M/3M3 BS EN 60721-3-4, classes 4M1/4M2/4M3
AH2	Medium severity	Usual industrial conditions Specially designed equipment or special arrangements	BS EN 60721-3-3, classes 3M4/3M5/3M6 BS EN 60721-3-4, classes 4M4/4M5/4M6
AH3	High severity	Industrial installations subject to severe conditions Specially designed equipment or special arrangements	BS EN 60721-3-3, classes 3M7/3M8 BS EN 60721-3-4, classes 4M7/4M8
AJ	*Other mechanical stresses*	Under consideration	
AK	*Presence of flora and/or mould growth*		
AK1	No hazard	No harmful hazard from flora and/or mould growth Normal[b]	BS EN 60721-3-3, class 3B1 BS EN 60721-3-4, class 4B1
AK2	Hazard	Harmful hazard from flora and/or mould growth The hazard depends on local conditions and the nature of flora. Distinction should be made between harmful growth of vegetation or conditions for promotion of mould growth Special protection, such as: – increased degree of protection (see AE) – special materials or protective coating of enclosures – arrangements to exclude flora from location	BS EN 60721-3-3, class 3B2 BS EN 60721-3-4, class 4B2
AL	*Presence of fauna*		
AL1	No hazard	No harmful hazard from fauna Normal[b]	BS EN 60721-3-3, class 3B1 BS EN 60721-3-4, class 4B1

A ENVIRONMENT *(cont.)*

Code	External influences	Characteristics required for selection and erection of equipment	Reference for information only
AL2	Hazard	Harmful hazard from fauna (insects, birds, small animals) The hazard depends on the nature of the fauna. Distinction should be made between: – presence of insects in harmful quantity or of an aggressive nature; – presence of small animals or birds in harmful quantity or of an aggressive nature Protection may include: – an appropriate degree of protection against penetration of foreign solid bodies (see AE) – sufficient mechanical resistance (see AG) – precautions to exclude fauna from the location (such as cleanliness, use of pesticides) – special equipment or protective coating of enclosures	BS EN 60721-3-3, class 3B2 BS EN 60721-3-4, class 4B2

A ENVIRONMENT *(cont.)*

Code	External influences	Characteristics required for selection and erection of equipment	Reference for information only
AM	*Electromagnetic, electrostatic, or ionizing influences* *Low-frequency electromagnetic phenomena (conducted or radiated)* *Harmonics, interharmonics*		
AM-1-1	Controlled level	Care should be taken that the controlled situation is not impaired	Lower than table 1 of BS EN 61000-2-2
AM-1-2	Normal level	Special measures in the design of the installation, e.g. filters	Complying with table 1 of BS EN 61000-2-2
AM-1-3	High level	Special measures in the design of the installation, e.g. filters	Locally higher than table 1 of BS EN 61000-2-2
	Signalling voltages		
AM-2-1	Controlled level	Possibly: blocking circuits	Lower than specified below
AM-2-2	Medium level	No additional requirement	IEC/TR 61000-2-1 and BS EN 61000-2-2
AM-2-3	High level	Appropriate measures	
	Voltage amplitude variations		
AM-3-1	Controlled level	e.g. controlled by UPS	
AM-3-2	Normal level	Compliance with BS 7671 Chapter 44	
AM-4	*Voltage unbalance*		Compliance with BS EN 61000-2-2
AM-5	*Power frequency variations*		±1 Hz according to BS EN 61000-2-2
AM-6	*Induced low-frequency voltages* No classification	Refer to BS 7671 Chapter 44 High withstand of signal and control systems of switchgear and controlgear	ITU-T (International Telecommunication Union)
AM-7	*Direct current in AC networks* No classification	Measures to limit their presence in level and time in the current-using equipment or their vicinity	
	Radiated magnetic fields		
AM-8-1	Medium level	Normal[b]	Level 2 of BS EN 61000-4-8
AM-8-2	High level	Protection by appropriate measures e.g. screening and/or separation	Level 4 of BS EN 61000-4-8

A ENVIRONMENT *(cont.)*

Code	External influences	Characteristics required for selection and erection of equipment	Reference for information only
	Electric fields		
AM-9-1	Negligible level	Normal[b]	
AM-9-2	Medium level		BS EN 61000-2-5
AM-9-3	High level		BS EN 61000-2-5
AM-9-4	Very high level		BS EN 61000-2-5
	High-frequency electromagnetic phenomena conducted, induced or radiated (continuous or transient)		
AM-21	*Induced oscillatory voltages or currents*		
	No classification	Normal[b]	BS EN 61000-4-6
	Conducted unidirectional transients of the nanosecond time scale		BS EN 61000-4-4
AM-22-1	Negligible level	Protective measures are necessary	Level 1
AM-22-2	Medium level	Protective measures are necessary	Level 2
AM-22-3	High level	Normal equipment	Level 3
AM-22-4	Very high level	High immunity equipment	Level 4
	Conducted unidirectional transients of microsecond to millisecond time scale		
AM-23-1	Controlled level	Impulse withstand of equipment and overvoltage protective means chosen taking into account the nominal supply voltage and the impulse withstand category according to BS 7671 Chapter 44	BS 7671 Chapter 44
AM-23-2	Medium level		BS 7671 Chapter 44
AM-23-3	High level		
	Conducted oscillatory transients		
AM-24-1	Medium level	Refer to BS EN 61000-4-12	BS EN 61000-4-12
AM-24-2	High level	Refer to BS EN 60255-26	BS EN 60255-26:2013
	Radiated high-frequency phenomena		BS EN 61000-4-3
AM-25-1	Negligible level		Level 1
AM-25-2	Medium level	Normal[b]	Level 2
AM-25-3	High level	Reinforced level	Level 3
	Electrostatic discharges		BS EN 61000-4-2
AM-31-1	Small level	Normal[b]	Level 1
AM-31-2	Medium level	Normal[b]	Level 2
AM-31-3	High level	Normal[b]	Level 3
AM-31-4	Very high level	Reinforced	Level 4
AM-41-1	*Ionization* No classification	Special protection such as: – Spacings from source – Interposition of screens, enclosure by special materials	

A ENVIRONMENT *(cont.)*

Code	External influences	Characteristics required for selection and erection of equipment	Reference for information only
AN	*Solar radiation*		
AN1	Low	Intensity ≤ 500 W/m² Normal[b]	BS EN 60721-3-3 class 3K1
AN2	Medium	500 W/m² < intensity ≤ 700 W/m² Appropriate arrangements must be made[c]	BS EN 60721-3-3 Classes 3K2 to 3K5
AN3	High	700 W/m² < intensity ≤ 1120 W/m² Appropriate arrangements must be made[c] Such arrangements could be: – material resistant to ultraviolet radiation – special colour coating – interposition of screens	BS EN 60721-3-3 Higher than class 3K5 BS EN 60721-3-4
AP	*Seismic effects*		
AP1	Negligible	Acceleration ≤ 30 Gal (1 Gal = 1 cm/s²) Normal[b]	
AP2	Low severity	30 Gal < acceleration ≤ 300 Gal Under consideration	
AP3	Medium severity	300 Gal < acceleration ≤ 600 Gal Under consideration	
AP4	High severity	600 Gal < acceleration Under consideration Vibration which may cause the destruction of the building is outside the classification Frequency is not taken into account in the classification; however, if the seismic wave resonates with the building, seismic effects must be specially considered. In general, the frequency of seismic acceleration is between 0 Hz and 10 Hz	
AQ	*Lightning*		
AQ1	Negligible	Normal[b]	
AQ2	Indirect exposure	In accordance with Section 443	
AQ3	Direct exposure	BS EN 62305-1	
AR	*Movement of air*		
AR1	Low	Speed ≤ 1 m/s Normal[b]	
AR2	Medium	1 m/s < speed ≤ 5 m/s Appropriate arrangements should be made[c]	
AR3	High	5 m/s < speed ≤ 10 m/s Appropriate arrangements should be made[c]	
AS	*Wind*		
AS1	Low	Speed ≤ 20 m/s Normal[b]	
AS2	Medium	20 m/s < speed ≤ 30 m/s Appropriate arrangements should be made[c]	
AS3	High	30 m/s < speed ≤ 50 m/s Appropriate arrangements should be made[c]	

B UTILIZATION:

Code	External influences	Characteristics required for selection and erection of equipment	Reference for information only
BA	**Capability of persons**		
BA1	Ordinary	Uninstructed persons Normal[b]	Inaccessibility of electrical equipment. Limitation of temperature of accessible surfaces
BA2	Children	Locations intended for presence of children e.g. nurseries, infant schools, etc. Equipment of degrees of protection equal to or greater than IP2XC. Inaccessibility of equipment with external surface temperature exceeding 60 °C	
BA3	Disabled	Persons not in command of all their physical and/or intellectual abilities (sick persons, old persons) According to the nature of the disability	
BA4	Instructed	Persons adequately advised or supervised by skilled persons to enable them to avoid dangers which electricity may create (operating and maintenance staff) Electrical operating areas	
BA5	Skilled	Equipment not having basic protection against direct contact with live parts admitted solely in locations which are accessible only to duly authorized persons with technical knowledge or sufficient experience to enable them to avoid danger which electricity may create (engineers and technicians) Closed electrical operating areas	
BB	**Electrical resistance of the human body** Under consideration		
BC	**Contact of persons with Earth potential**		
		Class of equipment according to BS EN 61140 I II III	
BC1	None	Persons in non-conducting situations A A A	
BC2	Occasional	Persons who do not in usual conditions make contact with extraneous-conductive-parts or stand on conducting surfaces A A A	
BC3	Frequent	Persons who are frequently in touch with extraneous-conductive-parts or stands on conducting services Locations with extraneous-conductive-parts, either numerous or of a large area Class of equipment according to BS EN 61140 0-01 I II III X A A A A Equipment permitted X Equipment prohibited Y Permitted if used as class 0	
BC4	Continuous	Persons who are immersed in water or in long term permanent contact with metallic surroundings and for whom the possibility of interrupting contact is limited Metallic surroundings such as boilers and tanks	Under consideration
BD	**Conditions of evacuation in an emergency**		
BD1	Low density / easy exit	Normal[b] Low density occupation, easy conditions of evacuation Buildings of normal or low height used for habitation	
BD2	Low density / difficult exit	Low density occupation, difficult conditions of evacuation High-rise buildings	
BD3	High density / easy exit	High density occupation, easy conditions of evacuation Locations open to the public (theatres, cinemas, departments stores, etc.)	
BD4	High density / difficult exit	High density occupation, difficult conditions of evacuation High-rise buildings open to the public (hotels, hospitals, etc.)	

Code	External influences	Characteristics required for selection and erection of equipment	Reference for information only
BE	**Nature of processed or stored materials**		
BE1	No significant risk	Normal[b]	
BE2	Fire risks	Manufacture, processing or storage of flammable materials including presence of dust Barns, wood-working shops, paper factories Equipment made of material retarding the spread of flame Arrangements such that a significant temperature rise or a spark within electrical equipment cannot initiate an external fire	Chapter 42 Chapter 52
BE3	Explosion risks	Processing or storage of explosive or low-flash-point materials including presence of explosive dusts Oil refineries, hydrocarbon stores Requirements for electrical apparatus for explosive atmospheres (see BS EN 60079),	Under consideration
BE4	Contamination risks	Presence of unprotected foodstuffs, pharmaceutics, and similar products without protection Foodstuff industries, kitchens: Certain precautions may be necessary, in the event of fault, to prevent processed materials being contaminated by electrical equipment, e.g. by broken lamps Appropriate arrangements, such as: protection against falling debris from broken lamps and other fragile objects screens against harmful radiation such as infrared or ultraviolet	Under consideration

C CONSTRUCTION OF BUILDINGS:

Code	External influences	Characteristics required for selection and erection of equipment	Reference for information only
CA	**Construction materials**		
CA1	Non-combustible	Normal	
CA2	Combustible	Buildings mainly constructed of combustible materials Wooden buildings Under consideration	HD 60364-4-42
CB	**Building design**		
CB1	Negligible risks	Normal	
CB2	Propagation of fire	Buildings of which the shape and dimensions facilitate the spread of fire (e.g. chimney effect) High-rise buildings. Forced ventilation systems Equipment made of material retarding the propagation of fire including fires not originating from the electrical installation. Fire barriers	HD 60364-4-42 HD 60364-5-52
CB3	Movement	Risk due to structural movement (e.g. displacement) Buildings of considerable length or erected on unstable ground Contraction or expansion joints in electrical wiring	Contraction or expansion joints (under consideration) HD 60364-5-52
CB4	Flexible or unstable	Structures which are weak or subject to movement (e.g. oscillation) Tents, air-support structures, false ceilings, removable partitions. Installations to be structurally self-supporting Under consideration	Flexible wiring HD 60364-5-52

APPENDIX 6 (Informative)

MODEL FORMS FOR CERTIFICATION AND REPORTING

Introduction

(i) The Electrical Installation Certificate required by Part 6 should be made out and signed or otherwise authenticated by a skilled person or persons in respect of the design, construction, inspection and testing of the work.

(ii) The Minor Electrical Installation Works Certificate required by Part 6 should be made out and signed or otherwise authenticated by a skilled person in respect of the design, construction, inspection and testing of the minor work.

(iii) The Electrical Installation Condition Report required by Part 6 should be made out and signed or otherwise authenticated by a skilled person or persons in respect of the inspection and testing of an existing installation.

(iv) Skilled persons will, as appropriate to their function under (i) (ii) and (iii) above, have a sound knowledge and experience relevant to the nature of the work undertaken and to the technical standards set out in BS 7671, be fully versed in the inspection and testing procedures contained in BS 7671 and employ adequate testing equipment.

(v) Electrical Installation Certificates will indicate the responsibility for design, construction, inspection and testing, whether in relation to new work or further work on an existing installation.

Where the design, construction, inspection and testing are the responsibility of one person a Certificate with a single-signature declaration in the form shown below may replace the multiple signatures section of the model form.

FOR DESIGN, CONSTRUCTION, INSPECTION & TESTING

I being the person responsible for the Design, Construction, Inspection & Testing of the electrical installation (as indicated by my signature below), particulars of which are described above, having exercised reasonable skill and care when carrying out the Design, Construction, Inspection & Testing, hereby CERTIFY that the said work for which I have been responsible is to the best of my knowledge and belief in accordance with BS 7671:2018, amended to ....(date) except for the departures, if any, detailed as follows.

(vi) A Minor Electrical Installation Works Certificate will indicate the responsibility for design, construction, inspection and testing of the work described on the certificate.

(vii) An Electrical Installation Condition Report will indicate the responsibility for the inspection and testing of an existing installation within the extent and limitations specified on the report.

(viii) Details of Schedules of Inspection and Schedules of Circuit Details and Test Results as required by Part 6 should be issued with the associated Electrical Installation Certificate or Electrical Installation Condition Report.

(ix) When making out and signing a form on behalf of a company or other business entity, individuals should state for whom they are acting.

(x) Additional forms may be required as clarification, if needed by ordinary persons, or in expansion, for larger or more complex installations.

ELECTRICAL INSTALLATION CERTIFICATE
(REQUIREMENTS FOR ELECTRICAL INSTALLATIONS - BS 7671)

Certificate No.: ....

DETAILS OF THE CLIENT

....

INSTALLATION ADDRESS

....
....

DESCRIPTION AND EXTENT OF THE INSTALLATION Description of installation:	New installation ☐
Extent of installation covered by this Certificate: (Use continuation sheet if necessary) See continuation sheet No:	Addition to an existing installation ☐ Alteration to an existing installation ☐

FOR DESIGN

I/We, being the person(s) responsible for the design of the electrical installation (as indicated by my/our signatures below), particulars of which are described above, having exercised reasonable skill and care when carrying out the design, hereby CERTIFY that the design work for which I/we have been responsible is to the best of my/our knowledge and belief in accordance with BS 7671:2018, amended to .... (date) except for the departures, if any, detailed as follows:

Details of departures from BS 7671 (Regulations 120.3, 133.1.3 and 133.5):

Details of permitted exceptions (Regulation 411.3.3). Where applicable, a suitable risk assessment(s) must be attached to this Certificate.

Risk assessment attached ☐

The extent of liability of the signatory or signatories is limited to the work described above as the subject of this Certificate.

For the DESIGN of the installation: **(Where there is mutual responsibility for the design)

Signature: .... Date: .... Name (IN BLOCK LETTERS): .... Designer No 1

Signature: .... Date: .... Name (IN BLOCK LETTERS): .... Designer No 2**

FOR CONSTRUCTION

I, being the person responsible for the construction of the electrical installation (as indicated by my signature below), particulars of which are described above, having exercised reasonable skill and care when carrying out the construction hereby CERTIFY that the construction work for which I have been responsible is to the best of my knowledge and belief in accordance with BS 7671:2018, amended to ....(date) except for the departures, if any, detailed as follows:

Details of departures from BS 7671 (Regulations 120.3 and 133.5):

The extent of liability of the signatory is limited to the work described above as the subject of this Certificate.

For CONSTRUCTION of the installation:

Signature: .... Date: .... Name (IN BLOCK LETTERS): .... Constructor

FOR INSPECTION & TESTING

I, being the person responsible for the inspection & testing of the electrical installation (as indicated by my signature below), particulars of which are described above, having exercised reasonable skill and care when carrying out the inspection & testing hereby CERTIFY that the work for which I have been responsible is to the best of my knowledge and belief in accordance with BS 7671:2018, amended to ....(date) except for the departures, if any, detailed as follows:

Details of departures from BS 7671 (Regulations 120.3 and 133.5):

The extent of liability of the signatory is limited to the work described above as the subject of this Certificate.

For INSPECTION AND TESTING of the installation:

Signature: .... Date: .... Name (IN BLOCK LETTERS): .... Inspector

NEXT INSPECTION

I/We, the designer(s), recommend that this installation is further inspected and tested after an interval of not more than .... years/months.

Certificate No.: ....

PARTICULARS OF SIGNATORIES TO THE ELECTRICAL INSTALLATION CERTIFICATE

Designer (No 1)
Name: .... Company: ....
Address: ....
.... Postcode: .... Tel No: ....

Designer (No 2) (if applicable)
Name: .... Company: ....
Address: ....
.... Postcode: .... Tel No: ....

Constructor
Name: .... Company: ....
Address: ....
.... Postcode: .... Tel No: ....

Inspector
Name: .... Company: ....
Address: ....
.... Postcode: .... Tel No: ....

SUPPLY CHARACTERISTICS AND EARTHING ARRANGEMENTS

Earthing arrangements	Number and Type of Live Conductors		Nature of Supply Parameters	Supply Protective Device
TN-C ☐	AC ☐	DC ☐	Nominal voltage, U / U_0[(1)] V	BS (EN) ....
TN-S ☐	1-phase, 2-wire ☐	2-wire ☐	Nominal frequency, f[(1)] .... Hz	Type ....
TN-C-S ☐	2-phase, 3-wire ☐	3-wire ☐	Prospective fault current, I_{pf}[(2)] .... kA	Rated current .... A
TT ☐	3-phase, 3-wire ☐	Other ☐	External earth fault loop impedance, Z_e[(2)] .... Ω	
IT ☐	3-phase, 4-wire ☐			
	Confirmation of supply polarity ☐		*(Note: (1) by enquiry (2) by enquiry or by measurement)*	

Other sources of supply (as detailed on attached schedule) ☐

PARTICULARS OF INSTALLATION REFERRED TO IN THE CERTIFICATE

Means of Earthing

Distributor's facility ☐

Installation earth electrode ☐

Maximum Demand

Maximum demand (load) .... kVA / Amps (Delete as appropriate)

Details of Installation Earth Electrode (where applicable)

Type (e.g. rod(s), tape etc) ....

Location ....

Electrode resistance to Earth .... Ω

Main Protective Conductors

Earthing conductor	Material .... csa .... mm^2	Connection / continuity verified ☐
Main protective bonding conductors	Material .... csa .... mm^2	Connection / continuity verified ☐

To water installation pipes ☐	To gas installation pipes ☐	To oil installation pipes ☐	To structural steel ☐
To lightning protection ☐	To other ☐ Specify		

Main switch / Switch-fuse / Circuit-breaker / RCD

		If RCD main switch
Location	Current rating .... A	RCD Type ....
....	Fuse / device rating or setting .... A	Rated residual operating current ($I_{\Delta n}$) .... mA
BS(EN)	Voltage rating V	Rated time delay .... ms
No of poles		Measured operating time .... ms

Schedule of Inspections

Item No.	Description	Outcome ✓ / N/A	Item No.	Description	Outcome ✓ / N/A
1.0	Condition of consumer's intake equipment (Visual inspection only)		8.0	Circuits (Distribution and Final)	
			9.0	Isolation and switching	
2.0	Parallel or switched alternative sources of supply		10.0	Current-using equipment (permanently connected)	
3.0	Protective measure: Automatic Disconnection of Supply (ADS)		11.0	Identification and notices	
4.0	Basic protection		12.0	Location(s) containing a bath or shower	
5.0	Protective measures other than ADS		13.0	Other special installations or locations	
6.0	Additional protection		14.0	Prosumer's low voltage electrical installation(s)	
7.0	Distribution equipment				

COMMENTS ON EXISTING INSTALLATION (in the case of an addition or alteration see Regulation 644.1.2):
....
....
....
....
....
....

SCHEDULES
This Certificate is valid only when..Schedules of Circuit Details and Test Results are attached.
(Enter quantities of schedules attached).

ELECTRICAL INSTALLATION CERTIFICATE

Notes for the person producing the Certificate:

1 The Electrical Installation Certificate is to be used for:

- the initial certification of a new installation or for an addition or alteration to an existing installation where new circuits have been introduced, or
- the replacement of a consumer unit/distribution board, or
- certifying for where there are multiple additions, or alterations or remedial works to the existing installation which do not extend to new circuits as an alternative to the issue of multiple Minor Electrical Installation Works Certificates.

It is not to be used for periodic inspection and testing, for which an Electrical Installation Condition Report should be used. For an addition or alteration which does not extend to the introduction of new circuits, a Minor Electrical Installation Works Certificate may be used.

The 'original' Certificate is to be issued to the person ordering the work (Regulation 644.4). A duplicate should be retained by the person issuing the certificate.

2 This Certificate is only valid if the Schedule of Inspections has been completed to confirm that all relevant inspections have been carried out and where accompanied by Schedule(s) of Circuit Details and Test Results.

3 The signatures appended are those of the persons authorized by the companies executing the work of design, construction, inspection and testing respectively. A signatory authorized to certify more than one category of work should sign in each of the appropriate places. (Where a single signature electrical installation certificate is used, the person authorized for executing the work of design, construction, inspection and testing shall sign the certificate.)

4 The time interval recommended before the first periodic inspection must be inserted.

The proposed date for the next inspection should take into consideration the frequency and quality of maintenance that the installation can reasonably be expected to receive during its intended life, and the period should be agreed between the designer, installer and other relevant parties.

5 The page numbers for the Schedule(s) of Circuit Details and Test Results should be indicated, together with the total number of pages associated with the certification provided.

6 The maximum prospective value of fault current (I_{pf}) recorded should be the greater of either the prospective value of short-circuit current or the prospective value of earth fault current.

ELECTRICAL INSTALLATION CERTIFICATE

GUIDANCE FOR RECIPIENTS (to be appended to the Certificate)

This safety Certificate has been issued to confirm that the electrical installation work to which it relates has been designed, constructed, inspected and tested in accordance with BS 7671.

You should have received an 'original' Certificate and the person that issued the Certificate should have retained a duplicate. If you were the person ordering the work, but not the owner of the installation, you should pass this Certificate, or a full copy of it including the schedules, immediately to the owner.

The 'original' Certificate should be retained in a safe place and be shown to any person inspecting or undertaking further work on the electrical installation in the future. If you later vacate the property, this Certificate will demonstrate to the new owner that the electrical installation complied with the requirements of BS 7671 at the time the Certificate was issued. The Construction (Design and Management) Regulations require that, for a project covered by those Regulations, a copy of this Certificate, together with schedules, is included in the project health and safety documentation.

For safety reasons, the electrical installation will need to be inspected at appropriate intervals by a skilled person or persons, competent in such work. The maximum time interval recommended before the next inspection is stated on Page 1 under 'NEXT INSPECTION'.

This Certificate is intended to be issued only for a new electrical installation or for new work associated with an addition or alteration to an existing installation. It should not have been issued for the inspection and testing of an existing electrical installation. An 'Electrical Installation Condition Report' should be issued for such an inspection.

This Certificate is only valid if the Schedule of Inspections has been completed to confirm that all relevant inspections have been carried out and where accompanied by Schedule(s) of Circuit Details and Test Results.

Where the installation includes a residual current device (RCD) it should be tested six-monthly by pressing the button marked 'T' or 'Test'. The device should switch off the supply and should then be switched on to restore the supply. If the device does not switch off the supply when the button is pressed, seek expert advice. **For safety reasons it is important that this instruction is followed.**

Where the installation includes an arc fault detection device (AFDD) having a manual test facility it should be tested six-monthly by pressing the test button. Where an AFDD has both a test button and automatic test function, manufacturer's instructions shall be followed with respect to test button operation.

Where the installation includes a surge protective device (SPD) the status indicator should be checked to confirm it is in operational condition in accordance with manufacturer's information. If the indication shows that the device is not operational, seek expert advice. **For safety reasons it is important that this instruction is followed.**

Where the installation includes alternative or additional sources of supply, warning notices should be found at the origin or meter position or, if remote from the origin, at the consumer unit or distribution board and at all points of isolation of all sources of supply.

MINOR ELECTRICAL INSTALLATION WORKS CERTIFICATE

Certificate No.: ....

(REQUIREMENTS FOR ELECTRICAL INSTALLATIONS - BS 7671)

To be used only for minor electrical work which does not include the provision of a new circuit

PART 1: Description of the minor works

1. Details of the Client .... Date minor works completed ....
2. Installation location/address ....
3. Description of the minor works ....
4. Details of any departures from BS 7671:2018 as amended to .... (date) for the circuit altered or extended (Regulation 120.3, 133.1.3 and 133.5).
   Details of permitted exceptions (Regulation 411.3.3). Where applicable, a suitable risk assessment(s) must be attached to this Certificate.
   Risk assessment attached ☐
   ....
5. **Comments on (including any defects observed in) the existing installation (Regulation 644.1.2):**
   ....

PART 2: Presence and adequacy of installation earthing and bonding arrangements (Regulation 132.16)

1. System earthing arrangement: TN-S ☐ TN-C-S ☐ TT ☐
2. Earth fault loop impedance at distribution board (Z_{db}) supplying the final circuit .... Ω
3. Presence of adequate main protective conductors:
   Earthing conductor ☐
   Main protective bonding conductor(s) to: Water ☐ Gas ☐ Oil ☐ Structural steel ☐ Other (Specify).............. ☐

PART 3: Circuit details

DB Reference No.: .... DB Location and type: ....

Circuit No.: .... Circuit description: .... Installation reference method ....

Number & size of conductors: Live .... mm^2 cpc .... mm^2

Circuit overcurrent protective device: BS (EN) .... Type .... Rating ....A

RCD: BS (EN) .... Type .... Rating ....A Rated residual operating current ($I_{\Delta n}$) ....mA

AFDD: BS (EN) .... Rating ....A

SPD: BS (EN) .... Type ....

PART 4: Test results for the altered or extended circuit (where relevant and practicable)

Protective conductor continuity: (R_1+ R_2) .... Ω or R_2 .... Ω

Continuity of ring final circuit conductors: L/L .... Ω N/N .... Ω cpc/cpc .... Ω

Insulation resistance: Test voltage........V Live - Live .... MΩ Live - Earth .... MΩ

Polarity satisfactory: ☐ Maximum measured earth fault loop impedance: Z_s .... Ω

RCD disconnection time at rated residual operating current ($I_{\Delta n}$) ....ms Satisfactory test button operation: ☐

AFDD satisfactory test button operation: ☐ NOTE: Not all AFDDs have a test button

SPD functionality confirmed: ☐ NOTE: Not all SPDs have visible functionality indication

PART 5: Declaration

I certify that the work covered by this certificate does not impair the safety of the existing installation and the work has been designed, constructed, inspected and tested in accordance with BS 7671:2018 amended to .... (date) and that to the best of my knowledge and belief, at the time of my inspection, complied with BS 7671 except as detailed in Part 1 above.

Name:	
For and on behalf of:	
Address:	Signature: ....
....	Position: ....
....	Date: ....

MINOR ELECTRICAL INSTALLATION WORKS CERTIFICATE

Notes for the person producing the Certificate:

The Minor Electrical Installation Works Certificate is intended to be used for additions and alterations to an installation that do not extend to the provision of a new circuit. Examples include the addition of socket-outlets or lighting points to an existing circuit, the relocation of a light switch etc. This Certificate may also be used for the replacement of equipment such as accessories or luminaires, but not for the replacement of distribution boards or similar items. Appropriate inspection and testing, however, should always be carried out irrespective of the extent of the work undertaken.

MINOR ELECTRICAL INSTALLATION WORKS CERTIFICATE

GUIDANCE FOR RECIPIENTS (to be appended to the Certificate)

This Certificate has been issued to confirm that the electrical installation work to which it relates has been designed, constructed, inspected and tested in accordance with BS 7671.

You should have received an 'original' Certificate and the person that issued the Certificate should have retained a duplicate. If you were the person ordering the work, but not the owner of the installation, you should pass this Certificate, or a copy of it, to the owner. A separate Certificate should have been received for each existing circuit on which minor works have been carried out. This Certificate is not appropriate if you requested the person that issued the Certificate to undertake more extensive installation work, for which you should have received an Electrical Installation Certificate.

The Certificate should be retained in a safe place and be shown to any person inspecting or undertaking further work on the electrical installation in the future. If you later vacate the property, this Certificate will demonstrate to the new owner that the minor electrical installation work carried out complied with the requirements of BS 7671 at the time the Certificate was issued.

For safety reasons, the electrical installation will need to be inspected at appropriate intervals by a skilled person or persons, competent in such work.

Where the installation includes a residual current device (RCD) it should be tested six-monthly by pressing the button marked ‘T’ or ‘Test’. The device should switch off the supply and should then be switched on to restore the supply. If the device does not switch off the supply when the button is pressed, seek expert advice. **For safety reasons it is important that this instruction is followed.**

Where the installation includes an arc fault detection device (AFDD) having a manual test facility it should be tested six-monthly by pressing the test button. Where an AFDD has both a test button and automatic test function, manufacturer’s instructions shall be followed with respect to test button operation.

Where the installation includes a surge protective device (SPD) the status indicator should be checked to confirm it is in operational condition in accordance with manufacturer’s information. If the indication shows that the device is not operational, seek expert advice. **For safety reasons it is important that this instruction is followed.**

Where the installation includes alternative or additional sources of supply, warning notices should be found at the origin or meter position or, if remote from the origin, at the consumer unit or distribution board and at all points of isolation of all sources of supply.

Example checklist of items requiring inspection during initial verification

This checklist provides examples of items to aid completion of the schedule of inspections in the electrical installation certificate.

The inspection should include a check of all relevant items in relation to the electrical installation. However, the list of items is not exhaustive and so this inspection checklist can be adjusted to reflect the particular installation.

Numbers in brackets are Regulation references.

Item No	DESCRIPTION
1.0	**CONDITION OF CONSUMER'S INTAKE EQUIPMENT (VISUAL INSPECTION ONLY)**
	Where inadequacies in the intake equipment are encountered, which may result in a dangerous or potentially dangerous situation, it is recommended that the person ordering the work informs the appropriate authority.
1.1	Consumer's isolator (where present)
1.2	Consumer's meter tails

2.0	**PARALLEL OR SWITCHED ALTERNATIVE SOURCES OF SUPPLY**
2.1	Presence of adequate arrangements where generator to operate as a switched alternative (551.6)
2.2	Dedicated earthing arrangement independent of that of the public supply (551.4.3.2.1)
2.3	Presence of adequate arrangements where generator to operate in parallel with the public supply system (551.7)
2.4	Correct connection of generator in parallel (551.7.2)
2.5	Compatibility of characteristics of means of generation (551.7.3)
2.6	Means to provide automatic disconnection of generator in the event of loss of public supply system or voltage or frequency deviation beyond declared values (551.7.4)
2.7	Means to prevent connection of generator in the event of loss of public supply system or voltage or frequency deviation beyond declared values (551.7.5)
2.8	Means to isolate generator from the public supply system (551.7.6)

3.0	**PROTECTIVE MEASURE: AUTOMATIC DISCONNECTION OF SUPPLY (ADS)**
3.1	Distributor's earthing arrangement (542.1.2.1; 542.1.2.2)
3.2	Installation earth electrode (where applicable) (542.1.2.3)
3.3	Earthing conductor and connections, including accessibility (542.3; 543.3.2)
3.4	Main protective bonding conductors and connections, including accessibility (411.3.1.2; 543.3.2; 544.1)
3.5	Provision of safety electrical earthing/bonding labels at all appropriate locations (514.13)
3.6	RCD(s) provided for fault protection (411.4.204; 411.5.3)
3.7	Provisions where automatic disconnection is not feasible (411.3.2.5)
3.8	FELV - requirements satisfied (411.7; 411.7.1)
3.9	RLV - requirements satisfied (411.8)

4.0	**BASIC PROTECTION**
4.1	Insulation of live parts (416.1)
4.2	Barriers or enclosures (416.2; 416.2.1)
4.3	Obstacles (Section 417; 417.2.1; 417.2.2)
4.4	Placing out of reach (Section 417; 417.3)

5.0	**PROTECTIVE MEASURES OTHER THAN ADS**
5.1	SELV (Section 414)
5.2	PELV (Section 414)
5.3	Double insulation i.e. Class II or equivalent equipment and associated circuits (Section 412)
5.4	Reinforced insulation i.e. Class II or equivalent equipment and associated circuits (Section 412)
5.5	Non-conducting location (418.1)
5.6	Earth-free local equipotential bonding (418.2)
5.7	Electrical separation (Section 413; 418.3)

Item No	DESCRIPTION
6.0	**ADDITIONAL PROTECTION**
6.1	RCDs not exceeding 30 mA as specified (415.1)
6.2	Supplementary bonding (Section 415; 415.2)

7.0	**DISTRIBUTION EQUIPMENT**
7.1	Adequacy of working space/accessibility to equipment (132.12; 513.1)
7.2	Security of fixing (134.1.1)
7.3	Insulation of live parts not damaged during erection (416.1)
7.4	Adequacy/security of barriers (416.2)
7.5	Suitability of enclosures for IP and fire ratings (416.2; 421.1.6; 421.1.201; 526.5)
7.6	Enclosures not damaged during installation (134.1.1)
7.7	Presence and effectiveness of obstacles (417.2)
7.8	Components are suitable according to manufacturers' assembly instructions or literature (536.4.203)
7.9	Presence of main switch(es), linked where required (462.1.201)
7.10	Isolators, for every circuit or group of circuits and all items of equipment (462.2)
7.11	Operation of main switch(es) (functional check) (643.10)
7.12	Manual operation of circuit-breakers, RCDs and AFDDs to prove functionality (643.10)
7.13	Confirmation that integral test button/switch causes RCD(s) to trip when operated (functional check) (643.10)
7.14	Confirmation overvoltage protection (SPDs) provided where specified (534.4.1.1)
7.15	Selection of protective device(s) and base(s); correct type and rating (411.3.2; 411.4; 411.5; 411.6; Sections 432, 433, 434, 537.1.1)
7.16	Single-pole protective devices in line conductors only (132.14.1; 530.3.3; 643.6)
7.17	Protection against mechanical damage where cables enter equipment (522.8.1; 522.8.5; 522.8.11)
7.18	Protection against electromagnetic effects where cables enter ferromagnetic enclosures (521.5.1)
7.19	Confirmation that ALL conductor connections, including connections to busbars, are correctly located in terminals and are tight and secure (526.1)

8.0	**CIRCUITS (Distribution and Final)**
8.1	Identification of conductors (514.3.1)
8.2	Conductors correctly identified by colour, lettering or numbering (Section 514)
8.3	Cables correctly supported throughout, with protection against abrasion (521.10.202; 522.8.5)
8.4	No basic insulation of a conductor visible outside enclosure (526.8)
8.5	Examination of cables for signs of mechanical damage during installation (522.6.1; 522.8.1; 522.8.3)
8.6	Examination of insulation of live parts, not damaged during erection (522.6.1; 522.8.1)
8.7	Non-sheathed cables protected by enclosure in conduit, ducting or trunking (521.10.1; 526.8)
8.8	Suitability of containment systems (including flexible conduit) (Section 522)
8.9	Correct temperature rating of cable insulation (522.1.1; Table 52.1)
8.10	Adequacy of cables for current-carrying capacity with regard for the type and nature of installation (Section 523)
8.11	Adequacy of overcurrent protective devices: type and fault current rating for fault protection (434.5)
8.12	Adequacy of RCDs: type and current rating (531.3.3)
8.13	Adequacy of AFDDs: current rating (532.6)
8.14	Presence, adequacy and correct termination of protective conductors (411.3.1.1; 543.1)
8.15	Coordination between conductors and overload protective devices (433.1; 533.2.1)
8.16	Wiring systems and cable installation methods/practices with regard to the type and nature of installation and external influences (Section 522)
8.17	Cables concealed under floors, above ceilings, in walls/partitions, adequately protected against damage (522.6.201; 522.6.202; 522.6.203; 522.6.204)

Item No	DESCRIPTION
8.18	Provision of additional protection by RCDs having rated residual operating current ($I_{\Delta n}$) not exceeding 30 mA (i) For all socket-outlets of rating 32 A or less, unless an exception is permitted (411.3.3) (ii) Supplies for mobile equipment not exceeding 32 A rating for use outdoors (411.3.3) (iii) For cables concealed in walls at a depth of less than 50 mm (522.6.202, 522.6.203) (iv) For cables concealed in walls/partitions containing metal parts regardless of depth (522.6.202, 522.6.203) (v) Final circuits supplying luminaires within domestic (household) premises (411.3.4)
8.19	Provision of fire barriers, sealing arrangements so as to minimize the spread of fire (Section 527)
8.20	Segregation/separation of Band I (ELV) and Band II (LV) circuits (528.1)
8.21	Cables segregated/separated from non-electrical services (528.3)
8.22	Termination of cables at enclosures (Section 526) (i) Connections under no undue strain (522.8.5, 526.6) (ii) No basic insulation of a conductor visible outside enclosure (526.8) (iii) Connections of live conductors adequately enclosed (526.5) (iv) Adequately connected at point of entry to enclosure (glands, bushes etc.) (522.8.5)
8.23	Suitability of circuit accessories for external influences (512.2)
8.24	Circuit accessories not damaged, securely fixed, correctly connected, suitable for external influences (134.101; 512.2; Section 526)
8.25	Single-pole devices for switching or protection in line conductors only (132.14.1; 530.3.3; 643.6)
8.26	Cables and conductors correctly connected, enclosed and with no undue mechanical strain (Section 526)
8.27	Adequacy of connections, including cpcs, within accessories and at fixed and stationary equipment (Section 526)

9.0	**ISOLATION AND SWITCHING**
9.1	Isolators (462; 537.2) (i) Presence and location of appropriate devices (Section 462; 537.2.7) (ii) Capable of being secured in the OFF position (537.2.4) (iii) Correct operation verified (functional check) (643.10) (iv) The installation, circuit or part thereof that will be isolated clearly identified by location and/or durable marking (537.2.7)
9.2	Switching off for mechanical maintenance (464; 537.3.2) (i) Presence of appropriate devices (464.1; 537.3.2) (ii) Acceptable location (537.3.2.4) (iii) Capable of being secured in the OFF position (464.2) (iv) Correct operation verified (functional check) (643.10)
9.3	Emergency switching off (Section 465; 537.3.3; 537.4) (i) Presence of appropriate devices (465.1; 537.3.3; 537.4) (ii) Readily accessible for operation where danger might occur (537.3.3.6) (iii) Correct operation verified (functional check) (643.10) (iv) Firefighter's switches (537.4)
9.4	Functional switching (463.1; 537.3.1) (i) Presence of appropriate devices (537.3.1.1; 537.3.1.2) (ii) Correct operation verified (functional check) (537.3.1.1; 537.3.1.2; 643.10) (iii) Functional switching, for control of parts of the installation and current-using equipment (463.1; 537.3.1)

Item No	DESCRIPTION
10.0	**CURRENT–USING EQUIPMENT (PERMANENTLY CONNECTED)**
10.1	Suitability of equipment in terms of IP and fire ratings (416.2; 421.1; 421.1.201; 526.5)
10.2	Enclosure not damaged/deteriorated during installation so as to impair safety (134.1.1)
10.3	Suitability for the environment and external influences (512.2)
10.4	Security of fixing (134.1.1)
10.5	Cable entry holes in ceilings above luminaires, sized or sealed so as to restrict the spread of fire (527.2)
10.6	Provision of undervoltage protection, where specified (Section 445)
10.7	Provision of overload protection, where specified (Section 433; 552.1)
10.8	Provision of overload and/or undervoltage protection e.g. for rotating machines, if required (Sections 445, 552)
10.9	Correct selection and installation of luminaires fitted (559.3)
10.10	Installed to minimize the build-up of heat and restrict the spread of fire (421.1.4, 559.4.1)
10.11	Adequacy of working space/accessibility to equipment (132.12, 513.1)

11.0	**IDENTIFICATION AND NOTICES**
11.1	Presence of RCD six-monthly test notice; where required (514.12.2)
11.2	AFDD six-monthly test notice; where required
11.3	Presence of diagrams, charts or schedules at or near each distribution board, where required (514.9.1)
11.4	Presence of alternative supply warning notice at or near (514.15) (i) The origin (ii) The meter position, if remote from origin (iii) The distribution board to which the alternative/additional sources are connected (iv) All points of isolation of ALL sources of supply
11.5	Presence of next inspection recommendation label (514.12.1)
11.6	Presence of other required labelling (Section 514)
11.7	Presence of labels to indicate the purpose of switchgear and protective devices (514.1.1; 514.8)
11.8	Warning notice posted in situation where live parts cannot be isolated by the operation of a single device (514.11.1; 537.1.2)
11.9	The circuit or part thereof to be disconnected clearly identified by location and/or durable marking (537.3.2.3; 537.3.2.4)
11.10	The installation, circuit or part thereof to be disconnected clearly identified by location and/or durable marking (537.3.3.6)

12.0	**LOCATION(S) CONTAINING A BATH OR SHOWER**
12.1	Additional protection for all low voltage (LV) circuits by RCD not exceeding 30 mA (701.411.3.3)
12.2	Where used as a protective measure, requirements for SELV or PELV met (701.414.4.5)
12.3	Shaver supply units comply with BS EN 61558-2-5 formerly BS 3535 (701.512.3)
12.4	Presence of supplementary bonding conductors, unless not required by BS 7671 (701.415.2)
12.5	Low voltage (e.g. 230 V) socket-outlets sited at least 2.5 m from zone 1 (701.512.3)
12.6	Suitability of equipment for external influences for installed location in terms of IP rating (701.512.2)
12.7	Suitability of accessories and controlgear etc. for a particular zone (701.512.3)
12.8	Suitability of current-using equipment for particular position within the location (701.55)

13.0	**OTHER SPECIAL INSTALLATIONS OR LOCATIONS**
13.1	Where the installation includes special installations or locations relating to sections of Part 7, additional inspection items should be added to the checklist.
13.2	List all other special installations or locations present, if any. (Record separately the results of particular inspections applied)

14.0	**PROSUMER'S LOW VOLTAGE ELECTRICAL INSTALLATION(S)**
14.1	Where the installation includes additional requirements and recommendations relating to Chapter 82, additional inspection items should be added to the checklist.

ELECTRICAL INSTALLATION CONDITION REPORT

Report No.: ....

SECTION A. DETAILS OF THE PERSON ORDERING THE REPORT

Name ....

Address ....

....

SECTION B. REASON FOR PRODUCING THIS REPORT ....

....

Date(s) on which inspection and testing was carried out ....

SECTION C. DETAILS OF THE INSTALLATION WHICH IS THE SUBJECT OF THIS REPORT

Occupier ....

Address ....

....

Description of premises

Residential ☐ Commercial ☐ Industrial ☐ Other (include brief description) ☐ ....

Estimated age of wiring system ....years

Evidence of additions / alterations? Yes ☐ No ☐ Not apparent ☐ If yes, estimate age ....years

Installation records available? (Regulation 651.1) Yes ☐ No ☐ Date of last inspection .... (date)

SECTION D. EXTENT AND LIMITATIONS OF INSPECTION AND TESTING

Extent of the electrical installation covered by this report ....

....

Agreed limitations including the reasons (see Regulation 653.2) ....

....

Agreed with: ....

Operational limitations including the reasons (see page no.............) ....

....

The inspection and testing detailed in this report and accompanying schedules have been carried out in accordance with BS 7671:2018 as amended to ....

It should be noted that cables concealed within trunking and conduits, under floors, in roof spaces, and generally within the fabric of the building or underground, have **not** been inspected unless specifically agreed between the client and inspector prior to the inspection. An inspection should be made within an accessible roof space housing other electrical equipment.

SECTION E. SUMMARY OF THE CONDITION OF THE INSTALLATION

General condition of the installation (in terms of electrical safety) ....

....

....

Overall assessment of the installation in terms of its suitability for continued use

SATISFACTORY / UNSATISFACTORY* (Delete as appropriate)

*An unsatisfactory assessment indicates that dangerous (code C1) and/or potentially dangerous (code C2) conditions have been identified.

SECTION F. RECOMMENDATIONS

Where the overall assessment of the suitability of the installation for continued use above is stated as UNSATISFACTORY, I / we recommend that any observations classified as *'Danger present'* (code C1) or *'Potentially dangerous'* (code C2) are acted upon as a matter of urgency.
Investigation without delay is recommended for observations identified as *'Further investigation required' (code FI).*
Observations classified as *'Improvement recommended'* (code C3) should be given due consideration.

Subject to the necessary remedial action being taken, I / We recommend that the installation is further inspected and tested by ....(date) for the following reasons..

SECTION G. DECLARATION

I/We, being the person(s) responsible for the inspection and testing of the electrical installation (as indicated by my/our signatures below), particulars of which are described above, having exercised reasonable skill and care when carrying out the inspection and testing, hereby declare that the information in this report, including the observations and the attached schedules, provides an accurate assessment of the condition of the electrical installation taking into account the stated extent and limitations in section D of this report.

Inspected and tested by:	**Report authorised for issue by:**
Name (Capitals)	Name (Capitals) ....
Signature	Signature ....
For/on behalf of	For/on behalf of ....
Position	Position ....
Address	Address ....
Date	Date ....

SECTION H. SCHEDULE(S)

............Inspection Schedule(s) and .... Schedule(s) of Circuit Details and Test Results are attached.

The attached schedule(s) are part of this document and this report is valid only when they are attached to it.

Report No.: ....

SECTION I. SUPPLY CHARACTERISTICS AND EARTHING ARRANGEMENTS				
Earthing arrangements	**Number and Type of Live Conductors**		**Nature of Supply Parameters**	**Supply Protective Device**
TN-C ☐ TN-S ☐ TN-C-S ☐ TT ☐ IT ☐	AC ☐ 1-phase, 2-wire ☐ 2-phase, 3-wire ☐ 3-phase, 3-wire ☐ 3-phase, 4-wire ☐	DC ☐ 2-wire ☐ 3-wire ☐ Other ☐	Nominal voltage, U / U_0(1) .... V Nominal frequency, f(1) .... Hz Prospective fault current, I_{pf}(2) .... kA External earth fault loop impedance, Z_e(2) .... Ω	BS (EN) .... Type .... Rated current .... A
	Confirmation of supply polarity ☐		*(Note: (1) by enquiry (2) by enquiry or by measurement)*	
Other sources of supply (as detailed on attached schedule) ☐				

SECTION J. PARTICULARS OF INSTALLATION REFERRED TO IN THE REPORT	
Means of Earthing	**Details of Installation Earth Electrode** (where applicable)
Distributor's facility ☐	Type ....
Installation earth electrode ☐	Location ....
	Resistance to Earth .... Ω

Main Protective Conductors

Earthing conductor	Material .... csa ....mm^2	Connection / continuity verified ☐
Main protective bonding conductors	Material .... csa ....mm^2	Connection / continuity verified ☐

To water installation pipes ☐	To gas installation pipes ☐	To oil installation pipes ☐	To structural steel ☐
To lightning protection ☐	To other ☐ Specify		

Main switch / Switch-fuse / Circuit-breaker / RCD

Location .... BS(EN) .... No of poles	Current rating .... A Fuse / device rating or setting .... A Voltage rating V	**If RCD main switch** RCD Type .... Rated residual operating current ($I_{\Delta n}$) ....mA Rated time delay ....ms Measured operating time ....ms

SECTION K. OBSERVATIONS

Referring to the attached inspection schedule(s) and schedule(s) of circuit details and test results, and subject to the limitations specified at the *Extent and limitations of inspection and testing* section

No remedial action is required ☐ The following observations are made ☐ (see below):

OBSERVATION(S) Include schedule reference, as appropriate	CLASSIFICATION CODE
....	....
....	....
....	....
....	....
....	....
....	....
....	....
....	....
....	....
....	....
....	....
....	....
....	....
....	....
....	....
....	....
....	....
....	....
....	....
....	....

One of the following codes, as appropriate, has been allocated to each of the observations made above to indicate to the person(s) responsible for the installation the degree of urgency for remedial action.

C1 – Danger present. Risk of injury. Immediate remedial action required
C2 – Potentially dangerous - urgent remedial action required
C3 – Improvement recommended
FI – Further investigation required without delay

CONDITION REPORT

Notes for the person producing the Report:

1 This Report should only be used for reporting on the condition of an existing electrical installation, and not for the replacement of a consumer unit/distribution board. An installation which was designed to an earlier edition of BS 7671 or the IEE Wiring Regulations and which does not fully comply with the current edition is not necessarily unsafe for continued use, or requires upgrading. Only damage, deterioration, defects, dangerous conditions and non-compliance with the requirements of BS 7671 or the IEE Wiring Regulations, which may give rise to danger, should be recorded.

2 The Report, normally comprising at least five pages, should include schedules of both the inspection and the test results. Additional pages may be necessary for other than a simple installation and for the 'Guidance for recipients'. The number of each page should be indicated, together with the total number of pages involved.

3 The reason for producing this Report, such as change of occupancy or landlord's periodic maintenance, should be identified in Section B.

4 Those elements of the installation that are covered by the Report and those that are not should be identified in Section D (Extent and limitations). These aspects should have been agreed with the person ordering the report and other interested parties before the inspection and testing commenced. Any operational limitations, such as inability to gain access to parts of the installation or an item of equipment, should also be recorded in Section D.

5 The maximum prospective value of fault current (I_{pf}) recorded should be the greater of either the prospective value of short-circuit current or the prospective value of earth fault current.

6 Where an installation has an alternative source of supply a further schedule of supply characteristics and earthing arrangements based upon Section I of this Report should be provided.

7 A summary of the condition of the installation in terms of safety should be clearly stated in Section E. Observations, if any, should be categorised in Section K using the coding C1 to C3 as appropriate. Any observation given a code C1 or C2 classification should result in the overall condition of the installation being reported as unsatisfactory.

8 Wherever practicable, **items classified as 'Danger present' (C1) should be made safe on discovery**. Where this is not possible the owner or user should be given written notification as a matter of urgency.

9 Where an observation requires further investigation (FI) because the inspection has revealed an apparent deficiency which could not, owing to the extent or limitations of the inspection, be fully identified and further investigation may reveal a code C1 or C2 item, this should be recorded within Section K, given the code FI and marked as unsatisfactory in Section E.

10 If the space available for observations in Section K is insufficient, additional pages should be provided as necessary.

11 The date by which the next Electrical Installation Condition Report is recommended should be given in Section F. The interval between inspections should take into account the requirements of Regulation 652.1 and the overall condition of the installation.

12 Any deficiencies with intake equipment should be reported to the person ordering the work.

CONDITION REPORT

GUIDANCE FOR RECIPIENTS

(to be appended to the Report)

This Report is an important and valuable document which should be retained for future reference.

1 The purpose of this Report is to confirm, so far as reasonably practicable, whether or not the electrical installation is in a satisfactory condition for continued service (see Section E). The Report should identify any damage, deterioration, defects and/or conditions which may give rise to danger (see Section K).

2 This Report is only valid if accompanied by the Inspection Schedule(s) and the Schedule(s) of Circuit Details and Test Results.

3 The person ordering the Report should have received the 'original' Report and the inspector should have retained a duplicate.

4 The 'original' Report should be retained in a safe place and be made available to any person inspecting or undertaking work on the electrical installation in the future. If the property is vacated, this Report will provide the new owner/occupier with details of the condition of the electrical installation at the time the Report was issued.

5 Section D (Extent and Limitations) should identify fully the extent of the installation covered by this Report and any limitations on the inspection and testing. The inspector should have agreed these aspects with the person ordering the Report and with other interested parties (licensing authority, insurance company, mortgage provider and the like) before the inspection was carried out.

6 Some operational limitations such as inability to gain access to parts of the installation or an item of equipment may have been encountered during the inspection. The inspector should have noted these in Section D.

7 For items classified in Section K as C1 ('Danger present'), **the safety of those using the installation is at risk**, and it is recommended that a skilled person or persons competent in electrical installation work undertakes the necessary remedial work immediately.

8 For items classified in Section K as C2 ('Potentially dangerous'), **the safety of those using the installation may be at risk** and it is recommended that a skilled person or persons competent in electrical installation work undertakes the necessary remedial work as a matter of urgency.

9 Where it has been stated in Section K that an observation requires further investigation (code FI) the inspection has revealed an apparent deficiency which may result in a code C1 or C2, and could not, due to the extent or limitations of the inspection, be fully identified. Such observations should be investigated without delay. A further examination of the installation will be necessary, to determine the nature and extent of the apparent deficiency (see Section F).

10 For safety reasons, the electrical installation should be re-inspected at appropriate intervals by a skilled person or persons, competent in such work . The recommended date by which the next inspection is due is stated in Section F of the Report under 'Recommendations'.

11 Where the installation includes a residual current device (RCD) it should be tested six-monthly by pressing the button marked 'T' or 'Test'. The device should switch off the supply and should then be switched on to restore the supply. If the device does not switch off the supply when the button is pressed, seek expert advice. **For safety reasons it is important that this instruction is followed.**

12 Where the installation includes an arc fault detection device (AFDD) having a manual test facility it should be tested six-monthly by pressing the test button. Where an AFDD has both a test button and automatic test function, manufacturer's instructions shall be followed with respect to test button operation.

13 Where the installation includes a surge protective device (SPD) the status indicator should be checked to confirm it is in operational condition in accordance with manufacturer's information. If the indication shows that the device is not operational, seek expert advice. For safety reasons it is important that this instruction is followed.

14 Where the installation includes alternative or additional sources of supply, warning notices should be found at the origin or meter position or, if remote from the origin, at the consumer unit or distribution board and at all points of isolation of all sources of supply.

CONDITION REPORT INSPECTION SCHEDULE

GUIDANCE FOR THE INSPECTOR

1. Section 1.0. Where inadequacies in the intake equipment are encountered the inspector should advise the person ordering the work to inform the appropriate authority.
2. The schedule is not exhaustive.
3. Numbers in brackets are regulation references to specified requirements.

EXAMPLES OF ITEMS REQUIRING INSPECTION

FOR AN ELECTRICAL INSTALLATION CONDITION REPORT

A visual inspection should first be made of the external condition of all electrical equipment which is not concealed.

Further detailed inspection, including partial dismantling of equipment as required, should be carried out as agreed with the person ordering the work. (651.2)

These examples are not exhaustive. The persons responsible for the periodic inspection of the installation should include the relevant items in relation to the electrical installation, the inspection schedule can be reduced or expanded depending on the requirements for the installation.

Numbers in brackets are Regulation references.

EXTERNAL CONDITION OF INTAKE EQUIPMENT (VISUAL INSPECTION ONLY)

Where inadequacies in intake equipment are encountered, it is recommended that the person ordering the report informs the appropriate authority.

- Service cable
- Service head
- Earthing arrangements
- Meter tails
- Metering equipment
- Isolator (where present)

PRESENCE OF ADEQUATE ARRANGEMENTS FOR PARALLEL OR SWITCHED ALTERNATIVE SOURCES

- Adequate arrangements where a generating set operates as a switched alternative to the public supply (551.6)
- Adequate arrangements where a generating set operates in parallel with the public supply (551.7)

AUTOMATIC DISCONNECTION OF SUPPLY

- Main earthing/bonding arrangements (411.3; Chap 54)
  1. Presence of distributor's earthing arrangement (542.1.2.1; 542.1.2.2), or presence of installation earth electrode arrangement (542.1.2.3)
  2. Adequacy of earthing conductor size (542.3; 543.1.1)
  3. Adequacy of earthing conductor connections (542.3.2)
  4. Accessibility of earthing conductor connections (543.3.2)
  5. Adequacy of main protective bonding conductor sizes (544.1)
  6. Adequacy and location of main protective bonding conductor connections (543.3.2; 544.1.2)
  7. Accessibility of all protective bonding connections (543.3.2)
  8. Provision of earthing/bonding labels at all appropriate locations (514.13)
- FELV - requirements satisfied (411.7; 411.7.1)

OTHER METHODS OF PROTECTION

(Where any of the methods listed below are employed details should be provided on separate sheets)

- Non-conducting location (418.1)
- Earth-free local equipotential bonding (418.2)
- Electrical separation (Section 413; 418.3)
- Double insulation (Section 412)
- Reinforced insulation (Section 412)

DISTRIBUTION EQUIPMENT

- Adequacy of working space/accessibility to equipment (132.12; 513.1)
- Security of fixing (134.1.1)
- Condition of insulation of live parts (416.1)
- Adequacy/security of barriers (416.2)
- Condition of enclosure(s) in terms of IP rating etc (416.2)
- Condition of enclosure(s) in terms of fire rating etc (421.1.6; 421.1.201; 526.5)
- Enclosure not damaged/deteriorated so as to impair safety (651.2)
- Presence and effectiveness of obstacles (417.2)
- Presence of main switch(es), linked where required (462.1; 462.1.201; 462.2)
- Operation of main switch(es) (functional check) (643.10)
- Manual operation of circuit-breakers, RCDs and AFDDs to prove functionality (643.10)
- Confirmation that integral test button/switch causes RCD(s) to trip when operated (functional check) (643.10)
- RCD(s) provided for fault protection – includes RCBOs (411.4.204; 411.5.2; 531.2)
- RCD(s) provided for additional protection/requirements, where required - includes RCBOs (411.3.3; 415.1)
- Presence of RCD six-monthly test notice, where required (514.12.2)
- Presence of diagrams, charts or schedules at or near equipment, where required (514.9.1)
- Presence of alternative supply warning notice at or near equipment, where required (514.15)
- Presence of next inspection recommendation label (514.12.1)
- Presence of other required labelling (please specify) (Section 514)
- Compatibility of protective devices, bases and other components; correct type and rating (no signs of unacceptable thermal damage, arcing or overheating) (411.3.2; 411.4; 411.5; 411.6; Sections 432, 433)
- Single-pole switching or protective devices in line conductors only (132.14.1; 530.3.3)
- Protection against mechanical damage where cables enter equipment (522.8.1; 522.8.5; 522.8.11)
- Protection against electromagnetic effects where cables enter ferromagnetic enclosures (521.5.1)

DISTRIBUTION CIRCUITS

- Identification of conductors (514.3.1)
- Cables correctly supported throughout their run (521.10.202; 522.8.5)
- Condition of insulation of live parts (416.1)
- Non-sheathed cables protected by enclosure in conduit, ducting or trunking (521.10.1)
- Suitability of containment systems for continued use (including flexible conduit) (Section 522)
- Cables correctly terminated in enclosures (Section 526)
- Confirmation that ALL conductor connections, including connections to busbars, are correctly located in terminals and are tight and secure (526.1)
- Examination of cables for signs of unacceptable thermal or mechanical damage/deterioration (421.1; 522.6)
- Adequacy of cables for current-carrying capacity with regard for the type and nature of installation (Section 523)
- Adequacy of protective devices: type and rated current for fault protection (411.3)
- Presence and adequacy of circuit protective conductors (411.3.1.1; 543.1)
- Coordination between conductors and overload protective devices (433.1; 533.2.1)
- Cable installation methods/practices with regard to the type and nature of installation and external influences (Section 522)

- Where exposed to direct sunlight, cable of a suitable type (522.11.1)
- Cables concealed under floors, above ceilings, in walls/partitions less than 50 mm from a surface, and in partitions containing metal parts
  1. installed in prescribed zones (see Section D. *Extent and limitations*) (522.6.202) or
  2. incorporating earthed armour or sheath, or run within earthed wiring system, or otherwise protected against mechanical damage by nails, screws and the like (see Section D. *Extent and limitations*) (522.6.204)
- Provision of fire barriers, sealing arrangements and protection against thermal effects (Section 527)
- Band II cables segregated/separated from Band I cables (528.1)
- Cables segregated/separated from non-electrical services (528.3)
- Condition of circuit accessories (651.2)
- Suitability of circuit accessories for external influences (512.2)
- Single-pole switching or protective devices in line conductors only (132.14.1; 530.3.3)
- Adequacy of connections, including cpcs, within accessories and to fixed and stationary equipment – identify/record numbers and locations of items inspected (Section 526)
- Presence, operation and correct location of appropriate devices for isolation and switching (Chapter 46; Section 537)
- General condition of wiring systems (651.2)
- Temperature rating of cable insulation (522.1.1; Table 52.1)

FINAL CIRCUITS

- Identification of conductors (514.3.1)
- Cables correctly supported throughout their run (521.10.202; 522.8.5)
- Condition of insulation of live parts (416.1)
- Non-sheathed cables protected by enclosure in conduit, ducting or trunking (521.10.1)
- Suitability of containment systems for continued use (including flexible conduit) (Section 522)
- Adequacy of cables for current-carrying capacity with regard for the type and nature of installation (Section 523)
- Adequacy of protective devices: type and rated current for fault protection (411.3)
- Presence and adequacy of circuit protective conductors (411.3.1.1; 543.1)
- Coordination between conductors and overload protective devices (433.1; 533.2.1) |
- Wiring system(s) appropriate for the type and nature of the installation and external influences (Section 522)
- Cables concealed under floors, above ceilings, in walls/partitions, adequately protected against damage (522.6.201; 522.6.202; 522.6.203; 522.6.204)
  1. installed in prescribed zones (see Section D. *Extent and limitations*) (522.6.202)
  2. incorporating earthed armour or sheath, or run within earthed wiring system, or otherwise protected against mechanical damage by nails, screws and the like (see Section D. *Extent and limitations*) (522.6.201; 522.6.204)
- Provision of additional protection by 30 mA RCD
  1. *for all socket-outlets of rating 32 A or less, unless an exception is permitted (411.3.3) |
  2. *for the supply of mobile equipment not exceeding 32 A rating for use outdoors (411.3.3)
  3. *for cables concealed in walls at a depth of less than 50 mm (522.6.202; 522.6.203)
  4. *for cables concealed in walls/partitions containing metal parts regardless of depth (522.6.203)
  5. *for final circuits supplying luminaires within domestic (household) premises (411.3.4)
- Provision of fire barriers, sealing arrangements and protection against thermal effects (Section 527)
- Band II cables segregated/separated from Band I cables (528.1)
- Cables segregated/separated from non-electrical services (528.3)
- Termination of cables at enclosures – identify/record numbers and locations of items inspected (Section 526)
  1. Connections under no undue strain (526.6)
  2. No basic insulation of a conductor visible outside enclosure (526.8)

3 Connections of live conductors adequately enclosed (526.5)

4 Adequately connected at point of entry to enclosure (glands, bushes etc.) (522.8.5)

- Condition of accessories including socket-outlets, switches and joint boxes (651.2)
- Suitability of accessories for external influences (512.2)
- Single-pole switching or protective devices in line conductors only (132.14.1; 530.3.3)

***Note**: Older installations designed prior to BS 7671:2018 may not have been provided with RCDs for additional protection.

ISOLATION AND SWITCHING

- Isolators (Sections 460; 537)

  1 Presence and condition of appropriate devices (Section 462; 537.2.7)

  2 Acceptable location – state if local or remote from equipment in question (Section 462; 537.2.7)

  3 Capable of being secured in the OFF position (462.3)

  4 Correct operation verified (643.10)

  5 Clearly identified by position and/or durable marking (537.2.6)

  6 Warning label posted in situations where live parts cannot be isolated by the operation of a single device (514.11.1; 537.1.2)
- Switching off for mechanical maintenance (Section 464; 537.3.2)

  1 Presence and condition of appropriate devices (464.1; 537.3.2)

  2 Acceptable location – state if local or remote from equipment in question (537.3.2.4)

  3 Capable of being secured in the OFF position (462.3)

  4 Correct operation verified (643.10)

  5 Clearly identified by position and/or durable marking (537.3.2.4)
- Emergency switching/stopping (Section 465; 537.3.3)

  1 Presence and condition of appropriate devices (Section 465; 537.3.3; 537.4)

  2 Readily accessible for operation where danger might occur (537.3.3.6)

  3 Correct operation verified (643.10)

  4 Clearly identified by position and/or durable marking (537.3.3.6)
- Functional switching (Section 463; 537.3.1)

  1 Presence and condition of appropriate devices (537.3.1.1; 537.3.1.2)

  2 Correct operation verified (537.3.1.1; 537.3.1.2)

CURRENT-USING EQUIPMENT (PERMANENTLY CONNECTED)

- Condition of equipment in terms of IP rating etc (416.2)
- Equipment does not constitute a fire hazard (Section 421)
- Enclosure not damaged/deteriorated so as to impair safety (134.1.1; 416.2; 512.2)
- Suitability for the environment and external influences (512.2)
- Security of fixing (134.1.1)
- Cable entry holes in ceiling above luminaires, sized or sealed so as to restrict the spread of fire: List number and location of luminaires inspected (separate page) (527.2)
- Recessed luminaires (downlighters)

  1 Correct type of lamps fitted (559.3.1)

  2 Installed to minimise build-up of heat by use of "fire rated" fittings, insulation displacement box or similar (421.1.2)

  3 No signs of overheating to surrounding building fabric (559.4.1)

  4 No signs of overheating to conductors/terminations (526.1)

PART 7 SPECIAL INSTALLATIONS OR LOCATIONS

- If any special installations or locations are present, list the particular inspections applied, and issue additional inspection schedule(s) on separate pages.

CONDITION REPORT INSPECTION SCHEDULE FOR RESIDENTIAL AND SIMILAR PREMISES WITH UP TO 100 A SUPPLY

NOTE: *This form is suitable for many types of smaller installation, not exclusively residential.*

The persons responsible for the periodic inspection of the installation should include the relevant items in relation to the electrical installation, the inspection schedule can be reduced or expanded depending on the requirements for the installation.

OUTCOMES	*Acceptable condition*	✓	*Unacceptable condition*	State **C1** or **C2**	*Improvement recommended*	State **C3**	*Further investigation*	**FI**	*Not verified*	**N/V**	*Limitation*	**LIM**	*Not applicable*	**N/A**

ITEM NO	DESCRIPTION	OUTCOME *(Use codes above. Provide additional comment where appropriate. C1, C2, C3 and FI coded items to be recorded in Section K of the Condition Report)*
1.0	**INTAKE EQUIPMENT (VISUAL INSPECTION ONLY)** An outcome against an item in this section, other than access to live parts, should not be used to determine the overall outcome.	
1.1	Distributor/supplier intake equipment • Service cable • Service head • Earthing arrangement • Meter tails • Metering equipment • Isolator (where present) **NOTE 1:** Where inadequacies in the intake equipment are encountered, which may result in a dangerous or potentially dangerous situation, the person ordering the work and/or dutyholder must be informed. It is strongly recommended that the person ordering the work informs the appropriate authority. **NOTE 2:** For this section only, where inadequacies are found, an 'X' should be put against the appropriate item and a comment made in Section K.	
	Person ordering work /dutyholder notified (Delete as appropriate)	Y / N/A
1.2	Consumer's isolator (where present)	
1.3	Consumer's meter tails	
2.0	**PRESENCE OF ADEQUATE ARRANGEMENTS FOR OTHER SOURCES SUCH AS MICROGENERATORS (551.6; 551.7)**	
3.0	**EARTHING / BONDING ARRANGEMENTS (411.3; Chap 54)**	
3.1	Presence and condition of distributor's earthing arrangement (542.1.2.1; 542.1.2.2)	
3.2	Presence and condition of earth electrode connection where applicable (542.1.2.3)	
3.3	Provision of earthing/bonding labels at all appropriate locations (514.13.1)	
3.4	Confirmation of earthing conductor size (542.3; 543.1.1)	
3.5	Accessibility and condition of earthing conductor at MET (543.3.2)	
3.6	Confirmation of main protective bonding conductor sizes (544.1)	
3.7	Condition and accessibility of main protective bonding conductor connections (543.3.2; 544.1.2)	
3.8	Accessibility and condition of other protective bonding connections (543.3.1; 543.3.2)	
4.0	**CONSUMER UNIT(S) / DISTRIBUTION BOARD(S)**	
4.1	Adequacy of working space/accessibility to consumer unit/distribution board (132.12; 513.1)	
4.2	Security of fixing (134.1.1)	
4.3	Condition of enclosure(s) in terms of IP rating etc (416.2)	
4.4	Condition of enclosure(s) in terms of fire rating etc (421.1.201; 526.5)	
4.5	Enclosure not damaged/deteriorated so as to impair safety (651.2)	
4.6	Presence of main linked switch (as required by 462.1.201)	
4.7	Operation of main switch (functional check) (643.10)	
4.8	Manual operation of circuit-breakers and RCDs to prove disconnection (643.10)	
4.9	Correct identification of circuit details and protective devices (514.8.1; 514.9.1)	
4.10	Presence of RCD six-monthly test notice, where required (514.12.2)	
4.11	Presence of alternative supply warning notice at or near consumer unit/distribution board (514.15)	
4.12	Presence of other required labelling (please specify) (Section 514)	
4.13	Compatibility of protective devices, bases and other components; correct type and rating (No signs of unacceptable thermal damage, arcing or overheating) (411.3.2; 411.4; 411.5; 411.6; Sections 432, 433)	
4.14	Single-pole switching or protective devices in line conductor only (132.14.1; 530.3.3)	
4.15	Protection against mechanical damage where cables enter consumer unit/distribution board (522.8.1; 522.8.5; 522.8.11)	
4.16	Protection against electromagnetic effects where cables enter consumer unit/distribution board/enclosures (521.5.1)	
4.17	RCD(s) provided for fault protection - includes RCBOs (411.4.204; 411.5.2; 531.2)	
4.18	RCD(s) provided for additional protection/requirements - includes RCBOs (411.3.3; 415.1)	
4.19	Confirmation of indication that SPD is functional (651.4)	

OUTCOMES	*Acceptable condition*	✓	*Unacceptable condition*	State **C1** or **C2**	*Improvement recommended*	State **C3**	*Further investigation*	**FI**	*Not verified*	**N/V**	*Limitation*	**LIM**	*Not applicable*	**N/A**

ITEM NO	DESCRIPTION	**OUTCOME** *(Use codes above. Provide additional comment where appropriate. C1, C2, C3 and FI coded items to be recorded in Section K of the Condition Report)*
4.20	Confirmation that ALL conductor connections, including connections to busbars, are correctly located in terminals and are tight and secure (526.1)	
4.21	Adequate arrangements where a generating set operates as a switched alternative to the public supply (551.6)	
4.22	Adequate arrangements where a generating set operates in parallel with the public supply (551.7)	

5.0	**FINAL CIRCUITS**	
5.1	Identification of conductors (514.3.1)	
5.2	Cables correctly supported throughout their run (521.10.202; 522.8.5)	
5.3	Condition of insulation of live parts (416.1)	
5.4	Non-sheathed cables protected by enclosure in conduit, ducting or trunking (521.10.1) • To include the integrity of conduit and trunking systems (metallic and plastic)	
5.5	Adequacy of cables for current-carrying capacity with regard for the type and nature of installation (Section 523)	
5.6	Coordination between conductors and overload protective devices (433.1; 533.2.1)	
5.7	Adequacy of protective devices: type and rated current for fault protection (411.3)	
5.8	Presence and adequacy of circuit protective conductors (411.3.1; Section 543)	
5.9	Wiring system(s) appropriate for the type and nature of the installation and external influences (Section 522)	
5.10	Concealed cables installed in prescribed zones (see Section D. Extent and limitations) (522.6.202)	
5.11	Cables concealed under floors, above ceilings or in walls/partitions, adequately protected against damage (see Section D. Extent and limitations) (522.6.204)	
5.12	Provision of additional requirements for protection by RCD not exceeding 30 mA: • for all socket-outlets of rating 32 A or less, unless an exception is permitted (411.3.3) • for the supply of mobile equipment not exceeding 32 A rating for use outdoors (411.3.3) • for cables concealed in walls at a depth of less than 50 mm (522.6.202; 522.6.203) • for cables concealed in walls/partitions containing metal parts regardless of depth (522.6.203) • Final circuits supplying luminaires within domestic (household) premises (411.3.4)	
5.13	Provision of fire barriers, sealing arrangements and protection against thermal effects (Section 527)	
5.14	Band II cables segregated/separated from Band I cables (528.1)	
5.15	Cables segregated/separated from communications cabling (528.2)	
5.16	Cables segregated/separated from non-electrical services (528.3)	
5.17	Termination of cables at enclosures - indicate extent of sampling in Section D of the report (Section 526) • Connections soundly made and under no undue strain (526.6) • No basic insulation of a conductor visible outside enclosure (526.8) • Connections of live conductors adequately enclosed (526.5) • Adequately connected at point of entry to enclosure (glands, bushes etc.) (522.8.5)	
5.18	Condition of accessories including socket-outlets, switches and joint boxes (651.2(v))	
5.19	Suitability of accessories for external influences (512.2)	
5.20	Adequacy of working space/accessibility to equipment (132.12; 513.1)	
5.21	Single-pole switching or protective devices in line conductors only (132.14.1; 530.3.3)	

6.0	**LOCATION(S) CONTAINING A BATH OR SHOWER**	
6.1	Additional protection for all low voltage (LV) circuits by RCD not exceeding 30 mA (701.411.3.3)	
6.2	Where used as a protective measure, requirements for SELV or PELV met (701.414.4.5)	
6.3	Shaver supply units comply with BS EN 61558-2-5 formerly BS 3535 (701.512.3)	
6.4	Presence of supplementary bonding conductors, unless not required by BS 7671:2018 (701.415.2)	
6.5	Low voltage (e.g. 230 V) socket-outlets sited at least 2.5 m from zone 1 (701.512.3)	
6.6	Suitability of equipment for external influences for installed location in terms of IP rating (701.512.2)	
6.7	Suitability of accessories and controlgear etc. for a particular zone (701.512.3)	

6.8	Suitability of current-using equipment for particular position within the location (701.55)	
7.0	**OTHER PART 7 SPECIAL INSTALLATIONS OR LOCATIONS**	
7.1	List all other special installations or locations present, if any. (Record separately the results of particular inspections applied.)	
8.0	**PROSUMER'S LOW VOLTAGE ELECTRICAL INSTALLATION(S)**	
8.1	Where the installation includes additional requirements and recommendations relating to Chapter 82, additional inspection items should be added to the checklist.	

Inspected by:

Name (Capitals) .... **Signature** .... **Date** ....

GENERIC SCHEDULE OF CIRCUIT DETAILS

Certificate/Report number:......................

Distribution board details

DB reference: ______ Location: ______ Supplied from: ______

Distribution circuit OCPD: BS (EN): ______ Type: ______ Rating/Setting: ______ A

SPD Details: Type(s)*: T1 ☐ T2 ☐ T3† ☐ N/A ☐

CIRCUIT DETAILS

		Conductor details					Overcurrent protective device					RCD			
					Number & size										
1 Circuit number	2 Circuit description	3 Type of wiring	4 Reference method‡	5 Number of points served	6 Live (mm^2)	7 cpc (mm^2)	8 BS (EN)	9 Type	10 Rating (A)	11 Breaking capacity (kA)	12 Maximum permitted Z_s (Ω)§	13 BS (EN)	14 Type	15 $I_{\Delta n}$ (mA)	16 Rating (A)

CODES FOR TYPES OF WIRING

A	B	C	D	E	F	G	H	O
Thermoplastic insulated/ sheathed cables	Thermoplastic cables in metallic conduit	Thermoplastic cables in non-metallic conduit	Thermoplastic cables in metallic trunking	Thermoplastic cables in non-metallic trunking	Thermoplastic SWA cables	Thermosetting SWA cables	Mineral insulated cables	Other - please state

* SPD Type. Where a combined T1 + T2 or T2 + T3 device is installed, indicate by ticking both Type boxes.

† Where a T3 SPD is installed to protect sensitive equipment, enter details in 'Remarks', column 31, of the Schedule of Test Results. (See Section 534 of BS 7671:2018+A2:2022.)

‡ See Table 4A2 of Appendix 4 of BS 7671:2018+A2:2022.

§ Where the maximum permitted earth fault loop impedance value stated in column 12 is taken from a source other than the tabulated values given in Chapter 41 of BS 7671:2018+A2:2022, state the source of the data in the appropriate cell for the circuit in the 'Remarks', column 31, of the Schedule of Test Results.

GENERIC SCHEDULE OF TEST RESULTS

Certificate/Report number:......................

Distribution board details

DB reference: .... Z_{db} .... Ω I_{pf} .... kA

Confirmed: Correct polarity ☐ Phase sequence ☐

SPD: Operational status confirmed¶ ☐ N/A ☐

Details of test instruments used (serial and/or asset numbers)

Multifunction: ....

Continuity: ....

Insulation resistance: ....

Earth fault loop impedance: ....

RCD: ....

Earth electrode resistance: ....

TEST RESULT DETAILS

	Continuity (Ω)					Insulation resistance				Z_s (Ω)	RCD		AFDD	
	Ring final circuit			$(R_1 + R_2)$ or R_2										**Remarks** Include details of circuits and/or installed equipment vulnerable to damage when testing (continue on a separate sheet if necessary)
17 Circuit number	18 r_1 (line)	19 r_n (neutral)	20 r_2 (cpc)	21 $(R_1 + R_2)$	22 R_2	23 Test voltage (V)	24 Live - Live (MΩ)	25 Live - Earth (MΩ)	26 Polarity#	27 Maximum measured	28 Disconnection time (ms)**	29 Test button operation	30 Manual test button operation††	31

Tested by name (Capitals): ....

Signature: .... Date: ....

¶ Not all SPDs have visible functionality indication.
# Where this schedule is issued with an Electrical Installation Condition Report, and incorrect polarity is identified, an 'X' should be entered.
** RCD effectiveness is verified using an alternating current test at rated residual operating current ($I_{\Delta n}$).
†† Not all AFDDs have a test button.

APPENDIX 7: Deleted by BS 7671:2018+A2:2022

APPENDIX 8 (Informative)

CURRENT-CARRYING CAPACITY AND VOLTAGE DROP FOR BUSBAR TRUNKING AND POWERTRACK SYSTEMS

1 Basis of current-carrying capacity

The current-carrying capacity (I_{nA}) of a busbar trunking or powertrack system relates to continuous loading and is declared by the manufacturer based on tests to BS EN 61439-6 (busbar trunking) or BS EN 61534 series of standards (Powertrack). The current-carrying capacity is designed to provide for satisfactory life of the system, subject to the thermal effects of carrying current for sustained periods in normal service.

Considerations affecting the choice of size of a busbar trunking or powertrack system include the requirements for protection against electric shock (see Chapter 41), protection against thermal effects (see Chapter 42), overcurrent protection (see Chapter 43 and sec 4 below) and voltage drop (see sec 5 below).

2 Rating factors for current-carrying capacity

The current-carrying capacity (I_{nA}) can be affected by the ambient temperature and the mounting conditions (for example the orientation of the conductors).

I_{nA} for Powertrack according to BS EN 61534 is defined as 'The rated current value assigned to a Powertrack system by the manufacturer and to which operation and performance characteristics are referred.'

I_{nA} for a busbar trunking system according to BS EN 61439-6 is defined as 'The rated current value declared by the manufacturer which can be carried without the temperature rise of various parts of the busbar trunking system exceeding specified limits under specified conditions.'

I_{nC} for a busbar trunking system according to BS EN 61439-6 is the value of the current that can be carried by a circuit loaded alone, under normal service conditions without the temperature rise of the various parts of the ASSEMBLY exceeding the limits specified.

Installation ambient temperature

For a busbar trunking system, if the ambient temperature exceeds 35 °C the rating factor k_{1A} to be applied is obtained from the manufacturer of the busbar trunking system ($k_{1A} = 1$ for 35 °C).

The effective current-carrying capacity (I_{nA}) at the new temperature is $k_{1A} \times I_{nA}$.

Mounting attitude

For a busbar trunking system the mounting factor k_{2C} to be applied is obtained from the manufacturer of the busbar trunking system.

The effective current-carrying capacity (I'_{nC}) under the new mounting conditions is $k_{2C} \times I_{nC}$.

In a typical installation, both factors may have to be taken into account and the effective current-carrying capacity (I'_{nC}) then becomes $k_{1C} \times k_{2C} \times I_{nC}$

where:

k_{1C} is a temperature factor , equal to 1 at an ambient air temperature of 35 °C

k_{2C} is a mounting factor, equal to 1 in the reference mounting conditions.

NOTE: For a powertrack system the effective current-carrying capacity is the rated current I_{nA} declared by the manufacturer in accordance with BS EN 61534 series under all normal conditions.

3 Effective current-carrying capacity

I_z must be not less than I_b, such that: $I_z \geq I_b$

where:

I_z is the effective current-carrying capacity of the busbar trunking (I'_{nC}) or powertrack system (I_{nA}) for continuous service under the particular installation conditions, and

I_b is the design current of the circuit.

4 Protection against overload current

The minimum operating current of the protective device should not exceed 1.45 I_z. Where the protective device is a fuse to BS 88 series or a circuit-breaker to either BS EN 60947-2 or BS EN 60898, this requirement is satisfied by selecting a value of I_z not less than I_n, where I_n is the rated current or current setting of the device protecting the circuit against overcurrent.

5 Voltage drop

The voltage drop (V_d) for the busbar trunking or powertrack system is obtained from the manufacturer. It is usually expressed as mV/ampere/metre, tabulated according to the value of the load-circuit power factor. The voltage drop given is calculated on the basis of a single load at the end of the run and, in this case, the total voltage drop = $V_d \times I_b \times L / 1000$ volts, where L is the length of run in metres.

In the case of an evenly distributed load (tapped off at intervals along the busbar trunking or powertrack system) then the voltage drop at the furthest tap-off point may be taken as 0.5 V_d, and is calculated by the above method.

In the case of an unevenly distributed load it will be necessary to calculate the voltage drop for each section between tap-off points and add them together to find the voltage drop at the furthest tap-off point.

APPENDIX 9 (Informative)

DEFINITIONS – MULTIPLE SOURCE, DC AND OTHER SYSTEMS

Fig 9 – Explanation of symbols used within Appendix 9

Symbol	Meaning
	Neutral conductor (N); midpoint conductor (M)
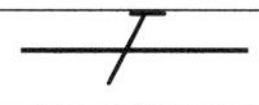	Protective conductor (PE)
	Combined protective and neutral conductor (PEN)

NOTE 1: The dotted lines indicate the parts of the system that are not covered by the scope of the Standard, whereas the solid lines indicate the part that is covered by the Standard.

NOTE 2: For private systems, the source and/or the distribution system may be considered as part of the installation within the meaning of BS 7671. For this case, the figures may be completely shown in solid lines.

Fig 9A – TN-C-S multiple source system with separate protective conductor and neutral conductor to current-using equipment

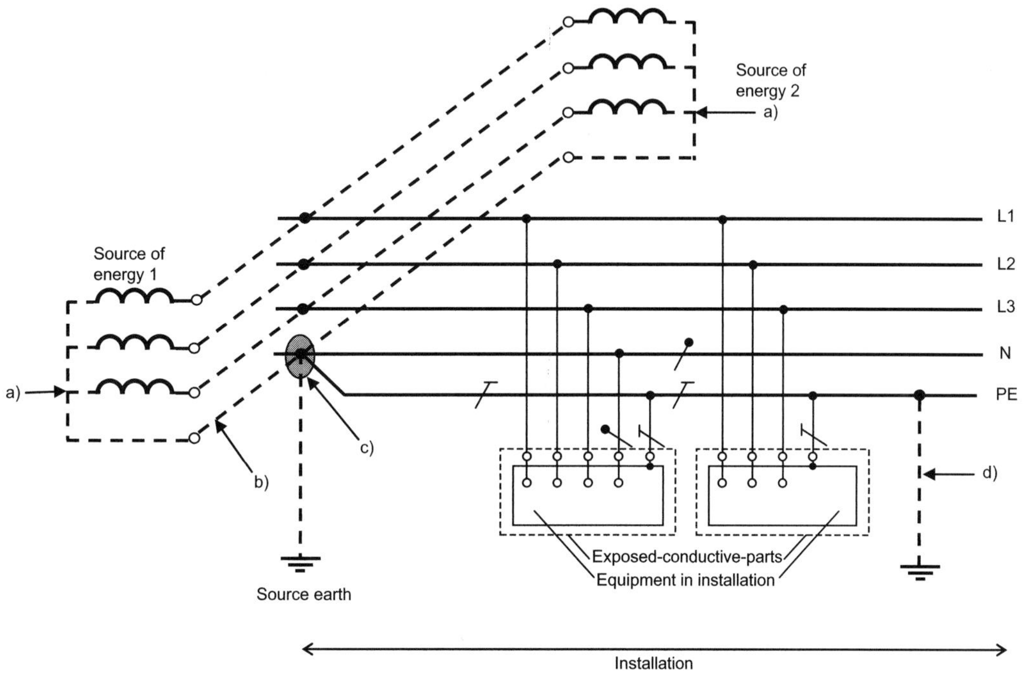

Fig 9B – TN multiple source system with protective conductor and no neutral conductor throughout the system for 2- or 3-phase load

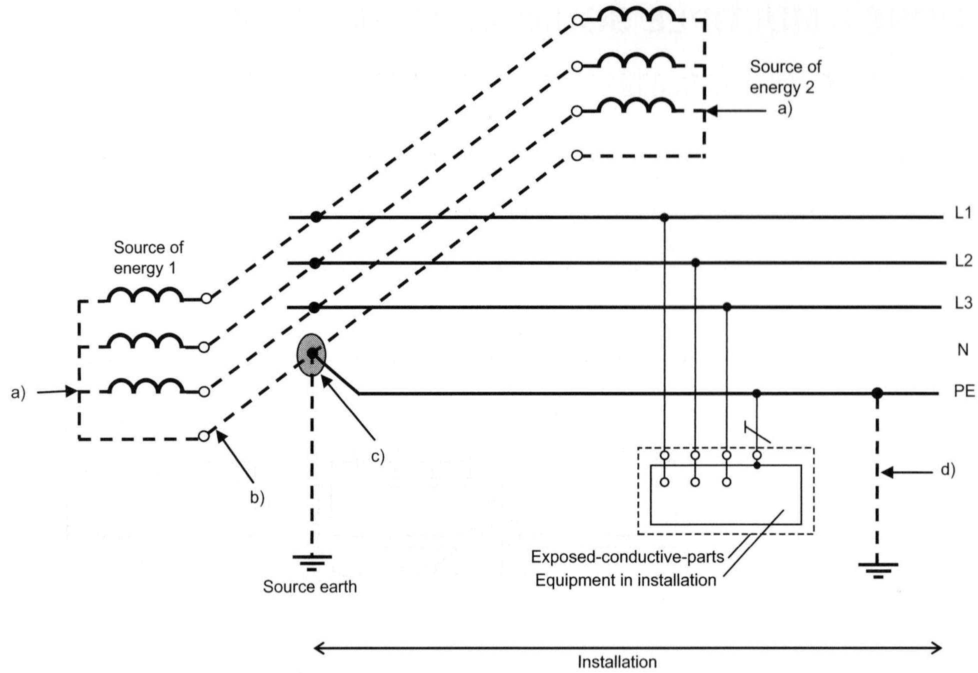

NOTES to Figures 9A and 9B

(a) No direct connection from either the transformer neutral point or the generator star point to Earth is permitted.

(b) The interconnection conductor between either the neutral points of the transformers or the generator star points is to be insulated. The function of this conductor is similar to a PEN; however, it must not be connected to current-using equipment.

(c) Only one connection between the interconnected neutral points of the sources and the PE is to be provided. This connection is to be located inside the main switchgear assembly.

(d) Additional earthing of the PE in the installation may be provided.

Fig 9C – IT system with exposed-conductive-parts earthed in groups or individually

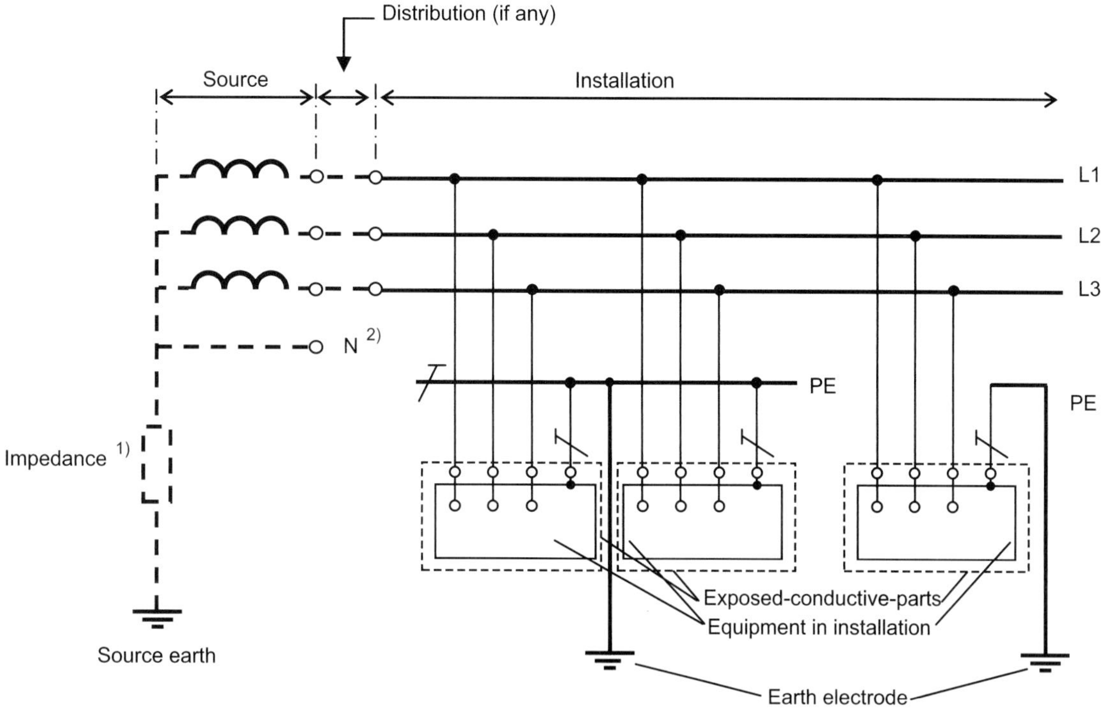

NOTES

Additional earthing of the PE in the installation may be provided.

(1) The system may be connected to Earth via a sufficiently high impedance.

(2) The neutral conductor may or may not be distributed.

Fig 9D – TN-S DC system with earthed line conductor L− separated from the protective conductor throughout the installation

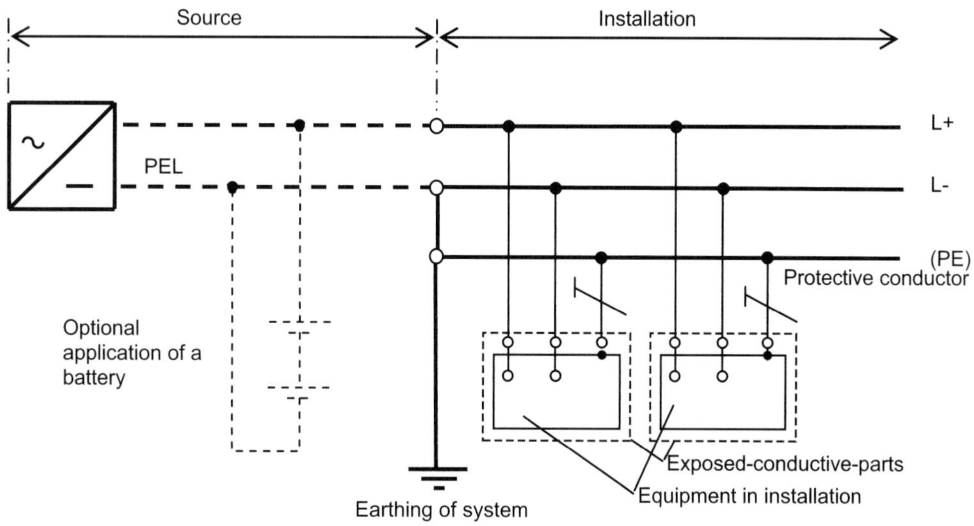

Fig 9E – TN-S DC system with earthed midpoint conductor M separated from the protective conductor throughout the installation

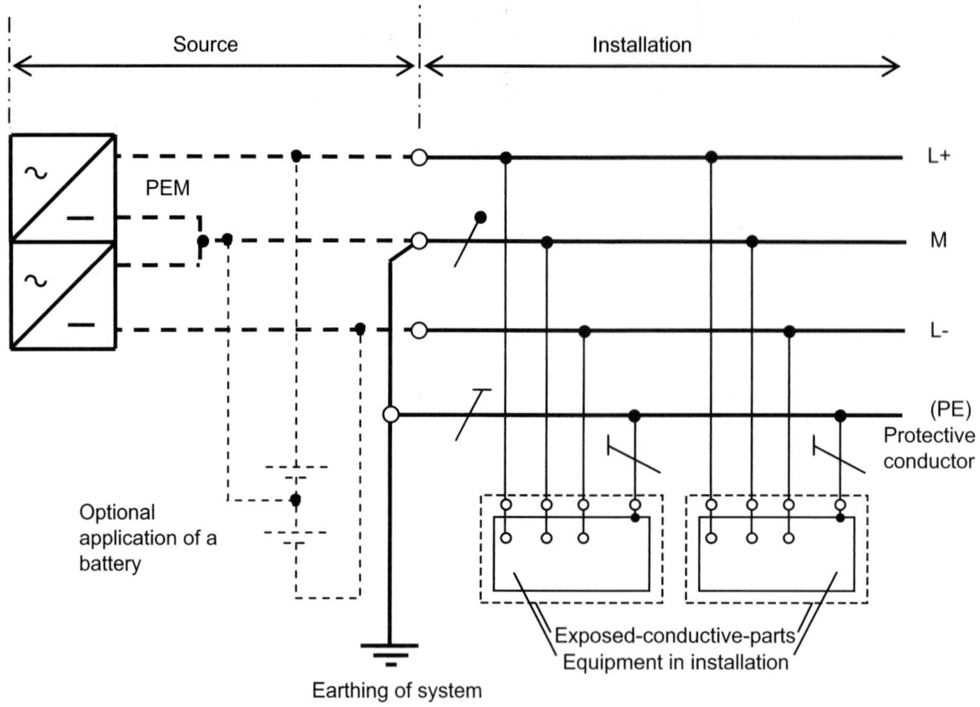

NOTE to Figures 9D and 9E

Additional earthing of the PE in the installation may be provided.

Fig 9F – TN-C DC system with earthed line conductor L– and protective conductor combined in one single conductor PEL throughout the installation

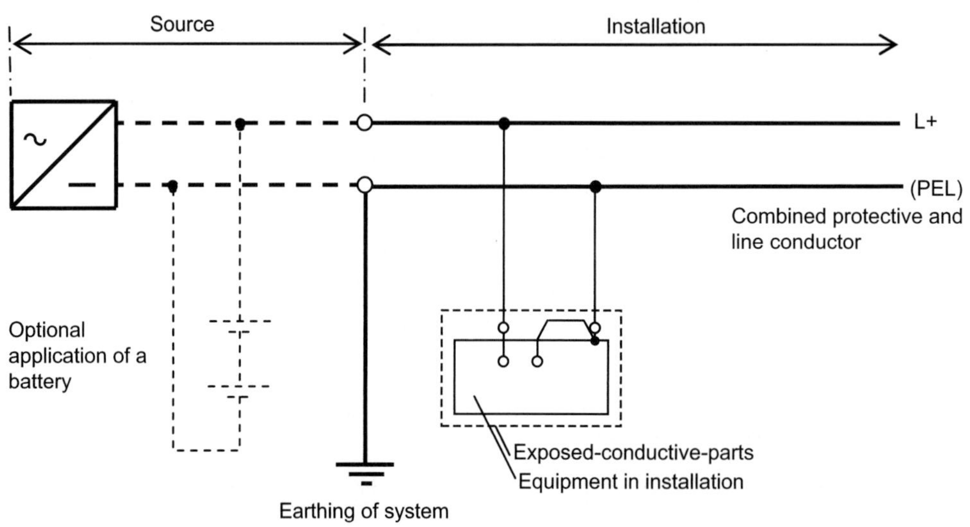

Fig 9G – TN-C DC system with earthed midpoint conductor M and protective conductor combined in one single conductor PEM throughout the installation

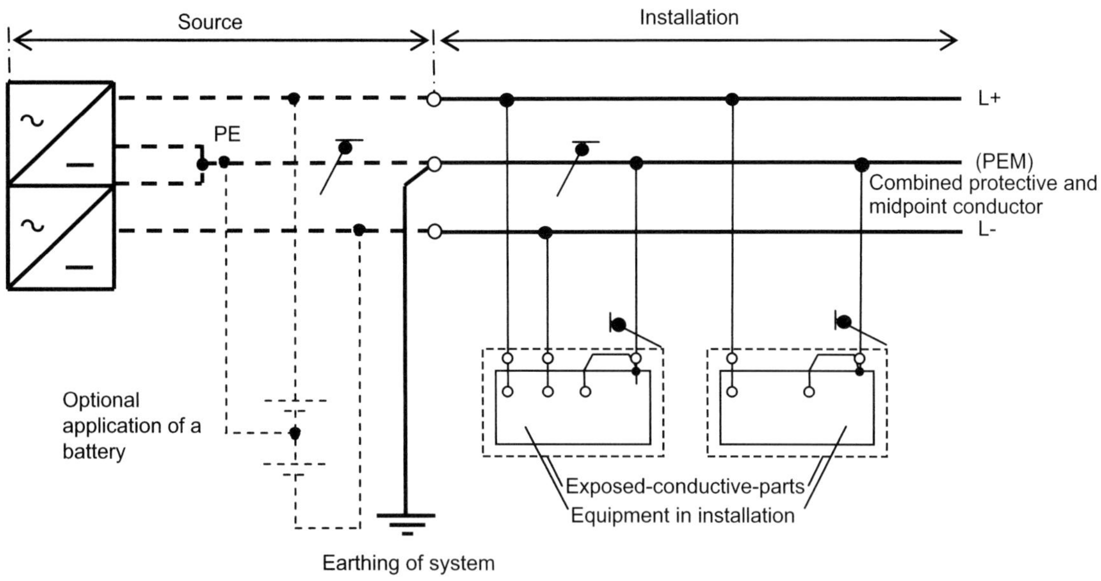

NOTE to Figures 9F and 9G

Additional earthing of the PEL or PEM in the installation may be provided.

Fig 9H – TN-C-S DC system with earthed line conductor L – and protective conductor combined in one single conductor PEL in a part of the installation

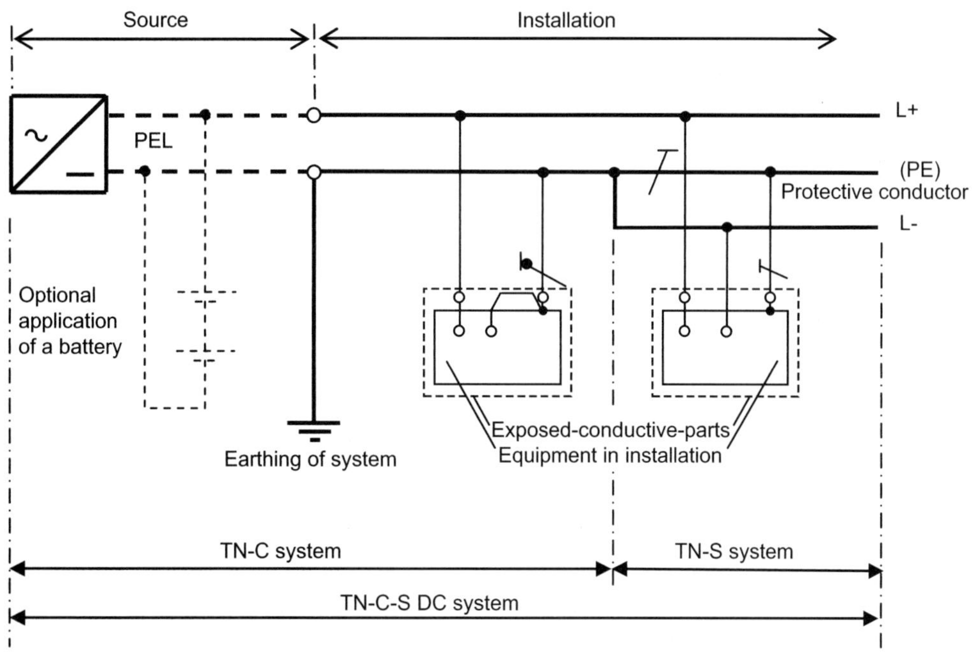

Fig 9I – TN-C-S DC system with earthed midpoint conductor M and protective conductor combined in one single conductor PEM in a part of the installation

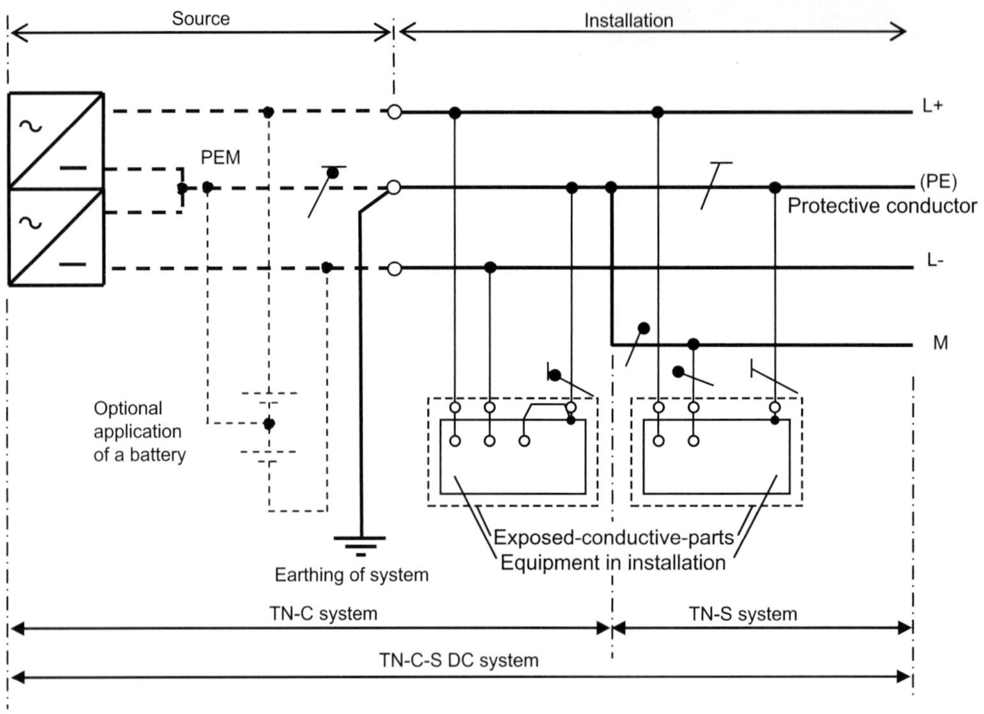

NOTES to Figures 9H and 9I

Additional earthing of the PE in the installation may be provided.

Regulation 8(4) of the Electricity Safety, Quality and Continuity Regulations (ESQCR) states that a consumer shall not combine the neutral and protective functions in a single conductor in his consumer's installation.

Fig 9J – TT DC system

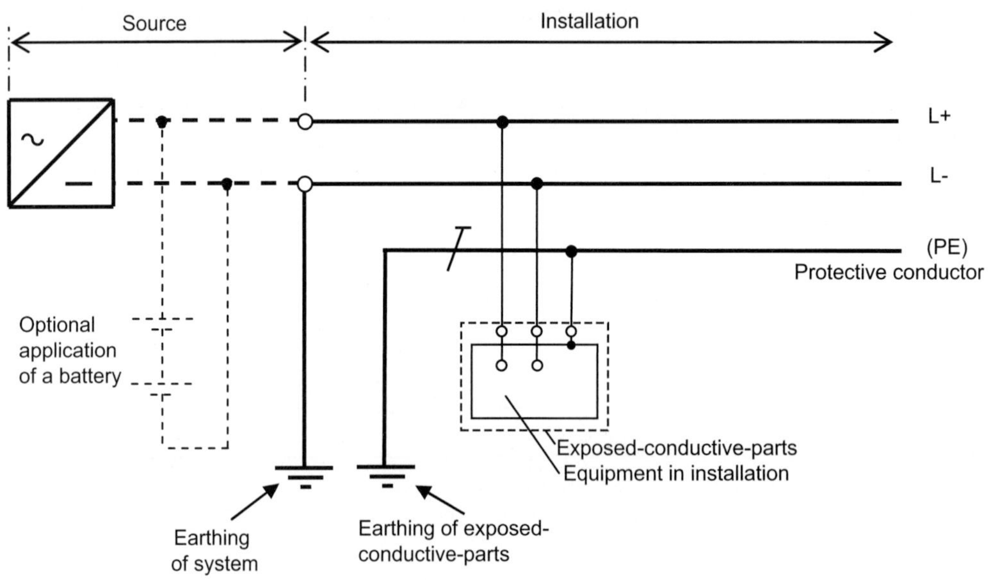

Fig 9K – TT DC system

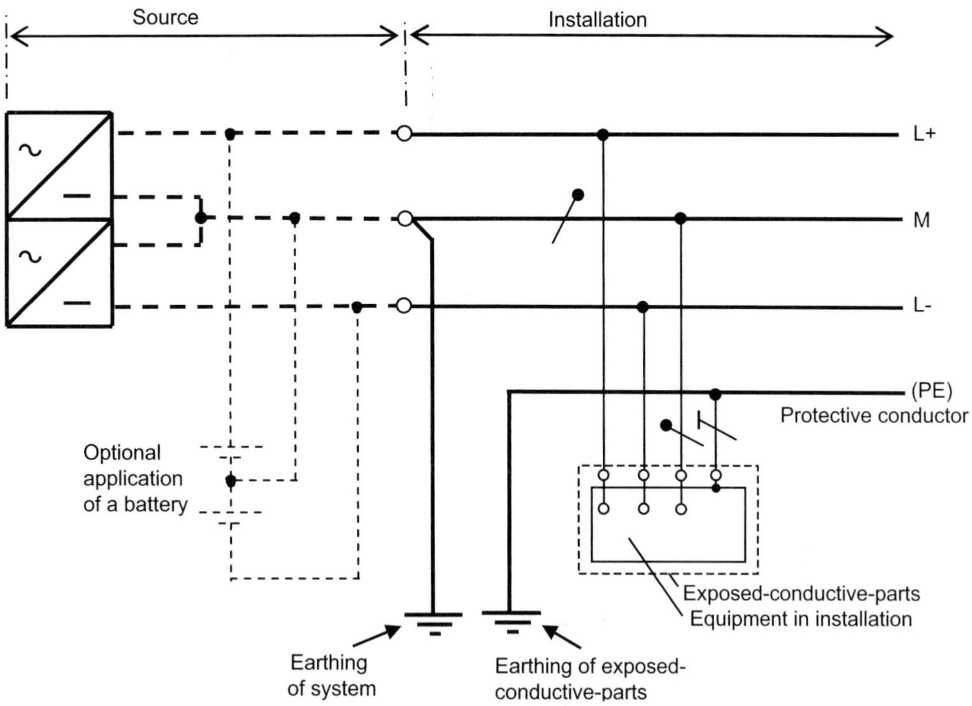

NOTE to Figures 9J and 9K

Additional earthing of the PE in the installation may be provided.

Fig 9L – IT DC system with earthed line conductor L− and protective conductor

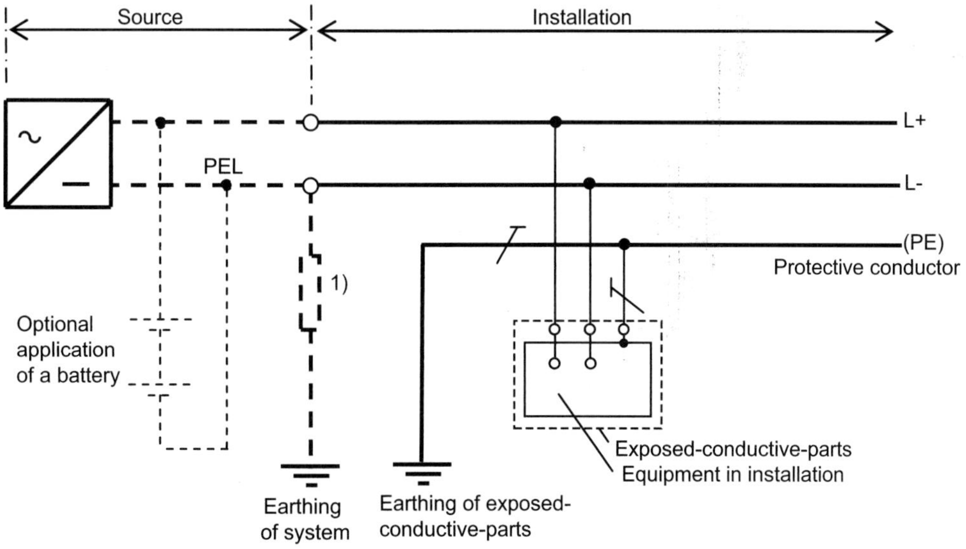

Fig 9M – IT DC system with earthed midpoint conductor M and protective conductor

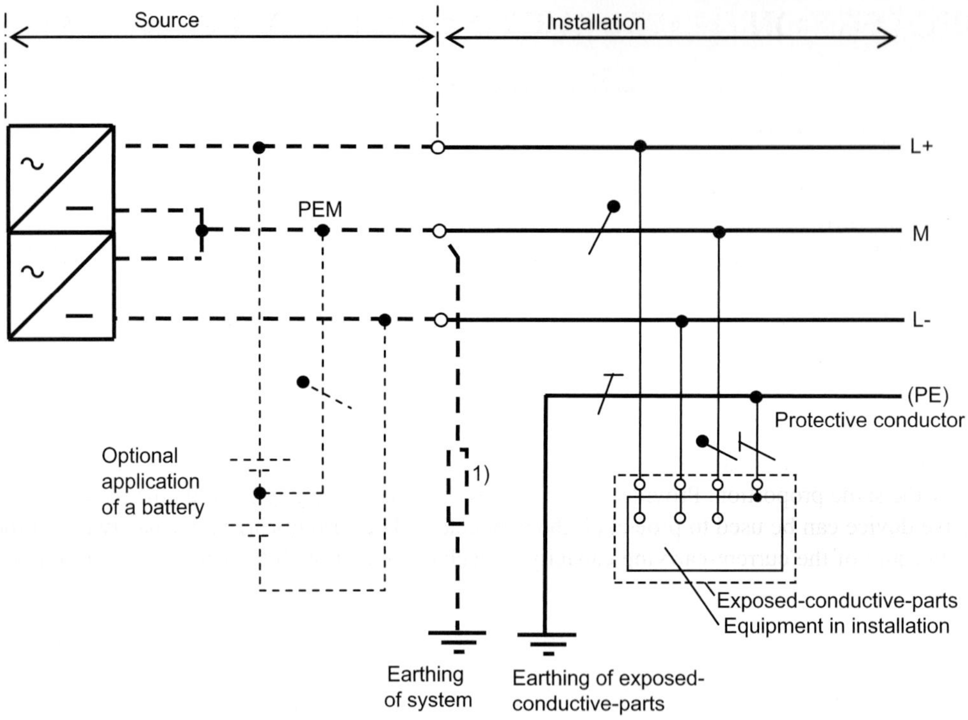

NOTES to Figures 9L and 9M

Additional earthing of the PE in the installation may be provided.

(1) The system may be connected to Earth via a sufficiently high impedance.

APPENDIX 10 (Informative)
PROTECTION OF CONDUCTORS IN PARALLEL AGAINST OVERCURRENT

1 INTRODUCTION

Overcurrent protection provided for conductors connected in parallel should provide adequate protection for all of the parallel conductors. For two conductors of the same cross-sectional area, conductor material, length and disposition arranged to carry substantially equal currents the requirements for overcurrent protection are straightforward. For more complex conductor arrangements, detailed consideration should be given to unequal current sharing between conductors and multiple fault current paths. This appendix gives guidance on the necessary considerations.

NOTE: A more detailed method for calculating the current between parallel conductors is given in BS 7769 (BS IEC 60287).

2 OVERLOAD PROTECTION OF CONDUCTORS IN PARALLEL

When an overload occurs in a circuit containing parallel conductors of multicore cables, the current in each conductor will increase by the same proportion. Provided that the current is shared equally between the parallel conductors, a single protective device can be used to protect all the conductors. The current-carrying capacity (I_z) of the parallel conductors is the sum of the current-carrying capacity of each conductor, with the appropriate grouping and other factors applied.

The current sharing between parallel cables is a function of the impedance of the cables. For large single-core cables the reactive component of the impedance is greater than the resistive component and will have a significant effect on the current sharing. The reactive component is influenced by the relative physical position of each cable. If, for example, a circuit consists of two large cables per phase, having the same length, construction and cross-sectional area and arranged in parallel with unfavourable relative position (i.e. cables of the same phase bunched together) the current sharing may be more like 70/30 rather than 50/50.

Where the difference in impedance between parallel conductors causes unequal current sharing, for example greater than 10 % difference, the design current and requirements for overload protection for each conductor should be considered individually.

The design current for each conductor can be calculated from the total load and the impedance of each conductor.

For a total of m conductors in parallel, the design current I_{bk} for conductor k is given by:

$$I_{bk} = \frac{I_b}{\left(\frac{Z_k}{Z_1} + \frac{Z_k}{Z_2} + \ldots + \frac{Z_k}{Z_{k-1}} + \frac{Z_k}{Z_k} + \frac{Z_k}{Z_{k+1}} + \ldots + \frac{Z_k}{Z_m}\right)}$$

where:

I_b is the current for which the circuit is designed

I_{bk} is the design current for conductor k

Z_k is the impedance of conductor k

Z_1, Z_2 and Z_m are the impedances of conductors 1, 2 and m respectively.

For parallel conductors up to and including 120 mm^2 cross-sectional area (csa) the design current I_{bk} for conductor k is given by:

$$I_{bk} = I_b \frac{S_k}{S_1 + S_2 + \ldots + S_m}$$

where:

$S_1, \ldots S_m$ is the csa of the conductors and

S_k is the csa of conductor k.

In the case of single-core cables, the impedance is a function of the relative positions of the cables as well as the design of the cable, for example, armoured or unarmoured. Methods for calculating the impedance are given in BS 7769 (BS IEC 60287). It is recommended that current sharing between parallel cables is verified by measurement.

The design current I_{bk} replaces I_b in Regulation 433.1.1 as follows:

$$I_{bk} \leq I_n \leq I_{zk}$$

The value used for I_z in Regulation 433.1.1 is either:

(i) the continuous current-carrying capacity of each conductor, I_{zk}, if an overload protective device is provided for each conductor (see Figure 10A), hence:

$$I_{bk} \leq I_{nk} \leq I_{zk}$$

or

(ii) the sum of the current-carrying capacities of all the conductors, ΣI_{zk}, if a single overload protective device is provided for the conductors in parallel (see Figure 10B), hence:

$$I_b \leq I_n \leq \Sigma I_{zk}$$

where:

I_{nk}	is the rated current of the protective device for conductor k
I_{zk}	is the continuous current-carrying capacity of conductor k
I_n	is the rated current of the protective device
ΣI_{zk}	is the sum of the continuous current-carrying capacities of the m conductors in parallel.

NOTE: For busbar systems, information should be obtained either from the manufacturer or from BS EN 61439-6.

For powertrack systems, information should be obtained either from the manufacturer or from BS EN 61534.

Fig 10A – Circuit in which an overload protective device is provided for each of the m conductors in parallel

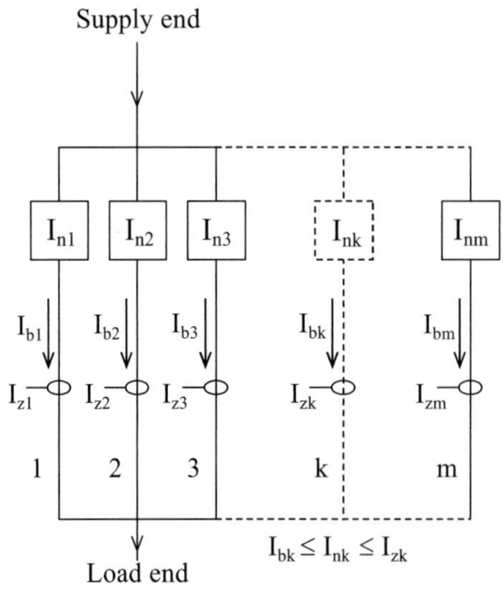

Fig 10B – Circuit in which a single overload protective device is provided for the m conductors in parallel

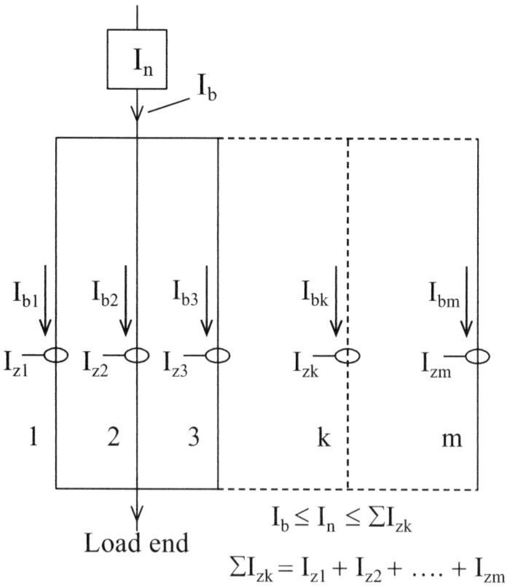

3 SHORT-CIRCUIT PROTECTION OF CONDUCTORS IN PARALLEL

Where conductors are connected in parallel, the possibility of a short-circuit within the parallel section should be considered.

If two conductors are connected in parallel and the operation of a single protective device may not be effective, then each conductor should have individual protection.

Where three or more conductors are connected in parallel then multiple fault current paths can occur and it may be necessary to provide short-circuit protection at both the supply and load ends of each parallel conductor. This situation is illustrated in Figures 10C and 10D.

Fig 10C – Current flow at the beginning of the fault

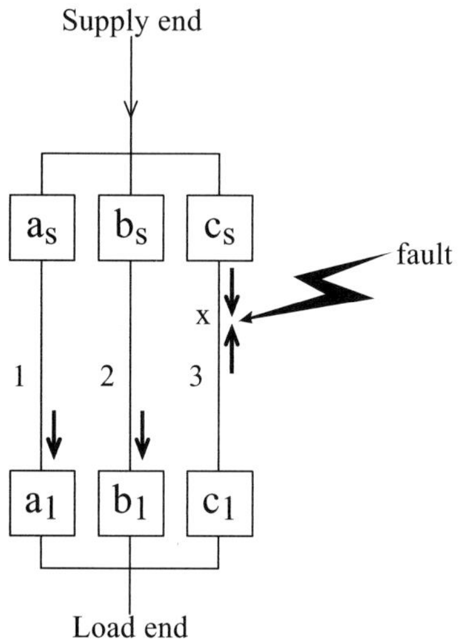

Fig 10D – Current flow after operation of the protective device c_s

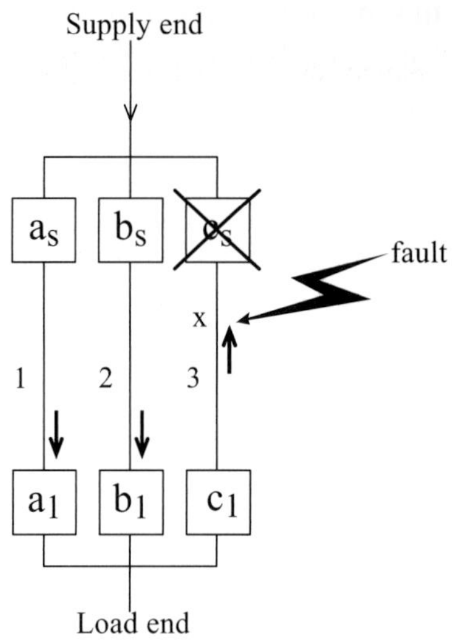

Figure 10C shows that, if a fault occurs in parallel conductor c at point x, the fault current will flow in conductors 1, 2 and 3. The magnitude of the fault current and the proportion of the fault current which flows through protective devices c_s and c_l will depend on the location of the fault. In this example it has been assumed that the highest proportion of the fault current will flow through protective device c_s. Figure 10D shows that, once c_s has operated, current will still flow to the fault at x via conductors 1 and 2. Because conductors 1 and 2 are in parallel, the current through protective devices a_s and b_s may not be sufficient for them to operate in the required time. If this is the case, the protective device c_l is necessary. It should be noted that the current flowing through c_l will be less than the current which caused c_s to operate. If the fault was close enough to c_l then c_l would operate first. The same situation would exist if a fault occurred in conductors 1 or 2, hence the protective devices a_l and b_l will be required.

The method of providing protective devices at both ends has two disadvantages. Firstly, if a fault at x is cleared by the operation of c_s and c_l then the circuit will continue to operate with the load being carried by conductors 1 and 2. Hence the fault and subsequent overloading of 1 and 2 may not be detected. Secondly, the fault at x may burn open-circuit at the c_l side leaving one side of the fault live and undetected.

An alternative method to providing protective devices at both ends would be to provide linked protective devices at the supply end (Figure 10E). This would prevent the continued operation of the circuit under fault conditions.

Fig 10E – Linked protective devices installed at the supply end of the parallel conductors

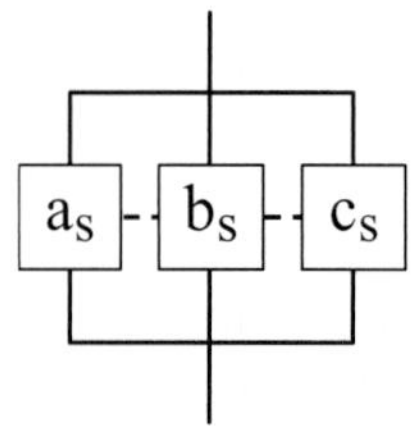

APPENDIX 11 (Informative)

WARNING AND USER INSTRUCTION LABELS

1 Introduction

Regulation 514.9.2 specifies harmonized standards to be used for user instructions and warning signs. The approach permits more flexibility in the sizing of appropriate labels for particular circumstances than previous editions of BS 7671, whilst accounting for an inclusive range of visual clarity and legibility.

This appendix provides particular guidance for the types of safety signs, warning signs and instructions required to be applied to electrical installations.

2 Text and graphical symbols

An easy to read sans serif font should be used. Examples include, but are not limited to, Arial, Calibri, Helvetica, and Verdana.

Graphical symbols should comply with appropriate standards, which include BS EN ISO 7010, IEC 60417, IEC 60617, and BS ISO 7001.

Sizes of text and graphical symbols should comply with BS EN IEC/IEEE 82079-1. Table 11A provides minimum sizes for labels relevant to electrical installations.

TABLE 11A – Minimum label text and symbol sizes to BS EN IEC/IEEE 82079-1 relevant to electrical installations

Label use	Recommended minimum height of graphical representations		Recommended minimum text size	
	Graphical symbols	Graphical safety signs	Dark on light background (e.g. black on white or black on yellow)	Low contrast (e.g. white on blue, white on red, white on black)
General labels	15 mm	15 mm	14 pt (4.9 mm)	16 pt (5.6 mm)
Labels for surfaces less than 10 cm^2	5 mm	10 mm	7 pt (2.5 mm)	12 pt (4.2 mm)

3 Safety signs, warnings and instructions provided on electronic displays

It is recommended that, where electronic displays for controls and indications relating to the electrical installation incorporate safety signs, warnings and instructions, these should also comply as relevant with BS EN ISO 9241-300, BS EN 60073 and BS EN 60447.

4 Examples of labels and signs required by BS 7671

NOTE 1: The examples in this appendix are provided for illustrative purposes. Alternative text and symbol sizes and layouts complying with BS EN IEC/IEEE 82079-1 are permitted (see item 2 of this appendix).

Fig 11A – Examples of the instruction notice required by Regulation 514.12.1

1. General Use (font size 14 pt)

> **IMPORTANT**
>
> This installation should be periodically inspected and tested and a report on its condition obtained, as prescribed in BS 7671 Requirements for Electrical Installations.
>
> Date of last inspection ....
>
> Recommended date of next inspection ....

Key:

■ Black (text)

□ White or yellow (background)

2. Use on surfaces less than 10 cm^2 (font size 7 pt)

> **IMPORTANT**
>
> This installation should be periodically inspected and tested and a report on its condition obtained, as prescribed in BS 7671 Requirements for Electrical Installations.
>
> Date of last inspection ....
>
> Recommended date of next inspection ....

Key:

■ Black (text)

□ White or yellow (background)

NOTE 2: If dates are to be printed with the label, rather than handwritten, space for handwritten dates is not necessary.

Fig 11B – Examples of the instruction notice required by Regulation 514.12.2

1. General use (font size 14 pt)

> **This installation, or part of it, is protected by a device which automatically switches off the supply if a fault develops. Test six-monthly by pressing the relevant test button(s) which should operate the device. Afterwards, manually switch on the device. If the device does not operate, or indicates a fault, seek expert advice.**

Key:

■ Black (text)

□ White or yellow (background)

2. Use on surfaces less than 10 cm^2 (font size 7 pt)

> **This installation, or part of it, is protected by a device which automatically switches off the supply if a fault develops. Test six-monthly by pressing the relevant test button(s) which should operate the device. Afterwards, manually switch on the device. If the device does not operate, or indicates a fault, seek expert advice.**

Key:

■ Black (text)

□ White or yellow (background)

Fig 11C – Example of the warning notice required by Regulation 514.13.1 complying with BS EN IEC/IEEE 82079-1

> **Safety Electrical Connection – Do Not Remove**

Key:

■ Black (text)

■ Yellow (background)

Fig 11D – Example of the warning notice required by Regulation 514.13.2 (font size 14 pt, graphical symbol optional)

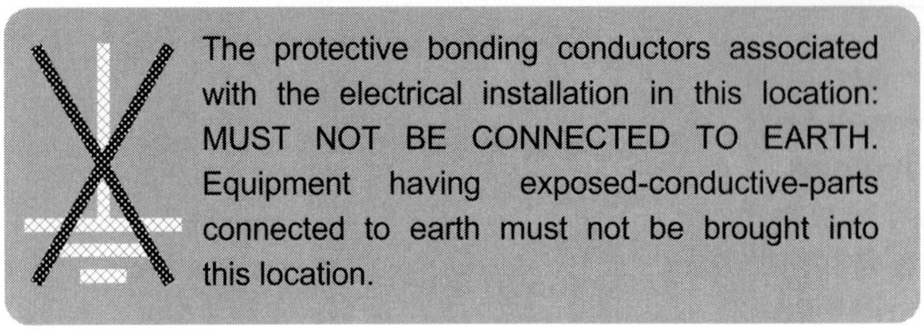

Key:

- Black (text)
- Yellow (background)
- Green (earth symbol)
- Red (cross)

Fig 11E – Example of the warning notice required by Regulation 514.15.1

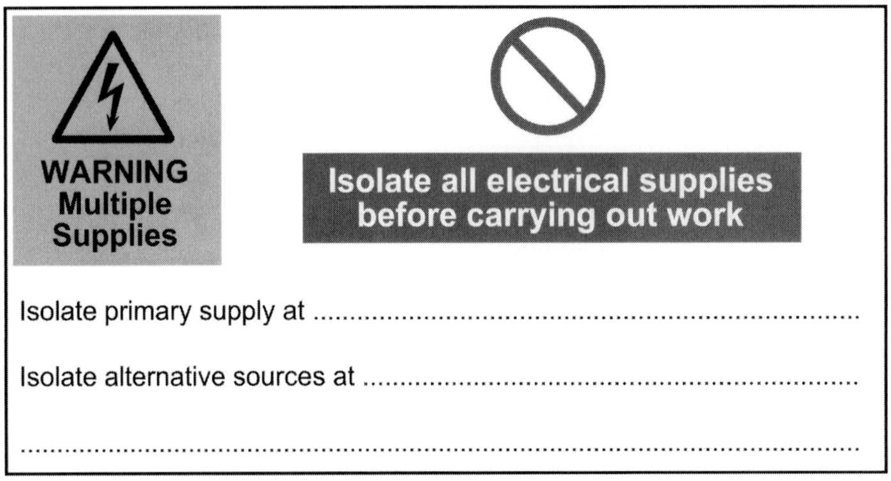

Key:

- Black (text, warning triangle and electric shock symbol)
- White (background and prohibition text)
- Yellow (background)
- Red (general prohibition sign and prohibition background text)

APPENDIX 12 NOT USED

APPENDIX 13 (Informative)

ESCAPE ROUTES AND FIRE PROTECTION

1 Cables in protected escape routes

Means of escape in a building is the provisions (i.e. the 'means') by which safe escape to a place of safety is provided. Often, particularly in smaller and simple buildings, this can be achieved simply by limiting the distance which has to be travelled to an exit from any place in the building. However, in larger and more complex buildings this is generally not possible and longer distances have to be travelled. It is then necessary to protect escapees during this longer period of travel.

This is achieved by providing *protected escape routes* (as defined) in which people escaping from the building are protected from the effects of fire by a specific fire and smoke-resisting structure. Such protected escape routes are composed of one or more of the following:

- Protected lobby
- Protected corridor
- Protected stairway.

These protected routes afford a safe passage through the building to an exit leading to a place of safety, usually the open air.

For safe escape, it is essential that as well as being enclosed in a fire and smoke-resisting construction these routes are essentially fire sterile. In other words, ideally they should not contain combustible material which itself could cause a hazard to the escaping occupants. Thus the amount of combustibles in a protected escape route is strictly controlled.

For this reason reference should be made to Regulation 422.2.1. Generally this means that cable within a protected route should be limited to that related to lighting, emergency lighting and fire detection and alarm. In some high hazard buildings such as chemical plant and nuclear buildings other alarm systems may also need to be included. Also in buildings such as hospitals with an evacuation strategy that does not involve protected escape routes a different approach may be justified. These different approaches should be justified in the fire safety strategy for the building.

Where expert fire safety advice has been obtained, it may be acceptable to install a minimal amount of cable in fully sealed non-combustible conduit as experience has demonstrated that this does not present an appreciable hazard. Ferrous metal, e.g. steel, is deemed to be an example of a non-combustible material. There is no restriction on the use of non-combustible cable (e.g. bare MICC).

Electrical accessories, such as socket-outlets, may be installed within a protected escape route provided that the structure's fire resistance is not compromised. It may be necessary to take additional precautions, for example the installation of internal or external fire resisting pads fitted in accordance with the accessory manufacturers' guidance. Refer also to Section 527.

Not all lobbies, corridors and stairways are protected escape routes (as defined). It is essential that available fire safety information on the building (fire safety plans/fire safety strategy/fire risk assessment, as available) be consulted to identify the protected escape routes. Where inadequate information exists, expert fire safety advice should be sought.

Cables installed in a fire resisting enclosure are deemed to be outside the protected escape route.

2 Fire sealing

Building elements

Where a wiring system passes through a building element that has specific fire-resistant properties, the ability of the element to resist the spread of fire will, in all likelihood, have been compromised. If a building element that has specific fire-resistant properties has been compromised, Regulation 527.2.1 of BS 7671 requires that the '... openings remaining after the passage of the wiring system shall be sealed according to the degree of fire-resistance (if any) prescribed for the respective element of building construction before penetration'. Guidance on sealing arrangements should be obtained from a competent person/authority at the design stage.

BS 7671 requires that the seal:

- resists the products of combustion to the same extent as the elements of building construction which have been penetrated, and
- provides the same degree of protection from water penetration as that required for the building construction element in which it has been installed, and
- is compatible with the material of the wiring system that it is in contact with, and
- permits thermal movement of the wiring system without reducing the sealing quality, and
- has adequate mechanical stability to withstand the stresses which may arise through damage to the support of the wiring system due to fire.

Internal sealing

Where a wiring system such as conduit, cable ducting, cable trunking, busbar or busbar trunking passes through a building element that has specific fire resisting properties and has an internal cross-sectional area greater than 710 mm^2, then it is to be internally sealed in accordance with the requirements of Regulation 527.2.3.

Whatever type of material is selected for internal sealing the installer should check that the sealant will not have a detrimental effect on the enclosed cables, such as chemical reaction on the sheath or insulation of the cables.

External sealing

It is likely that where a wiring system passes through a building element that has specific fire-resistant properties, the hole formed will be larger than that required for the wiring system. This surrounding area should also be sealed, such that the degree of fire resistance required for the element concerned is maintained. Reference should be made to Regulation 527.2.1

APPENDIX 14 (Informative)

DETERMINATION OF PROSPECTIVE FAULT CURRENT

Regulation 434.1 requires the prospective fault current to be determined at every relevant point of an installation. Relevant points are switchgear and protective devices that may have to operate and possibly disconnect a fault current. The devices have to be able to withstand the fault currents safely and protect downstream equipment from damage in the event of a fault.

Fault currents can occur in the event of insulation failure between live conductors and between line conductors and Earth.

Regulation 643.7.3.201 requires the prospective fault current to be measured, calculated or determined by another method, at the origin and at other relevant points in the installation.

In a three-phase installation the highest prospective fault current occurs with a simultaneous fault between all line conductors.

An approximation of the prospective fault current between line conductors can be determined by a measurement between a line conductor and neutral multiplied by $\sqrt{3}$.

An approximation of the prospective fault current due to a simultaneous short-circuit fault between all line conductors is determined by measurement between line and neutral multiplied by 2.

In a single-phase system the prospective fault current is the greater of either the fault current between the line conductor and neutral or the fault current between line conductor and Earth.

Other methods of determining prospective fault current are not precluded.

For installations with multiple sources of supply, measures should be taken to determine prospective fault current for all combinations of supply arrangements, so that the contribution made by any privately controlled embedded generation or uninterrupted supply arrangements is included.

In domestic (household) premises, where a consumer unit to BS EN 61439-3 is used and the maximum prospective fault current declared by the distributor is 16 kA, it is not necessary to measure or calculate prospective fault current at the origin of the supply.

When measuring prospective fault current, the person carrying out the testing should be aware of the danger present when connecting test instruments to live parts, and take suitable precautions as required by the Electricity at Work Regulations 1989 (EAWR). The HSE publication HSR25 provides guidance on this.

The measurement should always be made on the output terminals of a suitably rated protective device. If such a device is not present then a temporary one should be fitted. Measurement should never be made where overcurrent protection is not present between the point of connection and the supply transformer. Fused test leads alone do not meet this requirement.

APPENDIX 15 (Informative)

RING AND RADIAL FINAL CIRCUIT ARRANGEMENTS, REGULATION 433.1

This appendix sets out options for the design of ring and radial final circuits for household and similar premises in accordance with Regulation 433.1, using socket-outlets and fused connection units. It does not cover other aspects of the design of a circuit such as:

- Protection against electric shock, Chapter 41
- Protection against thermal effects, Chapter 42
- Protection against overcurrent, Chapter 43
- Selection and erection of equipment, Part 5.

Fig 15A – Ring final circuit arrangements, Regulation 433.1.204

The load current in any part of the circuit should be unlikely to exceed for long periods the current-carrying capacity of the cable (Regulation 433.1.204 refers). This can generally be achieved by:

(i) locating socket-outlets to provide reasonable sharing of the load around the ring

(ii) not supplying immersion heaters, comprehensive electric space heating or loads of a similar profile from the ring circuit

(iii) connecting cookers, ovens and hobs with a rated power exceeding 2 kW on their own dedicated radial circuit

(iv) taking account of the total floor area being served. (Historically, a limit of 100 m^2 has been adopted.)

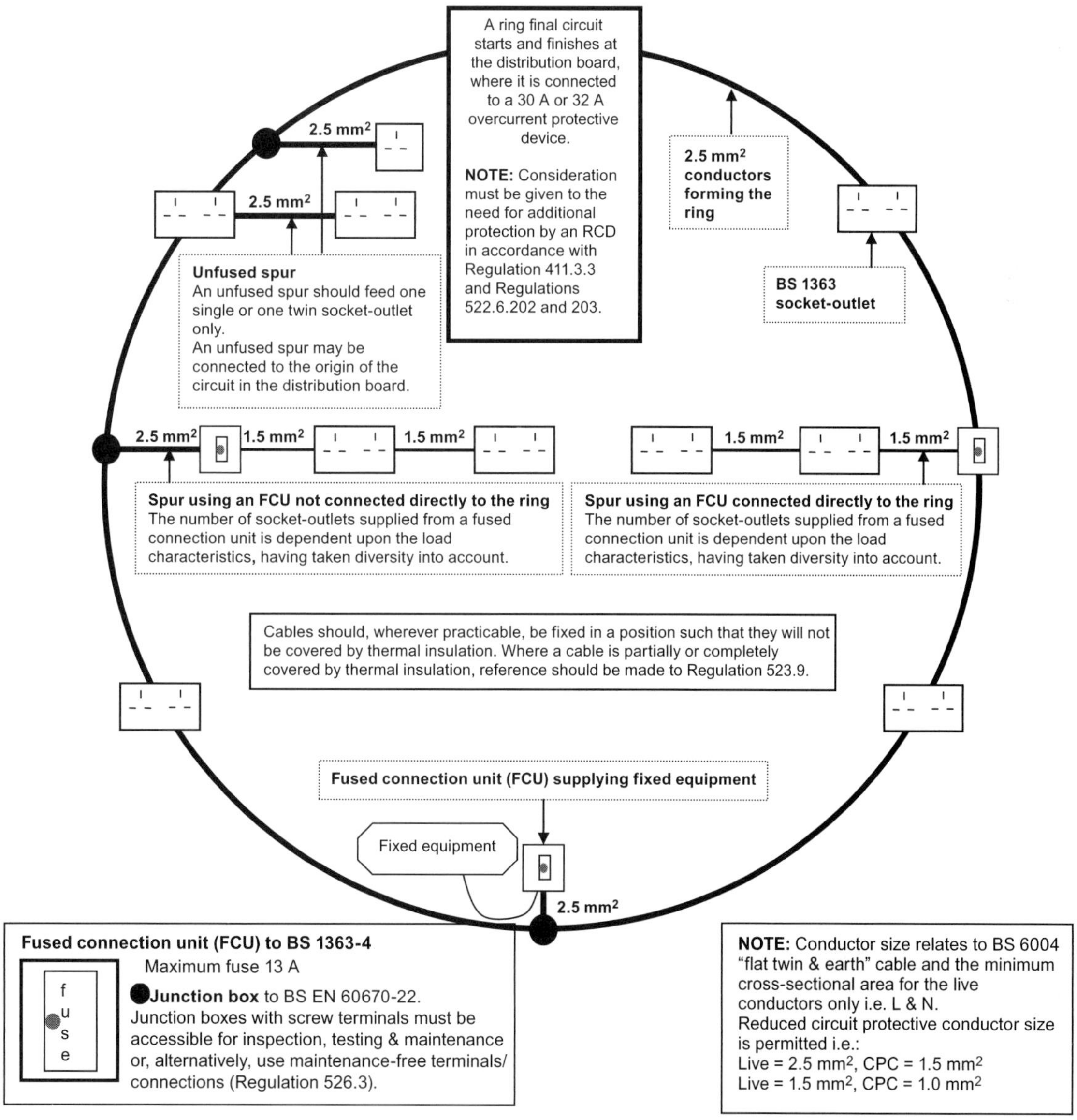

Fig 15B – Radial final circuit arrangements, Regulation 433.1

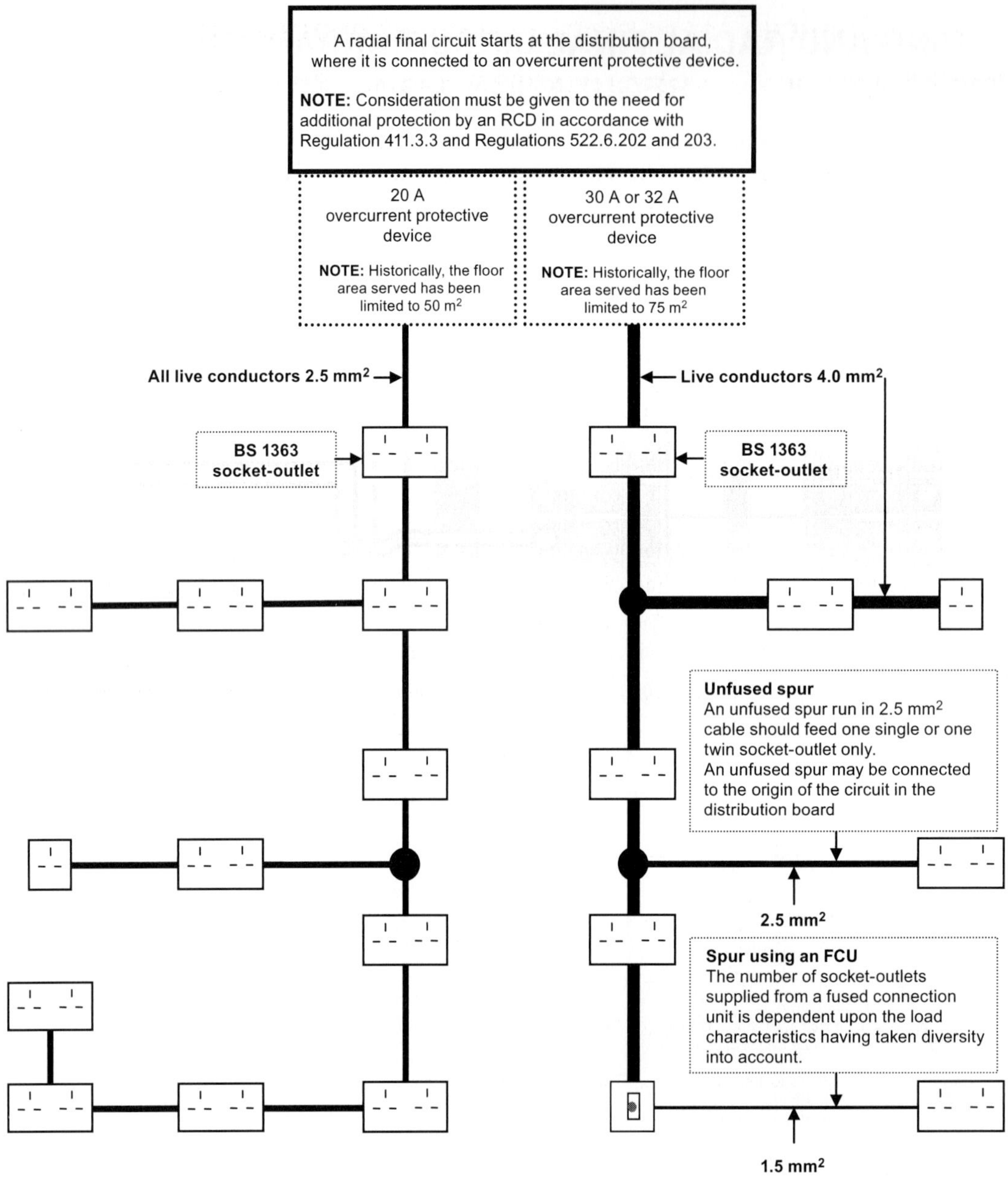

Cables should, wherever practicable, be fixed in a position such that they will not be covered by thermal insulation. Where a cable is partially or completely covered by thermal insulation, reference should be made to Regulation 523.9.

Fused connection unit (FCU) to BS 1363-4

f u s e

Maximum fuse 13 A

Junction box to BS EN 60670-22.
Junction boxes with screw terminals must be accessible for inspection, testing & maintenance or, alternatively, use maintenance-free terminals/connections (Regulation 526.3).

NOTE: Conductor size relates to BS 6004 "flat twin & earth" cable and the minimum cross-sectional area for the live conductors only i.e. L & N.
Reduced circuit protective conductor size is permitted i.e.:
Live = 4.0 mm^2, CPC = 1.5 mm^2
Live = 2.5 mm^2, CPC = 1.5 mm^2
Live = 1.5 mm^2, CPC = 1.0 mm^2

APPENDIX 16 (Informative)

Devices for protection against overvoltage

Typical installation of a surge protective device (SPD) in a power distribution board for a TN-S system.

Fig 16A1 – SPD connected to the first overcurrent protective device (OCPD) to the incoming supply

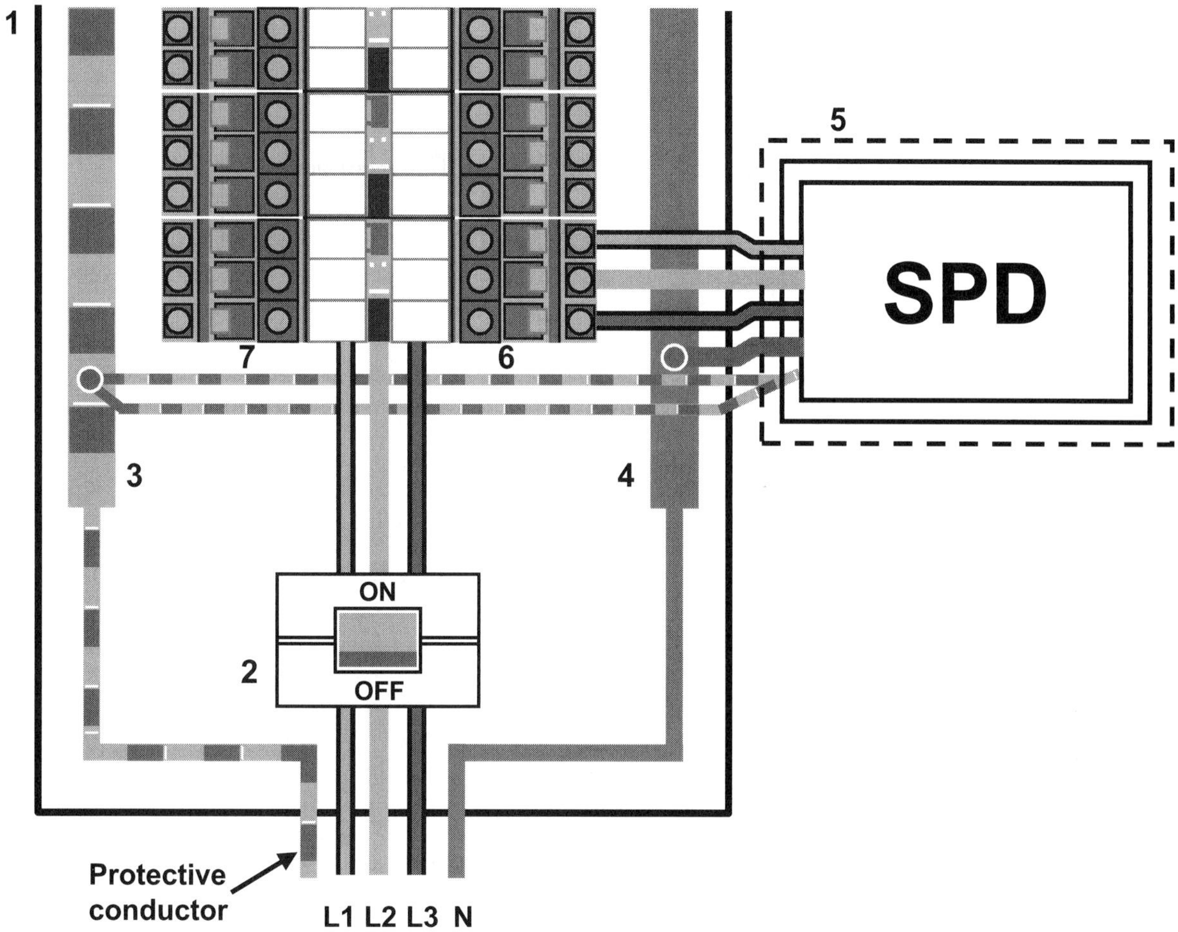

Key :

1 Distribution board

2 Main switch

3 Earthing bar

4 Neutral bar

5 Enclosure for SPD

6 First OCPD

7 Alternative first OCPD

NOTE: The OCPD provides a convenient means to protect the SPD and a means of isolation. As there is insufficient room within the distribution board the SPD is mounted in a separate enclosure for electrical safety. This enclosure is mounted directly alongside the distribution board so that the connecting leads are kept short. An additional local bonding connection is made to further minimize voltage drop on the connecting leads.

Installation of surge protective devices in TT systems, Connection Type 1 (CT 1)

Fig 16A2 – SPDs on the load side of an RCD [according to Regulation 534.4.6(a)]

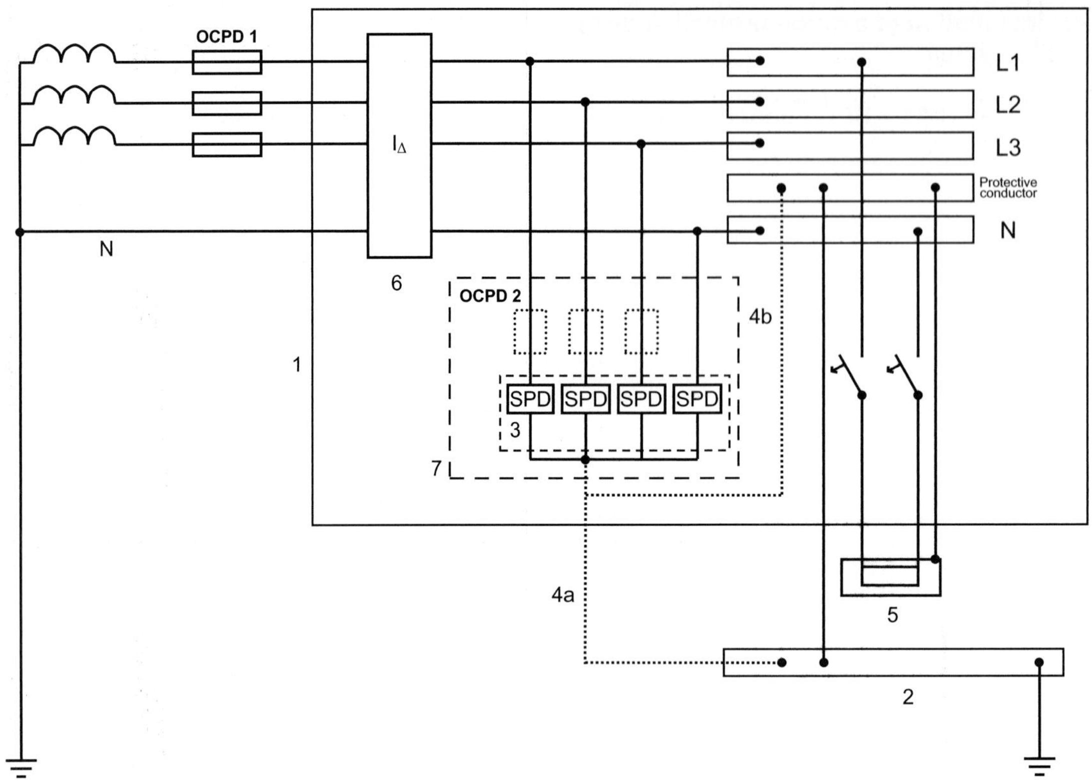

Key

1	Distribution board	OCPD 1	Overcurrent protective devices at the origin of the installation
2	Main earthing terminal or bar	OCPD 2	Overcurrent protective devices
3	Surge protective devices ensuring a protection level in accordance with overvoltage Category II		
4a, 4b	Earthing connection of surge protective devices, either 4a or 4b, whichever is the shorter route		
5	Current-using equipment		
6	Residual current protective device (RCD)		
7	SPD assembly		

Installation of surge protective devices in TT systems, Connection Type 2 (CT 2)

Fig 16A3 – SPDs on the supply side of an RCD [according to Regulation 534.4.6(b)]

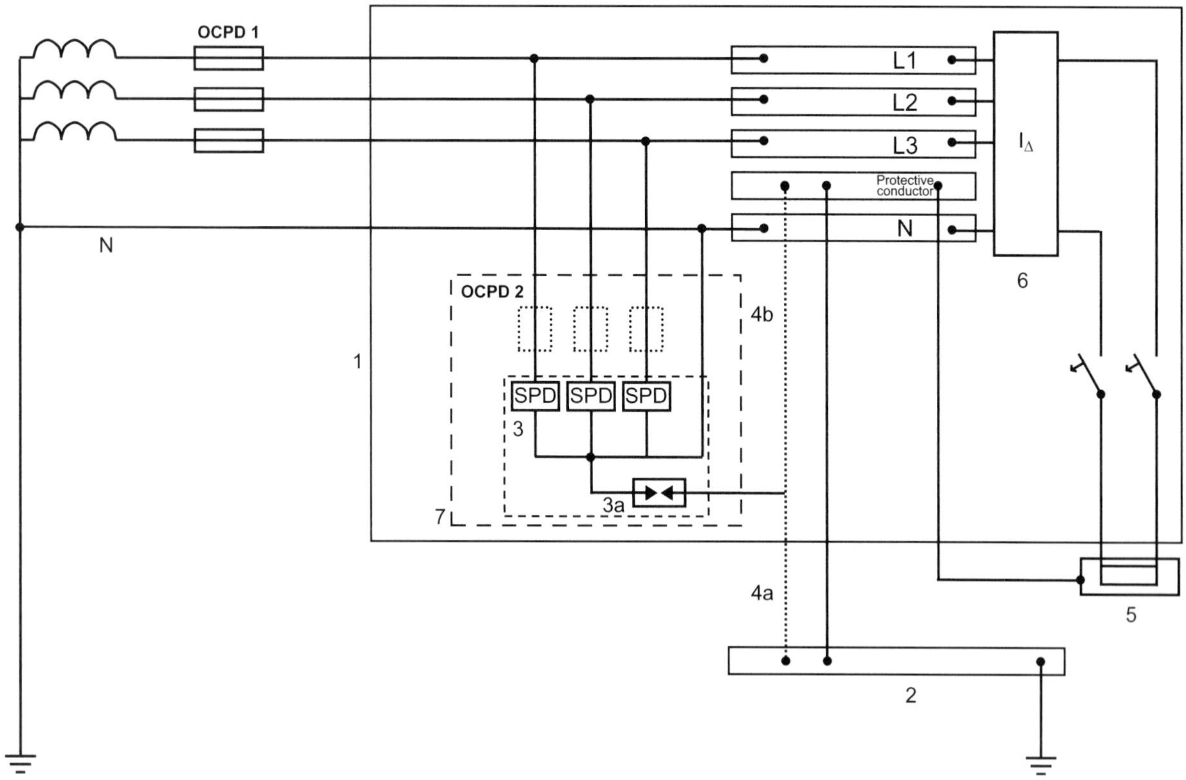

Key

1	Distribution board
2	Main earthing terminal or bar
3	Surge protective devices
3a	Surge protective device (SPDs 3 and 3a in series ensuring a protection level in accordance with overvoltage Category II)
4a, 4b	Earthing connection of surge protective devices, either 4a or 4b, whichever is the shorter route
5	Current-using equipment
6	Residual current protective device (RCD) installed downstream of the surge protective devices
7	SPD assembly
OCPD 1	Overcurrent protective devices at the origin of the installation
OCPD 2	Overcurrent protective devices

Installation of surge protective devices in IT systems

Fig 16A4 – SPDs on the load side of an RCD

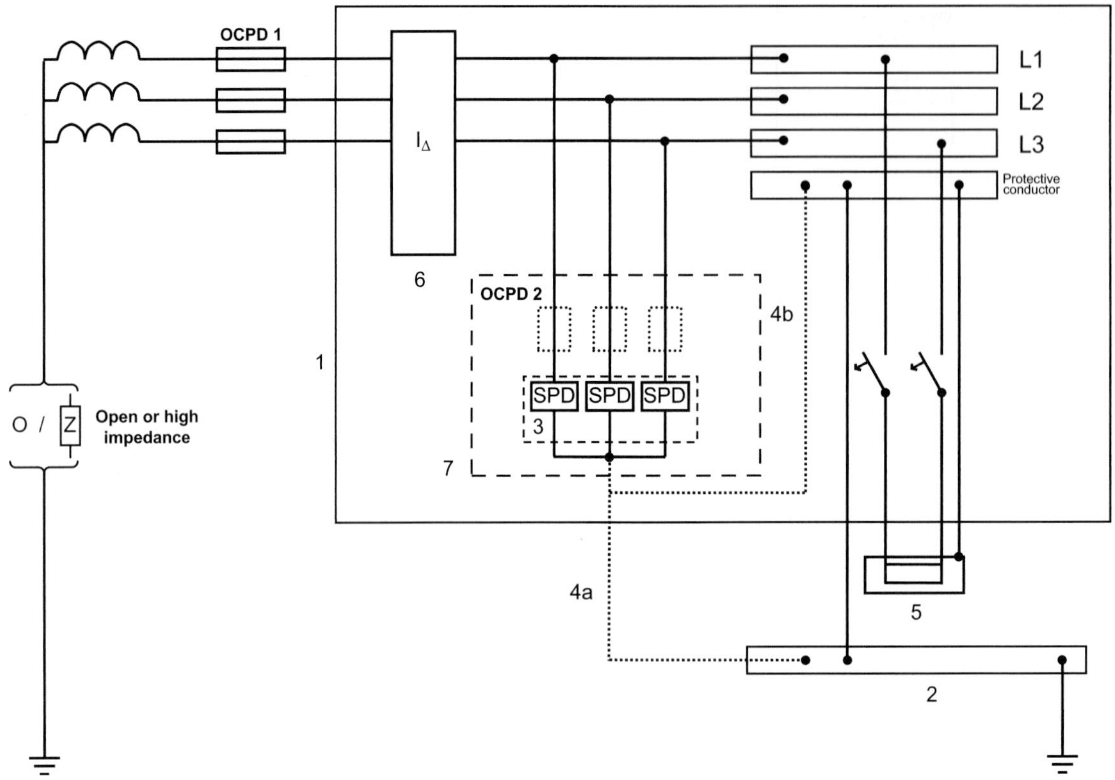

Key

1	Distribution board	OCPD 1	Overcurrent protective devices at the origin of the installation
2	Main earthing terminal or bar	OCPD 2	Overcurrent protective devices
3	Surge protective devices ensuring a protection level in accordance with overvoltage Category II		
4a, 4b	Earthing connection of surge protective devices, either 4a or 4b, whichever is the shorter route		
5	Current-using equipment		
6	Residual current protective device (RCD) installed upstream of the surge protective devices		
7	SPD assembly		

Installation of Types 1, 2 and 3 SPDs, for example in TN-C-S systems

Fig 16A5 – Installation example of Types 1, 2 and 3 coordinated SPDs

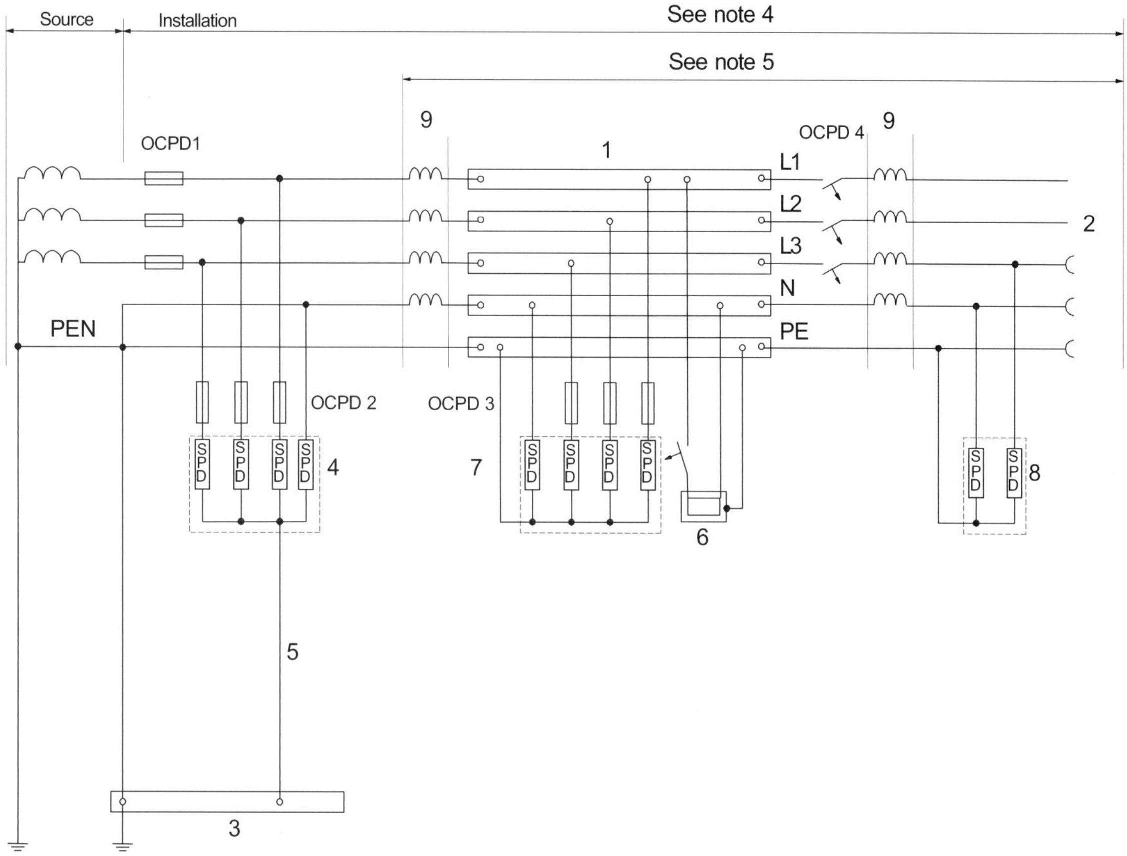

Key

1	Distribution board	7	Surge protective device, Type 2
2	Distribution outlet	8	Surge protective device, Type 2 or Type 3
3	Main earthing terminal or bar	9	Decoupling element or line length **NOTE:** If the cable length between the SPD types is short (refer to manufacturers' data), a decoupling element is employed to provide inductance for correct SPD coordination.
4	Surge protective device, Type 1	OCPD 1, 2, 3, 4	Overcurrent protective devices
5	Earthing connection (earthing conductor) of surge protective device		
6	Fixed equipment to be protected		

NOTE 1: For further information reference should be made to DD CLC/TS 61643-12.

NOTE 2: SPDs 4 and 7 (or 7 and 8) can be combined in a single SPD.

NOTE 3: SPDs may require additional modes of protection for sensitive equipment.

NOTE 4: BS EN 62305-4 covers the protection of electrical and electronic systems within structures against lightning.

NOTE 5: Section 443 of BS 7671 deals with the protection of electrical installations against transient overvoltages of atmospheric origin, transmitted by the supply distribution system, and against switching overvoltages generated by the equipment within the installation.

Typically, Type 1 SPDs are used at the origin of the installation, Type 2 SPDs are used at distribution boards and Type 3 SPDs are used near terminal equipment. Combined Type SPDs are classified with more than one Type, e.g. Type 1+2, Type 2+3. Type 1 SPDs are only used where there is a risk of direct lightning current.

The most important aspect in selecting an SPD is its limiting voltage performance (protection level U_p) during the expected surge event, and not the energy withstand (e.g. I_{imp}) which it can handle. An SPD with a low protection level will provide adequate protection of the equipment, while an SPD with a high energy withstand may only result in a longer operating life.

TABLE 16A – Information on SPD classification

SPD according to BS EN 62305	SPD according to BS EN 61643-11
SPD tested with I_{imp}	Type 1
SPD tested with I_n	Type 2
SPD tested with a combination wave	Type 3

SPD tested with Iimp (BS EN 62305-4)

SPDs which withstand the partial lightning current (with a typical waveform 10/350 μs) require a corresponding impulse test current I_{imp}.

NOTE 1: For power lines, a suitable test current I_{imp} is defined in the Class I test procedure of BS EN 61643-11.

SPD tested with In (BS EN 62305-4)

SPDs which withstand induced surge currents with a typical waveform 8/20 μs require a corresponding impulse test current I_{nspd}.

NOTE 2: For power lines a suitable test current I_{nspd} is defined in the Class II test procedure of BS EN 61643-11.

SPD tested with a combination wave (BS EN 62305-4)

SPDs that withstand induced surge currents with a typical waveform 8/20 μs and require a corresponding impulse test current I_{sc}.

NOTE 3: For power lines a suitable combination wave test is defined in the Class III test procedure of BS EN 61643-11, defining the open-circuit voltage U_{oc} 1.2/50 μs and the short-circuit current I_{sc} 8/20 μs of a 2 Ω combination wave generator.

APPENDIX 17 (Informative)

ENERGY EFFICIENCY

17.1 Scope

This appendix provides additional recommendations for the design and erection of electrical installations, including installations having local production and storage of energy, for optimizing the overall efficient use of electricity.

NOTE: On-site renewable energy sources and other local production sources do not of themselves increase the efficiency of an electrical installation. However, they do reduce the overall public electricity network losses as the consumption of the installation from the public supply is reduced. This may be considered an indirect energy efficiency measure. For installation of solar photovoltaic (PV) power supply systems, see Section 712 of BS 7671.

The recommendations within the scope of this appendix apply for new electrical installations and modification of existing installations. Much of this appendix will not apply to dwellings and similar installations. In dwellings and similar installations it is recommended that use is made of LED lighting. Reference should also be made to the Building Regulations - Part L in England and Wales, the Technical Handbook in Scotland and Technical Booklets F in Northern Ireland.

It is intended that this appendix will be developed into a section of Part 8 of BS 7671 in a future amendment.

17.2 Availability of electrical energy and user decision

The user specification will be the key input to the design of the energy efficiency management system.

The designer should take into account at least the following considerations:

(i) Selection of energy efficient appliances
(ii) Assignment of load priority for use as an input of the load optimization process (e.g. load shedding)
(iii) Intended use of the installation in providing an energy efficient design
(iv) Provision of a manual override facility which enables the user to take control from the automatic functions.

Energy efficiency management should be so designed that it does not reduce electrical supply availability and/or services or operation below the level desired by the user. The user of the electrical installation must be able to take the final decision over whether or not to operate a service at nominal value, or optimized value or not to operate it for a certain time.

NOTE: Examples are when someone is ill, the user may decide to heat the room at a higher temperature, even during peak consumption; when a company receives an urgent delivery order, the workshop may need to work at an unexpected hour.

17.3 Design requirements

The designer should take into account the following without losing the quality of service and the performance of the electrical installation:

(i) Load energy profile (active and passive)
(ii) Availability of local generation (solar, wind, generator, etc.)
(iii) Reduction of energy losses in the electrical installation
(iv) The tariff structure offered by the supplier of electrical energy
(v) The arrangement of the circuits with regard to energy efficiency (meshes, see 17.5)
(vi) The use of energy according to customer demand.

In order to benchmark the achievement of electrical energy efficiency measures, an overall energy efficiency assessment should be made.

17.4 Design recommendations

Determination of load profile

The main load demands within the installation have to be determined. The loads, together with their durations of operation, and/or an estimate of the annual load consumption of the main load demands (in kWh) should be identified.

Voltage drop

Consideration should be given to limiting the voltage drop within an installation to a level below that required by Regulation 525.202, to reduce the energy losses in the wiring systems.

Cross-sectional areas of conductors

Increasing the cross-sectional area of conductors will reduce the energy losses but will increase initial installation costs. The decision as to whether to do this should be made by assessing both the savings within a time scale and the additional cost due to the increased size. Practical constraints, such as size of terminations, will also affect the sizing of conductors.

NOTE: In some applications (particularly industrial), the most economical cross-sectional area of conductor may be several sizes larger than that required for the design current.

Power factor correction

Consideration should be given to improving the load power factor to reduce thermal losses in the wiring and tariff penalties.

Power factor correction can be made at the load level or centrally, depending on the type of application. The complexity of the issue requires careful consideration of each individual application.

17.5 Determination of zones and meshes

The installation should be divided into zones and meshes for the purpose of energy efficiency and analysis. A zone represents a floor area in square metres or a location where the electricity is used.

NOTE: Zones may correspond, for example, to:

- an industrial workshop
- a floor in building
- a space near windows or a space far from windows
- a room in a dwelling
- highway road lighting
- a hotel kitchen.

A Mesh is one or more circuits of the electrical installation for one or more zones including one or more services supplying a group of electrical equipment for the purpose of electrical energy efficiency. One circuit belongs to one mesh or a cluster of meshes.

Electrical management for energy efficiency is a system approach aiming to optimize the management of energy used for a specific service within a defined 'electrical mesh', taking into account all necessary information concerning the technical and economic approaches.

17.6 Determining the usages within the identified zones

The use of a particular circuit or zone should be clearly identified to enable accurate measurement and analysis of its energy consumption.

NOTE: Examples of different usages are:

- hot water production
- HVAC (cooling and heating)
- lighting
- motors
- appliances.

17.7 Energy efficiency and load management system

User requirements

User requirements are the first input to take into consideration. These requirements will be the key input to the design of the energy efficiency management system. Actions may include, amongst other things:

(i) Desk studies
(ii) Previous design considerations
(iii) Energy consumption records.

The designer, and/or end user should agree on the daily, weekly, monthly or yearly schedule for the periods when some services or applications shall be available or can be reduced or shut off. Identifying these applications and gathering them in a mesh are key from an energy usage point of view.

Requirements on the loads

The designer and installer must take into account the user decisions on selection of energy efficient appliances.

The user may give priority to the usage of the different loads as an input of the load optimization process (e.g. load shedding).

17.8 Inputs from loads, sensors and forecasts

Requirements on accuracy and measuring range

Provision must be made to allow the measurement and recording of energy consumption throughout the major parts of the installation, to provide for the management of that consumption. Accuracy of measurement should at least comply with the following:

- the meter at the origin of the loads must be accurate for billing purposes and can be used for the measurement of the efficiency of the whole installation
- at a lower level it may be necessary to provide measurement with an accuracy allowing sub-billing within the same entity. For example, a company such as a hotel may wish to charge the department for catering separately from the department in charge of entertainment
- at the lowest level of the final circuit directly powering loads it can be enough to provide information for following trends without precise needs for current to power conversion.

Measurement is a key parameter to determine the efficiency of the installation giving the user an awareness of energy consumption. Consequently, device accuracy and measuring range must be adapted to the intended use, as close as possible to the loads.

17.9 Loads

Loads should be classified regarding their user's acceptance of load shedding. Some loads, such as information technology equipment systems, computers and TV sets, are not suitable for load shedding. Others, such as heaters, refrigerators and electric vehicles, can accept a load shedding for up to a certain length of time without any impact on their service.

For each type of load, an acceptable duration of shedding in normal conditions should be determined.

NOTE 1: Examples of acceptable durations of shedding are 50 ms for a lamp and 15 minutes for a refrigerator or heater.

NOTE 2: Information on the ability of loads to accept, or not, load shedding and the corresponding duration(s) is useful.

Energy sensors

Energy-sensing devices should be of at least the same class as the energy performance and monitoring devices defined in Annex D of BS EN 61557-12.

Forecasts

Forecasts can be used as inputs to the energy efficiency management system, such as weather and occupancy forecasts.

Data logging

Historical data, where available, can be an input for making energy demand forecasts.

To achieve a design capable of delivering a high level of energy efficiency, all available energy consumption data should be taken into account.

17.10 Inputs from the supplies: energy availability and pricing, smart metering

The user should consider information concerning the energy availability and pricing which may vary with time.

Where the supply is a local source, the user should consider the minimum and/or the maximum available power and define the price of the corresponding energy based on the total cost of ownership including fixed and variable costs.

17.11 Impacts on the design of an electrical installation

The design of an electrical installation should consider energy efficiency at every stage, including the impact of different load demands, usage, zones and meshes.

The installation of fixed equipment for power metering and monitoring devices, control and energy management should be considered for new installations and future modifications.

Distribution switchboards should enable the segregation of circuits supplying each zone or each mesh as defined.

17.12 Information for the user: monitoring the electrical installation

The installation should be designed to enable the measurement of its total consumption in kWh for every hour of each day. This data, and the related cost of energy information, should be logged and stored for a minimum of one year and be accessible to the user.

NOTE: Multiple years of data can be useful for effective trend analysis.

In addition, by the use of sub-metering for example, the installation should be designed to enable the recording and saving of data for the consumption of individual or grouped loads totalling 97 % of the total load.

An energy efficiency management system comprises monitors for the whole electrical installation, including loads, local production and storage. It can manually (easiest cases) or automatically (most situations) monitor the electrical installation so as to allow optimization of the overall consumption of the system, taking into account the user requirements and the input parameters coming from the public electricity supply network, local electricity production and storage, the loads, sensors, and forecasts etc.

17.13 Efficiency measures for equipment

Motors and controls

An AC induction motor will consume more energy than it actually needs when operated at less than full-load conditions. This excess consumption of energy is dissipated by the motor in the form of heat. A better choice of motor and motor control will improve the global energy efficiency of the electric motor system.

Motors with a rated output of 0.75 kW – 375 kW must meet either the IE3 efficiency level or the IE2 level and be equipped with a variable speed drive. The energy efficiency classes are as follows:

- IE2 (High efficiency)
- IE3 (Premium efficiency)
- IE4 (Super premium efficiency).

NOTE 1: See BS EN 60034-30-1 Rotating electrical machines: Efficiency classes of line operated AC motors (IE code).

As about 95 % of the operating cost of a motor comes from its electrical energy consumption, adopting a higher energy efficiency class according to BS EN 60034-30-1, especially for high-duty applications, saves significant energy.

Consideration should be given to the use of motor starters, or other motor control devices such as variable speed drives, to achieve higher energy efficiency, particularly for efficient management of energy for intensive consumption applications (e.g. flow control of fans, pumps, air compressors).

Examples of aspects to be considered are:

- reducing electrical energy consumption
- optimizing the rated power
- reducing the inrush current
- reducing noise and vibration, in this way avoiding mechanical damage and failures within the air conditioning or heating system
- better control and better accuracy in achieving required flow and pressure.

NOTE 2: In industry, it is understood that 60 % of consumed electricity is used to turn motors and 63 % of this energy is used for applications such as pumps and fans.

Lighting

Lighting can represent a large amount of energy consumption in an electrical installation, depending upon the type of lamps and luminaires selected for each application. Lighting control is one of the easiest ways to improve energy efficiency. Therefore, careful consideration should be given to lighting control. The type of lamp, ballast switchgear and controlgear should be taken into consideration when applying lighting control. Building regulations have specific requirements on the selection of luminaire types for locations, the minimum required energy performance of luminaires and energy consumption per square metre of the installation.

NOTE: Solutions for lighting control can improve the energy efficiency by more than 50 %. These systems should be flexible and designed for the comfort of the users. The solutions can range from very small and local, such as with timer and occupancy sensors, up to sophisticated customized and centralized solutions that are part of complete building automation systems.

To operate lighting only when and where needed, permanent control of lighting may be implemented by using, for example:

- movement detectors
- dimming controls
- timed switches
- light-sensitive switches
- constant brightness controls.

Heating, ventilation and air conditioning (HVAC)

Consideration should be given to:

- the choice of HVAC equipment depending on the installation structure and usage
- the appropriate control system to optimize environment control (e.g. temperature, humidity, etc.) depending on the usage and occupancy of individual spaces.

NOTE: An example is a heating system controlled by a timer and thermostat monitoring the temperature threshold according to the expected occupancy.

17.14 Distribution system

Wiring systems

The cross-sectional areas of conductors may be optimized to reduce losses.

The electrical infrastructure may be optimized by locating the power source at an appropriate position to minimize circuit lengths.

The impact of thermal losses, off-load consumption and on-load energy consumption of equipment connected in the wiring system, e.g. switchgear and controlgear, power monitors and relays included in an electrical circuit, is negligible compared to the energy used in the load and in the energy transportation.

Power factor correction

Reduction of reactive current improves electrical energy efficiency by, for example, reducing thermal losses in wiring systems.

Where a reduction of reactive current is required, the optimized level of reactive current should be determined. This level generally depends on the public electricity supply contract.

In order to reduce reactive current, the following may be implemented:

- selection of current-using equipment with low reactive current component
- systems for compensation of reactive energy by using capacitors.

NOTE: Harmonic distortion rate and voltage impulse are important considerations for selecting capacitor banks. The provision of harmonic filters may need to be considered.

Energy

It is of prime importance, in terms of electrical energy efficiency, to ascertain by either calculation or measurement, the energy consumption of each item of current-using equipment.

Load profile

Measurement of the energy used is necessary to give a load profile. This should be over a period of a minimum of 24 h to give a reasonable estimate of load profile.

NOTE: The time period of measurement is typically from every 10 mins to 1 hour. The time period varies depending on the usage, zone and the sector of activity, and also the season (especially for lighting and HVAC).

Voltage drop

Voltage drop reduces energy efficiency of the electrical installation.

Where the voltage drop measurement is required, voltage measurements should be made on the current-using equipment and at the origin of the circuit serving the current-using equipment.

17.15 Periodic assessment

Where an electrical installation has been designed with energy efficiency measures, it may be appropriate periodically to confirm that such measures remain in line with any energy efficiency measures that exist.

Where a periodic assessment identifies the electrical installation efficiency is lower than the required value, an action plan is typically needed to achieve the required or desired electrical installation efficiency class.

Table of figures

Table of tables

INDEX

B

C

D

E

F

G

H

M

S

T

U

V

NOTES

NOTES

NOTES

NOTES

NOTES

NOTES

NOTES

NOTES

NOTES

NOTES

NOTES

NOTES

NOTES

NOTES

NOTES

NOTES